Studies in Systems, Decision and Control

Volume 305

The series "Studies in Systems, Decision and Control" (SSDC) covers both new developments and advances, as well as the state of the art, in the various areas of broadly perceived systems, decision-making, and control–quickly, up to date and with a high quality. The intent is to cover the theory, applications, and perspectives on the state of the art and future developments relevant to systems, decision-making, control, complex processes and related areas, as embedded in the fields of engineering, computer science, physics, economics, social and life sciences, as well as the paradigms and methodologies behind them. The series contains monographs, textbooks, lecture notes, and edited volumes in systems, decision-making, and control spanning the areas of Cyber-Physical Systems, Autonomous Systems, Sensor Networks, Control Systems, Energy Systems, Automotive Systems, Biological Systems, Vehicular Networking and Connected Vehicles, Aerospace Systems, Automation, Manufacturing, Smart Grids, Nonlinear Systems, Power Systems, Robotics, Social Systems, Economic Systems and other. Of particular value to both the contributors and the readership are the short publication timeframe and the worldwide distribution and exposure which enables both wide and rapid dissemination of research output.

** Indexing: The books of this series are submitted to ISI, SCOPUS, DBLP, Ulrichs, MathSciNet, Current Mathematical Publications, Mathematical Reviews, Zentralblatt Math: MetaPress and Springerlink.

More information about this series at http://www.springer.com/series/13304

George A. Anastassiou

Generalized Fractional Calculus

New Advancements and Applications

Springer

George A. Anastassiou
Department of Mathematical Sciences
University of Memphis
Memphis, TN, USA

ISSN 2198-4182 ISSN 2198-4190 (electronic)
Studies in Systems, Decision and Control
ISBN 978-3-030-56961-7 ISBN 978-3-030-56962-4 (eBook)
https://doi.org/10.1007/978-3-030-56962-4

This Springer imprint is published by the registered company Springer Nature Switzerland AG
The registered company address is: Gewerbestrasse 11, 6330 Cham, Switzerland

Dedicated to my wife Koula and to all doctors fighting Covid—19

Preface

Computational and Fractional Analysis nowadays are more and more in the center of mathematics and of other related sciences either by themselves because of their rapid development, which is based on very old foundations, or because they cover a great variety of applications in the real world. In this monograph, all presented is original work by the author given at a very general level to cover a maximum number of cases in various applications. We apply generalized fractional differentiation techniques of Caputo, Canavati, and conformable types to various kinds of inequalities such as Ostrowski, Grüss, Iyengar, Polya, Poincaré, Sobolev, Hilbert-Pachpatte, and Opial type inequalities. Some of these are extended to Banach space valued functions. Many of these are studying the average of a function as compared to values of the on hand function. These inequalities also have a great impact on numerical analysis, stochastics, and fractional differential equations. We continue with generalized fractional approximation by positive sublinear operators including the case of iterated one which derives from the newly presented Korovkin type inequalities. The linear case is extended to Banach space valued functions. It follows for the first time multivariate complex Korovkin quantitative approximation theory.

We also present M-fractional integral inequalities of Ostrowski and Polya types. The M-fractional derivative is based on Mittag-Leffler special function and possesses all nice properties of ordinary derivative.

Our results are weighted so they provide a great variety of cases and applications. The last eight chapters of the book deal for the first time with the quantitative fractional Korovkin type approximation of stochastic processes. We lay there the foundations of stochastic fractional calculus. We consider both Caputo and Conformable fractional directions and we derive regular and trigonometric results. Our positive linear operators can be expectation operator commutative or not. The case of non-commutative operators is especially elaborate and interesting.

This monograph is the natural and expected evolution of recent author's research work put in a book form for the first time. The presented approaches are original, and the chapters are self-contained and can be read independently. This monograph is suitable to be used in related graduate classes and research projects. We exhibit to the maximum our computational, estimating and approximation methods to all possible directions.

The motivation to write this monograph came by the following: various issues related to the modeling and analysis of ordinary and fractional order systems have gained an increased popularity, as witnessed by many books and volumes in Springer's program:

http://www.springer.com/gp/search?query=fractional&submit=Prze%C5%9Blij and the purpose of our book is to capture at a very general level a deeper formal analysis on some issues that are relevant to many areas for instance: decision making, complex processes, systems modeling and control, and related areas.

The above are deeply embedded in the fields of mathematics, engineering, computer science, physics, economics, social and life sciences.

The complete list of presented topics follows:

Mixed Caputo weighted-fractional Ostrowski type Inequalities,

Caputo weighted-fractional Ostrowski and Grüss inequalities for several functions,

Weighted fractional Iyengar type Inequalities in the Caputo direction,

Generalized Canavati type g-fractional Iyengar and Ostrowski type inequalities,

Generalized Canavati type g-fractional Polya type inequalities,

Caputo generalized weighted-fractional integral Inequalities,

Generalized weighted-fractional Approximation by Sublinear Operators,

Generalized g-iterated fractional Approximations by Sublinear Operators,

Generalized g-Fractional vector Representation Formula and integral Inequalities for Banach space valued functions,

Iterated g-Fractional vector Representation Formulae and Inequalities for Banach space valued functions,

Vectorial generalized g-fractional direct and iterated approximations by linear operators,

Quantitative Multivariate Complex Korovkin Theory,

M-fractional integral inequalities,

Foundation of Stochastic Fractional Calculus with Fractional Approximation of Stochastic Processes,

Trigonometric Fractional Approximation of Stochastic Processes,

Conformable Fractional Approximation of Stochastic Processes,

Trigonometric Conformable Fractional Quantitative Approximation of Stochastic Processes,

Commutative Caputo fractional Korovkin inequalities for Stochastic Processes,

Trigonometric Commutative Caputo fractional Korovkin theory for Stochastic Processes,

Commutative Conformable fractional Korovkin properties for Stochastic Processes,

Trigonometric Commutative Conformable fractional Korovkin properties for Stochastic Processes.

An extensive list of references is given per chapter.

The book's results are expected to find applications in many areas of pure and applied mathematics, especially in approximation theory, stochastic processes, inequalities and differential equations as they connect to fractionality. Thus this monograph is suitable for researchers, graduate students, and seminars of the above disciplines, also to be in all science and engineering libraries.

The preparation of the book took place during 2019–2020 at the University of Memphis and finished during the Coronavirus outbreak keeping alive and in control the author!

The author likes to thank Professor Alina Lupas of University of Oradea, Romania, for checking and reading the manuscript.

Memphis, TN, USA
June 2020

George A. Anastassiou

Contents

Chapter 1
Caputo ψ-Fractional Ostrowski Inequalities

Very general univariate mixed Caputo ψ-fractional Ostrowski type inequalities are presented. Estimates are with respect to $\|\cdot\|_p$, $1 \leq p \leq \infty$. We give also applications. This chapter relies on [4].

1.1 Introduction

In 1938, Ostrowski [5] proved the following important inequality.

Theorem 1.1 *Let $f : [a, b] \to \mathbb{R}$ be continuous on $[a, b]$ and differentiable on (a, b) whose derivative $f' : (a, b) \to \mathbb{R}$ is bounded on (a, b), i.e., $\|f'\|_\infty := \sup_{t \in (a,b)} |f'(t)| < +\infty$. Then*

$$\left| \frac{1}{b-a} \int_a^b f(t)\, dt - f(x) \right| \leq \left[\frac{1}{4} + \frac{\left(x - \frac{a+b}{2}\right)^2}{(b-a)^2} \right] \cdot (b-a) \|f'\|_\infty, \quad (1.1)$$

for any $x \in [a, b]$. The constant $\frac{1}{4}$ is the best possible.

Since then there has been a lot of activity around these inequalities with important applications to numerical analysis and probability.

In this chapter we are greatly motived and inspired by Theorem 1.1, see also [2]. Here we present various ψ-fractional Ostrowski type inequalities and we give interesting applications.

G. A. Anastassiou, *Generalized Fractional Calculus*, Studies in Systems, Decision and Control 305, https://doi.org/10.1007/978-3-030-56962-4_1

1.2 Background

Here we follow [1].

Let $\alpha > 0$, $[a,b] \subset \mathbb{R}$, $f : [a,b] \to \mathbb{R}$ which is integrable and $\psi \in C^1([a,b])$ an increasing function such that $\psi'(x) \neq 0$, for all $x \in [a,b]$. Consider $n = \lceil \alpha \rceil$, the ceiling of α. The left and right fractional integrals are defined, respectively, as follows:

$$I_{a+}^{\alpha,\psi} f(x) := \frac{1}{\Gamma(\alpha)} \int_a^x \psi'(t) (\psi(x) - \psi(t))^{\alpha-1} f(t) \, dt, \tag{1.2}$$

and

$$I_{b-}^{\alpha,\psi} f(x) := \frac{1}{\Gamma(\alpha)} \int_x^b \psi'(t) (\psi(t) - \psi(x))^{\alpha-1} f(t) \, dt, \tag{1.3}$$

for any $x \in [a,b]$, where Γ is the gamma function.

The following semigroup property is valid for fractional integrals: if $\alpha, \beta > 0$, then

$$I_{a+}^{\alpha,\psi} I_{a+}^{\beta,\psi} f(x) = I_{a+}^{\alpha+\beta,\psi} f(x), \text{ and } I_{b-}^{\alpha,\psi} I_{b-}^{\beta,\psi} f(x) = I_{b-}^{\alpha+\beta,\psi} f(x).$$

We mention

Definition 1.2 *([1])* Let $\alpha > 0$, $n \in \mathbb{N}$ such that $n = \lceil \alpha \rceil$, $[a,b] \subset \mathbb{R}$ and $f, \psi \in C^n([a,b])$ with ψ being increasing and $\psi'(x) \neq 0$, for all $x \in [a,b]$. The left ψ-Caputo fractional derivative of f of order α is given by

$${}^C D_{a+}^{\alpha,\psi} f(x) := I_{a+}^{n-\alpha,\psi} \left(\frac{1}{\psi'(x)} \frac{d}{dx} \right)^n f(x), \tag{1.4}$$

and the right ψ-Caputo fractional derivative of f is given by

$${}^C D_{b-}^{\alpha,\psi} f(x) := I_{b-}^{n-\alpha,\psi} \left(-\frac{1}{\psi'(x)} \frac{d}{dx} \right)^n f(x). \tag{1.5}$$

To simplify notation, we will use the symbol

$$f_\psi^{[n]}(x) := \left(\frac{1}{\psi'(x)} \frac{d}{dx} \right)^n f(x), \tag{1.6}$$

with $f_\psi^{[0]}(x) = f(x)$.

By the definition, when $\alpha = m \in \mathbb{N}$, we have

$$\begin{gathered} {}^C D_{a+}^{\alpha,\psi} f(x) = f_\psi^{[m]}(x) \\ \text{and} \\ {}^C D_{b-}^{\alpha,\psi} f(x) = (-1)^m f_\psi^{[m]}(x). \end{gathered} \tag{1.7}$$

If $\alpha \notin \mathbb{N}$, we have

$$ {}^{C}D_{a+}^{\alpha,\psi} f(x) = \frac{1}{\Gamma(n-\alpha)} \int_a^x \psi'(t) (\psi(x) - \psi(t))^{n-\alpha-1} f_\psi^{[n]}(t)\, dt, \tag{1.8} $$

and

$$ {}^{C}D_{b-}^{\alpha,\psi} f(x) = \frac{(-1)^n}{\Gamma(n-\alpha)} \int_x^b \psi'(t) (\psi(t) - \psi(x))^{n-\alpha-1} f_\psi^{[n]}(t)\, dt, \tag{1.9} $$

$\forall\, x \in [a, b]$.

In particular, when $\alpha \in (0, 1)$, we have

$$ \begin{aligned} &{}^{C}D_{a+}^{\alpha,\psi} f(x) = \tfrac{1}{\Gamma(1-\alpha)} \textstyle\int_a^x (\psi(x) - \psi(t))^{-\alpha} f'(t)\, dt, \\ &\text{and} \\ &{}^{C}D_{b-}^{\alpha,\psi} f(x) = \tfrac{-1}{\Gamma(1-\alpha)} \textstyle\int_x^b (\psi(t) - \psi(x))^{-\alpha} f'(t)\, dt \end{aligned} \tag{1.10} $$

$\forall\, x \in [a, b]$.

Clearly the above is a generalization of left and right Caputo fractional derivatives. For more see [1].

Still we need from [1] the following left and right fractional Taylor's formulae:

Theorem 1.3 ([1]) *Let $\alpha > 0$, $n \in \mathbb{N}$ such that $n = \lceil \alpha \rceil$, $[a, b] \subset \mathbb{R}$ and $f, \psi \in C^n([a, b])$ with ψ being increasing and $\psi'(x) \neq 0$, for all $x \in [a, b]$. Then, the left fractional Taylor formula follows,*

$$ f(x) = \sum_{k=0}^{n-1} \frac{f_\psi^{[k]}(a)}{k!} (\psi(x) - \psi(a))^k + I_{a+}^{\alpha,\psi}\, {}^{C}D_{a+}^{\alpha,\psi} f(x), \tag{1.11} $$

and the right fractional Taylor formula follows,

$$ f(x) = \sum_{k=0}^{n-1} (-1)^k \frac{f_\psi^{[k]}(b)}{k!} (\psi(b) - \psi(x))^k + I_{b-}^{\alpha,\psi}\, {}^{C}D_{b-}^{\alpha,\psi} f(x), \tag{1.12} $$

$\forall\, x \in [a, b]$.

In particular, given $\alpha \in (0, 1)$, we have

$$ \begin{aligned} &f(x) = f(a) + I_{a+}^{\alpha,\psi}\, {}^{C}D_{a+}^{\alpha,\psi} f(x), \\ &\textit{and} \\ &f(x) = f(b) + I_{b-}^{\alpha,\psi}\, {}^{C}D_{b-}^{\alpha,\psi} f(x), \end{aligned} \tag{1.13} $$

$\forall\, x \in [a, b]$.

Remark 1.4 For convenience we can rewrite (1.11)–(1.13) as follows:

$$f(x) = \sum_{k=0}^{n-1} \frac{f_{\psi}^{[k]}(a)}{k!} (\psi(x) - \psi(a))^k + \tag{1.14}$$

$$\frac{1}{\Gamma(\alpha)} \int_a^x \psi'(t) (\psi(x) - \psi(t))^{\alpha-1} \, {}^C D_{a+}^{\alpha,\psi} f(t) \, dt,$$

and

$$f(x) = \sum_{k=0}^{n-1} \frac{(-1)^k f_{\psi}^{[k]}(b)}{k!} (\psi(b) - \psi(x))^k + \tag{1.15}$$

$$\frac{1}{\Gamma(\alpha)} \int_x^b \psi'(t) (\psi(t) - \psi(x))^{\alpha-1} \, {}^C D_{b-}^{\alpha,\psi} f(t) \, dt,$$

$\forall \, x \in [a, b]$.

When $\alpha \in (0, 1)$, we get:

$$\begin{aligned} &f(x) = f(a) + \tfrac{1}{\Gamma(\alpha)} \int_a^x \psi'(t) (\psi(x) - \psi(t))^{\alpha-1} \, {}^C D_{a+}^{\alpha,\psi} f(t) \, dt, \\ &\text{and} \\ &f(x) = f(b) + \tfrac{1}{\Gamma(\alpha)} \int_x^b \psi'(t) (\psi(t) - \psi(x))^{\alpha-1} \, {}^C D_{b-}^{\alpha,\psi} f(t) \, dt, \end{aligned} \tag{1.16}$$

$\forall \, x \in [a, b]$.

Again from [1] we have the following:

Consider the norms $\|\cdot\|_{\infty} : C([a, b]) \to \mathbb{R}$ and $\|\cdot\|_{C_{\psi}^{[n]}} : C^n([a, b]) \to \mathbb{R}$, where $\|f\|_{C_{\psi}^{[n]}} := \sum_{k=0}^{n} \left\| f_{\psi}^{[k]} \right\|_{\infty}$.

We have

Theorem 1.5 ([1]) *The ψ-Caputo fractional derivatives are bounded operators. For all $\alpha > 0$ $(n = \lceil \alpha \rceil)$*

$$\left\| {}^C D_{a+}^{\alpha,\psi} \right\|_{\infty} \leq K \, \|f\|_{C_{\psi}^{[n]}} \tag{1.17}$$

and

$$\left\| {}^C D_{b-}^{\alpha,\psi} \right\|_{\infty} \leq K \, \|f\|_{C_{\psi}^{[n]}}, \tag{1.18}$$

where

$$K = \frac{(\psi(b) - \psi(a))^{n-\alpha}}{\Gamma(n + 1 - \alpha)} > 0. \tag{1.19}$$

1.3 Main Results

We present the following ψ-fractional Ostrowski type inequalities:

Theorem 1.6 *Let* $\alpha > 0$, $n \in \mathbb{N} : n = \lceil \alpha \rceil$, $[a, b] \subset \mathbb{R}$ *and* $f, \psi \in C^n([a, b])$ *with* ψ *being increasing and* $\psi'(x) \neq 0$, *for all* $x \in [a, b]$. *Let* $x_0 \in [a, b]$ *and assume that* $f_\psi^{[k]}(x_0) = 0$, *for* $k = 1, \ldots, n-1$. *Then*

$$\left| \frac{1}{(\psi(b) - \psi(a))} \int_a^b f(x) \psi'(x) dx - f(x_0) \right| \leq \tag{1.20}$$

$$\frac{1}{(\psi(b) - \psi(a)) \Gamma(\alpha + 2)} \left\{ (\psi(x_0) - \psi(a))^{\alpha+1} \left\| {}^C D_{x_0-}^{\alpha,\psi} f \right\|_{\infty,[a,x_0]} \right.$$

$$\left. + (\psi(b) - \psi(x_0))^{\alpha+1} \left\| {}^C D_{x_0+}^{\alpha,\psi} f \right\|_{\infty,[x_0,b]} \right\} \leq$$

$$\frac{1}{\Gamma(\alpha + 2)} \max \left\{ \left\| {}^C D_{x_0-}^{\alpha,\psi} f \right\|_{\infty,[a,x_0]}, \left\| {}^C D_{x_0+}^{\alpha,\psi} f \right\|_{\infty,[x_0,b]} \right\} (\psi(b) - \psi(a))^\alpha. \tag{1.21}$$

In case of $0 < \alpha \leq 1$, *(1.20)–(1.21) are still valid without any initial conditions.*

Proof By Theorem 1.3 we have that

$$f(x) - f(x_0) = \frac{1}{\Gamma(\alpha)} \int_{x_0}^x \psi'(t) (\psi(x) - \psi(t))^{\alpha-1} \, {}^C D_{x_0+}^{\alpha,\psi} f(t) \, dt, \tag{1.22}$$

$\forall \, x \in [x_0, b]$,
and

$$f(x) - f(x_0) = \frac{1}{\Gamma(\alpha)} \int_x^{x_0} \psi'(t) (\psi(t) - \psi(x))^{\alpha-1} \, {}^C D_{x_0-}^{\alpha,\psi} f(t) \, dt, \tag{1.23}$$

$\forall \, x \in [a, x_0]$.
Hence

$$|f(x) - f(x_0)| \leq \frac{1}{\Gamma(\alpha)} \int_{x_0}^x \psi'(t) (\psi(x) - \psi(t))^{\alpha-1} \left| {}^C D_{x_0+}^{\alpha,\psi} f(t) \right| dt \leq$$

$$\frac{1}{\Gamma(\alpha)} \left(\int_{x_0}^x \psi'(t) (\psi(x) - \psi(t))^{\alpha-1} dt \right) \left\| {}^C D_{x_0+}^{\alpha,\psi} f \right\|_{\infty,[x_0,b]} =$$

$$\frac{\left\| {}^C D_{x_0+}^{\alpha,\psi} f \right\|_{\infty,[x_0,b]}}{\Gamma(\alpha + 1)} (\psi(x) - \psi(x_0))^\alpha. \tag{1.24}$$

That is

$$|f(x)-f(x_0)| \leq \frac{\left\| {}^{C}D_{x_0+}^{\alpha,\psi} f \right\|_{\infty,[x_0,b]}}{\Gamma(\alpha+1)} (\psi(x)-\psi(x_0))^{\alpha}, \tag{1.25}$$

$\forall\, x \in [x_0, b]$.

Similarly, it holds

$$|f(x)-f(x_0)| \leq \frac{1}{\Gamma(\alpha)} \int_x^{x_0} \psi'(t)(\psi(t)-\psi(x))^{\alpha-1} \left| {}^{C}D_{x_0-}^{\alpha,\psi} f(t) \right| dt \leq$$

$$\frac{1}{\Gamma(\alpha)} \left(\int_x^{x_0} \psi'(t)(\psi(t)-\psi(x))^{\alpha-1} dt \right) \left\| {}^{C}D_{x_0-}^{\alpha,\psi} f \right\|_{\infty,[a,x_0]} =$$

$$\frac{\left\| {}^{C}D_{x_0-}^{\alpha,\psi} f \right\|_{\infty,[a,x_0]}}{\Gamma(\alpha+1)} (\psi(x_0)-\psi(x))^{\alpha}.$$

That is

$$|f(x)-f(x_0)| \leq \frac{\left\| {}^{C}D_{x_0-}^{\alpha,\psi} f \right\|_{\infty,[a,x_0]}}{\Gamma(\alpha+1)} (\psi(x_0)-\psi(x))^{\alpha}, \tag{1.26}$$

$\forall\, x \in [a, x_0]$.

We observe that

$$\left| \frac{1}{(\psi(b)-\psi(a))} \int_a^b f(x)\psi'(x)\,dx - f(x_0) \right| =$$

$$\left| \frac{1}{(\psi(b)-\psi(a))} \left[\int_a^b f(x)\psi'(x)\,dx - f(x_0)(\psi(b)-\psi(a)) \right] \right| =$$

$$\left| \frac{1}{(\psi(b)-\psi(a))} \left[\int_a^b f(x)\psi'(x)\,dx - \int_a^b f(x_0)\psi'(x)\,dx \right] \right| =$$

$$\left| \frac{1}{(\psi(b)-\psi(a))} \left[\int_a^b (f(x)-f(x_0))\psi'(x)\,dx \right] \right| \leq$$

$$\frac{1}{(\psi(b)-\psi(a))} \int_a^b |f(x)-f(x_0)|\,\psi'(x)\,dx = \tag{1.27}$$

$$\frac{1}{(\psi(b)-\psi(a))} \left\{ \int_a^{x_0} |f(x)-f(x_0)|\,\psi'(x)\,dx + \int_{x_0}^b |f(x)-f(x_0)|\,\psi'(x)\,dx \right\}$$

$$\overset{\text{(by (1.25), (1.26))}}{\leq}$$

$$\frac{1}{(\psi(b)-\psi(a))\,\Gamma(\alpha+1)}\left\{\left(\int_a^{x_0}(\psi(x_0)-\psi(x))^{\alpha}\psi'(x)\,dx\right)\left\|{}^C D_{x_0-}^{\alpha,\psi}f\right\|_{\infty,[a,x_0]}\right.$$

$$\left.+\left(\int_{x_0}^{b}(\psi(x)-\psi(x_0))^{\alpha}\psi'(x)\,dx\right)\left\|{}^C D_{x_0+}^{\alpha,\psi}f\right\|_{\infty,[x_0,b]}\right\}= \tag{1.28}$$

$$\frac{1}{(\psi(b)-\psi(a))\,\Gamma(\alpha+2)}\left\{\left\|{}^C D_{x_0-}^{\alpha,\psi}f\right\|_{\infty,[a,x_0]}(\psi(x_0)-\psi(a))^{\alpha+1}\right.$$

$$\left.+\left\|{}^C D_{x_0+}^{\alpha,\psi}f\right\|_{\infty,[x_0,b]}(\psi(b)-\psi(x_0))^{\alpha+1}\right\}\leq$$

$$\frac{1}{\Gamma(\alpha+2)}\max\left\{\left\|{}^C D_{x_0-}^{\alpha,\psi}f\right\|_{\infty,[a,x_0]},\left\|{}^C D_{x_0+}^{\alpha,\psi}f\right\|_{\infty,[x_0,b]}\right\}(\psi(b)-\psi(a))^{\alpha}. \tag{1.29}$$

■

We make

Remark 1.7 In our setting, clearly, it is $f_{\psi}^{[n]}\in C([a,b])$. Given $f\in C([a,b])$, by Theorem 4.10, p. 98 of [3], we get that $I_{a+}^{\alpha,\psi}f\in C([a,b])$, and by Theorem 4.11, p. 101 of [3], we get that $I_{b-}^{\alpha,\psi}f\in C([a,b])$.

Therefore, we obtain that ${}^C D_{a+}^{\alpha,\psi}f,\ {}^C D_{b-}^{\alpha,\psi}f\in C([a,b])$.

We continue with

Theorem 1.8 *All as in Theorem 1.6 and* $\alpha\geq 1$. *Then*

$$\left|\frac{1}{(\psi(b)-\psi(a))}\int_a^b f(x)\,\psi'(x)\,dx-f(x_0)\right|\leq \tag{1.30}$$

$$\frac{1}{(\psi(b)-\psi(a))\,\Gamma(\alpha+1)}\left\{(\psi(x_0)-\psi(a))^{\alpha}\left\|{}^C D_{x_0-}^{\alpha,\psi}f\right\|_{L_1([a,x_0],\psi)}\right.$$

$$\left.+(\psi(b)-\psi(x_0))^{\alpha}\left\|{}^C D_{x_0+}^{\alpha,\psi}f\right\|_{L_1([x_0,b],\psi)}\right\}\leq$$

$$\frac{1}{\Gamma(\alpha+1)}\max\left\{\left\|{}^C D_{x_0-}^{\alpha,\psi}f\right\|_{L_1([a,x_0],\psi)},\left\|{}^C D_{x_0+}^{\alpha,\psi}f\right\|_{L_1([x_0,b],\psi)}\right\}(\psi(b)-\psi(a))^{\alpha-1}. \tag{1.31}$$

Proof By (1.22) we obtain:

$$|f(x)-f(x_0)|\leq\frac{(\psi(x)-\psi(x_0))^{\alpha-1}}{\Gamma(\alpha)}\int_{x_0}^{x}\left|{}^C D_{x_0+}^{\alpha,\psi}f(t)\right|d\psi(t)\leq$$

$$\frac{(\psi(x)-\psi(x_0))^{\alpha-1}}{\Gamma(\alpha)}\int_{x_0}^{b}\left|{}^{C}D_{x_0+}^{\alpha,\psi}f(t)\right|d\psi(t)=$$

$$\frac{(\psi(x)-\psi(x_0))^{\alpha-1}}{\Gamma(\alpha)}\left\|{}^{C}D_{x_0+}^{\alpha,\psi}f\right\|_{L_1([x_0,b],\psi)}. \tag{1.32}$$

That is, we get

$$|f(x)-f(x_0)|\leq\frac{(\psi(x)-\psi(x_0))^{\alpha-1}}{\Gamma(\alpha)}\left\|{}^{C}D_{x_0+}^{\alpha,\psi}f\right\|_{L_1([x_0,b],\psi)}, \tag{1.33}$$

$\forall\, x\in[x_0,b]$.

Similarly, by (1.23), we get:

$$|f(x)-f(x_0)|\leq\frac{(\psi(x_0)-\psi(x))^{\alpha-1}}{\Gamma(\alpha)}\int_{x}^{x_0}\left|{}^{C}D_{x_0-}^{\alpha,\psi}f(t)\right|d\psi(t)\leq$$

$$\frac{(\psi(x_0)-\psi(x))^{\alpha-1}}{\Gamma(\alpha)}\int_{a}^{x_0}\left|{}^{C}D_{x_0-}^{\alpha,\psi}f(t)\right|d\psi(t)= \tag{1.34}$$

$$\frac{(\psi(x_0)-\psi(x))^{\alpha-1}}{\Gamma(\alpha)}\left\|{}^{C}D_{x_0-}^{\alpha,\psi}f\right\|_{L_1([a,x_0],\psi)}.$$

That is, we derive

$$|f(x)-f(x_0)|\leq\frac{(\psi(x_0)-\psi(x))^{\alpha-1}}{\Gamma(\alpha)}\left\|{}^{C}D_{x_0-}^{\alpha,\psi}f\right\|_{L_1([a,x_0],\psi)}, \tag{1.35}$$

$\forall\, x\in[a,x_0]$.

As in the proof of Theorem 1.6, we have

$$\left|\frac{1}{(\psi(b)-\psi(a))}\int_a^b f(x)\,\psi'(x)\,dx-f(x_0)\right|\leq$$

$$\frac{1}{(\psi(b)-\psi(a))}\left\{\int_a^{x_0}|f(x)-f(x_0)|\,\psi'(x)\,dx+\int_{x_0}^{b}|f(x)-f(x_0)|\,\psi'(x)\,dx\right\} \tag{1.36}$$

$$\overset{\text{(by (1.33), (1.35))}}{\leq}$$

$$\frac{1}{(\psi(b)-\psi(a))\,\Gamma(\alpha)}\left\{\left(\int_a^{x_0}(\psi(x_0)-\psi(x))^{\alpha-1}d\psi(x)\right)\left\|{}^{C}D_{x_0-}^{\alpha,\psi}f\right\|_{L_1([a,x_0],\psi)}\right.$$

$$+\left(\int_{x_0}^{b}\left(\psi\left(x\right)-\psi\left(x_0\right)\right)^{\alpha-1}d\psi\left(x\right)\right)\left\|{}^{C}D_{x_0+}^{\alpha,\psi}f\right\|_{L_1([x_0,b],\psi)}\Bigg\}=$$

$$\frac{1}{\left(\psi\left(b\right)-\psi\left(a\right)\right)\Gamma\left(\alpha+1\right)}\Bigg\{\left(\psi\left(x_0\right)-\psi\left(a\right)\right)^{\alpha}\left\|{}^{C}D_{x_0-}^{\alpha,\psi}f\right\|_{L_1([a,x_0],\psi)}$$

$$+\left(\psi\left(b\right)-\psi\left(x_0\right)\right)^{\alpha}\left\|{}^{C}D_{x_0+}^{\alpha,\psi}f\right\|_{L_1([x_0,b],\psi)}\Bigg\}\leq \tag{1.37}$$

$$\frac{1}{\Gamma\left(\alpha+1\right)}\max\left\{\left\|{}^{C}D_{x_0-}^{\alpha,\psi}f\right\|_{L_1([a,x_0],\psi)},\left\|{}^{C}D_{x_0+}^{\alpha,\psi}f\right\|_{L_1([x_0,b],\psi)}\right\}\left(\psi\left(b\right)-\psi\left(a\right)\right)^{\alpha-1}.$$

■

Next we present

Theorem 1.9 *All as in Theorem 1.6. Let* $p,q>1:\frac{1}{p}+\frac{1}{q}=1$, *and* $\alpha\geq 1$. *Then*

$$\left|\frac{1}{\left(\psi\left(b\right)-\psi\left(a\right)\right)}\int_{a}^{b}f\left(x\right)\psi'\left(x\right)dx-f\left(x_0\right)\right|\leq$$

$$\frac{1}{\left(\psi\left(b\right)-\psi\left(a\right)\right)\Gamma\left(\alpha\right)\left(p\left(\alpha-1\right)+1\right)^{\frac{1}{p}}\left(\alpha+\frac{1}{p}\right)} \tag{1.38}$$

$$\Bigg\{\left(\psi\left(x_0\right)-\psi\left(a\right)\right)^{\alpha+\frac{1}{p}}\left\|{}^{C}D_{x_0-}^{\alpha,\psi}f\right\|_{L_q([a,x_0],\psi)}+$$

$$\left(\psi\left(b\right)-\psi\left(x_0\right)\right)^{\alpha+\frac{1}{p}}\left\|{}^{C}D_{x_0+}^{\alpha,\psi}f\right\|_{L_q([x_0,b],\psi)}\Bigg\}\leq$$

$$\frac{1}{\Gamma\left(\alpha\right)\left(p\left(\alpha-1\right)+1\right)^{\frac{1}{p}}\left(\alpha+\frac{1}{p}\right)} \tag{1.39}$$

$$\max\left\{\left\|{}^{C}D_{x_0-}^{\alpha,\psi}f\right\|_{L_q([a,x_0],\psi)},\left\|{}^{C}D_{x_0+}^{\alpha,\psi}f\right\|_{L_q([x_0,b],\psi)}\right\}\left(\psi\left(b\right)-\psi\left(a\right)\right)^{\alpha-\frac{1}{q}}.$$

Proof By (1.22) and Hölder's inequality we have

$$\left|f\left(x\right)-f\left(x_0\right)\right|\leq\frac{1}{\Gamma\left(\alpha\right)}\int_{x_0}^{x}\left(\psi\left(x\right)-\psi\left(t\right)\right)^{\alpha-1}\left|{}^{C}D_{x_0+}^{\alpha,\psi}f\left(t\right)\right|d\psi\left(t\right)\leq$$

$$\frac{1}{\Gamma\left(\alpha\right)}\left(\int_{x_0}^{x}\left(\psi\left(x\right)-\psi\left(t\right)\right)^{p(\alpha-1)}d\psi\left(t\right)\right)^{\frac{1}{p}}\left(\int_{x_0}^{x}\left|{}^{C}D_{x_0+}^{\alpha,\psi}f\left(t\right)\right|^{q}d\psi\left(t\right)\right)^{\frac{1}{q}}\leq \tag{1.40}$$

$$\frac{1}{\Gamma(\alpha)}\frac{(\psi(x)-\psi(x_0))^{\alpha-1+\frac{1}{p}}}{(p(\alpha-1)+1)^{\frac{1}{p}}}\left\|^{C}D_{x_0+}^{\alpha,\psi}f\right\|_{L_q([x_0,b],\psi)}.$$

That is

$$|f(x)-f(x_0)|\leq\frac{1}{\Gamma(\alpha)}\frac{(\psi(x)-\psi(x_0))^{\alpha-1+\frac{1}{p}}}{(p(\alpha-1)+1)^{\frac{1}{p}}}\left\|^{C}D_{x_0+}^{\alpha,\psi}f\right\|_{L_q([x_0,b],\psi)},\tag{1.41}$$

$\forall\, x\in[x_0,b]$.

By (1.23) and Hölder's inequality we have

$$|f(x)-f(x_0)|\leq\frac{1}{\Gamma(\alpha)}\int_x^{x_0}(\psi(t)-\psi(x))^{\alpha-1}\left|^{C}D_{x_0-}^{\alpha,\psi}f(t)\right|d\psi(t)\leq$$

$$\frac{1}{\Gamma(\alpha)}\left(\int_x^{x_0}(\psi(t)-\psi(x))^{p(\alpha-1)}d\psi(t)\right)^{\frac{1}{p}}\left(\int_x^{x_0}\left|^{C}D_{x_0-}^{\alpha,\psi}f(t)\right|^{q}d\psi(t)\right)^{\frac{1}{q}}\leq\tag{1.42}$$

$$\frac{1}{\Gamma(\alpha)}\frac{(\psi(x_0)-\psi(x))^{\alpha-1+\frac{1}{p}}}{(p(\alpha-1)+1)^{\frac{1}{p}}}\left\|^{C}D_{x_0-}^{\alpha,\psi}f\right\|_{L_q([a,x_0],\psi)}.$$

That is

$$|f(x)-f(x_0)|\leq\frac{1}{\Gamma(\alpha)}\frac{(\psi(x_0)-\psi(x))^{\alpha-1+\frac{1}{p}}}{(p(\alpha-1)+1)^{\frac{1}{p}}}\left\|^{C}D_{x_0-}^{\alpha,\psi}f\right\|_{L_q([a,x_0],\psi)},\tag{1.43}$$

$\forall\, x\in[a,x_0]$.

As in the proof of Theorem 1.6, we have

$$\left|\frac{1}{(\psi(b)-\psi(a))}\int_a^b f(x)\psi'(x)\,dx-f(x_0)\right|\leq$$

$$\frac{1}{(\psi(b)-\psi(a))}\left\{\int_a^{x_0}|f(x)-f(x_0)|\psi'(x)\,dx+\int_{x_0}^b|f(x)-f(x_0)|\psi'(x)\,dx\right\}\tag{1.44}$$

$$\overset{\text{(by (1.41), (1.43))}}{\leq}\frac{1}{(\psi(b)-\psi(a))\Gamma(\alpha)(p(\alpha-1)+1)^{\frac{1}{p}}}$$

$$\left\{\left(\int_a^{x_0}(\psi(x_0)-\psi(x))^{\alpha-1+\frac{1}{p}}d\psi(x)\right)\left\|^{C}D_{x_0-}^{\alpha,\psi}f\right\|_{L_q([a,x_0],\psi)}\right.$$

$$\left.+\left(\int_{x_0}^b(\psi(x)-\psi(x_0))^{\alpha-1+\frac{1}{p}}d\psi(x)\right)\left\|^{C}D_{x_0+}^{\alpha,\psi}f\right\|_{L_q([x_0,b],\psi)}\right\}=$$

$$\frac{1}{\left(\psi(b)-\psi(a)\right)\Gamma(\alpha)\left(p(\alpha-1)+1\right)^{\frac{1}{p}}\left(\alpha+\frac{1}{p}\right)}$$

$$\left\{\left(\psi(x_0)-\psi(a)\right)^{\alpha+\frac{1}{p}}\left\|{}^{C}D_{x_0-}^{\alpha,\psi}f\right\|_{L_q([a,x_0],\psi)}+\right.$$

$$\left.\left(\psi(b)-\psi(x_0)\right)^{\alpha+\frac{1}{p}}\left\|{}^{C}D_{x_0+}^{\alpha,\psi}f\right\|_{L_q([x_0,b],\psi)}\right\}$$

$$\leq\frac{1}{\Gamma(\alpha)\left(p(\alpha-1)+1\right)^{\frac{1}{p}}\left(\alpha+\frac{1}{p}\right)} \tag{1.45}$$

$$\max\left\{\left\|{}^{C}D_{x_0-}^{\alpha,\psi}f\right\|_{L_q([a,x_0],\psi)},\left\|{}^{C}D_{x_0+}^{\alpha,\psi}f\right\|_{L_q([x_0,b],\psi)}\right\}\left(\psi(b)-\psi(a)\right)^{\alpha-\frac{1}{q}}.$$

■

Corollary 1.10 (to Theorem 1.9) *All as in Theorem 1.6. Here $p=q=2$ and $\alpha\geq 1$. Then*

$$\left|\frac{1}{\left(\psi(b)-\psi(a)\right)}\int_a^b f(x)\,\psi'(x)\,dx-f(x_0)\right|\leq$$

$$\frac{1}{\left(\psi(b)-\psi(a)\right)\Gamma(\alpha)\sqrt{(2\alpha-1)}\left(\alpha+\frac{1}{2}\right)} \tag{1.46}$$

$$\left\{\left(\psi(x_0)-\psi(a)\right)^{\alpha+\frac{1}{2}}\left\|{}^{C}D_{x_0-}^{\alpha,\psi}f\right\|_{L_2([a,x_0],\psi)}+\right.$$

$$\left.\left(\psi(b)-\psi(x_0)\right)^{\alpha+\frac{1}{2}}\left\|{}^{C}D_{x_0+}^{\alpha,\psi}f\right\|_{L_2([x_0,b],\psi)}\right\}$$

$$\leq\frac{1}{\Gamma(\alpha)\sqrt{(2\alpha-1)}\left(\alpha+\frac{1}{2}\right)} \tag{1.47}$$

$$\max\left\{\left\|{}^{C}D_{x_0-}^{\alpha,\psi}f\right\|_{L_2([a,x_0],\psi)},\left\|{}^{C}D_{x_0+}^{\alpha,\psi}f\right\|_{L_2([x_0,b],\psi)}\right\}\left(\psi(b)-\psi(a)\right)^{\alpha-\frac{1}{2}}.$$

Some applications of Theorem 1.6 follow.

In the case of $\psi(x)=e^x$ we get:

Proposition 1.11 *Let $\alpha>0$, $n\in\mathbb{N}:n=\lceil\alpha\rceil$, $[a,b]\subset\mathbb{R}$, $f\in C^n([a,b])$. Let $x_0\in[a,b]$ and assume that $f_{e^x}^{[k]}(x_0)=0$, $k=1,\ldots,n-1$. Then*

$$\left| \frac{1}{\left(e^b - e^a\right)} \int_a^b f(x) e^x dx - f(x_0) \right| \leq \frac{1}{\left(e^b - e^a\right) \Gamma(\alpha + 2)} \tag{1.48}$$

$$\left\{ \left(e^{x_0} - e^a\right)^{\alpha+1} \left\| {}^C D_{x_0-}^{\alpha, e^x} f \right\|_{\infty, [a, x_0]} + \left(e^b - e^{x_0}\right)^{\alpha+1} \left\| {}^C D_{x_0+}^{\alpha, e^x} f \right\|_{\infty, [x_0, b]} \right\} \leq$$

$$\frac{1}{\Gamma(\alpha+2)} \max \left\{ \left\| {}^C D_{x_0-}^{\alpha, e^x} f \right\|_{\infty, [a, x_0]}, \left\| {}^C D_{x_0+}^{\alpha, e^x} f \right\|_{\infty, [x_0, b]} \right\} \left(e^b - e^a\right)^{\alpha}. \tag{1.49}$$

In case of $0 < \alpha \leq 1$, *(1.48)–(1.49) are still valid without any initial conditions.*

In case of $\psi(x) = \ln x$ we obtain:

Proposition 1.12 *Let* $\alpha > 0$, $n \in \mathbb{N} : n = \lceil \alpha \rceil$, $[a, b] \subset (0, +\infty)$, $f \in C^n([a, b])$. *Let* $x_0 \in [a, b]$ *and assume that* $f_{\ln x}^{[k]}(x_0) = 0$, $k = 1, \ldots, n-1$. *Then*

$$\left| \frac{1}{\ln \frac{b}{a}} \int_a^b \frac{f(x)}{x} dx - f(x_0) \right| \leq \frac{1}{\left(\ln \frac{b}{a}\right) \Gamma(\alpha + 2)} \tag{1.50}$$

$$\left\{ \left(\ln \frac{x_0}{a}\right)^{\alpha+1} \left\| {}^C D_{x_0-}^{\alpha, \ln x} f \right\|_{\infty, [a, x_0]} + \left(\ln \frac{b}{x_0}\right)^{\alpha+1} \left\| {}^C D_{x_0+}^{\alpha, \ln x} f \right\|_{\infty, [x_0, b]} \right\} \leq$$

$$\frac{1}{\Gamma(\alpha+2)} \max \left\{ \left\| {}^C D_{x_0-}^{\alpha, \ln x} f \right\|_{\infty, [a, x_0]}, \left\| {}^C D_{x_0+}^{\alpha, \ln x} f \right\|_{\infty, [x_0, b]} \right\} \left(\ln \frac{b}{a}\right)^{\alpha}. \tag{1.51}$$

In case of $0 < \alpha \leq 1$, *(1.50)–(1.51) are still valid without any initial conditions.*

Note. Many other interesting examples of our theorems could follow but we skip this task.

References

1. Almeida, R.: A Caputo fractional derivative of a function with respect to another function. Commun. Nonlinear Sci. Numer. Simul. **44**, 460–481 (2017)
2. Anastassiou, G.A.: Ostrowski type inequalities. Proc. AMS **123**, 3775–3781 (1995)
3. Anastassiou, G.A.: Intelligent Computations: Abstract Fractional Calculus, Inequalities, Approximations. Springer, Heidelberg (2018)
4. Anastassiou, G.: Mixed Caputo ψ-fractional Ostrowski type inequalities (2019). Submitted
5. Ostrowski, A.: Über die Absolutabweichung einer differtentiebaren Funktion von ihrem Integralmittelwert. Comment. Math. Helv. **10**, 226–227 (1938)

Chapter 2
Caputo ψ-Fractional Ostrowski and Grüss Inequalities Involving Several Functions

Very general univariate mixed Caputo ψ-fractional Ostrowski and Grüss type inequalities for several functions are presented. Estimates are with respect to $\|\cdot\|_p$, $1 \leq p \leq \infty$. We give also applications. See also [4].

2.1 Introduction

In 1938, Ostrowski [6] proved the following important inequality.

Theorem 2.1 *Let $f : [a,b] \to \mathbb{R}$ be continuous on $[a,b]$ and differentiable on (a,b) whose derivative $f' : (a,b) \to \mathbb{R}$ is bounded on (a,b), i.e., $\|f'\|_\infty := \sup_{t\in(a,b)} |f'(t)| < +\infty$. Then*

$$\left| \frac{1}{b-a} \int_a^b f(t)\,dt - f(x) \right| \leq \left[\frac{1}{4} + \frac{\left(x - \frac{a+b}{2}\right)^2}{(b-a)^2} \right] \cdot (b-a) \|f'\|_\infty, \tag{2.1}$$

for any $x \in [a,b]$. The constant $\frac{1}{4}$ is the best possible.

Since then there has been a lot of activity around these inequalities with important applications to numerical analysis and probability.

In this chapter we are greatly motivated and inspired by Theorem 2.1, see also [2]. Here we present various ψ-fractional Ostrowski and Grüss type inequalities for several functions and we give interesting applications.

G. A. Anastassiou, *Generalized Fractional Calculus*, Studies in Systems, Decision and Control 305, https://doi.org/10.1007/978-3-030-56962-4_2

2.2 Background

Here we follow [1].

Let $\alpha > 0$, $[a, b] \subset \mathbb{R}$, $f : [a, b] \to \mathbb{R}$ which is integrable and $\psi \in C^1([a, b])$ an increasing function such that $\psi'(x) \neq 0$, for all $x \in [a, b]$. Consider $n = \lceil \alpha \rceil$, the ceiling of α. The left and right fractional integrals are defined, respectively, as follows:

$$I_{a+}^{\alpha,\psi} f(x) := \frac{1}{\Gamma(\alpha)} \int_a^x \psi'(t) (\psi(x) - \psi(t))^{\alpha-1} f(t) \, dt, \tag{2.2}$$

and

$$I_{b-}^{\alpha,\psi} f(x) := \frac{1}{\Gamma(\alpha)} \int_x^b \psi'(t) (\psi(t) - \psi(x))^{\alpha-1} f(t) \, dt, \tag{2.3}$$

for any $x \in [a, b]$, where Γ is the gamma function.

The following semigroup property is valid for fractional integrals: if $\alpha, \beta > 0$, then

$$I_{a+}^{\alpha,\psi} I_{a+}^{\beta,\psi} f(x) = I_{a+}^{\alpha+\beta,\psi} f(x), \text{ and } I_{b-}^{\alpha,\psi} I_{b-}^{\beta,\psi} f(x) = I_{b-}^{\alpha+\beta,\psi} f(x).$$

We mention

Definition 2.2 *([1])* Let $\alpha > 0$, $n \in \mathbb{N}$ such that $n = \lceil \alpha \rceil$, $[a, b] \subset \mathbb{R}$ and $f, \psi \in C^n([a, b])$ with ψ being increasing and $\psi'(x) \neq 0$, for all $x \in [a, b]$. The left ψ-Caputo fractional derivative of f of order α is given by

$$^C D_{a+}^{\alpha,\psi} f(x) := I_{a+}^{n-\alpha,\psi} \left(\frac{1}{\psi'(x)} \frac{d}{dx} \right)^n f(x), \tag{2.4}$$

and the right ψ-Caputo fractional derivative of f is given by

$$^C D_{b-}^{\alpha,\psi} f(x) := I_{b-}^{n-\alpha,\psi} \left(-\frac{1}{\psi'(x)} \frac{d}{dx} \right)^n f(x). \tag{2.5}$$

To simplify notation, we will use the symbol

$$f_{\psi}^{[n]}(x) := \left(\frac{1}{\psi'(x)} \frac{d}{dx} \right)^n f(x), \tag{2.6}$$

with $f_{\psi}^{[0]}(x) = f(x)$.

By the definition, when $\alpha = m \in \mathbb{N}$, we have

$$\begin{gathered} ^C D_{a+}^{\alpha,\psi} f(x) = f_{\psi}^{[m]}(x) \\ \text{and} \\ ^C D_{b-}^{\alpha,\psi} f(x) = (-1)^m f_{\psi}^{[m]}(x). \end{gathered} \tag{2.7}$$

If $\alpha \notin \mathbb{N}$, we have

$$ {}^{C}D_{a+}^{\alpha,\psi} f(x) = \frac{1}{\Gamma(n-\alpha)} \int_a^x \psi'(t)(\psi(x) - \psi(t))^{n-\alpha-1} f_{\psi}^{[n]}(t)\, dt, \tag{2.8} $$

and

$$ {}^{C}D_{b-}^{\alpha,\psi} f(x) = \frac{(-1)^n}{\Gamma(n-\alpha)} \int_x^b \psi'(t)(\psi(t) - \psi(x))^{n-\alpha-1} f_{\psi}^{[n]}(t)\, dt, \tag{2.9} $$

$\forall\, x \in [a, b]$.

In particular, when $\alpha \in (0, 1)$, we have

$$ \begin{aligned} &{}^{C}D_{a+}^{\alpha,\psi} f(x) = \tfrac{1}{\Gamma(1-\alpha)} \textstyle\int_a^x (\psi(x) - \psi(t))^{-\alpha} f'(t)\, dt, \\ &\text{and} \\ &{}^{C}D_{b-}^{\alpha,\psi} f(x) = \tfrac{-1}{\Gamma(1-\alpha)} \textstyle\int_x^b (\psi(t) - \psi(x))^{-\alpha} f'(t)\, dt \end{aligned} \tag{2.10} $$

$\forall\, x \in [a, b]$.

Clearly the above is a generalization of left and right Caputo fractional derivatives. For more see [1].

Still we need from [1] the following left and right fractional Taylor's formulae:

Theorem 2.3 ([1]) *Let $\alpha > 0$, $n \in \mathbb{N}$ such that $n = \lceil \alpha \rceil$, $[a, b] \subset \mathbb{R}$ and $f, \psi \in C^n([a, b])$ with ψ being increasing and $\psi'(x) \neq 0$, for all $x \in [a, b]$. Then, the left fractional Taylor formula follows,*

$$ f(x) = \sum_{k=0}^{n-1} \frac{f_{\psi}^{[k]}(a)}{k!} (\psi(x) - \psi(a))^k + I_{a+}^{\alpha,\psi}\, {}^{C}D_{a+}^{\alpha,\psi} f(x), \tag{2.11} $$

and the right fractional Taylor formula follows,

$$ f(x) = \sum_{k=0}^{n-1} (-1)^k \frac{f_{\psi}^{[k]}(b)}{k!} (\psi(b) - \psi(x))^k + I_{b-}^{\alpha,\psi}\, {}^{C}D_{b-}^{\alpha,\psi} f(x), \tag{2.12} $$

$\forall\, x \in [a, b]$.

In particular, given $\alpha \in (0, 1)$, we have

$$ \begin{aligned} &f(x) = f(a) + I_{a+}^{\alpha,\psi}\, {}^{C}D_{a+}^{\alpha,\psi} f(x), \\ &\textit{and} \\ &f(x) = f(b) + I_{b-}^{\alpha,\psi}\, {}^{C}D_{b-}^{\alpha,\psi} f(x), \end{aligned} \tag{2.13} $$

$\forall\, x \in [a, b]$.

Remark 2.4 For convenience we can rewrite (2.11)–(2.13) as follows:

$$f(x) = \sum_{k=0}^{n-1} \frac{f_{\psi}^{[k]}(a)}{k!} (\psi(x) - \psi(a))^k + \tag{2.14}$$

$$\frac{1}{\Gamma(\alpha)} \int_a^x \psi'(t) (\psi(x) - \psi(t))^{\alpha-1} \, {}^C D_{a+}^{\alpha,\psi} f(t) \, dt,$$

and

$$f(x) = \sum_{k=0}^{n-1} \frac{(-1)^k f_{\psi}^{[k]}(b)}{k!} (\psi(b) - \psi(x))^k + \tag{2.15}$$

$$\frac{1}{\Gamma(\alpha)} \int_x^b \psi'(t) (\psi(t) - \psi(x))^{\alpha-1} \, {}^C D_{b-}^{\alpha,\psi} f(t) \, dt,$$

$\forall \, x \in [a, b]$.

When $\alpha \in (0, 1)$, we get:

$$\begin{array}{l} f(x) = f(a) + \frac{1}{\Gamma(\alpha)} \int_a^x \psi'(t) (\psi(x) - \psi(t))^{\alpha-1} \, {}^C D_{a+}^{\alpha,\psi} f(t) \, dt, \\ \text{and} \\ f(x) = f(b) + \frac{1}{\Gamma(\alpha)} \int_x^b \psi'(t) (\psi(t) - \psi(x))^{\alpha-1} \, {}^C D_{b-}^{\alpha,\psi} f(t) \, dt, \end{array} \tag{2.16}$$

$\forall \, x \in [a, b]$.

Again from [1] we have the following:

Consider the norms $\|\cdot\|_{\infty} : C([a, b]) \to \mathbb{R}$ and $\|\cdot\|_{C_{\psi}^{[n]}} : C^n([a, b]) \to \mathbb{R}$, where $\|f\|_{C_{\psi}^{[n]}} := \sum_{k=0}^{n} \left\| f_{\psi}^{[k]} \right\|_{\infty}$.

We have

Theorem 2.5 ([1]) *The ψ-Caputo fractional derivatives are bounded operators. For all $\alpha > 0$ $(n = \lceil \alpha \rceil)$*

$$\left\| {}^C D_{a+}^{\alpha,\psi} \right\|_{\infty} \leq K \|f\|_{C_{\psi}^{[n]}} \tag{2.17}$$

and

$$\left\| {}^C D_{b-}^{\alpha,\psi} \right\|_{\infty} \leq K \|f\|_{C_{\psi}^{[n]}}, \tag{2.18}$$

where

$$K = \frac{(\psi(b) - \psi(a))^{n-\alpha}}{\Gamma(n + 1 - \alpha)} > 0. \tag{2.19}$$

2.3 Main Results

At first we present the following ψ-fractional Ostrowski type inequalities for several functions:

Theorem 2.6 *Let* $\alpha > 0$, $n \in \mathbb{N} : n = \lceil \alpha \rceil$, $[a, b] \subset \mathbb{R}$ *and* $f_i, \psi \in C^n([a, b])$, $i = 1, \ldots, r$; *with* ψ *being increasing and* $\psi'(x) \neq 0$, *for all* $x \in [a, b]$. *Let* $x_0 \in [a, b]$ *and assume that* $f_{i\psi}^{[k]}(x_0) = 0$, *for* $k = 1, \ldots, n-1$; $i = 1, \ldots, r$. *Set*

$$\Phi(f_1, \ldots, f_r)(x_0) := r \int_a^b \left(\prod_{k=1}^{r} f_k(x) \right) d\psi(x) - \tag{2.20}$$

$$\sum_{i=1}^{r} \left[f_i(x_0) \int_a^b \left(\prod_{\substack{j=1 \\ j \neq i}}^{r} f_j(x) \right) d\psi(x) \right].$$

Then

$$|\Phi(f_1, \ldots, f_r)(x_0)| \leq \sum_{i=1}^{r} \left[\left[\left\| {}^C D_{x_0-}^{\alpha,\psi} f_i \right\|_{\infty,[a,x_0]} I_{a+}^{\alpha+1,\psi} \left(\prod_{\substack{j=1 \\ j \neq i}}^{r} |f_j(x_0)| \right) \right] \right.$$

$$\left. + \left[\left\| {}^C D_{x_0+}^{\alpha,\psi} f_i \right\|_{\infty,[x_0,b]} I_{b-}^{\alpha+1,\psi} \left(\prod_{\substack{j=1 \\ j \neq i}}^{r} |f_j(x_0)| \right) \right] \right]. \tag{2.21}$$

If $0 < \alpha \leq 1$, *then (2.21) is valid without any initial conditions.*

Proof By Theorem 2.3 we have that

$$f_i(x) - f_i(x_0) = \frac{1}{\Gamma(\alpha)} \int_{x_0}^{x} \psi'(t) (\psi(x) - \psi(t))^{\alpha-1} \, {}^C D_{x_0+}^{\alpha,\psi} f_i(t) \, dt, \tag{2.22}$$

$\forall\, x \in [x_0, b]$,
and

$$f_i(x) - f_i(x_0) = \frac{1}{\Gamma(\alpha)} \int_{x}^{x_0} \psi'(t) (\psi(t) - \psi(x))^{\alpha-1} \, {}^C D_{x_0-}^{\alpha,\psi} f_i(t) \, dt, \tag{2.23}$$

$\forall\, x \in [a, x_0]$;
for all $i = 1, \ldots, r$.

Multiplying (2.22) and (2.23) by $\left(\prod_{\substack{j=1\\j\neq i}}^{r} f_j(x)\right)$ we obtain, respectively,

$$\prod_{k=1}^{r} f_k(x) - \left(\prod_{\substack{j=1\\j\neq i}}^{r} f_j(x)\right) f_i(x_0) =$$

$$\frac{\left(\prod_{\substack{j=1\\j\neq i}}^{r} f_j(x)\right)}{\Gamma(\alpha)} \int_{x_0}^{x} \psi'(t)(\psi(x)-\psi(t))^{\alpha-1}\, {}^{C}D_{x_0+}^{\alpha,\psi} f_i(t)\,dt, \tag{2.24}$$

$\forall\, x \in [x_0, b]$,

$$\prod_{k=1}^{r} f_k(x) - \left(\prod_{\substack{j=1\\j\neq i}}^{r} f_j(x)\right) f_i(x_0) =$$

$$\frac{\left(\prod_{\substack{j=1\\j\neq i}}^{r} f_j(x)\right)}{\Gamma(\alpha)} \int_{x}^{x_0} \psi'(t)(\psi(t)-\psi(x))^{\alpha-1}\, {}^{C}D_{x_0-}^{\alpha,\psi} f_i(t)\,dt, \tag{2.25}$$

$\forall\, x \in [a, x_0]$;

for all $i = 1, \ldots, r$.

Adding (2.24) and (2.25), separately, we obtain

$$r\left(\prod_{k=1}^{r} f_k(x)\right) - \sum_{i=1}^{r}\left[\left(\prod_{\substack{j=1\\j\neq i}}^{r} f_j(x)\right) f_i(x_0)\right] =$$

$$\frac{1}{\Gamma(\alpha)}\sum_{i=1}^{r}\left[\left(\prod_{\substack{j=1\\j\neq i}}^{r} f_j(x)\right)\int_{x_0}^{x} \psi'(t)(\psi(x)-\psi(t))^{\alpha-1}\, {}^{C}D_{x_0+}^{\alpha,\psi} f_i(t)\,dt\right], \tag{2.26}$$

$\forall\, x \in [x_0, b]$,

and

$$r\left(\prod_{k=1}^{r} f_k(x)\right) - \sum_{i=1}^{r}\left[\left(\prod_{\substack{j=1\\ j\neq i}}^{r} f_j(x)\right) f_i(x_0)\right] =$$

$$\frac{1}{\Gamma(\alpha)}\sum_{i=1}^{r}\left[\left(\prod_{\substack{j=1\\ j\neq i}}^{r} f_j(x)\right)\int_x^{x_0}\psi'(t)(\psi(t)-\psi(x))^{\alpha-1}\, {}^{C}D_{x_0-}^{\alpha,\psi} f_i(t)\,dt\right], \quad (2.27)$$

$\forall\, x \in [a, x_0]$.

Next we integrate (2.26) and (2.27) with respect to $\psi(x)$, $x \in [a, b]$. We have

$$r\int_{x_0}^{b}\left(\prod_{k=1}^{r} f_k(x)\right) d\psi(x) - \sum_{i=1}^{r}\left[f_i(x_0)\int_{x_0}^{b}\left(\prod_{\substack{j=1\\ j\neq i}}^{r} f_j(x)\right) d\psi(x)\right] =$$

$$\frac{1}{\Gamma(\alpha)}\sum_{i=1}^{r}\left[\int_{x_0}^{b}\left(\prod_{\substack{j=1\\ j\neq i}}^{r} f_j(x)\right)\left[\int_{x_0}^{x}\psi'(t)(\psi(x)-\psi(t))^{\alpha-1}\, {}^{C}D_{x_0+}^{\alpha,\psi} f_i(t)\,dt\right] d\psi(x)\right], \quad (2.28)$$

and

$$r\int_{a}^{x_0}\left(\prod_{k=1}^{r} f_k(x)\right) d\psi(x) - \sum_{i=1}^{r}\left[f_i(x_0)\int_{a}^{x_0}\left(\prod_{\substack{j=1\\ j\neq i}}^{r} f_j(x)\right) d\psi(x)\right] =$$

$$\frac{1}{\Gamma(\alpha)}\sum_{i=1}^{r}\left[\int_{a}^{x_0}\left(\prod_{\substack{j=1\\ j\neq i}}^{r} f_j(x)\right)[\int_{x}^{x_0}\psi'(t)(\psi(t)-\psi(x))^{\alpha-1}\, {}^{C}D_{x_0-}^{\alpha,\psi} f_i(t)\,dt] d\psi(x)\right]. \quad (2.29)$$

Adding (2.28) and (2.29) we derive the identity:

$$\Phi(f_1, \ldots, f_r)(x_0) := r\int_{a}^{b}\left(\prod_{k=1}^{r} f_k(x)\right) d\psi(x) -$$

$$\sum_{i=1}^{r}\left[f_i(x_0)\int_{a}^{b}\left(\prod_{\substack{j=1\\ j\neq i}}^{r} f_j(x)\right) d\psi(x)\right] =$$

$$\frac{1}{\Gamma(\alpha)}\sum_{i=1}^{r}\left[\left[\int_a^{x_0}\left(\prod_{\substack{j=1\\j\neq i}}^{r}f_j(x)\right)\left(\int_x^{x_0}\psi'(t)(\psi(t)-\psi(x))^{\alpha-1}\ {}^{C}D_{x_0-}^{\alpha,\psi}f_i(t)\,dt\right)d\psi(x)\right]\right. \tag{2.30}$$

$$\left.+\left[\int_{x_0}^{b}\left(\prod_{\substack{j=1\\j\neq i}}^{r}f_j(x)\right)\left(\int_{x_0}^{x}\psi'(t)(\psi(x)-\psi(t))^{\alpha-1}\ {}^{C}D_{x_0+}^{\alpha,\psi}f_i(t)\,dt\right)d\psi(x)\right]\right].$$

Hence it holds

$$|\Phi(f_1,\ldots,f_r)(x_0)|\leq$$

$$\frac{1}{\Gamma(\alpha)}\sum_{i=1}^{r}\left[\left[\int_a^{x_0}\left(\prod_{\substack{j=1\\j\neq i}}^{r}|f_j(x)|\right)\left(\int_x^{x_0}\psi'(t)(\psi(t)-\psi(x))^{\alpha-1}\left|{}^{C}D_{x_0-}^{\alpha,\psi}f_i(t)\right|dt\right)d\psi(x)\right]\right.$$

$$\left.+\left[\int_{x_0}^{b}\left(\prod_{\substack{j=1\\j\neq i}}^{r}|f_j(x)|\right)\left(\int_{x_0}^{x}\psi'(t)(\psi(x)-\psi(t))^{\alpha-1}\left|{}^{C}D_{x_0+}^{\alpha,\psi}f_i(t)\right|dt\right)d\psi(x)\right]\right]=:(*). \tag{2.31}$$

We observe that

$$(*)\leq\frac{1}{\Gamma(\alpha+1)}\sum_{i=1}^{r}\left[\left[\left\|{}^{C}D_{x_0-}^{\alpha,\psi}f_i\right\|_{\infty,[a,x_0]}\int_a^{x_0}\left(\prod_{\substack{j=1\\j\neq i}}^{r}|f_j(x)|\right)(\psi(x_0)-\psi(x))^{\alpha}\,d\psi(x)\right]\right.$$

$$\left.+\left[\left\|{}^{C}D_{x_0+}^{\alpha,\psi}f_i\right\|_{\infty,[x_0,b]}\int_{x_0}^{b}\left(\prod_{\substack{j=1\\j\neq i}}^{r}|f_j(x)|\right)(\psi(x)-\psi(x_0))^{\alpha}\,d\psi(x)\right]\right]= \tag{2.32}$$

$$\sum_{i=1}^{r}\left[\left[\left\|{}^{C}D_{x_0-}^{\alpha,\psi}f_i\right\|_{\infty,[a,x_0]}I_{a+}^{\alpha+1,\psi}\left(\prod_{\substack{j=1\\j\neq i}}^{r}|f_j(x_0)|\right)\right]+\right.$$

$$\left.\left[\left\|{}^{C}D_{x_0+}^{\alpha,\psi}f_i\right\|_{\infty,[x_0,b]}I_{b-}^{\alpha+1,\psi}\left(\prod_{\substack{j=1\\j\neq i}}^{r}|f_j(x_0)|\right)\right]\right].$$

By Theorem 4.10, p. 98 of [3], we get that $I_{a+}^{\alpha+1,\psi}\left(\prod_{\substack{j=1\\ j\neq i}}^{r}\left|f_j\right|\right)\in C([a,b])$ and so at any $x_0\in[a,b]$ is finite, $i=1,\dots,r$. Similarly, by Theorem 4.11, p. 101 of [3], we get that $I_{b-}^{\alpha+1,\psi}\left(\prod_{\substack{j=1\\ j\neq i}}^{r}\left|f_j\right|\right)\in C([a,b])$ and so at any $x_0\in[a,b]$ is finite, $i=1,\dots,r$. Arguing similarly, we get that ${}^{C}D_{a+}^{\alpha,\psi}f_i,\ {}^{C}D_{b-}^{\alpha,\psi}f_i\in C([a,b])$, for all $i=1,\dots,r$.

The theorem is proved. ■

We continue with

Theorem 2.7 *All as in Theorem 2.6 with $\alpha\geq 1$. Then*

$$\left|\Phi(f_1,\dots,f_r)(x_0)\right|\leq\sum_{i=1}^{r}\left[\left\|{}^{C}D_{x_0-}^{\alpha,\psi}f_i\right\|_{L_1([a,x_0],\psi)}I_{a+}^{\alpha,\psi}\left(\prod_{\substack{j=1\\ j\neq i}}^{r}\left|f_j(x_0)\right|\right)\right.$$

$$\left.+\left\|{}^{C}D_{x_0+}^{\alpha,\psi}f_i\right\|_{L_1([x_0,b],\psi)}I_{b-}^{\alpha,\psi}\left(\prod_{\substack{j=1\\ j\neq i}}^{r}\left|f_j(x_0)\right|\right)\right]. \tag{2.33}$$

Proof From (2.31) we get

$$(*)\leq\frac{1}{\Gamma(\alpha)}\sum_{i=1}^{r}\left[\left[\left\|{}^{C}D_{x_0-}^{\alpha,\psi}f_i\right\|_{L_1([a,x_0],\psi)}\right.\right.$$

$$\left.\left(\int_a^{x_0}\left(\prod_{\substack{j=1\\ j\neq i}}^{r}\left|f_j(x)\right|\right)(\psi(x_0)-\psi(x))^{\alpha-1}d\psi(x)\right)\right]+$$

$$\left.\left[\left\|{}^{C}D_{x_0+}^{\alpha,\psi}f_i\right\|_{L_1([x_0,b],\psi)}\left(\int_{x_0}^{b}\left(\prod_{\substack{j=1\\ j\neq i}}^{r}\left|f_j(x)\right|\right)(\psi(x)-\psi(x_0))^{\alpha-1}d\psi(x)\right)\right]\right] \tag{2.34}$$

$$=\sum_{i=1}^{r}\left[\left\|{}^{C}D_{x_0-}^{\alpha,\psi}f_i\right\|_{L_1([a,x_0],\psi)}I_{a+}^{\alpha,\psi}\left(\prod_{\substack{j=1\\ j\neq i}}^{r}\left|f_j(x_0)\right|\right)+\right.$$

$$\left\| {}^{C}D_{x_0+}^{\alpha,\psi} f_i \right\|_{L_1([x_0,b],\psi)} I_{b-}^{\alpha,\psi} \left(\prod_{\substack{j=1\\ j\neq i}}^{r} \left| f_j(x_0) \right| \right) \Bigg] , \tag{2.35}$$

proving the theorem. ∎

We continue with

Theorem 2.8 *All as in Theorem 2.6 with $p, q > 1 : \frac{1}{p} + \frac{1}{q} = 1$, $\alpha \geq 1$. Then*

$$\left| \Phi(f_1, \ldots, f_r)(x_0) \right| \leq \frac{\Gamma\left(\alpha + \frac{1}{p}\right)}{\Gamma(\alpha)\left(p(\alpha-1)+1\right)^{\frac{1}{p}}}$$

$$\sum_{i=1}^{r} \left[\left[\left\| {}^{C}D_{x_0-}^{\alpha,\psi} f_i \right\|_{L_q([a,x_0],\psi)} I_{a+}^{\alpha+\frac{1}{p},\psi} \left(\prod_{\substack{j=1\\ j\neq i}}^{r} \left| f_j(x_0) \right| \right) \right] \right. \tag{2.36}$$

$$\left. + \left[\left\| {}^{C}D_{x_0+}^{\alpha,\psi} f_i \right\|_{L_q([x_0,b],\psi)} I_{b-}^{\alpha+\frac{1}{p},\psi} \left(\prod_{\substack{j=1\\ j\neq i}}^{r} \left| f_j(x_0) \right| \right) \right] \right].$$

Proof From (2.31) we obtain

$$(*) \leq \frac{1}{\Gamma(\alpha)} \sum_{i=1}^{r} \left[\left[\int_a^{x_0} \left(\prod_{\substack{j=1\\ j\neq i}}^{r} \left| f_j(x) \right| \right) \right. \right.$$

$$\left(\int_x^{x_0} (\psi(t) - \psi(x))^{p(\alpha-1)} d\psi(t) \right)^{\frac{1}{p}} \left(\int_x^{x_0} \left| {}^{C}D_{x_0-}^{\alpha,\psi} f_i(t) \right|^q d\psi(t) \right)^{\frac{1}{q}} d\psi(x) \Bigg] +$$

$$\left[\int_{x_0}^{b} \left(\prod_{\substack{j=1\\ j\neq i}}^{r} \left| f_j(x) \right| \right) \left(\int_{x_0}^{x} (\psi(x) - \psi(t))^{p(\alpha-1)} d\psi(t) \right)^{\frac{1}{p}} \right. \tag{2.37}$$

$$\left. \left. \left(\int_{x_0}^{x} \left| {}^{C}D_{x_0+}^{\alpha,\psi} f_i(t) \right|^q d\psi(t) \right)^{\frac{1}{q}} d\psi(x) \right] \right] \leq \frac{1}{\left(p(\alpha-1)+1\right)^{\frac{1}{p}} \Gamma(\alpha)}$$

$$\sum_{i=1}^{r} \left[\left[\left\| {}^{C}D_{x_0-}^{\alpha,\psi} f_i \right\|_{L_q([a,x_0],\psi)} \int_a^{x_0} \left(\prod_{\substack{j=1\\ j\neq i}}^{r} \left| f_j(x) \right| \right) (\psi(x_0) - \psi(x))^{\alpha-1+\frac{1}{p}} d\psi(x) \right] \right.$$

$$+\left[\left\|{}^{C}D_{x_0+}^{\alpha,\psi}f_i\right\|_{L_q([x_0,b],\psi)}\int_{x_0}^{b}\left(\prod_{\substack{j=1\\j\neq i}}^{r}|f_j(x)|\right)(\psi(x)-\psi(x_0))^{\alpha-1+\frac{1}{p}}d\psi(x)\right]\Bigg] = \quad (2.38)$$

$$\frac{\Gamma\left(\alpha+\frac{1}{p}\right)}{\Gamma(\alpha)(p(\alpha-1)+1)^{\frac{1}{p}}}\sum_{i=1}^{r}\left[\left[\left\|{}^{C}D_{x_0-}^{\alpha,\psi}f_i\right\|_{L_q([a,x_0],\psi)}I_{a+}^{\alpha+\frac{1}{p},\psi}\left(\prod_{\substack{j=1\\j\neq i}}^{r}|f_j(x_0)|\right)\right]\right.$$

$$\left.+\left[\left\|{}^{C}D_{x_0+}^{\alpha,\psi}f_i\right\|_{L_q([x_0,b],\psi)}I_{b-}^{\alpha+\frac{1}{p},\psi}\left(\prod_{\substack{j=1\\j\neq i}}^{r}|f_j(x_0)|\right)\right]\right], \quad (2.39)$$

proving the theorem. ■

We mention as motivation for Grüss type inequalities the following:

Theorem 2.9 (1882, Čebyšev [5]) *Let $f, g : [a, b] \to \mathbb{R}$ be absolutely continuous functions with $f', g' \in L_\infty([a, b])$. Then*

$$\left|\frac{1}{b-a}\int_a^b f(x)g(x)dx - \left(\frac{1}{b-a}\int_a^b f(x)dx\right)\left(\frac{1}{b-a}\int_a^b g(x)dx\right)\right|$$

$$\leq \frac{1}{12}(b-a)^2\left\|f'\right\|_\infty\left\|g'\right\|_\infty. \quad (2.40)$$

The above integrals are assumed to exist.

Next follow ψ-Caputo fractional Grüss type inequalities for several functions.

Theorem 2.10 *Let $0 < \alpha \leq 1$, $[a, b] \subset \mathbb{R}$ and $f_i, \psi \in C^1([a, b])$, $i = 1, \ldots, r \in \mathbb{N} - \{1\}$; with ψ being increasing and $\psi'(x) \neq 0$, for all $x \in [a, b]$. Assume that $\sup_{x_0\in[a,b]}\left\|{}^{C}D_{x_0-}^{\alpha,\psi}f_i\right\|_{\infty,[a,x_0]} < \infty$, and $\sup_{x_0\in[a,b]}\left\|{}^{C}D_{x_0+}^{\alpha,\psi}f_i\right\|_{\infty,[x_0,b]} < \infty$, $i = 1, \ldots, r$. Set*

$$\Delta^\psi(f_1,\ldots,f_r) := r(\psi(b)-\psi(a))\left(\int_a^b\left(\prod_{k=1}^{r}f_k(x)\right)d\psi(x)\right) -$$

$$\sum_{i=1}^{r}\left[\left(\int_a^b f_i(x)d\psi(x)\right)\left(\int_a^b\left(\prod_{\substack{j=1\\j\neq i}}^{r}f_j(x)\right)d\psi(x)\right)\right]. \quad (2.41)$$

Then

$$\left|\Delta^\psi(f_1,\ldots,f_r)\right| \leq (\psi(b)-\psi(a))$$

$$\left\{ \sum_{i=1}^{r} \left[\left[\sup_{x_0 \in [a,b]} \left\| {}^{C}D_{x_0-}^{\alpha,\psi} f_i \right\|_{\infty,[a,x_0]} \sup_{x_0 \in [a,b]} I_{a+}^{\alpha+1,\psi} \left(\prod_{\substack{j=1 \\ j \neq i}}^{r} \left| f_j (x_0) \right| \right) \right] \right. \right.$$

$$\left. \left. + \left[\sup_{x_0 \in [a,b]} \left\| {}^{C}D_{x_0+}^{\alpha,\psi} f_i \right\|_{\infty,[x_0,b]} \sup_{x_0 \in [a,b]} I_{b-}^{\alpha+1,\psi} \left(\prod_{\substack{j=1 \\ j \neq i}}^{r} \left| f_j (x_0) \right| \right) \right] \right] \right\}. \tag{2.42}$$

Proof Here $0 < \alpha \leq 1$, i.e. $n = 1$. Then (2.30) is valid without any initial conditions. Clearly $\Phi (f_1, \ldots, f_r) \in C ([a, b])$. Thus, by integrating (2.30) against ψ we obtain

$$\Delta^{\psi} (f_1, \ldots, f_r) = \int_a^b \Phi (f_1, \ldots, f_r) (x_0) \, d\psi (x_0) =$$

$$r (\psi (b) - \psi (a)) \left(\int_a^b \left(\prod_{k=1}^{r} f_k (x) \right) d\psi (x) \right) -$$

$$\sum_{i=1}^{r} \left[\left(\int_a^b f_i (x) \, d\psi (x) \right) \left(\int_a^b \left(\prod_{\substack{j=1 \\ j \neq i}}^{r} f_j (x) \right) d\psi (x) \right) \right] =$$

$$\frac{1}{\Gamma (\alpha)} \int_a^b \left\{ \sum_{i=1}^{r} \left[\left[\int_a^{x_0} \left(\prod_{\substack{j=1 \\ j \neq i}}^{r} f_j (x) \right) \right. \right. \right.$$

$$\left. \left(\int_x^{x_0} \psi' (t) (\psi (t) - \psi (x))^{\alpha-1} \, {}^{C}D_{x_0-}^{\alpha,\psi} f_i (t) \, dt \right) d\psi (x) \right] + \tag{2.43}$$

$$\left. \left. \left[\int_{x_0}^b \left(\prod_{\substack{j=1 \\ j \neq i}}^{r} f_j (x) \right) \left(\int_{x_0}^x \psi' (t) (\psi (x) - \psi (t))^{\alpha-1} \, {}^{C}D_{x_0+}^{\alpha,\psi} f_i (t) \, dt \right) d\psi (x) \right] \right] \right\} d\psi (x_0).$$

Hence it holds

$$\left| \Delta^{\psi} (f_1, \ldots, f_r) \right| \leq \frac{1}{\Gamma (\alpha)} \int_a^b \left\{ \sum_{i=1}^{r} \left[\left[\int_a^{x_0} \left(\prod_{\substack{j=1 \\ j \neq i}}^{r} \left| f_j (x) \right| \right) \right. \right. \right.$$

$$\left(\int_x^{x_0} \psi'(t)(\psi(t)-\psi(x))^{\alpha-1}\left|{}^C D_{x_0-}^{\alpha,\psi} f_i(t)\right| dt\right) d\psi(x)\Bigg] + \left[\int_{x_0}^b \left(\prod_{\substack{j=1\\ j\neq i}}^r |f_j(x)|\right)\right.$$

$$\left.\left(\int_{x_0}^x \psi'(t)(\psi(x)-\psi(t))^{\alpha-1}\left|{}^C D_{x_0+}^{\alpha,\psi} f_i(t)\right| dt\right) d\psi(x)\right]\Bigg]\Bigg\} d\psi(x_0) =: (**) . \tag{2.44}$$

Using (2.21) we derive

$$(**) \leq (\psi(b)-\psi(a))$$

$$\Bigg\{\sum_{i=1}^r \Bigg[\Bigg[\sup_{x_0\in[a,b]}\left\|{}^C D_{x_0-}^{\alpha,\psi} f_i\right\|_{\infty,[a,x_0]} \sup_{x_0\in[a,b]} I_{a+}^{\alpha+1,\psi}\left(\prod_{\substack{j=1\\ j\neq i}}^r |f_j(x_0)|\right)\Bigg]$$

$$+\left[\sup_{x_0\in[a,b]}\left\|{}^C D_{x_0+}^{\alpha,\psi} f_i\right\|_{\infty,[x_0,b]} \sup_{x_0\in[a,b]} I_{b-}^{\alpha+1,\psi}\left(\prod_{\substack{j=1\\ j\neq i}}^r |f_j(x_0)|\right)\right]\Bigg]\Bigg\}, \tag{2.45}$$

proving the theorem. ∎

We make

Remark 2.11 Let $\alpha > 0, [a, b] \subset \mathbb{R}$, $f \in C([a, b])$ and $\psi \in C^1([a, b])$ an increasing function such that $\psi'(x) \neq 0$, for all $x \in [a, b]$. Let $x_0 \in [a, b]$. We observe the following

$$\left|I_{a+}^{\alpha,\psi} f(x_0)\right| \overset{(2.2)}{\leq} \frac{1}{\Gamma(\alpha)}\int_a^{x_0} \psi'(t)(\psi(x_0)-\psi(t))^{\alpha-1}|f(t)|\, dt \leq$$

$$\frac{\|f\|_{\infty,[a,x_0]}}{\Gamma(\alpha)}\int_a^{x_0}(\psi(x_0)-\psi(t))^{\alpha-1} d\psi(t) = \frac{\|f\|_{\infty,[a,x_0]}}{\Gamma(\alpha+1)}(\psi(x_0)-\psi(a))^{\alpha} . \tag{2.46}$$

That is

$$\left|I_{a+}^{\alpha,\psi} f(x_0)\right| \leq \frac{\|f\|_{\infty,[a,x_0]}}{\Gamma(\alpha+1)}(\psi(x_0)-\psi(a))^{\alpha} . \tag{2.47}$$

Similarly, we obtain

$$\left|I_{b-}^{\alpha,\psi} f(x_0)\right| \overset{(2.3)}{\leq} \frac{1}{\Gamma(\alpha)} \int_{x_0}^{b} \psi'(t)(\psi(t)-\psi(x_0))^{\alpha-1} |f(t)|\, dt \leq$$

$$\frac{\|f\|_{\infty,[x_0,b]}}{\Gamma(\alpha)} \int_{x_0}^{b} (\psi(t)-\psi(x_0))^{\alpha-1} d\psi(t) = \frac{\|f\|_{\infty,[x_0,b]}}{\Gamma(\alpha+1)} (\psi(b)-\psi(x_0))^{\alpha}. \tag{2.48}$$

That is

$$\left|I_{b-}^{\alpha,\psi} f(x_0)\right| \leq \frac{\|f\|_{\infty,[x_0,b]}}{\Gamma(\alpha+1)} (\psi(b)-\psi(x_0))^{\alpha}. \tag{2.49}$$

We make

Remark 2.12 Let $\alpha \geq 1$, the rest as in Remark 2.11. We observe that

$$\left|I_{a+}^{\alpha,\psi} f(x_0)\right| \overset{(2.2)}{\leq} \frac{1}{\Gamma(\alpha)} \int_{a}^{x_0} \psi'(t)(\psi(x_0)-\psi(t))^{\alpha-1} |f(t)|\, dt \leq$$

$$\frac{(\psi(x_0)-\psi(a))^{\alpha-1}}{\Gamma(\alpha)} \int_{a}^{x_0} |f(t)|\, d\psi(t) = \frac{(\psi(x_0)-\psi(a))^{\alpha-1}}{\Gamma(\alpha)} \|f\|_{L_1([a,x_0],\psi)}. \tag{2.50}$$

That is

$$\left|I_{a+}^{\alpha,\psi} f(x_0)\right| \leq \frac{(\psi(x_0)-\psi(a))^{\alpha-1}}{\Gamma(\alpha)} \|f\|_{L_1([a,x_0],\psi)}. \tag{2.51}$$

Similarly, we get

$$\left|I_{b-}^{\alpha,\psi} f(x_0)\right| \overset{(2.3)}{\leq} \frac{1}{\Gamma(\alpha)} \int_{x_0}^{b} \psi'(t)(\psi(t)-\psi(x_0))^{\alpha-1} |f(t)|\, dt \leq$$

$$\frac{(\psi(b)-\psi(x_0))^{\alpha-1}}{\Gamma(\alpha)} \int_{x_0}^{b} |f(t)|\, d\psi(t) = \frac{(\psi(b)-\psi(x_0))^{\alpha-1}}{\Gamma(\alpha)} \|f\|_{L_1([x_0,b],\psi)}. \tag{2.52}$$

That is

$$\left|I_{b-}^{\alpha,\psi} f(x_0)\right| \leq \frac{(\psi(b)-\psi(x_0))^{\alpha-1}}{\Gamma(\alpha)} \|f\|_{L_1([x_0,b],\psi)}. \tag{2.53}$$

Next, we simplify our main theorems:

Proposition 2.13 *All as in Theorem 2.6. Then*

$$|\Phi(f_1,\ldots,f_r)(x_0)| \leq \frac{1}{\Gamma(\alpha+2)}$$

$$\sum_{i=1}^{r}\left[\left[\left\|{}^{C}D_{x_0-}^{\alpha,\psi}f_i\right\|_{\infty,[a,x_0]}\left\|\prod_{\substack{j=1\\j\neq i}}^{r}f_j\right\|_{\infty,[a,x_0]}(\psi(x_0)-\psi(a))^{\alpha+1}\right]\right.$$

$$\left.+\left[\left\|{}^{C}D_{x_0+}^{\alpha,\psi}f_i\right\|_{\infty,[x_0,b]}\left\|\prod_{\substack{j=1\\j\neq i}}^{r}f_j\right\|_{\infty,[x_0,b]}(\psi(b)-\psi(x_0))^{\alpha+1}\right]\right]. \tag{2.54}$$

If $0<\alpha\leq 1$, *then (2.54) is valid without any initial conditions.*

Proof By (2.21), (2.47) and (2.49). ■

Next comes

Proposition 2.14 *All as in Theorem 2.6 with* $\alpha\geq 1$. *Then*

$$|\Phi(f_1,\ldots,f_r)(x_0)|\leq\frac{1}{\Gamma(\alpha)}$$

$$\sum_{i=1}^{r}\left[\left[\left\|{}^{C}D_{x_0-}^{\alpha,\psi}f_i\right\|_{L_1([a,x_0],\psi)}\left\|\prod_{\substack{j=1\\j\neq i}}^{r}f_j\right\|_{L_1([a,x_0],\psi)}(\psi(x_0)-\psi(a))^{\alpha-1}\right]\right.$$

$$\left.+\left[\left\|{}^{C}D_{x_0+}^{\alpha,\psi}f_i\right\|_{L_1([x_0,b],\psi)}\left\|\prod_{\substack{j=1\\j\neq i}}^{r}f_j\right\|_{L_1([x_0,b],\psi)}(\psi(b)-\psi(x_0))^{\alpha-1}\right]\right]. \tag{2.55}$$

Proof By (2.33), (2.51) and (2.53). ■

Next follows

Proposition 2.15 *All as in Theorem 2.6 with* $p,q>1:\frac{1}{p}+\frac{1}{q}=1$, $\alpha\geq 1$. *Then*

$$|\Phi(f_1,\ldots,f_r)(x_0)|\leq\frac{1}{\Gamma(\alpha)\left(\alpha+\frac{1}{p}\right)(p(\alpha-1)+1)^{\frac{1}{p}}}$$

$$\sum_{i=1}^{r}\left[\left[\left\|{}^{C}D_{x_0-}^{\alpha,\psi}f_i\right\|_{L_q([a,x_0],\psi)}\left\|\prod_{\substack{j=1\\j\neq i}}^{r}f_j\right\|_{\infty,[a,x_0]}(\psi(x_0)-\psi(a))^{\alpha+\frac{1}{p}}\right]\right.$$

$$+\left[\left\|{}^{C}D_{x_0+}^{\alpha,\psi}f_i\right\|_{L_q([x_0,b],\psi)}\left\|\prod_{\substack{j=1\\ j\neq i}}^{r}f_j\right\|_{\infty,[x_0,b]}(\psi(b)-\psi(x_0))^{\alpha+\frac{1}{p}}\right]\right]. \tag{2.56}$$

Proof By (2.36), (2.47) and (2.49). ∎

We continue with

Proposition 2.16 *All as in Theorem 2.10. Then*

$$\left|\Delta^{\psi}(f_1,\dots,f_r)\right|\leq\frac{(\psi(b)-\psi(a))^{\alpha+2}}{\Gamma(\alpha+2)}$$

$$\left\{\sum_{i=1}^{r}\left[\sup_{x_0\in[a,b]}\left\|{}^{C}D_{x_0-}^{\alpha,\psi}f_i\right\|_{\infty,[a,x_0]}+\sup_{x_0\in[a,b]}\left\|{}^{C}D_{x_0+}^{\alpha,\psi}f_i\right\|_{\infty,[x_0,b]}\right]\left\|\prod_{\substack{j=1\\ j\neq i}}^{r}f_j\right\|_{\infty,[a,b]}\right\}. \tag{2.57}$$

Proof By (2.42), (2.47) and (2.49). ∎

Next we make some applications of our main results.
We need

Remark 2.17 We have that $(r=2)$

$$\Phi(f_1,f_2)(x_0)=2\int_a^b f_1(x)f_2(x)d\psi(x)- \tag{2.58}$$

$$f_1(x_0)\int_a^b f_2(x)d\psi(x)-f_2(x_0)\int_a^b f_1(x)d\psi(x),$$

and $(r=3)$

$$\Phi(f_1,f_2,f_3)(x_0)=3\int_a^b f_1(x)f_2(x)f_3(x)d\psi(x)-f_1(x_0)\int_a^b f_2(x)f_3(x)d\psi(x) \tag{2.59}$$

$$-f_2(x_0)\int_a^b f_1(x)f_3(x)d\psi(x)-f_3(x_0)\int_a^b f_1(x)f_2(x)d\psi(x),$$

etc.

Furthermore we derive ($r = 2$)

$$\Delta^{\psi}(f_1, f_2) = 2\left[(\psi(b) - \psi(a))\left(\int_a^b f_1(x) f_2(x) d\psi(x)\right) - \left(\int_a^b f_1(x) d\psi(x)\right)\left(\int_a^b f_2(x) d\psi(x)\right)\right], \tag{2.60}$$

and ($r = 3$)

$$\Delta^{\psi}(f_1, f_2, f_3) = 3(\psi(b) - \psi(a))\left(\int_a^b f_1(x) f_2(x) f_3(x) d\psi(x)\right) - \left(\int_a^b f_1(x) d\psi(x)\right)\left(\int_a^b f_2(x) f_3(x) d\psi(x)\right) - \left(\int_a^b f_2(x) d\psi(x)\right)\left(\int_a^b f_1(x) f_3(x) d\psi(x)\right) - \left(\int_a^b f_3(x) d\psi(x)\right)\left(\int_a^b f_1(x) f_2(x) d\psi(x)\right), \tag{2.61}$$

etc.

We give the special cases of fractional Ostrowski type inequalities.

Proposition 2.18 *Let $\alpha > 0$, $n \in \mathbb{N} : n = \lceil \alpha \rceil$, $[a, b] \subset \mathbb{R}$ and $f_1, f_2 \in C^n([a, b])$. Let $x_0 \in [a, b]$ and assume that $f_{1e^x}^{[k]}(x_0) = f_{2e^x}^{[k]}(x_0) = 0$, for $k = 1, \ldots, n-1$. Then*

$$\left|2\int_a^b f_1(x) f_2(x) e^x dx - f_1(x_0)\int_a^b f_2(x) e^x dx - f_2(x_0)\int_a^b f_1(x) e^x dx\right| \leq$$

$$\frac{1}{\Gamma(\alpha+2)}\sum_{i=1}^{2}\left[\left[\left\|{}^{C}D_{x_0-}^{\alpha,e^x} f_i\right\|_{\infty,[a,x_0]}\left\|\prod_{\substack{j=1\\ j\neq i}}^{2} f_j\right\|_{\infty,[a,x_0]}\left(e^{x_0} - e^a\right)^{\alpha+1}\right] + \left[\left\|{}^{C}D_{x_0+}^{\alpha,e^x} f_i\right\|_{\infty,[x_0,b]}\left\|\prod_{\substack{j=1\\ j\neq i}}^{2} f_j\right\|_{\infty,[x_0,b]}\left(e^b - e^{x_0}\right)^{\alpha+1}\right]\right]. \tag{2.62}$$

If $0 < \alpha \leq 1$, then (2.62) is valid without any initial conditions.

Proof Case of $\psi(x) = e^x$, apply Proposition 2.13 for $r = 2$. ■

We continue with

Proposition 2.19 *Let $\alpha > 0$, $n \in \mathbb{N} : n = \lceil \alpha \rceil$, $[a, b] \subset (0, +\infty)$ and $f_1, f_2, f_3 \in C^n([a, b])$. Let $x_0 \in [a, b]$ and assume that $f_{i\,\ln x}^{[k]}(x_0) = 0$, for $k = 1, \ldots, n-1$; $i = 1, 2, 3$. Then*

$$\left| 3 \int_a^b \frac{f_1(x) f_2(x) f_3(x)}{x} dx - f_1(x_0) \int_a^b \frac{f_2(x) f_3(x)}{x} dx - \right.$$

$$\left. f_2(x_0) \int_a^b \frac{f_1(x) f_3(x)}{x} dx - f_3(x_0) \int_a^b \frac{f_1(x) f_2(x)}{x} dx \right| \leq$$

$$\frac{1}{\Gamma(\alpha+2)} \sum_{i=1}^{3} \left[\left[\left\| {}^C D_{x_0-}^{\alpha, \ln x} f_i \right\|_{\infty, [a, x_0]} \left\| \prod_{\substack{j=1 \\ j \neq i}}^{3} f_j \right\|_{\infty, [a, x_0]} \left(\ln \left(\frac{x_0}{a} \right) \right)^{\alpha+1} \right] \right.$$

$$\left. + \left[\left\| {}^C D_{x_0+}^{\alpha, \ln x} f_i \right\|_{\infty, [x_0, b]} \left\| \prod_{\substack{j=1 \\ j \neq i}}^{3} f_j \right\|_{\infty, [x_0, b]} \left(\ln \left(\frac{b}{x_0} \right) \right)^{\alpha+1} \right] \right]. \tag{2.63}$$

If $0 < \alpha \leq 1$, then (2.63) is valid without any initial conditions.

Proof Case of $\psi(x) = \ln x$, apply Proposition 2.13 for $r = 3$. ■

Next we present the special cases of fractional Grüss type inequality:

Proposition 2.20 *Let $0 < \alpha \leq 1$, $[a, b] \subset \mathbb{R}$ and $f_1, f_2 \in C^1([a, b])$. Assume that $\sup_{x_0 \in [a,b]} \left\| {}^C D_{x_0-}^{\alpha, e^x} f_i \right\|_{\infty, [a, x_0]} < \infty$, and $\sup_{x_0 \in [a,b]} \left\| {}^C D_{x_0+}^{\alpha, e^x} f_i \right\|_{\infty, [x_0, b]} < \infty$, $i = 1, 2$.*
Then

$$2 \left| \left(e^b - e^a\right) \left(\int_a^b f_1(x) f_2(x) e^x dx \right) - \left(\int_a^b f_1(x) e^x dx \right) \left(\int_a^b f_2(x) e^x dx \right) \right| \leq$$

$$\frac{\left(e^b - e^a\right)^{\alpha+2}}{\Gamma(\alpha+2)}$$

$$\left\{ \sum_{i=1}^{2} \left[\sup_{x_0 \in [a,b]} \left\| {}^C D_{x_0-}^{\alpha, e^x} f_i \right\|_{\infty, [a, x_0]} + \sup_{x_0 \in [a,b]} \left\| {}^C D_{x_0+}^{\alpha, e^x} f_i \right\|_{\infty, [x_0, b]} \right] \left\| \prod_{\substack{j=1 \\ j \neq i}}^{2} f_j \right\|_{\infty, [a, b]} \right\}. \tag{2.64}$$

Proof Apply Proposition 2.16, for $\psi(x) = e^x$, $r = 2$. ∎

We finish with another ψ-fractional Grüss type inequality:

Proposition 2.21 *Let* $0 < \alpha \leq 1$, $[a, b] \subset (0, \infty)$ *and* $f_i \in C^1([a, b])$, $i = 1, 2, 3$. *Assume that* $\sup_{x_0 \in [a,b]} \left\| {}^C D_{x_0-}^{\alpha, \ln x} f_i \right\|_{\infty, [a, x_0]} < \infty$, $\sup_{x_0 \in [a,b]} \left\| {}^C D_{x_0+}^{\alpha, \ln x} f_i \right\|_{\infty, [x_0, b]} < \infty$, $i = 1, 2, 3$. *Then*

$$\left| \ln\left(\frac{b}{a}\right)^3 \left(\int_a^b \frac{f_1(x) f_2(x) f_3(x)}{x} dx\right) - \left(\int_a^b \frac{f_1(x)}{x} dx\right)\left(\int_a^b \frac{f_2(x) f_3(x)}{x} dx\right) - \right.$$

$$\left. \left(\int_a^b \frac{f_2(x)}{x} dx\right)\left(\int_a^b \frac{f_1(x) f_3(x)}{x} dx\right) - \left(\int_a^b \frac{f_3(x)}{x} dx\right)\left(\int_a^b \frac{f_1(x) f_2(x)}{x} dx\right) \right| \quad (2.65)$$

$$\leq \frac{\left(\ln\left(\frac{b}{a}\right)\right)^{\alpha+2}}{\Gamma(\alpha+2)} \left\{ \sum_{i=1}^{3} \left[\sup_{x_0 \in [a,b]} \left\| {}^C D_{x_0-}^{\alpha, \ln x} f_i \right\|_{\infty, [a, x_0]} + \right.\right.$$

$$\left.\left. \sup_{x_0 \in [a,b]} \left\| {}^C D_{x_0+}^{\alpha, \ln x} f_i \right\|_{\infty, [x_0, b]} \right] \left\| \prod_{\substack{j=1 \\ j \neq 1}}^{3} f_j \right\|_{\infty, [a,b]} \right\}.$$

Proof By Proposition 2.16, for $\psi(x) = \ln x$, $r = 3$. ∎

References

1. Almeida, R.: A Caputo fractional derivative of a function with respect to another function. Commun. Nonlinear Sci. Numer. Simul. **44**, 460–481 (2017)
2. Anastassiou, G.A.: Ostrowski type inequalities. Proc. AMS **123**, 3775–3781 (1995)
3. Anastassiou, G.A.: Intelligent Computations: Abstract Fractional Calculus, Inequalities, Approximations. Springer, Heidelberg (2018)
4. Anastassiou, G.: Caputo ψ-fractional Ostrowski and Grüss inequalities for several functions. J. Comput. Anal. Appl. **29**(6), 1097–1114 (2021)
5. Čebyšev, P.L.: Sur les expressions approximatives des intégrales définies par les aures prises entre les mêmes limites. Proc. Math. Soc. Charkov **2**, 93–98 (1882)
6. Ostrowski, A.: Über die Absolutabweichung einer differtentiebaren Funktion von ihrem Integralmittelwert. Comment. Math. Helv. **10**, 226–227 (1938)

Chapter 3
Weighted Caputo Fractional Iyengar Type Inequalities

Here we present weighted fractional Iyengar type inequalities with respect to L_p norms, with $1 \leq p \leq \infty$. Our employed fractional calculus is of Caputo type defined with respect to another function. Our results provide quantitative estimates for the approximation of the Lebesgue-Stieljes integral of a function, based on its values over a finite set of points including at the endpoints of its interval of definition. Our method relies on the right and left generalized fractional Taylor's formulae. The iterated generalized fractional derivatives case is also studied. We give applications at the end. See also [3].

3.1 Introduction

We are motivated by the following famous Iyengar inequality (1938), [5].

Theorem 3.1 *Let f be a differentiable function on $[a, b]$ and $\left|f'(x)\right| \leq M$. Then*

$$\left|\int_a^b f(x)\,dx - \frac{1}{2}(b-a)(f(a)+f(b))\right| \leq \frac{M(b-a)^2}{4} - \frac{(f(b)-f(a))^2}{4M}. \tag{3.1}$$

We need

Definition 3.2 *([1])* Let $\alpha > 0$, $\lceil \alpha \rceil = n$, $\lceil \cdot \rceil$ the ceiling of the number. Here $g \in AC([a, b])$ (absolutely continuous functions) and strictly increasing. We assume that $\left(f \circ g^{-1}\right)^{(n)} \circ g \in L_\infty([a, b])$. We define the left generalized g-fractional derivative of f of order α as follows:

G. A. Anastassiou, *Generalized Fractional Calculus*, Studies in Systems, Decision and Control 305, https://doi.org/10.1007/978-3-030-56962-4_3

$$\left(D_{a+;g}^{\alpha}f\right)(x) := \frac{1}{\Gamma(n-\alpha)}\int_a^x (g(x)-g(t))^{n-\alpha-1} g'(t)\left(f \circ g^{-1}\right)^{(n)}(g(t))\,dt, \tag{3.2}$$

$x \geq a$.

If $\alpha \notin \mathbb{N}$, by [2], pp. 360–361, we have that $D_{a+;g}^{\alpha}f \in C([a,b])$.

We see that

$$\left(I_{a+;g}^{n-\alpha}\left(\left(f \circ g^{-1}\right)^{(n)} \circ g\right)\right)(x) = \left(D_{a+;g}^{\alpha}f\right)(x), \quad x \geq a. \tag{3.3}$$

We set

$$D_{a+;g}^{n}f(x) := \left(\left(f \circ g^{-1}\right)^{(n)} \circ g\right)(x), \tag{3.4}$$

$$D_{a+;g}^{0}f(x) = f(x), \quad \forall\, x \in [a,b]. \tag{3.5}$$

When $g = id$, then

$$D_{a+;g}^{\alpha}f = D_{a+;id}^{\alpha}f = D_{*a}^{\alpha}f, \tag{3.6}$$

the usual left Caputo fractional derivative.

We mention the following g-left fractional generalized Taylor's formula:

Theorem 3.3 *([1]) Let g be strictly increasing function and $g \in AC([a,b])$. We assume that $\left(f \circ g^{-1}\right) \in AC^n([g(a), g(b)])$, i.e. $\left(f \circ g^{-1}\right)^{(n-1)} \in AC([g(a), g(b)])$, where $\mathbb{N} \ni n = \lceil \alpha \rceil$, $\alpha > 0$. Also we assume that $\left(f \circ g^{-1}\right)^{(n)} \circ g \in L_{\infty}([a,b])$. Then*

$$f(x) = f(a) + \sum_{k=1}^{n-1} \frac{\left(f \circ g^{-1}\right)^{(k)}(g(a))}{k!}(g(x)-g(a))^k +$$

$$\frac{1}{\Gamma(\alpha)}\int_a^x (g(x)-g(t))^{\alpha-1} g'(t)\left(D_{a+;g}^{\alpha}f\right)(t)\,dt, \quad \forall\, x \in [a,b]. \tag{3.7}$$

Calling $R_n(a,x)$ the remainder of (3.7), we get that

$$R_n(a,x) = \frac{1}{\Gamma(\alpha)}\int_{g(a)}^{g(x)} (g(x)-z)^{\alpha-1}\left(\left(D_{a+;g}^{\alpha}f\right) \circ g^{-1}\right)(z)\,dz, \quad \forall\, x \in [a,b]. \tag{3.8}$$

We need

Definition 3.4 *([1])* Here $g \in AC([a,b])$ and strictly increasing. We assume that $\left(f \circ g^{-1}\right)^{(n)} \circ g \in L_{\infty}([a,b])$, where $N \ni n = \lceil \alpha \rceil, \alpha > 0$. We define the right generalized g-fractional derivative of f of order α as follows:

$$\left(D_{b-;g}^{\alpha}f\right)(x) := \frac{(-1)^n}{\Gamma(n-\alpha)}\int_x^b (g(t)-g(x))^{n-\alpha-1} g'(t)\left(f \circ g^{-1}\right)^{(n)}(g(t))\,dt, \tag{3.9}$$

all $x \in [a, b]$.

If $\alpha \notin \mathbb{N}$, by [2], p. 378, we get that $\left(D_{b-;g}^{\alpha} f\right) \in C([a, b])$.

We see that

$$I_{b-;g}^{n-\alpha}\left((-1)^n \left(f \circ g^{-1}\right)^{(n)} \circ g\right)(x) = \left(D_{b-;g}^{\alpha} f\right)(x), \ a \leq x \leq b. \tag{3.10}$$

We set

$$D_{b-;g}^{n} f(x) = (-1)^n \left(\left(f \circ g^{-1}\right)^{(n)} \circ g\right)(x), \tag{3.11}$$

$$D_{b-;g}^{0} f(x) = f(x), \ \forall x \in [a, b].$$

When $g = id$, then

$$D_{b-;g}^{\alpha} f(x) = D_{b-;id}^{\alpha} f(x) = D_{b-}^{\alpha} f, \tag{3.12}$$

the usual right Caputo fractional derivative.

We mention the g-right generalized fractional Taylor's formula:

Theorem 3.5 ([1]) *Let g be strictly increasing function and $g \in AC([a, b])$. We assume that $\left(f \circ g^{-1}\right) \in AC^n([g(a), g(b)])$, where $\mathbb{N} \ni n = \lceil \alpha \rceil$, $\alpha > 0$. Also we assume that $\left(f \circ g^{-1}\right)^{(n)} \circ g \in L_\infty([a, b])$. Then*

$$f(x) = f(b) + \sum_{k=1}^{n-1} \frac{\left(f \circ g^{-1}\right)^{(k)}(g(b))}{k!} (g(x) - g(b))^k +$$

$$\frac{1}{\Gamma(\alpha)} \int_x^b (g(t) - g(x))^{\alpha-1} g'(t) \left(D_{b-;g}^{\alpha} f\right)(t)\, dt, \ all\ a \leq x \leq b. \tag{3.13}$$

Calling $R_n(b, x)$ the remainder in (3.13), we get that

$$R_n(b, x) = \frac{1}{\Gamma(\alpha)} \int_{g(x)}^{g(b)} (z - g(x))^{\alpha-1} \left(\left(D_{b-;g}^{\alpha} f\right) \circ g^{-1}\right)(z)\, dz, \ \forall x \in [a, b]. \tag{3.14}$$

Denote by

$$D_{b-;g}^{n\alpha} := D_{b-;g}^{\alpha} D_{b-;g}^{\alpha} \dots D_{b-;g}^{\alpha} \ \ (n\text{-times}), n \in \mathbb{N}. \tag{3.15}$$

We mention the following g-right generalized modified Taylor's formula:

Theorem 3.6 ([1]) *Suppose that $F_k := D_{b-;g}^{k\alpha} f$, for $k = 0, 1, \dots, n+1$, fulfill: $F_k \circ g^{-1} \in AC([c, d])$, where $c = g(a)$, $d = g(b)$, and $\left(F_k \circ g^{-1}\right)' \circ g \in L_\infty([a, b])$, where $0 < \alpha \leq 1$. Then*

$$f(x)=\sum_{i=0}^{n}\frac{(g(b)-g(x))^{i\alpha}}{\Gamma(i\alpha+1)}\left(D_{b-;g}^{i\alpha}f\right)(b)+$$

$$\frac{1}{\Gamma((n+1)\alpha)}\int_{x}^{b}(g(t)-g(x))^{(n+1)\alpha-1}g'(t)\left(D_{b-;g}^{(n+1)\alpha}f\right)(t)\,dt= \tag{3.16}$$

$$\sum_{i=0}^{n}\frac{(g(b)-g(x))^{i\alpha}}{\Gamma(i\alpha+1)}\left(D_{b-;g}^{i\alpha}f\right)(b)+\frac{\left(D_{b-;g}^{(n+1)\alpha}f\right)(\psi_x)}{\Gamma((n+1)\alpha+1)}(g(b)-g(x))^{(n+1)\alpha}, \tag{3.17}$$

where $\psi_x\in[x,b]$, *any* $x\in[a,b]$.

Denote by

$$D_{a+;g}^{n\alpha}:=D_{a+;g}^{\alpha}D_{a+;g}^{\alpha}\dots D_{a+;g}^{\alpha}\ (n\text{-times}),\ n\in\mathbb{N}. \tag{3.18}$$

We mention the following g-left generalized modified Taylor's formula:

Theorem 3.7 ([1]) *Suppose that* $F_k:=D_{a+;g}^{k\alpha}f$, *for* $k=0,1,\dots,n+1$, *fulfill:* $F_k\circ g^{-1}\in AC([c,d])$, *where* $c=g(a)$, $d=g(b)$, *and* $\left(F_k\circ g^{-1}\right)'\circ g\in L_\infty([a,b])$, *where* $0<\alpha\leq1$. *Then*

$$f(x)=\sum_{i=0}^{n}\frac{(g(x)-g(a))^{i\alpha}}{\Gamma(i\alpha+1)}\left(D_{a+;g}^{i\alpha}f\right)(a)+ \tag{3.19}$$

$$\frac{1}{\Gamma((n+1)\alpha)}\int_{a}^{x}(g(x)-g(t))^{(n+1)\alpha-1}g'(t)\left(D_{a+;g}^{(n+1)\alpha}f\right)(t)\,dt=$$

$$\sum_{i=0}^{n}\frac{(g(x)-g(a))^{i\alpha}}{\Gamma(i\alpha+1)}\left(D_{a+;g}^{i\alpha}f\right)(a)+\frac{\left(D_{a+;g}^{(n+1)\alpha}f\right)(\psi_x)}{\Gamma((n+1)\alpha+1)}(g(x)-g(a))^{(n+1)\alpha}, \tag{3.20}$$

where $\psi_x\in[a,x]$, *any* $x\in[a,b]$.

Next we present generalized fractional Iyengar type inequalities.

3.2 Main Results

We present the following Caputo type generalized g-fractional Iyengar type inequality:

Theorem 3.8 *Let* g *be strictly increasing function and* $g\in AC([a,b])$. *We assume that* $\left(f\circ g^{-1}\right)\in AC^n([g(a),g(b)])$, *where* $\mathbb{N}\ni n=\lceil\alpha\rceil$, $\alpha>0$. *We also assume that* $\left(f\circ g^{-1}\right)^{(n)}\circ g\in L_\infty([a,b])$ *(clearly here it is* $f\in C([a,b])$*). Then*

(i)

$$\left|\int_a^b f(x)\,dg(x) - \sum_{k=0}^{n-1}\frac{1}{(k+1)!}\left[\left(f\circ g^{-1}\right)^{(k)}(g(a))(g(t)-g(a))^{k+1}\right.\right.$$

$$\left.\left. + (-1)^k\left(f\circ g^{-1}\right)^{(k)}(g(b))(g(b)-g(t))^{k+1}\right]\right| \leq$$

$$\frac{\max\left\{\left\|D_{a+;g}^{\alpha}f\right\|_{L_\infty([a,b])},\left\|D_{b-;g}^{\alpha}f\right\|_{L_\infty([a,b])}\right\}}{\Gamma(\alpha+2)}$$

$$\left[(g(t)-g(a))^{\alpha+1}+(g(b)-g(t))^{\alpha+1}\right], \tag{3.21}$$

$\forall\, t\in[a,b]$,

(ii) *at* $g(t)=\frac{g(a)+g(b)}{2}$, *the right hand side of (3.21) is minimized, and we get:*

$$\left|\int_a^b f(x)\,dg(x) - \sum_{k=0}^{n-1}\frac{1}{(k+1)!}\frac{(g(b)-g(a))^{k+1}}{2^{k+1}}\right.$$

$$\left.\left[\left(f\circ g^{-1}\right)^{(k)}(g(a)) + (-1)^k\left(f\circ g^{-1}\right)^{(k)}(g(b))\right]\right| \leq$$

$$\frac{\max\left\{\left\|D_{a+;g}^{\alpha}f\right\|_{L_\infty([a,b])},\left\|D_{b-;g}^{\alpha}f\right\|_{L_\infty([a,b])}\right\}}{\Gamma(\alpha+2)}\frac{(g(b)-g(a))^{\alpha+1}}{2^{\alpha}}, \tag{3.22}$$

(iii) *if* $\left(f\circ g^{-1}\right)^{(k)}(g(a)) = \left(f\circ g^{-1}\right)^{(k)}(g(b)) = 0$, *for* $k=0,1,\ldots,n-1$, *we obtain*

$$\left|\int_a^b f(x)\,dg(x)\right| \leq$$

$$\max\left\{\left\|D_{a+;g}^{\alpha}f\right\|_{L_\infty([a,b])},\left\|D_{b-;g}^{\alpha}f\right\|_{L_\infty([a,b])}\right\}\frac{(g(b)-g(a))^{\alpha+1}}{\Gamma(\alpha+2)\,2^{\alpha}}, \tag{3.23}$$

which is a sharp inequality,

(iv) *more generally, for* $j=0,1,2,\ldots,N\in\mathbb{N}$, *it holds*

$$\left|\int_a^b f(x)\,dg(x) - \sum_{k=0}^{n-1}\frac{1}{(k+1)!}\left(\frac{g(b)-g(a)}{N}\right)^{k+1}\right.$$

$$\left.\left[j^{k+1}\left(f\circ g^{-1}\right)^{(k)}(g(a)) + (-1)^k(N-j)^{k+1}\left(f\circ g^{-1}\right)^{(k)}(g(b))\right]\right| \leq$$

$$\frac{\max\left\{\left\|D_{a+;g}^{\alpha}f\right\|_{L_{\infty}([a,b])},\left\|D_{b-;g}^{\alpha}f\right\|_{L_{\infty}([a,b])}\right\}}{\Gamma(\alpha+2)}$$

$$\left(\frac{g(b)-g(a)}{N}\right)^{\alpha+1}\left[j^{\alpha+1}+(N-j)^{\alpha+1}\right], \tag{3.24}$$

(v) *if* $\left(f\circ g^{-1}\right)^{(k)}(g(a))=\left(f\circ g^{-1}\right)^{(k)}(g(b))=0$, *for* $k=1,\ldots,n-1$, *from (3.24) we obtain*

$$\left|\int_{a}^{b}f(x)\,dg(x)-\left(\frac{g(b)-g(a)}{N}\right)[jf(a)+(N-j)f(b)]\right|\leq$$

$$\frac{\max\left\{\left\|D_{a+;g}^{\alpha}f\right\|_{L_{\infty}([a,b])},\left\|D_{b-;g}^{\alpha}f\right\|_{L_{\infty}([a,b])}\right\}}{\Gamma(\alpha+2)}$$

$$\left(\frac{g(b)-g(a)}{N}\right)^{\alpha+1}\left[j^{\alpha+1}+(N-j)^{\alpha+1}\right], \tag{3.25}$$

$j=0,1,2,\ldots,N$,

(vi) *when* $N=2$, $j=1$, *(3.25) turns to*

$$\left|\int_{a}^{b}f(x)\,dg(x)-\left(\frac{g(b)-g(a)}{2}\right)(f(a)+f(b))\right|\leq$$

$$\frac{\max\left\{\left\|D_{a+;g}^{\alpha}f\right\|_{L_{\infty}([a,b])},\left\|D_{b-;g}^{\alpha}f\right\|_{L_{\infty}([a,b])}\right\}}{\Gamma(\alpha+2)}\frac{(g(b)-g(a))^{\alpha+1}}{2^{\alpha}}. \tag{3.26}$$

(vii) *when* $0<\alpha\leq 1$, *inequality (3.26) is again valid without any boundary conditions.*

Proof We have by (3.7) that

$$f(x)-\sum_{k=0}^{n-1}\frac{\left(f\circ g^{-1}\right)^{(k)}(g(a))}{k!}(g(x)-g(a))^{k}=$$

$$\frac{1}{\Gamma(\alpha)}\int_{a}^{x}(g(x)-g(t))^{\alpha-1}g'(t)\left(D_{a+;g}^{\alpha}f\right)(t)\,dt, \tag{3.27}$$

$\forall\, x\in[a,b]$.

Also by (3.13) we get

$$f(x) - \sum_{k=0}^{n-1} \frac{\left(f \circ g^{-1}\right)^{(k)}(g(b))}{k!} (g(x) - g(b))^k =$$

$$\frac{1}{\Gamma(\alpha)} \int_x^b (g(t) - g(x))^{\alpha-1} g'(t) \left(D_{b-;g}^{\alpha} f\right)(t) \, dt, \tag{3.28}$$

$\forall\, x \in [a, b]$.

By (3.27) we derive (by [6], p. 107)

$$\left| f(x) - \sum_{k=0}^{n-1} \frac{\left(f \circ g^{-1}\right)^{(k)}(g(a))}{k!} (g(x) - g(a))^k \right| \leq$$

$$\frac{\left\| D_{a+;g}^{\alpha} f \right\|_{L_\infty([a,b])}}{\Gamma(\alpha+1)} (g(x) - g(a))^{\alpha}, \tag{3.29}$$

and by (3.28) we obtain

$$\left| f(x) - \sum_{k=0}^{n-1} \frac{\left(f \circ g^{-1}\right)^{(k)}(g(b))}{k!} (g(x) - g(b))^k \right| \leq$$

$$\frac{\left\| D_{b-;g}^{\alpha} f \right\|_{L_\infty([a,b])}}{\Gamma(\alpha+1)} (g(b) - g(x))^{\alpha}, \tag{3.30}$$

$\forall\, x \in [a, b]$.

Call

$$\varphi_1 := \frac{\left\| D_{a+;g}^{\alpha} f \right\|_{L_\infty([a,b])}}{\Gamma(\alpha+1)}, \tag{3.31}$$

and

$$\varphi_2 := \frac{\left\| D_{b-;g}^{\alpha} f \right\|_{L_\infty([a,b])}}{\Gamma(\alpha+1)}. \tag{3.32}$$

Set

$$\varphi := \max\{\varphi_1, \varphi_2\}. \tag{3.33}$$

That is

$$\left| f(x) - \sum_{k=0}^{n-1} \frac{\left(f \circ g^{-1}\right)^{(k)}(g(a))}{k!} (g(x) - g(a))^k \right| \leq \varphi (g(x) - g(a))^{\alpha},$$

and

$$\left| f(x) - \sum_{k=0}^{n-1} \frac{\left(f \circ g^{-1}\right)^{(k)}(g(b))}{k!}(g(x) - g(b))^k \right| \leq \varphi (g(b) - g(x))^{\alpha}, \quad (3.34)$$

$\forall\, x \in [a, b]$.

Equivalently, we have

$$\sum_{k=0}^{n-1} \frac{\left(f \circ g^{-1}\right)^{(k)}(g(a))}{k!}(g(x) - g(a))^k - \varphi (g(x) - g(a))^{\alpha} \leq \quad (3.35)$$

$$f(x) \leq \sum_{k=0}^{n-1} \frac{\left(f \circ g^{-1}\right)^{(k)}(g(a))}{k!}(g(x) - g(a))^k + \varphi (g(x) - g(a))^{\alpha},$$

and

$$\sum_{k=0}^{n-1} \frac{\left(f \circ g^{-1}\right)^{(k)}(g(b))}{k!}(g(x) - g(b))^k - \varphi (g(b) - g(x))^{\alpha} \leq \quad (3.36)$$

$$f(x) \leq \sum_{k=0}^{n-1} \frac{\left(f \circ g^{-1}\right)^{(k)}(g(b))}{k!}(g(x) - g(b))^k + \varphi (g(b) - g(x))^{\alpha},$$

$\forall\, x \in [a, b]$.

Let any $t \in [a, b]$, then by integration against g over $[a, t]$ and $[t, b]$, respectively, we obtain

$$\sum_{k=0}^{n-1} \frac{\left(f \circ g^{-1}\right)^{(k)}(g(a))}{(k+1)!}(g(t) - g(a))^{k+1} - \frac{\varphi}{(\alpha+1)}(g(t) - g(a))^{\alpha+1}$$

$$\leq \int_a^t f(x)\, dg(x) \leq$$

$$\sum_{k=0}^{n-1} \frac{\left(f \circ g^{-1}\right)^{(k)}(g(a))}{(k+1)!}(g(t) - g(a))^{k+1} + \frac{\varphi}{(\alpha+1)}(g(t) - g(a))^{\alpha+1}, \quad (3.37)$$

and

$$-\sum_{k=0}^{n-1} \frac{\left(f \circ g^{-1}\right)^{(k)}(g(b))}{(k+1)!}(g(t) - g(b))^{k+1} - \frac{\varphi}{(\alpha+1)}(g(b) - g(t))^{\alpha+1}$$

$$\leq \int_t^b f(x)\,dg(x) \leq$$

$$-\sum_{k=0}^{n-1} \frac{\left(f \circ g^{-1}\right)^{(k)}(g(b))}{(k+1)!}(g(t)-g(b))^{k+1} + \frac{\varphi}{(\alpha+1)}(g(b)-g(t))^{\alpha+1}. \tag{3.38}$$

Adding (3.37) and (3.38), we obtain

$$\left\{\sum_{k=0}^{n-1} \frac{1}{(k+1)!}\left[\left(f \circ g^{-1}\right)^{(k)}(g(a))(g(t)-g(a))^{k+1} - \right.\right.$$

$$\left.\left.\left(f \circ g^{-1}\right)^{(k)}(g(b))(g(t)-g(b))^{k+1}\right]\right\} -$$

$$\frac{\varphi}{(\alpha+1)}\left[(g(t)-g(a))^{\alpha+1} + (g(b)-g(t))^{\alpha+1}\right]$$

$$\leq \int_a^b f(x)\,dg(x) \leq$$

$$\left\{\sum_{k=0}^{n-1} \frac{1}{(k+1)!}\left[\left(f \circ g^{-1}\right)^{(k)}(g(a))(g(t)-g(a))^{k+1} - \right.\right.$$

$$\left.\left.\left(f \circ g^{-1}\right)^{(k)}(g(b))(g(t)-g(b))^{k+1}\right]\right\} +$$

$$\frac{\varphi}{(\alpha+1)}\left[(g(t)-g(a))^{\alpha+1} + (g(b)-g(t))^{\alpha+1}\right], \tag{3.39}$$

$\forall\, t \in [a, b]$.

Consequently we derive:

$$\left|\int_a^b f(x)\,dg(x) - \sum_{k=0}^{n-1} \frac{1}{(k+1)!}\left[\left(f \circ g^{-1}\right)^{(k)}(g(a))(g(t)-g(a))^{k+1}\right.\right. \tag{3.40}$$

$$\left.\left. + (-1)^k \left(f \circ g^{-1}\right)^{(k)}(g(b))(g(b)-g(t))^{k+1}\right]\right| \leq$$

$$\frac{\varphi}{(\alpha+1)}\left[(g(t)-g(a))^{\alpha+1} + (g(b)-g(t))^{\alpha+1}\right],$$

$\forall\, t \in [a, b]$.

Let us consider

$$\theta(z) := (z - g(a))^{\alpha+1} + (g(b) - z)^{\alpha+1}, \forall z \in [g(a), g(b)].$$

That is

$$\theta(g(t)) = (g(t) - g(a))^{\alpha+1} + (g(b) - g(t))^{\alpha+1}, \forall t \in [a, b].$$

We have that

$$\theta'(z) = (\alpha+1)\left[(z - g(a))^{\alpha} - (g(b) - z)^{\alpha}\right] = 0,$$

giving $(z - g(a))^{\alpha} = (g(b) - z)^{\alpha}$ and $z - g(a) = g(b) - z$, that is $z = \frac{g(a)+g(b)}{2}$ the only critical number of θ. We have that $\theta(g(a)) = \theta(g(b)) = (g(b) - g(a))^{\alpha+1}$, and $\theta\left(\frac{g(a)+g(b)}{2}\right) = \frac{(g(b)-g(a))^{\alpha+1}}{2^{\alpha}}$, which is the minimum of θ over $[g(a), g(b)]$.

Consequently the right hand side of (3.40) is minimized when $g(t) = \frac{g(a)+g(b)}{2}$, with value $\frac{\varphi}{(\alpha+1)} \frac{(g(b)-g(a))^{\alpha+1}}{2^{\alpha}}$.

Assuming $\left(f \circ g^{-1}\right)^{(k)}(g(a)) = \left(f \circ g^{-1}\right)^{(k)}(g(b)) = 0$, for $k = 0, 1, \ldots, n-1$, then we obtain that

$$\left|\int_a^b f(x)\,dg(x)\right| \leq \frac{\varphi}{(\alpha+1)} \frac{(g(b) - g(a))^{\alpha+1}}{2^{\alpha}}, \tag{3.41}$$

which is a sharp inequality.

When $g(t) = \frac{g(a)+g(b)}{2}$, then (3.40) becomes

$$\left|\int_a^b f(x)\,dg(x) - \sum_{k=0}^{n-1} \frac{1}{(k+1)!} \frac{(g(b) - g(a))^{k+1}}{2^{k+1}}\right.$$

$$\left.\left[\left(f \circ g^{-1}\right)^{(k)}(g(a)) + (-1)^k \left(f \circ g^{-1}\right)^{(k)}(g(b))\right]\right| \leq$$

$$\frac{\varphi}{(\alpha+1)} \frac{(g(b) - g(a))^{\alpha+1}}{2^{\alpha}}. \tag{3.42}$$

Next let $N \in \mathbb{N}$, $j = 0, 1, 2, \ldots, N$ and $g(t_j) = g(a) + j\left(\frac{g(b)-g(a)}{N}\right)$, that is $g(t_0) = g(a)$, $g(t_1) = g(a) + \frac{(g(b)-g(a))}{N}, \ldots, g(t_N) = g(b)$.

Hence it holds

$$g(t_j) - g(a) = j\left(\frac{g(b) - g(a)}{N}\right), \quad g(b) - g(t_j) = (N - j)\left(\frac{g(b) - g(a)}{N}\right), \tag{3.43}$$

$j = 0, 1, 2, \ldots, N$.

We notice

$$\left(g\left(t_j\right)-g\left(a\right)\right)^{\alpha+1}+\left(g\left(b\right)-g\left(t_j\right)\right)^{\alpha+1}=$$

$$\left(\frac{g\left(b\right)-g\left(a\right)}{N}\right)^{\alpha+1}\left[j^{\alpha+1}+\left(N-j\right)^{\alpha+1}\right], \tag{3.44}$$

$j=0,1,2,\ldots,N$,

and (for $k=0,1,\ldots,n-1$)

$$\left[\left(f\circ g^{-1}\right)^{(k)}\left(g\left(a\right)\right)\left(g\left(t_j\right)-g\left(a\right)\right)^{k+1}+\right.$$

$$\left.(-1)^k\left(f\circ g^{-1}\right)^{(k)}\left(g\left(b\right)\right)\left(g\left(b\right)-g\left(t_j\right)\right)^{k+1}\right]=$$

$$\left(\frac{g\left(b\right)-g\left(a\right)}{N}\right)^{k+1}\left[\left(f\circ g^{-1}\right)^{(k)}\left(g\left(a\right)\right)j^{k+1}+\right.$$

$$\left.(-1)^k\left(f\circ g^{-1}\right)^{(k)}\left(g\left(b\right)\right)\left(N-j\right)^{k+1}\right], \tag{3.45}$$

$j=0,1,2,\ldots,N$.

By (3.40) we get

$$\left|\int_a^b f\left(x\right)dg\left(x\right)-\sum_{k=0}^{n-1}\frac{1}{(k+1)!}\left(\frac{g\left(b\right)-g\left(a\right)}{N}\right)^{k+1}\right.$$

$$\left.\left[\left(f\circ g^{-1}\right)^{(k)}\left(g\left(a\right)\right)j^{k+1}+(-1)^k\left(f\circ g^{-1}\right)^{(k)}\left(g\left(b\right)\right)\left(N-j\right)^{k+1}\right]\right|\leq$$

$$\left(\frac{\varphi}{\alpha+1}\right)\left(\frac{g\left(b\right)-g\left(a\right)}{N}\right)^{\alpha+1}\left[j^{\alpha+1}+\left(N-j\right)^{\alpha+1}\right], \tag{3.46}$$

$j=0,1,2,\ldots,N$.

If $\left(f\circ g^{-1}\right)^{(k)}\left(g\left(a\right)\right)=\left(f\circ g^{-1}\right)^{(k)}\left(g\left(b\right)\right)=0$, $k=1,\ldots,n-1$, then (3.46) becomes

$$\left|\int_a^b f\left(x\right)dg\left(x\right)-\left(\frac{g\left(b\right)-g\left(a\right)}{N}\right)\left[jf\left(a\right)+\left(N-j\right)f\left(b\right)\right]\right|\leq$$

$$\left(\frac{\varphi}{\alpha+1}\right)\left(\frac{g\left(b\right)-g\left(a\right)}{N}\right)^{\alpha+1}\left[j^{\alpha+1}+\left(N-j\right)^{\alpha+1}\right], \tag{3.47}$$

$j=0,1,2,\ldots,N$.

When $N=2$ and $j=1$, then (3.47) becomes

$$\left|\int_a^b f(x)\,dg(x) - \left(\frac{g(b)-g(a)}{2}\right)(f(a)+f(b))\right| \leq$$

$$\left(\frac{\varphi}{\alpha+1}\right)2\frac{(g(b)-g(a))^{\alpha+1}}{2^{\alpha+1}} = \left(\frac{\varphi}{\alpha+1}\right)\frac{(g(b)-g(a))^{\alpha+1}}{2^{\alpha}}. \tag{3.48}$$

Let $0 < \alpha \leq 1$, then $n = \lceil \alpha \rceil = 1$.

In that case, without any boundary conditions, we derive from (3.48) again that

$$\left|\int_a^b f(x)\,dg(x) - \left(\frac{g(b)-g(a)}{2}\right)(f(a)+f(b))\right| \leq$$

$$\left(\frac{\varphi}{\alpha+1}\right)\frac{(g(b)-g(a))^{\alpha+1}}{2^{\alpha}}. \tag{3.49}$$

We have proved theorem in all possible cases. ∎

Next we give modified g-fractional Iyengar type inequalities:

Theorem 3.9 *Let g be strictly increasing function and $g \in AC([a,b])$, and $f \in C([a,b])$. Let $0 < \alpha \leq 1$, and $F_k := D_{a+;g}^{k\alpha} f$, for $k = 0, 1, \ldots, n+1$; $n \in \mathbb{N}$. We assume that $F_k \circ g^{-1} \in AC([g(a), g(b)])$ and $\left(F_k \circ g^{-1}\right)' \circ g \in L_\infty([a,b])$. Also let $\overline{F_k} := D_{b-;g}^{k\alpha} f$, for $k = 0, 1, \ldots, n+1$, they fulfill $\overline{F_k} \circ g^{-1} \in AC([g(a), g(b)])$ and $\left(\overline{F_k} \circ g^{-1}\right)' \circ g \in L_\infty([a,b])$. Then*

(i)

$$\left|\int_a^b f(x)\,dg(x) - \left\{\sum_{i=0}^{n} \frac{1}{\Gamma(i\alpha+2)}\left[\left(D_{a+;g}^{i\alpha} f\right)(a)(g(t)-g(a))^{i\alpha+1}\right.\right.\right.$$

$$\left.\left.\left. + \left(D_{b-;g}^{i\alpha} f\right)(b)(g(b)-g(t))^{i\alpha+1}\right]\right\}\right| \leq$$

$$\frac{\max\left\{\left\|D_{a+;g}^{(n+1)\alpha} f\right\|_{\infty,[a,b]}, \left\|D_{b-;g}^{(n+1)\alpha} f\right\|_{\infty,[a,b]}\right\}}{\Gamma((n+1)\alpha+2)}$$

$$\left[(g(t)-g(a))^{(n+1)\alpha+1} + (g(b)-g(t))^{(n+1)\alpha+1}\right], \tag{3.50}$$

$\forall\, t \in [a,b]$,

(ii) at $g(t) = \frac{g(a)+g(b)}{2}$, the right hand side of (3.50) is minimized, and we get:

$$\left|\int_a^b f(x)\,dg(x) - \left\{\sum_{i=0}^{n} \frac{1}{\Gamma(i\alpha+2)} \frac{(g(b)-g(a))^{i\alpha+1}}{2^{i\alpha+1}}\right.\right.$$

$$\left[\left(D_{a+;g}^{i\alpha}f\right)(a)+\left(D_{b-;g}^{i\alpha}f\right)(b)\right]\Big\}\Big|\leq$$

$$\frac{\max\left\{\left\|D_{a+;g}^{(n+1)\alpha}f\right\|_{\infty,[a,b]},\left\|D_{b-;g}^{(n+1)\alpha}f\right\|_{\infty,[a,b]}\right\}}{\Gamma\left((n+1)\alpha+2\right)}\frac{\left(g\left(b\right)-g\left(a\right)\right)^{(n+1)\alpha+1}}{2^{(n+1)\alpha}}, \tag{3.51}$$

(iii) assuming $\left(D_{a+;g}^{i\alpha}f\right)(a)=\left(D_{b-;g}^{i\alpha}f\right)(b)=0$*, for* $i=0,1,\ldots,n$*, we obtain*

$$\left|\int_a^b f(x)\,dg(x)\right|\leq$$

$$\frac{\max\left\{\left\|D_{a+;g}^{(n+1)\alpha}f\right\|_{\infty,[a,b]},\left\|D_{b-;g}^{(n+1)\alpha}f\right\|_{\infty,[a,b]}\right\}}{\Gamma\left((n+1)\alpha+2\right)}\frac{\left(g\left(b\right)-g\left(a\right)\right)^{(n+1)\alpha+1}}{2^{(n+1)\alpha}}, \tag{3.52}$$

which is a sharp inequality,

(iv) more generally, for $j=0,1,2,\ldots,N\in\mathbb{N}$*, it holds*

$$\left|\int_a^b f(x)\,dg(x)-\left\{\sum_{i=0}^{n}\frac{1}{\Gamma(i\alpha+2)}\left(\frac{g(b)-g(a)}{N}\right)^{i\alpha+1}\right.\right.$$

$$\left.\left.\left[\left(D_{a+;g}^{i\alpha}f\right)(a)\,j^{i\alpha+1}+\left(D_{b-;g}^{i\alpha}f\right)(b)\,(N-j)^{i\alpha+1}\right]\right\}\right|\leq$$

$$\frac{\max\left\{\left\|D_{a+;g}^{(n+1)\alpha}f\right\|_{\infty,[a,b]},\left\|D_{b-;g}^{(n+1)\alpha}f\right\|_{\infty,[a,b]}\right\}}{\Gamma\left((n+1)\alpha+2\right)}$$

$$\left(\frac{g(b)-g(a)}{N}\right)^{(n+1)\alpha+1}\left[j^{(n+1)\alpha+1}+(N-j)^{(n+1)\alpha+1}\right], \tag{3.53}$$

(v) if $\left(D_{a+;g}^{i\alpha}f\right)(a)=\left(D_{b-;g}^{i\alpha}f\right)(b)=0$*, for* $i=1,\ldots,n$*, from (3.53) we get:*

$$\left|\int_a^b f(x)\,dg(x)-\left(\frac{g(b)-g(a)}{N}\right)\left[jf(a)+(N-j)\,f(b)\right]\right|\leq$$

$$\frac{\max\left\{\left\|D_{a+;g}^{(n+1)\alpha}f\right\|_{\infty,[a,b]},\left\|D_{b-;g}^{(n+1)\alpha}f\right\|_{\infty,[a,b]}\right\}}{\Gamma\left((n+1)\alpha+2\right)}$$

$$\left(\frac{g(b)-g(a)}{N}\right)^{(n+1)\alpha+1}\left[j^{(n+1)\alpha+1}+(N-j)^{(n+1)\alpha+1}\right], \tag{3.54}$$

for $j = 0, 1, 2, \ldots, N,$

(vi) *when* $N = 2$, $j = 1$, *(3.54) becomes*

$$\left| \int_a^b f(x)\, dg(x) - \left(\frac{g(b) - g(a)}{2} \right) (f(a) + f(b)) \right| \leq$$

$$\frac{\max \left\{ \left\| D_{a+;g}^{(n+1)\alpha} f \right\|_{\infty,[a,b]}, \left\| D_{b-;g}^{(n+1)\alpha} f \right\|_{\infty,[a,b]} \right\}}{\Gamma((n+1)\alpha + 2)} \frac{(g(b) - g(a))^{(n+1)\alpha+1}}{2^{(n+1)\alpha}}. \tag{3.55}$$

Proof We have by (3.19) that

$$f(x) = \sum_{i=0}^{n} \frac{(g(x) - g(a))^{i\alpha}}{\Gamma(i\alpha + 1)} \left(D_{a+;g}^{i\alpha} f \right)(a) +$$

$$\frac{1}{\Gamma((n+1)\alpha)} \int_a^x (g(x) - g(t))^{(n+1)\alpha - 1} g'(t) \left(D_{a+;g}^{(n+1)\alpha} f \right)(t)\, dt, \tag{3.56}$$

$\forall\, x \in [a, b]$.

Also by (3.16) we get

$$f(x) = \sum_{i=0}^{n} \frac{(g(b) - g(x))^{i\alpha}}{\Gamma(i\alpha + 1)} \left(D_{b-;g}^{i\alpha} f \right)(b) +$$

$$\frac{1}{\Gamma((n+1)\alpha)} \int_x^b (g(t) - g(x))^{(n+1)\alpha - 1} g'(t) \left(D_{b-;g}^{(n+1)\alpha} f \right)(t)\, dt, \tag{3.57}$$

$\forall\, x \in [a, b]$.

Clearly here it is $D_{a+;g}^{(n+1)\alpha} f,\ D_{b-;g}^{(n+1)\alpha} f \in C([a, b])$.

By (3.56) we derive (by [6], p. 107)

$$\left| f(x) - \sum_{i=0}^{n} \frac{(g(x) - g(a))^{i\alpha}}{\Gamma(i\alpha + 1)} \left(D_{a+;g}^{i\alpha} f \right)(a) \right| \leq$$

$$\left\| D_{a+;g}^{(n+1)\alpha} f \right\|_{\infty,[a,b]} \frac{(g(x) - g(a))^{(n+1)\alpha}}{\Gamma((n+1)\alpha + 1)}, \tag{3.58}$$

and by (3.57) we obtain

$$\left| f(x) - \sum_{i=0}^{n} \frac{(g(b) - g(x))^{i\alpha}}{\Gamma(i\alpha + 1)} \left(D_{b-;g}^{i\alpha} f \right)(b) \right| \leq$$

$$\left\| D_{b-;g}^{(n+1)\alpha} f \right\|_{\infty,[a,b]} \frac{(g(b) - g(x))^{(n+1)\alpha}}{\Gamma((n+1)\alpha + 1)}, \tag{3.59}$$

$\forall\, x \in [a, b]$.

Call

$$\gamma_1 := \frac{\left\| D_{a+;g}^{(n+1)\alpha} f \right\|_{\infty,[a,b]}}{\Gamma((n+1)\alpha + 1)}, \tag{3.60}$$

and

$$\gamma_2 := \frac{\left\| D_{b-;g}^{(n+1)\alpha} f \right\|_{\infty,[a,b]}}{\Gamma((n+1)\alpha + 1)}. \tag{3.61}$$

Set

$$\gamma := \max\{\gamma_1, \gamma_2\}. \tag{3.62}$$

That is

$$\left| f(x) - \sum_{i=0}^{n} \frac{(g(x) - g(a))^{i\alpha}}{\Gamma(i\alpha + 1)} \left(D_{a+;g}^{i\alpha} f\right)(a) \right| \leq \gamma (g(x) - g(a))^{(n+1)\alpha}, \tag{3.63}$$

and

$$\left| f(x) - \sum_{i=0}^{n} \frac{(g(b) - g(x))^{i\alpha}}{\Gamma(i\alpha + 1)} \left(D_{b-;g}^{i\alpha} f\right)(b) \right| \leq \gamma (g(b) - g(x))^{(n+1)\alpha}, \tag{3.64}$$

$\forall\, x \in [a, b]$.

Equivalently, we have

$$\sum_{i=0}^{n} \frac{(g(x) - g(a))^{i\alpha}}{\Gamma(i\alpha + 1)} \left(D_{a+;g}^{i\alpha} f\right)(a) - \gamma (g(x) - g(a))^{(n+1)\alpha} \leq f(x) \leq$$

$$\sum_{i=0}^{n} \frac{(g(x) - g(a))^{i\alpha}}{\Gamma(i\alpha + 1)} \left(D_{a+;g}^{i\alpha} f\right)(a) + \gamma (g(x) - g(a))^{(n+1)\alpha}, \tag{3.65}$$

and

$$\sum_{i=0}^{n} \frac{(g(b) - g(x))^{i\alpha}}{\Gamma(i\alpha + 1)} \left(D_{b-;g}^{i\alpha} f\right)(b) - \gamma (g(b) - g(x))^{(n+1)\alpha} \leq f(x) \leq$$

$$\sum_{i=0}^{n} \frac{(g(b) - g(x))^{i\alpha}}{\Gamma(i\alpha + 1)} \left(D_{b-;g}^{i\alpha} f\right)(b) + \gamma (g(b) - g(x))^{(n+1)\alpha}, \tag{3.66}$$

$\forall\, x \in [a, b]$.

Let any $t \in [a, b]$, then by integration against g over $[a, t]$ and $[t, b]$, respectively, we obtain

$$\sum_{i=0}^{n} \left(D_{a+;g}^{i\alpha} f\right)(a) \frac{(g(t) - g(a))^{i\alpha+1}}{\Gamma(i\alpha+2)} - \frac{\gamma}{((n+1)\alpha+1)} (g(t) - g(a))^{(n+1)\alpha+1}$$

$$\leq \int_a^t f(x)\, dg(x) \leq$$

$$\sum_{i=0}^{n} \left(D_{a+;g}^{i\alpha} f\right)(a) \frac{(g(t) - g(a))^{i\alpha+1}}{\Gamma(i\alpha+2)} + \frac{\gamma}{((n+1)\alpha+1)} (g(t) - g(a))^{(n+1)\alpha+1}, \tag{3.67}$$

and

$$\sum_{i=0}^{n} \frac{(g(b) - g(t))^{i\alpha+1}}{\Gamma(i\alpha+2)} \left(D_{b-;g}^{i\alpha} f\right)(b) - \frac{\gamma}{((n+1)\alpha+1)} (g(b) - g(t))^{(n+1)\alpha+1}$$

$$\leq \int_t^b f(x)\, dg(x) \leq$$

$$\sum_{i=0}^{n} \frac{(g(b) - g(t))^{i\alpha+1}}{\Gamma(i\alpha+2)} \left(D_{b-;g}^{i\alpha} f\right)(b) + \frac{\gamma}{((n+1)\alpha+1)} (g(b) - g(t))^{(n+1)\alpha+1}. \tag{3.68}$$

Adding (3.67) and (3.68), we obtain

$$\left\{ \sum_{i=0}^{n} \frac{1}{\Gamma(i\alpha+2)} \left[\left(D_{a+;g}^{i\alpha} f\right)(a) (g(t) - g(a))^{i\alpha+1} + \left(D_{b-;g}^{i\alpha} f\right)(b) (g(b) - g(t))^{i\alpha+1} \right] \right\}$$

$$- \frac{\gamma}{((n+1)\alpha+1)} \left[(g(t) - g(a))^{(n+1)\alpha+1} + (g(b) - g(t))^{(n+1)\alpha+1} \right]$$

$$\leq \int_a^b f(x)\, dg(x) \leq$$

$$\left\{ \sum_{i=0}^{n} \frac{1}{\Gamma(i\alpha+2)} \left[\left(D_{a+;g}^{i\alpha} f\right)(a) (g(t) - g(a))^{i\alpha+1} + \left(D_{b-;g}^{i\alpha} f\right)(b) (g(b) - g(t))^{i\alpha+1} \right] \right\}$$

$$+ \frac{\gamma}{((n+1)\alpha+1)} \left[(g(t) - g(a))^{(n+1)\alpha+1} + (g(b) - g(t))^{(n+1)\alpha+1} \right], \tag{3.69}$$

$\forall\, t \in [a, b]$.

Consequently, we derive:

$$\left|\int_a^b f(x)\,dg(x) - \left\{\sum_{i=0}^{n} \frac{1}{\Gamma(i\alpha+2)} \left[\left(D_{a+;g}^{i\alpha} f\right)(a)\,(g(t)-g(a))^{i\alpha+1}\right.\right.\right.$$

$$\left.\left.\left. + \left(D_{b-;g}^{i\alpha} f\right)(b)\,(g(b)-g(t))^{i\alpha+1}\right]\right\}\right| \leq$$

$$\frac{\gamma}{((n+1)\alpha+1)} \left[(g(t)-g(a))^{(n+1)\alpha+1} + (g(b)-g(t))^{(n+1)\alpha+1}\right], \tag{3.70}$$

$\forall\, t \in [a,b]$.

Let us consider

$$\phi(z) := (z-g(a))^{(n+1)\alpha+1} + (g(b)-z)^{(n+1)\alpha+1},$$

$\forall\, z \in [g(a), g(b)]$.

That is

$$\phi(g(t)) = (g(t)-g(a))^{(n+1)\alpha+1} + (g(b)-g(t))^{(n+1)\alpha+1},$$

$\forall\, t \in [a,b]$.

We have that

$$\phi'(z) = ((n+1)\alpha+1)\left[(z-g(a))^{(n+1)\alpha} - (g(b)-z)^{(n+1)\alpha}\right] = 0,$$

giving $(z-g(a))^{(n+1)\alpha} = (g(b)-z)^{(n+1)\alpha}$ and $z-g(a) = g(b)-z$, that is $z = \frac{g(a)+g(b)}{2}$ the only critical number of ϕ. We have that

$$\phi(g(a)) = \phi(g(b)) = (g(b)-g(a))^{(n+1)\alpha+1},$$

and

$$\phi\left(\frac{g(a)+g(b)}{2}\right) = \frac{(g(b)-g(a))^{(n+1)\alpha+1}}{2^{(n+1)\alpha}},$$

which is the minimum of ϕ over $[g(a), g(b)]$.

Consequently, the right hand side of (3.70) is minimized when $g(t) = \frac{g(a)+g(b)}{2}$, for some $t \in [a,b]$, with value $\frac{\gamma}{((n+1)\alpha+1)} \frac{(g(b)-g(a))^{(n+1)\alpha+1}}{2^{(n+1)\alpha}}$.

Assuming $\left(D_{a+;g}^{i\alpha} f\right)(a) = \left(D_{b-;g}^{i\alpha} f\right)(b) = 0$, $i = 0, 1, \ldots, n$, then we obtain that

$$\left|\int_a^b f(x)\,dg(x)\right| \leq \frac{\gamma}{((n+1)\alpha+1)} \frac{(g(b)-g(a))^{(n+1)\alpha+1}}{2^{(n+1)\alpha}}, \tag{3.71}$$

which is a sharp inequality.

When $g(t) = \frac{g(a)+g(b)}{2}$, then (3.70) becomes

$$\left| \int_a^b f(x) \, dg(x) - \left\{ \sum_{i=0}^{n} \frac{1}{\Gamma(i\alpha+2)} \frac{(g(b)-g(a))^{i\alpha+1}}{2^{i\alpha+1}} \right. \right.$$

$$\left. \left. \left[\left(D_{a+;g}^{i\alpha} f \right)(a) + \left(D_{b-;g}^{i\alpha} \right)(b) \right] \right\} \right| \leq$$

$$\frac{\gamma}{((n+1)\alpha+1)} \frac{(g(b)-g(a))^{(n+1)\alpha+1}}{2^{(n+1)\alpha}}. \tag{3.72}$$

Next let $N \in \mathbb{N}$, $j = 0, 1, 2, \ldots, N$ and $g(t_j) = g(a) + j\left(\frac{g(b)-g(a)}{N}\right)$, that is $g(t_0) = g(a)$, $g(t_1) = g(a) + \frac{(g(b)-g(a))}{N}, \ldots, g(t_N) = g(b)$.

Hence it holds

$$g(t_j) - g(a) = j\left(\frac{g(b)-g(a)}{N}\right), g(b) - g(t_j) = (N-j)\left(\frac{g(b)-g(a)}{N}\right), \tag{3.73}$$

$j = 0, 1, 2, \ldots, N.$

We notice

$$\left(g(t_j) - g(a)\right)^{(n+1)\alpha+1} + \left(g(b) - g(t_j)\right)^{(n+1)\alpha+1} =$$

$$\left(\frac{g(b)-g(a)}{N}\right)^{(n+1)\alpha+1} \left[j^{(n+1)\alpha+1} + (N-j)^{(n+1)\alpha+1} \right], \tag{3.74}$$

$j = 0, 1, 2, \ldots, N,$

and (for $i = 0, 1, \ldots, n$)

$$\left[\left(D_{a+;g}^{i\alpha} f \right)(a) \left(g(t_j) - g(a)\right)^{i\alpha+1} + \left(D_{b-;g}^{i\alpha} f \right)(b) \left(g(b) - g(t_j)\right)^{i\alpha+1} \right] =$$

$$\left(\frac{g(b)-g(a)}{N}\right)^{i\alpha+1} \left[\left(D_{a+;g}^{i\alpha} f \right)(a) \, j^{i\alpha+1} + \left(D_{b-;g}^{i\alpha} \right) f(b) (N-j)^{i\alpha+1} \right], \tag{3.75}$$

for $j = 0, 1, 2, \ldots, N$.

By (3.70) we get

$$\left| \int_a^b f(x) \, dg(x) - \left\{ \sum_{i=0}^{n} \frac{1}{\Gamma(i\alpha+2)} \left(\frac{g(b)-g(a)}{N}\right)^{i\alpha+1} \right. \right.$$

$$\left. \left. \left[\left(D_{a+;g}^{i\alpha} f \right)(a) \, j^{i\alpha+1} + \left(D_{b-;g}^{i\alpha} f \right)(b) (N-j)^{i\alpha+1} \right] \right\} \right| \leq$$

$$\frac{\gamma}{((n+1)\alpha+1)}\left(\frac{g(b)-g(a)}{N}\right)^{(n+1)\alpha+1}\left[j^{(n+1)\alpha+1}+(N-j)^{(n+1)\alpha+1}\right], \quad (3.76)$$

$j = 0, 1, 2, \ldots, N$.

If $\left(D_{a+;g}^{i\alpha}f\right)(a) = \left(D_{b-;g}^{i\alpha}f\right)(b) = 0$, $i = 1, \ldots, n$, then (3.76) becomes

$$\left|\int_a^b f(x)\,dg(x) - \left(\frac{g(b)-g(a)}{N}\right)[jf(a)+(N-j)f(b)]\right| \leq$$

$$\frac{\gamma}{((n+1)\alpha+1)}\left(\frac{g(b)-g(a)}{N}\right)^{(n+1)\alpha+1}\left[j^{(n+1)\alpha+1}+(N-j)^{(n+1)\alpha+1}\right], \quad (3.77)$$

$j = 0, 1, 2, \ldots, N$.

When $N = 2$ and $j = 1$, then (3.77) becomes

$$\left|\int_a^b f(x)\,dg(x) - \left(\frac{g(b)-g(a)}{2}\right)(f(a)+f(b))\right| \leq$$

$$\frac{\gamma}{((n+1)\alpha+1)}\frac{2(g(b)-g(a))^{(n+1)\alpha+1}}{2^{(n+1)\alpha+1}} =$$

$$\frac{\gamma}{((n+1)\alpha+1)}\frac{(g(b)-g(a))^{(n+1)\alpha+1}}{2^{(n+1)\alpha}}. \quad (3.78)$$

We have proved theorem in all possible cases. ■

We give L_1 variants of last theorems:

Theorem 3.10 *All as in Theorem 3.8 with $\alpha \geq 1$. If $\alpha = n \in \mathbb{N}$, we assume that $\left(f \circ g^{-1}\right)^{(n)} \circ g \in C([a,b])$. Then*

(i)

$$\left|\int_a^b f(x)\,dg(x) - \sum_{k=0}^{n-1}\frac{1}{(k+1)!}\left[\left(f\circ g^{-1}\right)^{(k)}(g(a))(g(t)-g(a))^{k+1}\right.\right.$$

$$\left.\left.+(-1)^k\left(f\circ g^{-1}\right)^{(k)}(g(b))(g(b)-g(t))^{k+1}\right]\right| \leq$$

$$\frac{\max\left\{\left\|D_{a+;g}^{\alpha}f\right\|_{L_1([a,b],g)}, \left\|D_{b-;g}^{\alpha}f\right\|_{L_1([a,b],g)}\right\}}{\Gamma(\alpha+1)}$$

$$\left[(g(t)-g(a))^{\alpha}+(g(b)-g(t))^{\alpha}\right], \quad (3.79)$$

$\forall\, t \in [a,b]$,

(ii) *at* $g(t) = \frac{g(a)+g(b)}{2}$, *the right hand side of (3.79) is minimized, and we get:*

$$\left| \int_a^b f(x)\, dg(x) - \sum_{k=0}^{n-1} \frac{1}{(k+1)!} \frac{(g(b)-g(a))^{k+1}}{2^{k+1}} \right.$$

$$\left. \left[\left(f \circ g^{-1}\right)^{(k)} (g(a)) + (-1)^k \left(f \circ g^{-1}\right)^{(k)} (g(b)) \right] \right| \leq$$

$$\frac{\max \left\{ \left\| D_{a+;g}^{\alpha} f \right\|_{L_1([a,b],g)}, \left\| D_{b-;g}^{\alpha} f \right\|_{L_1([a,b],g)} \right\}}{\Gamma(\alpha+1)} \frac{(g(b)-g(a))^{\alpha}}{2^{\alpha-1}}, \qquad (3.80)$$

(iii) *if* $\left(f \circ g^{-1}\right)^{(k)} (g(a)) = \left(f \circ g^{-1}\right)^{(k)} (g(b)) = 0$, *for* $k = 0, 1, \ldots, n-1$, *we obtain*

$$\left| \int_a^b f(x)\, dg(x) \right| \leq$$

$$\frac{\max \left\{ \left\| D_{a+;g}^{\alpha} f \right\|_{L_1([a,b],g)}, \left\| D_{b-;g}^{\alpha} f \right\|_{L_1([a,b],g)} \right\}}{\Gamma(\alpha+1)} \frac{(g(b)-g(a))^{\alpha}}{2^{\alpha-1}}, \qquad (3.81)$$

which is a sharp inequality,

(iv) *more generally, for* $j = 0, 1, 2, \ldots, N \in \mathbb{N}$, *it holds*

$$\left| \int_a^b f(x)\, dg(x) - \sum_{k=0}^{n-1} \frac{1}{(k+1)!} \left(\frac{g(b)-g(a)}{N} \right)^{k+1} \right.$$

$$\left. \left[j^{k+1} \left(f \circ g^{-1}\right)^{(k)} (g(a)) + (-1)^k (N-j)^{k+1} \left(f \circ g^{-1}\right)^{(k)} (g(b)) \right] \right| \leq$$

$$\frac{\max \left\{ \left\| D_{a+;g}^{\alpha} f \right\|_{L_1([a,b],g)}, \left\| D_{b-;g}^{\alpha} f \right\|_{L_1([a,b],g)} \right\}}{\Gamma(\alpha+1)}$$

$$\left(\frac{g(b)-g(a)}{N} \right)^{\alpha} \left[j^{\alpha} + (N-j)^{\alpha} \right], \qquad (3.82)$$

(v) *if* $\left(f \circ g^{-1}\right)^{(k)} (g(a)) = \left(f \circ g^{-1}\right)^{(k)} (g(b)) = 0$, *for* $k = 1, \ldots, n-1$, *from (3.82) we obtain*

$$\left| \int_a^b f(x)\, dg(x) - \left(\frac{g(b)-g(a)}{N} \right) [j f(a) + (N-j) f(b)] \right| \leq$$

$$\frac{\max\left\{\left\|D^{\alpha}_{a+;g}f\right\|_{L_1([a,b],g)},\left\|D^{\alpha}_{b-;g}f\right\|_{L_1([a,b],g)}\right\}}{\Gamma(\alpha+1)}$$

$$\left(\frac{g(b)-g(a)}{N}\right)^{\alpha}\left[j^{\alpha}+(N-j)^{\alpha}\right], \tag{3.83}$$

$j=0,1,2,\ldots,N,$

(vi) *when* $N=2$, $j=1$, *(3.83) turns to*

$$\left|\int_a^b f(x)\,dg(x)-\left(\frac{g(b)-g(a)}{2}\right)(f(a)+f(b))\right|\leq$$

$$\frac{\max\left\{\left\|D^{\alpha}_{a+;g}f\right\|_{L_1([a,b],g)},\left\|D^{\alpha}_{b-;g}f\right\|_{L_1([a,b],g)}\right\}}{\Gamma(\alpha+1)}\frac{(g(b)-g(a))^{\alpha}}{2^{\alpha-1}}. \tag{3.84}$$

Proof From (3.27) we get

$$\left|f(x)-\sum_{k=0}^{n-1}\frac{\left(f\circ g^{-1}\right)^{(k)}(g(a))}{k!}(g(x)-g(a))^{k}\right|\leq$$

$$\frac{1}{\Gamma(\alpha)}\int_a^x (g(x)-g(t))^{\alpha-1}g'(t)\left|\left(D^{\alpha}_{a+;g}f\right)(t)\right|dt\leq$$

$$\frac{(g(x)-g(a))^{\alpha-1}}{\Gamma(\alpha)}\int_a^x g'(t)\left|\left(D^{\alpha}_{a+;g}f\right)(t)\right|dt\leq \tag{3.85}$$

$$\frac{(g(x)-g(a))^{\alpha-1}}{\Gamma(\alpha)}\int_a^b g'(t)\left|\left(D^{\alpha}_{a+;g}f\right)(t)\right|dt=$$

$$\frac{(g(x)-g(a))^{\alpha-1}}{\Gamma(\alpha)}\int_a^b \left|\left(D^{\alpha}_{a+;g}f\right)(t)\right|dg(t)=$$

$$\frac{\left\|D^{\alpha}_{a+;g}f\right\|_{L_1([a,b],g)}}{\Gamma(\alpha)}(g(x)-g(a))^{\alpha-1},$$

$\forall\, x\in[a,b]$.

Similarly, from (3.28) we obtain

$$\left| f(x) - \sum_{k=0}^{n-1} \frac{\left(f \circ g^{-1}\right)^{(k)}(g(b))}{k!} (g(x) - g(b))^k \right| \leq$$

$$\frac{1}{\Gamma(\alpha)} \int_x^b (g(t) - g(x))^{\alpha-1} g'(t) \left| \left(D_{b-;g}^{\alpha} f\right)(t) \right| dt \leq$$

$$\frac{(g(b) - g(x))^{\alpha-1}}{\Gamma(\alpha)} \int_x^b \left| \left(D_{b-;g}^{\alpha} f\right)(t) \right| dg(t) \leq \tag{3.86}$$

$$\frac{\left\| D_{b-;g}^{\alpha} f \right\|_{L_1([a,b],g)}}{\Gamma(\alpha)} (g(b) - g(x))^{\alpha-1},$$

$\forall\, x \in [a, b]$.

Call

$$\delta := \max\left\{ \left\| D_{a+;g}^{\alpha} f \right\|_{L_1([a,b],g)}, \left\| D_{b-;g}^{\alpha} f \right\|_{L_1([a,b],g)} \right\}. \tag{3.87}$$

We have proved that

$$\left| f(x) - \sum_{k=0}^{n-1} \frac{\left(f \circ g^{-1}\right)^{(k)}(g(a))}{k!} (g(x) - g(a))^k \right| \leq$$

$$\frac{\delta}{\Gamma(\alpha)} (g(x) - g(a))^{\alpha-1}, \tag{3.88}$$

and

$$\left| f(x) - \sum_{k=0}^{n-1} \frac{\left(f \circ g^{-1}\right)^{(k)}(g(b))}{k!} (g(x) - g(b))^k \right| \leq$$

$$\frac{\delta}{\Gamma(\alpha)} (g(b) - g(x))^{\alpha-1}, \tag{3.89}$$

$\forall\, x \in [a, b]$.

The rest of the proof is as in Theorem 3.8. ■

It follows

Theorem 3.11 *All as in Theorem 3.9, with* $\frac{1}{n+1} \leq \alpha \leq 1$. *Call*

$$\rho := \max\left\{ \left\| D_{a+;g}^{(n+1)\alpha} f \right\|_{L_1([a,b],g)}, \left\| D_{b-;g}^{(n+1)\alpha} f \right\|_{L_1([a,b],g)} \right\}. \tag{3.90}$$

Then

(i)

$$\left|\int_a^b f(x)\,dg(x) - \left\{\sum_{i=0}^{n} \frac{1}{\Gamma(i\alpha+2)} \left[\left(D_{a+;g}^{i\alpha} f\right)(a)\,(g(t)-g(a))^{i\alpha+1}\right.\right.\right.$$

$$\left.\left.\left. + \left(D_{b-;g}^{i\alpha} f\right)(b)\,(g(b)-g(t))^{i\alpha+1}\right]\right\}\right| \leq$$

$$\frac{\rho}{\Gamma((n+1)\alpha+1)} \left[(g(t)-g(a))^{(n+1)\alpha} + (g(b)-g(t))^{(n+1)\alpha}\right], \tag{3.91}$$

$\forall\, t \in [a,b]$,

(ii) *at* $g(t) = \frac{g(a)+g(b)}{2}$, *the right hand side of (3.91) is minimized, and we get:*

$$\left|\int_a^b f(x)\,dg(x) - \left\{\sum_{i=0}^{n} \frac{1}{\Gamma(i\alpha+2)} \frac{(g(b)-g(a))^{i\alpha+1}}{2^{i\alpha+1}}\right.\right.$$

$$\left.\left.\left[\left(D_{a+;g}^{i\alpha} f\right)(a) + \left(D_{b-;g}^{i\alpha} f\right)(b)\right]\right\}\right| \leq$$

$$\frac{\rho}{\Gamma((n+1)\alpha+1)} \frac{(g(b)-g(a))^{(n+1)\alpha}}{2^{(n+1)\alpha-1}}, \tag{3.92}$$

(iii) *assuming* $\left(D_{a+;g}^{i\alpha} f\right)(a) = \left(D_{b-;g}^{i\alpha} f\right)(b) = 0$, $i = 0, 1, \ldots, n$, *we obtain*

$$\left|\int_a^b f(x)\,dg(x)\right| \leq$$

$$\frac{\rho}{\Gamma((n+1)\alpha+1)} \frac{(g(b)-g(a))^{(n+1)\alpha}}{2^{(n+1)\alpha-1}}, \tag{3.93}$$

which is a sharp inequality,

(iv) *more generally, for* $j = 0, 1, 2, \ldots, N \in \mathbb{N}$, *it holds*

$$\left|\int_a^b f(x)\,dg(x) - \left\{\sum_{i=0}^{n} \frac{1}{\Gamma(i\alpha+2)} \left(\frac{g(b)-g(a)}{N}\right)^{i\alpha+1}\right.\right.$$

$$\left.\left.\left[\left(D_{a+;g}^{i\alpha} f\right)(a)\, j^{i\alpha+1} + \left(D_{b-;g}^{i\alpha} f\right)(b)\,(N-j)^{i\alpha+1}\right]\right\}\right| \leq$$

$$\frac{\rho}{\Gamma((n+1)\alpha+1)} \left(\frac{g(b)-g(a)}{N}\right)^{(n+1)\alpha} \left[j^{(n+1)\alpha} + (N-j)^{(n+1)\alpha}\right], \tag{3.94}$$

(v) if $\left(D_{a+;g}^{i\alpha}f\right)(a)=\left(D_{b-;g}^{i\alpha}f\right)(b)=0$, $i=1,\ldots,n$, *from (3.94) we get:*

$$\left|\int_a^b f(x)\,dg(x)-\left(\frac{g(b)-g(a)}{N}\right)[jf(a)+(N-j)f(b)]\right|\leq$$

$$\frac{\rho}{\Gamma((n+1)\alpha+1)}\left(\frac{g(b)-g(a)}{N}\right)^{(n+1)\alpha}\left[j^{(n+1)\alpha}+(N-j)^{(n+1)\alpha}\right], \tag{3.95}$$

for $j=0,1,2,\ldots,N$,

(vi) when $N=2$ *and* $j=1$, *(3.95) becomes*

$$\left|\int_a^b f(x)\,dg(x)-\left(\frac{g(b)-g(a)}{2}\right)(f(a)+f(b))\right|\leq$$

$$\frac{\rho}{\Gamma((n+1)\alpha+1)}\frac{(g(b)-g(a))^{(n+1)\alpha}}{2^{(n+1)\alpha-1}}. \tag{3.96}$$

Proof By (3.56) we get

$$\left|f(x)-\sum_{i=0}^{n}\frac{(g(x)-g(a))^{i\alpha}}{\Gamma(i\alpha+1)}\left(D_{a+;g}^{i\alpha}f\right)(a)\right|\leq$$

$$\frac{1}{\Gamma((n+1)\alpha)}\int_a^x(g(x)-g(t))^{(n+1)\alpha-1}g'(t)\left|\left(D_{a+;g}^{(n+1)\alpha}f\right)(t)\right|dt\leq$$

$$\frac{(g(x)-g(a))^{(n+1)\alpha-1}}{\Gamma((n+1)\alpha)}\int_a^x g'(t)\left|\left(D_{a+;g}^{(n+1)\alpha}f\right)(t)\right|dt\leq \tag{3.97}$$

$$\frac{(g(x)-g(a))^{(n+1)\alpha-1}}{\Gamma((n+1)\alpha)}\int_a^b g'(t)\left|\left(D_{a+;g}^{(n+1)\alpha}f\right)(t)\right|dt=$$

$$\frac{(g(x)-g(a))^{(n+1)\alpha-1}}{\Gamma((n+1)\alpha)}\int_a^b\left|\left(D_{a+;g}^{(n+1)\alpha}f\right)(t)\right|dg(t)=$$

$$\frac{\left\|D_{a+;g}^{(n+1)\alpha}f\right\|_{L_1([a,b],g)}}{\Gamma((n+1)\alpha)}(g(x)-g(a))^{(n+1)\alpha-1},$$

$\forall\, x\in[a,b]$.

Similarly, from (3.57) we derive

$$\left| f(x) - \sum_{i=0}^{n} \frac{(g(b) - g(x))^{i\alpha}}{\Gamma(i\alpha + 1)} \left(D_{b-;g}^{i\alpha} f\right)(b) \right| \leq$$

$$\frac{1}{\Gamma((n+1)\alpha)} \int_x^b (g(t) - g(x))^{(n+1)\alpha - 1} g'(t) \left| \left(D_{b-;g}^{(n+1)\alpha} f\right)(t) \right| dt \leq$$

$$\frac{(g(b) - g(x))^{(n+1)\alpha - 1}}{\Gamma((n+1)\alpha)} \int_x^b \left| \left(D_{b-;g}^{(n+1)\alpha} f\right)(t) \right| dg(t) \leq \tag{3.98}$$

$$\frac{\left\| D_{b-;g}^{(n+1)\alpha} f \right\|_{L_1([a,b],g)}}{\Gamma((n+1)\alpha)} (g(b) - g(x))^{(n+1)\alpha - 1},$$

$\forall\, x \in [a, b]$.

We have proved that

$$\left| f(x) - \sum_{i=0}^{n} \frac{(g(x) - g(a))^{i\alpha}}{\Gamma(i\alpha + 1)} \left(D_{a+;g}^{i\alpha} f\right)(a) \right| \leq$$

$$\frac{\rho}{\Gamma((n+1)\alpha)} (g(x) - g(a))^{(n+1)\alpha - 1}, \tag{3.99}$$

and

$$\left| f(x) - \sum_{i=0}^{n} \frac{(g(b) - g(x))^{i\alpha}}{\Gamma(i\alpha + 1)} \left(D_{b-;g}^{i\alpha} f\right)(b) \right| \leq$$

$$\frac{\rho}{\Gamma((n+1)\alpha)} (g(b) - g(x))^{(n+1)\alpha - 1}, \tag{3.100}$$

$\forall\, x \in [a, b]$.

The rest of the proof is as in Theorem 3.9. ■

Next follow L_p variants of Theorems 3.8, 3.9.

Theorem 3.12 *All as in Theorem 3.8 with* $\alpha \geq 1$, *and* $p, q > 1 : \frac{1}{p} + \frac{1}{q} = 1$. *If* $\alpha = n \in \mathbb{N}$, *we assume that* $\left(f \circ g^{-1}\right)^{(n)} \circ g \in C([a, b])$. *Set*

$$\mu := \max \left\{ \left\| D_{a+;g}^{\alpha} f \right\|_{L_q([a,b],g)}, \left\| D_{b-;g}^{\alpha} f \right\|_{L_q([a,b],g)} \right\}. \tag{3.101}$$

Then

(i)

$$\left| \int_a^b f(x)\,dg(x) - \sum_{k=0}^{n-1} \frac{1}{(k+1)!} \left[\left(f \circ g^{-1}\right)^{(k)} (g(a)) (g(t) - g(a))^{k+1} \right. \right.$$

$$\left. \left. + (-1)^k \left(f \circ g^{-1}\right)^{(k)} (g(b)) (g(b) - g(t))^{k+1} \right] \right| \leq$$

$$\frac{\mu}{\Gamma(\alpha)\left(\alpha + \frac{1}{p}\right)(p(\alpha-1)+1)^{\frac{1}{p}}}$$

$$\left[(g(t) - g(a))^{\alpha + \frac{1}{p}} + (g(b) - g(t))^{\alpha + \frac{1}{p}} \right], \tag{3.102}$$

$\forall\, t \in [a, b]$,

(ii) *at* $g(t) = \frac{g(a)+g(b)}{2}$, *the right hand side of (3.102) is minimized, and we get:*

$$\left| \int_a^b f(x)\,dg(x) - \sum_{k=0}^{n-1} \frac{1}{(k+1)!} \frac{(g(b) - g(a))^{k+1}}{2^{k+1}} \right.$$

$$\left. \left[\left(f \circ g^{-1}\right)^{(k)} (g(a)) + (-1)^k \left(f \circ g^{-1}\right)^{(k)} (g(b)) \right] \right| \leq$$

$$\frac{\mu}{\Gamma(\alpha)\left(\alpha + \frac{1}{p}\right)(p(\alpha-1)+1)^{\frac{1}{p}}} \frac{(g(b) - g(a))^{\alpha + \frac{1}{p}}}{2^{\alpha - \frac{1}{q}}}, \tag{3.103}$$

(iii) *if* $\left(f \circ g^{-1}\right)^{(k)} (g(a)) = \left(f \circ g^{-1}\right)^{(k)} (g(b)) = 0$, *for* $k = 0, 1, \ldots, n-1$, *we obtain*

$$\left| \int_a^b f(x)\,dg(x) \right| \leq$$

$$\frac{\mu}{\Gamma(\alpha)\left(\alpha + \frac{1}{p}\right)(p(\alpha-1)+1)^{\frac{1}{p}}} \frac{(g(b) - g(a))^{\alpha + \frac{1}{p}}}{2^{\alpha - \frac{1}{q}}}, \tag{3.104}$$

which is a sharp inequality,

(iv) *more generally, for* $j = 0, 1, 2, \ldots, N \in \mathbb{N}$, *it holds*

$$\left| \int_a^b f(x)\,dg(x) - \sum_{k=0}^{n-1} \frac{1}{(k+1)!} \left(\frac{g(b) - g(a)}{N} \right)^{k+1} \right.$$

$$\left[j^{k+1}\left(f\circ g^{-1}\right)^{(k)}(g(a))+(-1)^{k}(N-j)^{k+1}\left(f\circ g^{-1}\right)^{(k)}(g(b))\right]\Bigg|\leq$$

$$\frac{\mu}{\Gamma(\alpha)\left(\alpha+\frac{1}{p}\right)(p(\alpha-1)+1)^{\frac{1}{p}}}\left(\frac{g(b)-g(a)}{N}\right)^{\alpha+\frac{1}{p}}\left[j^{\alpha+\frac{1}{p}}+(N-j)^{\alpha+\frac{1}{p}}\right], \tag{3.105}$$

(v) if $\left(f\circ g^{-1}\right)^{(k)}(g(a))=\left(f\circ g^{-1}\right)^{(k)}(g(b))=0$*, for* $k=1,\ldots,n-1$*, from (3.105) we obtain*

$$\left|\int_a^b f(x)\,dg(x)-\left(\frac{g(b)-g(a)}{N}\right)[jf(a)+(N-j)f(b)]\right|\leq$$

$$\frac{\mu}{\Gamma(\alpha)\left(\alpha+\frac{1}{p}\right)(p(\alpha-1)+1)^{\frac{1}{p}}}\left(\frac{g(b)-g(a)}{N}\right)^{\alpha+\frac{1}{p}}\left[j^{\alpha+\frac{1}{p}}+(N-j)^{\alpha+\frac{1}{p}}\right], \tag{3.106}$$

$j=0,1,2,\ldots,N,$

(vi) when $N=2$*,* $j=1$*, (3.106) turns to*

$$\left|\int_a^b f(x)\,dg(x)-\left(\frac{g(b)-g(a)}{2}\right)(f(a)+f(b))\right|\leq$$

$$\frac{\mu}{\Gamma(\alpha)\left(\alpha+\frac{1}{p}\right)(p(\alpha-1)+1)^{\frac{1}{p}}}\frac{(g(b)-g(a))^{\alpha+\frac{1}{p}}}{2^{\alpha-\frac{1}{q}}}. \tag{3.107}$$

Proof From (3.27) we get

$$\left|f(x)-\sum_{k=0}^{n-1}\frac{\left(f\circ g^{-1}\right)^{(k)}(g(a))}{k!}(g(x)-g(a))^{k}\right|\leq$$

$$\frac{1}{\Gamma(\alpha)}\int_a^x(g(x)-g(t))^{\alpha-1}g'(t)\left|\left(D_{a+;g}^{\alpha}f\right)(t)\right|dt=$$

(by [7], p. 439)

$$\frac{1}{\Gamma(\alpha)}\int_a^x(g(x)-g(t))^{\alpha-1}\left|\left(D_{a+;g}^{\alpha}f\right)(t)\right|dg(t)\leq \tag{3.108}$$

(by [4])

$$\frac{1}{\Gamma(\alpha)}\left(\int_a^x(g(x)-g(t))^{p(\alpha-1)}dg(t)\right)^{\frac{1}{p}}\left(\int_a^x\left|\left(D_{a+;g}^{\alpha}f\right)(t)\right|^{q}dg(t)\right)^{\frac{1}{q}}\leq$$

$$\frac{1}{\Gamma(\alpha)} \frac{(g(x)-g(a))^{\alpha-\frac{1}{q}}}{(p(\alpha-1)+1)^{\frac{1}{p}}} \left\| D_{a+;g}^{\alpha} f \right\|_{L_q([a,b],g)}.$$

That is

$$\left| f(x) - \sum_{k=0}^{n-1} \frac{\left(f \circ g^{-1}\right)^{(k)}(g(a))}{k!} (g(x)-g(a))^k \right| \leq$$

$$\frac{\left\| D_{a+;g}^{\alpha} f \right\|_{L_q([a,b],g)}}{\Gamma(\alpha)(p(\alpha-1)+1)^{\frac{1}{p}}} (g(x)-g(a))^{\alpha-\frac{1}{q}}, \tag{3.109}$$

$\forall\, x \in [a,b]$.

Similarly, from (3.28) we obtain

$$\left| f(x) - \sum_{k=0}^{n-1} \frac{\left(f \circ g^{-1}\right)^{(k)}(g(b))}{k!} (g(x)-g(b))^k \right| \leq$$

$$\frac{1}{\Gamma(\alpha)} \int_x^b (g(t)-g(x))^{\alpha-1} g'(t) \left| \left(D_{b-;g}^{\alpha} f\right)(t) \right| dt =$$

(by [7], p. 439)

$$\frac{1}{\Gamma(\alpha)} \int_x^b (g(t)-g(x))^{\alpha-1} \left| \left(D_{b-;g}^{\alpha} f\right)(t) \right| dg(t) \leq$$

(by [4])

$$\frac{1}{\Gamma(\alpha)} \left(\int_x^b (g(t)-g(x))^{p(\alpha-1)} dg(t) \right)^{\frac{1}{p}} \left(\int_x^b \left| \left(D_{b-;g}^{\alpha} f\right)(t) \right|^q dg(t) \right)^{\frac{1}{q}} \leq \tag{3.110}$$

$$\frac{1}{\Gamma(\alpha)} \frac{(g(b)-g(x))^{\alpha-\frac{1}{q}}}{(p(\alpha-1)+1)^{\frac{1}{p}}} \left\| D_{b-;g}^{\alpha} f \right\|_{L_q([a,b],g)}.$$

That is

$$\left| f(x) - \sum_{k=0}^{n-1} \frac{\left(f \circ g^{-1}\right)^{(k)}(g(b))}{k!} (g(x)-g(b))^k \right| \leq$$

$$\frac{\left\| D_{b-;g}^{\alpha} f \right\|_{L_q([a,b],g)}}{\Gamma(\alpha)(p(\alpha-1)+1)^{\frac{1}{p}}} (g(b)-g(x))^{\alpha-\frac{1}{q}}, \tag{3.111}$$

$\forall\, x \in [a,b]$.

We have proved that

$$\left| f(x) - \sum_{k=0}^{n-1} \frac{\left(f \circ g^{-1}\right)^{(k)} (g(a))}{k!} (g(x) - g(a))^k \right| \leq$$

$$\frac{\mu}{\Gamma(\alpha) (p(\alpha - 1) + 1)^{\frac{1}{p}}} (g(x) - g(a))^{\alpha - \frac{1}{q}}, \tag{3.112}$$

and

$$\left| f(x) - \sum_{k=0}^{n-1} \frac{\left(f \circ g^{-1}\right)^{(k)} (g(b))}{k!} (g(x) - g(b))^k \right| \leq$$

$$\frac{\mu}{\Gamma(\alpha) (p(\alpha - 1) + 1)^{\frac{1}{p}}} (g(b) - g(x))^{\alpha - \frac{1}{q}}, \tag{3.113}$$

$\forall\, x \in [a, b]$.

The rest of the proof is as in Theorem 3.8. ■

We continue with

Theorem 3.13 *All as in Theorem 3.9, with* $\frac{1}{n+1} \leq \alpha \leq 1$, *and* $p, q > 1 : \frac{1}{p} + \frac{1}{q} = 1$. *Set*

$$\theta := \max \left\{ \left\| D_{a+;g}^{(n+1)\alpha} f \right\|_{L_q([a,b],g)}, \left\| D_{b-;g}^{(n+1)\alpha} f \right\|_{L_q([a,b],g)} \right\}. \tag{3.114}$$

Then

(i)

$$\left| \int_a^b f(x)\, dg(x) - \left\{ \sum_{i=0}^{n} \frac{1}{\Gamma(i\alpha + 2)} \left[\left(D_{a+;g}^{i\alpha} f\right)(a) (g(t) - g(a))^{i\alpha+1} \right.\right.\right.$$

$$\left.\left.\left. + \left(D_{b-;g}^{i\alpha} f\right)(b) (g(b) - g(t))^{i\alpha+1} \right] \right\} \right| \leq$$

$$\frac{\theta}{\Gamma((n+1)\alpha) \left((n+1)\alpha + \frac{1}{p}\right) (p((n+1)\alpha - 1) + 1)^{\frac{1}{p}}}$$

$$\left[(g(t) - g(a))^{(n+1)\alpha + \frac{1}{p}} + (g(b) - g(t))^{(n+1)\alpha + \frac{1}{p}} \right], \tag{3.115}$$

$\forall\, t \in [a, b]$,

(ii) at $g(t) = \frac{g(a)+g(b)}{2}$, *the right hand side of (3.115) is minimized, and we get:*

$$\left| \int_a^b f(x)\, dg(x) - \left\{ \sum_{i=0}^{n} \frac{1}{\Gamma(i\alpha + 2)} \frac{(g(b) - g(a))^{i\alpha+1}}{2^{i\alpha+1}} \right.\right.$$

$$\left[\left(D_{a+;g}^{i\alpha}f\right)(a)+\left(D_{b-;g}^{i\alpha}f\right)(b)\right]\Bigg\}\Bigg|\leq$$

$$\frac{\theta}{\Gamma\left(\left(n+1\right)\alpha\right)\left(\left(n+1\right)\alpha+\frac{1}{p}\right)\left(p\left(\left(n+1\right)\alpha-1\right)+1\right)^{\frac{1}{p}}}\frac{\left(g\left(b\right)-g\left(a\right)\right)^{\left(n+1\right)\alpha+\frac{1}{p}}}{2^{\left(n+1\right)\alpha-\frac{1}{q}}},\tag{3.116}$$

(iii) assuming $\left(D_{a+;g}^{i\alpha}f\right)(a)=\left(D_{b-;g}^{i\alpha}f\right)(b)=0$, $i=0,1,\ldots,n$, *we obtain*

$$\left|\int_a^b f(x)\,dg(x)\right|\leq$$

$$\frac{\theta}{\Gamma\left(\left(n+1\right)\alpha\right)\left(\left(n+1\right)\alpha+\frac{1}{p}\right)\left(p\left(\left(n+1\right)\alpha-1\right)+1\right)^{\frac{1}{p}}}\frac{\left(g\left(b\right)-g\left(a\right)\right)^{\left(n+1\right)\alpha+\frac{1}{p}}}{2^{\left(n+1\right)\alpha-\frac{1}{q}}},\tag{3.117}$$

which is a sharp inequality,

(iv) more generally, for $j=0,1,2,\ldots,N\in\mathbb{N}$, *it holds*

$$\left|\int_a^b f(x)\,dg(x)-\left\{\sum_{i=0}^{n}\frac{1}{\Gamma(i\alpha+2)}\left(\frac{g(b)-g(a)}{N}\right)^{i\alpha+1}\right.\right.$$

$$\left.\left.\left[\left(D_{a+;g}^{i\alpha}f\right)(a)\,j^{i\alpha+1}+\left(D_{b-;g}^{i\alpha}f\right)(b)\,(N-j)^{i\alpha+1}\right]\right\}\right|\leq$$

$$\frac{\theta}{\Gamma\left(\left(n+1\right)\alpha\right)\left(\left(n+1\right)\alpha+\frac{1}{p}\right)\left(p\left(\left(n+1\right)\alpha-1\right)+1\right)^{\frac{1}{p}}}$$

$$\left(\frac{g(b)-g(a)}{N}\right)^{(n+1)\alpha+\frac{1}{p}}\left[j^{(n+1)\alpha+\frac{1}{p}}+(N-j)^{(n+1)\alpha+\frac{1}{p}}\right],\tag{3.118}$$

(v) if $\left(D_{a+;g}^{i\alpha}f\right)(a)=\left(D_{b-;g}^{i\alpha}f\right)(b)=0$, $i=1,\ldots,n$, *from (3.118) we get:*

$$\left|\int_a^b f(x)\,dg(x)-\left(\frac{g(b)-g(a)}{N}\right)\left[jf(a)+(N-j)\,f(b)\right]\right|\leq$$

$$\frac{\theta}{\Gamma\left(\left(n+1\right)\alpha\right)\left(\left(n+1\right)\alpha+\frac{1}{p}\right)\left(p\left(\left(n+1\right)\alpha-1\right)+1\right)^{\frac{1}{p}}}$$

$$\left(\frac{g(b)-g(a)}{N}\right)^{(n+1)\alpha+\frac{1}{p}}\left[j^{(n+1)\alpha+\frac{1}{p}}+(N-j)^{(n+1)\alpha+\frac{1}{p}}\right],\tag{3.119}$$

$j = 0, 1, 2, \ldots, N,$

(vi) when $N = 2$, $j = 1$, *(3.119) turns to*

$$\left| \int_a^b f(x)\, dg(x) - \left(\frac{g(b) - g(a)}{2} \right) (f(a) + f(b)) \right| \leq$$

$$\frac{\theta}{\Gamma((n+1)\alpha)\left((n+1)\alpha + \frac{1}{p}\right)(p((n+1)\alpha - 1) + 1)^{\frac{1}{p}}} \frac{(g(b) - g(a))^{(n+1)\alpha + \frac{1}{p}}}{2^{(n+1)\alpha - \frac{1}{q}}}. \tag{3.120}$$

Proof By (3.56) we get

$$\left| f(x) - \sum_{i=0}^{n} \frac{(g(x) - g(a))^{i\alpha}}{\Gamma(i\alpha + 1)} \left(D_{a+;g}^{i\alpha} f \right)(a) \right| \leq \tag{3.121}$$

$$\frac{1}{\Gamma((n+1)\alpha)} \int_a^x (g(x) - g(t))^{(n+1)\alpha - 1} g'(t) \left| \left(D_{a+;g}^{(n+1)\alpha} f \right)(t) \right| dt =$$

(by [7])

$$\frac{1}{\Gamma((n+1)\alpha)} \int_a^x (g(x) - g(t))^{(n+1)\alpha - 1} \left| \left(D_{a+;g}^{(n+1)\alpha} f \right)(t) \right| dg(t) \leq$$

(by [4])

$$\frac{1}{\Gamma((n+1)\alpha)} \left(\int_a^x (g(x) - g(t))^{p((n+1)\alpha - 1)} dg(t) \right)^{\frac{1}{p}}$$

$$\left(\int_a^x \left| \left(D_{a+;g}^{(n+1)\alpha} f \right)(t) \right|^q dg(t) \right)^{\frac{1}{q}} \leq$$

$$\frac{1}{\Gamma((n+1)\alpha)} \frac{(g(x) - g(a))^{\frac{p((n+1)\alpha - 1) + 1}{p}}}{(p((n+1)\alpha - 1) + 1)^{\frac{1}{p}}} \left\| D_{a+;g}^{(n+1)\alpha} f \right\|_{L_q([a,b],g)}.$$

That is

$$\left| f(x) - \sum_{i=0}^{n} \frac{(g(x) - g(a))^{i\alpha}}{\Gamma(i\alpha + 1)} \left(D_{a+;g}^{i\alpha} f \right)(a) \right| \leq$$

$$\frac{\left\| D_{a+;g}^{(n+1)\alpha} f \right\|_{L_q([a,b],g)}}{\Gamma((n+1)\alpha)(p((n+1)\alpha - 1) + 1)^{\frac{1}{p}}} (g(x) - g(a))^{(n+1)\alpha - \frac{1}{q}}, \tag{3.122}$$

$\forall\, x \in [a, b]$.

Similarly, from (3.57) we derive

$$\left| f(x) - \sum_{i=0}^{n} \frac{(g(b) - g(x))^{i\alpha}}{\Gamma(i\alpha + 1)} \left(D_{b-;g}^{i\alpha} f\right)(b) \right| \leq$$

$$\frac{1}{\Gamma((n+1)\alpha)} \int_x^b (g(t) - g(x))^{(n+1)\alpha - 1} g'(t) \left| \left(D_{b-;g}^{(n+1)\alpha} f\right)(t) \right| dt =$$

(by [7])

$$\frac{1}{\Gamma((n+1)\alpha)} \int_x^b (g(t) - g(x))^{(n+1)\alpha - 1} \left| \left(D_{b-;g}^{(n+1)\alpha} f\right)(t) \right| dg(t) \leq$$

(by [4])

$$\frac{1}{\Gamma((n+1)\alpha)} \left(\int_x^b (g(t) - g(x))^{p((n+1)\alpha - 1)} dg(t) \right)^{\frac{1}{p}}$$

$$\left(\int_x^b \left| \left(D_{b-;g}^{(n+1)\alpha} f\right)(t) \right|^q dg(t) \right)^{\frac{1}{q}} \leq \tag{3.123}$$

$$\frac{1}{\Gamma((n+1)\alpha)} \frac{(g(b) - g(x))^{\frac{p((n+1)\alpha - 1) + 1}{p}}}{(p((n+1)\alpha - 1) + 1)^{\frac{1}{p}}} \left\| D_{b-;g}^{(n+1)\alpha} f \right\|_{L_q([a,b],g)}.$$

That is

$$\left| f(x) - \sum_{i=0}^{n} \frac{(g(b) - g(x))^{i\alpha}}{\Gamma(i\alpha + 1)} \left(D_{b-;g}^{i\alpha} f\right)(b) \right| \leq$$

$$\frac{\left\| D_{b-;g}^{(n+1)\alpha} f \right\|_{L_q([a,b],g)}}{\Gamma((n+1)\alpha)(p((n+1)\alpha - 1) + 1)^{\frac{1}{p}}} (g(b) - g(x))^{(n+1)\alpha - \frac{1}{q}}, \tag{3.124}$$

$\forall\, x \in [a, b]$.

We have proved that

$$\left| f(x) - \sum_{i=0}^{n} \frac{(g(x) - g(a))^{i\alpha}}{\Gamma(i\alpha + 1)} \left(D_{a+;g}^{i\alpha} f\right)(a) \right| \leq$$

$$\frac{\theta}{\Gamma((n+1)\alpha)(p((n+1)\alpha - 1) + 1)^{\frac{1}{p}}} (g(x) - g(a))^{(n+1)\alpha - \frac{1}{q}}, \tag{3.125}$$

and

$$\left| f(x) - \sum_{i=0}^{n} \frac{(g(b) - g(x))^{i\alpha}}{\Gamma(i\alpha + 1)} \left(D_{b-;g}^{i\alpha} f \right)(b) \right| \leq$$

$$\frac{\theta}{\Gamma((n+1)\alpha)(p((n+1)\alpha - 1) + 1)^{\frac{1}{p}}} (g(b) - g(x))^{(n+1)\alpha - \frac{1}{q}}, \qquad (3.126)$$

$\forall\, x \in [a, b]$.

The rest of the proof is as in Theorem 3.9. ■

Applications follow:

Proposition 3.14 *We assume that* $(f \circ \ln x) \in AC^n([e^a, e^b])$, *where* $\mathbb{N} \ni n = \lceil \alpha \rceil$, $\alpha > 0$. *We also assume that* $(f \circ \ln x)^{(n)} \circ e^x \in L_\infty([a, b])$, $f \in C([a, b])$. *Set*

$$T_1 := \max \left\{ \left\| D_{a+;e^x}^{\alpha} f \right\|_{L_\infty([a,b])}, \left\| D_{b-;e^x}^{\alpha} f \right\|_{L_\infty([a,b])} \right\}. \qquad (3.127)$$

Then

(i)

$$\left| \int_a^b f(x) e^x dx - \sum_{k=0}^{n-1} \frac{1}{(k+1)!} \left[(f \circ \ln x)^{(k)} (e^a) \left(e^t - e^a \right)^{k+1} \right. \right.$$

$$\left. \left. (-1)^k (f \circ \ln x)^{(k)} \left(e^b \right) \left(e^b - e^t \right)^{k+1} \right] \right| \leq$$

$$\frac{T_1}{\Gamma(\alpha + 2)} \left[\left(e^t - e^a \right)^{\alpha+1} + \left(e^b - e^t \right)^{\alpha+1} \right], \qquad (3.128)$$

$\forall\, t \in [a, b]$,

(ii) *at* $t = \ln\left(\frac{e^a + e^b}{2}\right)$, *the right hand side of (3.128) is minimized, and we get:*

$$\left| \int_a^b f(x) e^x dx - \sum_{k=0}^{n-1} \frac{1}{(k+1)!} \frac{\left(e^b - e^a \right)^{k+1}}{2^{k+1}} \right.$$

$$\left. \left[(f \circ \ln x)^{(k)} (e^a) + (-1)^k (f \circ \ln x)^{(k)} \left(e^b \right) \right] \right| \leq$$

$$\frac{T_1}{\Gamma(\alpha + 2)} \frac{\left(e^b - e^a \right)^{\alpha+1}}{2^\alpha}, \qquad (3.129)$$

(iii) *if* $(f \circ \ln x)^{(k)} (e^a) = (f \circ \ln x)^{(k)} \left(e^b \right) = 0$, *for* $k = 0, 1, \ldots, n-1$, *we obtain*

$$\left| \int_a^b f(x) e^x dx \right| \leq T_1 \frac{\left(e^b - e^a \right)^{\alpha+1}}{\Gamma(\alpha + 2) 2^\alpha}, \qquad (3.130)$$

which is a sharp inequality,

(iv) *more generally, for $j = 0, 1, 2, \ldots, N \in \mathbb{N}$, it holds*

$$\left| \int_a^b f(x) e^x dx - \sum_{k=0}^{n-1} \frac{1}{(k+1)!} \left(\frac{e^b - e^a}{N} \right)^{k+1} \right.$$

$$\left. \left[j^{k+1} (f \circ \ln x)^{(k)} (e^a) + (-1)^k (N-j)^{k+1} (f \circ \ln x)^{(k)} (e^b) \right] \right| \leq$$

$$\frac{T_1}{\Gamma(\alpha+2)} \left(\frac{e^b - e^a}{N} \right)^{\alpha+1} \left[j^{\alpha+1} + (N-j)^{\alpha+1} \right], \tag{3.131}$$

(v) *if $(f \circ \ln x)^{(k)} (e^a) = (f \circ \ln x)^{(k)} (e^b) = 0$, for $k = 1, \ldots, n-1$, from (3.131) we obtain*

$$\left| \int_a^b f(x) e^x dx - \left(\frac{e^b - e^a}{N} \right) [j f(a) + (N-j) f(b)] \right| \leq$$

$$\frac{T_1}{\Gamma(\alpha+2)} \left(\frac{e^b - e^a}{N} \right)^{\alpha+1} \left[j^{\alpha+1} + (N-j)^{\alpha+1} \right], \tag{3.132}$$

$j = 0, 1, 2, \ldots, N$,

(vi) *when $N = 2$, $j = 1$, (3.132) turns to*

$$\left| \int_a^b f(x) e^x dx - \left(\frac{e^b - e^a}{2} \right) (f(a) + f(b)) \right| \leq$$

$$\frac{T_1}{\Gamma(\alpha+2)} \frac{\left(e^b - e^a \right)^{\alpha+1}}{2^\alpha}, \tag{3.133}$$

(vii) *when $0 < \alpha \leq 1$, inequality (3.133) is again valid without any boundary conditions.*

Proof By Theorem 3.8, for $g(x) = e^x$. ∎

We continue with

Proposition 3.15 *Here $f \in C([a,b])$, where $[a,b] \subset (0, +\infty)$. Let $0 < \alpha \leq 1$, and $G_k := D_{a+;\ln x}^{k\alpha} f$, for $k = 0, 1, \ldots, n+1$; $n \in \mathbb{N}$. We assume that $G_k \circ e^x \in AC([\ln a, \ln b])$ and $(G_k \circ e^x)' \circ \ln x \in L_\infty([a,b])$. Also let $\overline{G_k} := D_{b-;\ln x}^{k\alpha} f$, for $k = 0, 1, \ldots, n+1$, they fulfill $\overline{G_k} \circ e^x \in AC([\ln a, \ln b])$ and $\left(\overline{G_k} \circ e^x \right)' \circ \ln x \in L_\infty([a,b])$. Set*

$$T_2 := \max \left\{ \left\| D_{a+;\ln x}^{(n+1)\alpha} f \right\|_{\infty,[a,b]}, \left\| D_{b-;\ln x}^{(n+1)\alpha} f \right\|_{\infty,[a,b]} \right\}. \tag{3.134}$$

Then

(i)

$$\left|\int_a^b \frac{f(x)}{x}dx - \left\{\sum_{i=0}^{n} \frac{1}{\Gamma(i\alpha+2)}\left[\left(D_{a+;\ln x}^{i\alpha} f\right)(a)\left(\ln\frac{t}{a}\right)^{i\alpha+1}\right.\right.\right.$$

$$\left.\left.\left.+\left(D_{b-;\ln x}^{i\alpha} f\right)(b)\left(\ln\frac{b}{t}\right)^{i\alpha+1}\right]\right\}\right| \leq$$

$$\frac{T_2}{\Gamma((n+1)\alpha+2)}\left[\left(\ln\frac{t}{a}\right)^{(n+1)\alpha+1} + \left(\ln\frac{b}{t}\right)^{(n+1)\alpha+1}\right], \tag{3.135}$$

$\forall\, t \in [a, b]$,

(ii) *at* $t = e^{\left(\frac{\ln ab}{2}\right)}$, *the right hand side of (3.135) is minimized, and we get:*

$$\left|\int_a^b \frac{f(x)}{x}dx - \left\{\sum_{i=0}^{n} \frac{1}{\Gamma(i\alpha+2)} \frac{\left(\ln\frac{b}{a}\right)^{i\alpha+1}}{2^{i\alpha+1}}\right.\right.$$

$$\left.\left.\left[\left(D_{a+;\ln x}^{i\alpha} f\right)(a) + \left(D_{b-;\ln x}^{i\alpha} f\right)(b)\right]\right\}\right| \leq$$

$$\frac{T_2}{\Gamma((n+1)\alpha+2)} \frac{\left(\ln\frac{b}{a}\right)^{(n+1)\alpha+1}}{2^{(n+1)\alpha}}, \tag{3.136}$$

(iii) *assuming* $\left(D_{a+;\ln x}^{i\alpha} f\right)(a) = \left(D_{b-;\ln x}^{i\alpha} f\right)(b) = 0$, $i = 0, 1, \ldots, n$, *we obtain*

$$\left|\int_a^b \frac{f(x)}{x}dx\right| \leq \frac{T_2}{\Gamma((n+1)\alpha+2)} \frac{\left(\ln\frac{b}{a}\right)^{(n+1)\alpha+1}}{2^{(n+1)\alpha}}, \tag{3.137}$$

which is a sharp inequality,

(iv) *more generally, for* $j = 0, 1, 2, \ldots, N \in \mathbb{N}$, *it holds*

$$\left|\int_a^b \frac{f(x)}{x}dx - \left\{\sum_{i=0}^{n} \frac{1}{\Gamma(i\alpha+2)} \left(\frac{\ln\frac{b}{a}}{N}\right)^{i\alpha+1}\right.\right.$$

$$\left.\left.\left[\left(D_{a+;\ln x}^{i\alpha} f\right)(a)\, j^{i\alpha+1} + \left(D_{b-;\ln x}^{i\alpha} f\right)(b)\,(N-j)^{i\alpha+1}\right]\right\}\right| \leq$$

$$\frac{T_2}{\Gamma((n+1)\alpha+2)} \left(\frac{\ln\frac{b}{a}}{N}\right)^{(n+1)\alpha+1} \left[j^{(n+1)\alpha+1} + (N-j)^{(n+1)\alpha+1}\right], \tag{3.138}$$

(v) *if* $\left(D_{a+;\ln x}^{i\alpha} f\right)(a) = \left(D_{b-;\ln x}^{i\alpha} f\right)(b) = 0$, $i = 1, \ldots, n$, *from (3.138) we get:*

$$\left| \int_a^b \frac{f(x)}{x} dx - \left(\frac{\ln \frac{b}{a}}{N} \right) (jf(a) + (N-j) f(b)) \right| \leq$$

$$\frac{T_2}{\Gamma((n+1)\alpha + 2)} \left(\frac{\ln \frac{b}{a}}{N} \right)^{(n+1)\alpha+1} \left[j^{(n+1)\alpha+1} + (N-j)^{(n+1)\alpha+1} \right], \quad (3.139)$$

for $j = 0, 1, 2, \ldots, N$,

(vi) *if* $N = 2$ *and* $j = 1$, *(3.139) becomes*

$$\left| \int_a^b \frac{f(x)}{x} dx - \left(\frac{\ln \frac{b}{a}}{2} \right) (f(a) + f(b)) \right| \leq$$

$$\frac{T_2}{\Gamma((n+1)\alpha + 2)} \frac{\left(\ln \frac{b}{a}\right)^{(n+1)\alpha+1}}{2^{(n+1)\alpha}}. \quad (3.140)$$

Proof By Theorem 3.9, for $g(x) = \ln x$. ∎

We could give many other interesting applications that are based in our other theorems, due to lack of space we skip this task.

References

1. Anastassiou, G.: Advanced fractional Taylor's formulae
2. Anastassiou, G., Argyros, I.: Intelligent Numerical Methods: Applications to Fractional Calculus. Springer, Heidelberg (2016)
3. Anastassiou, G.: Weighted fractional Iyengar type Inequalities in the Caputo direction. Mathematics **7**, 1119 (2019). https://doi.org/10.3390/MATH7111119
4. Dragomir, S.S.: Inequalities for the Riemann-Stieljes integral of $(p, q) - H-$*Dominated integrators with applications*. Appl. Math. E-Notes **15**, 243–260 (2015). http://www.math.nthu.edu.tw/~amen
5. Iyengar, K.S.K.: Note on an inequality. Math. Student **6**, 75–76 (1938)
6. Royden, H.L.: Real Analysis, 2nd edn. MacMillan Publishing Co. Inc., New York (1968)
7. Royden, H.L., Fitzpatrick, P.M.: Real Analysis, 4th edn. Pearson, New York (2010)

Chapter 4
Generalized Canavati g-Fractional Iyengar and Ostrowski Inequalities

We present here generalized Canavati type g-fractional Iyengar and Ostrowski type inequalities. Our inequalities are with respect to all L_p norms: $1 \leq p \leq \infty$. We finish with applications. See also [6].

4.1 Background—I

We are motivated by the following famous Iyengar inequality (1938), [8].

Theorem 4.1 *Let f be a differentiable function on $[a, b]$ and $\left|f'(x)\right| \leq M$. Then*

$$\left|\int_a^b f(x)\,dx - \frac{1}{2}(b-a)(f(a)+f(b))\right| \leq \frac{M(b-a)^2}{4} - \frac{(f(b)-f(a))^2}{4M}. \tag{4.1}$$

We need the following fractional calculus background:

Let $\alpha > 0$, $m = [\alpha]$, ($[\cdot]$ is the integral part), $\beta = \alpha - m$, $0 < \beta < 1$, $f \in C([a, b])$, $[a, b] \subset \mathbb{R}$, $x \in [a, b]$. The gamma function Γ is given by $\Gamma(\alpha) = \int_0^\infty e^{-t} t^{\alpha-1} dt$. We define the left Riemann-Liouville integral [1, p. 24]

$$\left(J_\alpha^{a+} f\right)(x) = \frac{1}{\Gamma(\alpha)} \int_a^x (x-t)^{\alpha-1} f(t)\,dt, \tag{4.2}$$

$a \leq x \leq b$. We define the subspace $C_{a+}^\alpha([a, b])$ of $C^m([a, b])$:

$$C_{a+}^\alpha([a, b]) = \left\{f \in C^m([a, b]) : J_{1-\beta}^{a+} f^{(m)} \in C^1([a, b])\right\}. \tag{4.3}$$

G. A. Anastassiou, *Generalized Fractional Calculus*, Studies in Systems, Decision and Control 305, https://doi.org/10.1007/978-3-030-56962-4_4

For $f \in C_{a+}^{\alpha}([a,b])$, we define the left generalized α-fractional derivative of f over $[a,b]$ as

$$D_{a+}^{\alpha} f := \left(J_{1-\beta}^{a+} f^{(m)} \right)', \tag{4.4}$$

see [1, p. 24]. Canavati first in [7] introduced the above over [0, 1].

We have that $D_{a+}^{n} f = f^{(n)}$; $n \in \mathbb{N}$.

Notice that $D_{a+}^{\alpha} f \in C([a,b])$.

Furthermore we need:

Let again $\alpha > 0$, $m = [\alpha]$, $\beta = \alpha - m$, $f \in C([a,b])$, call the right Riemann-Liouville fractional integral operator by

$$\left(J_{b-}^{\alpha} f \right)(x) := \frac{1}{\Gamma(\alpha)} \int_x^b (t-x)^{\alpha-1} f(t)\, dt, \tag{4.5}$$

$x \in [a,b]$, see [2]. Define the subspace of functions

$$C_{b-}^{\alpha}([a,b]) = \left\{ f \in C^m([a,b]) : J_{b-}^{1-\beta} f^{(m)} \in C^1([a,b]) \right\}. \tag{4.6}$$

Define the right generalized α-fractional derivative of f over $[a,b]$ as

$$D_{b-}^{\alpha} f = (-1)^{m-1} \left(J_{b-}^{1-\beta} f^{(m)} \right)', \tag{4.7}$$

see [2]. We set $D_{b-}^{0} f = f$. We have $D_{b-}^{n} f = (-1)^n f^{(n)}$; $n \in \mathbb{N}$. Notice that $D_{b-}^{\alpha} f \in C([a,b])$.

From [5] we have the following Canavati fractional Iyengar type inequalities:

Theorem 4.2 *Let $\nu \geq 1$, $n = [\nu]$ and $f \in C_{a+}^{\nu}([a,b]) \cap C_{b-}^{\nu}([a,b])$. Then*
(i)

$$\left| \int_a^b f(x)\,dx - \sum_{k=0}^{n-1} \frac{1}{(k+1)!} \left[f^{(k)}(a)(t-a)^{k+1} + (-1)^k f^{(k)}(b)(b-t)^{k+1} \right] \right| \leq$$

$$\frac{\max\left\{ \left\| D_{a+}^{\nu} f \right\|_{\infty,([a,b])}, \left\| D_{b-}^{\nu} f \right\|_{\infty,([a,b])} \right\}}{\Gamma(\nu+2)} \left[(t-a)^{\nu+1} + (b-t)^{\nu+1} \right], \tag{4.8}$$

$\forall\, t \in [a,b]$,

(ii) at $t = \frac{a+b}{2}$, the right hand side of (4.8) is minimized, and we get:

$$\left| \int_a^b f(x)\,dx - \sum_{k=0}^{n-1} \frac{1}{(k+1)!} \frac{(b-a)^{k+1}}{2^{k+1}} \left[f^{(k)}(a) + (-1)^k f^{(k)}(b) \right] \right| \leq$$

$$\frac{\max\left\{ \left\| D_{a+}^{\nu} f \right\|_{\infty,([a,b])}, \left\| D_{b-}^{\nu} f \right\|_{\infty,([a,b])} \right\}}{\Gamma(\nu+2)} \frac{(b-a)^{\nu+1}}{2^{\nu}}, \tag{4.9}$$

(iii) if $f^{(k)}(a) = f^{(k)}(b) = 0$*, for all* $k = 0, 1, \ldots, n-1$*, we obtain*

$$\left| \int_a^b f(x)\,dx \right| \leq \frac{\max\left\{ \left\| D_{a+}^{\nu} f \right\|_{\infty,([a,b])}, \left\| D_{b-}^{\nu} f \right\|_{\infty,([a,b])} \right\}}{\Gamma(\nu+2)} \frac{(b-a)^{\nu+1}}{2^{\nu}}, \tag{4.10}$$

which is a sharp inequality,

(iv) more generally, for $j = 0, 1, 2, \ldots, N \in \mathbb{N}$*, it holds*

$$\left| \int_a^b f(x)\,dx - \sum_{k=0}^{n-1} \frac{1}{(k+1)!} \left(\frac{b-a}{N} \right)^{k+1} \left[j^{k+1} f^{(k)}(a) + (-1)^k (N-j)^{k+1} f^{(k)}(b) \right] \right|$$

$$\leq \frac{\max\left\{ \left\| D_{a+}^{\nu} f \right\|_{\infty,([a,b])}, \left\| D_{b-}^{\nu} f \right\|_{\infty,([a,b])} \right\}}{\Gamma(\nu+2)} \left(\frac{b-a}{N} \right)^{\nu+1} \left[j^{\nu+1} + (N-j)^{\nu+1} \right], \tag{4.11}$$

(v) if $f^{(k)}(a) = f^{(k)}(b) = 0$, $k = 1, \ldots, n-1$*, from (4.11) we get:*

$$\left| \int_a^b f(x)\,dx - \left(\frac{b-a}{N} \right) \left[j f(a) + (N-j) f(b) \right] \right| \leq$$

$$\frac{\max\left\{ \left\| D_{a+}^{\nu} f \right\|_{\infty,([a,b])}, \left\| D_{b-}^{\nu} f \right\|_{\infty,([a,b])} \right\}}{\Gamma(\nu+2)} \left(\frac{b-a}{N} \right)^{\nu+1} \left[j^{\nu+1} + (N-j)^{\nu+1} \right], \tag{4.12}$$

$j = 0, 1, 2, \ldots, N,$

(vi) when $N = 2$ *and* $j = 1$*, (4.12) turns to*

$$\left| \int_a^b f(x)\,dx - \left(\frac{b-a}{2} \right) (f(a) + f(b)) \right| \leq$$

$$\frac{\max\left\{ \left\| D_{a+}^{\nu} f \right\|_{\infty,([a,b])}, \left\| D_{b-}^{\nu} f \right\|_{\infty,([a,b])} \right\}}{\Gamma(\nu+2)} \frac{(b-a)^{\nu+1}}{2^{\nu}}. \tag{4.13}$$

We continue with L_1 estimates:

Theorem 4.3 ([5]) *Let* $\nu \geq 1$, $n = [\nu]$, *and* $f \in C_{a+}^{\nu}([a,b]) \cap C_{b-}^{\nu}([a,b])$. *Then*

(i)

$$\left| \int_a^b f(x)\,dx - \sum_{k=0}^{n-1} \frac{1}{(k+1)!} \left[f^{(k)}(a)(t-a)^{k+1} + (-1)^k f^{(k)}(b)(b-t)^{k+1} \right] \right| \leq$$

$$\frac{\max\left\{ \left\| D_{a+}^{\nu} f \right\|_{L_1([a,b])}, \left\| D_{b-}^{\nu} f \right\|_{L_1([a,b])} \right\}}{\Gamma(\nu+1)} \left[(t-a)^{\nu} + (b-t)^{\nu} \right], \tag{4.14}$$

$\forall\, t \in [a,b]$,

(ii) *when* $\nu = 1$, *from (4.14), we have*

$$\left| \int_a^b f(x)\,dx - \left[f(a)(t-a) + f(b)(b-t) \right] \right| \leq$$

$$\left\| f' \right\|_{L_1([a,b])} (b-a), \ \forall\, t \in [a,b], \tag{4.15}$$

(iii) *from (4.15), we obtain* ($\nu = 1$ *case*)

$$\left| \int_a^b f(x)\,dx - \left(\frac{b-a}{2} \right) (f(a) + f(b)) \right| \leq \left\| f' \right\|_{L_1([a,b])} (b-a), \tag{4.16}$$

(iv) *at* $t = \frac{a+b}{2}$, $\nu > 1$, *the right hand side of (4.14) is minimized, and we get:*

$$\left| \int_a^b f(x)\,dx - \sum_{k=0}^{n-1} \frac{1}{(k+1)!} \frac{(b-a)^{k+1}}{2^{k+1}} \left[f^{(k)}(a) + (-1)^k f^{(k)}(b) \right] \right| \leq$$

$$\frac{\max\left\{ \left\| D_{a+}^{\nu} f \right\|_{L_1([a,b])}, \left\| D_{b-}^{\nu} f \right\|_{L_1([a,b])} \right\}}{\Gamma(\nu+1)} \frac{(b-a)^{\nu}}{2^{\nu-1}}, \tag{4.17}$$

(v) *if* $f^{(k)}(a) = f^{(k)}(b) = 0$, *for all* $k = 0, 1, \ldots, n-1$; $\nu > 1$, *from (4.17), we obtain*

$$\left| \int_a^b f(x)\,dx \right| \leq \frac{\max\left\{ \left\| D_{a+}^{\nu} f \right\|_{L_1([a,b])}, \left\| D_{b-}^{\nu} f \right\|_{L_1([a,b])} \right\}}{\Gamma(\nu+1)} \frac{(b-a)^{\nu}}{2^{\nu-1}}, \tag{4.18}$$

which is a sharp inequality,

(vi) *more generally, for* $j = 0, 1, 2, \ldots, N \in \mathbb{N}$, *it holds*

$$\left| \int_a^b f(x)\,dx - \sum_{k=0}^{n-1} \frac{1}{(k+1)!} \left(\frac{b-a}{N} \right)^{k+1} \left[j^{k+1} f^{(k)}(a) + (-1)^k (N-j)^{k+1} f^{(k)}(b) \right] \right|$$

$$\leq \frac{\max \left\{ \left\| D_{a+}^{\nu} f \right\|_{L_1([a,b])}, \left\| D_{b-}^{\nu} f \right\|_{L_1([a,b])} \right\}}{\Gamma(\nu+1)} \left(\frac{b-a}{N} \right)^{\nu} \left[j^{\nu} + (N-j)^{\nu} \right], \quad (4.19)$$

(vii) *if* $f^{(k)}(a) = f^{(k)}(b) = 0$, $k = 1, \ldots, n-1$, *from (4.19) we get:*

$$\left| \int_a^b f(x)\,dx - \left(\frac{b-a}{N} \right) [j f(a) + (N-j) f(b)] \right| \leq$$

$$\frac{\max \left\{ \left\| D_{a+}^{\nu} f \right\|_{L_1([a,b])}, \left\| D_{b-}^{\nu} f \right\|_{L_1([a,b])} \right\}}{\Gamma(\nu+1)} \left(\frac{b-a}{N} \right)^{\nu} \left[j^{\nu} + (N-j)^{\nu} \right], \quad (4.20)$$

$j = 0, 1, 2, \ldots, N$,

(viii) *when* $N = 2$ *and* $j = 1$, *(4.20) turns to*

$$\left| \int_a^b f(x)\,dx - \frac{(b-a)}{2} (f(a) + f(b)) \right| \leq$$

$$\frac{\max \left\{ \left\| D_{a+}^{\nu} f \right\|_{L_1([a,b])}, \left\| D_{b-}^{\nu} f \right\|_{L_1([a,b])} \right\}}{\Gamma(\nu+1)} \frac{(b-a)^{\nu}}{2^{\nu-1}}, \quad (4.21)$$

We continue with L_p estimates:

Theorem 4.4 ([5]) *Let* $p, q > 1 : \frac{1}{p} + \frac{1}{q} = 1$, $\nu \geq 1$, $n = [\nu]$; $f \in C_{a+}^{\nu}([a,b]) \cap C_{b-}^{\nu}([a,b])$. *Then*

(i)

$$\left| \int_a^b f(x)\,dx - \sum_{k=0}^{n-1} \frac{1}{(k+1)!} \left[f^{(k)}(a)(t-a)^{k+1} + (-1)^k f^{(k)}(b)(b-t)^{k+1} \right] \right| \leq$$

$$\frac{\max \left\{ \left\| D_{a+}^{\nu} f \right\|_{L_q([a,b])}, \left\| D_{b-}^{\nu} f \right\|_{L_q([a,b])} \right\}}{\Gamma(\nu) \left(\nu + \frac{1}{p} \right) (p(\nu-1)+1)^{\frac{1}{p}}} \left[(t-a)^{\nu+\frac{1}{p}} + (b-t)^{\nu+\frac{1}{p}} \right], \quad (4.22)$$

$\forall\, t \in [a,b]$,

(ii) *at* $t = \frac{a+b}{2}$, *the right hand side of (4.22) is minimized, and we get:*

$$\left| \int_a^b f(x)\,dx - \sum_{k=0}^{n-1} \frac{1}{(k+1)!} \frac{(b-a)^{k+1}}{2^{k+1}} \left[f^{(k)}(a) + (-1)^k f^{(k)}(b) \right] \right| \leq$$

$$\frac{\max\left\{ \left\| D_{a+}^{\nu} f \right\|_{L_q([a,b])}, \left\| D_{b-}^{\nu} f \right\|_{L_q([a,b])} \right\}}{\Gamma(\nu)\left(\nu+\frac{1}{p}\right)(p(\nu-1)+1)^{\frac{1}{p}}} \frac{(b-a)^{\nu+\frac{1}{p}}}{2^{\nu-\frac{1}{q}}}, \tag{4.23}$$

(iii) if $f^{(k)}(a) = f^{(k)}(b) = 0$, *for all* $k = 0, 1, \ldots, n-1$, *we obtain*

$$\left| \int_a^b f(x)\,dx \right| \leq \frac{\max\left\{ \left\| D_{a+}^{\nu} f \right\|_{L_q([a,b])}, \left\| D_{b-}^{\nu} f \right\|_{L_q([a,b])} \right\}}{\Gamma(\nu)\left(\nu+\frac{1}{p}\right)(p(\nu-1)+1)^{\frac{1}{p}}} \frac{(b-a)^{\nu+\frac{1}{p}}}{2^{\nu-\frac{1}{q}}}, \tag{4.24}$$

which is a sharp inequality,

(iv) more generally, for $j = 0, 1, 2, \ldots, N \in \mathbb{N}$, *it holds*

$$\left| \int_a^b f(x)\,dx - \sum_{k=0}^{n-1} \frac{1}{(k+1)!} \left(\frac{b-a}{N}\right)^{k+1} \left[j^{k+1} f^{(k)}(a) + (-1)^k (N-j)^{k+1} f^{(k)}(b) \right] \right|$$

$$\leq \frac{\max\left\{ \left\| D_{a+}^{\nu} f \right\|_{L_q([a,b])}, \left\| D_{b-}^{\nu} f \right\|_{L_q([a,b])} \right\}}{\Gamma(\nu)\left(\nu+\frac{1}{p}\right)(p(\nu-1)+1)^{\frac{1}{p}}} \left(\frac{b-a}{N}\right)^{\nu+\frac{1}{p}} \left[j^{\nu+\frac{1}{p}} + (N-j)^{\nu+\frac{1}{p}} \right], \tag{4.25}$$

(v) if $f^{(k)}(a) = f^{(k)}(b) = 0$, $k = 1, \ldots, n-1$, *from (4.25) we get:*

$$\left| \int_a^b f(x)\,dx - \left(\frac{b-a}{N}\right) [j f(a) + (N-j) f(b)] \right| \leq$$

$$\frac{\max\left\{ \left\| D_{a+}^{\nu} f \right\|_{L_q([a,b])}, \left\| D_{b-}^{\nu} f \right\|_{L_q([a,b])} \right\}}{\Gamma(\nu)\left(\nu+\frac{1}{p}\right)(p(\nu-1)+1)^{\frac{1}{p}}} \left(\frac{b-a}{N}\right)^{\nu+\frac{1}{p}} \left[j^{\nu+\frac{1}{p}} + (N-j)^{\nu+\frac{1}{p}} \right], \tag{4.26}$$

for $j = 0, 1, 2, \ldots, N$,

(vi) when $N = 2$ *and* $j = 1$, *(4.26) turns to*

$$\left| \int_a^b f(x)\,dx - \left(\frac{b-a}{2}\right) (f(a) + f(b)) \right| \leq$$

$$\frac{\max\left\{ \left\| D_{a+}^{\nu} f \right\|_{L_q([a,b])}, \left\| D_{b-}^{\nu} f \right\|_{L_q([a,b])} \right\}}{\Gamma(\nu)\left(\nu+\frac{1}{p}\right)(p(\nu-1)+1)^{\frac{1}{p}}} \frac{(b-a)^{\nu+\frac{1}{p}}}{2^{\nu-\frac{1}{q}}}. \tag{4.27}$$

Next we follow [4].

Let $g : [a, b] \to \mathbb{R}$ be a strictly increasing function. Let $f \in C^n([a, b])$, $n \in \mathbb{N}$. Assume that $g \in C^1([a, b])$, and $g^{-1} \in C^n([a, b])$. Call $l := f \circ g^{-1} : [g(a), g(b)] \to \mathbb{R}$. It is clear that $l, l', \ldots, l^{(n)}$ are continuous functions from $[g(a), g(b)]$ into $f([a, b]) \subseteq \mathbb{R}$.

Let $\nu \geq 1$ such that $[\nu] = n$, $n \in \mathbb{N}$ as above, where $[\cdot]$ is the integral part of the number.

Clearly when $0 < \nu < 1$, $[\nu] = 0$. Next we follow [1, pp. 7–9].

(I) Let $h \in C([g(a), g(b)])$, we define the left Riemann-Liouville fractional integral as

$$\left(J_\nu^{z_0} h\right)(z) := \frac{1}{\Gamma(\nu)} \int_{z_0}^{z} (z - t)^{\nu - 1} h(t)\, dt, \tag{4.28}$$

for $g(a) \leq z_0 \leq z \leq g(b)$, where Γ is the gamma function; $\Gamma(\nu) = \int_0^\infty e^{-t} t^{\nu - 1} dt$.

We set $J_0^{z_0} h = h$.

Let $\alpha := \nu - [\nu]$ $(0 < \alpha < 1)$. We define the subspace $C_{g(x_0)}^{\nu}([g(a), g(b)])$ of $C^{[\nu]}([g(a), g(b)])$, where $x_0 \in [a, b]$:

$$C_{g(x_0)}^{\nu}([g(a), g(b)]) := \left\{h \in C^{[\nu]}([g(a), g(b)]) : J_{1-\alpha}^{g(x_0)} h^{([\nu])} \in C^1([g(x_0), g(b)])\right\}. \tag{4.29}$$

So let $h \in C_{g(x_0)}^{\nu}([g(a), g(b)])$; we define the left g-generalized fractional derivative of h of order ν, of Canavati type, over $[g(x_0), g(b)]$ as

$$D_{g(x_0)}^{\nu} h := \left(J_{1-\alpha}^{g(x_0)} h^{([\nu])}\right)'. \tag{4.30}$$

Clearly, for $h \in C_{g(x_0)}^{\nu}([g(a), g(b)])$, there exists

$$\left(D_{g(x_0)}^{\nu} h\right)(z) = \frac{1}{\Gamma(1-\alpha)} \frac{d}{dz} \int_{g(x_0)}^{z} (z - t)^{-\alpha} h^{([\nu])}(t)\, dt, \tag{4.31}$$

for all $g(x_0) \leq z \leq g(b)$.

In particular, when $f \circ g^{-1} \in C_{g(x_0)}^{\nu}([g(a), g(b)])$ we have that

$$\left(D_{g(x_0)}^{\nu}\left(f \circ g^{-1}\right)\right)(z) = \frac{1}{\Gamma(1-\alpha)} \frac{d}{dz} \int_{g(x_0)}^{z} (z - t)^{-\alpha} \left(f \circ g^{-1}\right)^{([\nu])}(t)\, dt, \tag{4.32}$$

for all $g(x_0) \leq z \leq g(b)$. We have $D_{g(x_0)}^{n}\left(f \circ g^{-1}\right) = \left(f \circ g^{-1}\right)^{(n)}$ and $D_{g(x_0)}^{0}\left(f \circ g^{-1}\right) = f \circ g^{-1}$.

(II) Next we follow [3, pp. 345–348].
Let $h \in C([g(a), g(b)])$, we define the right Riemann-Liouville fractional integral as

$$\left(J_{z_0-}^{\nu} h\right)(z) := \frac{1}{\Gamma(\nu)} \int_z^{z_0} (t-z)^{\nu-1} h(t)\, dt, \tag{4.33}$$

for $g(a) \leq z \leq z_0 \leq g(b)$. We set $J_{z_0-}^{0} h = h$.
Let $\alpha := \nu - [\nu]$ $(0 < \alpha < 1)$. We define the subspace $C_{g(x_0)-}^{\nu}([g(a), g(b)])$ of $C^{[\nu]}([g(a), g(b)])$, where $x_0 \in [a, b]$:

$$C_{g(x_0)-}^{\nu}([g(a), g(b)]) :=$$

$$\left\{h \in C^{[\nu]}([g(a), g(b)]) : J_{g(x_0)-}^{1-\alpha} h^{([\nu])} \in C^1([g(a), g(x_0)])\right\}. \tag{4.34}$$

So let $h \in C_{g(x_0)-}^{\nu}([g(a), g(b)])$; we define the right g-generalized fractional derivative of h of order ν, of Canavati type, over $[g(a), g(x_0)]$ as

$$D_{g(x_0)-}^{\nu} h := (-1)^{n-1} \left(J_{g(x_0)-}^{1-\alpha} h^{([\nu])}\right)'. \tag{4.35}$$

Clearly, for $h \in C_{g(x_0)-}^{\nu}([g(a), g(b)])$, there exists

$$\left(D_{g(x_0)-}^{\nu} h\right)(z) = \frac{(-1)^{n-1}}{\Gamma(1-\alpha)} \frac{d}{dz} \int_z^{g(x_0)} (t-z)^{-\alpha} h^{([\nu])}(t)\, dt, \tag{4.36}$$

for all $g(a) \leq z \leq g(x_0) \leq g(b)$.
In particular, when $f \circ g^{-1} \in C_{g(x_0)-}^{\nu}([g(a), g(b)])$ we have that

$$\left(D_{g(x_0)-}^{\nu}\left(f \circ g^{-1}\right)\right)(z) = \frac{(-1)^{n-1}}{\Gamma(1-\alpha)} \frac{d}{dz} \int_z^{g(x_0)} (t-z)^{-\alpha} \left(f \circ g^{-1}\right)^{([\nu])}(t)\, dt, \tag{4.37}$$

for all $g(a) \leq z \leq g(x_0) \leq g(b)$.
We get that

$$\left(D_{g(x_0)-}^{n}\left(f \circ g^{-1}\right)\right)(z) = (-1)^n \left(f \circ g^{-1}\right)^{(n)}(z) \tag{4.38}$$

and $\left(D_{g(x_0)-}^{0}\left(f \circ g^{-1}\right)\right)(z) = \left(f \circ g^{-1}\right)(z)$, all $z \in [g(a), g(x_0)]$.
Let g be strictly increasing and continuous over $[a, b]$, and $f \in C([a, b])$. We have that

$$\int_a^b f(x)\, dg(x) = \int_a^b \left(f \circ g^{-1}\right)(g(x))\, dg(x) = \int_{g(a)}^{g(b)} \left(f \circ g^{-1}\right)(z)\, dz. \tag{4.39}$$

Here it is $f \circ g^{-1} \in C([g(a), g(b)])$.

4.2 Main Results—I

Next we present generalized g-fractional Iyengar type inequalities:

Theorem 4.5 *Let $\nu \geq 1$ such that $[\nu] = n \in \mathbb{N}$, and $g : [a, b] \to \mathbb{R}$ be a strictly increasing function. Assume that $f \in C^n([a, b])$, $g \in C^1([a, b])$, $g^{-1} \in C^n([g(a), g(b)])$.*

Assume further that $f \circ g^{-1} \in C^{\nu}_{g(a)}([g(a), g(b)]) \cap C^{\nu}_{g(b)-}([g(a), g(b)])$. Set

$$M_1(f, g) := \max\left\{ \left\| D^{\nu}_{g(a)}\left(f \circ g^{-1}\right) \right\|_{\infty,[g(a),g(b)]}, \left\| D^{\nu}_{g(b)-}\left(f \circ g^{-1}\right) \right\|_{\infty,[g(a),g(b)]} \right\}. \tag{4.40}$$

Then

(i)

$$\left| \int_a^b f(x)\, dg(x) - \sum_{k=0}^{n-1} \frac{1}{(k+1)!} \left[\left(f \circ g^{-1}\right)^{(k)} (g(a))(z - g(a))^{k+1} \right.\right.$$

$$\left.\left. +(-1)^k \left(f \circ g^{-1}\right)^{(k)} (g(b))(g(b) - z)^{k+1} \right] \right| \leq$$

$$\frac{M_1(f, g)}{\Gamma(\nu+2)} \left[(z - g(a))^{\nu+1} + (g(b) - z)^{\nu+1} \right], \tag{4.41}$$

$\forall\, z \in [g(a), g(b)]$,

(ii) at $z = \frac{g(a)+g(b)}{2}$, the right hand side of (4.41) is minimized, and we get:

$$\left| \int_a^b f(x)\, dg(x) - \sum_{k=0}^{n-1} \frac{1}{(k+1)!} \frac{(g(b) - g(a))^{k+1}}{2^{k+1}} \right.$$

$$\left. \left[\left(f \circ g^{-1}\right)^{(k)} (g(a)) + (-1)^k \left(f \circ g^{-1}\right)^{(k)} (g(b)) \right] \right| \leq$$

$$\frac{M_1(f, g)}{\Gamma(\nu+2)} \frac{(g(b) - g(a))^{\nu+1}}{2^{\nu}}, \tag{4.42}$$

(iii) if $\left(f \circ g^{-1}\right)^{(k)} (g(a)) = \left(f \circ g^{-1}\right)^{(k)} (g(b)) = 0$, for all $k = 0, 1, \ldots, n-1$, we obtain

$$\left| \int_a^b f(x)\, dg(x) \right| \leq \frac{M_1(f, g)}{\Gamma(\nu+2)} \frac{(g(b) - g(a))^{\nu+1}}{2^{\nu}}, \tag{4.43}$$

which is a sharp inequality,

(iv) *more generally, for* $j = 0, 1, 2, \ldots, N \in \mathbb{N}$, *it holds*

$$\left| \int_a^b f(x)\, dg(x) - \sum_{k=0}^{n-1} \frac{1}{(k+1)!} \left(\frac{g(b) - g(a)}{N} \right)^{k+1} \right.$$

$$\left. \left[j^{k+1} \left(f \circ g^{-1} \right)^{(k)} (g(a)) + (-1)^k (N-j)^{k+1} \left(f \circ g^{-1} \right)^{(k)} (g(b)) \right] \right|$$

$$\leq \frac{M_1(f,g)}{\Gamma(\nu+2)} \left(\frac{g(b) - g(a)}{N} \right)^{\nu+1} \left[j^{\nu+1} + (N-j)^{\nu+1} \right], \tag{4.44}$$

(v) *if* $\left(f \circ g^{-1} \right)^{(k)} (g(a)) = \left(f \circ g^{-1} \right)^{(k)} (g(b)) = 0, \quad k = 1, \ldots, n-1,$ *from (4.44) we get:*

$$\left| \int_a^b f(x)\, dg(x) - \left(\frac{g(b) - g(a)}{N} \right) [jf(a) + (N-j) f(b)] \right| \leq$$

$$\frac{M_1(f,g)}{\Gamma(\nu+2)} \left(\frac{g(b) - g(a)}{N} \right)^{\nu+1} \left[j^{\nu+1} + (N-j)^{\nu+1} \right], \tag{4.45}$$

$j = 0, 1, 2, \ldots, N,$

(vi) *when* $N = 2$ *and* $j = 1$, *(4.45) turns to*

$$\left| \int_a^b f(x)\, dg(x) - \left(\frac{g(b) - g(a)}{2} \right) (f(a) + f(b)) \right| \leq$$

$$\frac{M_1(f,g)}{\Gamma(\nu+2)} \frac{(g(b) - g(a))^{\nu+1}}{2^\nu}. \tag{4.46}$$

Proof Apply Theorem 4.2 for $f \circ g^{-1}$ over $[g(a), g(b)]$ and take into account (4.39). ■

Next come L_1 estimates:

Theorem 4.6 *All as in Theorem 4.5. Set*

$$M_2(f,g) := \max \left\{ \left\| D_{g(a)}^\nu \left(f \circ g^{-1} \right) \right\|_{L_1[g(a), g(b)]}, \left\| D_{g(b)-}^\nu \left(f \circ g^{-1} \right) \right\|_{L_1[g(a), g(b)]} \right\}. \tag{4.47}$$

Then

(i)

$$\left|\int_a^b f(x)\,dg(x) - \sum_{k=0}^{n-1}\frac{1}{(k+1)!}\left[\left(f\circ g^{-1}\right)^{(k)}(g(a))\,(z-g(a))^{k+1}\right.\right.$$

$$\left.\left. +(-1)^k\left(f\circ g^{-1}\right)^{(k)}(g(b))\,(g(b)-z)^{k+1}\right]\right| \leq$$

$$\frac{M_2(f,g)}{\Gamma(\nu+1)}\left[(z-g(a))^{\nu}+(g(b)-z)^{\nu}\right], \tag{4.48}$$

$\forall\, z \in [g(a), g(b)]$,

(ii) *when* $\nu = 1$, *from (4.48), we have*

$$\left|\int_a^b f(x)\,dg(x) - [f(a)(z-g(a)) + f(b)(g(b)-z)]\right| \leq$$

$$\left\|\left(f\circ g^{-1}\right)'\right\|_{L_1([g(a),g(b)])}(g(b)-g(a)), \tag{4.49}$$

$\forall\, z \in [g(a), g(b)]$,

(iii) *from (4.49) we obtain* ($\nu = 1$ *case)*

$$\left|\int_a^b f(x)\,dg(x) - \left(\frac{g(b)-g(a)}{2}\right)(f(a)+f(b))\right| \leq$$

$$\left\|\left(f\circ g^{-1}\right)'\right\|_{L_1([g(a),g(b)])}(g(b)-g(a)), \tag{4.50}$$

(iv) *at* $z = \frac{g(a)+g(b)}{2}$, $\nu > 1$, *the right hand side of (4.48) is minimized, and we get:*

$$\left|\int_a^b f(x)\,dg(x) - \sum_{k=0}^{n-1}\frac{1}{(k+1)!}\frac{(g(b)-g(a))^{k+1}}{2^{k+1}}\right.$$

$$\left.\left[\left(f\circ g^{-1}\right)^{(k)}(g(a)) + (-1)^k\left(f\circ g^{-1}\right)^{(k)}(g(b))\right]\right| \leq$$

$$\frac{M_2(f,g)}{\Gamma(\nu+1)}\frac{(g(b)-g(a))^{\nu}}{2^{\nu-1}}, \tag{4.51}$$

(v) *if* $\left(f\circ g^{-1}\right)^{(k)}(g(a)) = \left(f\circ g^{-1}\right)^{(k)}(g(b)) = 0$, *for all* $k = 0, 1, \ldots, n-1$; $\nu > 1$, *from (4.51), we obtain*

$$\left|\int_a^b f(x)\,dg(x)\right| \leq \frac{M_2(f,g)}{\Gamma(\nu+1)} \frac{(g(b)-g(a))^{\nu}}{2^{\nu-1}}, \tag{4.52}$$

which is a sharp inequality,

(vi) more generally, for $j = 0, 1, 2, \ldots, N \in \mathbb{N}$*, it holds*

$$\left|\int_a^b f(x)\,dg(x) - \sum_{k=0}^{n-1} \frac{1}{(k+1)!} \left(\frac{g(b)-g(a)}{N}\right)^{k+1}\right.$$

$$\left.\left[j^{k+1}\left(f \circ g^{-1}\right)^{(k)}(g(a)) + (-1)^k (N-j)^{k+1} \left(f \circ g^{-1}\right)^{(k)}(g(b))\right]\right|$$

$$\leq \frac{M_2(f,g)}{\Gamma(\nu+1)} \left(\frac{g(b)-g(a)}{N}\right)^{\nu} \left[j^{\nu} + (N-j)^{\nu}\right], \tag{4.53}$$

(vii) if $\left(f \circ g^{-1}\right)^{(k)}(g(a)) = \left(f \circ g^{-1}\right)^{(k)}(g(b)) = 0, \quad k = 1, \ldots, n-1,$ *from (4.53) we get:*

$$\left|\int_a^b f(x)\,dg(x) - \left(\frac{g(b)-g(a)}{N}\right)[jf(a) + (N-j)f(b)]\right| \leq$$

$$\frac{M_2(f,g)}{\Gamma(\nu+1)} \left(\frac{g(b)-g(a)}{N}\right)^{\nu} \left[j^{\nu} + (N-j)^{\nu}\right], \tag{4.54}$$

$j = 0, 1, 2, \ldots, N,$

(viii) when $N = 2$ *and* $j = 1$*, (4.54) turns to*

$$\left|\int_a^b f(x)\,dg(x) - \left(\frac{g(b)-g(a)}{2}\right)(f(a) + f(b))\right| \leq$$

$$\frac{M_2(f,g)}{\Gamma(\nu+1)} \frac{(g(b)-g(a))^{\nu}}{2^{\nu-1}}. \tag{4.55}$$

Proof Application of Theorem 4.3 for $f \circ g^{-1}$ over $[g(a), g(b)]$ and take into account (4.39). ■

We continue with L_p estimates:

Theorem 4.7 *All as in Theorem 4.5. Let* $p, q > 1 : \frac{1}{p} + \frac{1}{q} = 1$*. Set*

$$M_3(f,g) := \max\left\{\left\|D_{g(a)}^{\nu}\left(f \circ g^{-1}\right)\right\|_{L_q[g(a),g(b)]}, \left\|D_{g(b)-}^{\nu}\left(f \circ g^{-1}\right)\right\|_{L_q[g(a),g(b)]}\right\}. \tag{4.56}$$

Then

(i)

$$\left|\int_a^b f(x)\,dg(x) - \sum_{k=0}^{n-1}\frac{1}{(k+1)!}\left[\left(f\circ g^{-1}\right)^{(k)}(g(a))(z-g(a))^{k+1}\right.\right.$$

$$\left.\left.+(-1)^k\left(f\circ g^{-1}\right)^{(k)}(g(b))(g(b)-z)^{k+1}\right]\right| \leq$$

$$\frac{M_3(f,g)}{\Gamma(\nu)\left(\nu+\frac{1}{p}\right)(p(\nu-1)+1)^{\frac{1}{p}}}\left[(z-g(a))^{\nu+\frac{1}{p}}+(g(b)-z)^{\nu+\frac{1}{p}}\right], \tag{4.57}$$

$\forall\, z\in[g(a),g(b)]$,

(ii) *at* $z=\frac{g(a)+g(b)}{2}$*, the right hand side of (4.57) is minimized, and we get:*

$$\left|\int_a^b f(x)\,dg(x) - \sum_{k=0}^{n-1}\frac{1}{(k+1)!}\frac{(g(b)-g(a))^{k+1}}{2^{k+1}}\right.$$

$$\left.\left[\left(f\circ g^{-1}\right)^{(k)}(g(a))+(-1)^k\left(f\circ g^{-1}\right)^{(k)}(g(b))\right]\right| \leq$$

$$\frac{M_3(f,g)}{\Gamma(\nu)\left(\nu+\frac{1}{p}\right)(p(\nu-1)+1)^{\frac{1}{p}}}\frac{(g(b)-g(a))^{\nu+\frac{1}{p}}}{2^{\nu-\frac{1}{q}}}, \tag{4.58}$$

(iii) *if* $\left(f\circ g^{-1}\right)^{(k)}(g(a))=\left(f\circ g^{-1}\right)^{(k)}(g(b))=0$*, for all* $k=0,1,\ldots,n-1$*, we obtain*

$$\left|\int_a^b f(x)\,dg(x)\right| \leq \frac{M_3(f,g)}{\Gamma(\nu)\left(\nu+\frac{1}{p}\right)(p(\nu-1)+1)^{\frac{1}{p}}}\frac{(g(b)-g(a))^{\nu+\frac{1}{p}}}{2^{\nu-\frac{1}{q}}}, \tag{4.59}$$

which is a sharp inequality,

(iv) *more generally, for* $j=0,1,2,\ldots,N\in\mathbb{N}$*, it holds*

$$\left|\int_a^b f(x)\,dg(x) - \sum_{k=0}^{n-1}\frac{1}{(k+1)!}\left(\frac{g(b)-g(a)}{N}\right)^{k+1}\right.$$

$$\left.\left[j^{k+1}\left(f\circ g^{-1}\right)^{(k)}(g(a))+(-1)^k(N-j)^{k+1}\left(f\circ g^{-1}\right)^{(k)}(g(b))\right]\right| \leq$$

$$\frac{M_3(f,g)}{\Gamma(\nu)\left(\nu+\frac{1}{p}\right)(p(\nu-1)+1)^{\frac{1}{p}}}\left(\frac{g(b)-g(a)}{N}\right)^{\nu+\frac{1}{p}}\left[j^{\nu+\frac{1}{p}}+(N-j)^{\nu+\frac{1}{p}}\right], \tag{4.60}$$

(v) *if* $\left(f \circ g^{-1}\right)^{(k)}(g(a)) = \left(f \circ g^{-1}\right)^{(k)}(g(b)) = 0, \quad k = 1, \ldots, n-1,$ *from* (4.60) *we get:*

$$\left| \int_a^b f(x)\, dg(x) - \left(\frac{g(b) - g(a)}{N}\right) [jf(a) + (N - j) f(b)] \right| \leq$$

$$\frac{M_3(f, g)}{\Gamma(\nu)\left(\nu + \frac{1}{p}\right)(p(\nu - 1) + 1)^{\frac{1}{p}}} \left(\frac{g(b) - g(a)}{N}\right)^{\nu + \frac{1}{p}} \left[j^{\nu + \frac{1}{p}} + (N - j)^{\nu + \frac{1}{p}}\right], \tag{4.61}$$

$j = 0, 1, 2, \ldots, N,$

(vi) *when* $N = 2$ *and* $j = 1$, *(4.61) turns to*

$$\left| \int_a^b f(x)\, dg(x) - \left(\frac{g(b) - g(a)}{2}\right)(f(a) + f(b)) \right| \leq$$

$$\frac{M_3(f, g)}{\Gamma(\nu)\left(\nu + \frac{1}{p}\right)(p(\nu - 1) + 1)^{\frac{1}{p}}} \frac{(g(b) - g(a))^{\nu + \frac{1}{p}}}{2^{\nu - \frac{1}{q}}}. \tag{4.62}$$

Proof Application of Theorem 4.4 to $f \circ g^{-1}$ over $[g(a), g(b)]$ and using (4.39). ■

4.3 Background—II

In 1938, Ostsrowski [9] proved the following important inequality.

Theorem 4.8 *Let* $f : [a, b] \to \mathbb{R}$ *be continuous on* $[a, b]$ *and differentiable on* (a, b) *whose derivative* $f' : (a, b) \to \mathbb{R}$ *is bounded on* (a, b), *i.e.,* $\|f'\|_\infty := \sup_{t \in (a,b)} |f'(t)| < +\infty$. *Then*

$$\left| \frac{1}{b - a} \int_a^b f(t)\, dt - f(x) \right| \leq \left[\frac{1}{4} + \frac{\left(x - \frac{a+b}{2}\right)^2}{(b - a)^2} \right] \cdot (b - a) \|f'\|_\infty, \tag{4.63}$$

for any $x \in [a, b]$. *The constant* $\frac{1}{4}$ *is the best possible.*

Since then there has been a lot of activity around these inequalities with important applications to numerical analysis and probability.

We mention and need the following left generalized g-fractional, of Canavati type, Taylor's formula:

Theorem 4.9 ([4]) *Let* $f \circ g^{-1} \in C^{\nu}_{g(x_0)}([g(a), g(b)])$, *where* $x_0 \in [a, b]$ *is fixed,* $\nu \geq 1$. *Then*

$$f(x) - f(x_0) = \sum_{k=1}^{[\nu]-1} \frac{\left(f \circ g^{-1}\right)^{(k)}(g(x_0))}{k!}(g(x) - g(x_0))^k +$$

$$\frac{1}{\Gamma(\nu)} \int_{g(x_0)}^{g(x)} (g(x) - t)^{\nu-1} \left(D_{g(x_0)}^{\nu}\left(f \circ g^{-1}\right)\right)(t)\,dt, \text{ all } x \in [a,b] : x \geq x_0. \tag{4.64}$$

We also mention and need the following right generalized g-fractional, of Canavati type, Taylor's formula:

Theorem 4.10 ([4]) *Let $f \circ g^{-1} \in C_{g(x_0)-}^{\nu}([g(a), g(b)])$, where $x_0 \in [a,b]$ is fixed, $\nu \geq 1$. Then*

$$f(x) - f(x_0) = \sum_{k=1}^{[\nu]-1} \frac{\left(f \circ g^{-1}\right)^{(k)}(g(x_0))}{k!}(g(x) - g(x_0))^k +$$

$$\frac{1}{\Gamma(\nu)} \int_{g(x)}^{g(x_0)} (t - g(x))^{\nu-1} \left(D_{g(x_0)-}^{\nu}\left(f \circ g^{-1}\right)\right)(t)\,dt, \text{ all } a \leq x \leq x_0. \tag{4.65}$$

4.4 Main Results—II

Next we present generalized g-fractional Ostrowski type inequalities:

Theorem 4.11 *Let $g : [a,b] \to \mathbb{R}$ be a strictly increasing function, $\nu \geq 1$, $[\nu] = n \in \mathbb{N}$, $f \in C^n([a,b])$. Assume that $g \in C^1([a,b])$, and $g^{-1} \in C^n([g(a), g(b)])$. For $x_0 \in [a,b]$, assume that $f \circ g^{-1} \in C_{g(x_0)}^{\nu}([g(a), g(b)])$ and $f \circ g^{-1} \in C_{g(x_0)-}^{\nu}([g(a), g(b)])$. Furthermore assume that $\left(f \circ g^{-1}\right)^{(k)}(g(x_0)) = 0$, all $k = 1, \ldots, n-1$. Then*

$$\left| \frac{1}{(g(b) - g(a))} \int_a^b f(x)\,dg(x) - f(x_0) \right| \leq$$

$$\frac{1}{(g(b) - g(a))} \Big\{ \left\| D_{g(x_0)-}^{\nu}\left(f \circ g^{-1}\right) \right\|_{\infty,[g(a),g(x_0)]} (g(x_0) - g(a))^{\nu+1} +$$

$$\left\| D_{g(x_0)}^{\nu}\left(f \circ g^{-1}\right) \right\|_{\infty,[g(x_0),g(b)]} (g(b) - g(x_0))^{\nu+1} \Big\} \leq \tag{4.66}$$

$$\frac{1}{\Gamma(\nu+2)} \max\Big\{ \left\| D_{g(x_0)-}^{\nu}\left(f \circ g^{-1}\right) \right\|_{\infty,[g(a),g(x_0)]},$$

$$\left\| D_{g(x_0)}^{\nu}\left(f \circ g^{-1}\right) \right\|_{\infty,[g(x_0),g(b)]} \Big\} (g(b) - g(a))^{\nu}.$$

Proof By Theorem 4.9, when $\left(f \circ g^{-1}\right)^{(k)}(g(x_0)) = 0$, for $k = 1, \ldots, n-1$, we get

$$f(x) - f(x_0) = \frac{1}{\Gamma(\nu)} \int_{g(x_0)}^{g(x)} (g(x) - t)^{\nu-1} \left(D_{g(x_0)}^{\nu}\left(f \circ g^{-1}\right)\right)(t)\, dt, \qquad (4.67)$$

$\forall\, x \in [x_0, b]$.

By Theorem 4.10, when $\left(f \circ g^{-1}\right)^{(k)}(g(x_0)) = 0$, for $k = 1, \ldots, n-1$, we get

$$f(x) - f(x_0) = \frac{1}{\Gamma(\nu)} \int_{g(x)}^{g(x_0)} (t - g(x))^{\nu-1} \left(D_{g(x_0)-}^{\nu}\left(f \circ g^{-1}\right)\right)(t)\, dt, \qquad (4.68)$$

$\forall\, x \in [a, x_0]$.

Hence

$$|f(x) - f(x_0)| \leq \frac{1}{\Gamma(\nu)} \int_{g(x_0)}^{g(x)} (g(x) - t)^{\nu-1} \left|\left(D_{g(x_0)}^{\nu}\left(f \circ g^{-1}\right)\right)(t)\right| dt$$

$$\leq \frac{1}{\Gamma(\nu)} \left(\int_{g(x_0)}^{g(x)} (g(x) - t)^{\nu-1}\, dt\right) \left\|D_{g(x_0)}^{\nu}\left(f \circ g^{-1}\right)\right\|_{\infty,[g(x_0),g(b)]}$$

$$= \frac{1}{\Gamma(\nu)} \frac{(g(x) - g(x_0))^{\nu}}{\nu} \left\|D_{g(x_0)}^{\nu}\left(f \circ g^{-1}\right)\right\|_{\infty,[g(x_0),g(b)]}$$

$$= \frac{(g(x) - g(x_0))^{\nu}}{\Gamma(\nu+1)} \left\|D_{g(x_0)}^{\nu}\left(f \circ g^{-1}\right)\right\|_{\infty,[g(x_0),g(b)]}, \qquad (4.69)$$

$\forall\, x \in [x_0, b]$.

That is

$$|f(x) - f(x_0)| \leq \frac{\left\|D_{g(x_0)}^{\nu}\left(f \circ g^{-1}\right)\right\|_{\infty,[g(x_0),g(b)]}}{\Gamma(\nu+1)} (g(x) - g(x_0))^{\nu}, \qquad (4.70)$$

$\forall\, x \in [x_0, b]$.

Similarly, it holds

$$|f(x) - f(x_0)| \leq \frac{1}{\Gamma(\nu)} \int_{g(x)}^{g(x_0)} (t - g(x))^{\nu-1} \left|\left(D_{g(x_0)-}^{\nu}\left(f \circ g^{-1}\right)\right)(t)\right| dt$$

$$\leq \frac{1}{\Gamma(\nu)} \left(\int_{g(x)}^{g(x_0)} (t - g(x))^{\nu-1}\, dt\right) \left\|D_{g(x_0)-}^{\nu}\left(f \circ g^{-1}\right)\right\|_{\infty,[g(a),g(x_0)]}$$

$$= \frac{1}{\Gamma(\nu+1)} (g(x_0) - g(x))^{\nu} \left\|D_{g(x_0)-}^{\nu}\left(f \circ g^{-1}\right)\right\|_{\infty,[g(a),g(x_0)]}, \qquad (4.71)$$

$\forall\, x \in [a, x_0]$.

That is

$$|f(x) - f(x_0)| \leq \frac{\left\| D^{\nu}_{g(x_0)-} \left(f \circ g^{-1}\right) \right\|_{\infty,[g(a),g(x_0)]}}{\Gamma(\nu+1)} (g(x_0) - g(x))^{\nu}, \quad (4.72)$$

$\forall\, x \in [a, x_0]$.

We observe that

$$\left| \frac{1}{(g(b) - g(a))} \int_a^b f(x)\, g'(x)\, dx - f(x_0) \right| =$$

$$\left| \frac{1}{(g(b) - g(a))} \int_a^b f(x)\, g'(x)\, dx - f(x_0)(g(b) - g(a)) \right| =$$

$$\left| \frac{1}{(g(b) - g(a))} \left[\int_a^b f(x)\, g'(x)\, dx - \int_a^b f(x_0)\, g'(x)\, dx \right] \right| =$$

$$\left| \frac{1}{(g(b) - g(a))} \left[\int_a^b (f(x) - f(x_0))\, g'(x)\, dx \right] \right| \leq \quad (4.73)$$

$$\frac{1}{(g(b) - g(a))} \int_a^b |f(x) - f(x_0)|\, g'(x)\, dx =$$

$$\frac{1}{(g(b) - g(a))} \left\{ \int_a^{x_0} |f(x) - f(x_0)|\, g'(x)\, dx + \int_{x_0}^b |f(x) - f(x_0)|\, g'(x)\, dx \right\}$$

$$\overset{\text{(by (4.70), (4.72))}}{\leq} \frac{1}{(g(b) - g(a))} \left\{ \frac{\left\| D^{\nu}_{g(x_0)-} \left(f \circ g^{-1}\right) \right\|_{\infty,[g(a),g(x_0)]}}{\Gamma(\nu+1)} \right.$$

$$\int_a^{x_0} (g(x_0) - g(x))^{\nu}\, g'(x)\, dx +$$

$$\left. \frac{\left\| D^{\nu}_{g(x_0)} \left(f \circ g^{-1}\right) \right\|_{\infty,[g(x_0),g(b)]}}{\Gamma(\nu+1)} \int_{x_0}^b (g(x) - g(x_0))^{\nu}\, g'(x)\, dx \right\} = \quad (4.74)$$

$$\frac{1}{(g(b) - g(a))\, \Gamma(\nu+2)} \left\{ \left\| D^{\nu}_{g(x_0)-} \left(f \circ g^{-1}\right) \right\|_{\infty,[g(a),g(x_0)]} (g(x_0) - g(a))^{\nu+1} \right.$$

$$\left. + \left\| D^{\nu}_{g(x_0)} \left(f \circ g^{-1}\right) \right\|_{\infty,[g(x_0),g(b)]} (g(b) - g(x_0))^{\nu+1} \right\} \leq$$

$$\frac{1}{\Gamma(\nu+2)} \max \left\{ \left\| D^{\nu}_{g(x_0)-} \left(f \circ g^{-1}\right) \right\|_{\infty,[g(a),g(x_0)]}, \right.$$

$$\left. \left\| D^{\nu}_{g(x_0)} \left(f \circ g^{-1}\right) \right\|_{\infty,[g(x_0),g(b)]} \right\} (g(b) - g(a))^{\nu}. \tag{4.75}$$

■

We continue with an L_1-estimate:

Theorem 4.12 *All as in Theorem 4.11. Then*

$$\left| \frac{1}{(g(b) - g(a))} \int_a^b f(x)\, dg(x) - f(x_0) \right| \leq$$

$$\frac{1}{(g(b) - g(a))\,\Gamma(\nu+1)} \left\{ \left\| D^{\nu}_{g(x_0)-} \left(f \circ g^{-1}\right) \right\|_{L_1([g(a),g(x_0)])} (g(x_0) - g(a))^{\nu} \right.$$

$$\left. + \left\| D^{\nu}_{g(x_0)} \left(f \circ g^{-1}\right) \right\|_{L_1[g(x_0),g(b)]} (g(b) - g(x_0))^{\nu} \right\} \leq$$

$$\frac{1}{\Gamma(\nu+1)} \max \left\{ \left\| D^{\nu}_{g(x_0)-} \left(f \circ g^{-1}\right) \right\|_{L_1([g(a),g(x_0)])}, \right.$$

$$\left. \left\| D^{\nu}_{g(x_0)} \left(f \circ g^{-1}\right) \right\|_{L_1([g(x_0),g(b)])} \right\} (g(b) - g(a))^{\nu-1}. \tag{4.76}$$

Proof Similar to the proof of Theorem 4.11. ■

We continue with an L_p-estimate:

Theorem 4.13 *All as in Theorem 4.11, and let* $p, q > 1 : \frac{1}{p} + \frac{1}{q} = 1$. *Then*

$$\left| \frac{1}{(g(b) - g(a))} \int_a^b f(x)\, dg(x) - f(x_0) \right| \leq$$

$$\frac{1}{(g(b) - g(a))\,\Gamma(\nu)\,(p(\nu-1)+1)^{\frac{1}{p}} \left(\nu + \frac{1}{p}\right)}$$

$$\left\{ (g(x_0) - g(a))^{\nu+\frac{1}{p}} \left\| D^{\nu}_{g(x_0)-} \left(f \circ g^{-1}\right) \right\|_{L_q([g(a),g(x_0)])} \right.$$

$$\left. + (g(b) - g(x_0))^{\nu+\frac{1}{p}} \left\| D^{\nu}_{g(x_0)} \left(f \circ g^{-1}\right) \right\|_{L_q[g(x_0),g(b)]} \right\} \leq$$

$$\frac{1}{\Gamma(\nu)(p(\nu-1)+1)^{\frac{1}{p}}\left(\nu+\frac{1}{p}\right)}\max\left\{\left\|D^{\nu}_{g(x_0)-}\left(f\circ g^{-1}\right)\right\|_{L_q([g(a),g(x_0)])},\right.$$

$$\left.\left\|D^{\nu}_{g(x_0)}\left(f\circ g^{-1}\right)\right\|_{L_q([g(x_0),g(b)])}\right\}(g(b)-g(a))^{\nu-\frac{1}{q}}. \tag{4.77}$$

Proof Similar to the proof of Theorem 4.11. ∎

Applications follow:

Proposition 4.14 *Let $\nu \geq 1$ such that $[\nu] = n \in \mathbb{N}$, and $f \in C^n([a,b])$, where $[a,b] \subset (0,+\infty)$. Assume that $f \circ \ln x \in C^{\nu}_{e^a}([e^a,e^b]) \cap C^{\nu}_{e^b-}([e^a,e^b])$. Set*

$$M_1\left(f,e^x\right):=\max\left\{\left\|D^{\nu}_{e^a}(f\circ\ln x)\right\|_{\infty,[e^a,e^b]},\left\|D^{\nu}_{e^b-}(f\circ\ln x)\right\|_{\infty,[e^a,e^b]}\right\}. \tag{4.78}$$

Then

(i)

$$\left|\int_a^b f(x)e^x dx-\sum_{k=0}^{n-1}\frac{1}{(k+1)!}\left[(f\circ\ln x)^{(k)}\left(e^a\right)\left(z-e^a\right)^{k+1}\right.\right.$$

$$\left.\left.+(-1)^k(f\circ\ln x)^{(k)}\left(e^b\right)\left(e^b-z\right)^{k+1}\right]\right|\leq$$

$$\frac{M_1(f,e^x)}{\Gamma(\nu+2)}\left[\left(z-e^a\right)^{\nu+1}+\left(e^b-z\right)^{\nu+1}\right], \tag{4.79}$$

$\forall z \in [e^a,e^b]$,

(ii) at $z=\frac{e^a+e^b}{2}$, the right hand side of (4.79) is minimized, and we get:

$$\left|\int_a^b f(x)e^x dx-\sum_{k=0}^{n-1}\frac{1}{(k+1)!}\frac{\left(e^b-e^a\right)^{k+1}}{2^{k+1}}\right.$$

$$\left.\left[(f\circ\ln x)^{(k)}\left(e^a\right)+(-1)^k(f\circ\ln x)^{(k)}\left(e^b\right)\right]\right|\leq$$

$$\frac{M_1(f,e^x)}{\Gamma(\nu+2)}\frac{\left(e^b-e^a\right)^{\nu+1}}{2^{\nu}}, \tag{4.80}$$

(iii) if $(f\circ\ln x)^{(k)}(e^a)=(f\circ\ln x)^{(k)}(e^b)=0$, for all $k=0,1,\ldots,n-1$, we obtain

$$\left|\int_a^b f(x)e^x dx\right|\leq\frac{M_1(f,e^x)}{\Gamma(\nu+2)}\frac{\left(e^b-e^a\right)^{\nu+1}}{2^{\nu}}, \tag{4.81}$$

which is a sharp inequality,

(iv) *more generally, for* $j = 0, 1, 2, \ldots, N \in \mathbb{N}$, *it holds*

$$\left| \int_a^b f(x) e^x dx - \sum_{k=0}^{n-1} \frac{1}{(k+1)!} \left(\frac{e^b - e^a}{N} \right)^{k+1} \right.$$

$$\left. \left[j^{k+1} (f \circ \ln x)^{(k)} (e^a) + (-1)^k (N-j)^{k+1} (f \circ \ln x)^{(k)} \left(e^b\right) \right] \right|$$

$$\leq \frac{M_1(f, e^x)}{\Gamma(\nu+2)} \left(\frac{e^b - e^a}{N} \right)^{\nu+1} \left[j^{\nu+1} + (N-j)^{\nu+1} \right], \tag{4.82}$$

(v) *if* $(f \circ \ln x)^{(k)} (e^a) = (f \circ \ln x)^{(k)} \left(e^b\right) = 0$, $k = 1, \ldots, n-1$, *from (4.82) we get:*

$$\left| \int_a^b f(x) e^x dx - \left(\frac{e^b - e^a}{N} \right) [j f(a) + (N-j) f(b)] \right| \leq$$

$$\frac{M_1(f, e^x)}{\Gamma(\nu+2)} \left(\frac{e^b - e^a}{N} \right)^{\nu+1} \left[j^{\nu+1} + (N-j)^{\nu+1} \right], \tag{4.83}$$

$j = 0, 1, 2, \ldots, N$,

(vi) *when* $N = 2$ *and* $j = 1$, *(4.83) turns to*

$$\left| \int_a^b f(x) e^x dx - \left(\frac{e^b - e^a}{2} \right) (f(a) + f(b)) \right| \leq$$

$$\frac{M_1(f, e^x)}{\Gamma(\nu+2)} \frac{\left(e^b - e^a\right)^{\nu+1}}{2^\nu}. \tag{4.84}$$

Proof By application of Theorem 4.5 for $g(x) = e^x$. ■

We finish with

Proposition 4.15 *Let* $\nu \geq 1$, $[\nu] = n \in \mathbb{N}$, $f \in C^n([a,b])$, $[a,b] \subset (0, +\infty)$. *For* $x_0 \in [a,b]$, *assume that* $f \circ e^x \in C^\nu_{\ln x_0}([\ln a, \ln b])$, *and* $f \circ e^x \in C^\nu_{\ln x_0 -}([\ln a, \ln b])$. *Furthermore assume that* $(f \circ e^x)^{(k)} (\ln x_0) = 0$, *all* $k = 1, \ldots, n-1$. *Then*

$$\left| \frac{1}{\left(\ln \frac{b}{a}\right)} \int_a^b \frac{f(x)}{x} dx - f(x_0) \right| \leq$$

$$\frac{1}{\left(\ln \frac{b}{a}\right) \Gamma(\nu+2)} \left\{ \left\| D^\nu_{\ln x_0 -} (f \circ e^x) \right\|_{\infty, [\ln a, \ln x_0]} \left(\ln \frac{x_0}{a} \right)^{\nu+1} + \right.$$

$$\left. \left\| D^\nu_{\ln x_0} (f \circ e^x) \right\|_{\infty, [\ln x_0, \ln b]} \left(\ln \frac{b}{x_0} \right)^{\nu+1} \right\} \leq \tag{4.85}$$

$$\frac{1}{\Gamma(\nu+2)} \max \left\{ \left\| D^{\nu}_{\ln x_0 -} \left(f \circ e^x \right) \right\|_{\infty, [\ln a, \ln x_0]}, \right.$$

$$\left. \left\| D^{\nu}_{\ln x_0} \left(f \circ e^x \right) \right\|_{\infty, [\ln x_0, \ln b]} \right\} \left(\ln \frac{b}{a} \right)^{\nu}.$$

Proof By Theorem 4.11, for $g(x) = \ln x$. ■

References

1. Anastassiou, G.A.: Fractional Differentiation Inequalities. Research Monograph. Springer, New York (2009)
2. Anastassiou, G.A.: On right fractional calculus. Chaos Solitons Fractals **42**, 365–376 (2009)
3. Anastassiou, G.A.: Inteligent Mathematics: Computational Analysis. Springer, Heidelberg (2011)
4. Anastassiou, G.A.: Generalized Canavati type Fractional Taylor's formulae. J. Comput. Anal. Appl. **21**(7), 1205–1212 (2016)
5. Anastassiou, G.A.: Canavati fractional Iyengar type Inequalities. Analele Universitatii Oradea, Fasc. Matematica, Tom **XXVI**(1), 141–151 (2019)
6. Anastassiou, G.: Generalized Canavati type g-fractional Iyengar and Ostrowski type inequalities. Ann. Commun. Math. **2**(2), 57–72 (2019)
7. Canavati, J.A.: The Riemann-Liouville Integral. Nieuw Archief Voor Wiskunde **5**(1), 53–75 (1987)
8. Iyengar, K.S.K.: Note on a inequality. Math. Stud. **6**, 75–76 (1938)
9. Ostrowski, A.: Über die Absolutabweichung einer differtentiebaren Funktion von ihrem Integralmittelwert. Comment. Math. Helv. **10**, 226–227 (1938)

Chapter 5
Generalized Canavati g-Fractional Polya Inequalities

We present here generalized Canavati type g-fractional Polya type inequalities. We cover also the iterated case. Our inequalities are with respect to all L_p norms: $1 \leq p \leq \infty$. We finish with applications. See also [6].

5.1 Introduction

We are motivated by the following famous Polya's integral inequality, see [12], [13, p. 62], [14] and [15, p. 83].

Theorem 5.1 *Let $f(x)$ be differentiable and not identically a constant on $[a,b]$ with $f(a) = f(b) = 0$. Then there exists at least one point $\xi \in [a,b]$ such that*

$$\left|f'(\xi)\right| > \frac{4}{(b-a)^2} \int_a^b f(x)\,dx. \tag{5.1}$$

We need the following fractional calculus background:

Let $\alpha > 0, m = [\alpha], [\cdot]$ is the integral part, $\beta = \alpha - m, 0 < \beta < 1, f \in C([a,b])$, $[a,b] \subset \mathbb{R}$, $x \in [a,b]$. The gamma function Γ is given by $\Gamma(\alpha) = \int_0^\infty e^{-t} t^{\alpha-1} dt$. We define the left Riemann–Liouville integral

$$\left(J_\alpha^{a+} f\right)(x) = \frac{1}{\Gamma(\alpha)} \int_a^x (x-t)^{\alpha-1} f(t)\,dt, \tag{5.2}$$

$a \leq x \leq b$. We define the subspace $C_{a+}^\alpha([a,b])$ of $C^m([a,b])$:

$$C_{a+}^\alpha([a,b]) = \left\{f \in C^m([a,b]) : J_{1-\beta}^{a+} f^{(m)} \in C^1([a,b])\right\}. \tag{5.3}$$

G. A. Anastassiou, *Generalized Fractional Calculus*, Studies in Systems, Decision and Control 305, https://doi.org/10.1007/978-3-030-56962-4_5

For $f \in C_{a+}^{\alpha}([a,b])$, we define the left generalized α-fractional derivative of f over $[a,b]$ as

$$D_{a+}^{\alpha} f := \left(J_{1-\beta}^{a+} f^{(m)}\right)', \tag{5.4}$$

see [1, p. 24]. Canavati first in [7] introduced the above over $[0,1]$.

Notice that $D_{a+}^{\alpha} f \in C([a,b])$.

Furthermore we need:

Let again $\alpha > 0$, $m = [\alpha]$, $\beta = \alpha - m$, $f \in C([a,b])$, call the right Riemann–Liouville fractional integral operator by

$$\left(J_{b-}^{\alpha} f\right)(x) := \frac{1}{\Gamma(\alpha)} \int_x^b (t-x)^{\alpha-1} f(t)\,dt, \tag{5.5}$$

$x \in [a,b]$, see also [2, 8–10, 16]. Define the subspace of functions

$$C_{b-}^{\alpha}([a,b]) = \left\{ f \in C^m([a,b]) : J_{b-}^{1-\beta} f^{(m)} \in C^1([a,b]) \right\}. \tag{5.6}$$

Define the right generalized α-fractional derivative of f over $[a,b]$ as

$$D_{b-}^{\alpha} f = (-1)^{m-1} \left(J_{b-}^{1-\beta} f^{(m)}\right)', \tag{5.7}$$

see [2]. We set $D_{b-}^{0} f = f$. Notice that $D_{b-}^{\alpha} f \in C([a,b])$.

We are also motivated by the following fractional Polya type integral inequality without any boundary conditions.

Theorem 5.2 ([4, pp. 1–7]) *Let $0 < \alpha < 1$, $f \in C([a,b])$. Assume $f \in C_{a+}^{\alpha}\left(\left[a, \frac{a+b}{2}\right]\right)$ and $f \in C_{b-}^{\alpha}\left(\left[\frac{a+b}{2}, b\right]\right)$. Set*

$$M(f) = \max\left\{ \left\| D_{a+}^{\alpha} f \right\|_{\infty,\left[a,\frac{a+b}{2}\right]}, \left\| D_{b-}^{\alpha} f \right\|_{\infty,\left[\frac{a+b}{2},b\right]} \right\}. \tag{5.8}$$

Then

$$\left| \int_a^b f(x)\,dx \right| \leq \int_a^b |f(x)|\,dx \leq M(f) \frac{(b-a)^{\alpha+1}}{\Gamma(\alpha+2)\,2^{\alpha}}. \tag{5.9}$$

Inequality (5.9) is sharp, namely it is attained by

$$f_*(x) = \left\{ \begin{array}{l} (x-a)^{\alpha}, \ x \in \left[a, \frac{a+b}{2}\right] \\ (b-x)^{\alpha}, \ x \in \left[\frac{a+b}{2}, b\right] \end{array} \right\}, \quad 0 < \alpha < 1. \tag{5.10}$$

Clearly here non zero constant functions f are excluded.

In this chapter we present very general g-fractional Polya type inequalities.

5.2 Background

Here we follow [5].

Let $g : [a, b] \to \mathbb{R}$ be a strictly increasing function. Let $f \in C^n([a, b])$, $n \in \mathbb{N}$. Assume that $g \in C^1([a, b])$, and $g^{-1} \in C^n([g(a), g(b)])$. Call $l := f \circ g^{-1} : [g(a), g(b)] \to \mathbb{R}$. It is clear that $l, l', \ldots, l^{(n)}$ are continuous functions from $[g(a), g(b)]$ into $f([a, b]) \subseteq \mathbb{R}$.

Let $\nu \geq 1$ such that $[\nu] = n$, $n \in \mathbb{N}$ as above, where $[\cdot]$ is the integral part of the number.

Clearly when $0 < \nu < 1$, $[\nu] = 0$ and $n = 0$. Next we follow [1, pp. 7–9].

(I) Let $h \in C([g(a), g(b)])$, we define the left Riemann–Liouville fractional integral as

$$\left(J_{\nu}^{z_0} h\right)(z) := \frac{1}{\Gamma(\nu)} \int_{z_0}^{z} (z-t)^{\nu-1} h(t)\, dt, \tag{5.11}$$

for $g(a) \leq z_0 \leq z \leq g(b)$, where Γ is the gamma function; $\Gamma(\nu) = \int_0^{\infty} e^{-t} t^{\nu-1} dt$.

We set $J_0^{z_0} h = h$.

Let $\alpha := \nu - [\nu]$ $(0 < \alpha < 1)$. We define the subspace $C_{g(x_0)}^{\nu}([g(a), g(b)])$ of $C^{[\nu]}([g(a), g(b)])$, where $x_0 \in [a, b]$:

$$C_{g(x_0)}^{\nu}([g(a), g(b)]) := \left\{h \in C^{[\nu]}([g(a), g(b)]) : J_{1-\alpha}^{g(x_0)} h^{([\nu])} \in C^1([g(x_0), g(b)])\right\}. \tag{5.12}$$

So let $h \in C_{g(x_0)}^{\nu}([g(a), g(b)])$; we define the left g-generalized fractional derivative of h of order ν, of Canavati type, over $[g(x_0), g(b)]$ as

$$D_{g(x_0)}^{\nu} h := \left(J_{1-\alpha}^{g(x_0)} h^{([\nu])}\right)'. \tag{5.13}$$

Clearly, for $h \in C_{g(x_0)}^{\nu}([g(a), g(b)])$, there exists

$$\left(D_{g(x_0)}^{\nu} h\right)(z) = \frac{1}{\Gamma(1-\alpha)} \frac{d}{dz} \int_{g(x_0)}^{z} (z-t)^{-\alpha} h^{([\nu])}(t)\, dt, \tag{5.14}$$

for all $g(x_0) \leq z \leq g(b)$.

In particular, when $f \circ g^{-1} \in C_{g(x_0)}^{\nu}([g(a), g(b)])$ we have that

$$\left(D_{g(x_0)}^{\nu}\left(f \circ g^{-1}\right)\right)(z) = \frac{1}{\Gamma(1-\alpha)} \frac{d}{dz} \int_{g(x_0)}^{z} (z-t)^{-\alpha} \left(f \circ g^{-1}\right)^{([\nu])}(t)\, dt, \tag{5.15}$$

for all $g(x_0) \leq z \leq g(b)$. We have $D_{g(x_0)}^{n}\left(f \circ g^{-1}\right) = \left(f \circ g^{-1}\right)^{(n)}$ and $D_{g(x_0)}^{0}\left(f \circ g^{-1}\right) = f \circ g^{-1}$.

We mention and need the following left generalized g-fractional, of Canavati type, Taylor's formula:

Theorem 5.3 *Let* $f \circ g^{-1} \in C^{\nu}_{g(x_0)}([g(a), g(b)])$, *where* $x_0 \in [a,b]$ *is fixed.*

(i) if $\nu \geq 1$, *then*

$$f(x) - f(x_0) = \sum_{k=1}^{[\nu]-1} \frac{\left(f \circ g^{-1}\right)^{(k)}(g(x_0))}{k!}(g(x) - g(x_0))^k +$$

$$\frac{1}{\Gamma(\nu)} \int_{g(x_0)}^{g(x)} (g(x) - t)^{\nu-1} \left(D^{\nu}_{g(x_0)}\left(f \circ g^{-1}\right)\right)(t)\, dt, \quad \textit{all } x \in [a,b] : x \geq x_0, \tag{5.16}$$

(ii) if $0 < \nu < 1$, *we get*

$$f(x) = \frac{1}{\Gamma(\nu)} \int_{g(x_0)}^{g(x)} (g(x) - t)^{\nu-1} \left(D^{\nu}_{g(x_0)}\left(f \circ g^{-1}\right)\right)(t)\, dt, \quad \textit{all } x \in [a,b] : x \geq x_0. \tag{5.17}$$

By the change of variable method, see [11], we may rewrite the remainder of (5.16), (5.17), as

$$\frac{1}{\Gamma(\nu)} \int_{g(x_0)}^{g(x)} (g(x) - t)^{\nu-1} \left(D^{\nu}_{g(x_0)}\left(f \circ g^{-1}\right)\right)(t)\, dt = \tag{5.18}$$

$$\frac{1}{\Gamma(\nu)} \int_{x_0}^{x} (g(x) - g(s))^{\nu-1} \left(D^{\nu}_{g(x_0)}\left(f \circ g^{-1}\right)\right)(g(s))\, g'(s)\, ds,$$

all $x \in [a,b] : x \geq x_0$.

We may rewrite (5.17) as follows:
if $0 < \nu < 1$, we have

$$f(x) = \left(J^{g(x_0)}_{\nu}\left(D^{\nu}_{g(x_0)}\left(f \circ g^{-1}\right)\right)\right)(g(x)), \tag{5.19}$$

all $x \in [a,b] : x \geq x_0$.

(II) Next we follow [3, pp. 345–348].

Let $h \in C([g(a), g(b)])$, we define the right Riemann–Liouville fractional integral as

$$\left(J^{\nu}_{z_0-}h\right)(z) := \frac{1}{\Gamma(\nu)} \int_{z}^{z_0} (t - z)^{\nu-1} h(t)\, dt, \tag{5.20}$$

for $g(a) \leq z \leq z_0 \leq g(b)$. We set $J^0_{z_0-}h = h$.

Let $\alpha := \nu - [\nu]$ $(0 < \alpha < 1)$. We define the subspace $C^{\nu}_{g(x_0)-}([g(a), g(b)])$ of $C^{[\nu]}([g(a), g(b)])$, where $x_0 \in [a,b]$:

$$C^{\nu}_{g(x_0)-}([g(a), g(b)]) :=$$

$$\left\{h \in C^{[\nu]}\left([g(a), g(b)]\right) : J_{g(x_0)-}^{1-\alpha} h^{([\nu])} \in C^1\left([g(a), g(x_0)]\right)\right\}. \tag{5.21}$$

So let $h \in C_{g(x_0)-}^{\nu}([g(a), g(b)])$; we define the right g-generalized fractional derivative of h of order ν, of Canavati type, over $[g(a), g(x_0)]$ as

$$D_{g(x_0)-}^{\nu} h := (-1)^{n-1}\left(J_{g(x_0)-}^{1-\alpha} h^{([\nu])}\right)'. \tag{5.22}$$

Clearly, for $h \in C_{g(x_0)-}^{\nu}([g(a), g(b)])$, there exists

$$\left(D_{g(x_0)-}^{\nu} h\right)(z) = \frac{(-1)^{n-1}}{\Gamma(1-\alpha)} \frac{d}{dz} \int_z^{g(x_0)} (t-z)^{-\alpha} h^{([\nu])}(t)\, dt, \tag{5.23}$$

for all $g(a) \le z \le g(x_0) \le g(b)$.

In particular, when $f \circ g^{-1} \in C_{g(x_0)-}^{\nu}([g(a), g(b)])$ we have that

$$\left(D_{g(x_0)-}^{\nu}\left(f \circ g^{-1}\right)\right)(z) = \frac{(-1)^{n-1}}{\Gamma(1-\alpha)} \frac{d}{dz} \int_z^{g(x_0)} (t-z)^{-\alpha}\left(f \circ g^{-1}\right)^{([\nu])}(t)\, dt, \tag{5.24}$$

for all $g(a) \le z \le g(x_0) \le g(b)$.

We get that

$$\left(D_{g(x_0)-}^{n}\left(f \circ g^{-1}\right)\right)(z) = (-1)^n\left(f \circ g^{-1}\right)^{(n)}(z) \tag{5.25}$$

and $\left(D_{g(x_0)-}^{0}\left(f \circ g^{-1}\right)\right)(z) = \left(f \circ g^{-1}\right)(z)$, all $z \in [g(a), g(x_0)]$.

We mention and need the following right generalized g-fractional, of Canavati type, Taylor's formula:

Theorem 5.4 *Let $f \circ g^{-1} \in C_{g(x_0)-}^{\nu}([g(a), g(b)])$, where $x_0 \in [a, b]$ is fixed.*
(i) if $\nu \ge 1$, then

$$f(x) - f(x_0) = \sum_{k=1}^{[\nu]-1} \frac{\left(f \circ g^{-1}\right)^{(k)}(g(x_0))}{k!}(g(x) - g(x_0))^k +$$

$$\frac{1}{\Gamma(\nu)} \int_{g(x)}^{g(x_0)} (t - g(x))^{\nu-1}\left(D_{g(x_0)-}^{\nu}\left(f \circ g^{-1}\right)\right)(t)\, dt, \quad \text{all } a \le x \le x_0, \tag{5.26}$$

(ii) if $0 < \nu < 1$, we get

$$f(x) = \frac{1}{\Gamma(\nu)} \int_{g(x)}^{g(x_0)} (t - g(x))^{\nu-1}\left(D_{g(x_0)-}^{\nu}\left(f \circ g^{-1}\right)\right)(t)\, dt, \quad \text{all } a \le x \le x_0. \tag{5.27}$$

By change of variable, see [11], we may rewrite the remainder of (5.26), (5.27), as

$$\frac{1}{\Gamma(\nu)}\int_{g(x)}^{g(x_0)}(t-g(x))^{\nu-1}\left(D_{g(x_0)-}^{\nu}\left(f\circ g^{-1}\right)\right)(t)\,dt= \tag{5.28}$$

$$\frac{1}{\Gamma(\nu)}\int_{x}^{x_0}(g(s)-g(x))^{\nu-1}\left(D_{g(x_0)-}^{\nu}\left(f\circ g^{-1}\right)\right)(g(s))\,g'(s)\,ds,$$

all $a\le x\le x_0$.

We may rewrite (5.27) as follows:

if $0<\nu<1$, we have

$$f(x)=\left(J_{g(x_0)-}^{\nu}\left(D_{g(x_0)-}^{\nu}\left(f\circ g^{-1}\right)\right)\right)(g(x)), \tag{5.29}$$

all $a\le x\le x_0\le b$.

(III) Denote by

$$D_{g(x_0)}^{m\nu}=D_{g(x_0)}^{\nu}D_{g(x_0)}^{\nu}\ldots D_{g(x_0)}^{\nu}\ \ (m\text{-times}),\ m\in\mathbb{N}. \tag{5.30}$$

Also denote by

$$J_{m\nu}^{g(x_0)}=J_{\nu}^{g(x_0)}J_{\nu}^{g(x_0)}\ldots J_{\nu}^{g(x_0)}\ \ (m\text{-times}),\ m\in\mathbb{N}. \tag{5.31}$$

We call $D_{g(x_0)}^{m\nu}$ an iterated fractional derivative.

We mention and need the following g-modified and generalized left fractional Taylor's formula of Canavati type:

Theorem 5.5 *Let* $0<\nu<1$. *Assume that* $\left(D_{g(x_0)}^{i\nu}\left(f\circ g^{-1}\right)\right)\in C_{g(x_0)}^{\nu}([g(a),g(b)])$, $x_0\in[a,b]$, *for* $i=0,1,\ldots,m$. *Assume also that* $\left(D_{g(x_0)}^{(m+1)\nu}\left(f\circ g^{-1}\right)\right)\in C([g(x_0),g(b)])$. *Then*

$$f(x)=\frac{1}{\Gamma((m+1)\nu)}\int_{g(x_0)}^{g(x)}(g(x)-z)^{(m+1)\nu-1}\left(D_{g(x_0)}^{(m+1)\nu}\left(f\circ g^{-1}\right)\right)(z)\,dz$$

$$=\frac{1}{\Gamma((m+1)\nu)}\int_{x_0}^{x}(g(x)-g(s))^{(m+1)\nu-1}\left(D_{g(x_0)}^{(m+1)\nu}\left(f\circ g^{-1}\right)\right)(g(s))\,g'(s)\,ds, \tag{5.32}$$

all $x_0\le x\le b$.

(IV) Denote by

$$D_{g(x_0)-}^{m\nu}=D_{g(x_0)-}^{\nu}D_{g(x_0)-}^{\nu}\ldots D_{g(x_0)-}^{\nu}\ \ (m\text{-times}),\ m\in\mathbb{N}. \tag{5.33}$$

Also denote by

$$J_{g(x_0)-}^{m\nu} = J_{g(x_0)-}^{\nu} J_{g(x_0)-}^{\nu} \ldots J_{g(x_0)-}^{\nu} \quad (m\text{-times}),\ m \in \mathbb{N}. \tag{5.34}$$

We call $D_{g(x_0)-}^{m\nu}$ an iterated fractional derivative.

We mention and need the following g-modified and generalized right fractional Taylor's formula of Canavati type:

Theorem 5.6 *Let $0 < \nu < 1$. Assume that $\left(D_{g(x_0)-}^{i\nu}\left(f \circ g^{-1}\right)\right) \in C_{g(x_0)-}^{\nu}([g(a), g(b)])$, $x_0 \in [a, b]$, for all $i = 0, 1, \ldots, m$. Assume also that $\left(D_{g(x_0)-}^{(m+1)\nu}\left(f \circ g^{-1}\right)\right) \in C([g(a), g(x_0)])$. Then*

$$f(x) = \frac{1}{\Gamma((m+1)\nu)} \int_{g(x)}^{g(x_0)} (z - g(x))^{(m+1)\nu-1} \left(D_{g(x_0)-}^{(m+1)\nu}\left(f \circ g^{-1}\right)\right)(z)\, dz$$

$$= \frac{1}{\Gamma((m+1)\nu)} \int_{x}^{x_0} (g(s) - g(x))^{(m+1)\nu-1} \left(D_{g(x_0)-}^{(m+1)\nu}\left(f \circ g^{-1}\right)\right)(g(s))\, g'(s)\, ds, \tag{5.35}$$

all $a \le x \le x_0 \le b$.

In what follows next it is based on this background.

5.3 Main Results

We present the following generalized g-fractional Polya type integral inequalities without any boundary conditions.

Theorem 5.7 *Let $0 < \nu < 1$, $f \in C([a, b])$, $g : [a, b] \to \mathbb{R}$ be strictly increasing and $g \in C^1([a, b])$. Assume $f \circ g^{-1} \in C_{g(a)}^{\nu}\left(\left[g(a), \frac{g(a)+g(b)}{2}\right]\right)$, and $f \circ g^{-1} \in C_{g(b)-}^{\nu}\left(\left[\frac{g(a)+g(b)}{2}, g(b)\right]\right)$. Set*

$$M(f, g) := \max\left\{ \left\| D_{g(a)}^{\nu}\left(f \circ g^{-1}\right)\right\|_{\infty, \left[g(a), \frac{g(a)+g(b)}{2}\right]}, \left\| D_{g(b)-}^{\nu}\left(f \circ g^{-1}\right)\right\|_{\infty, \left[\frac{g(a)+g(b)}{2}, g(b)\right]} \right\}. \tag{5.36}$$

Then

$$\left| \int_a^b f\, dg \right| \le \int_a^b |f|\, dg \le M(f, g) \frac{(g(b) - g(a))^{\nu+1}}{\Gamma(\nu+2)\, 2^{\nu}}. \tag{5.37}$$

Inequality (5.37) is sharp, namely it is attained by f_* such that

$$\left(f_{*}\circ g^{-1}\right)(z)=\left\{\begin{array}{l}(z-g(a))^{\nu},\ z\in\left[g(a),\frac{g(a)+g(b)}{2}\right]\\ (g(b)-z)^{\nu},\ z\in\left[\frac{g(a)+g(b)}{2},g(b)\right]\end{array}\right\},\quad 0<\nu<1. \tag{5.38}$$

Clearly here non zero constant functions $f\circ g^{-1}$ are excluded.

Proof Notice that

$$\left|\int_{a}^{b}fdg\right|=\left|\int_{a}^{b}fg'dx\right|\leq\int_{a}^{b}|f|\,g'dx=\int_{a}^{b}|f|\,dg.$$

Let $x_0\in[a,b]$ be such that $g(x_0)=\frac{g(a)+g(b)}{2}$, that is $x_0=g^{-1}\left(\frac{g(a)+g(b)}{2}\right)$.

Let $f\circ g^{-1}\in C_{g(a)}^{\nu}\left(\left[g(a),\frac{g(a)+g(b)}{2}\right]\right)$, $0<\nu<1$. By Theorem 5.3 we have that

$$f(x)=\frac{1}{\Gamma(\nu)}\int_{g(a)}^{g(x)}(g(x)-t)^{\nu-1}\left(D_{g(a)}^{\nu}\left(f\circ g^{-1}\right)\right)(t)\,dt, \tag{5.39}$$

all $x\in[a,x_0]$.

Assume also $f\circ g^{-1}\in C_{g(b)-}^{\nu}\left(\left[\frac{g(a)+g(b)}{2},g(b)\right]\right)$, by Theorem 5.4 we have that

$$f(x)=\frac{1}{\Gamma(\nu)}\int_{g(x)}^{g(b)}(t-g(x))^{\nu-1}\left(D_{g(b)-}^{\nu}\left(f\circ g^{-1}\right)\right)(t)\,dt, \tag{5.40}$$

all $x\in[x_0,b]$.

By (5.39) we get

$$|f(x)|\leq\frac{1}{\Gamma(\nu)}\left(\int_{g(a)}^{g(x)}(g(x)-t)^{\nu-1}dt\right)\left\|D_{g(a)}^{\nu}\left(f\circ g^{-1}\right)\right\|_{\infty,\left[g(a),\frac{g(a)+g(b)}{2}\right]}$$

$$=\left\|D_{g(a)}^{\nu}\left(f\circ g^{-1}\right)\right\|_{\infty,\left[g(a),\frac{g(a)+g(b)}{2}\right]}\frac{(g(x)-g(a))^{\nu}}{\Gamma(\nu+1)}. \tag{5.41}$$

That is

$$|f(x)|\leq\left\|D_{g(a)}^{\nu}\left(f\circ g^{-1}\right)\right\|_{\infty,\left[g(a),\frac{g(a)+g(b)}{2}\right]}\frac{(g(x)-g(a))^{\nu}}{\Gamma(\nu+1)}, \tag{5.42}$$

for any $x\in[a,x_0]$.

Similarly, by (5.40) we get

$$|f(x)|\leq\frac{1}{\Gamma(\nu)}\left(\int_{g(x)}^{g(b)}(t-g(x))^{\nu-1}dt\right)\left\|D_{g(b)-}^{\nu}\left(f\circ g^{-1}\right)\right\|_{\infty,\left[\frac{g(a)+g(b)}{2},g(b)\right]}$$

$$= \frac{(g(b) - g(x))^{\nu}}{\Gamma(\nu+1)} \left\| D^{\nu}_{g(b)-} \left(f \circ g^{-1} \right) \right\|_{\infty, \left[\frac{g(a)+g(b)}{2}, g(b) \right]}. \tag{5.43}$$

That is

$$|f(x)| \leq \left\| D^{\nu}_{g(b)-} \left(f \circ g^{-1} \right) \right\|_{\infty, \left[\frac{g(a)+g(b)}{2}, g(b) \right]} \frac{(g(b) - g(x))^{\nu}}{\Gamma(\nu+1)}, \tag{5.44}$$

for any $x \in [x_0, b]$.

Thus, we have

$$\left| \int_a^b f dg \right| \leq \int_a^b |f| \, dg = \int_a^b |f| \, g' dx =$$

$$\int_a^{x_0} |f(x)| \, g'(x) \, dx + \int_{x_0}^b |f(x)| \, g'(x) \, dx \overset{\text{(by (5.42), (5.44))}}{\leq}$$

$$\frac{\left\| D^{\nu}_{g(a)} \left(f \circ g^{-1} \right) \right\|_{\infty, \left[g(a), \frac{g(a)+g(b)}{2} \right]}}{\Gamma(\nu+1)} \int_a^{x_0} (g(x) - g(a))^{\nu} g'(x) \, dx + \tag{5.45}$$

$$\frac{\left\| D^{\nu}_{g(b)-} \left(f \circ g^{-1} \right) \right\|_{\infty, \left[\frac{g(a)+g(b)}{2}, g(b) \right]}}{\Gamma(\nu+1)} \int_{x_0}^b (g(b) - g(x))^{\nu} g'(x) \, dx =$$

$$\frac{\left\| D^{\nu}_{g(a)} \left(f \circ g^{-1} \right) \right\|_{\infty, \left[g(a), \frac{g(a)+g(b)}{2} \right]}}{\Gamma(\nu+1)} \frac{(g(x_0) - g(a))^{\nu+1}}{(\nu+1)} + \tag{5.46}$$

$$\frac{\left\| D^{\nu}_{g(b)-} \left(f \circ g^{-1} \right) \right\|_{\infty, \left[\frac{g(a)+g(b)}{2}, g(b) \right]}}{\Gamma(\nu+1)} \frac{(g(b) - g(x_0))^{\nu+1}}{(\nu+1)} =$$

$$\frac{1}{\Gamma(\nu+2)} \left[\left\| D^{\nu}_{g(a)} \left(f \circ g^{-1} \right) \right\|_{\infty, \left[g(a), \frac{g(a)+g(b)}{2} \right]} \left(\frac{g(b) - g(a)}{2} \right)^{\nu+1} + \right.$$

$$\left. \left\| D^{\nu}_{g(b)-} \left(f \circ g^{-1} \right) \right\|_{\infty, \left[\frac{g(a)+g(b)}{2}, g(b) \right]} \left(\frac{g(b) - g(a)}{2} \right)^{\nu+1} \right] =$$

$$\frac{1}{\Gamma(\nu+2)} \frac{\left(g(b) - g(a)^{\nu+1} \right)}{2^{\nu+1}} \left[\left\| D^{\nu}_{g(a)} \left(f \circ g^{-1} \right) \right\|_{\infty, \left[g(a), \frac{g(a)+g(b)}{2} \right]} \right. \tag{5.47}$$

$$+ \left\| D_{g(b)-}^{\nu} \left(f \circ g^{-1} \right) \right\|_{\infty, \left[\frac{g(a)+g(b)}{2}, g(b) \right]} \Bigg] \leq$$

$$\frac{M(f,g)}{\Gamma(\nu+2)} \frac{(g(b)-g(a))^{\nu+1}}{2^{\nu}}. \tag{5.48}$$

We have proved inequality (5.37). Next we prove sharpness of (5.37).

Here $[\nu] = 0$ and $\alpha = \nu$. We see that

$$\left(J_{1-\alpha}^{g(a)} \left(f_* \circ g^{-1} \right) \right)(z) = \frac{1}{\Gamma(1-\alpha)} \int_{g(a)}^{z} (z-t)^{-\alpha} \left(f_* \circ g^{-1} \right)(t)\, dt$$

$$= \frac{1}{\Gamma(1-\alpha)} \int_{g(a)}^{z} (z-t)^{(1-\alpha)-1} (t-g(a))^{(\nu+1)-1} dt = \tag{5.49}$$

(by [17, p. 256])

$$\frac{1}{\Gamma(1-\alpha)} \frac{\Gamma(1-\alpha)\Gamma(\nu+1)}{\Gamma(2)} (z-g(a)) = \Gamma(\nu+1)(z-g(a)).$$

Therefore

$$\left(D_{g(a)}^{\nu} \left(f_* \circ g^{-1} \right) \right)(z) = \Gamma(\nu+1), \text{ for all } z \in \left[g(a), \frac{g(a)+g(b)}{2} \right]. \tag{5.50}$$

That is

$$\left\| D_{g(a)}^{\nu} \left(f \circ g^{-1} \right) \right\|_{\infty, \left[g(a), \frac{g(a)+g(b)}{2} \right]} = \Gamma(\nu+1). \tag{5.51}$$

Furthermore we have

$$-\left(J_{g(b)-}^{1-\alpha} \left(f_* \circ g^{-1} \right) \right)(z) = \frac{-1}{\Gamma(1-\alpha)} \int_{z}^{g(b)} (t-z)^{-\alpha} (g(b)-t)^{\nu} dt =$$

$$-\frac{1}{\Gamma(1-\alpha)} \int_{z}^{g(b)} (g(b)-t)^{(\nu+1)-1} (t-z)^{(1-\alpha)-1} dt =$$

(by [17, p. 256])

$$-\frac{1}{\Gamma(1-\alpha)} \frac{\Gamma(\nu+1)\Gamma(1-\alpha)}{\Gamma(2)} (g(b)-z) = \Gamma(\nu+1)(z-g(b)), \tag{5.52}$$

for all $z \in \left[\frac{g(a)+g(b)}{2}, g(b) \right]$.

Hence

$$\left(D_{g(b)-}^{\nu} \left(f_* \circ g^{-1} \right) \right)(z) = \Gamma(\nu+1), \tag{5.53}$$

and

$$\left\| D^{\nu}_{g(b)-}\left(f_e \circ g^{-1}\right)\right\|_{\infty,\left[\frac{g(a)+g(b)}{2},g(b)\right]} = \Gamma\left(\nu+1\right). \tag{5.54}$$

Consequently we get that

$$M\left(f_*, g\right) = \Gamma\left(\nu+1\right). \tag{5.55}$$

Applying f_* into (5.37) we obtain:

$$\text{R.H.S (5.37) for } f_* = \frac{\Gamma\left(\nu+1\right)}{\Gamma\left(\nu+2\right)}\frac{\left(g\left(b\right)-g\left(a\right)\right)^{\nu+1}}{2^{\nu}} = \frac{\left(g\left(b\right)-g\left(a\right)\right)^{\nu+1}}{\left(\nu+1\right)2^{\nu}}. \tag{5.56}$$

We get the same result when we apply f_* in the

$$\text{L.H.S (5.37)} = \left|\int_a^b f_*\left(x\right)dg\left(x\right)\right| =$$

$$\left|\int_a^{x_0} f_*\left(x\right)dg\left(x\right) + \int_{x_0}^b f_*\left(x\right)dg\left(x\right)\right| =$$

$$\left|\int_a^{x_0}\left(f_*\circ g^{-1}\right)\left(g\left(x\right)\right)dg\left(x\right) + \int_{x_0}^b\left(f_*\circ g^{-1}\right)\left(g\left(x\right)\right)dg\left(x\right)\right| =$$

$$\left|\int_{g(a)}^{g(x_0)}\left(f_*\circ g^{-1}\right)\left(z\right)dz + \int_{g(x_0)}^{g(b)}\left(f_*\circ g^{-1}\right)\left(z\right)dz\right| = \tag{5.57}$$

$$\left|\int_{g(a)}^{g(x_0)}\left(z-g\left(a\right)\right)^{\nu}dz + \int_{g(x_0)}^{g(b)}\left(g\left(b\right)-z\right)^{\nu}dz\right| =$$

$$\frac{\left(g\left(x_0\right)-g\left(a\right)\right)^{\nu+1}}{\nu+1} + \frac{\left(g\left(b\right)-g\left(x_0\right)\right)^{\nu+1}}{\nu+1} =$$

$$\frac{2}{\nu+1}\frac{\left(g\left(b\right)-g\left(a\right)\right)^{\nu+1}}{2^{\nu+1}} = \frac{\left(g\left(b\right)-g\left(a\right)\right)^{\nu+1}}{\left(\nu+1\right)2^{\nu}}. \tag{5.58}$$

So equality holds in (5.37) for f_*. Furthermore we notice that

$$\left(f_*\circ g^{-1}\right)\left(\left(\frac{g\left(a\right)+g\left(b\right)}{2}\right)_-\right) = \left(f_*\circ g^{-1}\right)\left(\left(\frac{g\left(a\right)+g\left(b\right)}{2}\right)_+\right) \tag{5.59}$$

$$= \left(\frac{g\left(b\right)-g\left(a\right)}{2}\right)^{\nu},$$

thus $f_*\circ g^{-1}$ is continuous on $\left[g\left(a\right), g\left(b\right)\right]$ and hence $f_*\in C\left(\left[a,b\right]\right)$.

Sharpness of (5.37) has been proved. The theorem is proved completely. ∎

Remark 5.8 When $\nu \geq 1$, thus $n = [\nu] \geq 1$, and by assuming that $\left(f \circ g^{-1}\right)^{(k)} (g(a)) = \left(f \circ g^{-1}\right)^{(k)} (g(b)) = 0$, for $k = 0, 1, \ldots, n-1$, we can prove inequality (5.37) again. Here also $f \in C^n([a, b])$ and $g^{-1} \in C^n([g(a), g(b)])$.

The function $\left(f \circ g^{-1}\right)^{(n)}$ cannot be a constant different than zero, equivalently, $f \circ g^{-1}$ cannot be a non-trivial polynomial of degree n.

We continue with a fractional iterated Polya type inequality:

Theorem 5.9 *Let $0 < \nu < 1$, $f \in C([a, b])$, $g : [a, b] \to \mathbb{R}$ be strictly increasing and $g \in C^1([a, b])$. Assume that $\left(D^{i\nu}_{g(a)}\left(f \circ g^{-1}\right)\right) \in C^{\nu}_{g(a)}\left(\left[g(a), \frac{g(a)+g(b)}{2}\right]\right)$, for $i = 0, 1, \ldots, m \in \mathbb{N}$, and $\left(D^{(m+1)\nu}_{g(a)}\left(f \circ g^{-1}\right)\right) \in C\left(\left[g(a), \frac{g(a)+g(b)}{2}\right]\right)$. Also assume that $\left(D^{i\nu}_{g(b)-}\left(f \circ g^{-1}\right)\right) \in C^{\nu}_{g(b)-}\left(\left[\frac{g(a)+g(b)}{2}, g(b)\right]\right)$, for $i = 0, 1, \ldots, m \in \mathbb{N}$, and $\left(D^{(m+1)\nu}_{g(b)-}\left(f \circ g^{-1}\right)\right) \in C\left(\left[\frac{g(a)+g(b)}{2}, g(b)\right]\right)$. Set*

$$M^*(f, g) := \max\left\{\left\|D^{(m+1)\nu}_{g(a)}\left(f \circ g^{-1}\right)\right\|_{\infty,\left[g(a), \frac{g(a)+g(b)}{2}\right]},\right.$$

$$\left.\left\|D^{(m+1)\nu}_{g(b)-}\left(f \circ g^{-1}\right)\right\|_{\infty,\left[\frac{g(a)+g(b)}{2}, g(b)\right]}\right\}. \tag{5.60}$$

Then

$$\left|\int_a^b f dg\right| \leq \int_a^b |f| dg \leq M^*(f, g) \frac{(g(b) - g(a))^{(m+1)\nu+1}}{\Gamma((m+1)\nu+2)\, 2^{(m+1)\nu}}. \tag{5.61}$$

Proof By Theorem 5.5 we have

$$f(x) = \frac{1}{\Gamma((m+1)\nu)} \int_{g(a)}^{g(x)} (g(x) - z)^{(m+1)\nu-1} \left(D^{(m+1)\nu}_{g(a)}\left(f \circ g^{-1}\right)\right)(z)\, dz, \tag{5.62}$$

all $x \in [a, x_0]$, where $x_0 := g^{-1}\left(\frac{g(a)+g(b)}{2}\right)$.

By Theorem 5.6 we have

$$f(x) = \frac{1}{\Gamma((m+1)\nu)} \int_{g(x)}^{g(b)} (z - g(x))^{(m+1)\nu-1} \left(D^{(m+1)\nu}_{g(b)-}\left(f \circ g^{-1}\right)\right)(z)\, dz, \tag{5.63}$$

all $x \in [x_0, b]$.

By (5.62) we get

$$|f(x)| \leq \frac{1}{\Gamma((m+1)\nu)}$$

$$\left(\int_{g(a)}^{g(x)} (g(x) - z)^{(m+1)\nu - 1} dz\right) \left\| D_{g(a)}^{(m+1)\nu} \left(f \circ g^{-1}\right) \right\|_{\infty, \left[g(a), \frac{g(a)+g(b)}{2}\right]}$$

$$= \frac{(g(x) - g(a))^{(m+1)\nu}}{\Gamma((m+1)\nu + 1)} \left\| D_{g(a)}^{(m+1)\nu} \left(f \circ g^{-1}\right) \right\|_{\infty, \left[g(a), \frac{g(a)+g(b)}{2}\right]}, \quad (5.64)$$

for all $x \in [a, x_0]$.

Similarly, by (5.63), we obtain

$$|f(x)| \leq \frac{1}{\Gamma((m+1)\nu)}$$

$$\left(\int_{g(x)}^{g(b)} (z - g(x))^{(m+1)\nu - 1} dz\right) \left\| D_{g(b)-}^{(m+1)\nu} \left(f \circ g^{-1}\right) \right\|_{\infty, \left[\frac{g(a)+g(b)}{2}, g(b)\right]}$$

$$= \frac{(g(b) - g(x))^{(m+1)\nu}}{\Gamma((m+1)\nu + 1)} \left\| D_{g(b)-}^{(m+1)\nu} \left(f \circ g^{-1}\right) \right\|_{\infty, \left[\frac{g(a)+g(b)}{2}, g(b)\right]}, \quad (5.65)$$

for all $x \in [x_0, b]$.

Thus, we have

$$\left|\int_a^b f dg\right| \leq \int_a^b |f| dg = \int_a^b |f| g' dx =$$

$$\int_a^{x_0} |f(x)| g'(x) dx + \int_{x_0}^b |f(x)| g'(x) dx \overset{\text{(by (5.64), (5.65))}}{\leq}$$

$$\frac{\left\| D_{g(a)}^{(m+1)\nu} \left(f \circ g^{-1}\right) \right\|_{\infty, \left[g(a), \frac{g(a)+g(b)}{2}\right]}}{\Gamma((m+1)\nu + 1)} \int_a^{x_0} (g(x) - g(a))^{(m+1)\nu} g'(x) dx + \quad (5.66)$$

$$\frac{\left\| D_{g(b)-}^{(m+1)\nu} \left(f \circ g^{-1}\right) \right\|_{\infty, \left[\frac{g(a)+g(b)}{2}, g(b)\right]}}{\Gamma((m+1)\nu + 1)} \int_{x_0}^b (g(b) - g(x))^{(m+1)\nu} g'(x) dx =$$

$$\frac{1}{\Gamma((m+1)\nu + 2)} \left[\left\| D_{g(a)}^{(m+1)\nu} \left(f \circ g^{-1}\right) \right\|_{\infty, \left[g(a), \frac{g(a)+g(b)}{2}\right]} (g(x_0) - g(a))^{(m+1)\nu + 1}\right.$$

$$\left. + \left\| D_{g(b)-}^{(m+1)\nu} \left(f \circ g^{-1}\right) \right\|_{\infty, \left[\frac{g(a)+g(b)}{2}, g(b)\right]} (g(b) - g(x_0))^{(m+1)\nu + 1}\right] =$$

(by $g(x_0) = \frac{g(a)+g(b)}{2}$)

$$\frac{1}{\Gamma\left((m+1)\nu+2\right)}\frac{(g(b)-g(a))^{(m+1)\nu+1}}{2^{(m+1)\nu+1}}\left[\left\|D_{g(a)}^{(m+1)\nu}\left(f\circ g^{-1}\right)\right\|_{\infty,\left[g(a),\frac{g(a)+g(b)}{2}\right]}\right.$$
$$\left.+\left\|D_{g(b)-}^{(m+1)\nu}\left(f\circ g^{-1}\right)\right\|_{\infty,\left[\frac{g(a)+g(b)}{2},g(b)\right]}\right]\leq \tag{5.67}$$

$$\frac{M^*(f,g)}{\Gamma\left((m+1)\nu+2\right)}\frac{(g(b)-g(a))^{(m+1)\nu+1}}{2^{(m+1)\nu}}, \tag{5.68}$$

proving inequality (5.61). ∎

Next we give an L_1 generalized g-fractional Polya inequality:

Theorem 5.10 *Let $\nu\geq 1$, $n=[\nu]$, and $f\in C^n([a,b])$, $g:[a,b]\to\mathbb{R}$ be a strictly increasing function. Assume that $g\in C^1([a,b])$ and $g^{-1}\in C^n([g(a),g(b)])$. Also assume $f\circ g^{-1}\in C_{g(a)}^{\nu}\left(\left[g(a),\frac{g(a)+g(b)}{2}\right]\right)$, and $f\circ g^{-1}\in C_{g(b)-}^{\nu}\left(\left[\frac{g(a)+g(b)}{2},g(b)\right]\right)$, and that $\left(f\circ g^{-1}\right)^{(k)}(g(a))=\left(f\circ g^{-1}\right)^{(k)}(g(b))=0$, for $k=0,1,\ldots,n-1$. Set*

$$M_1(f,g):=\max\left\{\left\|D_{g(a)}^{\nu}\left(f\circ g^{-1}\right)\right\|_{L_1\left(\left[g(a),\frac{g(a)+g(b)}{2}\right]\right)},\right.$$
$$\left.\left\|D_{g(b)-}^{\nu}\left(f\circ g^{-1}\right)\right\|_{L_1\left(\left[\frac{g(a)+g(b)}{2},g(b)\right]\right)}\right\}. \tag{5.69}$$

Then

$$\left|\int_a^b f\,dg\right|\leq\int_a^b|f|\,dg\leq M_1(f,g)\frac{(g(b)-g(a))^{\nu}}{\Gamma(\nu+1)\,2^{\nu-1}}. \tag{5.70}$$

Here $f\circ g^{-1}$ cannot be a non-trivial polynomial of degree n.

Proof Let $f\circ g^{-1}\in C_{g(a)}^{\nu}\left(\left[g(a),\frac{g(a)+g(b)}{2}\right]\right)$, $\nu\geq 1$, and $\left(f\circ g^{-1}\right)^{(k)}(g(a))=0$, for $k=0,1,\ldots,n-1$, $n=[\nu]$. By Theorem 5.3 we have that

$$f(x)=\frac{1}{\Gamma(\nu)}\int_{g(a)}^{g(x)}(g(x)-t)^{\nu-1}\left(D_{g(a)}^{\nu}\left(f\circ g^{-1}\right)\right)(t)\,dt, \tag{5.71}$$

all $x\in[a,x_0]$, where $x_0:=g^{-1}\left(\frac{g(a)+g(b)}{2}\right)$.

Let $f\circ g^{-1}\in C_{g(b)-}^{\nu}\left(\left[\frac{g(a)+g(b)}{2},g(b)\right]\right)$, and $\left(f\circ g^{-1}\right)^{(k)}(g(b))=0$, for $k=0,1,\ldots,n-1$. By Theorem 5.4 we have that

$$f(x)=\frac{1}{\Gamma(\nu)}\int_{g(x)}^{g(b)}(t-g(x))^{\nu-1}\left(D_{g(b)-}^{\nu}\left(f\circ g^{-1}\right)\right)(t)\,dt, \tag{5.72}$$

all $x \in [x_0, b]$.

By (5.71) we have

$$|f(x)| \leq \frac{1}{\Gamma(\nu)} \int_{g(a)}^{g(x)} (g(x) - t)^{\nu-1} \left| \left(D_{g(a)}^{\nu} \left(f \circ g^{-1} \right) \right)(t) \right| dt \leq$$

$$\frac{(g(x) - g(a))^{\nu-1}}{\Gamma(\nu)} \int_{g(a)}^{g(x)} \left| \left(D_{g(a)}^{\nu} \left(f \circ g^{-1} \right) \right)(t) \right| dt \leq \tag{5.73}$$

$$\frac{(g(x) - g(a))^{\nu-1}}{\Gamma(\nu)} \left\| D_{g(a)}^{\nu} \left(f \circ g^{-1} \right) \right\|_{L_1\left(\left[g(a), \frac{g(a)+g(b)}{2}\right]\right)},$$

for all $x \in [a, x_0]$.

Similarly, by (5.72) we get

$$|f(x)| \leq \frac{1}{\Gamma(\nu)} \int_{g(x)}^{g(b)} (t - g(x))^{\nu-1} \left| \left(D_{g(b)-}^{\nu} \left(f \circ g^{-1} \right) \right)(t) \right| dt \leq$$

$$\frac{(g(b) - g(x))^{\nu-1}}{\Gamma(\nu)} \int_{g(x)}^{g(b)} \left| \left(D_{g(b)-}^{\nu} \left(f \circ g^{-1} \right) \right)(t) \right| dt \leq$$

$$\frac{(g(b) - g(x))^{\nu-1}}{\Gamma(\nu)} \left\| D_{g(b)-}^{\nu} \left(f \circ g^{-1} \right) \right\|_{L_1\left(\left[\frac{g(a)+g(b)}{2}, g(b)\right]\right)}, \tag{5.74}$$

for all $x \in [x_0, b]$.

Thus, we have

$$\left| \int_a^b f dg \right| \leq \int_a^b |f| dg = \int_a^b |f| g' dx =$$

$$\int_a^{x_0} |f(x)| g'(x) dx + \int_{x_0}^b |f(x)| g'(x) dx \overset{\text{(by (5.73), (5.74))}}{\leq}$$

$$\frac{\left\| D_{g(a)}^{\nu} \left(f \circ g^{-1} \right) \right\|_{L_1\left(\left[g(a), \frac{g(a)+g(b)}{2}\right]\right)}}{\Gamma(\nu)} \int_a^{x_0} (g(x) - g(a))^{\nu-1} g'(x) dx +$$

$$\frac{\left\| D_{g(b)-}^{\nu} \left(f \circ g^{-1} \right) \right\|_{L_1\left(\left[\frac{g(a)+g(b)}{2}, g(b)\right]\right)}}{\Gamma(\nu)} \int_{x_0}^b (g(b) - g(x))^{\nu-1} g'(x) dx =$$

$$\frac{\left\| D_{g(a)}^{\nu}\left(f \circ g^{-1}\right)\right\|_{L_1\left(\left[g(a), \frac{g(a)+g(b)}{2}\right]\right)}}{\Gamma(\nu+1)}(g(x_0)-g(a))^{\nu}+$$

$$\frac{\left\| D_{g(b)-}^{\nu}\left(f \circ g^{-1}\right)\right\|_{L_1\left(\left[\frac{g(a)+g(b)}{2}, g(b)\right]\right)}}{\Gamma(\nu+1)}(g(b)-g(x_0))^{\nu}= \tag{5.75}$$

$$\frac{1}{\Gamma(\nu+1)}\frac{(g(b)-g(a))^{\nu}}{2^{\nu}}\left[\left\| D_{g(a)}^{\nu}\left(f \circ g^{-1}\right)\right\|_{L_1\left(\left[g(a), \frac{g(a)+g(b)}{2}\right]\right)}\right.$$

$$\left.+\left\| D_{g(b)-}^{\nu}\left(f \circ g^{-1}\right)\right\|_{L_1\left(\left[\frac{g(a)+g(b)}{2}, g(b)\right]\right)}\right] \leq$$

$$\frac{M_1(f,g)}{\Gamma(\nu+1)}\frac{(g(b)-g(a))^{\nu}}{2^{\nu-1}}, \tag{5.76}$$

proving the claim. ■

We continue with a L_q generalized g-fractional Polya type inequality:

Theorem 5.11 *Let $p, q > 1 : \frac{1}{p}+\frac{1}{q}=1$, $\nu > \frac{1}{q}$, $n=[\nu]$, $f \in C^n([a,b])$, $g : [a,b] \to \mathbb{R}$ be a strictly increasing function. Assume that $g \in C^1([a,b])$ and $g^{-1} \in C^n([g(a), g(b)])$. Also assume $f \circ g^{-1} \in C_{g(a)}^{\nu}\left(\left[g(a), \frac{g(a)+g(b)}{2}\right]\right)$, and $f \circ g^{-1} \in C_{g(b)-}^{\nu}\left(\left[\frac{g(a)+g(b)}{2}, g(b)\right]\right)$, and that $\left(f \circ g^{-1}\right)^{(k)}(g(a)) = \left(f \circ g^{-1}\right)^{(k)}(g(b)) = 0$, for $k=0,1,\ldots,n-1$. When $\frac{1}{q} < \nu < 1$, the last boundary conditions are void. Set*

$$M_q(f,g) := \max\left\{\left\| D_{g(a)}^{\nu}\left(f \circ g^{-1}\right)\right\|_{L_q\left(\left[g(a), \frac{g(a)+g(b)}{2}\right]\right)},\right.$$

$$\left.\left\| D_{g(b)-}^{\nu}\left(f \circ g^{-1}\right)\right\|_{L_q\left(\left[\frac{g(a)+g(b)}{2}, g(b)\right]\right)}\right\}. \tag{5.77}$$

Then

$$\left|\int_a^b f dg\right| \leq \int_a^b |f|\, dg \leq \frac{M_q(f,g)}{2^{\nu-\frac{1}{q}}\Gamma(\nu)(p(\nu-1)+1)^{\frac{1}{p}}\left(\nu+\frac{1}{p}\right)}(g(b)-g(a))^{\nu+\frac{1}{p}}. \tag{5.78}$$

Again here $f \circ g^{-1}$ cannot be a non-trivial polynomial of degree n.

Proof Let $f \circ g^{-1} \in C_{g(a)}^{\nu}\left(\left[g(a), \frac{g(a)+g(b)}{2}\right]\right)$, $\nu > 0$, and $\left(f \circ g^{-1}\right)^{(k)}(g(a)) = 0$, for $k=0,1,\ldots,n-1$, $n=[\nu]$. By Theorem 5.3 we have that

$$f(x) = \frac{1}{\Gamma(\nu)} \int_{g(a)}^{g(x)} (g(x) - t)^{\nu-1} \left(D_{g(a)}^{\nu} \left(f \circ g^{-1}\right)\right)(t)\, dt, \tag{5.79}$$

all $x \in [a, x_0]$, where $x_0 := g^{-1}\left(\frac{g(a)+g(b)}{2}\right)$.

Let $f \circ g^{-1} \in C_{g(b)-}^{\nu}\left(\left[\frac{g(a)+g(b)}{2}, g(b)\right]\right)$, and $\left(f \circ g^{-1}\right)^{(k)}(g(b)) = 0$, for $k = 0, 1, \ldots, n-1$. By Theorem 5.4 we have that

$$f(x) = \frac{1}{\Gamma(\nu)} \int_{g(x)}^{g(b)} (t - g(x))^{\nu-1} \left(D_{g(b)-}^{\nu} \left(f \circ g^{-1}\right)\right)(t)\, dt, \tag{5.80}$$

all $x \in [x_0, b]$.

By (5.79) we have

$$|f(x)| \leq \frac{1}{\Gamma(\nu)} \int_{g(a)}^{g(x)} (g(x) - t)^{\nu-1} \left|\left(D_{g(a)}^{\nu} \left(f \circ g^{-1}\right)\right)(t)\right| dt \leq$$

$$\frac{1}{\Gamma(\nu)} \left(\int_{g(a)}^{g(x)} (g(x) - t)^{p(\nu-1)}\, dt\right)^{\frac{1}{p}} \left(\int_{g(a)}^{g(x)} \left|\left(D_{g(a)}^{\nu} \left(f \circ g^{-1}\right)\right)(t)\right|^q dt\right)^{\frac{1}{q}} \leq$$

$$\frac{1}{\Gamma(\nu)} \frac{(g(x) - g(a))^{\nu-1+\frac{1}{p}}}{(p(\nu-1)+1)^{\frac{1}{p}}} \left\|D_{g(a)}^{\nu} \left(f \circ g^{-1}\right)\right\|_{L_q\left(\left[g(a), \frac{g(a)+g(b)}{2}\right]\right)}, \tag{5.81}$$

for all $x \in [a, x_0]$.

Similarly, by (5.80) we get

$$|f(x)| \leq \frac{1}{\Gamma(\nu)} \int_{g(x)}^{g(b)} (t - g(x))^{\nu-1} \left|\left(D_{g(b)-}^{\nu} \left(f \circ g^{-1}\right)\right)(t)\right| dt \leq$$

$$\frac{1}{\Gamma(\nu)} \left(\int_{g(x)}^{g(b)} (t - g(x))^{p(\nu-1)}\, dt\right)^{\frac{1}{p}} \left(\int_{g(x)}^{g(b)} \left|\left(D_{g(b)-}^{\nu} \left(f \circ g^{-1}\right)\right)(t)\right|^q dt\right)^{\frac{1}{q}} \leq$$

$$\frac{1}{\Gamma(\nu)} \frac{(g(b) - g(x))^{\nu-1+\frac{1}{p}}}{(p(\nu-1)+1)^{\frac{1}{p}}} \left\|D_{g(b)-}^{\nu} \left(f \circ g^{-1}\right)\right\|_{L_q\left(\left[\frac{g(a)+g(b)}{2}, g(b)\right]\right)}, \tag{5.82}$$

for all $x \in [x_0, b]$.

Thus, we have

$$\left|\int_a^b f dg\right| \leq \int_a^b |f|\, dg =$$

$$\int_a^{x_0} |f(x)|\, g'(x)\, dx + \int_{x_0}^b |f(x)|\, g'(x)\, dx \overset{\text{(by (5.81), (5.82))}}{\leq} \frac{1}{\Gamma(\nu)(p(\nu-1)+1)^{\frac{1}{p}}}$$

$$\left[\left\| D^{\nu}_{g(a)} \left(f \circ g^{-1} \right) \right\|_{L_q\left(\left[g(a), \frac{g(a)+g(b)}{2} \right] \right)} \int_a^{x_0} (g(x) - g(a))^{\nu - 1 + \frac{1}{p}} g'(x) \, dx + \right.$$

$$\left. \left\| D^{\nu}_{g(b)-} \left(f \circ g^{-1} \right) \right\|_{L_q\left(\left[\frac{g(a)+g(b)}{2}, g(b) \right] \right)} \int_{x_0}^{b} (g(b) - g(x))^{\nu - 1 + \frac{1}{p}} g'(x) \, dx \right] = \tag{5.83}$$

$$\frac{1}{\Gamma(\nu)(p(\nu-1)+1)^{\frac{1}{p}} \left(\nu + \frac{1}{p} \right)} \left[\left\| D^{\nu}_{g(a)} \left(f \circ g^{-1} \right) \right\|_{L_q\left(\left[g(a), \frac{g(a)+g(b)}{2} \right] \right)} (g(x_0) - g(a))^{\nu + \frac{1}{p}} \right.$$

$$\left. + \left\| D^{\nu}_{g(b)-} \left(f \circ g^{-1} \right) \right\|_{L_q\left(\left[\frac{g(a)+g(b)}{2}, g(b) \right] \right)} (g(b) - g(x_0))^{\nu + \frac{1}{p}} \right] =$$

$$\frac{1}{\Gamma(\nu)(p(\nu-1)+1)^{\frac{1}{p}} \left(\nu + \frac{1}{p} \right)} \frac{(g(b) - g(a))^{\nu + \frac{1}{p}}}{2^{\nu + \frac{1}{p}}}$$

$$\left[\left\| D^{\nu}_{g(a)} \left(f \circ g^{-1} \right) \right\|_{L_q\left(\left[g(a), \frac{g(a)+g(b)}{2} \right] \right)} + \left\| D^{\nu}_{g(b)-} \left(f \circ g^{-1} \right) \right\|_{L_q\left(\left[\frac{g(a)+g(b)}{2}, g(b) \right] \right)} \right]$$

$$\leq \frac{M_q(f, g)}{\Gamma(\nu)(p(\nu-1)+1)^{\frac{1}{p}} \left(\nu + \frac{1}{p} \right)} \frac{(g(b) - g(a))^{\nu + \frac{1}{p}}}{2^{\nu - \frac{1}{q}}}, \tag{5.84}$$

proving the claim. ■

Combining facts from Theorems 5.7, 5.10, 5.11 and Remark 5.8, we get:

Theorem 5.12 *Let* $p, q > 1 : \frac{1}{p} + \frac{1}{q} = 1$, $\nu \geq 1$, $n = [\nu]$, $f \in C^n([a, b])$, $g : [a, b] \to \mathbb{R}$ *be a strictly increasing function. Assume that* $g \in C^1([a, b])$ *and* $g^{-1} \in C^n([g(a), g(b)])$. *Also assume* $f \circ g^{-1} \in C^{\nu}_{g(a)}\left(\left[g(a), \frac{g(a)+g(b)}{2} \right] \right)$, *and* $f \circ g^{-1} \in C^{\nu}_{g(b)-}\left(\left[\frac{g(a)+g(b)}{2}, g(b) \right] \right)$, *and that* $\left(f \circ g^{-1} \right)^{(k)}(g(a)) = \left(f \circ g^{-1} \right)^{(k)}(g(b)) = 0$, *for* $k = 0, 1, \ldots, n-1$. *Then*

$$\left| \int_a^b f dg \right| \leq \int_a^b |f| \, dg \leq \tag{5.85}$$

$$\min \left\{ M(f, g) \frac{(g(b) - g(a))^{\nu+1}}{\Gamma(\nu+2) 2^{\nu}}, M_1(f, g) \frac{(g(b) - g(a))^{\nu}}{\Gamma(\nu+1) 2^{\nu-1}}, \right.$$

$$\left. \frac{M_q(f, g)}{2^{\nu - \frac{1}{q}} \Gamma(\nu)(p(\nu-1)+1)^{\frac{1}{p}} \left(\nu + \frac{1}{p} \right)} (g(b) - g(a))^{\nu + \frac{1}{p}} \right\},$$

where $M(f, g)$ *as in (5.36),* $M_1(f, g)$ *as in (5.69) and* $M_q(f, g)$ *as in (5.77).*

Above $f \circ g^{-1}$ *cannot be a non-trivial polynomial of degree* n.

Corollary 5.13 *Here all as in Theorem 5.12. Then*

$$\left|\frac{1}{g(b)-g(a)}\int_a^b f dg\right| \leq \frac{1}{g(b)-g(a)}\int_a^b |f| dg \leq$$

$$\min\left\{M(f,g)\frac{(g(b)-g(a))^{\nu}}{\Gamma(\nu+2)2^{\nu}}, M_1(f,g)\frac{(g(b)-g(a))^{\nu-1}}{\Gamma(\nu+1)2^{\nu-1}},\right.$$

$$\left.\frac{M_q(f,g)}{2^{\nu-\frac{1}{q}}\Gamma(\nu)(p(\nu-1)+1)^{\frac{1}{p}}\left(\nu+\frac{1}{p}\right)}(g(b)-g(a))^{\nu-\frac{1}{q}}\right\}. \tag{5.86}$$

We continue with an L_1 iterated fractional Polya type inequality:

Theorem 5.14 *All as in Theorem 5.9, with* $\frac{1}{m+1} \leq \nu < 1$. *Set*

$$M_1^*(f,g) := \max\left\{\left\|D_{g(a)}^{(m+1)\nu}\left(f\circ g^{-1}\right)\right\|_{L_1\left(\left[g(a),\frac{g(a)+g(b)}{2}\right]\right)},\right.$$

$$\left.\left\|D_{g(b)-}^{(m+1)\nu}\left(f\circ g^{-1}\right)\right\|_{L_1\left(\left[\frac{g(a)+g(b)}{2},g(b)\right]\right)}\right\}. \tag{5.87}$$

Then

$$\left|\int_a^b f dg\right| \leq \int_a^b |f| dg \leq M_1^*(f,g)\frac{(g(b)-g(a))^{(m+1)\nu}}{\Gamma((m+1)\nu+1)2^{(m+1)\nu-1}}. \tag{5.88}$$

Proof Similar to Theorem 5.10. ∎

We continue with an L_p iterated fractional Polya type inequality:

Theorem 5.15 *All as in Theorem 5.9. Let* $p,q>1: \frac{1}{p}+\frac{1}{q}=1$, *such that* $\frac{1}{(m+1)q} < \nu < 1$. *Set*

$$M_q^*(f,g) := \max\left\{\left\|D_{g(a)}^{(m+1)\nu}\left(f\circ g^{-1}\right)\right\|_{L_q\left(\left[g(a),\frac{g(a)+g(b)}{2}\right]\right)},\right.$$

$$\left.\left\|D_{g(b)-}^{(m+1)\nu}\left(f\circ g^{-1}\right)\right\|_{L_q\left(\left[\frac{g(a)+g(b)}{2},g(b)\right]\right)}\right\}. \tag{5.89}$$

Then

$$\left|\int_a^b f dg\right| \leq \int_a^b |f| dg \leq$$

$$\frac{M_q^*(f,g)(g(b)-g(a))^{(m+1)\nu+\frac{1}{p}}}{2^{(m+1)\nu-\frac{1}{q}}\Gamma((m+1)\nu)(p((m+1)\nu-1)+1)^{\frac{1}{p}}\left((m+1)\nu+\frac{1}{p}\right)}. \tag{5.90}$$

Proof Similar to Theorem 5.11. ■

Applications follow:

Proposition 5.16 *Let* $0<\nu<1$, $f\in C([a,b])$. *Assume* $f\circ\ln x\in C_{e^a}^{\nu}\left(\left[e^a,\frac{e^a+e^b}{2}\right]\right)$, *and* $f\circ\ln x\in C_{e^b-}^{\nu}\left(\left[\frac{e^a+e^b}{2},e^b\right]\right)$. *Set*

$$M(f,e^x):=\max\left\{\left\|D_{e^a}^{\nu}(f\circ\ln)\right\|_{\infty,\left[e^a,\frac{e^a+e^b}{2}\right]},\right.$$

$$\left.\left\|D_{e^b-}^{\nu}(f\circ\ln)\right\|_{\infty,\left[\frac{e^a+e^b}{2},e^b\right]}\right\}. \tag{5.91}$$

Then

$$\left|\int_a^b f(x)e^x dx\right|\le\int_a^b|f(x)|e^x dx\le M(f,e^x)\frac{\left(e^b-e^a\right)^{\nu+1}}{\Gamma(\nu+2)2^{\nu}}. \tag{5.92}$$

Inequality (5.92) is sharp, namely it is attained by f_* *such that*

$$(f_*\circ\ln x)(z)=\left\{\begin{array}{l}(z-e^a)^{\nu},\ z\in\left[e^a,\frac{e^a+e^b}{2}\right]\\ \left(e^b-z\right)^{\nu},\ z\in\left[\frac{e^a+e^b}{2},e^b\right]\end{array}\right\},\quad 0<\nu<1. \tag{5.93}$$

Clearly here non zero constant functions $f\circ\ln x$ *are excluded.*

Proof By Theorem 5.7. ■

We continue with

Proposition 5.17 *Here* $[a,b]\subset(0,+\infty)$, $0<\nu<1$, $f\in C([a,b])$. *Assume that* $\left(D_{\ln a}^{i\nu}(f\circ e^x)\right)\in C_{\ln a}^{\nu}\left(\left[\ln a,\frac{\ln(ab)}{2}\right]\right)$, *for* $i=0,1,\ldots,m\in\mathbb{N}$, *and* $\left(D_{\ln a}^{(m+1)\nu}(f\circ e^x)\right)\in C\left(\left[\ln a,\left(\frac{\ln ab}{2}\right)\right]\right)$. *Also assume that* $\left(D_{\ln b-}^{i\nu}(f\circ e^x)\right)\in C_{\ln b-}^{\nu}\left(\left[\frac{\ln(ab)}{2},\ln b\right]\right)$, *for* $i=0,1,\ldots,m\in\mathbb{N}$, *and* $\left(D_{\ln b-}^{(m+1)\nu}(f\circ e^x)\right)\in C\left(\left[\frac{\ln(ab)}{2},\ln b\right]\right)$. *Set*

$$M^*(f,\ln x):=\max\left\{\left\|D_{\ln a}^{(m+1)\nu}(f\circ e^x)\right\|_{\infty,\left[\ln a,\frac{\ln(ab)}{2}\right]},\right.$$

$$\left.\left\|D_{\ln b-}^{(m+1)\nu}(f\circ e^x)\right\|_{\infty,\left[\frac{\ln(ab)}{2},\ln b\right]}\right\}. \tag{5.94}$$

Then

$$\left|\int_a^b \frac{f(x)}{x}dx\right| \leq \int_a^b \frac{|f(x)|}{x}dx \leq M^*(f, \ln x)\frac{\left(\ln \frac{b}{a}\right)^{(m+1)\nu+1}}{\Gamma((m+1)\nu+2)\,2^{(m+1)\nu}}. \tag{5.95}$$

Proof By Theorem 5.9. ∎

We can have many other interesting applications but we stop here.

References

1. Anastassiou, G.A.: Fractional Differentiation Inequalities. Research Monograph. Springer, New York (2009)
2. Anastassiou, G.A.: On right fractional calculus. Chaos, Solitons and Fractals **42**, 365–376 (2009)
3. Anastassiou, G.A.: Inteligent Mathematics: Computational Analysis. Springer, Heidelberg (2011)
4. Anastassiou, G.A.: Intelligent Comparisons: Analytic Inequalities. Springer, Heidelberg (2016)
5. Anastassiou, G.A.: Generalized Canavati type fractional Taylor's formulae. J. Comput. Anal. Appl. **21**(7), 1205–1212 (2016)
6. Anastassiou, G.: Generalized Canavati type g-fractional Polya type inequalities. Int. J. Appl. Math. **32**(5), 857–880 (2019)
7. Canavati, J.A.: The Riemann-Liouville integral. Nieuw Archief Voor Wiskunde **5**(1), 53–75 (1987)
8. El-Sayed, A.M.A., Gaber, M.: On the finite caputo and finite Riesz derivatives. Electron. J. Theor. Phys. **3**(12), 81–95 (2006)
9. Frederico, G.S., Torres, D.F.M.: Fractional optimal control in the sense of caputo and the fractional Noether's theorem. Int. Math. Forum **3**(10), 479–493 (2008)
10. Gorenflo, R., Mainardi, F.: Essentials of fractional calculus. Maphysto Center (2000). http://www.maphysto.dk/oldpages/events/LevyCAC2000/MainardiNotes/fm2k0a.ps
11. Jia, R.-Q.: Absolutely continuous functions (Chap. 3). https://www.ualberta.ca/126rjia/Math418/Notes/Chap.3.pdf
12. Polya, G.: Ein mittelwertsats für Funktionen mehrerer Veränderlichen. Tohoku Math. J. **19**, 1–3 (1921)
13. Polya, G., Szegö, G.: Aufgaben und Lehrsärtze aus der Analysis, vol. I. Springer, Berlin (1925). (German)
14. Polya, G., Szegö, G.: Problems and Theorems in Analysis. Classics in Mathematics, vol. I. Springer, Berlin (1972)
15. Polya, G., Szegö, G.: Problems and Theorems in Analysis, vol. I, Chinese edn. (1984)
16. Samko, S.G., Kilbas, A.A., Marichev, O.I.: Fractional Integrals and Derivatives, Theory and Applications. Gordon and Breach, Amsterdam (1993) [English translation from the Russian, Integrals and Derivatives of Fractional Order and Some of Their Applications (Nauka i Tekhnika, Minsk, 1987)]
17. Whittaker, E.T., Watson, G.N.: A Course in Modern Analysis. Cambridge University Press, Cambridge (1927)

Chapter 6
Caputo Generalized ψ-Fractional Integral Type Inequalities

Very general univariate and multivariate Caputo ψ-fractional integral inequalities of Poincaré, Sobolev and Hilbert–Pachpatte types are presented. Estimates are with respect to $\|\cdot\|_p$, $1 \leq p < \infty$. Applications are given. See also [5].

6.1 Background

This chapter among others is motivated by the famous Poincaré inequality: see [1]. Given a bounded domain $\Omega \subset \mathbb{R}^n$, $n \in \mathbb{N}$, it holds that

$$\|u\|_{L^p(\Omega)} \leq C \|\nabla u\|_{L^p(\Omega)} \tag{6.1}$$

for functions u with vanishing mean value over Ω, $1 \leq p \leq \infty$, under very general assumptions on Ω, where $\|\nabla u\|_{L^p(\Omega)}$ is defined as the L^p-norm of the Euclidean norm of ∇u.

Especially in [1] it is proven for a convex domain $\Omega \subset \mathbb{R}^n$ with diameter d that

$$\|u\|_{L^1(\Omega)} \leq \frac{d}{2} \|\nabla u\|_{L^1(\Omega)} \tag{6.2}$$

for any u with zero mean value on Ω, with the constant $\frac{1}{2}$ also being optimal.

Also this chapter is motivated by the famous Sobolev-type inequality (see [6, p. 263]), the Gagliardo-Nirenberg-Sobolev inequality: Assume $1 \leq p < n$. Then there exists a constant C, depending only on p and n, such that

G. A. Anastassiou, *Generalized Fractional Calculus*, Studies in Systems, Decision and Control 305, https://doi.org/10.1007/978-3-030-56962-4_6

$$\|u\|_{L^{p^*}(\mathbb{R}^n)} \leq C \|Du\|_{L^p(\mathbb{R}^n)}, \tag{6.3}$$

for all $u \in C_c^1(\mathbb{R}^n)$.

Here $p^* := np/n - p$, $p^* > p$, and Du is the gradient of u.

We are also motivated by the following result [6, p. 265]: estimates for Sobolev space $W^{1,p}$, $1 \leq p < n$: let U be a bounded open subset of $\mathbb{R}^n$, and suppose the boundary ∂U is C^1. Assume $1 \leq p < n$, and $u \in W^{1,p}(U)$. Then $u \in L^{p^*}(U)$, with the estimate

$$\|u\|_{L^{p^*}(U)} \leq C \|u\|_{W^{1,p}(U)}, \tag{6.4}$$

the constant C depending only on p, n, and U.

The next Hilbert–Pachpatte inequality is a main motivation here:

Theorem 6.1 (Pachpatte [7, Theorem 1]). *Let $n \geq 1$ and $0 \leq k \leq n-1$ be integers. Let $u \in C^n([0,x])$ and $v \in C^n([0,y])$, where $x > 0$, $y > 0$, and let $u^{(j)}(0) = v^{(j)}(0) = 0$ for $j \in \{0, \ldots, n-1\}$. Then*

$$\int_0^x \int_0^y \frac{\left|u^{(k)}(s)\right| \left|v^{(k)}(t)\right|}{s^{2n-2k-1} + t^{2n-2k-1}} ds dt \leq M(n,k,x,y)$$

$$\left(\int_0^x (x-s)\left(u^{(n)}(s)\right)^2 ds\right)^{\frac{1}{2}} \left(\int_0^y (y-t)\left(v^{(n)}(t)\right)^2 dt\right)^{\frac{1}{2}}, \tag{6.5}$$

where

$$M(n,k,x,y) := \frac{1}{2} \frac{\sqrt{xy}}{[(n-k-1)!]^2 (2n-2k-1)}.$$

Next we follow [2].

Let $\alpha > 0$, $[a,b] \subset \mathbb{R}$, $f : [a,b] \to \mathbb{R}$ which is integrable and $\psi \in C^1([a,b])$ an increasing function such that $\psi'(x) \neq 0$, for all $x \in [a,b]$. Consider $n = \lceil \alpha \rceil$, the ceiling of α. The left and right fractional integrals are defined, respectively, as follows:

$$I_{a+}^{\alpha,\psi} f(x) := \frac{1}{\Gamma(\alpha)} \int_a^x \psi'(t) (\psi(x) - \psi(t))^{\alpha-1} f(t) dt, \tag{6.6}$$

and

$$I_{b-}^{\alpha,\psi} f(x) := \frac{1}{\Gamma(\alpha)} \int_x^b \psi'(t) (\psi(t) - \psi(x))^{\alpha-1} f(t) dt, \tag{6.7}$$

for any $x \in [a,b]$, where Γ is the gamma function.

The following semigroup property is valid for fractional integrals: if $\alpha, \beta > 0$, then

$$I_{a+}^{\alpha,\psi} I_{a+}^{\beta,\psi} f(x) = I_{a+}^{\alpha+\beta,\psi} f(x), \text{ and } I_{b-}^{\alpha,\psi} I_{b-}^{\beta,\psi} f(x) = I_{b-}^{\alpha+\beta,\psi} f(x).$$

We mention

Definition 6.2 *([2])* Let $\alpha > 0$, $n \in \mathbb{N}$ such that $n = \lceil \alpha \rceil$, $[a, b] \subset \mathbb{R}$ and $f, \psi \in C^n([a, b])$ with ψ being increasing and $\psi'(x) \neq 0$, for all $x \in [a, b]$. The left ψ-Caputo fractional derivative of f of order α is given by

$$ {}^{C}D_{a+}^{\alpha,\psi} f(x) := I_{a+}^{n-\alpha,\psi} \left(\frac{1}{\psi'(x)} \frac{d}{dx} \right)^n f(x), \tag{6.8} $$

and the right ψ-Caputo fractional derivative of f is given by

$$ {}^{C}D_{b-}^{\alpha,\psi} f(x) := I_{b-}^{n-\alpha,\psi} \left(-\frac{1}{\psi'(x)} \frac{d}{dx} \right)^n f(x). \tag{6.9} $$

To simplify notation, we will use the symbol

$$ f_{\psi}^{[n]}(x) := \left(\frac{1}{\psi'(x)} \frac{d}{dx} \right)^n f(x), \tag{6.10} $$

with $f_{\psi}^{[0]}(x) = f(x)$.

By the definition, when $\alpha = m \in \mathbb{N}$, we have

$$ \begin{aligned} &{}^{C}D_{a+}^{\alpha,\psi} f(x) = f_{\psi}^{[m]}(x) \\ &\text{and} \\ &{}^{C}D_{b-}^{\alpha,\psi} f(x) = (-1)^m f_{\psi}^{[m]}(x). \end{aligned} \tag{6.11} $$

If $\alpha \notin \mathbb{N}$, we have

$$ {}^{C}D_{a+}^{\alpha,\psi} f(x) = \frac{1}{\Gamma(n-\alpha)} \int_a^x \psi'(t) (\psi(x) - \psi(t))^{n-\alpha-1} f_{\psi}^{[n]}(t)\, dt, \tag{6.12} $$

and

$$ {}^{C}D_{b-}^{\alpha,\psi} f(x) = \frac{(-1)^n}{\Gamma(n-\alpha)} \int_x^b \psi'(t) (\psi(t) - \psi(x))^{n-\alpha-1} f_{\psi}^{[n]}(t)\, dt, \tag{6.13} $$

$\forall\, x \in [a, b]$.

In particular, when $\alpha \in (0, 1)$, we have

$$ \begin{aligned} &{}^{C}D_{a+}^{\alpha,\psi} f(x) = \tfrac{1}{\Gamma(1-\alpha)} \int_a^x (\psi(x) - \psi(t))^{-\alpha} f'(t)\, dt, \\ &\text{and} \\ &{}^{C}D_{b-}^{\alpha,\psi} f(x) = \tfrac{-1}{\Gamma(1-\alpha)} \int_x^b (\psi(t) - \psi(x))^{-\alpha} f'(t)\, dt, \end{aligned} \tag{6.14} $$

$\forall\, x \in [a, b]$.

Clearly the above is a generalization of left and right Caputo fractional derivatives. For more see [2].

Still we need from [2] the following left and right fractional Taylor's formulae:

Theorem 6.3 ([2]) *Let $\alpha > 0$, $n \in \mathbb{N}$ such that $n = \lceil \alpha \rceil$, $[a, b] \subset \mathbb{R}$ and $f, \psi \in C^n([a, b])$ with ψ being increasing and $\psi'(x) \neq 0$, for all $x \in [a, b]$. Then, the left fractional Taylor formula follows,*

$$f(x) = \sum_{k=0}^{n-1} \frac{f_{\psi}^{[k]}(a)}{k!} (\psi(x) - \psi(a))^k + I_{a+}^{\alpha,\psi}\ {}^{C}D_{a+}^{\alpha,\psi} f(x), \tag{6.15}$$

and the right fractional Taylor formula follows,

$$f(x) = \sum_{k=0}^{n-1} (-1)^k \frac{f_{\psi}^{[k]}(b)}{k!} (\psi(b) - \psi(x))^k + I_{b-}^{\alpha,\psi}\ {}^{C}D_{b-}^{\alpha,\psi} f(x), \tag{6.16}$$

$\forall\, x \in [a, b]$.

In particular, given $\alpha \in (0, 1)$, we have

$$\begin{aligned} & f(x) = f(a) + I_{a+}^{\alpha,\psi}\ {}^{C}D_{a+}^{\alpha,\psi} f(x), \\ & \text{and} \\ & f(x) = f(b) + I_{b-}^{\alpha,\psi}\ {}^{C}D_{b-}^{\alpha,\psi} f(x), \end{aligned} \tag{6.17}$$

$\forall\, x \in [a, b]$.

Remark 6.4 For convenience we can rewrite (6.15)–(6.17) as follows:

$$f(x) = \sum_{k=0}^{n-1} \frac{f_{\psi}^{[k]}(a)}{k!} (\psi(x) - \psi(a))^k + \tag{6.18}$$

$$\frac{1}{\Gamma(\alpha)} \int_a^x \psi'(t) (\psi(x) - \psi(t))^{\alpha-1}\ {}^{C}D_{a+}^{\alpha,\psi} f(t)\, dt,$$

and

$$f(x) = \sum_{k=0}^{n-1} \frac{(-1)^k f_{\psi}^{[k]}(b)}{k!} (\psi(b) - \psi(x))^k + \tag{6.19}$$

$$\frac{1}{\Gamma(\alpha)} \int_x^b \psi'(t) (\psi(t) - \psi(x))^{\alpha-1}\ {}^{C}D_{b-}^{\alpha,\psi} f(t)\, dt,$$

$\forall\, x \in [a, b]$.

When $\alpha \in (0, 1)$, we get:

$$f(x) = f(a) + \frac{1}{\Gamma(\alpha)} \int_a^x \psi'(t)(\psi(x) - \psi(t))^{\alpha-1} \, {}^C D_{a+}^{\alpha,\psi} f(t) \, dt$$
and
$$f(x) = f(b) + \frac{1}{\Gamma(\alpha)} \int_x^b \psi'(t)(\psi(t) - \psi(x))^{\alpha-1} \, {}^C D_{b-}^{\alpha,\psi} f(t) \, dt \tag{6.20}$$

$\forall\, x \in [a, b]$.

Again from [2] we have the following:

Consider the norms $\|\cdot\|_\infty : C([a, b]) \to \mathbb{R}$ and $\|\cdot\|_{C_\psi^{[n]}} : C^n([a, b]) \to \mathbb{R}$, where

$\|f\|_{C_\psi^{[n]}} := \sum_{k=0}^{n} \left\| f_\psi^{[k]} \right\|_\infty$.

We have

Theorem 6.5 ([2]) *The ψ-Caputo fractional derivatives are bounded operators. For all $\alpha > 0$ ($n = \lceil \alpha \rceil$)*

$$\left\| {}^C D_{a+}^{\alpha,\psi} \right\|_\infty \leq K \|f\|_{C_\psi^{[n]}} \tag{6.21}$$

and

$$\left\| {}^C D_{b-}^{\alpha,\psi} \right\|_\infty \leq K \|f\|_{C_\psi^{[n]}}, \tag{6.22}$$

where

$$K = \frac{(\psi(b) - \psi(a))^{n-\alpha}}{\Gamma(n + 1 - \alpha)} > 0. \tag{6.23}$$

We make

Remark 6.6 In our setting, clearly, it is $f_\psi^{[n]} \in C([a, b])$. Given $f \in C([a, b])$, by Theorem 4.10, p. 98 of [4], we get that $I_{a+}^{\alpha,\psi} f \in C([a, b])$, and by Theorem 4.11, p. 101 of [4], we get that $I_{b-}^{\alpha,\psi} f \in C([a, b])$.

Therefore, we obtain that ${}^C D_{a+}^{\alpha,\psi} f, {}^C D_{b-}^{\alpha,\psi} f \in C([a, b])$.

6.2 Main Results

We present a generalized left fractional Poincaré type inequality:

Theorem 6.7 *Let $p, q > 1$ such that $\frac{1}{p} + \frac{1}{q} = 1$, and $\alpha > \frac{1}{q}$, $n \in \mathbb{N}$ such that $n = \lceil \alpha \rceil$, $[a, b] \subset \mathbb{R}$ and $f, \psi \in C^n([a, b])$ with ψ being strictly increasing and $\psi'(x) \neq 0$, for all $x \in [a, b]$. Assume that $f_\psi^{[k]}(a) = 0$, for $k = 0, 1, \ldots, n - 1$. Then*

$$\|f\|_{q,[a,b]} \leq \frac{(\psi(b) - \psi(a))^{\alpha - \frac{1}{q}} (b - a)^{\frac{1}{q}}}{\Gamma(\alpha)(p(\alpha - 1) + 1)^{\frac{1}{p}}} \left\| \left({}^C D_{a+}^{\alpha,\psi} f \right) \circ \psi^{-1} \right\|_{q,[\psi(a),\psi(b)]}. \tag{6.24}$$

Proof By (6.18), since $f_{\psi}^{[k]}(a) = 0$, for $k = 0, 1, \ldots, n-1$, we have

$$f(x) = \frac{1}{\Gamma(\alpha)} \int_a^x \psi'(t) (\psi(x) - \psi(t))^{\alpha-1} \, {}^C D_{a+}^{\alpha,\psi} f(t) \, dt, \tag{6.25}$$

$\forall \, x \in [a, b]$.

Hence

$$|f(x)| \leq \frac{1}{\Gamma(\alpha)} \int_a^x \psi'(t) (\psi(x) - \psi(t))^{\alpha-1} \left| {}^C D_{a+}^{\alpha,\psi} f(t) \right| dt =$$

$$\frac{1}{\Gamma(\alpha)} \int_a^x (\psi(x) - \psi(t))^{\alpha-1} \left| {}^C D_{a+}^{\alpha,\psi} f(t) \right| d\psi(t) = \tag{6.26}$$

(by ψ being strictly increasing)

$$\frac{1}{\Gamma(\alpha)} \int_{\psi(a)}^{\psi(x)} (\psi(x) - z)^{\alpha-1} \left| \left({}^C D_{a+}^{\alpha,\psi} f \right) \left(\psi^{-1}(z) \right) \right| dz \leq$$

$$\frac{1}{\Gamma(\alpha)} \left(\int_{\psi(a)}^{\psi(x)} (\psi(x) - z)^{p(\alpha-1)} dz \right)^{\frac{1}{p}} \left(\int_{\psi(a)}^{\psi(x)} \left| \left({}^C D_{a+}^{\alpha,\psi} f \right) \left(\psi^{-1}(z) \right) \right|^q dz \right)^{\frac{1}{q}} \leq$$

$$\frac{1}{\Gamma(\alpha)} \frac{(\psi(x) - \psi(a))^{\frac{p(\alpha-1)+1}{p}}}{(p(\alpha-1)+1)^{\frac{1}{p}}} \left\| \left({}^C D_{a+}^{\alpha,\psi} f \right) \circ \psi^{-1} \right\|_{q,[\psi(a),\psi(b)]}. \tag{6.27}$$

That is

$$|f(x)| \leq \frac{(\psi(x) - \psi(a))^{\frac{p(\alpha-1)+1}{p}}}{\Gamma(\alpha)(p(\alpha-1)+1)^{\frac{1}{p}}} \left\| \left({}^C D_{a+}^{\alpha,\psi} f \right) \circ \psi^{-1} \right\|_{q,[\psi(a),\psi(b)]}, \tag{6.28}$$

$\forall \, x \in [a, b]$.

Thus, we have

$$|f(x)|^q \leq \frac{(\psi(x) - \psi(a))^{(p(\alpha-1)+1)\frac{q}{p}}}{(\Gamma(\alpha))^q (p(\alpha-1)+1)^{\frac{q}{p}}} \left\| \left({}^C D_{a+}^{\alpha,\psi} f \right) \circ \psi^{-1} \right\|_{q,[\psi(a),\psi(b)]}^q, \tag{6.29}$$

$\forall \, x \in [a, b]$.

Therefore it holds

$$\int_a^b |f(x)|^q \, dx \leq \frac{(\psi(b) - \psi(a))^{(p(\alpha-1)+1)\frac{q}{p}} (b-a)}{(\Gamma(\alpha))^q (p(\alpha-1)+1)^{\frac{q}{p}}}$$

$$\left\| \left(^{C}D_{a+}^{\alpha,\psi} f \right) \circ \psi^{-1} \right\|_{q,[\psi(a),\psi(b)]}^{q}, \tag{6.30}$$

and

$$\left(\int_a^b |f(x)|^q \, dx \right)^{\frac{1}{q}} \leq \frac{(\psi(b) - \psi(a))^{\alpha - \frac{1}{q}} (b-a)^{\frac{1}{q}}}{\Gamma(\alpha) (p(\alpha - 1) + 1)^{\frac{1}{p}}}$$

$$\left\| \left(^{C}D_{a+}^{\alpha,\psi} f \right) \circ \psi^{-1} \right\|_{q,[\psi(a),\psi(b)]}, \tag{6.31}$$

proving the claim. ■

Next we give a generalized right fractional Poincaré type inequality:

Theorem 6.8 *Let $p, q > 1$ such that $\frac{1}{p} + \frac{1}{q} = 1$, and $\alpha > \frac{1}{q}$, $n \in \mathbb{N}$ such that $n = \lceil \alpha \rceil$, $[a,b] \subset \mathbb{R}$ and $f, \psi \in C^n([a,b])$ with ψ being strictly increasing and $\psi'(x) \neq 0$, for all $x \in [a,b]$. Assume that $f_{\psi}^{[k]}(b) = 0$, for $k = 0, 1, \ldots, n-1$. Then*

$$\|f\|_{q,[a,b]} \leq \frac{(\psi(b) - \psi(a))^{\alpha - \frac{1}{q}} (b-a)^{\frac{1}{q}}}{\Gamma(\alpha) (p(\alpha - 1) + 1)^{\frac{1}{p}}} \left\| \left(^{C}D_{b-}^{\alpha,\psi} f \right) \circ \psi^{-1} \right\|_{q,[\psi(a),\psi(b)]}. \tag{6.32}$$

Proof By (6.19), since $f_{\psi}^{[k]}(b) = 0$, for $k = 0, 1, \ldots, n-1$, we have

$$f(x) = \frac{1}{\Gamma(\alpha)} \int_x^b \psi'(t) (\psi(t) - \psi(x))^{\alpha - 1} \, {^{C}D_{b-}^{\alpha,\psi}} f(t) \, dt, \tag{6.33}$$

$\forall \, x \in [a,b]$.

Hence

$$|f(x)| \leq \frac{1}{\Gamma(\alpha)} \int_x^b \psi'(t) (\psi(t) - \psi(x))^{\alpha - 1} \left| {^{C}D_{b-}^{\alpha,\psi}} f(t) \right| dt =$$

$$\frac{1}{\Gamma(\alpha)} \int_x^b (\psi(t) - \psi(x))^{\alpha - 1} \left| {^{C}D_{b-}^{\alpha,\psi}} f(t) \right| d\psi(t) = \tag{6.34}$$

(by ψ being strictly increasing)

$$\frac{1}{\Gamma(\alpha)} \int_{\psi(x)}^{\psi(b)} (z - \psi(x))^{\alpha - 1} \left| \left(^{C}D_{b-}^{\alpha,\psi} f \right) \left(\psi^{-1}(z) \right) \right| dz \leq$$

$$\frac{1}{\Gamma(\alpha)} \left(\int_{\psi(x)}^{\psi(b)} (z - \psi(x))^{p(\alpha - 1)} dz \right)^{\frac{1}{p}} \left(\int_{\psi(x)}^{\psi(b)} \left| \left(^{C}D_{b-}^{\alpha,\psi} f \right) \left(\psi^{-1}(z) \right) \right|^q dz \right)^{\frac{1}{q}} \leq$$

$$\frac{1}{\Gamma(\alpha)}\frac{(\psi(b)-\psi(x))^{\frac{p(\alpha-1)+1}{p}}}{(p(\alpha-1)+1)^{\frac{1}{p}}}\left\|\left({}^{C}D_{b-}^{\alpha,\psi}f\right)\circ\psi^{-1}\right\|_{q,[\psi(a),\psi(b)]}. \quad (6.35)$$

That is

$$|f(x)|\leq\frac{(\psi(b)-\psi(x))^{\frac{p(\alpha-1)+1}{p}}}{\Gamma(\alpha)(p(\alpha-1)+1)^{\frac{1}{p}}}\left\|\left({}^{C}D_{b-}^{\alpha,\psi}f\right)\circ\psi^{-1}\right\|_{q,[\psi(a),\psi(b)]}, \quad (6.36)$$

$\forall\, x\in[a,b]$.

Thus, we have

$$|f(x)|^{q}\leq\frac{(\psi(b)-\psi(x))^{(p(\alpha-1)+1)\frac{q}{p}}}{(\Gamma(\alpha))^{q}(p(\alpha-1)+1)^{\frac{q}{p}}}\left\|\left({}^{C}D_{b-}^{\alpha,\psi}f\right)\circ\psi^{-1}\right\|_{q,[\psi(a),\psi(b)]}^{q}, \quad (6.37)$$

$\forall\, x\in[a,b]$.

Therefore it holds

$$\int_{a}^{b}|f(x)|^{q}dx\leq\frac{(\psi(b)-\psi(a))^{(p(\alpha-1)+1)\frac{q}{p}}(b-a)}{(\Gamma(\alpha))^{q}(p(\alpha-1)+1)^{\frac{q}{p}}}$$

$$\left\|\left({}^{C}D_{b-}^{\alpha,\psi}f\right)\circ\psi^{-1}\right\|_{q,[\psi(a),\psi(b)]}^{q}, \quad (6.38)$$

and

$$\left(\int_{a}^{b}|f(x)|^{q}dx\right)^{\frac{1}{q}}\leq\frac{(\psi(b)-\psi(a))^{\alpha-\frac{1}{q}}(b-a)^{\frac{1}{q}}}{\Gamma(\alpha)(p(\alpha-1)+1)^{\frac{1}{p}}}$$

$$\left\|\left({}^{C}D_{b-}^{\alpha,\psi}f\right)\circ\psi^{-1}\right\|_{q,[\psi(a),\psi(b)]}, \quad (6.39)$$

proving the claim. ■

We continue with a generalized left fractional Sobolev type inequality:

Theorem 6.9 *All as in Theorem 6.7, $r>0$. Then*

$$\|f\|_{r,[a,b]}\leq\frac{(\psi(b)-\psi(a))^{\alpha-\frac{1}{q}}(b-a)^{\frac{1}{r}}}{\Gamma(\alpha)(p(\alpha-1)+1)^{\frac{1}{p}}}\left\|\left({}^{C}D_{a+}^{\alpha,\psi}f\right)\circ\psi^{-1}\right\|_{q,[\psi(a),\psi(b)]}. \quad (6.40)$$

Proof From (6.28) we have

$$|f(x)|^{r}\leq\frac{(\psi(x)-\psi(a))^{\left(\frac{p(\alpha-1)+1}{p}\right)r}}{(\Gamma(\alpha))^{r}(p(\alpha-1)+1)^{\frac{r}{p}}}\left\|\left({}^{C}D_{a+}^{\alpha,\psi}f\right)\circ\psi^{-1}\right\|_{q,[\psi(a),\psi(b)]}^{r}, \quad (6.41)$$

$\forall\, x \in [a, b]$.

Hence we have

$$\int_a^b |f(x)|^r\, dx \le \frac{(\psi(b)-\psi(a))^{\left(\frac{p(\alpha-1)+1}{p}\right)r}(b-a)}{(\Gamma(\alpha))^r (p(\alpha-1)+1)^{\frac{r}{p}}} \left\| \left({}^{C}D_{a+}^{\alpha,\psi} f\right) \circ \psi^{-1} \right\|_{q,[\psi(a),\psi(b)]}^{r}, \tag{6.42}$$

That is

$$\left(\int_a^b |f(x)|^r\, dx\right)^{\frac{1}{r}} \le \frac{(\psi(b)-\psi(a))^{\frac{p(\alpha-1)+1}{p}}(b-a)^{\frac{1}{r}}}{\Gamma(\alpha)(p(\alpha-1)+1)^{\frac{1}{p}}} \left\| \left({}^{C}D_{a+}^{\alpha,\psi} f\right) \circ \psi^{-1} \right\|_{q,[\psi(a),\psi(b)]}, \tag{6.43}$$

proving the claim. ■

We continue with a generalized right fractional Sobolev type inequality:

Theorem 6.10 *All as in Theorem 6.8, $r > 0$. Then*

$$\|f\|_{r,[a,b]} \le \frac{(\psi(b)-\psi(a))^{\alpha-\frac{1}{q}}(b-a)^{\frac{1}{r}}}{\Gamma(\alpha)(p(\alpha-1)+1)^{\frac{1}{p}}} \left\| \left({}^{C}D_{b-}^{\alpha,\psi} f\right) \circ \psi^{-1} \right\|_{q,[\psi(a),\psi(b)]}. \tag{6.44}$$

Proof From (6.36) we get that

$$|f(x)|^r \le \frac{(\psi(b)-\psi(x))^{\left(\frac{p(\alpha-1)+1}{p}\right)r}}{(\Gamma(\alpha))^r (p(\alpha-1)+1)^{\frac{r}{p}}} \left\| \left({}^{C}D_{b-}^{\alpha,\psi} f\right) \circ \psi^{-1} \right\|_{q,[\psi(a),\psi(b)]}^{r}, \tag{6.45}$$

$\forall\, x \in [a, b]$.

Hence we have

$$\int_a^b |f(x)|^r\, dx \le \frac{(\psi(b)-\psi(a))^{\left(\frac{p(\alpha-1)+1}{p}\right)r}(b-a)}{(\Gamma(\alpha))^r (p(\alpha-1)+1)^{\frac{r}{p}}} \left\| \left({}^{C}D_{b-}^{\alpha,\psi} f\right) \circ \psi^{-1} \right\|_{q,[\psi(a),\psi(b)]}^{r}, \tag{6.46}$$

That is

$$\left(\int_a^b |f(x)|^r\, dx\right)^{\frac{1}{r}} \le \frac{(\psi(b)-\psi(a))^{\frac{p(\alpha-1)+1}{p}}(b-a)^{\frac{1}{r}}}{\Gamma(\alpha)(p(\alpha-1)+1)^{\frac{1}{p}}} \left\| \left({}^{C}D_{b-}^{\alpha,\psi} f\right) \circ \psi^{-1} \right\|_{q,[\psi(a),\psi(b)]}, \tag{6.47}$$

proving the claim. ■

Next comes a generalized left fractional Hilbert–Pachpatte inequality:

Theorem 6.11 *Let $p, q > 1$ such that $\frac{1}{p} + \frac{1}{q} = 1$, $\alpha_1 > \frac{1}{q}$, $n_1 \in \mathbb{N} : n_1 = \lceil \alpha_1 \rceil$; and $\alpha_2 > \frac{1}{p}$, $n_2 \in \mathbb{N} : n_2 = \lceil \alpha_2 \rceil$; $[a_i, b_i] \subset \mathbb{R}$, for $i = 1, 2$. Let $f_i, \psi_i \in C^{n_i}([a_i, b_i])$ with ψ_i being strictly increasing and $\psi_i'(x_i) \neq 0$, for all $x_i \in [a_i, b_i]$, for $i = 1, 2$. Assume that $f_{i\psi_i}^{[k_i]}(a_i) = 0$, for $k_i = 0, 1, \ldots, n_i - 1$; $i = 1, 2$. Then*

$$\int_{a_1}^{b_1}\int_{a_2}^{b_2}\frac{|f_1(x_1)|\,|f_2(x_2)|\,dx_1dx_2}{\left[\frac{(\psi_1(x_1)-\psi_1(a_1))^{p(\alpha_1-1)+1}}{p}+\frac{(\psi_2(x_2)-\psi_2(a_2))^{q(\alpha_2-1)+1}}{q}\right]}\leq$$

$$\frac{(b_1-a_1)(b_2-a_2)}{\Gamma(\alpha_1)\Gamma(\alpha_2)(p(\alpha_1-1)+1)^{\frac{1}{p}}(q(\alpha_2-1)+1)^{\frac{1}{q}}}$$

$$\left\|\left({}^{C}D_{a_1+}^{\alpha_1,\psi_1}f_1\right)\circ\psi_1^{-1}\right\|_{q,[\psi_1(a_1),\psi_1(b_1)]}\left\|\left({}^{C}D_{a_2+}^{\alpha_2,\psi_2}f_2\right)\circ\psi_2^{-1}\right\|_{p,[\psi_2(a_2),\psi_2(b_2)]}. \tag{6.48}$$

Proof Since $f_{i\psi_i}^{[k_i]}(a_i)=0$, for $k_i=0,1,\ldots,n_i-1; i=1,2$, by (6.18) we get that:

$$f_i(x_i)=\frac{1}{\Gamma(\alpha_i)}\int_{a_i}^{x_i}\psi_i'(t_i)(\psi_i(x_i)-\psi_i(t_i))^{\alpha_i-1}\,{}^{C}D_{a_i+}^{\alpha_i,\psi_i}f_i(t_i)\,dt_i, \tag{6.49}$$

$\forall\, x_i\in[a_i,b_i]$, $i=1,2$.

Hence it holds

$$|f_i(x_i)|\leq\frac{1}{\Gamma(\alpha_i)}\int_{a_i}^{x_i}\psi_i'(t_i)(\psi_i(x_i)-\psi_i(t_i))^{\alpha_i-1}\left|{}^{C}D_{a_i+}^{\alpha_i,\psi_i}f_i(t_i)\right|dt_i, \tag{6.50}$$

$\forall\, x_i\in[a_i,b_i]$, $i=1,2$.

Let $p,q>1:\frac{1}{p}+\frac{1}{q}=1$, then

$$|f_1(x_1)|\leq\frac{1}{\Gamma(\alpha_1)}\int_{a_1}^{x_1}\psi_1'(t_1)(\psi_1(x_1)-\psi_1(t_1))^{\alpha_1-1}\left|{}^{C}D_{a_1+}^{\alpha_1,\psi_1}f_1(t_1)\right|dt_1=$$

$$\frac{1}{\Gamma(\alpha_1)}\int_{a_1}^{x_1}(\psi_1(x_1)-\psi_1(t_1))^{\alpha_1-1}\left|{}^{C}D_{a_1+}^{\alpha_1,\psi_1}f_1(t_1)\right|d\psi_1(t_1)= \tag{6.51}$$

$$\frac{1}{\Gamma(\alpha_1)}\int_{\psi_1(a_1)}^{\psi_1(x_1)}(\psi_1(x_1)-z_1)^{\alpha_1-1}\left|\left({}^{C}D_{a_1+}^{\alpha_1,\psi_1}f_1\right)\left(\psi_1^{-1}(z_1)\right)\right|dz_1\leq$$

$$\frac{1}{\Gamma(\alpha_1)}\left(\int_{\psi_1(a_1)}^{\psi_1(x_1)}(\psi_1(x_1)-z_1)^{p(\alpha_1-1)}dz_1\right)^{\frac{1}{p}}$$

$$\left(\int_{\psi_1(a_1)}^{\psi_1(x_1)}\left|\left({}^{C}D_{a_1+}^{\alpha_1,\psi_1}f_1\right)\left(\psi_1^{-1}(z_1)\right)\right|^{q}dz_1\right)^{\frac{1}{q}}\leq$$

$$\frac{1}{\Gamma(\alpha_1)}\frac{(\psi_1(x_1)-\psi_1(a_1))^{\frac{p(\alpha_1-1)+1}{p}}}{(p(\alpha_1-1)+1)^{\frac{1}{p}}}\left\|\left({}^{C}D_{a_1+}^{\alpha_1,\psi_1}f_1\right)\circ\psi_1^{-1}\right\|_{q,[\psi_1(a_1),\psi_1(b_1)]}. \tag{6.52}$$

That is

$$|f_1(x_1)| \leq \frac{(\psi_1(x_1) - \psi_1(a_1))^{\frac{p(\alpha_1-1)+1}{p}}}{\Gamma(\alpha_1)(p(\alpha_1-1)+1)^{\frac{1}{p}}} \left\| \left({}^{C}D_{a_1+}^{\alpha_1,\psi_1} f_1\right) \circ \psi_1^{-1} \right\|_{q,[\psi_1(a_1),\psi_1(b_1)]}, \tag{6.53}$$

$\forall\, x_1 \in [a_1, b_1]$, where $\alpha_1 > \frac{1}{q}$.

Similarly, by assuming $\alpha_2 > \frac{1}{p}$, we get

$$|f_2(x_2)| \leq \frac{(\psi_2(x_2) - \psi_2(a_2))^{\frac{q(\alpha_2-1)+1}{q}}}{\Gamma(\alpha_2)(q(\alpha_2-1)+1)^{\frac{1}{q}}} \left\| \left({}^{C}D_{a_2+}^{\alpha_2,\psi_2} f_2\right) \circ \psi_2^{-1} \right\|_{p,[\psi_2(a_2),\psi_2(b_2)]}, \tag{6.54}$$

$\forall\, x_2 \in [a_2, b_2]$.

Hence we have (by (6.53) and (6.54) multiplication)

$$|f_1(x_1)|\,|f_2(x_2)| \leq \frac{1}{\Gamma(\alpha_1)\Gamma(\alpha_2)(p(\alpha_1-1)+1)^{\frac{1}{p}}(q(\alpha_2-1)+1)^{\frac{1}{q}}}$$

$$(\psi_1(x_1) - \psi_1(a_1))^{\frac{p(\alpha_1-1)+1}{p}} (\psi_2(x_2) - \psi_2(a_2))^{\frac{q(\alpha_2-1)+1}{q}}$$

$$\left\| \left({}^{C}D_{a_1+}^{\alpha_1,\psi_1} f_1\right) \circ \psi_1^{-1} \right\|_{q,[\psi_1(a_1),\psi_1(b_1)]} \left\| \left({}^{C}D_{a_2+}^{\alpha_2,\psi_2} f_2\right) \circ \psi_2^{-1} \right\|_{p,[\psi_2(a_2),\psi_2(b_2)]} \tag{6.55}$$

(by using Young's inequality for $a, b \geq 0$, $a^{\frac{1}{p}} b^{\frac{1}{q}} \leq \frac{a}{p} + \frac{b}{q}$)

$$\leq \frac{1}{\Gamma(\alpha_1)\Gamma(\alpha_2)(p(\alpha_1-1)+1)^{\frac{1}{p}}(q(\alpha_2-1)+1)^{\frac{1}{q}}}$$

$$\left[\frac{(\psi_1(x_1) - \psi_1(a_1))^{p(\alpha_1-1)+1}}{p} + \frac{(\psi_2(x_2) - \psi_2(a_2))^{q(\alpha_2-1)+1}}{q}\right]$$

$$\left\| \left({}^{C}D_{a_1+}^{\alpha_1,\psi_1} f_1\right) \circ \psi_1^{-1} \right\|_{q,[\psi_1(a_1),\psi_1(b_1)]} \left\| \left({}^{C}D_{a_2+}^{\alpha_2,\psi_2} f_2\right) \circ \psi_2^{-1} \right\|_{p,[\psi_2(a_2),\psi_2(b_2)]} \tag{6.56}$$

$\forall\, x_i \in [a_i, b_i]$, $i = 1, 2$.

Therefore we can write

$$\frac{|f_1(x_1)|\,|f_2(x_2)|}{\left[\frac{(\psi_1(x_1)-\psi_1(a_1))^{p(\alpha_1-1)+1}}{p} + \frac{(\psi_2(x_2)-\psi_2(a_2))^{q(\alpha_2-1)+1}}{q}\right]} \leq$$

$$\frac{\left\| \left({}^{C}D_{a_1+}^{\alpha_1,\psi_1} f_1\right) \circ \psi_1^{-1} \right\|_{q,[\psi_1(a_1),\psi_1(b_1)]} \left\| \left({}^{C}D_{a_2+}^{\alpha_2,\psi_2} f_2\right) \circ \psi_2^{-1} \right\|_{p,[\psi_2(a_2),\psi_2(b_2)]}}{\Gamma(\alpha_1)\Gamma(\alpha_2)(p(\alpha_1-1)+1)^{\frac{1}{p}}(q(\alpha_2-1)+1)^{\frac{1}{q}}}, \tag{6.57}$$

$\forall\, x_i \in [a_i, b_i]$, $i = 1, 2$.

The denominator in the left hand side of (6.57) can be zero only when both $x_1 = a_1$ and $x_2 = a_2$.

By integrating (6.57) over $[a_1, b_1] \times [a_2, b_2]$ we obtain (6.48). ■

Next comes a generalized right fractional Hilbert–Pachpatte inequality:

Theorem 6.12 *Let $p, q > 1$ such that $\frac{1}{p} + \frac{1}{q} = 1$, $\alpha_1 > \frac{1}{q}$, $n_1 \in \mathbb{N} : n_1 = \lceil \alpha_1 \rceil$; and $\alpha_2 > \frac{1}{p}$, $n_2 \in \mathbb{N} : n_2 = \lceil \alpha_2 \rceil$; $[a_i, b_i] \subset \mathbb{R}$, for $i = 1, 2$. Let $f_i, \psi_i \in C^{n_i}([a_i, b_i])$ with ψ_i being strictly increasing and $\psi_i'(x_i) \neq 0$, for all $x_i \in [a_i, b_i]$, for $i = 1, 2$. Assume that $f_{i\psi_i}^{[k_i]}(b_i) = 0$, for $k_i = 0, 1, \dots, n_i - 1$; $i = 1, 2$. Then*

$$\int_{a_1}^{b_1} \int_{a_2}^{b_2} \frac{|f_1(x_1)| |f_2(x_2)| dx_1 dx_2}{\left[\frac{(\psi_1(b_1) - \psi_1(x_1))^{p(\alpha_1 - 1) + 1}}{p} + \frac{(\psi_2(b_2) - \psi_2(x_2))^{q(\alpha_2 - 1) + 1}}{q}\right]} \leq$$

$$\frac{(b_1 - a_1)(b_2 - a_2)}{\Gamma(\alpha_1) \Gamma(\alpha_2) (p(\alpha_1 - 1) + 1)^{\frac{1}{p}} (q(\alpha_2 - 1) + 1)^{\frac{1}{q}}}$$

$$\left\| \left({}^C D_{b_1-}^{\alpha_1, \psi_1} f_1 \right) \circ \psi_1^{-1} \right\|_{q, [\psi_1(a_1), \psi_1(b_1)]} \left\| \left({}^C D_{b_2-}^{\alpha_2, \psi_2} f_2 \right) \circ \psi_2^{-1} \right\|_{p, [\psi_2(a_2), \psi_2(b_2)]}. \tag{6.58}$$

Proof Similar to the proof of Theorem 6.11, but now using (6.19). We omit the details. ■

Preparing to give multivariate results we make

Remark 6.13 Let $N \geq 2$, $S^{N-1} := \{x \in \mathbb{R}^N : |x| = 1\}$ the unit sphere on $\mathbb{R}^N$, where $|\cdot|$ stands for the Euclidean norm in $\mathbb{R}^N$. Also denote the ball

$$B(0, R) := \{x \in \mathbb{R}^N : |x| < R\} \subseteq \mathbb{R}^N, \ R > 0,$$

and the spherical shell

$$A := B(0, R_2) - \overline{B(0, R_1)}, \ 0 < R_1 < R_2.$$

For the following see [8, pp. 149–150] and [9, pp. 87–88].

For $x \in \mathbb{R}^N - \{0\}$ we can write uniquely $x = r\omega$, where $r = |x| > 0$, and $\omega = \frac{x}{r} \in S^{N-1}$, $|\omega| = 1$.

Clearly here

$$\mathbb{R}^N - \{0\} = (0, \infty) \times S^{N-1},$$

and

$$\overline{A} = [R_1, R_2] \times S^{N-1}.$$

In the sequel the following related theorem will be used:

Theorem 6.14 *([3], p. 458) Let $f : A \to \mathbb{R}$ be a Lebesgue integrable function. Then*

$$\int_A f(x)\,dx = \int_{S^{N-1}} \left(\int_{R_1}^{R_2} f(r\omega)\, r^{N-1} dr \right) d\omega. \tag{6.59}$$

So we are able to write an integral in polar form using the polar coordinates (r, ω). We need

Definition 6.15 Let $\alpha > 0$, $n \in \mathbb{N}$ such that $n = \lceil \alpha \rceil$, $[R_1, R_2] \subset \mathbb{R}$; $R_2 > R_1 > 0$, and $f(\cdot\omega), \psi \in C^n([R_1, R_2])$ for any $\omega \in S^{N-1}$, with ψ being increasing and $\psi'(r) \neq 0$, for all $r \in [R_1, R_2]$.

Denote by

$$f_\psi^{[n]}(r\omega) := \left(\frac{1}{\psi'(r)} \frac{d}{dr} \right)^n f(r\omega), \tag{6.60}$$

with $f_\psi^{[0]}(r\omega) = f(r\omega)$, $\forall\, \omega \in S^{N-1}$.

The left ψ-Caputo fractional radial derivative of $f : A \to \mathbb{R}$ of order α is given by: ($x \in A$ is written as $x = r\omega$) for $\alpha \notin \mathbb{N}$, we have

$$^C D_{R_1+}^{\alpha,\psi} f(x) = {}^C D_{R_1+}^{\alpha,\psi} f(r\omega) :=$$

$$\frac{1}{\Gamma(n-\alpha)} \int_{R_1}^{r} \psi'(t) (\psi(r) - \psi(t))^{n-\alpha-1} f_\psi^{[n]}(t\omega)\, dt, \tag{6.61}$$

for $\alpha = m \in \mathbb{N}$, we have

$$^C D_{R_1+}^{\alpha,\psi} f(x) = {}^C D_{R_1+}^{\alpha,\psi} f(r\omega) = f_\psi^{[m]}(x) = f_\psi^{[m]}(r\omega). \tag{6.62}$$

And the right ψ-Caputo fractional radial derivative of $f : A \to \mathbb{R}$ of order α is given by:

$$^C D_{R_2-}^{\alpha,\psi} f(x) = {}^C D_{R_2-}^{\alpha,\psi} f(r\omega) :=$$

$$\frac{(-1)^n}{\Gamma(n-\alpha)} \int_{r}^{R_2} \psi'(t) (\psi(t) - \psi(r))^{n-\alpha-1} f_\psi^{[n]}(t\omega)\, dt, \tag{6.63}$$

for $\alpha = m \in \mathbb{N}$, we have

$$^C D_{R_2-}^{\alpha,\psi} f(x) = {}^C D_{R_2-}^{\alpha,\psi} f(r\omega) = (-1)^m f_\psi^{[m]}(x) = (-1)^m f_\psi^{[m]}(r\omega). \tag{6.64}$$

We set

$$^{C}D_{R_1+}^{0,\psi}f = ^{C}D_{R_2-}^{0,\psi}f := f.$$

Next we present a generalized left fractional Poincaré type inequality on the shell A.

Theorem 6.16 *Let $p, q > 1$ such that $\frac{1}{p}+\frac{1}{q}=1$, and $\alpha > \frac{1}{q}$, $n \in \mathbb{N}$ such that $n = \lceil \alpha \rceil$. Let $f \in C^n\left(\overline{A}\right)$, $\psi \in C^n([R_1, R_2])$ with ψ being strictly increasing and $\psi'(r) \neq 0$, for all $r \in [R_1, R_2]$. Assume that $f_{\psi}^{[k]}(R_1\omega) = 0$, for $k = 0, 1, \ldots, n-1$; $\forall\, \omega \in S^{N-1}$, and $^{C}D_{R_1+}^{\alpha,\psi}f \in C\left(\overline{A}\right)$. Then*

$$\|f\|_{q,A} \leq \frac{\|\psi'\|_{\infty,[R_1,R_2]}^{\frac{1}{q}}\left(\psi(R_2)-\psi(R_1)\right)^{\alpha-\frac{1}{q}}(R_2-R_1)^{\frac{1}{q}}}{\Gamma(\alpha)\left(p(\alpha-1)+1\right)^{\frac{1}{p}}}\left(\frac{R_2}{R_1}\right)^{\frac{N-1}{q}}\left\|^{C}D_{R_1+}^{\alpha,\psi}f\right\|_{q,A}. \tag{6.65}$$

Proof By (6.18), since $f_{\psi}^{[k]}(R_1\omega) = 0$, for $k = 0, 1, \ldots, n-1$, $\forall\, \omega \in S^{N-1}$, we have

$$f(r\omega) = \frac{1}{\Gamma(\alpha)}\int_{R_1}^{r}\psi'(t)\left(\psi(r)-\psi(t)\right)^{\alpha-1}\, {}^{C}D_{R_1+}^{\alpha,\psi}f(t\omega)\,dt, \tag{6.66}$$

$\forall\, r \in [R_1, R_2]$, $\forall\, \omega \in S^{N-1}$.

Hence

$$|f(r\omega)| \leq \frac{1}{\Gamma(\alpha)}\int_{R_1}^{r}\psi'(t)\left(\psi(r)-\psi(t)\right)^{\alpha-1}\left|^{C}D_{R_1+}^{\alpha,\psi}f(t\omega)\right|dt =$$

$$\frac{1}{\Gamma(\alpha)}\int_{R_1}^{r}\left(\psi(r)-\psi(t)\right)^{\alpha-1}\left|^{C}D_{R_1+}^{\alpha,\psi}f(t\omega)\right|d\psi(t) = \tag{6.67}$$

(by ψ being strictly increasing)

$$\frac{1}{\Gamma(\alpha)}\int_{\psi(R_1)}^{\psi(r)}\left(\psi(r)-z\right)^{\alpha-1}\left|\left(^{C}D_{R_1+}^{\alpha,\psi}f\right)\left(\psi^{-1}(z)\,\omega\right)\right|dz \leq$$

$$\frac{1}{\Gamma(\alpha)}\left(\int_{\psi(R_1)}^{\psi(r)}\left(\psi(r)-z\right)^{p(\alpha-1)}dz\right)^{\frac{1}{p}}\left(\int_{\psi(R_1)}^{\psi(r)}\left|\left(^{C}D_{R_1+}^{\alpha,\psi}f\right)\left(\psi^{-1}(z)\,\omega\right)\right|^{q}dz\right)^{\frac{1}{q}} \leq \tag{6.68}$$

$$\frac{1}{\Gamma(\alpha)}\frac{\left(\psi(r)-\psi(R_1)\right)^{\frac{p(\alpha-1)+1}{p}}}{\left(p(\alpha-1)+1\right)^{\frac{1}{p}}}\left(\int_{\psi(R_1)}^{\psi(R_2)}\left|\left(^{C}D_{R_1+}^{\alpha,\psi}f\right)\left(\psi^{-1}(z)\,\omega\right)\right|^{q}dz\right)^{\frac{1}{q}}.$$

That is

$$|f(r\omega)| \leq \frac{(\psi(r)-\psi(R_1))^{\frac{p(\alpha-1)+1}{p}}}{\Gamma(\alpha)(p(\alpha-1)+1)^{\frac{1}{p}}} \left(\int_{\psi(R_1)}^{\psi(R_2)} \left| \left({}^C D_{R_1+}^{\alpha,\psi} f \right) \left(\psi^{-1}(z)\omega \right) \right|^q dz \right)^{\frac{1}{q}}, \tag{6.69}$$

$\forall\, r \in [R_1, R_2]$, $\forall\, \omega \in S^{N-1}$.

Thus, we have

$$|f(r\omega)|^q \leq \frac{(\psi(r)-\psi(R_1))^{(p(\alpha-1)+1)\frac{q}{p}}}{(\Gamma(\alpha))^q (p(\alpha-1)+1)^{\frac{q}{p}}} \left(\int_{\psi(R_1)}^{\psi(R_2)} \left| \left({}^C D_{R_1+}^{\alpha,\psi} f \right) \left(\psi^{-1}(z)\omega \right) \right|^q dz \right), \tag{6.70}$$

$\forall\, r \in [R_1, R_2]$, $\forall\, \omega \in S^{N-1}$.

Therefore it holds

$$\int_{R_1}^{R_2} |f(r\omega)|^q dr \leq \frac{(\psi(R_2)-\psi(R_1))^{(p(\alpha-1)+1)\frac{q}{p}} (R_2-R_1)}{(\Gamma(\alpha))^q (p(\alpha-1)+1)^{\frac{q}{p}}}$$

$$\left(\int_{\psi(R_1)}^{\psi(R_2)} \left| \left({}^C D_{R_1+}^{\alpha,\psi} f \right) \left(\psi^{-1}(z)\omega \right) \right|^q dz \right) = \tag{6.71}$$

$$\frac{(\psi(R_2)-\psi(R_1))^{(p(\alpha-1)+1)\frac{q}{p}} (R_2-R_1)}{(\Gamma(\alpha))^q (p(\alpha-1)+1)^{\frac{q}{p}}} \left(\int_{R_1}^{R_2} \left| \left({}^C D_{R_1+}^{\alpha,\psi} f \right) (t\omega) \right|^q \psi'(t)\, dt \right) \leq$$

$$\frac{\|\psi'\|_{\infty,[R_1,R_2]} (\psi(R_2)-\psi(R_1))^{(p(\alpha-1)+1)\frac{q}{p}} (R_2-R_1)}{(\Gamma(\alpha))^q (p(\alpha-1)+1)^{\frac{q}{p}}} \tag{6.72}$$

$$\left(\int_{R_1}^{R_2} \left| \left({}^C D_{R_1+}^{\alpha,\psi} f \right) (r\omega) \right|^q dr \right),$$

$\forall\, \omega \in S^{N-1}$.

We have proved that

$$\int_{R_1}^{R_2} |f(r\omega)|^q dr \leq \frac{\|\psi'\|_{\infty,[R_1,R_2]} (\psi(R_2)-\psi(R_1))^{(p(\alpha-1)+1)\frac{q}{p}} (R_2-R_1)}{(\Gamma(\alpha))^q (p(\alpha-1)+1)^{\frac{q}{p}}}$$

$$\left(\int_{R_1}^{R_2} \left| \left({}^C D_{R_1+}^{\alpha,\psi} f \right) (r\omega) \right|^q dr \right), \tag{6.73}$$

$\forall\, \omega \in S^{N-1}$.

Because $R_1 \leq r \leq R_2$, $N \geq 2$, we have that $R_1^{N-1} \leq r^{N-1} \leq R_2^{N-1}$ and $R_2^{1-N} \leq r^{1-N} \leq R_1^{1-N}$.

By (6.73) we get

$$R_2^{1-N}\int_{R_1}^{R_2} r^{N-1}\left|f(r\omega)\right|^q dr \le \int_{R_1}^{R_2} r^{1-N}r^{N-1}\left|f(r\omega)\right|^q dr = \int_{R_1}^{R_2}\left|f(r\omega)\right|^q dr \le$$

$$\frac{\left\|\psi'\right\|_{\infty,[R_1,R_2]}\left(\psi(R_2)-\psi(R_1)\right)^{(p(\alpha-1)+1)\frac{q}{p}}(R_2-R_1)}{(\Gamma(\alpha))^q\,(p(\alpha-1)+1)^{\frac{q}{p}}}$$

$$\left(\int_{R_1}^{R_2} r^{1-N}r^{N-1}\left|\left({}^{C}D_{R_1+}^{\alpha,\psi}f\right)(r\omega)\right|^q dr\right) \le \tag{6.74}$$

$$\frac{\left\|\psi'\right\|_{\infty,[R_1,R_2]}\left(\psi(R_2)-\psi(R_1)\right)^{(p(\alpha-1)+1)\frac{q}{p}}(R_2-R_1)\,R_1^{1-N}}{(\Gamma(\alpha))^q\,(p(\alpha-1)+1)^{\frac{q}{p}}}$$

$$\left(\int_{R_1}^{R_2} r^{N-1}\left|\left({}^{C}D_{R_1+}^{\alpha,\psi}f\right)(r\omega)\right|^q dr\right),$$

$\forall\,\omega\in S^{N-1}$.

We have proved that

$$\int_{R_1}^{R_2}\left|f(r\omega)\right|^q r^{N-1}dr \le$$

$$\frac{\left\|\psi'\right\|_{\infty,[R_1,R_2]}\left(\psi(R_2)-\psi(R_1)\right)^{(p(\alpha-1)+1)\frac{q}{p}}(R_2-R_1)}{(\Gamma(\alpha))^q\,(p(\alpha-1)+1)^{\frac{q}{p}}}\left(\frac{R_2}{R_1}\right)^{N-1}$$

$$\left(\int_{R_1}^{R_2}\left|\left({}^{C}D_{R_1+}^{\alpha,\psi}f\right)(r\omega)\right|^q r^{N-1}dr\right), \tag{6.75}$$

$\forall\,\omega\in S^{N-1}$.

Then by integration of (6.75) on S^{N-1} we obtain

$$\int_{S^{N-1}}\left(\int_{R_1}^{R_2}\left|f(r\omega)\right|^q r^{N-1}dr\right)d\omega \le$$

$$\frac{\left\|\psi'\right\|_{\infty,[R_1,R_2]}\left(\psi(R_2)-\psi(R_1)\right)^{(p(\alpha-1)+1)\frac{q}{p}}(R_2-R_1)}{(\Gamma(\alpha))^q\,(p(\alpha-1)+1)^{\frac{q}{p}}}\left(\frac{R_2}{R_1}\right)^{N-1}$$

$$\int_{S^{N-1}}\left(\int_{R_1}^{R_2}\left|\left({}^{C}D_{R_1+}^{\alpha,\psi}f\right)(r\omega)\right|^q r^{N-1}dr\right)d\omega. \tag{6.76}$$

That is (by (6.59))

$$\int_A\left|f(x)\right|^q dx \le$$

$$\frac{\left\|\psi'\right\|_{\infty,[R_1,R_2]} \left(\psi\left(R_2\right)-\psi\left(R_1\right)\right)^{(p(\alpha-1)+1)\frac{q}{p}} \left(R_2-R_1\right)}{\left(\Gamma\left(\alpha\right)\right)^q \left(p\left(\alpha-1\right)+1\right)^{\frac{q}{p}}} \left(\frac{R_2}{R_1}\right)^{N-1}$$

$$\int_A \left|\left({}^C D^{\alpha,\psi}_{R_1+} f\right)(x)\right|^q dx. \tag{6.77}$$

Consequently we get

$$\left(\int_A \left|f\left(x\right)\right|^q dx\right)^{\frac{1}{q}} \leq$$

$$\frac{\left\|\psi'\right\|^{\frac{1}{q}}_{\infty,[R_1,R_2]} \left(\psi\left(R_2\right)-\psi\left(R_1\right)\right)^{\alpha-\frac{1}{q}} \left(R_2-R_1\right)^{\frac{1}{q}}}{\Gamma\left(\alpha\right) \left(p\left(\alpha-1\right)+1\right)^{\frac{1}{p}}} \left(\frac{R_2}{R_1}\right)^{\frac{N-1}{q}}$$

$$\left(\int_A \left|\left({}^C D^{\alpha,\psi}_{R_1+} f\right)(x)\right|^q dx\right)^{\frac{1}{q}}, \tag{6.78}$$

proving the claim. ∎

It follows a generalized right fractional Poincaré type inequality on the shell A.

Theorem 6.17 *Let $p, q > 1$ such that $\frac{1}{p}+\frac{1}{q}=1$, and $\alpha > \frac{1}{q}$, $n \in \mathbb{N}$ such that $n = \lceil\alpha\rceil$. Let $f \in C^n\left(\overline{A}\right)$, $\psi \in C^n\left([R_1, R_2]\right)$ with ψ being strictly increasing and $\psi'(r) \neq 0$, for all $r \in [R_1, R_2]$. Assume that $f^{[k]}_{\psi}\left(R_2\omega\right) = 0$, for $k = 0, 1, \ldots, n-1$; $\forall\, \omega \in S^{N-1}$, and ${}^C D^{\alpha,\psi}_{R_2-} f \in C\left(\overline{A}\right)$. Then*

$$\left\|f\right\|_{q,A} \leq \frac{\left\|\psi'\right\|^{\frac{1}{q}}_{\infty,[R_1,R_2]} \left(\psi\left(R_2\right)-\psi\left(R_1\right)\right)^{\alpha-\frac{1}{q}} \left(R_2-R_1\right)^{\frac{1}{q}}}{\Gamma\left(\alpha\right) \left(p\left(\alpha-1\right)+1\right)^{\frac{1}{p}}} \left(\frac{R_2}{R_1}\right)^{\frac{N-1}{q}} \left\|{}^C D^{\alpha,\psi}_{R_2-} f\right\|_{q,A}. \tag{6.79}$$

Proof Similar to the proof of Theorem 6.16, now using (6.19). We skip the details. ∎

Fractional Poincaré L_1 inequalities come next

Theorem 6.18 *Let $\alpha \geq 1$, $n \in \mathbb{N} : n = \lceil\alpha\rceil$, $[a, b] \subset \mathbb{R}$ and $f, \psi \in C^n\left([a, b]\right)$ with ψ being strictly increasing and $\psi'(x) \neq 0$, for all $x \in [a, b]$. Assume that $f^{[k]}_{\psi}(a) = 0$, for $k = 0, 1, \ldots, n-1$. Then*

$$\left\|f\right\|_{L_1([a,b])} \leq \frac{\left(\psi\left(b\right)-\psi\left(a\right)\right)^{\alpha-1} \left\|\psi'\right\|_{\infty,[a,b]} \left(b-a\right)}{\Gamma\left(\alpha\right)} \left\|{}^C D^{\alpha,\psi}_{a+} f\right\|_{L_1([a,b])}. \tag{6.80}$$

Proof By (6.18), since $f^{[k]}_{\psi}(a) = 0$, for $k = 0, 1, \ldots, n-1$, we have

$$f(x) = \frac{1}{\Gamma(\alpha)} \int_a^x \psi'(t)(\psi(x) - \psi(t))^{\alpha-1} \, {}^C D_{a+}^{\alpha,\psi} f(t) \, dt, \tag{6.81}$$

$\forall x \in [a, b]$.

Hence

$$|f(x)| \leq \frac{1}{\Gamma(\alpha)} \int_a^x \psi'(t)(\psi(x) - \psi(t))^{\alpha-1} \left| {}^C D_{a+}^{\alpha,\psi} f(t) \right| dt \leq$$

$$\frac{\|\psi'\|_{\infty,[a,b]} (\psi(b) - \psi(a))^{\alpha-1}}{\Gamma(\alpha)} \int_a^x \left| {}^C D_{a+}^{\alpha,\psi} f(t) \right| dt \leq$$

$$\frac{\|\psi'\|_{\infty,[a,b]} (\psi(b) - \psi(a))^{\alpha-1}}{\Gamma(\alpha)} \int_a^b \left| {}^C D_{a+}^{\alpha,\psi} f(t) \right| dt, \tag{6.82}$$

$\forall x \in [a, b]$.

Therefore it holds

$$\int_a^b |f(x)| \, dx \leq \frac{\|\psi'\|_{\infty,[a,b]} (\psi(b) - \psi(a))^{\alpha-1} (b-a)}{\Gamma(\alpha)} \int_a^b \left| {}^C D_{a+}^{\alpha,\psi} f(x) \right| dx, \tag{6.83}$$

proving the claim. ∎

Theorem 6.19 *Let* $\alpha \geq 1$, $n \in \mathbb{N} : n = \lceil \alpha \rceil$, $[a, b] \subset \mathbb{R}$ *and* $f, \psi \in C^n([a, b])$ *with* ψ *being strictly increasing and* $\psi'(x) \neq 0$, *for all* $x \in [a, b]$. *Assume that* $f_\psi^{[k]}(b) = 0$, *for* $k = 0, 1, \ldots, n-1$. *Then*

$$\|f\|_{L_1([a,b])} \leq \frac{(\psi(b) - \psi(a))^{\alpha-1} \|\psi'\|_{\infty,[a,b]} (b-a)}{\Gamma(\alpha)} \left\| {}^C D_{b-}^{\alpha,\psi} f \right\|_{L_1([a,b])}. \tag{6.84}$$

Proof Similar to the proof of Theorem 6.18. ∎

Next come multivariate on the shell A L_1-fractional Poincaré type inequalities.

Theorem 6.20 *Let* $\alpha \geq 1$, $n \in \mathbb{N}$ *such that* $n = \lceil \alpha \rceil$. *Let* $f \in C^n(\overline{A})$, $\psi \in C^n([R_1, R_2])$ *with* ψ *being strictly increasing and* $\psi'(r) \neq 0$, *for all* $r \in [R_1, R_2]$. *Assume that* $f_\psi^{[k]}(R_1\omega) = 0$, *for* $k = 0, 1, \ldots, n-1$; $\forall \omega \in S^{N-1}$, *and* ${}^C D_{R_1+}^{\alpha,\psi} f \in C(\overline{A})$. *Then*

$$\|f\|_{L_1(A)} \leq \frac{(\psi(R_2) - \psi(R_1))^{\alpha-1} \|\psi'\|_{\infty,[R_1,R_2]} (R_2 - R_1)}{\Gamma(\alpha)} \left(\frac{R_2}{R_1}\right)^{N-1} \left\| {}^C D_{R_1+}^{\alpha,\psi} f \right\|_{L_1(A)}. \tag{6.85}$$

Proof By (6.83) we get:

$$R_2^{1-N} \int_{R_1}^{R_2} r^{N-1} |f(r\omega)| \, dr \leq \int_{R_1}^{R_2} r^{1-N} r^{N-1} f(r\omega) \, dr = \int_{R_1}^{R_2} |f(r\omega)| \, dr \leq$$

$$\frac{\left\|\psi'\right\|_{\infty,[R_1,R_2]}\left(\psi\left(R_2\right)-\psi\left(R_1\right)\right)^{\alpha-1}\left(R_2-R_1\right)}{\Gamma\left(\alpha\right)}$$

$$\int_{R_1}^{R_2} r^{1-N} r^{N-1}\left|\left({}^{C}D_{R_1+}^{\alpha,\psi}f\right)(r\omega)\right| dr \leq \tag{6.86}$$

$$\frac{\left\|\psi'\right\|_{\infty,[R_1,R_2]}\left(\psi\left(R_2\right)-\psi\left(R_1\right)\right)^{\alpha-1}\left(R_2-R_1\right)R_1^{1-N}}{\Gamma\left(\alpha\right)}$$

$$\int_{R_1}^{R_2} r^{N-1}\left|\left({}^{C}D_{R_1+}^{\alpha,\psi}f\right)(r\omega)\right| dr,$$

$\forall\, \omega \in S^{N-1}$.

Therefore it holds

$$\int_{R_1}^{R_2} \left|f\left(r\omega\right)\right| r^{N-1} dr \leq$$

$$\frac{\left\|\psi'\right\|_{\infty,[R_1,R_2]}\left(\psi\left(R_2\right)-\psi\left(R_1\right)\right)^{\alpha-1}\left(R_2-R_1\right)}{\Gamma\left(\alpha\right)}\left(\frac{R_2}{R_1}\right)^{N-1}$$

$$\int_{R_1}^{R_2} \left|\left({}^{C}D_{R_1+}^{\alpha,\psi}f\right)(r\omega)\right| r^{N-1} dr, \tag{6.87}$$

$\forall\, \omega \in S^{N-1}$.

Thus we obtain

$$\int_{S^{N-1}}\left(\int_{R_1}^{R_2} \left|f\left(r\omega\right)\right| r^{N-1} dr\right) d\omega \leq$$

$$\frac{\left\|\psi'\right\|_{\infty,[R_1,R_2]}\left(\psi\left(R_2\right)-\psi\left(R_1\right)\right)^{\alpha-1}\left(R_2-R_1\right)}{\Gamma\left(\alpha\right)}\left(\frac{R_2}{R_1}\right)^{N-1}$$

$$\int_{S^{N-1}}\left(\int_{R_1}^{R_2} \left|\left({}^{C}D_{R_1+}^{\alpha,\psi}f\right)(r\omega)\right| r^{N-1} dr\right) d\omega, \tag{6.88}$$

which means (by using (6.59)) that

$$\int_{A} \left|f\left(x\right)\right| dx \leq$$

$$\frac{\left\|\psi'\right\|_{\infty,[R_1,R_2]}\left(\psi\left(R_2\right)-\psi\left(R_1\right)\right)^{\alpha-1}\left(R_2-R_1\right)}{\Gamma\left(\alpha\right)}\left(\frac{R_2}{R_1}\right)^{N-1}$$

$$\int_A \left| {}^C D_{R_1+}^{\alpha,\psi} f(x) \right| dx, \tag{6.89}$$

proving the claim. ∎

Theorem 6.21 *Let $\alpha \geq 1$, $n \in \mathbb{N}$ such that $n = \lceil \alpha \rceil$. Let $f \in C^n(\overline{A})$, $\psi \in C^n([R_1, R_2])$ with ψ being strictly increasing and $\psi'(r) \neq 0$, for all $r \in [R_1, R_2]$. Assume that $f_{\psi}^{[k]}(R_2\omega) = 0$, for $k = 0, 1, \ldots, n-1$; $\forall\, \omega \in S^{N-1}$, and ${}^C D_{R_2-}^{\alpha,\psi} f \in C(\overline{A})$. Then*

$$\|f\|_{L_1(A)} \leq \frac{(\psi(R_2) - \psi(R_1))^{\alpha-1} \|\psi'\|_{\infty,[R_1,R_2]} (R_2 - R_1)}{\Gamma(\alpha)} \left(\frac{R_2}{R_1}\right)^{N-1} \left\| {}^C D_{R_2-}^{\alpha,\psi} f \right\|_{L_1(A)}. \tag{6.90}$$

Proof As similar to the proof of Theorem 6.20 is omitted. ∎

Applications follow:

Proposition 6.22 *Let $p, q > 1$ such that $\frac{1}{p} + \frac{1}{q} = 1$, $\alpha_1 > \frac{1}{q}$, $n_1 \in \mathbb{N} : n_1 = \lceil \alpha_1 \rceil$; and $\alpha_2 > \frac{1}{p}$, $n_2 \in \mathbb{N} : n_2 = \lceil \alpha_2 \rceil$; $[a_1, b_1] \subset \mathbb{R}$, $[a_2, b_2] \subset (0, \infty)$. Let $f_i \in C^{n_i}([a_i, b_i])$, $i = 1, 2$. Assume that $f_{1e^{x_1}}^{[k_1]}(a_1) = 0$, for $k_1 = 0, 1, \ldots, n_1 - 1$; $f_{2\ln x_2}^{[k_2]}(a_2) = 0$, for $k_2 = 0, 1, \ldots, n_2 - 1$. Then*

$$\int_{a_1}^{b_1} \int_{a_2}^{b_2} \frac{|f_1(x_1)| \, |f_2(x_2)| \, dx_1 dx_2}{\left[\frac{(e^{x_1} - e^{a_1})^{p(\alpha_1 - 1)+1}}{p} + \frac{\left(\ln\left(\frac{x_2}{a_2}\right)\right)^{q(\alpha_2 - 1)+1}}{q} \right]} \leq$$

$$\frac{(b_1 - a_1)(b_2 - a_2)}{\Gamma(\alpha_1)\Gamma(\alpha_2)(p(\alpha_1 - 1) + 1)^{\frac{1}{p}} (q(\alpha_2 - 1) + 1)^{\frac{1}{q}}}$$

$$\left\| \left({}^C D_{a_1+}^{\alpha_1, e^{x_1}} f_1 \right) \circ \ln x_1 \right\|_{q,[e^{a_1}, e^{b_1}]} \left\| \left({}^C D_{a_2+}^{\alpha_2, \ln x_2} f_2 \right) \circ e^{x_2} \right\|_{q,[\ln a_2, \ln b_2]}. \tag{6.91}$$

Proof By Theorem 6.11, for $\psi_1(x_1) = e^{x_1}$ and $\psi_2(x_2) = \ln x_2$. ∎

Proposition 6.23 *Let $p, q > 1$ such that $\frac{1}{p} + \frac{1}{q} = 1$, and $\alpha > \frac{1}{q}$, $n \in \mathbb{N} : n = \lceil \alpha \rceil$. Let $f \in C^n(\overline{A})$ and assume that $f_{e^r}^{[k]}(R_1\omega) = 0$, for $k = 0, 1, \ldots, n-1$; $\forall\, \omega \in S^{N-1}$, and ${}^C D_{R_1+}^{\alpha,e^r} f \in C(\overline{A})$. Then*

$$\|f\|_{q,A} \leq \frac{e^{\frac{R_2}{q}} \left(e^{R_2} - e^{R_1}\right)^{\alpha - \frac{1}{q}} (R_2 - R_1)^{\frac{1}{q}}}{\Gamma(\alpha)(p(\alpha - 1) + 1)^{\frac{1}{p}}} \left(\frac{R_2}{R_1}\right)^{\frac{N-1}{q}} \left\| {}^C D_{R_1+}^{\alpha,e^r} f \right\|_{q,A}. \tag{6.92}$$

Proof By Theorem 6.16, for $\psi(r) = e^r$, $r \in [R_1, R_2]$. ∎

References

1. Acosta, G., Durán, R.G.: An optimal Poincaré inequality in L' for convex domains. Proc. AMS **132**(1), 195–202 (2003)
2. Almeida, R.: A Caputo fractional derivative of a function with respect to another function. Commun. Nonlinear Sci. Numer. Simul. **44**, 460–481 (2017)
3. Anastassiou, G.A.: Fractional Differentiation Inequalities. Research Monograph. Springer, New York (2009)
4. Anastassiou, G.A.: Intelligent Computations: Abstract Fractional Calculus, Inequalities, Approximations. Springer, Heidelberg (2018)
5. Anastassiou, G.: Caputo generalized ψ-fractional integral inequalities (2019). Submitted
6. Evans, L.C.: Partial Differential Equations. Graduate studies in Mathematics, vol. 19. American Mathematical Society, Providence, R.I. (1998)
7. Pachpatte, B.G.: Inequalities similar to the integral analogue of Hilbert's inequalities. Tamkang J. Math. **30**(1), 139–146 (1999)
8. Rudin, W.: Real and Complex Analysis, International Student edn. Mc Graw Hill, London (1970)
9. Stroock, D.: A Concise Introduction to the Theory of Integration, 3rd edn. Birkäuser, Boston (1999)

Chapter 7
Generalized ψ-Fractional Quantitative Approximation by Sublinear Operators

Here we consider the approximation of functions by sublinear positive operators with applications to several Max-Product operators under generalized fractional differentiability. Our study is based on our generalized fractional results about positive sublinear operators. We derive Jackson type inequalities under iterated initial conditions. So our approach is quantitative by producing inequalities with their right hand sides involving the modulus of continuity of generalized fractional derivative of the function under approximation. See also [4].

7.1 Background

Here we study the approximation properties of positive sublinear operators converging the unit operator related to the generalized fractional smoothness of the involved function under approximation. This is a quantitative study using inequalities.

We need some background:

We mention

Definition 7.1 Here $C_+([a,b]) := \{f : [a,b] \to \mathbb{R}_+, \text{ continuous functions}\}$. Let $L_N : C_+([a,b]) \to C_+([a,b])$, operators, $\forall\, N \in \mathbb{N}$, such that

(i)

$$L_N(\alpha f) = \alpha L_N(f), \ \forall\, \alpha \geq 0, \ \forall\, f \in C_+([a,b]), \tag{7.1}$$

(ii) if $f, g \in C_+([a,b]) : f \leq g$, then

$$L_N(f) \leq L_N(g), \ \forall\, N \in \mathbb{N}, \tag{7.2}$$

(iii)

$$L_N(f+g) \leq L_N(f) + L_N(g), \ \ \forall\, f, g \in C_+([a,b]). \tag{7.3}$$

G. A. Anastassiou, *Generalized Fractional Calculus*, Studies in Systems, Decision and Control 305, https://doi.org/10.1007/978-3-030-56962-4_7

We call $\{L_N\}_{N\in\mathbb{N}}$ positive sublinear operators.

We need a Hölder's type inequality, see next:

Theorem 7.2 (see [2, p. 6]) *Let $L : C_+([a,b]) \to C_+([a,b])$, be a positive sublinear operator and $f, g \in C_+([a,b])$, furthermore let $p, q > 1 : \frac{1}{p} + \frac{1}{q} = 1$. Assume that $L\left((f(\cdot))^p\right)(s_*)$, $L\left((g(\cdot))^q\right)(s_*) > 0$ for some $s_* \in [a,b]$. Then*

$$L(f(\cdot)g(\cdot))(s_*) \le \left(L\left((f(\cdot))^p\right)(s_*)\right)^{\frac{1}{p}} \left(L\left((g(\cdot))^q\right)(s_*)\right)^{\frac{1}{q}}. \tag{7.4}$$

We make

Remark 7.3 By [5, p. 17], we get: let $f, g \in C_+([a,b])$, then

$$|L_N(f)(x) - L_N(g)(x)| \le L_N(|f - g|)(x), \quad \forall\, x \in [a,b]. \tag{7.5}$$

Furthermore, we also have that

$$|L_N(f)(x) - f(x)| \le L_N(|f(\cdot) - f(x)|)(x) + |f(x)|\,|L_N(e_0)(x) - 1|, \tag{7.6}$$

$\forall\, x \in [a,b]$; $e_0(t) = 1$.

From now on we assume that $L_N(1) = 1$. Hence it holds

$$|L_N(f)(x) - f(x)| \le L_N(|f(\cdot) - f(x)|)(x), \quad \forall\, x \in [a,b]. \tag{7.7}$$

In [5, p. 10], the authors introduced the basic Max-product Bernstein operators,

$$B_N^{(M)}(f)(x) = \frac{\bigvee_{k=0}^{N} p_{N,k}(x) f\left(\frac{k}{N}\right)}{\bigvee_{k=0}^{N} p_{N,k}(x)}, \quad N \in \mathbb{N}, \tag{7.8}$$

where $\bigvee$ stands for maximum, and $p_{N,k}(x) = \binom{N}{k} x^k (1-x)^{N-k}$ and $f : [0,1] \to \mathbb{R}_+ = [0,\infty)$.

These are nonlinear and piecewise rational operators, in particular they are positive sublinear operators.

The authors in [5] studied similar such nonlinear operators such as: the Max-product Favard–Szász–Mirakjan operators and their truncated version, the Max-product Baskakov operators and their truncated version, also many other similar specific operators. These Max-product operators tend to converge faster to the on hand function.

So we mention from [5, p. 30], that for $f : [0,1] \to \mathbb{R}_+$ continuous, we have the estimate

$$\left| B_N^{(M)}(f)(x) - f(x) \right| \leq 12\omega_1 \left(f, \frac{1}{\sqrt{N+1}} \right), \quad \text{for all } N \in \mathbb{N},\ x \in [0,1], \tag{7.9}$$

Also from [5, p. 36], we mention that for $f : [0,1] \to \mathbb{R}_+$ being concave function we get that

$$\left| B_N^{(M)}(f)(x) - f(x) \right| \leq 2\omega_1 \left(f, \frac{1}{N} \right), \quad \text{for all } x \in [0,1], \tag{7.10}$$

a much faster convergence.

Above ω_1 is the first modulus of continuity. Inequalities (7.9) and (7.10) motivate our work.

Some fractional background follows:

Next we use [1].

Let $\alpha > 0$, $[a,b] \subset \mathbb{R}$, $f : [a,b] \to \mathbb{R}$ which is integrable and $\psi \in C^1([a,b])$ an increasing function such that $\psi'(x) \neq 0$, for all $x \in [a,b]$. Consider $n = \lceil \alpha \rceil$, the ceiling of α. The left and right fractional integrals are defined, respectively, as follows:

$$I_{a+}^{\alpha,\psi} f(x) := \frac{1}{\Gamma(\alpha)} \int_a^x \psi'(t) (\psi(x) - \psi(t))^{\alpha-1} f(t)\, dt, \tag{7.11}$$

and

$$I_{b-}^{\alpha,\psi} f(x) := \frac{1}{\Gamma(\alpha)} \int_x^b \psi'(t) (\psi(t) - \psi(x))^{\alpha-1} f(t)\, dt, \tag{7.12}$$

for any $x \in [a,b]$, where Γ is the gamma function.

The following semigroup property is valid for fractional integrals: if $\alpha, \beta > 0$, then

$$I_{a+}^{\alpha,\psi} I_{a+}^{\beta,\psi} f(x) = I_{a+}^{\alpha+\beta,\psi} f(x), \text{ and } I_{b-}^{\alpha,\psi} I_{b-}^{\beta,\psi} f(x) = I_{b-}^{\alpha+\beta,\psi} f(x).$$

We mention

Definition 7.4 *([1])* Let $\alpha > 0$, $n \in \mathbb{N}$ such that $n = \lceil \alpha \rceil$, $[a,b] \subset \mathbb{R}$ and $f, \psi \in C^n([a,b])$ with ψ being increasing and $\psi'(x) \neq 0$, for all $x \in [a,b]$. The left ψ-Caputo fractional derivative of f of order α is given by

$${}^C D_{a+}^{\alpha,\psi} f(x) := I_{a+}^{n-\alpha,\psi} \left(\frac{1}{\psi'(x)} \frac{d}{dx} \right)^n f(x), \tag{7.13}$$

and the right ψ-Caputo fractional derivative of f is given by

$${}^C D_{b-}^{\alpha,\psi} f(x) := I_{b-}^{n-\alpha,\psi} \left(-\frac{1}{\psi'(x)} \frac{d}{dx} \right)^n f(x). \tag{7.14}$$

To simplify notation, we will use the symbol

$$f_{\psi}^{[n]}(x) := \left(\frac{1}{\psi'(x)}\frac{d}{dx}\right)^{n} f(x), \tag{7.15}$$

with $f_{\psi}^{[0]}(x) = f(x)$.

By the definition, when $\alpha = m \in \mathbb{N}$, we have

$$\begin{aligned} &{}^{C}D_{a+}^{\alpha,\psi} f(x) = f_{\psi}^{[m]}(x) \\ &\text{and} \\ &{}^{C}D_{b-}^{\alpha,\psi} f(x) = (-1)^{m} f_{\psi}^{[m]}(x). \end{aligned} \tag{7.16}$$

If $\alpha \notin \mathbb{N}$, we have

$${}^{C}D_{a+}^{\alpha,\psi} f(x) = \frac{1}{\Gamma(n-\alpha)} \int_{a}^{x} \psi'(t) (\psi(x) - \psi(t))^{n-\alpha-1} f_{\psi}^{[n]}(t)\, dt, \tag{7.17}$$

and

$${}^{C}D_{b-}^{\alpha,\psi} f(x) = \frac{(-1)^{n}}{\Gamma(n-\alpha)} \int_{x}^{b} \psi'(t) (\psi(t) - \psi(x))^{n-\alpha-1} f_{\psi}^{[n]}(t)\, dt, \tag{7.18}$$

$\forall\, x \in [a,b]$.

In particular, when $\alpha \in (0,1)$, we have

$$\begin{aligned} &{}^{C}D_{a+}^{\alpha,\psi} f(x) = \tfrac{1}{\Gamma(1-\alpha)} \textstyle\int_{a}^{x} (\psi(x) - \psi(t))^{-\alpha} f'(t)\, dt, \\ &\text{and} \\ &{}^{C}D_{b-}^{\alpha,\psi} f(x) = \tfrac{-1}{\Gamma(1-\alpha)} \textstyle\int_{x}^{b} (\psi(t) - \psi(x))^{-\alpha} f'(t)\, dt \end{aligned} \tag{7.19}$$

$\forall\, x \in [a,b]$.

Clearly the above is a generalization of left and right Caputo fractional derivatives. For more see [1].

Still we need from [1] the following left and right fractional Taylor's formulae:

Theorem 7.5 ([1]) *Let $\alpha > 0$, $n \in \mathbb{N}$ such that $n = \lceil \alpha \rceil$, $[a,b] \subset \mathbb{R}$ and $f, \psi \in C^{n}([a,b])$ with ψ being increasing and $\psi'(x) \neq 0$, for all $x \in [a,b]$. Then, the left fractional Taylor formula follows,*

$$f(x) = \sum_{k=0}^{n-1} \frac{f_{\psi}^{[k]}(a)}{k!} (\psi(x) - \psi(a))^{k} + I_{a+}^{\alpha,\psi}\, {}^{C}D_{a+}^{\alpha,\psi} f(x), \tag{7.20}$$

and the right fractional Taylor formula follows,

$$f(x) = \sum_{k=0}^{n-1} (-1)^k \frac{f_{\psi}^{[k]}(b)}{k!} (\psi(b) - \psi(x))^k + I_{b-}^{\alpha,\psi} \, {}^C D_{b-}^{\alpha,\psi} f(x), \tag{7.21}$$

$\forall \, x \in [a, b]$.

In particular, given $\alpha \in (0, 1)$, we have

$$\begin{array}{l} f(x) = f(a) + I_{a+}^{\alpha,\psi} \, {}^C D_{a+}^{\alpha,\psi} f(x), \\ \text{and} \\ f(x) = f(b) + I_{b-}^{\alpha,\psi} \, {}^C D_{b-}^{\alpha,\psi} f(x), \end{array} \tag{7.22}$$

$\forall \, x \in [a, b]$.

Remark 7.6 For convenience we can rewrite (7.20)–(7.22) as follows:

$$f(x) = \sum_{k=0}^{n-1} \frac{f_{\psi}^{[k]}(a)}{k!} (\psi(x) - \psi(a))^k + \tag{7.23}$$

$$\frac{1}{\Gamma(\alpha)} \int_a^x \psi'(t) (\psi(x) - \psi(t))^{\alpha-1} \, {}^C D_{a+}^{\alpha,\psi} f(t) \, dt,$$

and

$$f(x) = \sum_{k=0}^{n-1} \frac{(-1)^k f_{\psi}^{[k]}(b)}{k!} (\psi(b) - \psi(x))^k + \tag{7.24}$$

$$\frac{1}{\Gamma(\alpha)} \int_x^b \psi'(t) (\psi(t) - \psi(x))^{\alpha-1} \, {}^C D_{b-}^{\alpha,\psi} f(t) \, dt,$$

$\forall \, x \in [a, b]$.

When $\alpha \in (0, 1)$, we get:

$$\begin{array}{l} f(x) = f(a) + \frac{1}{\Gamma(\alpha)} \int_a^x \psi'(t) (\psi(x) - \psi(t))^{\alpha-1} \, {}^C D_{a+}^{\alpha,\psi} f(t) \, dt, \\ \text{and} \\ f(x) = f(b) + \frac{1}{\Gamma(\alpha)} \int_x^b \psi'(t) (\psi(t) - \psi(x))^{\alpha-1} \, {}^C D_{b-}^{\alpha,\psi} f(t) \, dt, \end{array} \tag{7.25}$$

$\forall \, x \in [a, b]$.

Again from [1] we have the following:

Consider the norms $\|\cdot\|_{\infty} : C([a, b]) \to \mathbb{R}$ and $\|\cdot\|_{C_{\psi}^{[n]}} : C^n([a, b]) \to \mathbb{R}$, where

$\|f\|_{C_{\psi}^{[n]}} := \sum_{k=0}^{n} \left\| f_{\psi}^{[k]} \right\|_{\infty}$.

We have

Theorem 7.7 ([1]) *The ψ-Caputo fractional derivatives are bounded operators. For all $\alpha > 0$ $(n = \lceil \alpha \rceil)$*

$$\left\| {}^{C}D_{a+}^{\alpha,\psi} \right\|_{\infty} \leq K \|f\|_{C_{\psi}^{[n]}} \tag{7.26}$$

and

$$\left\| {}^{C}D_{b-}^{\alpha,\psi} \right\|_{\infty} \leq K \|f\|_{C_{\psi}^{[n]}}, \tag{7.27}$$

where

$$K = \frac{(\psi(b) - \psi(a))^{n-\alpha}}{\Gamma(n+1-\alpha)} > 0. \tag{7.28}$$

We make

Remark 7.8 In our setting, clearly, it is $f_{\psi}^{(n)} \in C([a,b])$. Given $f \in C([a,b])$, by Theorem 4.10, p. 98 of [3], we get that $I_{a+}^{\alpha,\psi} f \in C([a,b])$, and by Theorem 4.11, p. 101 of [3], we get that $I_{b-}^{\alpha,\psi} f \in C([a,b])$.

Therefore, we obtain that ${}^{C}D_{a+}^{\alpha,\psi}$, ${}^{C}D_{b-}^{\alpha,\psi} \in C([a,b])$.

We mention

Definition 7.9 Let $f \in C([a,b])$, we define

$$\omega_1(f,\delta) := \sup_{\substack{x,y \in [a,b]: \\ |x-y| \leq \delta}} |f(x) - f(y)|, \text{ where } 0 < \delta \leq b - a.$$

7.2 Main Results

We need

Definition 7.10 Let $\left({}^{C}D_{x_0}^{\alpha,\psi} f\right) \circ \psi^{-1}$ denote any of $\left({}^{C}D_{x_0+}^{\alpha,\psi} f\right) \circ \psi^{-1}$, $\left({}^{C}D_{x_0-}^{\alpha,\psi} f\right) \circ \psi^{-1}$, and $\delta > 0$. We set

$$\omega_1\left(\left({}^{C}D_{x_0}^{\alpha,\psi} f\right) \circ \psi^{-1}, \delta\right) := \tag{7.29}$$

$$\max\left\{\omega_1\left(\left({}^{C}D_{x_0-}^{\alpha,\psi} f\right) \circ \psi^{-1}, \delta\right)_{[\psi(a),\psi(x_0)]}, \omega_1\left(\left({}^{C}D_{x_0+}^{\alpha,\psi} f\right) \circ \psi^{-1}, \delta\right)_{[\psi(x_0),\psi(b)]}\right\},$$

where $x_0 \in [a,b]$. Here the moduli of continuity are considered over $[\psi(a), \psi(x_0)]$ and $[\psi(x_0), \psi(b)]$, respectively.

We present

Theorem 7.11 *Let $\alpha > 0$, $\alpha \notin \mathbb{N}$, $n \in \mathbb{N}$ such that $n = \lceil \alpha \rceil$, $[a,b] \subset \mathbb{R}$, and $f, \psi \in C^n([a,b])$ with ψ being strictly increasing. Let a fixed $x_0 \in [a,b]$ such that $f_{\psi}^{[k]}(x_0) = 0$, for $k = 1, \ldots, n-1$. Then*

$$|f(x) - f(x_0)| \leq \frac{\omega_1\left(\left({}^C D_{x_0}^{\alpha,\psi} f\right) \circ \psi^{-1}, \delta\right)}{\Gamma(\alpha+1)}$$

$$\left[|\psi(x) - \psi(x_0)|^{\alpha} + \frac{|\psi(x) - \psi(x_0)|^{\alpha+1}}{\delta(\alpha+1)}\right], \tag{7.30}$$

$\delta > 0$, $\forall\, x \in [a, b]$.

If $0 < \alpha < 1$, *then we do not need any initial conditions.*

Proof By (7.23) we have

$$f(x) = \sum_{k=0}^{n-1} \frac{f_{\psi}^{[k]}(x_0)}{k!} (\psi(x) - \psi(x_0))^k + \tag{7.31}$$

$$\frac{1}{\Gamma(\alpha)} \int_{x_0}^{x} \psi'(t) (\psi(x) - \psi(t))^{\alpha-1}\ {}^C D_{x_0+}^{\alpha,\psi} f(t)\, dt,$$

$\forall\, x \in [x_0, b]$.

And, by (7.24) we have

$$f(x) = \sum_{k=0}^{n-1} \frac{(-1)^k f_{\psi}^{[k]}(x_0)}{k!} (\psi(x_0) - \psi(x))^k + \tag{7.32}$$

$$\frac{1}{\Gamma(\alpha)} \int_{x}^{x_0} \psi'(t) (\psi(t) - \psi(x))^{\alpha-1}\ {}^C D_{x_0-}^{\alpha,\psi} f(t)\, dt,$$

$\forall\, x \in [a, x_0]$.

By the assumption $f_{\psi}^{[k]}(x_0) = 0$, $k = 1, \ldots, n-1$, we get

$$f(x) - f(x_0) = \frac{1}{\Gamma(\alpha)} \int_{x_0}^{x} \psi'(t) (\psi(x) - \psi(t))^{\alpha-1}\ {}^C D_{x_0+}^{\alpha,\psi} f(t)\, dt, \tag{7.33}$$

$\forall\, x \in [x_0, b]$, and

$$f(x) - f(x_0) = \frac{1}{\Gamma(\alpha)} \int_{x}^{x_0} \psi'(t) (\psi(t) - \psi(x))^{\alpha-1}\ {}^C D_{x_0-}^{\alpha,\psi} f(t)\, dt, \tag{7.34}$$

$\forall\, x \in [a, x_0]$.

When $0 < \alpha < 1$, i.e. $n = 1$, then (7.33), (7.34) are valid without any initial condition.

We have that $f_{\psi}^{[k]} \in C([a, b])$, for all $k = 0, 1, \ldots, n$.

Here $\alpha \notin \mathbb{N}$ and we have

$$ {}^{C}D_{x_0+}^{\alpha,\psi} f(x) = \frac{1}{\Gamma(n-\alpha)} \int_{x_0}^{x} \psi'(t) (\psi(x) - \psi(t))^{n-\alpha-1} f_{\psi}^{[n]}(t) \, dt, \tag{7.35} $$

$\forall\, x \in [x_0, b]$, and

$$ {}^{C}D_{x_0-}^{\alpha,\psi} f(x) = \frac{(-1)^n}{\Gamma(n-\alpha)} \int_{x}^{x_0} \psi'(t) (\psi(t) - \psi(x))^{n-\alpha-1} f_{\psi}^{[n]}(t) \, dt, \tag{7.36} $$

$\forall\, x \in [a, x_0]$.

Thus, we derive

$$ \left| {}^{C}D_{x_0+}^{\alpha,\psi} f(x) \right| \leq \frac{1}{\Gamma(n-\alpha)} \int_{x_0}^{x} \psi'(t) (\psi(x) - \psi(t))^{n-\alpha-1} \left| f_{\psi}^{[n]}(t) \right| dt \leq $$

$$ \frac{\left\| f_{\psi}^{[n]} \right\|_{\infty,[a,b]}}{\Gamma(n-\alpha)} \int_{x_0}^{x} (\psi(x) - \psi(t))^{n-\alpha-1} \, d\psi(t) = \frac{\left\| f_{\psi}^{[n]} \right\|_{\infty,[a,b]}}{\Gamma(n-\alpha+1)} (\psi(x) - \psi(x_0))^{n-\alpha}. \tag{7.37} $$

We found that

$$ \left| {}^{C}D_{x_0+}^{\alpha,\psi} f(x) \right| \leq \frac{\left\| f_{\psi}^{[n]} \right\|_{\infty,[a,b]}}{\Gamma(n-\alpha+1)} (\psi(x) - \psi(x_0))^{n-\alpha}, \tag{7.38} $$

$\forall\, x \in [x_0, b]$.

Clearly, it holds ${}^{C}D_{x_0+}^{\alpha,\psi} f(x_0) = 0$.

Similarly, it holds

$$ \left| {}^{C}D_{x_0-}^{\alpha,\psi} f(x) \right| \leq \frac{1}{\Gamma(n-\alpha)} \int_{x}^{x_0} \psi'(t) (\psi(t) - \psi(x))^{n-\alpha-1} \left| f_{\psi}^{[n]}(t) \right| dt \leq $$

$$ \frac{\left\| f_{\psi}^{[n]} \right\|_{\infty,[a,b]}}{\Gamma(n-\alpha)} \int_{x}^{x_0} (\psi(t) - \psi(x))^{n-\alpha-1} \, d\psi(t) = \frac{\left\| f_{\psi}^{[n]} \right\|_{\infty,[a,b]}}{\Gamma(n-\alpha+1)} (\psi(x_0) - \psi(x))^{n-\alpha}. \tag{7.39} $$

That is

$$ \left| {}^{C}D_{x_0-}^{\alpha,\psi} f(x) \right| \leq \frac{\left\| f_{\psi}^{[n]} \right\|_{\infty,[a,b]}}{\Gamma(n-\alpha+1)} (\psi(x_0) - \psi(x))^{n-\alpha}, \tag{7.40} $$

$\forall\, x \in [a, x_0]$.

Clearly, it holds ${}^{C}D_{x_0-}^{\alpha,\psi} f(x_0) = 0$.

Therefore, we can write

$$f(x) - f(x_0) = \frac{1}{\Gamma(\alpha)} \int_{x_0}^{x} \psi'(t) (\psi(x) - \psi(t))^{\alpha-1} \left({}^C D_{x_0+}^{\alpha,\psi} f(t) - {}^C D_{x_0+}^{\alpha,\psi} f(x_0) \right) dt, \tag{7.41}$$

$\forall\, x \in [x_0, b]$, and

$$f(x) - f(x_0) = \frac{1}{\Gamma(\alpha)} \int_{x}^{x_0} \psi'(t) (\psi(t) - \psi(x))^{\alpha-1} \left({}^C D_{x_0-}^{\alpha,\psi} f(t) - {}^C D_{x_0-}^{\alpha,\psi} f(x_0) \right) dt, \tag{7.42}$$

$\forall\, x \in [a, x_0]$.

We can rewrite (7.41) and (7.42), as follows (by $z := \psi(t)$):

$$f(x) - f(x_0) = \frac{1}{\Gamma(\alpha)} \int_{\psi(x_0)}^{\psi(x)} (\psi(x) - z)^{\alpha-1}$$

$$\left(\left(\left({}^C D_{x_0+}^{\alpha,\psi} f \right) \circ \psi^{-1} \right)(z) - \left(\left({}^C D_{x_0+}^{\alpha,\psi} f \right) \circ \psi^{-1} \right)(\psi(x_0)) \right) dz, \tag{7.43}$$

$\forall\, x \in [x_0, b]$, and

$$f(x) - f(x_0) = \frac{1}{\Gamma(\alpha)} \int_{\psi(x)}^{\psi(x_0)} (z - \psi(x))^{\alpha-1}$$

$$\left(\left(\left({}^C D_{x_0-}^{\alpha,\psi} f \right) \circ \psi^{-1} \right)(z) - \left(\left({}^C D_{x_0-}^{\alpha,\psi} f \right) \circ \psi^{-1} \right)(\psi(x_0)) \right) dz, \tag{7.44}$$

$\forall\, x \in [a, x_0]$.

We have that ($x_0 \le x \le b$)

$$|f(x) - f(x_0)| \overset{(7.43)}{=} \frac{1}{\Gamma(\alpha)} \int_{\psi(x_0)}^{\psi(x)} (\psi(x) - z)^{\alpha-1}$$

$$\left| \left(\left({}^C D_{x_0+}^{\alpha,\psi} f \right) \circ \psi^{-1} \right)(z) - \left(\left({}^C D_{x_0+}^{\alpha,\psi} f \right) \circ \psi^{-1} \right)(\psi(x_0)) \right| dz \le$$

($\delta_1 > 0$)

$$\frac{1}{\Gamma(\alpha)} \int_{\psi(x_0)}^{\psi(x)} (\psi(x) - z)^{\alpha-1} \omega_1 \left(\left({}^C D_{x_0+}^{\alpha,\psi} f \right) \circ \psi^{-1}, \frac{\delta_1 |z - \psi(x_0)|}{\delta_1} \right)_{[\psi(x_0), \psi(b)]} dz \le$$

$$\frac{\omega_1 \left(\left({}^C D_{x_0+}^{\alpha,\psi} f \right) \circ \psi^{-1}, \delta_1 \right)_{[\psi(x_0), \psi(b)]}}{\Gamma(\alpha)} \int_{\psi(x_0)}^{\psi(x)} (\psi(x) - z)^{\alpha-1} \left(1 + \frac{(z - \psi(x_0))}{\delta_1} \right) dz \tag{7.45}$$

$$= \frac{\omega_1\left(\left({}^{C}D_{x_0+}^{\alpha,\psi}f\right)\circ\psi^{-1},\delta_1\right)_{[\psi(x_0),\psi(b)]}}{\Gamma(\alpha)}$$

$$\left[\frac{(\psi(x)-\psi(x_0))^{\alpha}}{\alpha}+\frac{1}{\delta_1}\int_{\psi(x_0)}^{\psi(x)}(\psi(x)-z)^{\alpha-1}(z-\psi(x_0))^{2-1}dz\right]=$$

$$\frac{\omega_1\left(\left({}^{C}D_{x_0+}^{\alpha,\psi}f\right)\circ\psi^{-1},\delta_1\right)_{[\psi(x_0),\psi(b)]}}{\Gamma(\alpha)}$$

$$\left[\frac{(\psi(x)-\psi(x_0))^{\alpha}}{\alpha}+\frac{1}{\delta_1}\frac{\Gamma(\alpha)\Gamma(2)}{\Gamma(\alpha+2)}(\psi(x)-\psi(x_0))^{\alpha+1}\right]=$$

$$\frac{\omega_1\left(\left({}^{C}D_{x_0+}^{\alpha,\psi}f\right)\circ\psi^{-1},\delta_1\right)_{[\psi(x_0),\psi(b)]}}{\Gamma(\alpha)}$$

$$\left[\frac{(\psi(x)-\psi(x_0))^{\alpha}}{\alpha}+\frac{1}{\delta_1}\frac{1}{(\alpha+1)\alpha}(\psi(x)-\psi(x_0))^{\alpha+1}\right], \tag{7.46}$$

$\forall\, x\in[x_0,b]$.

We have proved that

$$|f(x)-f(x_0)|\leq\frac{\omega_1\left(\left({}^{C}D_{x_0+}^{\alpha,\psi}f\right)\circ\psi^{-1},\delta_1\right)_{[\psi(x_0),\psi(b)]}}{\Gamma(\alpha+1)}$$

$$\left[(\psi(x)-\psi(x_0))^{\alpha}+\frac{(\psi(x)-\psi(x_0))^{\alpha+1}}{\delta_1(\alpha+1)}\right], \tag{7.47}$$

$\forall\, x\in[x_0,b]$, $\delta_1>0$.

By (7.44) we obtain

$$|f(x)-f(x_0)|\leq\frac{1}{\Gamma(\alpha)}\int_{\psi(x)}^{\psi(x_0)}(z-\psi(x))^{\alpha-1}$$

$$\left|\left(\left({}^{C}D_{x_0-}^{\alpha,\psi}f\right)\circ\psi^{-1}\right)(z)-\left(\left({}^{C}D_{x_0-}^{\alpha,\psi}f\right)\circ\psi^{-1}\right)(\psi(x_0))\right|dz\leq$$

$(\delta_2>0)$

$$\frac{1}{\Gamma(\alpha)}\int_{\psi(x)}^{\psi(x_0)}(z-\psi(x))^{\alpha-1}\omega_1\left(\left({}^{C}D_{x_0-}^{\alpha,\psi}f\right)\circ\psi^{-1},\frac{\delta_2(\psi(x_0)-z)}{\delta_2}\right)_{[\psi(a),\psi(x_0)]}dz\leq$$

$$\frac{\omega_1\left(\left({}^C D^{\alpha,\psi}_{x_0-} f\right)\circ\psi^{-1},\delta_2\right)_{[\psi(a),\psi(x_0)]}}{\Gamma(\alpha)}\int_{\psi(x)}^{\psi(x_0)}(z-\psi(x))^{\alpha-1}\left(1+\frac{(\psi(x_0)-z)}{\delta_2}\right)dz \tag{7.48}$$

$$=\frac{\omega_1\left(\left({}^C D^{\alpha,\psi}_{x_0-} f\right)\circ\psi^{-1},\delta_2\right)_{[\psi(a),\psi(x_0)]}}{\Gamma(\alpha)}$$

$$\left[\frac{(\psi(x_0)-\psi(x))^{\alpha}}{\alpha}+\frac{1}{\delta_2}\int_{\psi(x)}^{\psi(x_0)}(\psi(x_0)-z)^{2-1}(z-\psi(x))^{\alpha-1}dz\right]=$$

$$\frac{\omega_1\left(\left({}^C D^{\alpha,\psi}_{x_0-} f\right)\circ\psi^{-1},\delta_2\right)_{[\psi(a),\psi(x_0)]}}{\Gamma(\alpha)}$$

$$\left[\frac{(\psi(x_0)-\psi(x))^{\alpha}}{\alpha}+\frac{1}{\delta_2}\frac{\Gamma(2)\Gamma(\alpha)}{\Gamma(\alpha+2)}(\psi(x_0)-\psi(x))^{\alpha+1}\right]=$$

$$\frac{\omega_1\left(\left({}^C D^{\alpha,\psi}_{x_0-} f\right)\circ\psi^{-1},\delta_2\right)_{[\psi(a),\psi(x_0)]}}{\Gamma(\alpha)}$$

$$\left[\frac{(\psi(x_0)-\psi(x))^{\alpha}}{\alpha}+\frac{1}{\delta_2}\frac{1}{\alpha(\alpha+1)}(\psi(x_0)-\psi(x))^{\alpha+1}\right]= \tag{7.49}$$

$$\frac{\omega_1\left(\left({}^C D^{\alpha,\psi}_{x_0-} f\right)\circ\psi^{-1},\delta_2\right)_{[\psi(a),\psi(x_0)]}}{\Gamma(\alpha+1)}$$

$$\left[(\psi(x_0)-\psi(x))^{\alpha}+\frac{1}{\delta_2(\alpha+1)}(\psi(x_0)-\psi(x))^{\alpha+1}\right], \tag{7.50}$$

$\forall\, x\in[a,x_0]$.

In conclusion, we derive

$$|f(x)-f(x_0)|\leq\frac{\omega_1\left(\left({}^C D^{\alpha,\psi}_{x_0-} f\right)\circ\psi^{-1},\delta_2\right)_{[\psi(a),\psi(x_0)]}}{\Gamma(\alpha+1)}$$

$$\left[(\psi(x_0)-\psi(x))^{\alpha}+\frac{(\psi(x_0)-\psi(x))^{\alpha+1}}{\delta_2(\alpha+1)}\right], \tag{7.51}$$

$\forall\, x\in[a,x_0]$, $\delta_2>0$.

Choosing $\delta=\delta_1=\delta_2>0$, by (7.47) and (7.51), we get (7.30). ■

Corollary 7.12 *All as in Theorem 7.11. Then*

$$|f(\cdot) - f(x_0)| \leq \frac{\omega_1\left(\left({}^C D_{x_0}^{\alpha,\psi} f\right) \circ \psi^{-1}, \delta\right)}{\Gamma(\alpha+1)}$$

$$\left[|\psi(\cdot) - \psi(x_0)|^{\alpha} + \frac{|\psi(\cdot) - \psi(x_0)|^{\alpha+1}}{\delta(\alpha+1)}\right], \tag{7.52}$$

true over $[a, b]$, $\delta > 0$.

Our first result on the convergence of positive sublinear operators comes next.

Theorem 7.13 *Let* $\alpha > 0$, $\alpha \notin \mathbb{N}$, $n \in \mathbb{N}$ *such that* $n = \lceil \alpha \rceil$, $[a, b] \subset \mathbb{R}$, *and* $f \in C_+([a, b])$, $f, \psi \in C^n([a, b])$ *with* ψ *being strictly increasing. Let* $x_0 \in [a, b]$ *be fixed such that* $f_{\psi}^{[k]}(x_0) = 0$, $k = 1, \ldots, n-1$. *Let* $L_N : C_+([a, b]) \to C_+([a, b])$, $\forall\, N \in \mathbb{N}$, *be positive sublinear operators, such that* $L_N(1) = 1$, $\forall\, N \in \mathbb{N}$. *Then*

$$|L_N(f)(x_0) - f(x_0)| \leq \frac{\omega_1\left(\left({}^C D_{x_0}^{\alpha,\psi} f\right) \circ \psi^{-1}, \delta\right)}{\Gamma(\alpha+1)}$$

$$\left[L_N\left(|\psi(\cdot) - \psi(x_0)|^{\alpha}\right)(x_0) + \frac{L_N\left(|\psi(\cdot) - \psi(x_0)|^{\alpha+1}\right)(x_0)}{\delta(\alpha+1)}\right], \tag{7.53}$$

where $\delta > 0$, $\forall\, N \in \mathbb{N}$.

In particular (7.53) is true for $\alpha > 1$, $\alpha \notin \mathbb{N}$.

Proof By (7.7) and (7.52) we have

$$|L_N(f)(x_0) - f(x_0)| \leq L_N(|f(\cdot) - f(x_0)|)(x_0)$$

$$\leq \frac{\omega_1\left(\left({}^C D_{x_0}^{\alpha,\psi} f\right) \circ \psi^{-1}, \delta\right)}{\Gamma(\alpha+1)}$$

$$\left[L_N\left(|\psi(\cdot) - \psi(x_0)|^{\alpha}\right)(x_0) + \frac{L_N\left(|\psi(\cdot) - \psi(x_0)|^{\alpha+1}\right)(x_0)}{\delta(\alpha+1)}\right], \quad \delta > 0, \tag{7.54}$$

proving the claim. ■

Corollary 7.14 (to Theorem 7.13) *When* $0 < \alpha < 1$, *then (7.53) is valid for any* $x_0 \in [a, b]$, *and without any initial conditions.*

We give

Theorem 7.15 *All as in Theorem 7.13, and assume that* $L_N\left(|\psi(\cdot)-\psi(x_0)|^{\alpha+1}\right)(x_0)>0$, $\forall\, N\in\mathbb{N}$. *Then*

$$|L_N(f)(x_0)-f(x_0)|\leq\frac{(\alpha+2)}{\Gamma(\alpha+2)}$$

$$\omega_1\left(\left({}^{C}D_{x_0}^{\alpha,\psi}f\right)\circ\psi^{-1},\left(L_N\left(|\psi(\cdot)-\psi(x_0)|^{\alpha+1}\right)(x_0)\right)^{\frac{1}{\alpha+1}}\right)$$

$$\left(L_N\left(|\psi(\cdot)-\psi(x_0)|^{\alpha+1}\right)(x_0)\right)^{\frac{\alpha}{\alpha+1}}, \tag{7.55}$$

$\forall\, N\in\mathbb{N}$.

If $0<\alpha<1$, *then (7.55) is valid without any initial conditions and true for any* $x_0\in[a,b]$.

Proof By Theorem 7.2, see (7.4), we get

$$L_N\left(|\psi(\cdot)-\psi(x_0)|^{\alpha}\right)(x_0)\leq\left(L_N\left(|\psi(\cdot)-\psi(x_0)|^{\alpha+1}\right)(x_0)\right)^{\frac{\alpha}{\alpha+1}}. \tag{7.56}$$

Choose

$$\delta:=\left(L_N\left(|\psi(\cdot)-\psi(x_0)|^{\alpha+1}\right)(x_0)\right)^{\frac{1}{\alpha+1}}>0, \tag{7.57}$$

i.e.

$$\delta^{\alpha+1}=L_N\left(|\psi(\cdot)-\psi(x_0)|^{\alpha+1}\right)(x_0).$$

By (7.53) we obtain

$$|L_N(f)(x_0)-f(x_0)|\leq$$

$$\frac{\omega_1\left(\left({}^{C}D_{x_0}^{\alpha,\psi}f\right)\circ\psi^{-1},\left(L_N\left(|\psi(\cdot)-\psi(x_0)|^{\alpha+1}\right)(x_0)\right)^{\frac{1}{\alpha+1}}\right)}{\Gamma(\alpha+1)}$$

$$\left[\left(L_N\left(|\psi(\cdot)-\psi(x_0)|^{\alpha+1}\right)(x_0)\right)^{\frac{\alpha}{\alpha+1}}+\frac{\delta^{\alpha+1}}{\delta(\alpha+1)}\right]=$$

$$\frac{\omega_1\left(\left({}^{C}D_{x_0}^{\alpha,\psi}f\right)\circ\psi^{-1},\left(L_N\left(|\psi(\cdot)-\psi(x_0)|^{\alpha+1}\right)(x_0)\right)^{\frac{1}{\alpha+1}}\right)}{\Gamma(\alpha+1)}$$

$$\left[\left(L_N\left(|\psi(\cdot)-\psi(x_0)|^{\alpha+1}\right)(x_0)\right)^{\frac{\alpha}{\alpha+1}}+\frac{\left(L_N\left(|\psi(\cdot)-\psi(x_0)|^{\alpha+1}\right)(x_0)\right)^{\frac{\alpha}{\alpha+1}}}{(\alpha+1)}\right]= \tag{7.58}$$

$$\frac{\omega_1\left(\left({}^C D_{x_0}^{\alpha,\psi} f\right)\circ\psi^{-1}, \left(L_N\left(\left|\psi(\cdot)-\psi(x_0)\right|^{\alpha+1}\right)(x_0)\right)^{\frac{1}{\alpha+1}}\right)}{\Gamma(\alpha+1)}$$

$$\left(L_N\left(\left|\psi(\cdot)-\psi(x_0)\right|^{\alpha+1}\right)(x_0)\right)^{\frac{\alpha}{\alpha+1}}\left(1+\frac{1}{\alpha+1}\right)=$$

$$\frac{\omega_1\left(\left({}^C D_{x_0}^{\alpha,\psi} f\right)\circ\psi^{-1}, \left(L_N\left(\left|\psi(\cdot)-\psi(x_0)\right|^{\alpha+1}\right)(x_0)\right)^{\frac{1}{\alpha+1}}\right)(\alpha+2)}{\Gamma(\alpha+2)} \tag{7.59}$$

$$\left(L_N\left(\left|\psi(\cdot)-\psi(x_0)\right|^{\alpha+1}\right)(x_0)\right)^{\frac{\alpha}{\alpha+1}},$$

proving (7.55). ■

Remark 7.16 In the terms and assumptions of Theorem 7.15:
let $\left(L_N\left(\left|\psi(\cdot)-\psi(x_0)\right|^{\alpha+1}\right)(x_0)\right)\to 0$, then $L_N(f)(x_0)\to f(x_0)$, as $N\to+\infty$.

We need

Corollary 7.17 *All as in Theorem 7.13, and assume that* $L_N\left(\left|\cdot-x_0\right|^{\alpha+1}\right)(x_0)>0$, $\forall\, N\in\mathbb{N}$. *Then*

$$\left|L_N(f)(x_0)-f(x_0)\right|\leq\frac{(\alpha+2)}{\Gamma(\alpha+2)}$$

$$\omega_1\left(\left({}^C D_{x_0}^{\alpha,\psi} f\right)\circ\psi^{-1}, \left\|\psi'\right\|_{\infty,[a,b]}\left(L_N\left(\left|\cdot-x_0\right|^{\alpha+1}\right)(x_0)\right)^{\frac{1}{\alpha+1}}\right)$$

$$\left\|\psi'\right\|_{\infty,[a,b]}^{\alpha}\left(L_N\left(\left|\cdot-x_0\right|^{\alpha+1}\right)(x_0)\right)^{\frac{\alpha}{\alpha+1}}, \tag{7.60}$$

$\forall\, N\in\mathbb{N}$.

Proof Use of Theorem 7.15 (7.55), ω_1 and L_N properties, and that $\left|\psi(x)-\psi(x_0)\right|\leq\left\|\psi'\right\|_{\infty,[a,b]}\left|x-x_0\right|$, along with $\left|\psi(x)-\psi(x_0)\right|\geq\left(\inf_{x\in[a,b]}\left|\psi'(x)\right|\right)\left|x-x_0\right|$, $\forall\, x, x_0\in[a,b]$, given that $\psi'(x)\neq 0$, $\forall\, x\in[a,b]$. Clearly then (7.60) is valid, and $L_N\left(\left|\psi(\cdot)-\psi(x_0)\right|^{\alpha+1}\right)(x_0)>0$. ■

7.3 Applications

(I) Case $0<\alpha<1$.
Here we apply Corollary 7.17 to well known Max-product operators.
We make

Remark 7.18 The Max-product Bernstein operators $B_N^{(M)}(f)(x)$ are defined by (7.8), see also [5, p. 10]; here $f : [0, 1] \to \mathbb{R}_+$ is a continuous function.

We have $B_N^{(M)}(1) = 1$, and

$$B_N^{(M)}(|\cdot - x|)(x) \leq \frac{6}{\sqrt{N+1}}, \ \forall\, x \in [0, 1], \forall\, N \in \mathbb{N}, \tag{7.61}$$

see [5, p. 31].

$B_N^{(M)}$ are positive sublinear operators and thus they possess the monotonicity property, also since $|\cdot - x| \leq 1$, then $|\cdot - x|^{\beta} \leq 1, \forall\, x \in [0, 1], \forall\, \beta > 0$.

Therefore it holds

$$B_N^{(M)}\left(|\cdot - x|^{1+\beta}\right)(x) \leq \frac{6}{\sqrt{N+1}}, \ \forall\, x \in [0, 1], \forall\, N \in \mathbb{N}, \forall\, \beta > 0. \tag{7.62}$$

Furthermore, clearly it holds that

$$B_N^{(M)}\left(|\cdot - x|^{1+\beta}\right)(x) > 0, \forall\, N \in \mathbb{N}, \forall\, \beta \geq 0 \text{ and any } x \in (0, 1). \tag{7.63}$$

The operator $B_N^{(M)}$ maps $C_+([0, 1])$ into itself.

We present

Theorem 7.19 *Let* $0 < \alpha < 1$, $f \in C_+([0, 1])$, $f, \psi \in C^1([0, 1])$ *with* ψ *being strictly increasing. Then*

$$\left|B_N^{(M)}(f)(x) - f(x)\right| \leq \frac{(\alpha+2)}{\Gamma(\alpha+2)}$$

$$\omega_1\left(\left({}^{C}D_x^{\alpha,\psi} f\right) \circ \psi^{-1}, \|\psi'\|_{\infty,[0,1]}\left(\frac{6}{\sqrt{N+1}}\right)^{\frac{1}{\alpha+1}}\right)$$

$$\|\psi'\|_{\infty,[0,1]}^{\alpha}\left(\frac{6}{\sqrt{N+1}}\right)^{\frac{\alpha}{\alpha+1}}, \ \forall\, N \in \mathbb{N}, \tag{7.64}$$

$\forall\, x \in (0, 1)$.

As $N \to +\infty$, *we get* $B_N^{(M)}(f)(x) \to f(x)$, *for any* $x \in (0, 1)$.

Proof By Corollary 7.17. ∎

We continue with

Remark 7.20 The truncated Favard–Szász–Mirakjan operators are given by

$$T_N^{(M)}(f)(x)=\frac{\bigvee_{k=0}^{N}s_{N,k}(x)\,f\left(\frac{k}{N}\right)}{\bigvee_{k=0}^{N}s_{N,k}(x)},\quad x\in[0,1],\ N\in\mathbb{N},\ f\in C_{+}([0,1]),\tag{7.65}$$

$s_{N,k}(x)=\frac{(Nx)^k}{k!}$, see also [5, p. 11].

By [5, pp. 178–179], we get that

$$T_N^{(M)}(|\cdot-x|)(x)\leq\frac{3}{\sqrt{N}},\quad\forall\,x\in[0,1],\ \forall\,N\in\mathbb{N}.\tag{7.66}$$

Clearly it holds

$$T_N^{(M)}\left(|\cdot-x|^{1+\beta}\right)(x)\leq\frac{3}{\sqrt{N}},\quad\forall\,x\in[0,1],\ \forall\,N\in\mathbb{N},\forall\,\beta>0.\tag{7.67}$$

The operators $T_N^{(M)}$ are positive sublinear operators mapping $C_{+}([0,1])$ into itself, with $T_N^{(M)}(1)=1$.

Furthermore it holds

$$T_N^{(M)}\left(|\cdot-x|^{\lambda}\right)(x)=\frac{\bigvee_{k=0}^{N}\frac{(Nx)^k}{k!}\left|\frac{k}{N}-x\right|^{\lambda}}{\bigvee_{k=0}^{N}\frac{(Nx)^k}{k!}}>0,\quad\forall\,x\in(0,1],\ \forall\,\lambda\geq1,\forall\,N\in\mathbb{N}.\tag{7.68}$$

We give

Theorem 7.21 *All as in Theorem 7.19. Then*

$$\left|T_N^{(M)}(f)(x)-f(x)\right|\leq\frac{(\alpha+2)}{\Gamma(\alpha+2)}$$

$$\omega_1\left(\left({}^{C}D_x^{\alpha,\psi}f\right)\circ\psi^{-1},\|\psi'\|_{\infty,[0,1]}\left(\frac{3}{\sqrt{N}}\right)^{\frac{1}{\alpha+1}}\right)$$

$$\|\psi'\|_{\infty,[0,1]}^{\alpha}\left(\frac{3}{\sqrt{N}}\right)^{\frac{\alpha}{\alpha+1}},\quad\forall\,N\in\mathbb{N},\tag{7.69}$$

$\forall\,x\in(0,1]$.

As $N\to+\infty$, we get $T_N^{(M)}(f)(x)\to f(x)$, for any $x\in(0,1]$.

Proof By Corollary 7.17. ∎

We continue with

Remark 7.22 Next we study the truncated Max-product Baskakov operators (see [5, p. 11])

$$U_N^{(M)}(f)(x) = \frac{\bigvee_{k=0}^{N} b_{N,k}(x) f\left(\frac{k}{N}\right)}{\bigvee_{k=0}^{N} b_{N,k}(x)}, \quad x \in [0,1], \; f \in C_+([0,1]), \; N \in \mathbb{N}, \tag{7.70}$$

where

$$b_{N,k}(x) = \binom{N+k-1}{k} \frac{x^k}{(1+x)^{N+k}}. \tag{7.71}$$

From [5, pp. 217–218], we get ($x \in [0,1]$)

$$\left(U_N^{(M)}(|\cdot - x|)\right)(x) \leq \frac{2\sqrt{3}\left(\sqrt{2}+2\right)}{\sqrt{N+1}}, \quad N \geq 2, N \in \mathbb{N}. \tag{7.72}$$

Let $\lambda \geq 1$, clearly then it holds

$$\left(U_N^{(M)}\left(|\cdot - x|^{\lambda}\right)\right)(x) \leq \frac{2\sqrt{3}\left(\sqrt{2}+2\right)}{\sqrt{N+1}}, \quad \forall\, N \geq 2, N \in \mathbb{N}. \tag{7.73}$$

Also it holds $U_N^{(M)}(1) = 1$, and $U_N^{(M)}$ are positive sublinear operators from $C_+([0,1])$ into itself. Furthermore it holds

$$U_N^{(M)}\left(|\cdot - x|^{\lambda}\right)(x) > 0, \quad \forall\, x \in (0,1], \; \forall\, \lambda \geq 1, \forall\, N \in \mathbb{N}. \tag{7.74}$$

We present

Theorem 7.23 *All as in Theorem 7.19. Then*

$$\left|U_N^{(M)}(f)(x) - f(x)\right| \leq \frac{(\alpha+2)}{\Gamma(\alpha+2)}$$

$$\omega_1\left(\left({}^{C}D_{x}^{\alpha,\psi} f\right) \circ \psi^{-1}, \|\psi'\|_{\infty,[0,1]} \left(\frac{2\sqrt{3}\left(\sqrt{2}+2\right)}{\sqrt{N+1}}\right)^{\frac{1}{\alpha+1}}\right)$$

$$\|\psi'\|_{\infty,[0,1]}^{\alpha} \left(\frac{2\sqrt{3}\left(\sqrt{2}+2\right)}{\sqrt{N+1}}\right)^{\frac{\alpha}{\alpha+1}}, \quad \forall\, N \in \mathbb{N} - \{1\}, \tag{7.75}$$

$\forall\, x \in (0,1]$.

As $N \to +\infty$, we get $U_N^{(M)}(f)(x) \to f(x)$, for any $x \in (0,1]$.

Proof By Corollary 7.17. ∎

We give

Remark 7.24 Here we study the Max-product Meyer-Köning and Zeller operators (see [5, p. 11]) defined by

$$Z_N^{(M)}(f)(x) = \frac{\bigvee_{k=0}^{\infty} s_{N,k}(x) f\left(\frac{k}{N+k}\right)}{\bigvee_{k=0}^{\infty} s_{N,k}(x)}, \quad \forall\, N \in \mathbb{N},\, f \in C_+([0,1]), \tag{7.76}$$

$s_{N,k}(x) = \binom{N+k}{k} x^k$, $x \in [0,1]$.

By [5, p. 253], we get that

$$Z_N^{(M)}(|\cdot - x|)(x) \leq \frac{8\left(1+\sqrt{5}\right)}{3} \frac{\sqrt{x}(1-x)}{\sqrt{N}}, \quad \forall\, x \in [0,1], \ \forall\, N \geq 4,\, N \in \mathbb{N}. \tag{7.77}$$

As before we get that (for $\lambda \geq 1$)

$$Z_N^{(M)}\left(|\cdot - x|^{\lambda}\right)(x) \leq \frac{8\left(1+\sqrt{5}\right)}{3} \frac{\sqrt{x}(1-x)}{\sqrt{N}} := \rho(x), \tag{7.78}$$

$\forall\, x \in [0,1]$, $N \geq 4$, $N \in \mathbb{N}$.

Also it holds $Z_N^{(M)}(1) = 1$, and $Z_N^{(M)}$ are positive sublinear operators from $C_+([0,1])$ into itself. Also it holds

$$Z_N^{(M)}\left(|\cdot - x|^{\lambda}\right)(x) > 0, \quad \forall\, x \in (0,1), \ \forall\, \lambda \geq 1, \forall\, N \in \mathbb{N}. \tag{7.79}$$

We give

Theorem 7.25 *All as in Theorem 7.19. Then*

$$\left|Z_N^{(M)}(f)(x) - f(x)\right| \leq \frac{(\alpha+2)}{\Gamma(\alpha+2)}$$

$$\omega_1\left(\left({}^{C}D_x^{\alpha,\psi} f\right) \circ \psi^{-1}, \left\|\psi'\right\|_{\infty,[0,1]} (\rho(x))^{\frac{1}{\alpha+1}}\right)$$

$$\left\|\psi'\right\|_{\infty,[0,1]}^{\alpha} (\rho(x))^{\frac{\alpha}{\alpha+1}}, \quad \forall\, N \geq 4,\, N \in \mathbb{N}, \tag{7.80}$$

$\forall\, x \in (0,1)$.

As $N \to +\infty$, we get $Z_N^{(M)}(f)(x) \to f(x)$, for any $x \in (0,1)$.

Proof By Corollary 7.17. ■

Corollary 7.26 *All as in Theorem 7.19, for $\alpha = \frac{1}{2}$. Then*

$$\left| B_N^{(M)}(f)(x) - f(x) \right| \leq \sqrt{\|\psi'\|_{\infty,[0,1]}} \left(\frac{10}{3\sqrt{\pi}} \right) \left(\frac{\sqrt[3]{6}}{\sqrt[6]{N+1}} \right)$$

$$\omega_1 \left(\left({}^C D_x^{\frac{1}{2},\psi} f \right) \circ \psi^{-1}, \left\| \psi' \right\|_{\infty,[0,1]} \frac{\sqrt[3]{36}}{\sqrt[3]{N+1}} \right), \tag{7.81}$$

$\forall\, x \in (0,1)$, $\forall\, N \in \mathbb{N}$.

Proof By Theorem 7.19. ■

(II) Case of $\alpha > 1$, $\alpha \notin \mathbb{N}$.
We need

Corollary 7.27 (to Theorem 7.13) *Same assumptions as in Theorem 7.13. Then*

$$|L_N(f)(x_0) - f(x_0)| \leq \frac{\omega_1 \left(\left({}^C D_{x_0}^{\alpha,\psi} f \right) \circ \psi^{-1}, \delta \right)}{\Gamma(\alpha+1)}$$

$$\left\| \psi' \right\|_{\infty,[a,b]}^{\alpha} \left[L_N\left(|\cdot - x_0|^{\alpha} \right)(x_0) + \frac{\left\| \psi' \right\|_{\infty,[a,b]} L_N\left(|\cdot - x_0|^{\alpha+1} \right)(x_0)}{\delta(\alpha+1)} \right], \tag{7.82}$$

where $\delta > 0$, $\forall\, N \in \mathbb{N}$.

Proof By Theorem 7.13, and $|\psi(x) - \psi(x_0)| \leq \left\| \psi' \right\|_{\infty,[a,b]} |x - x_0|$, $\forall\, x, x_0 \in [a,b]$ and the properties of L_N. ■

We apply Corollary 7.27 to well-known Max-product operators.
We present

Theorem 7.28 *Let $\alpha > 1$, $\alpha \notin \mathbb{N}$, $n \in \mathbb{N}$ such that $n = \lceil \alpha \rceil$, and $f \in C_+([0,1])$, $f, \psi \in C^n([0,1])$ with ψ being strictly increasing. For a fixed $x \in [0,1]$ assume that $f_{\psi}^{[k]}(x) = 0$, $k = 1, \ldots, n-1$. Then*

$$\left| B_N^{(M)}(f)(x) - f(x) \right| \leq \frac{\omega_1 \left(\left({}^C D_x^{\alpha,\psi} f \right) \circ \psi^{-1}, \left(\frac{6}{\sqrt{N+1}} \right)^{\frac{1}{\alpha+1}} \right)}{\Gamma(\alpha+1)} \tag{7.83}$$

$$\left\| \psi' \right\|_{\infty,[0,1]}^{\alpha} \left[\frac{6}{\sqrt{N+1}} + \frac{\left\| \psi' \right\|_{\infty,[0,1]}}{(\alpha+1)} \left(\frac{6}{\sqrt{N+1}} \right)^{\frac{\alpha}{\alpha+1}} \right],$$

$\forall\, N \in \mathbb{N}$.
We get $\lim_{N \to +\infty} B_N^{(M)}(f)(x) = f(x)$.

Proof By (7.82) we get

$$\left|B_N^{(M)}(f)(x) - f(x)\right| \leq \frac{\omega_1\left(\left({}^C D_x^{\alpha,\psi} f\right) \circ \psi^{-1}, \delta\right)}{\Gamma(\alpha+1)}$$

$$\|\psi'\|_{\infty,[0,1]}^{\alpha}\left[B_N^{(M)}\left(|\cdot - x|^{\alpha}\right)(x) + \frac{\|\psi'\|_{\infty,[0,1]} B_N^{(M)}\left(|\cdot - x|^{\alpha+1}\right)(x)}{\delta(\alpha+1)}\right] \overset{(7.62)}{\leq}$$

$$\frac{\omega_1\left(\left({}^C D_x^{\alpha,\psi} f\right) \circ \psi^{-1}, \delta\right)}{\Gamma(\alpha+1)} \|\psi'\|_{\infty,[0,1]}^{\alpha} \tag{7.84}$$

$$\left[\frac{6}{\sqrt{N+1}} + \frac{\|\psi'\|_{\infty,[0,1]} \frac{6}{\sqrt{N+1}}}{\delta(\alpha+1)}\right] =: (*).$$

Choose $\delta = \left(\frac{6}{\sqrt{N+1}}\right)^{\frac{1}{\alpha+1}}$, then $\delta^{\alpha+1} = \frac{6}{\sqrt{N+1}}$, and apply it to (7.84).

We get that

$$(*) = \frac{\omega_1\left(\left({}^C D_x^{\alpha,\psi} f\right) \circ \psi^{-1}, \left(\frac{6}{\sqrt{N+1}}\right)^{\frac{1}{\alpha+1}}\right)}{\Gamma(\alpha+1)} \|\psi'\|_{\infty,[0,1]}^{\alpha}$$

$$\left[\frac{6}{\sqrt{N+1}} + \frac{\|\psi'\|_{\infty,[0,1]} \left(\frac{6}{\sqrt{N+1}}\right)^{\frac{\alpha}{\alpha+1}}}{(\alpha+1)}\right], \tag{7.85}$$

$\forall\, N \in \mathbb{N}$, proving the claim. ■

Next comes

Theorem 7.29 *All as in Theorem 7.28. Then*

$$\left|T_N^{(M)}(f)(x) - f(x)\right| \leq \frac{\omega_1\left(\left({}^C D_x^{\alpha,\psi} f\right) \circ \psi^{-1}, \left(\frac{3}{\sqrt{N}}\right)^{\frac{1}{\alpha+1}}\right)}{\Gamma(\alpha+1)}$$

$$\|\psi'\|_{\infty,[0,1]}^{\alpha}\left[\frac{3}{\sqrt{N}} + \frac{\|\psi'\|_{\infty,[0,1]}}{(\alpha+1)}\left(\frac{3}{\sqrt{N}}\right)^{\frac{\alpha}{\alpha+1}}\right], \tag{7.86}$$

$\forall\, N \in \mathbb{N}$.

We get $\lim_{N \to +\infty} T_N^{(M)}(f)(x) = f(x)$.

Proof Similar to Theorem 7.28. ∎

Next comes

Theorem 7.30 *All as in Theorem 7.28. Then*

$$\left| U_N^{(M)}(f)(x) - f(x) \right| \leq \frac{\omega_1\left(\left({}^C D_x^{\alpha,\psi} f\right)\circ\psi^{-1}, \left(\frac{2\sqrt{3}(\sqrt{2}+2)}{\sqrt{N+1}}\right)^{\frac{1}{\alpha+1}}\right)}{\Gamma(\alpha+1)}$$

$$\left\|\psi'\right\|_{\infty,[0,1]}^{\alpha}\left[\frac{2\sqrt{3}\left(\sqrt{2}+2\right)}{\sqrt{N+1}} + \frac{\left\|\psi'\right\|_{\infty,[0,1]}}{(\alpha+1)}\left(\frac{2\sqrt{3}\left(\sqrt{2}+2\right)}{\sqrt{N+1}}\right)^{\frac{\alpha}{\alpha+1}}\right], \tag{7.87}$$

$\forall\, N \in \mathbb{N} - \{1\}$.

We get $\lim_{N\to+\infty} U_N^{(M)}(f)(x) = f(x)$.

Proof Similar to Theorem 7.28. ∎

We finish with

Theorem 7.31 *All as in Theorem 7.28. Then*

$$\left| Z_N^{(M)}(f)(x) - f(x) \right| \leq \frac{\omega_1\left(\left({}^C D_x^{\alpha,\psi} f\right)\circ\psi^{-1}, (\rho(x))^{\frac{1}{\alpha+1}}\right)}{\Gamma(\alpha+1)}$$

$$\left\|\psi'\right\|_{\infty,[0,1]}^{\alpha}\left[\rho(x) + \frac{\left\|\psi'\right\|_{\infty,[0,1]}}{(\alpha+1)}(\rho(x))^{\frac{\alpha}{\alpha+1}}\right], \tag{7.88}$$

$\forall\, N \in \mathbb{N},\ N \geq 4$.

We get $\lim_{N\to+\infty} Z_N^{(M)}(f)(x) = f(x)$, *where* $\rho(x)$ *is as in (7.78).*

Proof Similar to Theorem 7.28. ∎

References

1. Almeida, R.: A Caputo fractional derivative of a function with respect to another function. Commun. Nonlinear Sci. Numer. Simul. **44**, 460–481 (2017)
2. Anastassiou, G.A.: Nonlinearity: Ordinary and Fractional Approximations by Sublinear and Max-Product Operators. Springer, Heidelberg (2018)
3. Anastassiou, G.A.: Intelligent Computations: Abstract Fractional Calculus, Inequalities, Approximations. Springer, Heidelberg (2018)
4. Anastassiou, G.: Generalized ψ-fractional approximation by sublinear operators (2019). Submitted
5. Bede, B., Coroianu, L., Gal, S.: Approximation by Max-Product type Operators. Springer, Heidelberg (2016)

Chapter 8
Generalized g-Iterated Fractional Quantitative Approximation By Sublinear Operators

Here we study the approximation of functions by sublinear positive operators with applications to several Max-Product operators under generalized g-iterated fractional differentiability. Our work is based on our generalized g-iterated fractional results about positive sublinear operators. We produce Jackson type inequalities under iterated initial conditions. So our approach is quantitative by deriving inequalities with their right hand sides involving the modulus of continuity of generalized g-iterated fractional derivative of the function under approximation. See also [3].

8.1 Background

Here we study the approximation properties of positive sublinear operators converging the unit operator related to the iterated generalized fractional smoothness of the involved function under approximation. This is a quantitative study using inequalities.

We need some background:

We mention

Definition 8.1 Here $C_+([a,b]) := \{f : [a,b] \to \mathbb{R}_+, \text{ continuous functions}\}$. Let $L_N : C_+([a,b]) \to C_+([a,b])$, operators, $\forall N \in \mathbb{N}$, such that

(i)

$$L_N(\alpha f) = \alpha L_N(f), \ \forall \alpha \geq 0, \ \forall f \in C_+([a,b]), \tag{8.1}$$

(ii) if $f, g \in C_+([a,b]) : f \leq g$, then

$$L_N(f) \leq L_N(g), \ \forall N \in \mathbb{N}, \tag{8.2}$$

G. A. Anastassiou, *Generalized Fractional Calculus*, Studies in Systems, Decision and Control 305, https://doi.org/10.1007/978-3-030-56962-4_8

(iii)

$$L_N(f+g) \le L_N(f) + L_N(g), \ \forall f, g \in C_+([a,b]). \tag{8.3}$$

We call $\{L_N\}_{N \in \mathbb{N}}$ positive sublinear operators.

We need a Hölder's type inequality, see next:

Theorem 8.2 (see [2, p. 6]) *Let $L : C_+([a,b]) \to C_+([a,b])$, be a positive sublinear operator and $f, g \in C_+([a,b])$, furthermore let $p, q > 1 : \frac{1}{p} + \frac{1}{q} = 1$. Assume that $L\left((f(\cdot))^p\right)(s_*), L\left((g(\cdot))^q\right)(s_*) > 0$ for some $s_* \in [a,b]$. Then*

$$L(f(\cdot)g(\cdot))(s_*) \le \left(L\left((f(\cdot))^p\right)(s_*)\right)^{\frac{1}{p}} \left(L\left((g(\cdot))^q\right)(s_*)\right)^{\frac{1}{q}}. \tag{8.4}$$

We make

Remark 8.3 By [5, p. 17], we get: let $f, g \in C_+([a,b])$, then

$$|L_N(f)(x) - L_N(g)(x)| \le L_N(|f-g|)(x), \ \forall x \in [a,b]. \tag{8.5}$$

Furthermore, we also have that

$$|L_N(f)(x) - f(x)| \le L_N(|f(\cdot) - f(x)|)(x) + |f(x)| \, |L_N(e_0)(x) - 1|, \tag{8.6}$$

$\forall x \in [a,b]$; $e_0(t) = 1$.

From now on we assume that $L_N(1) = 1$. Hence it holds

$$|L_N(f)(x) - f(x)| \le L_N(|f(\cdot) - f(x)|)(x), \ \forall x \in [a,b]. \tag{8.7}$$

In [5, p. 10], the authors introduced the basic Max-product Bernstein operators,

$$B_N^{(M)}(f)(x) = \frac{\bigvee_{k=0}^{N} p_{N,k}(x) f\left(\frac{k}{N}\right)}{\bigvee_{k=0}^{N} p_{N,k}(x)}, \ N \in \mathbb{N}, \tag{8.8}$$

where $\bigvee$ stands for maximum, and $p_{N,k}(x) = \binom{N}{k} x^k (1-x)^{N-k}$ and $f : [0,1] \to \mathbb{R}_+ = [0, \infty)$.

These are nonlinear and piecewise rational operators, in particular they are positive sublinear operators.

The authors in [5] studied similar such nonlinear operators such as: the Max-product Favard–Szász–Mirakjan operators and their truncated version, the Max-product Baskakov operators and their truncated version, also many other similar specific operators. These Max-product operators tend to converge faster to the on hand function.

So we mention from [5, p. 30], that for $f : [0, 1] \to \mathbb{R}_+$ continuous, we have the estimate

$$\left| B_N^{(M)} (f)(x) - f(x) \right| \leq 12\omega_1 \left(f, \frac{1}{\sqrt{N+1}} \right), \quad \text{for all } N \in \mathbb{N}, \ x \in [0, 1], \tag{8.9}$$

Also from [5, p. 36], we mention that for $f : [0, 1] \to \mathbb{R}_+$ being concave function we get that

$$\left| B_N^{(M)} (f)(x) - f(x) \right| \leq 2\omega_1 \left(f, \frac{1}{N} \right), \quad \text{for all } x \in [0, 1], \tag{8.10}$$

a much faster convergence.

Above ω_1 is the first modulus of continuity. Inequalities (8.9) and (8.10) motivate our work.

We mention

Definition 8.4 Let $f \in C([a, b])$, we define

$$\omega_1(f, \delta) := \sup_{\substack{x, y \in [a,b]: \\ |x-y| \leq \delta}} |f(x) - f(y)|, \text{ where } 0 < \delta \leq b - a.$$

We need

Definition 8.5 *([1])* Let $\alpha > 0$, $\lceil \alpha \rceil = m$, $\lceil \cdot \rceil$ the ceiling of the number. Here $g \in AC([a, b])$ (absolutely continuous functions) and strictly increasing. We assume that $\left(f \circ g^{-1} \right)^{(m)} \circ g \in L_\infty([a, b])$. We define the left generalized g-fractional derivative of f of order α as follows:

$$\left(D_{a+;g}^{\alpha} f \right)(x) := \frac{1}{\Gamma(m-\alpha)} \int_a^x (g(x) - g(t))^{m-\alpha-1} g'(t) \left(f \circ g^{-1} \right)^{(m)} (g(t)) \, dt, \tag{8.11}$$

all $x \in [a, b]$.

If $\alpha \notin \mathbb{N}$, by [4, pp. 360–361], we have that $D_{a+;g}^{\alpha} f \in C([a, b])$.

We see that

$$\left(I_{a+;g}^{m-\alpha} \left(\left(f \circ g^{-1} \right)^{(m)} \circ g \right) \right)(x) = \left(D_{a+;g}^{\alpha} f \right)(x), \ x \geq a. \tag{8.12}$$

We set

$$D_{a+;g}^{m} f(x) := \left(\left(f \circ g^{-1} \right)^{(m)} \circ g \right)(x), \tag{8.13}$$

$$D_{a+;g}^{0} f(x) = f(x), \ \forall x \in [a, b]. \tag{8.14}$$

When $g = id$, then

$$D_{a+;g}^{\alpha} f = D_{a+;id}^{\alpha} f = D_{*a}^{\alpha} f, \tag{8.15}$$

the usual left Caputo fractional derivative.

We need

Definition 8.6 ([1]) Here $g \in AC([a,b])$ and strictly increasing. We assume that $\left(f \circ g^{-1}\right)^{(m)} \circ g \in L_{\infty}([a,b])$, where $N \ni m = \lceil \alpha \rceil$, $\alpha > 0$. We define the right generalized g-fractional derivative of f of order α as follows:

$$\left(D_{b-;g}^{\alpha} f\right)(x) := \frac{(-1)^m}{\Gamma(m-\alpha)} \int_x^b (g(t) - g(x))^{m-\alpha-1} g'(t) \left(f \circ g^{-1}\right)^{(m)} (g(t))\, dt, \tag{8.16}$$

all $x \in [a,b]$.

If $\alpha \notin \mathbb{N}$, by [4, p. 378], we get that $\left(D_{b-;g}^{\alpha} f\right) \in C([a,b])$.

We see that

$$I_{b-;g}^{m-\alpha}\left((-1)^m \left(f \circ g^{-1}\right)^{(m)} \circ g\right)(x) = \left(D_{b-;g}^{\alpha} f\right)(x), \ a \le x \le b. \tag{8.17}$$

We set

$$D_{b-;g}^{m} f(x) = (-1)^m \left(\left(f \circ g^{-1}\right)^{(m)} \circ g\right)(x), \tag{8.18}$$

$$D_{b-;g}^{0} f(x) = f(x), \ \forall x \in [a,b].$$

When $g = id$, then

$$D_{b-;g}^{\alpha} f(x) = D_{b-;id}^{\alpha} f(x) = D_{b-}^{\alpha} f, \tag{8.19}$$

the usual right Caputo fractional derivative.

Denote by

$$D_{b-;g}^{n\alpha} := D_{b-;g}^{\alpha} D_{b-;g}^{\alpha} \dots D_{b-;g}^{\alpha} \ \ (n\text{-times}),\ n \in \mathbb{N}. \tag{8.20}$$

We mention the following g-right generalized modified Taylor's formula:

Theorem 8.7 ([1]) *Suppose that $F_k := D_{b-;g}^{k\alpha} f$, for $k = 0, 1, \dots, n+1$, fulfill: $F_k \circ g^{-1} \in AC([c,d])$, where $c = g(a)$, $d = g(b)$, and $\left(F_k \circ g^{-1}\right)' \circ g \in L_{\infty}([a,b])$, where $0 < \alpha \le 1$. Then*

$$f(x) = \sum_{i=0}^{n} \frac{(g(b) - g(x))^{i\alpha}}{\Gamma(i\alpha + 1)} \left(D_{b-;g}^{i\alpha} f\right)(b) +$$

$$\frac{1}{\Gamma((n+1)\alpha)} \int_x^b (g(t) - g(x))^{(n+1)\alpha - 1} g'(t) \left(D_{b-;g}^{(n+1)\alpha} f\right)(t)\, dt = \tag{8.21}$$

$$\sum_{i=0}^{n} \frac{(g(b) - g(x))^{i\alpha}}{\Gamma(i\alpha + 1)} \left(D_{b-;g}^{i\alpha} f\right)(b) + \frac{\left(D_{b-;g}^{(n+1)\alpha} f\right)(\psi_x)}{\Gamma((n+1)\alpha + 1)} (g(b) - g(x))^{(n+1)\alpha}, \tag{8.22}$$

where $\psi_x \in [x, b]$, *any* $x \in [a, b]$.

Denote by

$$D_{a+;g}^{n\alpha} := D_{a+;g}^{\alpha} D_{a+;g}^{\alpha} \dots D_{a+;g}^{\alpha} \text{ (n-times)}, n \in \mathbb{N}. \tag{8.23}$$

We mention the following g-left generalized modified Taylor's formula:

Theorem 8.8 ([1]) *Suppose that* $F_k^* := D_{a+;g}^{k\alpha} f$, *for* $k = 0, 1, \dots, n+1$, *fulfill:* $F_k^* \circ g^{-1} \in AC([c, d])$, *where* $c = g(a)$, $d = g(b)$, *and* $\left(F_k^* \circ g^{-1}\right)' \circ g \in L_\infty([a, b])$, *where* $0 < \alpha \leq 1$. *Then*

$$f(x) = \sum_{i=0}^{n} \frac{(g(x) - g(a))^{i\alpha}}{\Gamma(i\alpha + 1)} \left(D_{a+;g}^{i\alpha} f\right)(a) + \tag{8.24}$$

$$\frac{1}{\Gamma((n+1)\alpha)} \int_a^x (g(x) - g(t))^{(n+1)\alpha - 1} g'(t) \left(D_{a+;g}^{(n+1)\alpha} f\right)(t)\, dt =$$

$$\sum_{i=0}^{n} \frac{(g(x) - g(a))^{i\alpha}}{\Gamma(i\alpha + 1)} \left(D_{a+;g}^{i\alpha} f\right)(a) + \frac{\left(D_{a+;g}^{(n+1)\alpha} f\right)(\psi_x)}{\Gamma((n+1)\alpha + 1)} (g(x) - g(a))^{(n+1)\alpha}, \tag{8.25}$$

where $\psi_x \in [a, x]$, *any* $x \in [a, b]$.

We make

Remark 8.9 Here we restrict ourselves to $0 < \alpha < 1$. In this case

$$\left(D_{a+;g}^{\alpha} f\right)(x) = \frac{1}{\Gamma(1-\alpha)} \int_a^x (g(x) - g(t))^{-\alpha} g'(t) \left(f \circ g^{-1}\right)'(g(t))\, dt, \tag{8.26}$$

and

$$\left(D_{b-;g}^{\alpha} f\right)(x) = \frac{-1}{\Gamma(1-\alpha)} \int_x^b (g(t) - g(x))^{-\alpha} g'(t) \left(f \circ g^{-1}\right)'(g(t))\, dt, \tag{8.27}$$

$\forall\, x \in [a, b]$.

We observe that

$$\left|D_{a+;g}^{\alpha} f(x)\right| \leq \frac{1}{\Gamma(1-\alpha)} \int_a^x (g(x) - g(t))^{-\alpha} g'(t) \left|\left(f \circ g^{-1}\right)'(g(t))\right| dt$$

$$\leq \frac{\left\| \left(f \circ g^{-1} \right)' \circ g \right\|_{\infty,[a,b]}}{\Gamma (1-\alpha)} \int_a^x (g(x) - g(t))^{-\alpha} g'(t) \, dt$$

(by [6, p. 107])

$$= \frac{\left\| \left(f \circ g^{-1} \right)' \circ g \right\|_{\infty,[a,b]}}{\Gamma (1-\alpha)} \frac{(g(x) - g(a))^{1-\alpha}}{1-\alpha}$$

$$= \frac{\left\| \left(f \circ g^{-1} \right)' \circ g \right\|_{\infty,[a,b]}}{\Gamma (2-\alpha)} (g(x) - g(a))^{1-\alpha}. \tag{8.28}$$

That is

$$\left| \left(D_{a+;g}^{\alpha} f \right)(x) \right| \leq \frac{\left\| \left(f \circ g^{-1} \right)' \circ g \right\|_{\infty,[a,b]}}{\Gamma (2-\alpha)} (g(x) - g(a))^{1-\alpha} < +\infty, \tag{8.29}$$

$\forall\, x \in [a, b]$.

Clearly it holds $\left(D_{a+;g}^{\alpha} f \right)(a) = 0$.

I.e.

$$\left(D_{x_0+;g}^{\alpha} f \right)(x_0) = 0, \ \forall\, x_0 \in [a, b]. \tag{8.30}$$

We define

$$\left(D_{x_0+;g}^{\alpha} f \right)(x) = 0, \ \text{for } a \leq x < x_0. \tag{8.31}$$

Similarly we get (by [6, p. 107])

$$\left| D_{b-;g}^{\alpha} f(x) \right| \leq \frac{1}{\Gamma (1-\alpha)} \int_x^b (g(t) - g(x))^{-\alpha} g'(t) \left| \left(f \circ g^{-1} \right)' (g(t)) \right| dt$$

$$\leq \frac{\left\| \left(f \circ g^{-1} \right)' \circ g \right\|_{\infty,[a,b]}}{\Gamma (1-\alpha)} \int_x^b (g(t) - g(x))^{-\alpha} g'(t) \, dt$$

$$= \frac{\left\| \left(f \circ g^{-1} \right)' \circ g \right\|_{\infty,[a,b]}}{\Gamma (2-\alpha)} (g(b) - g(x))^{1-\alpha}. \tag{8.32}$$

That is

$$\left| D_{b-;g}^{\alpha} f(x) \right| \leq \frac{\left\| \left(f \circ g^{-1} \right)' \circ g \right\|_{\infty,[a,b]}}{\Gamma (2-\alpha)} (g(b) - g(x))^{1-\alpha} < +\infty, \tag{8.33}$$

$\forall\, x \in [a, b]$.

Clearly it holds $\left(D_{b-;g}^{\alpha}f\right)(b)=0.$
In particular we have

$$\left(D_{x_0-;g}^{\alpha}f\right)(x_0)=0,\ \forall\, x_0\in[a,b]. \tag{8.34}$$

We define

$$\left(D_{x_0-;g}^{\alpha}f\right)(x)=0,\ \text{ for } x_0<x\leq b. \tag{8.35}$$

From Theorem 8.7 we have

Theorem 8.10 *Let* $0<\alpha<1$, $g\in C^1([a,b])$ *and strictly increasing, so that* $g^{-1}\in C^1([g(a),g(b)])$. *Assume that* $F_k:=D_{x_0-;g}^{k\alpha}f$, *for* $k=0,1,\ldots,n+1$, $n\in\mathbb{N}$ *fulfill:* $F_k\in C^1([a,x_0])$, *where* $x_0\in[a,b]$ *is fixed. Then*

$$f(x)-f(x_0)=\sum_{i=2}^{n}\frac{(g(x_0)-g(x))^{i\alpha}}{\Gamma(i\alpha+1)}\left(D_{x_0-;g}^{i\alpha}f\right)(x_0)+$$

$$\frac{1}{\Gamma((n+1)\alpha)}\int_x^{x_0}(g(t)-g(x))^{(n+1)\alpha-1}g'(t)\left(D_{x_0-;g}^{(n+1)\alpha}f\right)(t)\,dt, \tag{8.36}$$

$\forall\, x\in[a,x_0]$.

From Theorem 8.8 we have

Theorem 8.11 *Let* $0<\alpha<1$, $g\in C^1([a,b])$ *and strictly increasing, so that* $g^{-1}\in C^1([g(a),g(b)])$. *Assume that* $F_k^*:=D_{x_0+;g}^{k\alpha}f$, *for* $k=0,1,\ldots,n+1$, $n\in\mathbb{N}$ *fulfill:* $F_k^*\in C^1([x_0,b])$, *where* $x_0\in[a,b]$ *is fixed. Then*

$$f(x)-f(x_0)=\sum_{i=2}^{n}\frac{(g(x)-g(x_0))^{i\alpha}}{\Gamma(i\alpha+1)}\left(D_{x_0+;g}^{i\alpha}f\right)(x_0)+$$

$$\frac{1}{\Gamma((n+1)\alpha)}\int_{x_0}^{x}(g(x)-g(t))^{(n+1)\alpha-1}g'(t)\left(D_{x_0+;g}^{(n+1)\alpha}f\right)(t)\,dt, \tag{8.37}$$

$\forall\, x\in[x_0,b]$.

We make

Remark 8.12 All as in Theorem 8.10 and $\left(D_{x_0-;g}^{i\alpha}f\right)(x_0)=0$, for $i=2,\ldots,n\in\mathbb{N}$, then we get

$$f(x)-f(x_0)=\frac{1}{\Gamma((n+1)\alpha)}\int_x^{x_0}(g(t)-g(x))^{(n+1)\alpha-1}g'(t)\left(D_{x_0-;g}^{(n+1)\alpha}f\right)(t)\,dt, \tag{8.38}$$

$\forall\, x\in[a,x_0]$.

Next, all as in Theorem 8.11 and $\left(D_{x_0+;g}^{i\alpha}f\right)(x_0) = 0$, for $i = 2, \ldots, n \in \mathbb{N}$, then we get

$$f(x) - f(x_0) = \frac{1}{\Gamma((n+1)\alpha)} \int_{x_0}^{x} (g(x) - g(t))^{(n+1)\alpha-1} g'(t) \left(D_{x_0+;g}^{(n+1)\alpha} f\right)(t)\, dt, \tag{8.39}$$

$\forall\, x \in [x_0, b]$.

We need

Definition 8.13 Let $D_{x_0;g}^{(n+1)\alpha} f$ denote any of $D_{x_0-;g}^{(n+1)\alpha} f$, $D_{x_0+;g}^{(n+1)\alpha} f$, $n \in \mathbb{N}$ and $\delta > 0$. We set

$$\omega_1\left(\left(D_{x_0;g}^{(n+1)\alpha} f\right) \circ g^{-1}, \delta\right) :=$$

$$\max\left\{\omega_1\left(\left(D_{x_0-;g}^{(n+1)\alpha} f\right) \circ g^{-1}, \delta\right)_{[g(a),g(x_0)]}, \omega_1\left(\left(D_{x_0+;g}^{(n+1)\alpha} f\right) \circ g^{-1}, \delta\right)_{[g(x_0),g(b)]}\right\}, \tag{8.40}$$

where $x_0 \in [a, b]$. Here the moduli of continuity are considered over $[g(a), g(x_0)]$ and $[g(x_0), g(b)]$, respectively.

8.2 Main Results

We will use the following important result.

Theorem 8.14 *Let $0 < \alpha < 1$, $f, g \in C^1([a, b])$, g is strictly increasing and $g^{-1} \in C^1([g(a), g(b)])$. Assume that $D_{x_0-;g}^{k\alpha} f \in C^1([a, x_0])$ and $D_{x_0+;g}^{k\alpha} f \in C^1([x_0, b])$, for $k = 1, \ldots, n+1$; where $x_0 \in [a, b]$ is fixed. Further assume that $\left(D_{x_0\pm;g}^{i\alpha} f\right)(x_0) = 0$, $i = 2, \ldots, n+1$. Then*

$$|f(x) - f(x_0)| \leq \frac{\omega_1\left(\left(D_{x_0;g}^{(n+1)\alpha} f\right) \circ g^{-1}, \delta\right)}{\Gamma((n+1)\alpha + 1)} \tag{8.41}$$

$$\left[|g(x) - g(x_0)|^{(n+1)\alpha} + \frac{|g(x) - g(x_0)|^{(n+1)\alpha+1}}{\delta((n+1)\alpha + 1)}\right],$$

$\forall\, x \in [a, b]$, $\delta > 0$.

Proof By $\left(D_{x_0+;g}^{i\alpha} f\right)(x_0) = 0$, for $i = 2, \ldots, n+1$, we have

$$f(x) - f(x_0) = \frac{1}{\Gamma((n+1)\alpha)} \int_{x_0}^{x} (g(x) - g(t))^{(n+1)\alpha-1} g'(t)$$

$$\left[\left(D_{x_0+;g}^{(n+1)\alpha} f\right)(t) - \left(D_{x_0+;g}^{(n+1)\alpha} f\right)(x_0)\right] dt, \tag{8.42}$$

$\forall\, x \in [x_0, b]$.

Hence ($z := g(t)$)

$$f(x) - f(x_0) = \frac{1}{\Gamma((n+1)\alpha)} \int_{g(x_0)}^{g(x)} (g(x) - z)^{(n+1)\alpha - 1}$$

$$\left[\left(\left(D_{x_0+;g}^{(n+1)\alpha} f\right) \circ g^{-1}\right)(z) - \left(\left(D_{x_0+;g}^{(n+1)\alpha} f\right) \circ g^{-1}\right)(g(x_0))\right] dz, \tag{8.43}$$

$\forall\, x \in [x_0, b]$.

By $\left(D_{x_0-;g}^{i\alpha} f\right)(x_0) = 0$, for $i = 2, \ldots, n+1 \in \mathbb{N}$, we have

$$f(x) - f(x_0) = \frac{1}{\Gamma((n+1)\alpha)} \int_x^{x_0} (g(t) - g(x))^{(n+1)\alpha - 1} g'(t)$$

$$\left[\left(D_{x_0-;g}^{(n+1)\alpha} f\right)(t) - \left(D_{x_0-;g}^{(n+1)\alpha} f\right)(x_0)\right] dt, \tag{8.44}$$

$\forall\, x \in [a, x_0]$.

Hence ($z := g(t)$)

$$f(x) - f(x_0) = \frac{1}{\Gamma((n+1)\alpha)} \int_{g(x)}^{g(x_0)} (z - g(x))^{(n+1)\alpha - 1}$$

$$\left[\left(\left(D_{x_0-;g}^{(n+1)\alpha} f\right) \circ g^{-1}\right)(z) - \left(\left(D_{x_0-;g}^{(n+1)\alpha} f\right) \circ g^{-1}\right)(g(x_0))\right] dz, \tag{8.45}$$

$\forall\, x \in [a, x_0]$.

We have that ($x_0 \leq x \leq b$)

$$|f(x) - f(x_0)| \leq \frac{1}{\Gamma((n+1)\alpha)} \int_{g(x_0)}^{g(x)} (g(x) - z)^{(n+1)\alpha - 1}$$

$$\left|\left(\left(D_{x_0+;g}^{(n+1)\alpha} f\right) \circ g^{-1}\right)(z) - \left(\left(D_{x_0+;g}^{(n+1)\alpha} f\right) \circ g^{-1}\right)(g(x_0))\right| dz \underset{(\delta_1 > 0)}{\leq}$$

$$\frac{1}{\Gamma((n+1)\alpha)} \int_{g(x_0)}^{g(x)} (g(x) - z)^{(n+1)\alpha - 1} \tag{8.46}$$

$$\omega_1\left(\left(D_{x_0+;g}^{(n+1)\alpha} f\right) \circ g^{-1}, \frac{\delta_1 |z - g(x_0)|}{\delta_1}\right)_{[g(x_0), g(b)]} dz \leq$$

$$\frac{\omega_1\left(\left(D_{x_0+;g}^{(n+1)\alpha} f\right)\circ g^{-1}, \delta_1\right)_{[g(x_0),g(b)]}}{\Gamma\left((n+1)\alpha\right)}$$

$$\int_{g(x_0)}^{g(x)} (g(x)-z)^{(n+1)\alpha-1}\left(1+\frac{(z-g(x_0))}{\delta_1}\right)dz =$$

$$\frac{\omega_1\left(\left(D_{x_0+;g}^{(n+1)\alpha} f\right)\circ g^{-1}, \delta_1\right)_{[g(x_0),g(b)]}}{\Gamma\left((n+1)\alpha\right)}$$

$$\left[\frac{(g(x)-g(x_0))^{(n+1)\alpha}}{(n+1)\alpha}+\frac{1}{\delta_1}\int_{g(x_0)}^{g(x)} (g(x)-z)^{(n+1)\alpha-1}(z-g(x_0))^{2-1}dz\right] = \tag{8.47}$$

$$\frac{\omega_1\left(\left(D_{x_0+;g}^{(n+1)\alpha} f\right)\circ g^{-1}, \delta_1\right)_{[g(x_0),g(b)]}}{\Gamma\left((n+1)\alpha\right)}$$

$$\left[\frac{(g(x)-g(x_0))^{(n+1)\alpha}}{(n+1)\alpha}+\frac{1}{\delta_1}\frac{\Gamma\left((n+1)\alpha\right)\Gamma(2)}{\Gamma\left((n+1)\alpha+2\right)}(g(x)-g(x_0))^{(n+1)\alpha+1}\right] =$$

$$\frac{\omega_1\left(\left(D_{x_0+;g}^{(n+1)\alpha} f\right)\circ g^{-1}, \delta_1\right)_{[g(x_0),g(b)]}}{\Gamma\left((n+1)\alpha\right)} \tag{8.48}$$

$$\left[\frac{(g(x)-g(x_0))^{(n+1)\alpha}}{(n+1)\alpha}+\frac{1}{\delta_1}\frac{(g(x)-g(x_0))^{(n+1)\alpha+1}}{(n+1)\alpha\left((n+1)\alpha+1\right)}\right].$$

We have proved that

$$|f(x)-f(x_0)| \leq \frac{\omega_1\left(\left(D_{x_0+;g}^{(n+1)\alpha} f\right)\circ g^{-1}, \delta_1\right)_{[g(x_0),g(b)]}}{\Gamma\left((n+1)\alpha+1\right)} \tag{8.49}$$

$$\left[(g(x)-g(x_0))^{(n+1)\alpha}+\frac{(g(x)-g(x_0))^{(n+1)\alpha+1}}{\delta_1\left((n+1)\alpha+1\right)}\right],$$

$\forall\, x \in [x_0, b]$, $\delta_1 > 0$.

We have that $(a \leq x \leq x_0)$

$$|f(x)-f(x_0)| \overset{(8.44)}{\leq} \frac{1}{\Gamma\left((n+1)\alpha\right)}\int_{g(x)}^{g(x_0)} (z-g(x))^{(n+1)\alpha-1}$$

$$\left|\left(\left(D_{x_0-;g}^{(n+1)\alpha}f\right)\circ g^{-1}\right)(z)-\left(\left(D_{x_0-;g}^{(n+1)\alpha}f\right)\circ g^{-1}\right)(g(x_0))\right|dz \underset{(\delta_2>0)}{\leq}$$

$$\frac{1}{\Gamma((n+1)\alpha)}\int_{g(x)}^{g(x_0)}(z-g(x))^{(n+1)\alpha-1}$$

$$\omega_1\left(\left(D_{x_0-;g}^{(n+1)\alpha}f\right)\circ g^{-1},\frac{\delta_2\left|z-g(x_0)\right|}{\delta_2}\right)_{[g(a),g(x_0)]}dz\leq \tag{8.50}$$

$$\frac{\omega_1\left(\left(D_{x_0-;g}^{(n+1)\alpha}f\right)\circ g^{-1},\delta_2\right)_{[g(a),g(x_0)]}}{\Gamma((n+1)\alpha)}$$

$$\int_{g(x)}^{g(x_0)}(z-g(x))^{(n+1)\alpha-1}\left(1+\frac{g(x_0)-z}{\delta_2}\right)dz=$$

$$\frac{\omega_1\left(\left(D_{x_0-;g}^{(n+1)\alpha}f\right)\circ g^{-1},\delta_2\right)_{[g(a),g(x_0)]}}{\Gamma((n+1)\alpha)}$$

$$\left[\frac{(g(x_0)-g(x))^{(n+1)\alpha}}{(n+1)\alpha}+\frac{1}{\delta_2}\int_{g(x)}^{g(x_0)}(g(x_0)-z)^{2-1}(z-g(x))^{(n+1)\alpha-1}dz\right]= \tag{8.51}$$

$$\frac{\omega_1\left(\left(D_{x_0-;g}^{(n+1)\alpha}f\right)\circ g^{-1},\delta_2\right)_{[g(a),g(x_0)]}}{\Gamma((n+1)\alpha)}$$

$$\left[\frac{(g(x_0)-g(x))^{(n+1)\alpha}}{(n+1)\alpha}+\frac{1}{\delta_2}\frac{\Gamma(2)\Gamma((n+1)\alpha)}{\Gamma((n+1)\alpha+2)}(g(x_0)-g(x))^{(n+1)\alpha+1}\right]=$$

$$\frac{\omega_1\left(\left(D_{x_0-;g}^{(n+1)\alpha}f\right)\circ g^{-1},\delta_2\right)_{[g(a),g(x_0)]}}{\Gamma((n+1)\alpha)}$$

$$\left[\frac{(g(x_0)-g(x))^{(n+1)\alpha}}{(n+1)\alpha}+\frac{1}{\delta_2}\frac{(g(x_0)-g(x))^{(n+1)\alpha+1}}{(n+1)\alpha((n+1)\alpha+1)}\right]= \tag{8.52}$$

$$\frac{\omega_1\left(\left(D_{x_0-;g}^{(n+1)\alpha}f\right)\circ g^{-1},\delta_2\right)_{[g(a),g(x_0)]}}{\Gamma((n+1)\alpha+1)}$$

$$\left[(g(x_0)-g(x))^{(n+1)\alpha}+\frac{(g(x_0)-g(x))^{(n+1)\alpha+1}}{\delta_2((n+1)\alpha+1)}\right].$$

We have proved that

$$|f(x)-f(x_0)| \leq \frac{\omega_1\left(\left(D_{x_0-;g}^{(n+1)\alpha}f\right)\circ g^{-1},\delta_2\right)_{[g(a),g(x_0)]}}{\Gamma((n+1)\alpha+1)}$$

$$\left[(g(x_0)-g(x))^{(n+1)\alpha}+\frac{(g(x_0)-g(x))^{(n+1)\alpha+1}}{\delta_2((n+1)\alpha+1)}\right], \tag{8.53}$$

$\forall\, x \in [a, x_0]$, $\delta_2 > 0$.

By (8.49) and (8.53), setting $\delta = \delta_1 = \delta_2 > 0$, we derive (8.41). ■

We state

Corollary 8.15 (to Theorem 8.14) *All as in Theorem 8.14. Then*

$$|f(\cdot)-f(x_0)| \leq \frac{\omega_1\left(\left(D_{x_0;g}^{(n+1)\alpha}f\right)\circ g^{-1},\delta\right)}{\Gamma((n+1)\alpha+1)}$$

$$\left[|g(\cdot)-g(x_0)|^{(n+1)\alpha}+\frac{|g(\cdot)-g(x_0)|^{(n+1)\alpha+1}}{\delta((n+1)\alpha+1)}\right], \tag{8.54}$$

$\delta > 0$, *true over* $[a, b]$.

Our first result on the approximation by positive sublinear operators follows:

Theorem 8.16 *Let* $0 < \alpha < 1$, $f \in C_+([a, b])$, $f, g \in C^1([a, b])$, *g is strictly increasing and* $g^{-1} \in C^1([g(a), g(b)])$. *Assume that* $D_{x_0-;g}^{k\alpha}f \in C^1([a, x_0])$ *and* $D_{x_0+;g}^{k\alpha}f \in C^1([x_0, b])$, *for* $k = 1, \ldots, n+1$; *where* $x_0 \in [a, b]$ *is fixed. Further assume that* $\left(D_{x_0\pm;g}^{i\alpha}f\right)(x_0) = 0$, $i = 2, \ldots, n+1$. *Let* $L_N : C_+([a, b]) \to C_+([a, b])$, $\forall\, N \in \mathbb{N}$, *be positive sublinear operators, such that* $L_N(1) = 1$, $\forall\, N \in \mathbb{N}$. *Then*

$$|L_N(f)(x_0)-f(x_0)| \leq \frac{\omega_1\left(\left(D_{x_0;g}^{(n+1)\alpha}f\right)\circ g^{-1},\delta\right)}{\Gamma((n+1)\alpha+1)}$$

$$\left[L_N\left(|g(\cdot)-g(x_0)|^{(n+1)\alpha}\right)(x_0)+\frac{L_N\left(|g(\cdot)-g(x_0)|^{(n+1)\alpha+1}\right)(x_0)}{\delta((n+1)\alpha+1)}\right], \tag{8.55}$$

$\forall\, N \in \mathbb{N}$, *where* $\delta > 0$.

Proof By (8.7) and (8.54) we have

$$|L_N(f)(x_0)-f(x)| \leq L_N(|f(\cdot)-f(x_0)|)(x_0)$$

$$\leq \frac{\omega_1\left(\left(D_{x_0;g}^{(n+1)\alpha}f\right)\circ g^{-1},\delta\right)}{\Gamma\left((n+1)\alpha+1\right)}$$

$$\left[L_N\left(|g(\cdot)-g(x_0)|^{(n+1)\alpha}\right)(x_0)+\frac{L_N\left(|g(\cdot)-g(x_0)|^{(n+1)\alpha+1}\right)(x_0)}{\delta\left((n+1)\alpha+1\right)}\right],\ \delta>0, \tag{8.56}$$

proving the claim. ■

We give

Theorem 8.17 *All as in Theorem 8.16, and assume that* $L_N\left(|g(\cdot)-g(x_0)|^{(n+1)\alpha+1}\right)(x_0)>0$, $\forall\ N\in\mathbb{N}$. *Then*

$$|L_N(f)(x_0)-f(x)|\leq\frac{((n+1)\alpha+2)}{\Gamma\left((n+1)\alpha+2\right)}$$

$$\omega_1\left(\left(D_{x_0;g}^{(n+1)\alpha}f\right)\circ g^{-1},\left(L_N\left(|g(\cdot)-g(x_0)|^{(n+1)\alpha+1}\right)(x_0)\right)^{\frac{1}{(n+1)\alpha+1}}\right)$$

$$\left(L_N\left(|g(\cdot)-g(x_0)|^{(n+1)\alpha+1}\right)(x_0)\right)^{\frac{(n+1)\alpha}{(n+1)\alpha+1}}, \tag{8.57}$$

$\forall\ N\in\mathbb{N}$.

Proof By Theorem 8.2, see (8.4), we get

$$L_N\left(|g(\cdot)-g(x_0)|^{(n+1)\alpha}\right)(x_0)\leq\left(L_N\left(|g(\cdot)-g(x_0)|^{(n+1)\alpha+1}\right)(x_0)\right)^{\frac{(n+1)\alpha}{(n+1)\alpha+1}}. \tag{8.58}$$

Choose

$$\delta:=\left(L_N\left(|g(\cdot)-g(x_0)|^{(n+1)\alpha+1}\right)(x_0)\right)^{\frac{1}{(n+1)\alpha+1}}>0, \tag{8.59}$$

i.e.

$$\delta^{(n+1)\alpha+1}=L_N\left(|g(\cdot)-g(x_0)|^{(n+1)\alpha+1}\right)(x_0).$$

By (8.55) we obtain

$$|L_N(f)(x_0)-f(x_0)|\leq$$

$$\frac{\omega_1\left(\left(D_{x_0;g}^{(n+1)\alpha}f\right)\circ g^{-1},\left(L_N\left(|g(\cdot)-g(x_0)|^{(n+1)\alpha+1}\right)(x_0)\right)^{\frac{1}{(n+1)\alpha+1}}\right)}{\Gamma\left((n+1)\alpha+1\right)}$$

$$\left[\left(L_N\left(|g(\cdot)-g(x_0)|^{(n+1)\alpha+1}\right)(x_0)\right)^{\frac{(n+1)\alpha}{(n+1)\alpha+1}}+\frac{\delta^{(n+1)\alpha+1}}{\delta\left((n+1)\alpha+1\right)}\right]=$$

$$\frac{\omega_1\left(\left(D_{x_0;g}^{(n+1)\alpha}f\right)\circ g^{-1},\left(L_N\left(|g(\cdot)-g(x_0)|^{(n+1)\alpha+1}\right)(x_0)\right)^{\frac{1}{(n+1)\alpha+1}}\right)}{\Gamma\left((n+1)\alpha+1\right)} \tag{8.60}$$

$$\left[\left(L_N\left(|g(\cdot)-g(x_0)|^{(n+1)\alpha+1}\right)(x_0)\right)^{\frac{(n+1)\alpha}{(n+1)\alpha+1}}+\right.$$

$$\left.\frac{\left(L_N\left(|g(\cdot)-g(x_0)|^{(n+1)\alpha+1}\right)(x_0)\right)^{\frac{(n+1)\alpha}{(n+1)\alpha+1}}}{((n+1)\alpha+1)}\right]=$$

$$\frac{\omega_1\left(\left(D_{x_0;g}^{(n+1)\alpha}f\right)\circ g^{-1},\left(L_N\left(|g(\cdot)-g(x_0)|^{(n+1)\alpha+1}\right)(x_0)\right)^{\frac{1}{(n+1)\alpha+1}}\right)}{\Gamma((n+1)\alpha+1)}$$

$$\left(L_N\left(|g(\cdot)-g(x_0)|^{(n+1)\alpha+1}\right)(x_0)\right)^{\frac{(n+1)\alpha}{(n+1)\alpha+1}}\left[1+\frac{1}{(n+1)\alpha+1}\right]= \tag{8.61}$$

$$\frac{((n+1)\alpha+2)}{\Gamma((n+1)\alpha+2)}\omega_1\left(\left(D_{x_0;g}^{(n+1)\alpha}f\right)\circ g^{-1},\left(L_N\left(|g(\cdot)-g(x_0)|^{(n+1)\alpha+1}\right)(x_0)\right)^{\frac{1}{(n+1)\alpha+1}}\right)$$

$$\left(L_N\left(|g(\cdot)-g(x_0)|^{(n+1)\alpha+1}\right)(x_0)\right)^{\frac{(n+1)\alpha}{(n+1)\alpha+1}},$$

proving (8.57). ■

We make

Remark 8.18 In the terms and assumptions of Theorem 8.17: let $\left(L_N\left(|g(\cdot)-g(x_0)|^{(n+1)\alpha+1}\right)(x_0)\right)\to 0$, then $L_N(f)(x_0)\to f(x_0)$, as $N\to+\infty$.

We need

Corollary 8.19 *All as in Theorem 8.16, and assume that* $L_N\left(|\cdot-x_0|^{(n+1)\alpha+1}\right)(x_0)>0$, $\forall\, N\in\mathbb{N}$. *Then*

$$|L_N(f)(x_0)-f(x_0)|\le\frac{((n+1)\alpha+2)}{\Gamma((n+1)\alpha+2)}$$

$$\omega_1\left(\left(D_{x_0;g}^{(n+1)\alpha}f\right)\circ g^{-1},\left\|g'\right\|_{\infty,[a,b]}\left(L_N\left(|\cdot-x_0|^{(n+1)\alpha+1}\right)(x_0)\right)^{\frac{1}{(n+1)\alpha+1}}\right)$$

$$\left\|g'\right\|_{\infty,[a,b]}^{(n+1)\alpha}\left(L_N\left(|\cdot-x_0|^{(n+1)\alpha+1}\right)(x_0)\right)^{\frac{(n+1)\alpha}{(n+1)\alpha+1}}, \tag{8.62}$$

$\forall\, N\in\mathbb{N}$.

Proof By Theorem 8.17 (8.57), ω_1 and L_N properties, and that $|g(x)-g(x_0)|\le\left\|g'\right\|_{\infty,[a,b]}|x-x_0|$, along with $|g(x)-g(x_0)|\ge\left(\inf_{x\in[a,b]}\left|g'(x)\right|\right)|x-x_0|$, $\forall\, x,x_0\in[a,b]$, by $g'(x)>0$, $\forall\, x\in[a,b]$. Clearly then (8.62) is valid, and $L_N\left(|g(\cdot)-g(x_0)|^{(n+1)\alpha+1}\right)(x_0)>0$, $\forall\, N\in\mathbb{N}$. ■

8.3 Applications

(I) Case $0 < (n+1)\alpha < 1$.

Here we apply Corollary 8.19 to well known Max-product operators.

We make

Remark 8.20 The Max-product Bernstein operators $B_N^{(M)}(f)(x)$ are defined by (8.8), see also [5, p. 10]; here $f : [0,1] \to \mathbb{R}_+$ is a continuous function.

We have $B_N^{(M)}(1) = 1$, and

$$B_N^{(M)}(|\cdot - x|)(x) \leq \frac{6}{\sqrt{N+1}}, \ \forall x \in [0,1], \forall N \in \mathbb{N}, \tag{8.63}$$

see [5, p. 31].

$B_N^{(M)}$ are positive sublinear operators and thus they possess the monotonicity property, also since $|\cdot - x| \leq 1$, then $|\cdot - x|^{\beta} \leq 1, \forall x \in [0,1], \forall \beta > 0$.

Therefore it holds

$$B_N^{(M)}\left(|\cdot - x|^{1+\beta}\right)(x) \leq \frac{6}{\sqrt{N+1}}, \ \forall x \in [0,1], \forall N \in \mathbb{N}, \forall \beta > 0. \tag{8.64}$$

Furthermore, clearly it holds that

$$B_N^{(M)}\left(|\cdot - x|^{1+\beta}\right)(x) > 0, \forall N \in \mathbb{N}, \forall \beta \geq 0 \text{ and any } x \in (0,1). \tag{8.65}$$

The operator $B_N^{(M)}$ maps $C_+([0,1])$ into itself.

We present

Theorem 8.21 *Let $0 < \alpha < 1$, $f \in C_+([0,1])$, $f, g \in C^1([0,1])$, g is strictly increasing and $g^{-1} \in C^1([g(0), g(1)])$. Assume that $D_{x-;g}^{k\alpha} f \in C^1([0,x])$ and $D_{x+;g}^{k\alpha} f \in C^1([x,1])$, for $k = 1, \ldots, n+1$; where $x_0 \in (0,1)$ is fixed. Further assume that $\left(D_{x\pm;g}^{i\alpha} f\right)(x) = 0$, $i = 2, \ldots, n+1$. Then*

$$\left|B_N^{(M)}(f)(x) - f(x)\right| \leq \frac{((n+1)\alpha + 2)}{\Gamma((n+1)\alpha + 2)}$$

$$\omega_1\left(\left(D_{x;g}^{(n+1)\alpha} f\right) \circ g^{-1}, \|g'\|_{\infty,[0,1]}\left(\frac{6}{\sqrt{N+1}}\right)^{\frac{1}{(n+1)\alpha+1}}\right) \tag{8.66}$$

$$\|g'\|_{\infty,[0,1]}^{(n+1)\alpha}\left(\frac{6}{\sqrt{N+1}}\right)^{\frac{(n+1)\alpha}{(n+1)\alpha+1}}, \ \forall N \in \mathbb{N}.$$

As $N \to +\infty$, we get $\left(B_N^{(M)}(f)\right)(x) \to f(x)$.

Proof By Corollary 8.19. ■

We continue with

Remark 8.22 The truncated Favard–Szász–Mirakjan operators are given by

$$T_N^{(M)}(f)(x) = \frac{\bigvee_{k=0}^{N} s_{N,k}(x) f\left(\frac{k}{N}\right)}{\bigvee_{k=0}^{N} s_{N,k}(x)}, \ x \in [0,1], \ N \in \mathbb{N}, \ f \in C_+([0,1]), \tag{8.67}$$

$s_{N,k}(x) = \frac{(Nx)^k}{k!}$, see also [5, p. 11].

By [5, pp. 178–179], we get that

$$T_N^{(M)}(|\cdot - x|)(x) \leq \frac{3}{\sqrt{N}}, \ \forall x \in [0,1], \ \forall N \in \mathbb{N}. \tag{8.68}$$

Clearly it holds

$$T_N^{(M)}\left(|\cdot - x|^{1+\beta}\right)(x) \leq \frac{3}{\sqrt{N}}, \ \forall x \in [0,1], \ \forall N \in \mathbb{N}, \forall \beta > 0. \tag{8.69}$$

The operators $T_N^{(M)}$ are positive sublinear operators mapping $C_+([0,1])$ into itself, with $T_N^{(M)}(1) = 1$.

Furthermore it holds

$$T_N^{(M)}\left(|\cdot - x|^{\lambda}\right)(x) = \frac{\bigvee_{k=0}^{N} \frac{(Nx)^k}{k!} \left|\frac{k}{N} - x\right|^{\lambda}}{\bigvee_{k=0}^{N} \frac{(Nx)^k}{k!}} > 0, \ \forall x \in (0,1], \ \forall \lambda \geq 1, \forall N \in \mathbb{N}. \tag{8.70}$$

We give

Theorem 8.23 *All as in Theorem 8.21, but now* $x \in (0,1]$. *Then*

$$\left|T_N^{(M)}(f)(x) - f(x)\right| \leq \frac{((n+1)\alpha + 2)}{\Gamma((n+1)\alpha + 2)}$$

$$\omega_1\left(\left(D_{x;g}^{(n+1)\alpha} f\right) \circ g^{-1}, \left\|g'\right\|_{\infty,[0,1]} \left(\frac{3}{\sqrt{N}}\right)^{\frac{1}{(n+1)\alpha+1}}\right)$$

$$\left\|g'\right\|_{\infty,[0,1]}^{(n+1)\alpha} \left(\frac{3}{\sqrt{N}}\right)^{\frac{(n+1)\alpha}{(n+1)\alpha+1}}, \ \forall N \in \mathbb{N}. \tag{8.71}$$

As $N \to +\infty$, *we get* $\left(T_N^{(M)}(f)\right)(x) \to f(x)$.

Proof By Corollary 8.19. ∎

We continue with

Remark 8.24 Next we study the truncated Max-product Baskakov operators (see [5, p. 11])

$$U_N^{(M)}(f)(x) = \frac{\bigvee_{k=0}^{N} b_{N,k}(x) f\left(\frac{k}{N}\right)}{\bigvee_{k=0}^{N} b_{N,k}(x)}, \ x \in [0,1], \ f \in C_+([0,1]), \ N \in \mathbb{N}, \tag{8.72}$$

where

$$b_{N,k}(x) = \binom{N+k-1}{k} \frac{x^k}{(1+x)^{N+k}}. \tag{8.73}$$

From [5, pp. 217–218], we get ($x \in [0,1]$)

$$\left(U_N^{(M)}(|\cdot - x|)\right)(x) \leq \frac{2\sqrt{3}\left(\sqrt{2}+2\right)}{\sqrt{N+1}}, \ N \geq 2, N \in \mathbb{N}. \tag{8.74}$$

Let $\lambda \geq 1$, clearly then it holds

$$\left(U_N^{(M)}\left(|\cdot - x|^{\lambda}\right)\right)(x) \leq \frac{2\sqrt{3}\left(\sqrt{2}+2\right)}{\sqrt{N+1}}, \ \forall N \geq 2, N \in \mathbb{N}. \tag{8.75}$$

Also it holds $U_N^{(M)}(1) = 1$, and $U_N^{(M)}$ are positive sublinear operators from $C_+([0,1])$ into itself. Furthermore it holds

$$U_N^{(M)}\left(|\cdot - x|^{\lambda}\right)(x) > 0, \ \forall x \in (0,1], \ \forall \lambda \geq 1, \forall N \in \mathbb{N}. \tag{8.76}$$

We present

Theorem 8.25 *Here all as in Theorem 8.23. Then*

$$\left|U_N^{(M)}(f)(x) - f(x)\right| \leq \frac{((n+1)\alpha+2)}{\Gamma((n+1)\alpha+2)}$$

$$\omega_1\left(\left(D_{x;g}^{(n+1)\alpha} f\right) \circ g^{-1}, \left\|g'\right\|_{\infty,[0,1]} \left(\frac{2\sqrt{3}\left(\sqrt{2}+2\right)}{\sqrt{N+1}}\right)^{\frac{1}{(n+1)\alpha+1}}\right)$$

$$\left\|g'\right\|_{\infty,[0,1]}^{(n+1)\alpha} \left(\frac{2\sqrt{3}\left(\sqrt{2}+2\right)}{\sqrt{N+1}}\right)^{\frac{(n+1)\alpha}{(n+1)\alpha+1}}, \ \forall N \geq 2, \ N \in \mathbb{N}. \tag{8.77}$$

As $N \to +\infty$, *we get* $U_N^{(M)}(f)(x) \to f(x)$.

Proof By Corollary 8.19. ■

We give

Remark 8.26 Here we study the Max-product Meyer-Köning and Zeller operators (see [5, p. 11]) defined by

$$Z_N^{(M)}(f)(x) = \frac{\bigvee_{k=0}^{\infty} s_{N,k}(x) f\left(\frac{k}{N+k}\right)}{\bigvee_{k=0}^{\infty} s_{N,k}(x)}, \forall N \in \mathbb{N}, f \in C_+([0,1]), \tag{8.78}$$

$s_{N,k}(x) = \binom{N+k}{k} x^k$, $x \in [0,1]$.

By [5, p. 253], we get that

$$Z_N^{(M)}(|\cdot - x|)(x) \leq \frac{8\left(1+\sqrt{5}\right)}{3} \frac{\sqrt{x}(1-x)}{\sqrt{N}}, \forall x \in [0,1], \forall N \geq 4, N \in \mathbb{N}. \tag{8.79}$$

As before we get that (for $\lambda \geq 1$)

$$Z_N^{(M)}\left(|\cdot - x|^{\lambda}\right)(x) \leq \frac{8\left(1+\sqrt{5}\right)}{3} \frac{\sqrt{x}(1-x)}{\sqrt{N}} := \rho(x), \tag{8.80}$$

$\forall x \in [0,1]$, $N \geq 4$, $N \in \mathbb{N}$.

Also it holds $Z_N^{(M)}(1) = 1$, and $Z_N^{(M)}$ are positive sublinear operators from $C_+([0,1])$ into itself. Also it holds

$$Z_N^{(M)}\left(|\cdot - x|^{\lambda}\right)(x) > 0, \forall x \in (0,1), \forall \lambda \geq 1, \forall N \in \mathbb{N}. \tag{8.81}$$

We give

Theorem 8.27 *Here all as in Theorem 8.21. Then*

$$\left|Z_N^{(M)}(f)(x) - f(x)\right| \leq \frac{((n+1)\alpha+2)}{\Gamma((n+1)\alpha+2)}$$

$$\omega_1\left(\left(D_{x;g}^{(n+1)\alpha} f\right) \circ g^{-1}, \|g'\|_{\infty,[0,1]} (\rho(x))^{\frac{1}{(n+1)\alpha+1}}\right)$$

$$\|g'\|_{\infty,[0,1]}^{(n+1)\alpha} (\rho(x))^{\frac{(n+1)\alpha}{(n+1)\alpha+1}}, \forall N \geq 4, N \in \mathbb{N}. \tag{8.82}$$

As $N \to +\infty$, we get $Z_N^{(M)}(f)(x) \to f(x)$.

Proof By Corollary 8.19. ■

(II) Case of $(n+1)\alpha > 1$.

We need

Corollary 8.28 *All as in Theorem 8.16. Then*

$$\left|L_N(f)(x_0) - f(x_0)\right| \leq \frac{\omega_1\left(\left(D_{x_0;g}^{(n+1)\alpha} f\right) \circ g^{-1}, \delta\right)}{\Gamma((n+1)\alpha+1)}$$

$$\left\|g'\right\|_{\infty,[a,b]}^{(n+1)\alpha} \left[L_N\left(\left|\cdot - x_0\right|^{(n+1)\alpha}\right)(x_0) + \frac{\left\|g'\right\|_{\infty,[a,b]} L_N\left(\left|\cdot - x_0\right|^{(n+1)\alpha+1}\right)(x_0)}{\delta((n+1)\alpha+1)}\right], \tag{8.83}$$

$\forall N \in \mathbb{N}$, *where* $\delta > 0$.

Proof By Theorem 8.16, and $|g(x) - g(x_0)| \leq \left\|g'\right\|_{\infty,[a,b]} |x - x_0|, \forall x, x_0 \in [a,b]$ and the properties of L_N. ∎

Applications follow:
We give

Theorem 8.29 *Let* $0 < \alpha < 1$, *and* $(n+1)\alpha > 1$, $n \in \mathbb{N}$, $f \in C_+([0,1])$, $f, g \in C^1([0,1])$, g *is strictly increasing and* $g^{-1} \in C^1([g(0), g(1)])$. *Assume that* $D_{x_0-;g}^{k\alpha} f \in C^1([0, x_0])$ *and* $D_{x_0+;g}^{k\alpha} f \in C^1([x_0, 1])$, *for* $k = 1, \ldots, n+1$; *where* $x_0 \in [0,1]$ *is fixed. Further assume that* $\left(D_{x_0\pm;g}^{i\alpha} f\right)(x_0) = 0$, $i = 2, \ldots, n+1$. *Then*

$$\left|B_N^{(M)}(f)(x_0) - f(x_0)\right| \leq \frac{\omega_1\left(\left(D_{x_0;g}^{(n+1)\alpha} f\right) \circ g^{-1}, \left(\frac{6}{\sqrt{N+1}}\right)^{\frac{1}{(n+1)\alpha+1}}\right)}{\Gamma(\alpha+1)} \tag{8.84}$$

$$\left\|g'\right\|_{\infty,[0,1]}^{(n+1)\alpha} \left[\frac{6}{\sqrt{N+1}} + \frac{\left\|g'\right\|_{\infty,[0,1]} \left(\frac{6}{\sqrt{N+1}}\right)^{\frac{(n+1)\alpha}{(n+1)\alpha+1}}}{((n+1)\alpha+1)}\right],$$

$\forall N \in \mathbb{N}$.

As $N \to +\infty$, *we get* $B_N^{(M)}(f)(x_0) \to f(x_0)$.

Proof By (8.83) we have

$$\left|B_N^{(M)}(f)(x_0) - f(x_0)\right| \leq \frac{\omega_1\left(\left(D_{x_0;g}^{(n+1)\alpha} f\right) \circ g^{-1}, \delta\right)}{\Gamma((n+1)\alpha+1)} \left\|g'\right\|_{\infty,[0,1]}^{(n+1)\alpha}$$

$$\left[B_N^{(M)}\left(\left|\cdot - x\right|^{(n+1)\alpha}\right)(x_0) + \frac{\left\|g'\right\|_{\infty,[0,1]} B_N^{(M)}\left(\left|\cdot - x_0\right|^{(n+1)\alpha+1}\right)(x_0)}{\delta((n+1)\alpha+1)}\right] \overset{(8.64)}{\leq}$$

$$\frac{\omega_1\left(\left(D_{x_0;g}^{(n+1)\alpha} f\right) \circ g^{-1}, \delta\right)}{\Gamma((n+1)\alpha+1)} \left\|g'\right\|_{\infty,[0,1]}^{(n+1)\alpha}$$

$$\left[\frac{6}{\sqrt{N+1}}+\frac{\left\|g'\right\|_{\infty,[0,1]}\frac{6}{\sqrt{N+1}}}{\delta\left((n+1)\alpha+1\right)}\right]=:(*)\,. \tag{8.85}$$

Choose $\delta=\left(\frac{6}{\sqrt{N+1}}\right)^{\frac{1}{(n+1)\alpha+1}}$, then $\delta^{(n+1)\alpha+1}=\frac{6}{\sqrt{N+1}}$, and apply it to (8.85).

We get that

$$(*)=\frac{\omega_1\left(\left(D_{x_0;g}^{(n+1)\alpha}f\right)\circ g^{-1},\left(\frac{6}{\sqrt{N+1}}\right)^{\frac{1}{(n+1)\alpha+1}}\right)}{\Gamma\left((n+1)\alpha+1\right)}\left\|g'\right\|_{\infty,[0,1]}^{(n+1)\alpha}$$

$$\left[\frac{6}{\sqrt{N+1}}+\frac{\left\|g'\right\|_{\infty,[0,1]}\delta^{(n+1)\alpha+1}}{\delta\left((n+1)\alpha+1\right)}\right]= \tag{8.86}$$

$$\frac{\omega_1\left(\left(D_{x_0;g}^{(n+1)\alpha}f\right)\circ g^{-1},\left(\frac{6}{\sqrt{N+1}}\right)^{\frac{1}{(n+1)\alpha+1}}\right)}{\Gamma\left((n+1)\alpha+1\right)}\left\|g'\right\|_{\infty,[0,1]}^{(n+1)\alpha}$$

$$\left[\frac{6}{\sqrt{N+1}}+\frac{\left\|g'\right\|_{\infty,[0,1]}\left(\frac{6}{\sqrt{N+1}}\right)^{\frac{(n+1)\alpha}{(n+1)\alpha+1}}}{\left((n+1)\alpha+1\right)}\right],$$

$\forall\, N\in\mathbb{N}$, proving the claim. ∎

Next comes

Theorem 8.30 *All as in Theorem 8.29. Then*

$$\left|T_N^{(M)}(f)(x_0)-f(x_0)\right|\leq\frac{\omega_1\left(\left(D_{x_0;g}^{(n+1)\alpha}f\right)\circ g^{-1},\left(\frac{3}{\sqrt{N}}\right)^{\frac{1}{(n+1)\alpha+1}}\right)}{\Gamma\left((n+1)\alpha+1\right)}$$

$$\left\|g'\right\|_{\infty,[0,1]}^{(n+1)\alpha}\left[\frac{3}{\sqrt{N}}+\frac{\left\|g'\right\|_{\infty,[0,1]}}{\left((n+1)\alpha+1\right)}\left(\frac{3}{\sqrt{N}}\right)^{\frac{(n+1)\alpha}{(n+1)\alpha+1}}\right], \tag{8.87}$$

$\forall\, N\in\mathbb{N}$.

We get $\lim_{N\to+\infty}T_N^{(M)}(f)(x_0)=f(x_0)$.

Proof Similar to Theorem 8.29. ∎

Next follows:

Theorem 8.31 *All as in Theorem 8.29. Then*

$$\left|U_N^{(M)}(f)(x_0) - f(x_0)\right| \leq$$

$$\frac{\omega_1\left(\left(D_{x_0;g}^{(n+1)\alpha} f\right) \circ g^{-1}, \left(\frac{2\sqrt{3}(\sqrt{2}+2)}{\sqrt{N+1}}\right)^{\frac{1}{(n+1)\alpha+1}}\right) \|g'\|_{\infty,[0,1]}^{(n+1)\alpha}}{\Gamma((n+1)\alpha+1)}$$

$$\left[\frac{2\sqrt{3}\left(\sqrt{2}+2\right)}{\sqrt{N+1}} + \frac{\|g'\|_{\infty,[0,1]}}{((n+1)\alpha+1)}\left(\frac{2\sqrt{3}\left(\sqrt{2}+2\right)}{\sqrt{N+1}}\right)^{\frac{(n+1)\alpha}{(n+1)\alpha+1}}\right], \tag{8.88}$$

$\forall\, N \geq 2$, $N \in \mathbb{N}$.

We get $\lim_{N\to+\infty} U_N^{(M)}(f)(x_0) = f(x_0)$.

Proof Similar to Theorem 8.29. ∎

We continue with

Theorem 8.32 *All as in Theorem 8.29. Then*

$$\left|Z_N^{(M)}(f)(x_0) - f(x_0)\right| \leq \frac{\omega_1\left(\left(D_{x_0;g}^{(n+1)\alpha} f\right) \circ g^{-1}, (\rho(x))^{\frac{1}{(n+1)\alpha+1}}\right) \|g'\|_{\infty,[0,1]}^{(n+1)\alpha}}{\Gamma((n+1)\alpha+1)}$$

$$\left[\rho(x) + \frac{\|g'\|_{\infty,[0,1]}}{((n+1)\alpha+1)}(\rho(x))^{\frac{(n+1)\alpha}{(n+1)\alpha+1}}\right], \tag{8.89}$$

$\forall\, N \in \mathbb{N}$, $N \geq 4$.

We get $\lim_{N\to+\infty} Z_N^{(M)}(f)(x_0) = f(x_0)$, *where* $\rho(x)$ *is as in (8.80).*

Proof Similar to Theorem 8.29. ∎

We finish with

Corollary 8.33 (to Theorem 8.29) *Let* $0 < \alpha < 1$, *and* $(n+1)\alpha > 1$, $n \in \mathbb{N}$, $f \in C_+([0,1])$ *and* $f \in C^1([0,1])$. *Assume that* $D^{k\alpha}_{x_0-;e^x} f \in C^1([0,x_0])$ *and* $D^{k\alpha}_{x_0+;e^x} f \in C^1([x_0,1])$, *for* $k = 1, \ldots, n+1$; *where* $x_0 \in [0,1]$ *is fixed. Further assume that* $\left(D^{i\alpha}_{x_0\pm;e^x} f\right)(x_0) = 0$, $i = 2, \ldots, n+1$. *Then*

$$\left|B_N^{(M)}(f)(x_0) - f(x_0)\right| \leq \frac{\omega_1\left(\left(D^{(n+1)\alpha}_{x_0;e^x} f\right) \circ \ln t, \left(\frac{6}{\sqrt{N+1}}\right)^{\frac{1}{(n+1)\alpha+1}}\right) e^{(n+1)\alpha}}{\Gamma((n+1)\alpha+1)}$$

$$\left[\frac{6}{\sqrt{N+1}} + \frac{e\left(\frac{6}{\sqrt{N+1}}\right)^{\frac{(n+1)\alpha}{(n+1)\alpha+1}}}{((n+1)\alpha+1)}\right], \; \forall N \in \mathbb{N}. \tag{8.90}$$

It holds $\lim_{N \to +\infty} B_N^{(M)}(f)(x_0) = f(x_0)$.

Proof By Theorem 8.29 and $g(x) = e^x$. ■

References

1. Anastassiou, G.: Advanced fractional Taylor's formulae. J. Comput. Anal. Appl. **21**(7), 1185–1204 (2016)
2. Anastassiou, G.A.: Nonlinearity: Ordinary and Fractional Approximations by Sublinear and Max-Product Operators. Springer, Heidelberg (2018)
3. Anastassiou, G.: Generalized *g-iterated fractional approximations by sublinear operators*. Appl. Math. (2020). Accepted
4. Anastassiou, G., Argyros, I.: Intelligent Numerical Methods: Applications to Fractional Calculus. Springer, Heidelberg (2016)
5. Bede, B., Coroianu, L., Gal, S.: Approximation by Max-Product type Operators. Springer, Heidelberg (2016)
6. Royden, H.L.: Real Analysis, 2nd edn. Macmillan, New York (1968)

Chapter 9
Generalized g-Fractional Vector Representation Formula And Bochner Integral Type Inequalities for Banach Space Valued Functions

Here we give a very general fractional Bochner integral representation formula for Banach space valued functions. We derive generalized left and right fractional Opial type inequalities, fractional Ostrowski type inequalities and fractional Grüss type inequalities. All these inequalities are very general having in their background Bochner type integrals. See also [4].

9.1 Background

We need

Definition 9.1 *([2]) Let $[a, b] \subset \mathbb{R}$, $(X, \|\cdot\|)$ a Banach space, $g \in C^1([a, b])$ and increasing, $f \in C([a, b], X)$, $\nu > 0$.*

We define the left Riemann–Liouville generalized fractional Bochner integral operator

$$\left(I_{a+;g}^{\nu} f\right)(x) := \frac{1}{\Gamma(\nu)} \int_a^x (g(x) - g(z))^{\nu-1} g'(z) f(z) \, dz, \tag{9.1}$$

$\forall\, x \in [a, b]$, where Γ is the gamma function.

The last integral is of Bochner type. Since $f \in C([a, b], X)$, then $f \in L_{\infty}([a, b], X)$. By [2] we get that $I_{a+;g}^{\nu} f \in C([a, b], X)$. Above we set $I_{a+;g}^{0} f := f$ and see that $\left(I_{a+;g}^{\nu} f\right)(a) = 0$.

When g is the identity function id, we get that $I_{a+;id}^{\nu} = I_{a+}^{\nu}$, the ordinary left Riemann–Liouville fractional integral

G. A. Anastassiou, *Generalized Fractional Calculus*, Studies in Systems, Decision and Control 305, https://doi.org/10.1007/978-3-030-56962-4_9

$$\left(I_{a+}^{\nu}f\right)(x)=\frac{1}{\Gamma(\nu)}\int_{a}^{x}(x-t)^{\nu-1}f(t)\,dt, \tag{9.2}$$

$\forall\, x\in[a,b]$, $\left(I_{a+}^{\nu}f\right)(a)=0$.

We need

Theorem 9.2 ([2]) *Let $\mu,\nu>0$ and $f\in C([a,b],X)$. Then*

$$I_{a+;g}^{\mu}I_{a+;g}^{\nu}f=I_{a+;g}^{\mu+\nu}f=I_{a+;g}^{\nu}I_{a+;g}^{\mu}f. \tag{9.3}$$

We need

Definition 9.3 *([2])* Let $[a,b]\subset\mathbb{R}$, $(X,\|\cdot\|)$ a Banach space, $g\in C^{1}([a,b])$ and increasing, $f\in C([a,b],X)$, $\nu>0$.

We define the right Riemann–Liouville generalized fractional Bochner integral operator

$$\left(I_{b-;g}^{\nu}f\right)(x):=\frac{1}{\Gamma(\nu)}\int_{x}^{b}(g(z)-g(x))^{\nu-1}g'(z)f(z)\,dz, \tag{9.4}$$

$\forall\, x\in[a,b]$, where Γ is the gamma function.

The last integral is of Bochner type. Since $f\in C([a,b],X)$, then $f\in L_{\infty}([a,b],X)$. By [2] we get that $I_{b-;g}^{\nu}f\in C([a,b],X)$. Above we set $I_{b-;g}^{0}f:=f$ and see that $\left(I_{b-;g}^{\nu}f\right)(b)=0$.

When g is the identity function id, we get that $I_{b-;id}^{\nu}=I_{b-}^{\nu}$, the ordinary right Riemann–Liouville fractional integral

$$\left(I_{b-}^{\nu}f\right)(x)=\frac{1}{\Gamma(\nu)}\int_{x}^{b}(t-x)^{\nu-1}f(t)\,dt, \tag{9.5}$$

$\forall\, x\in[a,b]$, with $\left(I_{b-}^{\nu}f\right)(b)=0$.

We need

Theorem 9.4 ([2]) *Let $\mu,\nu>0$ and $f\in C([a,b],X)$. Then*

$$I_{b-;g}^{\mu}I_{b-;g}^{\nu}f=I_{b-;g}^{\mu+\nu}f=I_{b-;g}^{\nu}I_{b-;g}^{\mu}f. \tag{9.6}$$

We will use

Definition 9.5 *([2])* Let $\alpha>0$, $\lceil\alpha\rceil=n$, $\lceil\cdot\rceil$ the ceiling of the number. Let $f\in C^{n}([a,b],X)$, where $[a,b]\subset\mathbb{R}$, and $(X,\|\cdot\|)$ is a Banach space. Let $g\in C^{1}([a,b])$, strictly increasing, such that $g^{-1}\in C^{n}([g(a),g(b)])$.

We define the left generalized g-fractional derivative X-valued of f of order α as follows:

$$\left(D_{a+;g}^{\alpha}f\right)(x) := \frac{1}{\Gamma(n-\alpha)} \int_a^x (g(x)-g(t))^{n-\alpha-1} g'(t) \left(f \circ g^{-1}\right)^{(n)} (g(t))\, dt, \tag{9.7}$$

$\forall\, x \in [a, b]$. The last integral is of Bochner type.

If $\alpha \notin \mathbb{N}$, by [2], we have that $\left(D_{a+;g}^{\alpha}f\right) \in C([a, b], X)$.

We see that

$$\left(I_{a+;g}^{n-\alpha}\left(\left(f \circ g^{-1}\right)^{(n)} \circ g\right)\right)(x) = \left(D_{a+;g}^{\alpha}f\right)(x),\ \forall\, x \in [a, b]. \tag{9.8}$$

We set

$$D_{a+;g}^{n}f(x) := \left(\left(f \circ g^{-1}\right)^{(n)} \circ g\right)(x) \in C([a, b], X),\ n \in \mathbb{N}, \tag{9.9}$$

$$D_{a+;g}^{0}f(x) = f(x),\ \forall\, x \in [a, b].$$

When $g = id$, then

$$D_{a+;g}^{\alpha}f = D_{a+;id}^{\alpha}f = D_{*a}^{\alpha}f, \tag{9.10}$$

the usual left X-valued Caputo fractional derivative, see [3].

We will use

Definition 9.6 *([2])* Let $\alpha > 0$, $\lceil\alpha\rceil = n$, $\lceil\cdot\rceil$ the ceiling of the number. Let $f \in C^n([a, b], X)$, where $[a, b] \subset \mathbb{R}$, and $(X, \|\cdot\|)$ is a Banach space. Let $g \in C^1([a, b])$, strictly increasing, such that $g^{-1} \in C^n([g(a), g(b)])$.

We define the right generalized g-fractional derivative X-valued of f of order α as follows:

$$\left(D_{b-;g}^{\alpha}f\right)(x) := \frac{(-1)^n}{\Gamma(n-\alpha)} \int_x^b (g(t)-g(x))^{n-\alpha-1} g'(t) \left(f \circ g^{-1}\right)^{(n)} (g(t))\, dt, \tag{9.11}$$

$\forall\, x \in [a, b]$. The last integral is of Bochner type.

If $\alpha \notin \mathbb{N}$, by [2], we have that $\left(D_{b-;g}^{\alpha}f\right) \in C([a, b], X)$.

We see that

$$I_{b-;g}^{n-\alpha}\left((-1)^n \left(f \circ g^{-1}\right)^{(n)} \circ g\right)(x) = \left(D_{b-;g}^{\alpha}f\right)(x),\ a \le x \le b. \tag{9.12}$$

We set

$$D_{b-;g}^{n}f(x) := (-1)^n \left(\left(f \circ g^{-1}\right)^{n} \circ g\right)(x) \in C([a, b], X),\ n \in \mathbb{N}, \tag{9.13}$$

$$D_{b-;g}^{0} f(x) := f(x), \forall x \in [a,b].$$

When $g = id$, then

$$D_{b-;g}^{\alpha} f(x) = D_{b-;id}^{\alpha} f(x) = D_{b-}^{\alpha} f, \tag{9.14}$$

the usual right X-valued Caputo fractional derivative, see [3].

We make

Remark 9.7 All as in Definition 9.5. We have (by Theorem 2.5, p. 7, [6])

$$\left\| \left(D_{a+;g}^{\alpha} f \right)(x) \right\| \leq \frac{1}{\Gamma(n-\alpha)} \int_{a}^{x} (g(x) - g(t))^{n-\alpha-1} g'(t) \left\| \left(f \circ g^{-1} \right)^{(n)} (g(t)) \right\| dt$$

$$\leq \frac{\left\| \left(f \circ g^{-1} \right)^{(n)} \circ g \right\|_{\infty,[a,b]}}{\Gamma(n-\alpha)} \int_{g(a)}^{g(x)} (g(x) - g(t))^{n-\alpha-1} dg(t) =$$

$$\frac{\left\| \left(f \circ g^{-1} \right)^{(n)} \circ g \right\|_{\infty,[a,b]}}{\Gamma(n-\alpha+1)} (g(x) - g(a))^{n-\alpha}. \tag{9.15}$$

That is

$$\left\| \left(D_{a+;g}^{\alpha} f \right)(x) \right\| \leq \frac{\left\| \left(f \circ g^{-1} \right)^{(n)} \circ g \right\|_{\infty,[a,b]}}{\Gamma(n-\alpha+1)} (g(x) - g(a))^{n-\alpha}, \tag{9.16}$$

$\forall x \in [a,b]$.

If $\alpha \notin \mathbb{N}$, then $\left(D_{a+;g}^{\alpha} f \right)(a) = 0$.

Similarly, by Definition 9.6 we derive

$$\left\| \left(D_{b-;g}^{\alpha} f \right)(x) \right\| \leq \frac{1}{\Gamma(n-\alpha)} \int_{x}^{b} (g(t) - g(x))^{n-\alpha-1} g'(t) \left\| \left(f \circ g^{-1} \right)^{(n)} (g(t)) \right\| dt$$

$$\leq \frac{\left\| \left(f \circ g^{-1} \right)^{(n)} \circ g \right\|_{\infty,[a,b]}}{\Gamma(n-\alpha)} \int_{g(x)}^{g(b)} (g(t) - g(x))^{n-\alpha-1} dg(t) =$$

$$\frac{\left\| \left(f \circ g^{-1} \right)^{(n)} \circ g \right\|_{\infty,[a,b]}}{\Gamma(n-\alpha+1)} (g(b) - g(x))^{n-\alpha}. \tag{9.17}$$

That is

$$\left\| \left(D_{b-;g}^{\alpha} f \right)(x) \right\| \leq \frac{\left\| \left(f \circ g^{-1} \right)^{(n)} \circ g \right\|_{\infty,[a,b]}}{\Gamma(n-\alpha+1)} (g(b) - g(x))^{n-\alpha}, \tag{9.18}$$

$\forall x \in [a,b]$.

If $\alpha \notin \mathbb{N}$, then $\left(D_{b-;g}^{\alpha} f\right)(b) = 0$.

Notation 9.8 *We denote by*

$$D_{a+;g}^{n\alpha} := D_{a+;g}^{\alpha} D_{a+;g}^{\alpha} \ldots D_{a+;g}^{\alpha} \text{ (n times)}, n \in \mathbb{N}, \tag{9.19}$$

$$I_{a+;g}^{n\alpha} := I_{a+;g}^{\alpha} I_{a+;g}^{\alpha} \ldots I_{a+;g}^{\alpha}, \tag{9.20}$$

$$D_{b-;g}^{n\alpha} := D_{b-;g}^{\alpha} D_{b-;g}^{\alpha} \ldots D_{b-;g}^{\alpha}, \tag{9.21}$$

and

$$I_{b-;g}^{n\alpha} := I_{b-;g}^{\alpha} I_{b-;g}^{\alpha} \ldots I_{b-;g}^{\alpha}, \tag{9.22}$$

(n times), $n \in \mathbb{N}$.

We are motivated by the following generalized fractional Ostrowski type inequality:

Theorem 9.9 ([2]) *Let $g \in C^1([a,b])$ and strictly increasing, such that $g^{-1} \in C^1([g(a), g(b)])$, and $0 < \alpha < 1$, $n \in \mathbb{N}$, $f \in C^1([a,b], X)$, where $(X, \|\cdot\|)$ is a Banach space. Let $x_0 \in [a,b]$ be fixed. Assume that $F_k^{x_0} := D_{x_0-;g}^{k\alpha} f$, for $k = 1, \ldots, n$, fulfill $F_k^{x_0} \in C^1([a, x_0], X)$ and $\left(D_{x_0-;g}^{i\alpha} f\right)(x_0) = 0$, $i = 2, \ldots, n$.*

Similarly, we assume that $G_k^{x_0} := D_{x_0+;g}^{k\alpha} f$, for $k = 1, \ldots, n$, fulfill $G_k^{x_0} \in C^1([x_0, b], X)$ and $\left(D_{x_0+;g}^{i\alpha} f\right)(x_0) = 0$, $i = 2, \ldots, n$.

Then

$$\left\| \frac{1}{b-a} \int_a^b f(x)\, dx - f(x_0) \right\| \leq \frac{1}{(b-a)\, \Gamma((n+1)\alpha + 1)} \cdot$$

$$\left\{ (g(b) - g(x_0))^{(n+1)\alpha} (b - x_0) \left\| D_{x_0+;g}^{(n+1)\alpha} f \right\|_{\infty,[x_0,b]} + \right.$$

$$\left. (g(x_0) - g(a))^{(n+1)\alpha} (x_0 - a) \left\| D_{x_0-;g}^{(n+1)\alpha} f \right\|_{\infty,[a,x_0]} \right\}. \tag{9.23}$$

In this work we will present several generalized fractional Bochner integral inequalities.

We mention the following g-left generalized X-valued Taylor's formula:

Theorem 9.10 ([2]) *Let $\alpha > 0$, $n = \lceil \alpha \rceil$, and $f \in C^n([a,b], X)$, where $[a,b] \subset \mathbb{R}$ and $(X, \|\cdot\|)$ is a Banach space. Let $g \in C^1([a,b])$, strictly increasing, such that $g^{-1} \in C^n([g(a), g(b)])$. Then*

$$f(x) = f(a) + \sum_{i=1}^{n-1} \frac{(g(x) - g(a))^i}{i!} \left(f \circ g^{-1}\right)^{(i)} (g(a)) +$$

$$\frac{1}{\Gamma(\alpha)}\int_a^x (g(x)-g(t))^{\alpha-1} g'(t)\left(D_{a+;g}^{\alpha} f\right)(t)\,dt =$$

$$f(a)+\sum_{i=1}^{n-1}\frac{(g(x)-g(a))^i}{i!}\left(f\circ g^{-1}\right)^{(i)}(g(a))+ \tag{9.24}$$

$$\frac{1}{\Gamma(\alpha)}\int_{g(a)}^{g(x)} (g(x)-z)^{\alpha-1}\left(\left(D_{a+;g}^{\alpha} f\right)\circ g^{-1}\right)(z)\,dz,\ \forall x\in[a,b].$$

We mention the following g-right generalized X-valued Taylor's formula:

Theorem 9.11 ([2]) *Let $\alpha>0$, $n=\lceil\alpha\rceil$, and $f\in C^n([a,b],X)$, where $[a,b]\subset\mathbb{R}$ and $(X,\|\cdot\|)$ is a Banach space. Let $g\in C^1([a,b])$, strictly increasing, such that $g^{-1}\in C^n([g(a),g(b)])$. Then*

$$f(x)=f(b)+\sum_{i=1}^{n-1}\frac{(g(x)-g(b))^i}{i!}\left(f\circ g^{-1}\right)^{(i)}(g(b))+$$

$$\frac{1}{\Gamma(\alpha)}\int_x^b (g(t)-g(x))^{\alpha-1} g'(t)\left(D_{b-;g}^{\alpha} f\right)(t)\,dt =$$

$$f(b)+\sum_{i=1}^{n-1}\frac{(g(x)-g(b))^i}{i!}\left(f\circ g^{-1}\right)^{(i)}(g(b))+ \tag{9.25}$$

$$\frac{1}{\Gamma(\alpha)}\int_{g(x)}^{g(b)} (z-g(x))^{\alpha-1}\left(\left(D_{b-;g}^{\alpha} f\right)\circ g^{-1}\right)(z)\,dz,\ \forall x\in[a,b].$$

For the Bochner integral excellent resources are [5, 7, 8] and [1, pp. 422–428].

9.2 Main Results

We give the following representation formula:

Theorem 9.12 *All as in Theorem 9.10. Then*

$$f(y)=\frac{1}{b-a}\int_a^b f(x)\,dx-\sum_{k=1}^{n-1}\frac{\left(f\circ g^{-1}\right)^{(k)}(g(y))}{k!\,(b-a)}\int_a^b (g(x)-g(y))^k\,dx+R_1(y), \tag{9.26}$$

for any $y\in[a,b]$, where

$$R_1(y)=-\frac{1}{\Gamma(\alpha)(b-a)}$$

$$\left[\int_a^b \chi_{[a,y)}(x)\left(\int_x^y |g(x)-g(t)|^{\alpha-1} g'(t)\left(D_{y-;g}^{\alpha} f\right)(t)\,dt\right)dx\right.$$

$$\left.\int_a^b \chi_{[y,b]}(x)\left(\int_y^x |g(x)-g(t)|^{\alpha-1} g'(t)\left(D_{y+;g}^{\alpha} f\right)(t)\,dt\right)dx\right]. \tag{9.27}$$

here χ_A stands for the characteristic function set A, where A is an arbitrary set.

One may write also that

$$R_1(y) = -\frac{1}{\Gamma(\alpha)(b-a)}\left[\int_a^y \left(\int_x^y (g(t)-g(x))^{\alpha-1} g'(t)\left(D_{y-;g}^{\alpha} f\right)(t)\,dt\right)dx\right.$$

$$\left.+\int_y^b \left(\int_y^x (g(x)-g(t))^{\alpha-1} g'(t)\left(D_{y+;g}^{\alpha} f\right)(t)\,dt\right)dx\right], \tag{9.28}$$

for any $y \in [a,b]$.

Putting things together, one has

$$f(y) = \frac{1}{b-a}\int_a^b f(x)\,dx - \sum_{k=1}^{n-1}\frac{\left(f\circ g^{-1}\right)^{(k)}(g(y))}{k!(b-a)}\int_a^b (g(x)-g(y))^k\,dx$$

$$-\frac{1}{\Gamma(\alpha)(b-a)}\left[\int_a^b \chi_{[a,y)}(x)\left(\int_x^y |g(x)-g(t)|^{\alpha-1} g'(t)\left(D_{y-;g}^{\alpha} f\right)(t)\,dt\right)dx\right.$$

$$\left.\int_a^b \chi_{[y,b]}(x)\left(\int_y^x |g(x)-g(t)|^{\alpha-1} g'(t)\left(D_{y+;g}^{\alpha} f\right)(t)\,dt\right)dx\right]. \tag{9.29}$$

In particular, one has

$$f(y) - \frac{1}{b-a}\int_a^b f(x)\,dx + \sum_{k=1}^{n-1}\frac{\left(f\circ g^{-1}\right)^{(k)}(g(y))}{k!(b-a)}\int_a^b (g(x)-g(y))^k\,dx = R_1(y), \tag{9.30}$$

for any $y \in [a,b]$.

Proof Here $x, y \in [a,b]$. We keep y as fixed.

By Theorem 9.10 we get:

$$f(x) = f(y) + \sum_{k=1}^{n-1}\frac{\left(f\circ g^{-1}\right)^{(k)}(g(y))}{k!}(g(x)-g(y))^k + \tag{9.31}$$

$$\frac{1}{\Gamma(\alpha)}\int_y^x (g(x)-g(t))^{\alpha-1} g'(t)\left(D_{y+;g}^{\alpha} f\right)(t)\,dt, \quad \text{for any } x \geq y.$$

By Theorem 9.11 we get:

$$f(x) = f(y) + \sum_{k=1}^{n-1} \frac{\left(f \circ g^{-1}\right)^{(k)}(g(y))}{k!} (g(x) - g(y))^k + \tag{9.32}$$

$$\frac{1}{\Gamma(\alpha)} \int_x^y (g(t) - g(x))^{\alpha-1} g'(t) \left(D_{y-;g}^{\alpha} f\right)(t)\, dt, \quad \text{for any } x \leq y.$$

By (9.31), (9.32) we notice that

$$\int_a^b f(x)\, dx = \int_a^y f(x)\, dx + \int_y^b f(x)\, dx = \tag{9.33}$$

$$\int_a^y f(y)\, dx + \sum_{k=1}^{n-1} \frac{\left(f \circ g^{-1}\right)^{(k)}(g(y))}{k!} \int_a^y (g(x) - g(y))^k\, dx +$$

$$\frac{1}{\Gamma(\alpha)} \int_a^y \left(\int_x^y (g(t) - g(x))^{\alpha-1} g'(t) \left(D_{y-;g}^{\alpha} f\right)(t)\, dt \right) dx +$$

$$\int_y^b f(y)\, dx + \sum_{k=1}^{n-1} \frac{\left(f \circ g^{-1}\right)^{(k)}(g(y))}{k!} \int_y^b (g(x) - g(y))^k\, dx +$$

$$\frac{1}{\Gamma(\alpha)} \int_y^b \left(\int_y^x (g(x) - g(t))^{\alpha-1} g'(t) \left(D_{y+;g}^{\alpha} f\right)(t)\, dt \right) dx.$$

Hence it holds

$$\frac{1}{b-a} \int_a^b f(x)\, dx = f(y) + \sum_{k=1}^{n-1} \frac{\left(f \circ g^{-1}\right)^{(k)}(g(y))}{k!\,(b-a)} \int_a^b (g(x) - g(y))^k\, dx +$$

$$\frac{1}{\Gamma(\alpha)(b-a)} \left[\int_a^y \left(\int_x^y |g(x) - g(t)|^{\alpha-1} g'(t) \left(D_{y-;g}^{\alpha} f\right)(t)\, dt \right) dx + \right. \tag{9.34}$$

$$\left. \int_y^b \left(\int_y^x |g(x) - g(t)|^{\alpha-1} g'(t) \left(D_{y+;g}^{\alpha} f\right)(t)\, dt \right) dx \right].$$

Therefore we obtain

$$f(y) = \frac{1}{b-a} \int_a^b f(x)\, dx - \sum_{k=1}^{n-1} \frac{\left(f \circ g^{-1}\right)^{(k)}(g(y))}{k!\,(b-a)} \int_a^b (g(x) - g(y))^k\, dx - \tag{9.35}$$

$$\frac{1}{\Gamma(\alpha)(b-a)}\left[\int_a^y\left(\int_x^y |g(x)-g(t)|^{\alpha-1} g'(t)\left(D_{y-;g}^{\alpha} f\right)(t)\,dt\right)dx+\right.$$

$$\left.\int_y^b\left(\int_y^x |g(x)-g(t)|^{\alpha-1} g'(t)\left(D_{y+;g}^{\alpha} f\right)(t)\,dt\right)dx\right].$$

Hence the remainder

$$R_1(y) := -\frac{1}{\Gamma(\alpha)(b-a)}\left[\int_a^y\left(\int_x^y |g(x)-g(t)|^{\alpha-1} g'(t)\left(D_{y-;g}^{\alpha} f\right)(t)\,dt\right)dx\right.$$

$$\left.+\int_y^b\left(\int_y^x |g(x)-g(t)|^{\alpha-1} g'(t)\left(D_{y+;g}^{\alpha} f\right)(t)\,dt\right)dx\right] =$$

$$-\frac{1}{\Gamma(\alpha)(b-a)}\left[\int_a^b \chi_{[a,y)}(x)\left(\int_x^y |g(x)-g(t)|^{\alpha-1} g'(t)\left(D_{y-;g}^{\alpha} f\right)(t)\,dt\right)dx\right.$$

$$\left.+\int_a^b \chi_{[y,b]}(x)\left(\int_y^x |g(x)-g(t)|^{\alpha-1} g'(t)\left(D_{y+;g}^{\alpha} f\right)(t)\,dt\right)dx\right]. \tag{9.36}$$

The theorem is proved. ■

Next we present a left fractional Opial type inequality:

Theorem 9.13 *All as in Theorem 9.10. Additionally assume that* $\alpha \geq 1$, $g \in C^1([a,b])$, *and* $\left(f \circ g^{-1}\right)^{(k)}(g(a)) = 0$, *for* $k = 0, 1, \ldots, n-1$. *Let* $p, q > 1 : \frac{1}{p} + \frac{1}{q} = 1$. *Then*

$$\int_a^x \|f(w)\| \left\|\left(D_{a+;g}^{\alpha} f\right)(w)\right\| g'(w)\,dw \leq \frac{1}{\Gamma(\alpha)\, 2^{\frac{1}{q}}}\cdot \tag{9.37}$$

$$\left(\int_a^x\left(\int_a^w (g(w)-g(t))^{p(\alpha-1)}\,dt\right)dw\right)^{\frac{1}{p}}\left(\int_a^x (g'(w))^q \left\|\left(D_{a+;g}^{\alpha} f\right)(w)\right\|^q dw\right)^{\frac{2}{q}},$$

$\forall\, x \in [a,b]$.

Proof By Theorem 9.10, we have that

$$f(x) = \frac{1}{\Gamma(\alpha)}\int_a^x (g(x)-g(t))^{\alpha-1} g'(t)\left(D_{a+;g}^{\alpha} f\right)(t)\,dt, \quad \forall\, x \in [a,b]. \tag{9.38}$$

Then, by Hölder's inequality we obtain,

$$\|f(x)\| \leq \frac{1}{\Gamma(\alpha)}\left(\int_a^x (g(x)-g(t))^{p(\alpha-1)}\,dt\right)^{\frac{1}{p}}\left(\int_a^x (g'(t))^q \left\|\left(D_{a+;g}^{\alpha} f\right)(t)\right\|^q dt\right)^{\frac{1}{q}}. \tag{9.39}$$

Call

$$z(x) := \int_a^x (g'(t))^q \left\| \left(D_{a+;g}^{\alpha} f\right)(t) \right\|^q dt, \tag{9.40}$$

$z(a) = 0$.

Thus

$$z'(x) = (g'(x))^q \left\| \left(D_{a+;g}^{\alpha} f\right)(x) \right\|^q \geq 0, \tag{9.41}$$

and

$$\left(z'(x)\right)^{\frac{1}{q}} = g'(x) \left\| \left(D_{a+;g}^{\alpha} f\right)(x) \right\| \geq 0, \ \forall x \in [a, b]. \tag{9.42}$$

Consequently, we get

$$\|f(w)\| g'(w) \left\| \left(D_{a+;g}^{\alpha} f\right)(w) \right\| \leq \tag{9.43}$$

$$\frac{1}{\Gamma(\alpha)} \left(\int_a^w (g(w) - g(t))^{p(\alpha-1)} dt\right)^{\frac{1}{p}} \left(z(w) z'(w)\right)^{\frac{1}{q}}, \quad \forall w \in [a, b].$$

Then

$$\int_a^x \|f(w)\| \left\| \left(D_{a+;g}^{\alpha} f\right)(w) \right\| g'(w) \, dw \leq \tag{9.44}$$

$$\frac{1}{\Gamma(\alpha)} \int_a^x \left(\int_a^w (g(w) - g(t))^{p(\alpha-1)} dt\right)^{\frac{1}{p}} \left(z(w) z'(w)\right)^{\frac{1}{q}} dw \leq$$

$$\frac{1}{\Gamma(\alpha)} \left(\int_a^x \left(\int_a^w (g(w) - g(t))^{p(\alpha-1)} dt\right) dw\right)^{\frac{1}{p}} \left(\int_a^x z(w) z'(w) \, dw\right)^{\frac{1}{q}} = \tag{9.45}$$

$$\frac{1}{\Gamma(\alpha)} \left(\int_a^x \left(\int_a^w (g(w) - g(t))^{p(\alpha-1)} dt\right) dw\right)^{\frac{1}{p}} \left(\frac{z^2(x)}{2}\right)^{\frac{1}{q}} =$$

$$\frac{1}{\Gamma(\alpha)} \left(\int_a^x \left(\int_a^w (g(w) - g(t))^{p(\alpha-1)} dt\right) dw\right)^{\frac{1}{p}} \cdot$$

$$\left(\int_a^x (g'(t))^q \left\| \left(D_{a+;g}^{\alpha} f\right)(t) \right\|^q dt\right)^{\frac{2}{q}} \cdot 2^{-\frac{1}{q}}. \tag{9.46}$$

The theorem is proved. ■

We also give a right fractional Opial type inequality:

Theorem 9.14 *All as in Theorem 9.11. Additionally assume that* $\alpha \geq 1$, $g \in C^1([a, b])$, *and* $\left(f \circ g^{-1}\right)^{(k)}(g(b)) = 0$, $k = 0, 1, \ldots, n-1$. *Let* $p, q > 1 : \frac{1}{p} + \frac{1}{q} = 1$. *Then*

$$\int_x^b \|f(w)\| \left\| \left(D_{b-;g}^{\alpha} f\right)(w) \right\| g'(w)\, dw \leq \frac{1}{2^{\frac{1}{q}} \Gamma(\alpha)} \cdot \tag{9.47}$$

$$\left(\int_x^b \left(\int_w^b (g(t) - g(w))^{p(\alpha-1)} dt \right) dw \right)^{\frac{1}{p}} \left(\int_x^b (g'(w))^q \left\| \left(D_{b-;g}^{\alpha} f\right)(w) \right\|^q dw \right)^{\frac{2}{q}},$$

all $a \leq x \leq b$.

Proof By Theorem 9.11, we have that

$$f(x) = \frac{1}{\Gamma(\alpha)} \int_x^b (g(t) - g(x))^{\alpha-1} g'(t) \left(D_{b-;g}^{\alpha} f\right)(t)\, dt, \quad \text{all } a \leq x \leq b. \tag{9.48}$$

Then, by Hölder's inequality we obtain,

$$\|f(x)\| \leq \frac{1}{\Gamma(\alpha)} \left(\int_x^b (g(t) - g(x))^{p(\alpha-1)} dt \right)^{\frac{1}{p}} \left(\int_x^b (g'(t))^q \left\| \left(D_{b-;g}^{\alpha} f\right)(t) \right\|^q dt \right)^{\frac{1}{q}}. \tag{9.49}$$

Call

$$z(x) := \int_x^b (g'(t))^q \left\| \left(D_{b-;g}^{\alpha} f\right)(t) \right\|^q dt, \tag{9.50}$$

$z(b) = 0$.

Hence

$$z'(x) = -(g'(x))^q \left\| \left(D_{b-;g}^{\alpha} f\right)(x) \right\|^q \leq 0, \tag{9.51}$$

and

$$-z'(x) = (g'(x))^q \left\| \left(D_{b-;g}^{\alpha} f\right)(x) \right\|^q \geq 0, \tag{9.52}$$

and

$$(-z'(x))^{\frac{1}{q}} = g'(x) \left\| \left(D_{b-;g}^{\alpha} f\right)(x) \right\| \geq 0, \ \forall x \in [a, b]. \tag{9.53}$$

Consequently, we get

$$\|f(w)\| g'(w) \left\| \left(D_{b-;g}^{\alpha} f\right)(w) \right\| \leq$$

$$\frac{1}{\Gamma(\alpha)} \left(\int_w^b (g(t) - g(w))^{p(\alpha-1)} dt \right)^{\frac{1}{p}} \left(z(w) \left(-z'(w) \right) \right)^{\frac{1}{q}}, \ \forall w \in [a, b]. \tag{9.54}$$

Then

$$\int_x^b \|f(w)\| \left\| \left(D_{b-;g}^{\alpha} f\right)(w) \right\| g'(w)\, dw \leq \tag{9.55}$$

$$\frac{1}{\Gamma(\alpha)} \int_x^b \left(\int_w^b (g(t) - g(w))^{p(\alpha-1)} dt \right)^{\frac{1}{p}} \left(-z(w) z'(w) \right)^{\frac{1}{q}} dw \leq$$

$$\frac{1}{\Gamma(\alpha)} \left(\int_x^b \left(\int_w^b (g(t) - g(w))^{p(\alpha-1)} dt \right) dw \right)^{\frac{1}{p}} \left(-\int_x^b z(w) z'(w) dw \right)^{\frac{1}{q}} = \tag{9.56}$$

$$\frac{1}{\Gamma(\alpha)} \left(\int_x^b \left(\int_w^b (g(t) - g(w))^{p(\alpha-1)} dt \right) dw \right)^{\frac{1}{p}} \left(\frac{z^2(x)}{2} \right)^{\frac{1}{q}} =$$

$$\frac{1}{2^{\frac{1}{q}} \Gamma(\alpha)} \left(\int_x^b \left(\int_w^b (g(t) - g(w))^{p(\alpha-1)} dt \right) dw \right)^{\frac{1}{p}} \cdot$$

$$\left(\int_x^b (g'(t))^q \left\| \left(D_{b-;g}^{\alpha} f \right)(t) \right\|^q dt \right)^{\frac{2}{q}}. \tag{9.57}$$

The theorem is proved. ■

Two extreme fractional Opial type inequalities follow (case $p = 1, q = \infty$).

Theorem 9.15 *All as in Theorem 9.10. Assume that* $\left(f \circ g^{-1}\right)^{(k)} (g(a)) = 0$, $k = 0, 1, \ldots, n-1$. *Then*

$$\int_a^x \|f(w)\| \left\| D_{a+;g}^{\alpha} f(w) \right\| dw \leq \frac{\left\| D_{a+;g}^{\alpha} f \right\|_{\infty}^2}{\Gamma(\alpha+1)} \left(\int_a^x (g(w) - g(a))^{\alpha} dw \right), \tag{9.58}$$

all $a \leq x \leq b$.

Proof For any $w \in [a, b]$, we have that

$$f(x) = \frac{1}{\Gamma(\alpha)} \int_a^w (g(w) - g(t))^{\alpha-1} g'(t) \left(D_{a+;g}^{\alpha} f \right)(t) dt, \tag{9.59}$$

and

$$\|f(x)\| \leq \frac{1}{\Gamma(\alpha)} \left(\int_a^w (g(w) - g(t))^{\alpha-1} g'(t) dt \right) \left\| D_{a+;g}^{\alpha} f \right\|_{\infty}$$

$$= \frac{\left\| D_{a+;g}^{\alpha} f \right\|_{\infty}}{\Gamma(\alpha+1)} (g(w) - g(a))^{\alpha}. \tag{9.60}$$

Hence we obtain

$$\|f(w)\| \left\| D_{a+;g}^{\alpha} f(w) \right\| \leq \frac{\left\| D_{a+;g}^{\alpha} f \right\|_{\infty}^2}{\Gamma(\alpha+1)} (g(w) - g(a))^{\alpha}. \tag{9.61}$$

Integrating (9.61) over $[a, x]$ we derive (9.58). ■

Theorem 9.16 *All as in Theorem 9.11. Assume that* $\left(f \circ g^{-1}\right)^{(k)}(g(b)) = 0$, $k = 0, 1, \ldots, n-1$. *Then*

$$\int_x^b \|f(w)\| \left\|D_{b-;g}^{\alpha} f(w)\right\| dw \leq \frac{\left\|D_{b-;g}^{\alpha} f\right\|_{\infty}^2}{\Gamma(\alpha+1)} \left(\int_x^b (g(b) - g(w))^{\alpha} dw\right), \tag{9.62}$$

all $a \leq x \leq b$.

Proof For any $w \in [a, b]$, we have

$$f(x) = \frac{1}{\Gamma(\alpha)} \int_w^b (g(t) - g(w))^{\alpha-1} g'(t) \left(D_{b-;g}^{\alpha} f\right)(t)\, dt, \tag{9.63}$$

and

$$\|f(x)\| \leq \frac{1}{\Gamma(\alpha)} \left(\int_w^b (g(t) - g(w))^{\alpha-1} g'(t)\, dt\right) \left\|D_{b-;g}^{\alpha} f\right\|_{\infty}$$

$$= \frac{\left\|D_{b-;g}^{\alpha} f\right\|_{\infty}}{\Gamma(\alpha+1)} (g(b) - g(w))^{\alpha}. \tag{9.64}$$

Hence we obtain

$$\|f(w)\| \left\|D_{b-;g}^{\alpha} f(w)\right\| \leq \frac{\left\|D_{b-;g}^{\alpha} f\right\|_{\infty}^2}{\Gamma(\alpha+1)} (g(b) - g(w))^{\alpha}. \tag{9.65}$$

Integrating (9.65) over $[x, b]$ we derive (9.62). ∎

Next we present three fractional Ostrowski type inequalities:

Theorem 9.17 *All as in Theorem 9.10. Then*

$$\left\| f(y) - \frac{1}{b-a} \int_a^b f(x)\, dx + \sum_{k=1}^{n-1} \frac{\left(f \circ g^{-1}\right)^{(k)}(g(y))}{k!\,(b-a)} \int_a^b (g(x) - g(y))^k dx \right\| \leq \tag{9.66}$$

$$\frac{1}{\Gamma(\alpha+1)(b-a)}\cdot$$

$$\left[(g(y) - g(a))^{\alpha} (y-a) \left\|D_{y-;g}^{\alpha} f\right\|_{\infty} + (g(b) - g(y))^{\alpha} (b-y) \left\|D_{y+;g}^{\alpha} f\right\|_{\infty}\right],$$

$\forall\, y \in [a, b]$.

Proof Define

$$\left(D^{\alpha}_{y+;g}f\right)(t)=0,\ \text{for } t<y,\quad \text{and} \quad \left(D^{\alpha}_{y-;g}f\right)(t)=0,\ \text{for } t>y. \tag{9.67}$$

Notice for $0<\alpha\notin\mathbb{N}$ by Remark 9.7 we have

$$\left(D^{\alpha}_{a+;g}f\right)(a)=0. \tag{9.68}$$

Similarly it holds ($0<\alpha\notin\mathbb{N}$) by Remark 9.7 that

$$\left(D^{\alpha}_{b-;g}f\right)(b)=0. \tag{9.69}$$

Thus

$$\left(D^{\alpha}_{y+;g}f\right)(y)=0,\quad \left(D^{\alpha}_{y-;g}f\right)(y)=0, \tag{9.70}$$

$0<\alpha\notin\mathbb{N}$, any $y\in[a,b]$.

We observe that

$$\|R_1(y)\| \overset{(9.28)}{\leq} \frac{1}{\Gamma(\alpha)(b-a)}\left[\left(\int_a^y\left(\int_x^y (g(t)-g(x))^{\alpha-1}g'(t)\,dt\right)dx\right)\left\|D^{\alpha}_{y-;g}f\right\|_{\infty}\right. \tag{9.71}$$

$$\left.+\left(\int_y^b\left(\int_y^x (g(x)-g(t))^{\alpha-1}g'(t)\,dt\right)dx\right)\left\|D^{\alpha}_{y+;g}f\right\|_{\infty}\right]=$$

$$\frac{1}{\Gamma(\alpha)(b-a)}\left[\left(\int_a^y\frac{(g(y)-g(x))^{\alpha}}{\alpha}dx\right)\left\|D^{\alpha}_{y-;g}f\right\|_{\infty}\right.$$

$$\left.+\left(\int_y^b\frac{(g(x)-g(y))^{\alpha}}{\alpha}dx\right)\left\|D^{\alpha}_{y+;g}f\right\|_{\infty}\right]\leq$$

$$\frac{1}{\Gamma(\alpha+1)(b-a)}\left[(g(y)-g(a))^{\alpha}(y-a)\left\|D^{\alpha}_{y-;g}f\right\|_{\infty}+\right.$$

$$\left.(g(b)-g(y))^{\alpha}(b-y)\left\|D^{\alpha}_{y+;g}f\right\|_{\infty}\right]. \tag{9.72}$$

We have proved that

$$\|R_1(y)\|\leq\frac{1}{\Gamma(\alpha+1)(b-a)}\left[(g(y)-g(a))^{\alpha}(y-a)\left\|D^{\alpha}_{y-;g}f\right\|_{\infty}\right. \tag{9.73}$$

$$\left.+(g(b)-g(y))^{\alpha}(b-y)\left\|D^{\alpha}_{y+;g}f\right\|_{\infty}\right],$$

any $y\in[a,b]$.

We have established the theorem. ∎

Theorem 9.18 *All as in Theorem 9.10. Here we take $\alpha \geq 1$. Then*

$$\left\| f(y) - \frac{1}{b-a}\int_a^b f(x)\,dx + \sum_{k=1}^{n-1} \frac{\left(f\circ g^{-1}\right)^{(k)}(g(y))}{k!\,(b-a)} \int_a^b (g(x)-g(y))^k\,dx \right\| \leq$$

$$\frac{1}{\Gamma(\alpha)(b-a)} \Big[\left\| \left(D_{y-;g}^{\alpha} f\right)\circ g^{-1} \right\|_{1,[g(a),g(y)]} (y-a)(g(y)-g(a))^{\alpha-1}$$

$$+ \left\| \left(D_{y+;g}^{\alpha} f\right)\circ g^{-1} \right\|_{1,[g(y),g(b)]} (b-y)(g(b)-g(y))^{\alpha-1} \Big], \tag{9.74}$$

$\forall\, y \in [a,b]$.

Proof We can rewrite

$$R_1(y) = -\frac{1}{\Gamma(\alpha)(b-a)} \left[\int_a^y \left(\int_{g(x)}^{g(y)} (z-g(x))^{\alpha-1} \left(\left(D_{y-;g}^{\alpha} f\right)\circ g^{-1}\right)(z)\,dz \right) dx \right. \tag{9.75}$$

$$\left. + \int_y^b \left(\int_{g(y)}^{g(x)} (g(x)-z)^{\alpha-1} \left(\left(D_{y+;g}^{\alpha} f\right)\circ g^{-1}\right)(z)\,dz \right) dx \right].$$

We assumed $\alpha \geq 1$, then

$$\|R_1(y)\| \leq \frac{1}{\Gamma(\alpha)(b-a)} \cdot$$

$$\left[\int_a^y \left(\int_{g(x)}^{g(y)} (z-g(x))^{\alpha-1} \left\| \left(\left(D_{y-;g}^{\alpha} f\right)\circ g^{-1}\right)(z) \right\| dz \right) dx \right. \tag{9.76}$$

$$\left. + \int_y^b \left(\int_{g(y)}^{g(x)} (g(x)-z)^{\alpha-1} \left\| \left(\left(D_{y+;g}^{\alpha} f\right)\circ g^{-1}\right)(z) \right\| dz \right) dx \right] \leq$$

$$\frac{1}{\Gamma(\alpha)(b-a)} \left[\left(\int_a^y \left(\int_{g(x)}^{g(y)} \left\| \left(\left(D_{y-;g}^{\alpha} f\right)\circ g^{-1}\right)(z) \right\| dz \right) dx \right) (g(y)-g(a))^{\alpha-1} \right.$$

$$\left. + \left(\int_y^b \left(\int_{g(y)}^{g(x)} \left\| \left(\left(D_{y+;g}^{\alpha} f\right)\circ g^{-1}\right)(z) \right\| dz \right) dx \right) (g(b)-g(y))^{\alpha-1} \right] \leq$$

$$\frac{1}{\Gamma(\alpha)(b-a)} \Big[\left\| \left(D_{y-;g}^{\alpha} f\right)\circ g^{-1} \right\|_{1,[g(a),g(y)]} (y-a)(g(y)-g(a))^{\alpha-1} \tag{9.77}$$

$$+\left\|\left(D_{y+;g}^{\alpha}f\right)\circ g^{-1}\right\|_{1,[g(y),g(b)]}(b-y)(g(b)-g(y))^{\alpha-1}\Big].$$

So when $\alpha \geq 1$, we obtained

$$\|R_1(y)\| \leq \frac{1}{\Gamma(\alpha)(b-a)}\Big[\left\|\left(D_{y-;g}^{\alpha}f\right)\circ g^{-1}\right\|_{1,[g(a),g(y)]}(y-a)(g(y)-g(a))^{\alpha-1}$$
$$+\left\|\left(D_{y+;g}^{\alpha}f\right)\circ g^{-1}\right\|_{1,[g(y),g(b)]}(b-y)(g(b)-g(y))^{\alpha-1}\Big]. \tag{9.78}$$

Clearly here g^{-1} is continuous, thus $\left(D_{y-;g}^{\alpha}f\right)\circ g^{-1} \in C([g(a),g(y)],X)$, and $\left(D_{y+;g}^{\alpha}f\right)\circ g^{-1} \in C([g(y),g(b)],X)$. Therefore

$$\left\|\left(D_{y-;g}^{\alpha}f\right)\circ g^{-1}\right\|_{1,[g(a),g(y)]},\ \left\|\left(D_{y+;g}^{\alpha}f\right)\circ g^{-1}\right\|_{1,[g(y),g(b)]} < \infty. \tag{9.79}$$

The proof of the theorem now is complete. ∎

Theorem 9.19 *All as in Theorem 9.10. Let $p,q>1:\frac{1}{p}+\frac{1}{q}=1$, $\alpha>\frac{1}{q}$. Then*

$$\left\|f(y)-\frac{1}{b-a}\int_a^b f(x)\,dx+\sum_{k=1}^{n-1}\frac{\left(f\circ g^{-1}\right)^{(k)}(g(y))}{k!(b-a)}\int_a^b (g(x)-g(y))^k\,dx\right\|$$
$$\leq \frac{1}{\Gamma(\alpha)(b-a)(p(\alpha-1)+1)^{\frac{1}{p}}}\cdot \tag{9.80}$$
$$\Big[(g(y)-g(a))^{\alpha-1+\frac{1}{p}}(y-a)\left\|\left(D_{y-;g}^{\alpha}f\right)\circ g^{-1}\right\|_{q,[g(a),g(y)]}$$
$$+(g(b)-g(y))^{\alpha-1+\frac{1}{p}}(b-y)\left\|\left(D_{y+;g}^{\alpha}f\right)\circ g^{-1}\right\|_{q,[g(y),g(b)]}\Big],$$

$\forall\, y \in [a,b]$.

Proof Here we use (9.75).

We get that

$$\|R_1(y)\| \leq \frac{1}{\Gamma(\alpha)(b-a)}\Big[\int_a^y\left(\int_{g(x)}^{g(y)}(z-g(x))^{p(\alpha-1)}dz\right)^{\frac{1}{p}}\cdot$$
$$\left(\int_{g(x)}^{g(y)}\left\|\left(\left(D_{y-;g}^{\alpha}f\right)\circ g^{-1}\right)(z)\right\|^q dz\right)^{\frac{1}{q}}dx+\int_y^b\left(\int_{g(y)}^{g(x)}(g(x)-z)^{p(\alpha-1)}dz\right)^{\frac{1}{p}}\cdot$$
$$\left(\int_{g(y)}^{g(x)}\left\|\left(\left(D_{y+;g}^{\alpha}f\right)\circ g^{-1}\right)(z)\right\|^q dz\right)^{\frac{1}{q}}dx\Big] \leq$$

$$\frac{1}{\Gamma(\alpha)(b-a)}\left[\left(\int_a^y \frac{(g(y)-g(x))^{(\alpha-1)+\frac{1}{p}}}{(p(\alpha-1)+1)^{\frac{1}{p}}}dx\right)\left\|\left(D_{y-;g}^{\alpha}f\right)\circ g^{-1}\right\|_{q,[g(a),g(y)]}\right.$$

$$\left.+\left(\int_a^y \frac{(g(x)-g(y))^{(\alpha-1)+\frac{1}{p}}}{(p(\alpha-1)+1)^{\frac{1}{p}}}dx\right)\left\|\left(D_{y+;g}^{\alpha}f\right)\circ g^{-1}\right\|_{q,[g(y),g(b)]}\right]. \tag{9.81}$$

(here it is $\alpha-1+\frac{1}{p}>0$)

Hence it holds

$$\|R_1(y)\| \leq \frac{1}{\Gamma(\alpha)(b-a)(p(\alpha-1)+1)^{\frac{1}{p}}}\cdot \tag{9.82}$$

$$\left[(g(y)-g(a))^{\alpha-1+\frac{1}{p}}(y-a)\left\|\left(D_{y-;g}^{\alpha}f\right)\circ g^{-1}\right\|_{q,[g(a),g(y)]}+\right.$$

$$\left.(g(b)-g(y))^{\alpha-1+\frac{1}{p}}(b-y)\left\|\left(D_{y+;g}^{\alpha}f\right)\circ g^{-1}\right\|_{q,[g(y),g(b)]}\right].$$

Clearly here

$$\left\|\left(D_{y-;g}^{\alpha}f\right)\circ g^{-1}\right\|_{q,[g(a),g(y)]}, \left\|\left(D_{y+;g}^{\alpha}f\right)\circ g^{-1}\right\|_{q,[g(y),g(b)]} < \infty.$$

We have proved the theorem. ■

Next we give some fractional Grüss type inequalities:

Theorem 9.20 *Let f, h as in Theorem 9.10. Here $R_1(y)$ will be renamed as $R_1(f,y)$, so we can consider $R_1(h,y)$. Then*

(1)

$$\Delta_n(f,h) := \frac{1}{b-a}\int_a^b f(x)h(x)\,dx - \frac{\left(\int_a^b f(x)\,dx\right)\left(\int_a^b h(x)\,dx\right)}{(b-a)^2}+$$

$$\frac{1}{2(b-a)^2}\sum_{k=1}^{n-1}\frac{1}{k!}\left[\int_a^b\left(\int_a^b\left(h(y)\left(f\circ g^{-1}\right)^{(k)}(g(y))+\right.\right.\right.$$

$$\left.\left.\left.f(y)\left(h\circ g^{-1}\right)^{(k)}(g(y))\right)(g(x)-g(y))^k\,dx\right)dy\right]=$$

$$\frac{1}{2(b-a)}\left[\int_a^b (h(y)R_1(f,y)+f(y)R_1(h,y))\,dy\right] =: K_n(f,h), \tag{9.83}$$

(2) it holds

$$\|\Delta_n(f,h)\| \leq \frac{(g(b)-g(a))^{\alpha}}{2\Gamma(\alpha+1)} \left[\|h\|_{\infty} \left(\sup_{y\in[a,b]} \left(\left\| D^{\alpha}_{y-;g} f \right\|_{\infty} + \left\| D^{\alpha}_{y+;g} f \right\|_{\infty} \right) \right) \right.$$

$$\left. + \|f\|_{\infty} \left(\sup_{y\in[a,b]} \left(\left\| D^{\alpha}_{y-;g} h \right\|_{\infty} + \left\| D^{\alpha}_{y+;g} h \right\|_{\infty} \right) \right) \right], \quad (9.84)$$

(3) if $\alpha \geq 1$, we get:

$$\|\Delta_n(f,h)\| \leq \frac{1}{2\Gamma(\alpha)(b-a)} (g(b)-g(a))^{\alpha-1} \cdot \quad (9.85)$$

$$\left\{ \|h\|_1 \left(\sup_{y\in[a,b]} \left(\left\| \left(D^{\alpha}_{y-;g} f \right) \circ g^{-1} \right\|_{1,[g(a),g(b)]} + \left\| \left(D^{\alpha}_{y+;g} f \right) \circ g^{-1} \right\|_{1,[g(a),g(b)]} \right) \right) + \right.$$

$$\left. \|f\|_1 \left(\sup_{y\in[a,b]} \left(\left\| \left(D^{\alpha}_{y-;g} h \right) \circ g^{-1} \right\|_{1,[g(a),g(b)]} + \left\| \left(D^{\alpha}_{y+;g} h \right) \circ g^{-1} \right\|_{1,[g(a),g(b)]} \right) \right) \right\},$$

(4) if $p,q>1 : \frac{1}{p}+\frac{1}{q}=1$, $\alpha > \frac{1}{q}$, we get:

$$\|\Delta_n(f,h)\| \leq \frac{(g(b)-g(a))^{\alpha-1+\frac{1}{p}}}{2\Gamma(\alpha)(p(\alpha-1)+1)^{\frac{1}{p}}} \cdot \quad (9.86)$$

$$\left\{ \|h\|_{\infty} \left(\sup_{y\in[a,b]} \left(\left\| \left(D^{\alpha}_{y-;g} f \right) \circ g^{-1} \right\|_{q,[g(a),g(b)]} + \left\| \left(D^{\alpha}_{y+;g} f \right) \circ g^{-1} \right\|_{q,[g(a),g(b)]} \right) \right) + \right.$$

$$\left. \|f\|_{\infty} \left(\sup_{y\in[a,b]} \left(\left\| \left(D^{\alpha}_{y-;g} h \right) \circ g^{-1} \right\|_{q,[g(a),g(b)]} + \left\| \left(D^{\alpha}_{y+;g} h \right) \circ g^{-1} \right\|_{q,[g(a),g(b)]} \right) \right) \right\}.$$

All right hand sides of (9.84)–(9.86) are finite.

Proof By Theorem 9.10 we have

$$h(y) f(y) = \frac{h(y)}{b-a} \int_a^b f(x)\,dx -$$

$$\sum_{k=1}^{n-1} \frac{h(y)\left(f\circ g^{-1}\right)^{(k)}(g(y))}{k!(b-a)} \int_a^b (g(x)-g(y))^k\,dx + h(y) R_1(f,y), \quad (9.87)$$

and

$$f(y)h(y) = \frac{f(y)}{b-a}\int_a^b h(x)\,dx -$$

$$\sum_{k=1}^{n-1}\frac{f(y)\left(h\circ g^{-1}\right)^{(k)}(g(y))}{k!(b-a)}\int_a^b (g(x)-g(y))^k\,dx + f(y)R_1(h,y), \quad (9.88)$$

$\forall\, y \in [a,b]$.

Then integrating (9.87) we find

$$\int_a^b h(y)f(y)\,dy = \frac{\left(\int_a^b h(y)\,dy\right)}{b-a}\left(\int_a^b f(x)\,dx\right) -$$

$$\sum_{k=1}^{n-1}\frac{1}{k!(b-a)}\int_a^b\int_a^b h(y)\left(f\circ g^{-1}\right)^{(k)}(g(y))(g(x)-g(y))^k\,dxdy$$

$$+\int_a^b h(y)R_1(f,y)\,dy, \quad (9.89)$$

and integrating (9.88) we obtain

$$\int_a^b f(y)h(y)\,dy = \frac{\left(\int_a^b f(y)\,dy\right)\left(\int_a^b h(x)\,dx\right)}{b-a} -$$

$$\sum_{k=1}^{n-1}\frac{1}{k!(b-a)}\int_a^b\int_a^b f(y)\left(h\circ g^{-1}\right)^{(k)}(g(y))(g(x)-g(y))^k\,dxdy$$

$$+\int_a^b f(y)R_1(h,y)\,dy. \quad (9.90)$$

Adding the last two equalities (9.89) and (9.90), we get:

$$2\int_a^b f(x)h(x)\,dx = \frac{2\left(\int_a^b f(x)\,dx\right)\left(\int_a^b h(x)\,dx\right)}{b-a} -$$

$$\sum_{k=1}^{n-1}\frac{1}{k!(b-a)}\left[\left[\int_a^b\int_a^b h(y)\left(f\circ g^{-1}\right)^{(k)}(g(y)) + f(y)\left(h\circ g^{-1}\right)^{(k)}(g(y))\right]\cdot\right.$$

$$\left.(g(x)-g(y))^k\,dxdy\right] + \int_a^b \left(h(y)R_1(f,y) + f(y)R_1(h,y)\right)dy. \quad (9.91)$$

Divide the last (9.91) by $2(b-a)$ to obtain (9.83).

Then, we upper bound $K_n(f,h)$ using Theorems 9.17–9.19, to obtain (9.84)–(9.86), respectively.

We use also that a norm is a continuous function. The theorem is proved. ■

We make

Remark 9.21 *(in support of the proof of Theorem* 9.20) Let $\alpha > 0$, $\alpha \notin \mathbb{N}$, $\lceil \alpha \rceil = n$. We have

$$\left(D_{y+;g}^{\alpha} f\right)(x) = \frac{1}{\Gamma(n-\alpha)} \int_y^x (g(x) - g(t))^{n-\alpha-1} g'(t) \left(f \circ g^{-1}\right)^{(n)} (g(t))\, dt, \tag{9.92}$$

$\forall\, x \in [y,b]$, and

$$\left(D_{y-;g}^{\alpha} f\right)(x) = \frac{(-1)^n}{\Gamma(n-\alpha)} \int_x^y (g(t) - g(x))^{n-\alpha-1} g'(t) \left(f \circ g^{-1}\right)^{(n)} (g(t))\, dt, \tag{9.93}$$

$\forall\, x \in [a,y]$, both are Bochner type integrals.

By change of variables for Bochner integrals, see [7, Lemma B. 4.10] and [8, p. 158], we get:

$$\left(D_{y+;g}^{\alpha} f\right)(x) = \frac{1}{\Gamma(n-\alpha)} \int_{g(y)}^{g(x)} (g(x) - z)^{n-\alpha-1} \left(f \circ g^{-1}\right)^{(n)} (z)\, dz =$$

$$\left(D_{g(y)+}^{\alpha} \left(f \circ g^{-1}\right)\right)(g(x)),\ \forall\, x \in [y,b], \tag{9.94}$$

and

$$\left(D_{y-;g}^{\alpha} f\right)(x) = \frac{(-1)^n}{\Gamma(n-\alpha)} \int_{g(x)}^{g(y)} (z - g(x))^{n-\alpha-1} \left(f \circ g^{-1}\right)^{(n)} (z)\, dz =$$

$$\left(D_{g(y)-}^{\alpha} \left(f \circ g^{-1}\right)\right)(g(x)),\ \forall\, x \in [a,y]. \tag{9.95}$$

Here $D_{g(y)+}^{\alpha}$, $D_{g(y)-}^{\alpha}$ are the left and right X-valued Caputo fractional differentiation operators.

Fix $w : w \geq x_0 \geq y_0$; $w, x_0, y_0 \in [a,b]$, then $g(w) \geq g(x_0) \geq g(y_0)$. Hence

$$\left\| \left(D_{y_0+;g}^{\alpha} f\right)(w) - \left(D_{x_0+;g}^{\alpha} f\right)(w) \right\| =$$

$$\left\| \left(D_{g(y_0)+}^{\alpha} \left(f \circ g^{-1}\right)\right)(g(w)) - \left(D_{g(x_0)+}^{\alpha} \left(f \circ g^{-1}\right)\right)(g(w)) \right\| =$$

$$\frac{1}{\Gamma(n-\alpha)} \left\| \int_{g(y_0)}^{g(x_0)} (g(w) - z)^{n-\alpha-1} \left(f \circ g^{-1}\right)^{(n)} (z)\, dz \right\| \leq \tag{9.96}$$

$$\frac{1}{\Gamma(n-\alpha)}\int_{g(y_0)}^{g(x_0)}(g(w)-z)^{n-\alpha-1}\left\|\left(f\circ g^{-1}\right)^{(n)}(z)\right\|dz\leq$$

$$\frac{\left\|\left(f\circ g^{-1}\right)^{(n)}\right\|_{\infty,[g(a),g(b)]}}{\Gamma(n-\alpha)}\int_{g(y_0)}^{g(x_0)}(g(w)-z)^{n-\alpha-1}dz=$$

$$\frac{\left\|\left(f\circ g^{-1}\right)^{(n)}\right\|_{\infty,[g(a),g(b)]}}{\Gamma(n-\alpha+1)}\left[(g(y_0)-z)^{n-\alpha}-(g(x_0)-z)^{n-\alpha}\right]\to 0,$$

as $y_0\to x_0$, then $g(y_0)\to g(x_0)$, proving continuity of $\left(D^{\alpha}_{g(y)+}\left(f\circ g^{-1}\right)\right)(g(x))$ with respect to $g(y)$, and of course continuity of $\left(D^{\alpha}_{y+;g}f\right)(x)$ in $y\in[a,b]$.

Similarly, it is proved that $\left(D^{\alpha}_{y-;g}f\right)(x)$ is continuous in $y\in[a,b]$, the proof is omitted.

Remark 9.22 Some examples for g follow:

$$\begin{aligned}
&g(x)=e^{x},\ x\in[a,b]\subset\mathbb{R},\\
&g(x)=\sin x,\\
&g(x)=\tan x,\\
\text{where } x\in&\left[-\frac{\pi}{2}+\varepsilon,\frac{\pi}{2}-\varepsilon\right],\text{ where }\varepsilon>0\text{ small.}
\end{aligned}$$

Indeed, the above examples of g are strictly increasing and continuous functions.

One can apply all of our results here for the above specific choices of g. We choose to omit this job.

References

1. Aliprantis, C.D., Border, K.C.: Infinite Dimensional Analysis. Springer, New York (2006)
2. Anastassiou, G.A.: Principles of general fractional analysis for Banach space valued functions. Bull. Allahabad Math. Soc. **32**(1), 71–145 (2017)
3. Anastassiou, G.: Intelligent Computations: Abstract Fractional Calculus, Inequalities, Approximations. Springer, Heidelberg (2018)
4. Anastassiou, G.: Generalized g-fractional vector representation formula and integral Inequalities for Banach space valued functions. J. Comput. Anal. Appl. **29**(6), 1063–1081 (2021)
5. Bochner integral. Encyclopedia of Mathematics. http://www.encyclopediaofmath.org/index.php?title=Bochner_integral&oldid=38659
6. Kreuter, M.: Sobolev space of vector-valued functions. Ulm University, Master thesis in Mathematics, Ulm, Germany (2015)
7. Mikkola, K.: Appendix B integration and differentiation in Banach spaces. http://math.aalto.fi/~kmikkola/research/thesis/contents/thesisb.pdf
8. Mikusinski, J.: The Bochner Integral. Academic, New York (1978)

Chapter 10
Iterated g-Fractional Vector Bochner Integral Representation Formulae and Inequalities for Banach Space Valued Functions

Here we present very general iterated fractional Bochner integral representation formulae for Banach space valued functions. Based on these we derive generalized and iterated left and right: fractional Poincaré type inequalities, fractional Opial type inequalities and fractional Hilbert–Pachpatte inequalities. All these inequalities are very general having in their background Bochner type integrals. See also [4].

10.1 Background

We need

Definition 10.1 *([2])* Let $[a,b] \subset \mathbb{R}$, $(X, \|\cdot\|)$ a Banach space, $g \in C^1([a,b])$ and increasing, $f \in C([a,b], X)$, $\nu > 0$.

We define the left Riemann–Liouville generalized fractional Bochner integral operator

$$\left(I_{a+;g}^{\nu} f\right)(x) := \frac{1}{\Gamma(\nu)} \int_a^x (g(x) - g(z))^{\nu-1} g'(z) f(z) \, dz, \tag{10.1}$$

$\forall\, x \in [a,b]$, where Γ is the gamma function.

The last integral is of Bochner type. Since $f \in C([a,b], X)$, then $f \in L_\infty([a,b], X)$. By [2] we get that $I_{a+;g}^{\nu} f \in C([a,b], X)$. Above we set $I_{a+;g}^{0} f := f$ and see that $\left(I_{a+;g}^{\nu} f\right)(a) = 0$.

When g is the identity function id, we get that $I_{a+;id}^{\nu} = I_{a+}^{\nu}$, the ordinary left Riemann–Liouville fractional integral

$$\left(I_{a+}^{\nu} f\right)(x) = \frac{1}{\Gamma(\nu)} \int_a^x (x-t)^{\nu-1} f(t) \, dt, \tag{10.2}$$

$\forall\, x \in [a,b]$, $\left(I_{a+}^{\nu} f\right)(a) = 0$.

G. A. Anastassiou, *Generalized Fractional Calculus*, Studies in Systems, Decision and Control 305, https://doi.org/10.1007/978-3-030-56962-4_10

We need

Theorem 10.2 ([2]) *Let* $\mu, \nu > 0$ *and* $f \in C([a,b], X)$. *Then*

$$I^{\mu}_{a+;g} I^{\nu}_{a+;g} f = I^{\mu+\nu}_{a+;g} f = I^{\nu}_{a+;g} I^{\mu}_{a+;g} f. \tag{10.3}$$

We need

Definition 10.3 *([2])* Let $[a,b] \subset \mathbb{R}$, $(X, \|\cdot\|)$ a Banach space, $g \in C^1([a,b])$ and increasing, $f \in C([a,b], X)$, $\nu > 0$.

We define the right Riemann–Liouville generalized fractional Bochner integral operator

$$\left(I^{\nu}_{b-;g} f\right)(x) := \frac{1}{\Gamma(\nu)} \int_x^b (g(z) - g(x))^{\nu-1} g'(z) f(z) dz, \tag{10.4}$$

$\forall\, x \in [a,b]$, where Γ is the gamma function.

The last integral is of Bochner type. Since $f \in C([a,b], X)$, then $f \in L_{\infty}([a,b], X)$. By [2] we get that $I^{\nu}_{b-;g} f \in C([a,b], X)$. Above we set $I^{0}_{b-;g} f := f$ and see that $\left(I^{\nu}_{b-;g} f\right)(b) = 0$.

When g is the identity function id, we get that $I^{\nu}_{b-;id} = I^{\nu}_{b-}$, the ordinary right Riemann–Liouville fractional integral

$$\left(I^{\nu}_{b-} f\right)(x) = \frac{1}{\Gamma(\nu)} \int_x^b (t - x)^{\nu-1} f(t) dt, \tag{10.5}$$

$\forall\, x \in [a,b]$, with $\left(I^{\nu}_{b-} f\right)(b) = 0$.

We need

Theorem 10.4 ([2]) *Let* $\mu, \nu > 0$ *and* $f \in C([a,b], X)$. *Then*

$$I^{\mu}_{b-;g} I^{\nu}_{b-;g} f = I^{\mu+\nu}_{b-;g} f = I^{\nu}_{b-;g} I^{\mu}_{b-;g} f. \tag{10.6}$$

We will use

Definition 10.5 *([2])* Let $\alpha > 0$, $\lceil \alpha \rceil = n$, $\lceil \cdot \rceil$ the ceiling of the number. Let $f \in C^n([a,b], X)$, where $[a,b] \subset \mathbb{R}$, and $(X, \|\cdot\|)$ is a Banach space. Let $g \in C^1([a,b])$, strictly increasing, such that $g^{-1} \in C^n([g(a), g(b)])$.

We define the left generalized g-fractional derivative X-valued of f of order α as follows:

$$\left(D^{\alpha}_{a+;g} f\right)(x) := \frac{1}{\Gamma(n-\alpha)} \int_a^x (g(x) - g(t))^{n-\alpha-1} g'(t) \left(f \circ g^{-1}\right)^{(n)} (g(t)) dt, \tag{10.7}$$

$\forall\, x \in [a,b]$. The last integral is of Bochner type.

If $\alpha \notin \mathbb{N}$, by [2], we have that $\left(D_{a+;g}^{\alpha} f\right) \in C([a,b], X)$.

We see that

$$\left(I_{a+;g}^{n-\alpha}\left(\left(f \circ g^{-1}\right)^{(n)} \circ g\right)\right)(x) = \left(D_{a+;g}^{\alpha} f\right)(x), \ \forall x \in [a,b]. \tag{10.8}$$

We set

$$D_{a+;g}^{n} f(x) := \left(\left(f \circ g^{-1}\right)^{(n)} \circ g\right)(x) \in C([a,b], X), \ n \in \mathbb{N}, \tag{10.9}$$

$$D_{a+;g}^{0} f(x) = f(x), \ \forall x \in [a,b].$$

When $g = id$, then

$$D_{a+;g}^{\alpha} f = D_{a+;id}^{\alpha} f = D_{*a}^{\alpha} f, \tag{10.10}$$

the usual left X-valued Caputo fractional derivative, see [3].

We will use

Definition 10.6 *([2])* Let $\alpha > 0$, $\lceil \alpha \rceil = n$, $\lceil \cdot \rceil$ the ceiling of the number. Let $f \in C^{n}([a,b], X)$, where $[a,b] \subset \mathbb{R}$, and $(X, \|\cdot\|)$ is a Banach space. Let $g \in C^{1}([a,b])$, strictly increasing, such that $g^{-1} \in C^{n}([g(a), g(b)])$.

We define the right generalized g-fractional derivative X-valued of f of order α as follows:

$$\left(D_{b-;g}^{\alpha} f\right)(x) := \frac{(-1)^{n}}{\Gamma(n-\alpha)} \int_{x}^{b} (g(t) - g(x))^{n-\alpha-1} g'(t) \left(f \circ g^{-1}\right)^{(n)} (g(t))\, dt, \tag{10.11}$$

$\forall x \in [a,b]$. The last integral is of Bochner type.

If $\alpha \notin \mathbb{N}$, by [2], we have that $\left(D_{b-;g}^{\alpha} f\right) \in C([a,b], X)$.

We see that

$$I_{b-;g}^{n-\alpha}\left((-1)^{n}\left(f \circ g^{-1}\right)^{(n)} \circ g\right)(x) = \left(D_{b-;g}^{\alpha} f\right)(x), \ a \leq x \leq b. \tag{10.12}$$

We set

$$D_{b-;g}^{n} f(x) := (-1)^{n}\left(\left(f \circ g^{-1}\right)^{n} \circ g\right)(x) \in C([a,b], X), \ n \in \mathbb{N}, \tag{10.13}$$

$$D_{b-;g}^{0} f(x) := f(x), \ \forall x \in [a,b].$$

When $g = id$, then

$$D_{b-;g}^{\alpha} f(x) = D_{b-;id}^{\alpha} f(x) = D_{b-}^{\alpha} f, \tag{10.14}$$

the usual right X-valued Caputo fractional derivative, see [3].

We make

Remark 10.7 All as in Definition 10.5. We have (by [6, Theorem 2.5, p. 7])

$$\left\| \left(D_{a+;g}^{\alpha} f \right) (x) \right\| \leq \frac{1}{\Gamma (n-\alpha)} \int_{a}^{x} (g(x) - g(t))^{n-\alpha-1} g'(t) \left\| \left(f \circ g^{-1} \right)^{(n)} (g(t)) \right\| dt$$

$$\leq \frac{\left\| \left(f \circ g^{-1} \right)^{(n)} \circ g \right\|_{\infty,[a,b]}}{\Gamma (n-\alpha)} \int_{g(a)}^{g(x)} (g(x) - g(t))^{n-\alpha-1} dg(t) =$$

$$\frac{\left\| \left(f \circ g^{-1} \right)^{(n)} \circ g \right\|_{\infty,[a,b]}}{\Gamma (n-\alpha+1)} (g(x) - g(a))^{n-\alpha}. \tag{10.15}$$

That is

$$\left\| \left(D_{a+;g}^{\alpha} f \right) (x) \right\| \leq \frac{\left\| \left(f \circ g^{-1} \right)^{(n)} \circ g \right\|_{\infty,[a,b]}}{\Gamma (n-\alpha+1)} (g(x) - g(a))^{n-\alpha}, \tag{10.16}$$

$\forall\, x \in [a, b]$.

If $\alpha \notin \mathbb{N}$, then $\left(D_{a+;g}^{\alpha} f \right)(a) = 0$.

Similarly, by Definition 10.6 we derive

$$\left\| \left(D_{b-;g}^{\alpha} f \right) (x) \right\| \leq \frac{1}{\Gamma (n-\alpha)} \int_{x}^{b} (g(t) - g(x))^{n-\alpha-1} g'(t) \left\| \left(f \circ g^{-1} \right)^{(n)} (g(t)) \right\| dt$$

$$\leq \frac{\left\| \left(f \circ g^{-1} \right)^{(n)} \circ g \right\|_{\infty,[a,b]}}{\Gamma (n-\alpha)} \int_{g(x)}^{g(b)} (g(t) - g(x))^{n-\alpha-1} dg(t) =$$

$$\frac{\left\| \left(f \circ g^{-1} \right)^{(n)} \circ g \right\|_{\infty,[a,b]}}{\Gamma (n-\alpha+1)} (g(b) - g(x))^{n-\alpha}. \tag{10.17}$$

That is

$$\left\| \left(D_{b-;g}^{\alpha} f \right) (x) \right\| \leq \frac{\left\| \left(f \circ g^{-1} \right)^{(n)} \circ g \right\|_{\infty,[a,b]}}{\Gamma (n-\alpha+1)} (g(b) - g(x))^{n-\alpha}, \tag{10.18}$$

$\forall\, x \in [a, b]$.

If $\alpha \notin \mathbb{N}$, then $\left(D_{b-;g}^{\alpha} f \right)(b) = 0$.

Notation 10.8 *We denote by*

$$D_{a+;g}^{n\alpha} := D_{a+;g}^{\alpha} D_{a+;g}^{\alpha} \ldots D_{a+;g}^{\alpha} \text{ (n times)}, \ n \in \mathbb{N}, \tag{10.19}$$

$$I_{a+;g}^{n\alpha} := I_{a+;g}^{\alpha} I_{a+;g}^{\alpha} \dots I_{a+;g}^{\alpha}, \tag{10.20}$$

$$D_{b-;g}^{n\alpha} := D_{b-;g}^{\alpha} D_{b-;g}^{\alpha} \dots D_{b-;g}^{\alpha}, \tag{10.21}$$

and

$$I_{b-;g}^{n\alpha} := I_{b-;g}^{\alpha} I_{b-;g}^{\alpha} \dots I_{b-;g}^{\alpha}, \tag{10.22}$$

(n times), $n \in \mathbb{N}$.

We are motivated by the following generalized fractional Ostrowski type inequality:

Theorem 10.9 ([2]) *Let* $g \in C^1([a,b])$ *and strictly increasing, such that* $g^{-1} \in C^1([g(a), g(b)])$, *and* $0 < \alpha < 1$, $n \in \mathbb{N}$, $f \in C^1([a,b], X)$, *where* $(X, \|\cdot\|)$ *is a Banach space. Let* $x_0 \in [a,b]$ *be fixed. Assume that* $F_k^{x_0} := D_{x_0-;g}^{k\alpha} f$, *for* $k = 1, \dots, n$, *fulfill* $F_k^{x_0} \in C^1([a, x_0], X)$ *and* $\left(D_{x_0-;g}^{i\alpha} f\right)(x_0) = 0$, $i = 2, \dots, n$.

Similarly, we assume that $G_k^{x_0} := D_{x_0+;g}^{k\alpha} f$, *for* $k = 1, \dots, n$, *fulfill* $G_k^{x_0} \in C^1([x_0, b], X)$ *and* $\left(D_{x_0+;g}^{i\alpha} f\right)(x_0) = 0$, $i = 2, \dots, n$.

Then

$$\left\| \frac{1}{b-a} \int_a^b f(x)\,dx - f(x_0) \right\| \leq \frac{1}{(b-a)\,\Gamma((n+1)\alpha+1)} \cdot$$

$$\left\{ (g(b) - g(x_0))^{(n+1)\alpha} (b - x_0) \left\| D_{x_0+;g}^{(n+1)\alpha} f \right\|_{\infty,[x_0,b]} + \right.$$

$$\left. (g(x_0) - g(a))^{(n+1)\alpha} (x_0 - a) \left\| D_{x_0-;g}^{(n+1)\alpha} f \right\|_{\infty,[a,x_0]} \right\}. \tag{10.23}$$

In this work we will present several generalized fractional Bochner integral inequalities.

We mention the following g-left generalized modified X-valued Taylor's formula.

Theorem 10.10 ([2]) *Let* $0 < \alpha \leq 1$, $n \in \mathbb{N}$, $f \in C^1([a,b], X)$, $g \in C^1([a,b])$, *strictly increasing, such that* $g^{-1} \in C^1([g(a), g(b)])$. *Let* $F_k := D_{a+;g}^{k\alpha} f$, $k = 1, \dots, n$, *that fulfill* $F_k \in C^1([a,b], X)$.

Then

$$f(x) = \sum_{i=0}^{n} \frac{(g(x) - g(a))^{i\alpha}}{\Gamma(i\alpha + 1)} \left(D_{a+;g}^{i\alpha} f\right)(a) +$$

$$\frac{1}{\Gamma((n+1)\alpha)} \int_a^x (g(x) - g(t))^{(n+1)\alpha - 1} g'(t) \left(D_{a+;g}^{(n+1)\alpha} f\right)(t)\,dt, \tag{10.24}$$

$\forall\, x \in [a,b]$.

Clearly here $D_{a+;g}^{(n+1)\alpha} f \in C([a,b], X)$.

We also mention the following g-right generalized modified X-valued Taylor's formula.

Theorem 10.11 ([2]) *Let* $f \in C^1([a,b],X)$, $g \in C^1([a,b])$, *strictly increasing, such that* $g^{-1} \in C^1([g(a),g(b)])$. *Suppose that* $F_k := D_{b-;g}^{k\alpha} f$, $k = 1,\ldots,n$, *fulfill* $F_k \in C^1([a,b],X)$, *where* $0 < \alpha \leq 1$, $n \in \mathbb{N}$.

Then

$$f(x) = \sum_{i=0}^{n} \frac{(g(b) - g(x))^{i\alpha}}{\Gamma(i\alpha+1)} \left(D_{b-;g}^{i\alpha} f\right)(b) +$$

$$\frac{1}{\Gamma((n+1)\alpha)} \int_x^b (g(t) - g(x))^{(n+1)\alpha-1} g'(t) \left(D_{b-;g}^{(n+1)\alpha} f\right)(t)\, dt, \tag{10.25}$$

$\forall\, x \in [a,b]$.

Clearly here $D_{b-;g}^{(n+1)\alpha} f \in C([a,b],X)$.

For differentiation of functions from real numbers to normed linear spaces the definition is the same as for the real valued functions, however the limit and convergence is in the norm of linear space $(X, \|\cdot\|)$.

We state

Corollary 10.12 (to Theorem 10.10) *Let* $0 < \alpha < 1$, $n \in \mathbb{N}$, $f \in C^1([a,b],X)$, $g \in C^1([a,b])$, *strictly increasing, such that* $g^{-1} \in C^1([g(a),g(b)])$. *Assume* $D_{a+;g}^{k\alpha} f \in C^1([a,b],X)$, $k = 1,\ldots,n$, *and* $\left(D_{a+;g}^{i\alpha} f\right)(a) = 0$, $i = 0,2,3,\ldots,n$.

Then

$$f(x) = \frac{1}{\Gamma((n+1)\alpha)} \int_a^x (g(x) - g(t))^{(n+1)\alpha-1} g'(t) \left(D_{a+;g}^{(n+1)\alpha} f\right)(t)\, dt, \tag{10.26}$$

$\forall\, x \in [a,b]$.

We state

Corollary 10.13 (to Theorem 10.11) *Let* $f \in C^1([a,b],X)$, $g \in C^1([a,b])$, *strictly increasing, such that* $g^{-1} \in C^1([g(a),g(b)])$. *Suppose that* $D_{b-;g}^{k\alpha} f \in C^1([a,b],X)$, *for* $k = 1,\ldots,n$; *where* $0 < \alpha < 1$, $n \in \mathbb{N}$. *We further assume that* $\left(D_{b-;g}^{i\alpha} f\right)(b) = 0$, $i = 0,2,3,\ldots,n$.

Then

$$f(x) = \frac{1}{\Gamma((n+1)\alpha)} \int_x^b (g(t) - g(x))^{(n+1)\alpha-1} g'(t) \left(D_{b-;g}^{(n+1)\alpha} f\right)(t)\, dt, \tag{10.27}$$

$\forall\, x \in [a,b]$.

For the Bochner integral excellent resources are [5, 8] and [1, pp. 422–428].

10.2 Main Results

We give the following representation formula:

Theorem 10.14 *All as in Corollary 10.12. Let $\gamma > 0$ with $\lceil \gamma \rceil = m < n+1$, such that $m < (n+1)\alpha$ (i.e. $\frac{m}{n+1} < \alpha < 1$). Then*

$$\left(D_{a+;g}^{\gamma} f \right)(x) = \tag{10.28}$$

$$\frac{1}{\Gamma((n+1)\alpha - \gamma)} \int_a^x (g(x) - g(t))^{(n+1)\alpha - \gamma - 1} g'(t) \left(D_{a+;g}^{(n+1)\alpha} f \right)(t) \, dt,$$

$\forall\, x \in [a,b]$ and $\left(D_{a+;g}^{\gamma} f \right) \in C([a,b], X)$.

Proof By Corollary 10.12 we have

$$f(x) = \frac{1}{\Gamma((n+1)\alpha)} \int_a^x (g(x) - g(t))^{(n+1)\alpha - 1} g'(t) \left(D_{a+;g}^{(n+1)\alpha} f \right)(t) \, dt, \tag{10.29}$$

$\forall\, x \in [a,b]$.

We can write

$$f(x) = \frac{1}{\Gamma((n+1)\alpha)} \int_a^x (g(x) - g(t))^{(n+1)\alpha - 1} \left(D_{a+;g}^{(n+1)\alpha} f \right)(t) \, dg(t) =$$

(set $z := g(t)$, $a \le t \le b$)

$$\frac{1}{\Gamma((n+1)\alpha)} \int_{g(a)}^{g(x)} (g(x) - z)^{(n+1)\alpha - 1} \left(\left(D_{a+;g}^{(n+1)\alpha} f \right) \circ g^{-1} \right)(z) \, dz, \tag{10.30}$$

$\forall\, x \in [a,b]$.

Hence it holds ($y = g(x)$)

$$\left(f \circ g^{-1} \right)(y) = \frac{1}{\Gamma((n+1)\alpha)} \int_{g(a)}^{y} (y - z)^{(n+1)\alpha - 1} \left(\left(D_{a+;g}^{(n+1)\alpha} f \right) \circ g^{-1} \right)(z) \, dz, \tag{10.31}$$

$\forall\, y \in [g(a), g(b)]$.

By assuming $(n+1)\alpha - 1 > 0$ (equivalently, $\alpha > \frac{1}{n+1}$) and by [2], we get:

$$\left(\left(f \circ g^{-1} \right)(y) \right)' = \frac{(n+1)\alpha - 1}{\Gamma((n+1)\alpha)} \int_{g(a)}^{y} (y - z)^{(n+1)\alpha - 2} \left(\left(D_{a+;g}^{(n+1)\alpha} f \right) \circ g^{-1} \right)(z) \, dz, \tag{10.32}$$

$\forall\, y \in [g(a), g(b)]$.

If $(n+1)\alpha - 2 > 0$ (equivalently, $\alpha > \frac{2}{n+1}$), we get

$$\left(\left(f \circ g^{-1}\right)(y)\right)'' = \frac{((n+1)\alpha - 1)((n+1)\alpha - 2)}{\Gamma((n+1)\alpha)}$$

$$\int_{g(a)}^{y} (y-z)^{(n+1)\alpha - 3} \left(\left(D_{a+;g}^{(n+1)\alpha} f\right) \circ g^{-1}\right)(z)\, dz, \tag{10.33}$$

$\forall\, y \in [g(a), g(b)]$.

In general, if $(n+1)\alpha - m > 0$ (equivalently, $\alpha > \frac{m}{n+1}$) we get that there exists

$$\left(f \circ g^{-1}\right)^{(m)}(y) = \frac{\prod_{j=1}^{m} ((n+1)\alpha - j)}{\Gamma((n+1)\alpha)}$$

$$\int_{g(a)}^{y} (y-z)^{(n+1)\alpha - m - 1} \left(\left(D_{a+;g}^{(n+1)\alpha} f\right) \circ g^{-1}\right)(z)\, dz, \tag{10.34}$$

$\forall\, y \in [g(a), g(b)]$.

By [2] we get $\left(f \circ g^{-1}\right)^{(m)} \in C([g(a), g(b)])$.

By (10.2) we have that

$$\left(f \circ g^{-1}\right)^{(m)}(y) = \frac{\prod_{j=1}^{m} ((n+1)\alpha - j)\,\Gamma((n+1)\alpha - m)}{\Gamma((n+1)\alpha)}$$

$$I_{g(a)+}^{(n+1)\alpha - m} \left(\left(D_{a+;g}^{(n+1)\alpha} f\right) \circ g^{-1}\right)(y) = I_{g(a)+}^{(n+1)\alpha - m} \left(\left(D_{a+;g}^{(n+1)\alpha} f\right) \circ g^{-1}\right)(y). \tag{10.35}$$

That is

$$\left(f \circ g^{-1}\right)^{(m)}(y) = I_{g(a)+}^{(n+1)\alpha - m} \left(\left(D_{a+;g}^{(n+1)\alpha} f\right) \circ g^{-1}\right)(y), \tag{10.36}$$

$\forall\, y \in [g(a), g(b)]$, and

$$\left(f \circ g^{-1}\right)^{(m)}(g(x)) = I_{g(a)+}^{(n+1)\alpha - m} \left(\left(D_{a+;g}^{(n+1)\alpha} f\right) \circ g^{-1}\right)(g(x)), \tag{10.37}$$

$\forall\, x \in [a, b]$.

Clearly, it holds (see [7, Lemma B.4.10] and [8, p. 158])

$$\left(f \circ g^{-1}\right)^{(m)}(g(x)) = \left(I_{a+;g}^{(n+1)\alpha - m} \left(D_{a+;g}^{(n+1)\alpha} f\right)\right)(x), \tag{10.38}$$

$\forall\, x \in [a, b]$.

Let $\gamma > 0$ with $\lceil \gamma \rceil = m < n+1$, such that $m < (n+1)\alpha$ (equivalently, $\alpha > \frac{m}{n+1}$).

We have that (case of $\gamma < m$)

$$\left(D_{a+;g}^{\gamma} f \right)(x) \overset{(10.8)}{=} \left(I_{a+;g}^{m-\gamma} \left(\left(f \circ g^{-1} \right)^{(m)} \circ g \right) \right)(x) \overset{(10.38)}{=}$$

$$I_{a+;g}^{m-\gamma} \left(I_{a+;g}^{(n+1)\alpha - m} \left(D_{a+;g}^{(n+1)\alpha} f \right) \right)(x) = \left(I_{a+}^{(n+1)\alpha-\gamma} \left(D_{a+;g}^{(n+1)\alpha} f \right) \right)(x), \quad (10.39)$$

(by (10.3)).

We have proved that

$$\left(D_{a+;g}^{\gamma} f \right)(x) = \left(I_{a+}^{(n+1)\alpha-\gamma} \left(D_{a+;g}^{(n+1)\alpha} f \right) \right)(x), \quad (10.40)$$

$\forall\, x \in [a,b]$, which is continuous, by [2]. ■

We continue with

Theorem 10.15 *All as in Corollary 10.12 and let $\gamma > 0$ with $\lceil \gamma \rceil = m$. When $\frac{\gamma+m}{n+1} < \alpha < 1$, we get that*

$$\left(D_{a+;g}^{2\gamma} f \right)(x) = \frac{1}{\Gamma\left((n+1)\alpha - 2\gamma \right)}$$

$$\int_a^x \left(g(x) - g(t) \right)^{(n+1)\alpha - 2\gamma - 1} g'(t) \left(D_{a+;g}^{(n+1)\alpha} f \right)(t)\, dt, \quad (10.41)$$

$\forall\, x \in [a,b]$ *and* $\left(D_{a+;g}^{2\gamma} f \right) \in C\left([a,b], X \right)$.

Proof Call $\lambda := (n+1)\alpha - \gamma - 1$, i.e. $\lambda + 1 = (n+1)\alpha - \gamma$, and call $\delta := (n+1)\alpha$. Then we can rewrite (10.28) as

$$\left(D_{a+;g}^{\gamma} f \right)(x) = \frac{1}{\Gamma(\lambda+1)} \int_a^x \left(g(x) - g(t) \right)^{\lambda} g'(t) \left(D_{a+;g}^{\delta} f \right)(t)\, dt, \quad (10.42)$$

$\forall\, x \in [a,b]$.

That is ($z = g(t)$, $a \le t \le b$)

$$\left(D_{a+;g}^{\gamma} f \right)(x) = \frac{1}{\Gamma(\lambda+1)} \int_{g(a)}^{g(x)} \left(g(x) - z \right)^{\lambda} \left(\left(D_{a+;g}^{\delta} f \right) \circ g^{-1} \right)(z)\, dz, \quad (10.43)$$

$\forall\, x \in [a,b]$.

Hence it holds ($y = g(x)$)

$$\left(\left(D_{a+;g}^{\gamma} f \right) \circ g^{-1} \right)(y) = \frac{1}{\Gamma(\lambda+1)} \int_{g(a)}^{y} (y - z)^{\lambda} \left(\left(D_{a+;g}^{\delta} f \right) \circ g^{-1} \right)(z)\, dz, \quad (10.44)$$

$\forall\, y \in [g(a), g(b)]$.

If $\lambda > 0$, then

$$\left(\left(D^{\gamma}_{a+;g} f\right) \circ g^{-1}\right)'(y) = \frac{\lambda}{\Gamma(\lambda+1)} \int_{g(a)}^{y} (y-z)^{\lambda-1} \left(\left(D^{\delta}_{a+;g} f\right) \circ g^{-1}\right)(z)\, dz, \tag{10.45}$$

$\forall\, y \in [g(a), g(b)]$.

If $\lambda - 1 > 0$, then

$$\left(\left(D^{\gamma}_{a+;g} f\right) \circ g^{-1}\right)''(y) = \frac{\lambda(\lambda-1)}{\Gamma(\lambda+1)} \int_{g(a)}^{y} (y-z)^{\lambda-2} \left(\left(D^{\delta}_{a+;g} f\right) \circ g^{-1}\right)(z)\, dz, \tag{10.46}$$

$\forall\, y \in [g(a), g(b)]$.

If $\lambda - 2 > 0$, then

$$\left(\left(D^{\gamma}_{a+;g} f\right) \circ g^{-1}\right)'''(y) = \frac{\lambda(\lambda-1)(\lambda-2)}{\Gamma(\lambda+1)} \int_{g(a)}^{y} (y-z)^{\lambda-3} \left(\left(D^{\delta}_{a+;g} f\right) \circ g^{-1}\right)(z)\, dz, \tag{10.47}$$

$\forall\, y \in [g(a), g(b)]$, etc.

In general, if $\lambda - m + 1 > 0$, then

$$\left(\left(D^{\gamma}_{a+;g} f\right) \circ g^{-1}\right)^{(m)}(y) = \frac{\lambda(\lambda-1)(\lambda-2)\ldots(\lambda-m+1)}{\Gamma(\lambda+1)}$$

$$\int_{g(a)}^{y} (y-z)^{(\lambda-m+1)-1} \left(\left(D^{\delta}_{a+;g} f\right) \circ g^{-1}\right)(z)\, dz =$$

$$\frac{\lambda(\lambda-1)(\lambda-2)\ldots(\lambda-m+1)\, I^{(\lambda-m+1)}_{g(\alpha)+}\left(\left(D^{\delta}_{a+;g} f\right) \circ g^{-1}\right)(y)}{\Gamma(\lambda+1)}$$

$$= I^{(\lambda-m+1)}_{g(a)+}\left(\left(D^{\delta}_{a+;g} f\right) \circ g^{-1}\right)(y), \tag{10.48}$$

$\forall\, y \in [g(a), g(b)]$.

That is, if $\lambda - m + 1 > 0$, we have

$$\left(\left(D^{\gamma}_{a+;g} f\right) \circ g^{-1}\right)^{(m)}(y) = I^{(\lambda-m+1)}_{g(a)+}\left(\left(D^{\delta}_{a+;g} f\right) \circ g^{-1}\right)(y), \tag{10.49}$$

$\forall\, y \in [g(a), g(b)]$.

Hence

$$\left(\left(D^{\gamma}_{a+;g} f\right) \circ g^{-1}\right)^{(m)}(g(x)) = I^{(\lambda-m+1)}_{g(a)+}\left(\left(D^{(n+1)\alpha}_{a+;g} f\right) \circ g^{-1}\right)(g(x)), \tag{10.50}$$

$\forall\, x \in [a, b]$.

Clearly, it holds by change of variables

$$\left(\left(D_{a+;g}^{\gamma}f\right)\circ g^{-1}\right)^{(m)}(g(x))=\left(I_{a+}^{(\lambda-m+1)}\left(D_{a+;g}^{(n+1)\alpha}f\right)\right)(x), \tag{10.51}$$

$\forall\, x\in[a,b]$, under $\lambda-m+1>0$.

We see that

$$\left(D_{a+;g}^{2\gamma}f\right)(x)=\left(D_{a+;g}^{\gamma}\left(D_{a+;g}^{\gamma}f\right)\right)(x)=$$

$$\left(I_{a+;g}^{m-\gamma}\left(\left(\left(D_{a+;g}^{\gamma}f\right)\circ g^{-1}\right)^{(m)}\circ g\right)\right)(x)\overset{(10.51)}{=}\left(I_{a+;g}^{m-\gamma}\left(I_{a+;g}^{(\lambda-m+1)}\left(D_{a+;g}^{(n+1)\alpha}f\right)\right)\right)(x)$$

$$=\left(I_{a+;g}^{\lambda-\gamma+1}\left(D_{a+;g}^{(n+1)\alpha}f\right)\right)(x)=\left(I_{a+;g}^{(n+1)\alpha-\gamma-1-\gamma+1}\left(D_{a+;g}^{(n+1)\alpha}f\right)\right)(x) \tag{10.52}$$

$$=\left(I_{a+;g}^{(n+1)\alpha-2\gamma}\left(D_{a+;g}^{(n+1)\alpha}f\right)\right)(x),$$

$\forall\, x\in[a,b]$.

We have derived that

$$\left(D_{a+;g}^{2\gamma}f\right)(x)=\left(I_{a+;g}^{(n+1)\alpha-2\gamma}\left(D_{a+;g}^{(n+1)\alpha}f\right)\right)(x), \tag{10.53}$$

$\forall\, x\in[a,b]$, under the condition $\frac{\gamma+m}{n+1}<\alpha<1$. ■

Similarly to the last two theorems, it holds

Theorem 10.16 *All as in Corollary 10.12 and let $\gamma>0$ with $\lceil\gamma\rceil=m$. When $\frac{m+2\gamma}{n+1}<\alpha<1$, we obtain that*

$$\left(D_{a+;g}^{3\gamma}f\right)(x)=\frac{1}{\Gamma((n+1)\alpha-3\gamma)}$$

$$\int_a^x(g(x)-g(t))^{(n+1)\alpha-3\gamma-1}g'(t)\left(D_{a+;g}^{(n+1)\alpha}f\right)(t)\,dt, \tag{10.54}$$

$\forall\, x\in[a,b]$ and $\left(D_{a+;g}^{3\gamma}f\right)\in C([a,b],X)$.

In general, we derive the iterated left fractional derivative representation formula:

Theorem 10.17 *All as in Corollary 10.12 and let $\gamma>0$ with $\lceil\gamma\rceil=m$. When $\frac{m+(k-1)\gamma}{n+1}<\alpha<1$, $k\in\mathbb{N}$, we obtain that*

$$\left(D_{a+;g}^{k\gamma}f\right)(x)=\frac{1}{\Gamma((n+1)\alpha-k\gamma)}$$

$$\int_a^x (g(x) - g(t))^{(n+1)\alpha - k\gamma - 1} g'(t) \left(D_{a+;g}^{(n+1)\alpha} f\right)(t)\, dt, \tag{10.55}$$

$\forall\, x \in [a, b]$ *and* $\left(D_{a+;g}^{k\gamma} f\right) \in C([a, b], X)$.

Similarly, it holds the iterated right fractional derivative representation formula:

Theorem 10.18 *All as in Corollary 10.13 and let* $\gamma > 0$ *with* $\lceil \gamma \rceil = m$. *Here we assume* $\frac{m+(k-1)\gamma}{n+1} < \alpha < 1$, $k \in \mathbb{N}$. *Then*

$$\left(D_{b-;g}^{k\gamma} f\right)(x) = \frac{1}{\Gamma((n+1)\alpha - k\gamma)}$$

$$\int_x^b (g(t) - g(x))^{(n+1)\alpha - k\gamma - 1} g'(t) \left(D_{b-;g}^{(n+1)\alpha} f\right)(t)\, dt, \tag{10.56}$$

$\forall\, x \in [a, b]$ *and* $\left(D_{b-;g}^{k\gamma} f\right) \in C([a, b], X)$.

It follows a left generalized and iterated fractional Poincaré type inequality:

Theorem 10.19 *Here all as in Theorem 10.17. Let* $p, q > 1 : \frac{1}{p} + \frac{1}{q} = 1$. *We further assume that* ($k \in \mathbb{N}$)

$$1 > \alpha > \max\left(\frac{m + (k-1)\gamma}{n+1}, \frac{k\gamma q + 1}{(n+1) q}\right). \tag{10.57}$$

Then

$$\left\| D_{a+;g}^{k\gamma} f \right\|_{q,[a,b]} \leq$$

$$\frac{(g(b) - g(a))^{((n+1)\alpha - k\gamma - 1) + \frac{1}{p}} (b-a)^{\frac{1}{q}}}{\Gamma((n+1)\alpha - k\gamma) \left(p((n+1)\alpha - k\gamma - 1) + 1\right)^{\frac{1}{p}}} \left\| \left(D_{a+;g}^{(n+1)\alpha} f\right) \circ g^{-1} \right\|_{q,[g(a),g(b)]}. \tag{10.58}$$

Proof We use (10.55). We observe that

$$\left\| \left(D_{a+;g}^{k\gamma} f\right)(x) \right\| \leq \frac{1}{\Gamma((n+1)\alpha - k\gamma)}\cdot \tag{10.59}$$

$$\int_a^x (g(x) - g(t))^{(n+1)\alpha - k\gamma - 1} g'(t) \left\| \left(D_{a+;g}^{(n+1)\alpha} f\right)(t) \right\| dt =$$

$$\frac{1}{\Gamma((n+1)\alpha - k\gamma)} \int_{g(a)}^{g(x)} (g(x) - z)^{(n+1)\alpha - k\gamma - 1} \left\| \left(D_{a+;g}^{(n+1)\alpha} f\right)\left(g^{-1}(z)\right) \right\| dz$$

$$\leq \frac{1}{\Gamma((n+1)\alpha - k\gamma)} \left(\int_{g(a)}^{g(x)} (g(x) - z)^{p((n+1)\alpha - k\gamma - 1)} dz \right)^{\frac{1}{p}} \tag{10.60}$$

$$\cdot \left(\int_{g(a)}^{g(b)} \left\| \left(D_{a+;g}^{(n+1)\alpha} f \right) \left(g^{-1}(z) \right) \right\|^q dz \right)^{\frac{1}{q}}$$

$$= \frac{(g(x) - g(a))^{\frac{p((n+1)\alpha - k\gamma - 1)+1}{p}}}{\Gamma((n+1)\alpha - k\gamma)(p((n+1)\alpha - k\gamma - 1) + 1)^{\frac{1}{p}}} \left\| \left(D_{a+;g}^{(n+1)\alpha} f \right) \circ g^{-1} \right\|_{q,[g(a),g(b)]}.$$

Thus we have

$$\left\| \left(D_{a+;g}^{k\gamma} f \right)(x) \right\|^q \leq \frac{(g(x) - g(a))^{(p((n+1)\alpha - k\gamma - 1)+1)\frac{q}{p}}}{\Gamma((n+1)\alpha - k\gamma)^q (p((n+1)\alpha - k\gamma - 1) + 1)^{\frac{q}{p}}} \cdot \left\| \left(D_{a+;g}^{(n+1)\alpha} f \right) \circ g^{-1} \right\|_{q,[g(a),g(b)]}^q. \tag{10.61}$$

Therefore it holds

$$\int_a^b \left\| \left(D_{a+;g}^{k\gamma} f \right)(x) \right\|^q dx \leq \frac{(g(x) - g(a))^{(p((n+1)\alpha - k\gamma - 1)+1)\frac{q}{p}} (b-a)}{\Gamma((n+1)\alpha - k\gamma)^q (p((n+1)\alpha - k\gamma - 1) + 1)^{\frac{q}{p}}} \cdot \left\| \left(D_{a+;g}^{(n+1)\alpha} f \right) \circ g^{-1} \right\|_{q,[g(a),g(b)]}^q, \tag{10.62}$$

proving the claim. ■

It follows a right generalized and iterated fractional Poincaré type inequality:

Theorem 10.20 *Here all as in Theorem 10.18. Let $p, q > 1 : \frac{1}{p} + \frac{1}{q} = 1$. We further assume that ($k \in \mathbb{N}$)*

$$1 > \alpha > \max\left(\frac{m + (k-1)\gamma}{n+1}, \frac{k\gamma q + 1}{(n+1)q} \right). \tag{10.63}$$

Then

$$\left\| D_{b-;g}^{k\gamma} f \right\|_{q,[a,b]} \leq$$

$$\frac{(g(b) - g(a))^{((n+1)\alpha - k\gamma - 1) + \frac{1}{p}} (b-a)^{\frac{1}{q}}}{\Gamma((n+1)\alpha - k\gamma)(p((n+1)\alpha - k\gamma - 1) + 1)^{\frac{1}{p}}} \left\| \left(D_{b-;g}^{(n+1)\alpha} f \right) \circ g^{-1} \right\|_{q,[g(a),g(b)]}. \tag{10.64}$$

Proof As similar to Theorem 10.19 is omitted. ■

Next comes a left generalized and iterated fractional Opial type inequality:

Theorem 10.21 *All here as in Theorem 10.17. Let $p, q > 1 : \frac{1}{p} + \frac{1}{q} = 1$. We further assume that ($k \in \mathbb{N}$)*

$$1 > \alpha > \max \left(\frac{m + (k-1)\gamma}{n+1}, \frac{k\gamma q + 1}{(n+1)q} \right). \tag{10.65}$$

Then

$$\int_{g(a)}^{y} \left\| \left(\left(D_{a+;g}^{k\gamma} f \right) \circ g^{-1} \right)(w) \right\| \left\| \left(\left(D_{a+;g}^{(n+1)\alpha} f \right) \circ g^{-1} \right)(w) \right\| dw \le$$

$$\frac{(y - g(a))^{((n+1)\alpha - k\gamma - 1) + \frac{2}{p}}}{2^{\frac{1}{q}} \Gamma((n+1)\alpha - k\gamma) \left[(p((n+1)\alpha - k\gamma - 1) + 1)(p((n+1)\alpha - k\gamma - 1) + 2) \right]^{\frac{1}{p}}} \cdot$$

$$\left(\int_{g(a)}^{y} \left\| \left(\left(D_{a+;g}^{(n+1)\alpha} f \right) \circ g^{-1} \right)(w) \right\|^{q} dw \right)^{\frac{2}{q}}, \quad \forall\, y \in [g(a), g(b)]. \tag{10.66}$$

Proof We use (10.55). We observe that

$$\left\| \left(D_{a+;g}^{k\gamma} f \right)(x) \right\| \le \frac{1}{\Gamma((n+1)\alpha - k\gamma)} \cdot \tag{10.67}$$

$$\int_{a}^{x} (g(x) - g(t))^{(n+1)\alpha - k\gamma - 1} g'(t) \left\| \left(D_{a+;g}^{(n+1)\alpha} f \right)(t) \right\| dt$$

$$= \frac{1}{\Gamma((n+1)\alpha - k\gamma)} \int_{g(a)}^{g(x)} (g(x) - z)^{(n+1)\alpha - k\gamma - 1} \left\| \left(D_{a+;g}^{(n+1)\alpha} f \right)\left(g^{-1}(z) \right) \right\| dz$$

$$\le \frac{1}{\Gamma((n+1)\alpha - k\gamma)} \left(\int_{g(a)}^{g(x)} (g(x) - z)^{p((n+1)\alpha - k\gamma - 1)} dz \right)^{\frac{1}{p}} \cdot \tag{10.68}$$

$$\left(\int_{g(a)}^{g(x)} \left\| \left(D_{a+;g}^{(n+1)\alpha} f \right)\left(g^{-1}(z) \right) \right\|^{q} dz \right)^{\frac{1}{q}}$$

$$= \frac{1}{\Gamma((n+1)\alpha - k\gamma)} \frac{(g(x) - g(a))^{\frac{p((n+1)\alpha - k\gamma - 1) + 1}{p}}}{(p((n+1)\alpha - k\gamma - 1) + 1)^{\frac{1}{p}}} \cdot$$

$$\left(\int_{g(a)}^{g(x)} \left\| \left(D_{a+;g}^{(n+1)\alpha} f \right)\left(g^{-1}(z) \right) \right\|^{q} dz \right)^{\frac{1}{q}}. \tag{10.69}$$

That is

$$\left\| \left(D_{a+;g}^{k\gamma} f \right) (x) \right\| \leq \frac{(g(x) - g(a))^{\frac{p((n+1)\alpha - k\gamma - 1)+1}{p}}}{\Gamma((n+1)\alpha - k\gamma)(p((n+1)\alpha - k\gamma - 1) + 1)^{\frac{1}{p}}} \cdot$$

$$\left(\int_{g(a)}^{g(x)} \left\| \left(D_{a+;g}^{(n+1)\alpha} f \right) \left(g^{-1}(z) \right) \right\|^q dz \right)^{\frac{1}{q}}, \tag{10.70}$$

$\forall\, x \in [a, b]$.

Call

$$y = g(x),$$

then

$$x = g^{-1}(y). \tag{10.71}$$

Hence it holds

$$\left\| \left(D_{a+;g}^{k\gamma} f \right) \left(g^{-1}(y) \right) \right\| \leq \frac{(y - g(a))^{\frac{p((n+1)\alpha - k\gamma - 1)+1}{p}}}{\Gamma((n+1)\alpha - k\gamma)(p((n+1)\alpha - k\gamma - 1) + 1)^{\frac{1}{p}}}$$

$$\cdot \left(\int_{g(a)}^{y} \left\| \left(D_{a+;g}^{(n+1)\alpha} f \right) \left(g^{-1}(z) \right) \right\|^q dz \right)^{\frac{1}{q}}, \tag{10.72}$$

$\forall\, y \in [g(a), g(b)]$.

Call

$$\eta(y) := \int_{g(a)}^{y} \left\| \left(D_{a+;g}^{(n+1)\alpha} f \right) \left(g^{-1}(z) \right) \right\|^q dz, \tag{10.73}$$

and

$$\eta(g(a)) = 0.$$

Thus

$$\eta'(y) = \left\| \left(D_{a+;g}^{(n+1)\alpha} f \right) \left(g^{-1}(y) \right) \right\|^q \geq 0, \tag{10.74}$$

and

$$\left(\eta'(y)\right)^{\frac{1}{q}} = \left\| \left(D_{a+;g}^{(n+1)\alpha} f \right) \left(g^{-1}(y) \right) \right\| \geq 0, \tag{10.75}$$

$\forall\, y \in [g(a), g(b)]$.

Consequently, we get

$$\left\| \left(D_{a+;g}^{k\gamma} f \right) \left(g^{-1}(w) \right) \right\| \left\| \left(D_{a+;g}^{(n+1)\alpha} f \right) \left(g^{-1}(w) \right) \right\| \leq$$

$$\frac{(w-g(a))^{\frac{p((n+1)\alpha-k\gamma-1)+1}{p}}}{\Gamma((n+1)\alpha-k\gamma)(p((n+1)\alpha-k\gamma-1)+1)^{\frac{1}{p}}}\left(\eta(w)\eta'(w)\right)^{\frac{1}{q}}, \quad (10.76)$$

$\forall\, w \in [g(a), g(b)]$.

Then it holds

$$\int_{g(a)}^{y}\left\|\left(D_{a+;g}^{k\gamma}f\right)\left(g^{-1}(w)\right)\right\|\left\|\left(D_{a+;g}^{(n+1)\alpha}f\right)\left(g^{-1}(w)\right)\right\|dw \leq \quad (10.77)$$

$$\frac{1}{\Gamma((n+1)\alpha-k\gamma)(p((n+1)\alpha-k\gamma-1)+1)^{\frac{1}{p}}}\cdot$$

$$\int_{g(a)}^{y}(w-g(a))^{\frac{p((n+1)\alpha-k\gamma-1)+1}{p}}\left(\eta(w)\eta'(w)\right)^{\frac{1}{q}}dw$$

$$\leq\frac{1}{\Gamma((n+1)\alpha-k\gamma)(p((n+1)\alpha-k\gamma-1)+1)^{\frac{1}{p}}}\cdot \quad (10.78)$$

$$\left(\int_{g(a)}^{y}(w-g(a))^{(p((n+1)\alpha-k\gamma-1)+1)}dw\right)^{\frac{1}{p}}\left(\int_{g(a)}^{y}\eta(w)\eta'(w)\,dw\right)^{\frac{1}{q}}=$$

$$\frac{(y-g(a))^{\frac{p((n+1)\alpha-k\gamma-1)+2}{p}}}{\Gamma((n+1)\alpha-k\gamma)\left[(p((n+1)\alpha-k\gamma-1)+1)(p((n+1)\alpha-k\gamma-1)+2)\right]^{\frac{1}{p}}}$$

$$\cdot\left(\frac{\eta^{2}(y)}{2}\right)^{\frac{1}{q}}= \quad (10.79)$$

$$\frac{(y-g(a))^{((n+1)\alpha-k\gamma-1)+\frac{2}{p}}}{2^{\frac{1}{q}}\Gamma((n+1)\alpha-k\gamma)\left[(p((n+1)\alpha-k\gamma-1)+1)(p((n+1)\alpha-k\gamma-1)+2)\right]^{\frac{1}{p}}}$$

$$\cdot\left(\int_{g(a)}^{y}\left\|\left(\left(D_{a+;g}^{(n+1)\alpha}f\right)\circ g^{-1}\right)(w)\right\|^{q}dw\right)^{\frac{2}{q}}, \quad (10.80)$$

$\forall\, y \in [g(a), g(b)]$, proving the claim. ■

Also we give a right generalized and iterated fractional Opial type inequality:

Theorem 10.22 *All here as in Theorem 10.18. Let $p, q > 1 : \frac{1}{p}+\frac{1}{q}=1$. We further assume that ($k \in \mathbb{N}$)*

$$1>\alpha>\max\left(\frac{m+(k-1)\gamma}{n+1},\frac{k\gamma q+1}{(n+1)q}\right). \quad (10.81)$$

Then

$$\int_y^{g(b)} \left\| \left(\left(D_{b-;g}^{k\gamma} f \right) \circ g^{-1} \right)(w) \right\| \left\| \left(\left(D_{b-;g}^{(n+1)\alpha} f \right) \circ g^{-1} \right)(w) \right\| dw \leq$$

$$\frac{(g(b) - y)^{((n+1)\alpha - k\gamma - 1) + \frac{2}{p}}}{2^{\frac{1}{q}} \Gamma((n+1)\alpha - k\gamma) \left[(p((n+1)\alpha - k\gamma - 1) + 1)(p((n+1)\alpha - k\gamma - 1) + 2) \right]^{\frac{1}{p}}} \cdot$$

$$\left(\int_y^{g(b)} \left\| \left(\left(D_{b-;g}^{(n+1)\alpha} f \right) \circ g^{-1} \right)(w) \right\|^q dw \right)^{\frac{2}{q}}, \quad \forall\, y \in [g(a), g(b)]. \tag{10.82}$$

Proof It is omitted as similar to Theorem 10.21. ■

Next we present a left generalized and iterated fractional Hilbert–Pachpatte type inequality:

Theorem 10.23 *Here $i = 1, 2$. Let $a_i, b_i \in \mathbb{R}$, $a_i < b_i$, $0 < \alpha_i < 1$, as in (10.83), (10.84), and $g_i \in C^1([a_i, b_i])$ that are strictly increasing, $f_i \in C^1([a_i, b_i], X)$, $g_i^{-1} \in C^1([g_i(a_i), g_i(b_i)])$. Assume that $D_{a_i+;g_i}^{k_i\alpha_i} f_i \in C^1([a_i, b_i], X)$, $k_i = 1, \ldots, n_i$, and $\left(D_{a_i+;g_i}^{\lambda_i\alpha_i} f_i \right)(a_i) = 0$, $\lambda_i = 0, 2, 3, \ldots, n_i$, where $n_i \in \mathbb{N}$. Let $\gamma_i > 0$ with $\lceil \gamma_i \rceil = m_i$, $k_i \in \mathbb{N}$, $p, q > 1 : \frac{1}{p} + \frac{1}{q} = 1$. We further assume:*

$$1 > \alpha_1 > \max\left(\frac{m_1 + (k_1 - 1)\gamma_1}{n_1 + 1}, \frac{k_1\gamma_1 q + 1}{(n_1 + 1) q} \right), \tag{10.83}$$

and

$$1 > \alpha_2 > \max\left(\frac{m_2 + (k_2 - 1)\gamma_2}{n_2 + 1}, \frac{k_2\gamma_2 p + 1}{(n_2 + 1) p} \right). \tag{10.84}$$

Then

$$\int_{g_1(a_1)}^{g_1(b_1)} \int_{g_2(a_2)}^{g_2(b_2)} \frac{\left\| \left(D_{a_1+;g_1}^{k_1\gamma_1} f_1 \right)\left(g_1^{-1}(y_1) \right) \right\| \left\| \left(D_{a_2+;g_2}^{k_2\gamma_2} f_2 \right)\left(g_2^{-1}(y_2) \right) \right\|}{\left[\frac{(y_1 - g_1(a_1))^{p((n_1+1)\alpha_1 - k_1\gamma_1 - 1) + 1}}{p(p((n_1+1)\alpha_1 - k_1\gamma_1 - 1) + 1)} + \frac{(y_2 - g_2(a_2))^{q((n_2+1)\alpha_2 - k_2\gamma_2 - 1) + 1}}{q(q((n_2+1)\alpha_2 - k_2\gamma_2 - 1) + 1)} \right]} dy_1 dy_2 \tag{10.85}$$

$$\leq \frac{(g_1(b_1) - g_1(a_1))(g_2(b_2) - g_2(a_2))}{\Gamma((n_1+1)\alpha_1 - k_1\gamma_1)\Gamma((n_2+1)\alpha_2 - k_2\gamma_2)} \cdot$$

$$\left(\int_{g_1(a_1)}^{g_1(b_1)} \left\| \left(D_{a_1+;g_1}^{(n_1+1)\alpha_1} f_1 \right)\left(g_1^{-1}(z_1) \right) \right\|^q dz_1 \right)^{\frac{1}{q}} \cdot$$

$$\left(\int_{g_2(a_2)}^{g_2(b_2)} \left\| (D_{a_2+;g_2}^{(n_2+1)\alpha_2} f_2)\left(g_2^{-1}(z_2) \right) \right\|^p dz_2 \right)^{\frac{1}{p}}.$$

Proof As in the proof of Theorem 10.21, we obtain:

$$\left\| \left(D_{a_1+;g_1}^{k_1\gamma_1} f_1 \right) \left(g_1^{-1}(y_1) \right) \right\| \leq$$

$$\frac{(y_1 - g_1(a_1))^{\frac{p((n_1+1)\alpha_1 - k_1\gamma_1 - 1)+1}{p}}}{\Gamma((n_1+1)\alpha_1 - k_1\gamma_1)(p((n_1+1)\alpha_1 - k_1\gamma_1 - 1)+1)^{\frac{1}{p}}} \cdot \tag{10.86}$$

$$\left(\int_{g_1(a_1)}^{y_1} \left\| \left(D_{a_1+;g_1}^{(n_1+1)\alpha_1} f_1 \right) \left(g_1^{-1}(z_1) \right) \right\|^q dz_1 \right)^{\frac{1}{q}},$$

$\forall\, y_1 \in [g_1(a_1), g_1(b_1)]$.

Also we obtain:

$$\left\| \left(D_{a_2+;g_2}^{k_2\gamma_2} f_2 \right) \left(g_2^{-1}(y_2) \right) \right\| \leq$$

$$\frac{(y_2 - g_2(a_2))^{\frac{q((n_2+1)\alpha_2 - k_2\gamma_2 - 1)+1}{q}}}{\Gamma((n_2+1)\alpha_2 - k_2\gamma_2)(q((n_2+1)\alpha_2 - k_2\gamma_2 - 1)+1)^{\frac{1}{q}}} \cdot$$

$$\left(\int_{g_2(a_2)}^{y_2} \left\| \left(D_{a_2+;g_2}^{(n_2+1)\alpha_2} f_2 \right) \left(g_2^{-1}(z_2) \right) \right\|^p dz_2 \right)^{\frac{1}{p}}, \tag{10.87}$$

$\forall\, y_2 \in [g_2(a_2), g_2(b_2)]$.

Multiplying (10.86) and (10.87), we get:

$$\left\| \left(D_{a_1+;g_1}^{k_1\gamma_1} f_1 \right) \left(g_1^{-1}(y_1) \right) \right\| \left\| \left(D_{a_2+;g_2}^{k_2\gamma_2} f_2 \right) \left(g_2^{-1}(y_2) \right) \right\| \leq$$

$$\frac{1}{\Gamma((n_1+1)\alpha_1 - k_1\gamma_1)\,\Gamma((n_2+1)\alpha_2 - k_2\gamma_2)} \cdot$$

$$\left(\frac{(y_1 - g_1(a_1))^{p((n_1+1)\alpha_1 - k_1\gamma_1 - 1)+1}}{(p((n_1+1)\alpha_1 - k_1\gamma_1 - 1)+1)} \right)^{\frac{1}{p}} \left(\frac{(y_2 - g_2(a_2))^{q((n_2+1)\alpha_2 - k_2\gamma_2 - 1)+1}}{(q((n_2+1)\alpha_2 - k_2\gamma_2 - 1)+1)} \right)^{\frac{1}{q}} \cdot \tag{10.88}$$

$$\left(\int_{g_1(a_1)}^{y_1} \left\| \left(D_{a_1+;g_1}^{(n_1+1)\alpha_1} f_1 \right) \left(g_1^{-1}(z_1) \right) \right\|^q dz_1 \right)^{\frac{1}{q}} \cdot$$

$$\left(\int_{g_2(a_2)}^{y_2} \left\| \left(D_{a_2+;g_2}^{(n_2+1)\alpha_2} f_2 \right) \left(g_2^{-1}(z_2) \right) \right\|^p dz_2 \right)^{\frac{1}{p}}$$

(using Young's inequality for $a, b \geq 0$, $a^{\frac{1}{p}} b^{\frac{1}{q}} \leq \frac{a}{p} + \frac{b}{q}$)

$$\leq \frac{1}{\Gamma\left(\left(n_1+1\right)\alpha_1-k_1\gamma_1\right)\Gamma\left(\left(n_2+1\right)\alpha_2-k_2\gamma_2\right)}\cdot$$

$$\left[\left(\frac{\left(y_1-g_1\left(a_1\right)\right)^{p\left(\left(n_1+1\right)\alpha_1-k_1\gamma_1-1\right)+1}}{p\left(p\left(\left(n_1+1\right)\alpha_1-k_1\gamma_1-1\right)+1\right)}\right)+\left(\frac{\left(y_2-g_2\left(a_2\right)\right)^{q\left(\left(n_2+1\right)\alpha_2-k_2\gamma_2-1\right)+1}}{q\left(q\left(\left(n_2+1\right)\alpha_2-k_2\gamma_2-1\right)+1\right)}\right)\right] \tag{10.89}$$

$$\cdot\left(\int_{g_1(a_1)}^{y_1}\left\|\left(D_{a_1+;g_1}^{(n_1+1)\alpha_1}f_1\right)\left(g_1^{-1}\left(z_1\right)\right)\right\|^q dz_1\right)^{\frac{1}{q}}$$

$$\cdot\left(\int_{g_2(a_2)}^{y_2}\left\|\left(D_{a_2+;g_2}^{(n_2+1)\alpha_2}f_2\right)\left(g_2^{-1}\left(z_2\right)\right)\right\|^p dz_2\right)^{\frac{1}{p}}.$$

So far we have

$$\frac{\left\|\left(D_{a_1+;g_1}^{k_1\gamma_1}f_1\right)\left(g_1^{-1}\left(y_1\right)\right)\right\|\left\|\left(D_{a_2+;g_2}^{k_2\gamma_2}f_2\right)\left(g_2^{-1}\left(y_2\right)\right)\right\|}{\left[\left(\frac{(y_1-g_1(a_1))^{p((n_1+1)\alpha_1-k_1\gamma_1-1)+1}}{p(p((n_1+1)\alpha_1-k_1\gamma_1-1)+1)}\right)+\left(\frac{(y_2-g_2(a_2))^{q((n_2+1)\alpha_2-k_2\gamma_2-1)+1}}{q(q((n_2+1)\alpha_2-k_2\gamma_2-1)+1)}\right)\right]}\leq \tag{10.90}$$

$$\frac{1}{\Gamma\left(\left(n_1+1\right)\alpha_1-k_1\gamma_1\right)\Gamma\left(\left(n_2+1\right)\alpha_2-k_2\gamma_2\right)}\cdot$$

$$\left(\int_{g_1(a_1)}^{y_1}\left\|\left(D_{a_1+;g_1}^{(n_1+1)\alpha_1}f_1\right)\left(g_1^{-1}\left(z_1\right)\right)\right\|^q dz_1\right)^{\frac{1}{q}}\cdot$$

$$\left(\int_{g_2(a_2)}^{y_2}\left\|\left(D_{a_2+;g_2}^{(n_2+1)\alpha_2}f_2\right)\left(g_2^{-1}\left(z_2\right)\right)\right\|^p dz_2\right)^{\frac{1}{p}}.$$

The denominator in (10.90) can be zero only when $y_1 = g_1(a_1)$ and $y_2 = g_2(a_2)$. Therefore we obtain

$$\int_{g_1(a_1)}^{g_1(b_1)}\int_{g_2(a_2)}^{g_2(b_2)}\frac{\left\|\left(D_{a_1+;g_1}^{k_1\gamma_1}f_1\right)\left(g_1^{-1}\left(y_1\right)\right)\right\|\left\|\left(D_{a_2+;g_2}^{k_2\gamma_2}f_2\right)\left(g_2^{-1}\left(y_2\right)\right)\right\|}{\left[\left(\frac{(y_1-g_1(a_1))^{p((n_1+1)\alpha_1-k_1\gamma_1-1)+1}}{p(p((n_1+1)\alpha_1-k_1\gamma_1-1)+1)}\right)+\left(\frac{(y_2-g_2(a_2))^{q((n_2+1)\alpha_2-k_2\gamma_2-1)+1}}{q(q((n_2+1)\alpha_2-k_2\gamma_2-1)+1)}\right)\right]}dy_1dy_2 \tag{10.91}$$

$$\leq \frac{1}{\Gamma\left(\left(n_1+1\right)\alpha_1-k_1\gamma_1\right)\Gamma\left(\left(n_2+1\right)\alpha_2-k_2\gamma_2\right)}\cdot$$

$$\left(\int_{g_1(a_1)}^{g_1(b_1)}\left(\int_{g_1(a_1)}^{y_1}\left\|\left(D_{a_1+;g_1}^{(n_1+1)\alpha_1}f_1\right)\left(g_1^{-1}\left(z_1\right)\right)\right\|^q dz_1\right)^{\frac{1}{q}}dy_1\right)\cdot$$

$$\left(\int_{g_2(a_2)}^{g_2(b_2)}\left(\int_{g_2(a_2)}^{y_2}\left\|\left(D_{a_2+;g_2}^{(n_2+1)\alpha_2}f_2\right)\left(g_2^{-1}(z_2)\right)\right\|^p dz_2\right)^{\frac{1}{p}}dy_2\right)$$

$$\leq \frac{1}{\Gamma\left((n_1+1)\alpha_1-k_1\gamma_1\right)\Gamma\left((n_2+1)\alpha_2-k_2\gamma_2\right)}.$$

$$\left(\int_{g_1(a_1)}^{g_1(b_1)}\left(\int_{g_1(a_1)}^{g_1(b_1)}\left\|\left(D_{a_1+;g_1}^{(n_1+1)\alpha_1}f_1\right)\left(g_1^{-1}(z_1)\right)\right\|^q dz_1\right)^{\frac{1}{q}}dy_1\right)\cdot$$

$$\left(\int_{g_2(a_2)}^{g_2(b_2)}\left(\int_{g_2(a_2)}^{g_2(b_2)}\left\|\left(D_{a_2+;g_2}^{(n_2+1)\alpha_2}f_2\right)\left(g_2^{-1}(z_2)\right)\right\|^p dz_2\right)^{\frac{1}{p}}dy_2\right) \tag{10.92}$$

$$= \frac{\left(g_1(b_1)-g_1(a_1)\right)\left(g_2(b_2)-g_2(a_2)\right)}{\Gamma\left((n_1+1)\alpha_1-k_1\gamma_1\right)\Gamma\left((n_2+1)\alpha_2-k_2\gamma_2\right)}.$$

$$\left(\int_{g_1(a_1)}^{g_1(b_1)}\left\|\left(D_{a_1+;g_1}^{(n_1+1)\alpha_1}f_1\right)\left(g_1^{-1}(z_1)\right)\right\|^q dz_1\right)^{\frac{1}{q}}\cdot \tag{10.93}$$

$$\left(\int_{g_2(a_2)}^{g_2(b_2)}\left\|\left(D_{a_2+;g_2}^{(n_2+1)\alpha_2}f_2\right)\left(g_2^{-1}(z_2)\right)\right\|^p dz_2\right)^{\frac{1}{p}}.$$

The theo is proved. ■

Finally we present a right generalized and iterated fractional Hilbert–Pachpatte type inequality:

Theorem 10.24 *Here $i=1,2$. Let $a_i,b_i\in\mathbb{R}$, $a_i<b_i$, $0<\alpha_i<1$, as in (10.94), (10.95), and $g_i\in C^1([a_i,b_i])$ that are strictly increasing, $f_i\in C^1([a_i,b_i],X)$, $g_i^{-1}\in C^1([g_i(a_i),g_i(b_i)])$. Assume that $D_{b_i-;g_i}^{k_i\alpha_i}f_i\in C^1([a_i,b_i],X)$, $k_i=1,\ldots,n_i$, and $\left(D_{b_i-;g_i}^{\lambda_i\alpha_i}f_i\right)(b_i)=0$, $\lambda_i=0,2,3,\ldots,n_i$, where $n_i\in\mathbb{N}$. Let $\gamma_i>0$ with $\lceil\gamma_i\rceil=m_i$, $k_i\in\mathbb{N}$, $p,q>1:\frac{1}{p}+\frac{1}{q}=1$. We further assume:*

$$1>\alpha_1>\max\left(\frac{m_1+(k_1-1)\gamma_1}{n_1+1},\frac{k_1\gamma_1 q+1}{(n_1+1)q}\right). \tag{10.94}$$

and

$$1>\alpha_2>\max\left(\frac{m_2+(k_2-1)\gamma_2}{n_2+1},\frac{k_2\gamma_2 p+1}{(n_2+1)p}\right). \tag{10.95}$$

Then

$$\int_{g_1(a_1)}^{g_1(b_1)} \int_{g_2(a_2)}^{g_2(b_2)} \frac{\left\| \left(D_{b_1-;g_1}^{k_1\gamma_1} f_1 \right) \left(g_1^{-1}(y_1) \right) \right\| \left\| \left(D_{b_2-;g_2}^{k_2\gamma_2} f_2 \right) \left(g_2^{-1}(y_2) \right) \right\|}{\left[\frac{(g_1(b_1)-y_1)^{p((n_1+1)\alpha_1-k_1\gamma_1-1)+1}}{p(p((n_1+1)\alpha_1-k_1\gamma_1-1)+1)} + \frac{(g_2(b_2)-y_2)^{q((n_2+1)\alpha_2-k_2\gamma_2-1)+1}}{q(q((n_2+1)\alpha_2-k_2\gamma_2-1)+1)} \right]} dy_1 dy_2 \tag{10.96}$$

$$\leq \frac{(g_1(b_1) - g_1(a_1))(g_2(b_2) - g_2(a_2))}{\Gamma((n_1+1)\alpha_1 - k_1\gamma_1)\Gamma((n_2+1)\alpha_2 - k_2\gamma_2)}.$$

$$\left(\int_{g_1(a_1)}^{g_1(b_1)} \left\| \left(D_{b_1-;g_1}^{(n_1+1)\alpha_1} f_1 \right) \left(g_1^{-1}(z_1) \right) \right\|^q dz_1 \right)^{\frac{1}{q}} \cdot$$

$$\left(\int_{g_2(a_2)}^{g_2(b_2)} \left\| \left(D_{b_2-;g_2}^{(n_2+1)\alpha_2} f_2 \right) \left(g_2^{-1}(z_2) \right) \right\|^p dz_2 \right)^{\frac{1}{p}}.$$

Proof It is similar to Theorem 10.23, thus it is omitted. ∎

Remark 10.25 Some examples for g follow:

$$g(x) = e^x, \ x \in [a, b] \subset \mathbb{R},$$
$$g(x) = \sin x,$$
$$g(x) = \tan x,$$
$$\text{where } x \in \left[-\frac{\pi}{2} + \varepsilon, \frac{\pi}{2} - \varepsilon \right], \text{ where } \varepsilon > 0 \text{ small.}$$

Indeed, the above examples of g are strictly increasing and continuous functions.

One can apply all of our results here for the above specific choices of g. We choose to omit this job.

References

1. Aliprantis, C.D., Border, K.C.: Infinite Dimensional Analysis. Springer, New York (2006)
2. Anastassiou, G.A.: Principles of general fractional analysis for Banach space valued functions. Bull. Allahabad Math. Soc. **32**(1), 71–145 (2017)
3. Anastassiou, G.: Intelligent Computations: Abstract Fractional Calculus, Inequalities, Approximations. Springer, Heidelberg (2018)
4. Anastassiou, G.: Iterated *g-fractional vector representation formulae and inequalities for Banach space valued functions*. Issues Anal. **9**(27)(1), 3–26 (2020)
5. Bochner integral. Encyclopedia of Mathematics. http://www.encyclopediaofmath.org/index.php?title=Bochner_integral&oldid=38659
6. Kreuter, M.: Sobolev space of vector-valued functions. Ulm University, Master thesis in Mathematics, Ulm, Germany (2015)
7. Mikkola, K.: Appendix B integration and differentiation in Banach spaces. http://math.aalto.fi/~kmikkola/research/thesis/contents/thesisb.pdf
8. Mikusinski, J.: The Bochner Integral. Academic, New York (1978)

Chapter 11
Vectorial Generalized g-Fractional Direct and Iterated Quantitative Approximation by Linear Operators

In this work we consider quantitatively with rates the convergence of sequences of linear operators applied on Banach space valued functions. The results are pointwise estimates with rates. To prove our main results we use an elegant and natural boundedness property of our linear operators by their companion positive linear operators. Our inequalities are generalized g-direct and iterated fractional involving the right and left vector Caputo type generalized g-direct and iterated fractional derivatives, built in vector moduli of continuity. We treat wide and general classes of Banach space valued functions. We give applications to vectorial Bernstein operators. See also [6].

11.1 Motivation

Let $(X, \|\cdot\|)$ be a Banach space, $N \in \mathbb{N}$. Consider $g \in C([0,1])$ and the classic Bernstein polynomials

$$\left(\widetilde{B}_N g\right)(t) = \sum_{k=0}^{N} g\left(\frac{k}{N}\right)\binom{N}{k} t^k (1-t)^{N-k}, \ \forall\, t \in [0,1]. \tag{11.1}$$

Let al.so $f \in C([0,1], X)$ and define the vector valued in X Bernstein linear operators

$$(B_N f)(t) = \sum_{k=0}^{N} f\left(\frac{k}{N}\right)\binom{N}{k} t^k (1-t)^{N-k}, \ \forall\, t \in [0,1]. \tag{11.2}$$

That is $(B_N f)(t) \in X$.

Clearly here $\|f\| \in C([0,1])$.

G. A. Anastassiou, *Generalized Fractional Calculus*, Studies in Systems, Decision and Control 305, https://doi.org/10.1007/978-3-030-56962-4_11

We notice that

$$\|(B_N f)(t)\| \leq \sum_{k=0}^{N} \left\| f\left(\frac{k}{N}\right) \right\| \binom{N}{k} t^k (1-t)^{N-k} = \left(\widetilde{B}_N (\|f\|)\right)(t), \quad (11.3)$$

$\forall\, t \in [0, 1]$.

The property

$$\|(B_N f)(t)\| \leq \left(\widetilde{B}_N (\|f\|)\right)(t), \ \forall t \in [0, 1], \quad (11.4)$$

is shared by almost all summation/integration similar operators and motivates our work here.

If $f(x) = c \in X$ the constant function, then

$$(B_N c) = c. \quad (11.5)$$

If $g \in C([0, 1])$ and $c \in X$, then $cg \in C([0, 1], X)$ and

$$(B_N (cg)) = c\widetilde{B}_N (g). \quad (11.6)$$

Again (11.5), (11.6) are fulfilled by many summation/integration operators.

In fact here (11.6) implies (11.5), when $g \equiv 1$.

The above can be generalized from $[0, 1]$ to any interval $[a, b] \subset \mathbb{R}$. All this discussion motivates us to consider the following situation.

Let $L_N : C([a, b], X) \hookrightarrow C([a, b], X)$, $(X, \|\cdot\|)$ a Banach space, L_N is a linear operator, $\forall\, N \in \mathbb{N}$, $x_0 \in [a, b]$. Let also $\widetilde{L}_N : C([a, b]) \hookrightarrow C([a, b])$, a sequence of positive linear operators, $\forall\, N \in \mathbb{N}$.

We assume that

$$\|(L_N (f))(x_0)\| \leq \left(\widetilde{L}_N (\|f\|)\right)(x_0), \quad (11.7)$$

$\forall\, N \in \mathbb{N}, \forall\, x_0 \in [a, b], \forall\, f \in C([a, b], X)$.

When $g \in C([a, b])$, $c \in X$, we assume that

$$(L_N (cg)) = c\widetilde{L}_N (g). \quad (11.8)$$

The special case of

$$\widetilde{L}_N (1) = 1, \quad (11.9)$$

implies

$$L_N (c) = c, \ \forall\, c \in X. \quad (11.10)$$

We call $\widetilde{L}_N$ the companion operator of L_N.

Based on the above fundamental properties we study the fractional approximation properties of the sequence of linear operators $\{L_N\}_{N\in\mathbb{N}}$, i.e. their fractional convergence to the unit operator. No kind of positivity property of $\{L_N\}_{N\in\mathbb{N}}$ is assumed.

11.2 Background

We need

Definition 11.1 ([3]) Let $[a,b]\subset\mathbb{R}$, $(X,\|\cdot\|)$ a Banach space, $g\in C^1([a,b])$ and increasing, $f\in C([a,b],X)$, $\nu>0$.

We define the left Riemann–Liouville generalized fractional Bochner integral operator

$$\left(I_{a+;g}^{\nu}f\right)(x):=\frac{1}{\Gamma(\nu)}\int_a^x(g(x)-g(z))^{\nu-1}g'(z)f(z)\,dz, \tag{11.11}$$

$\forall\, x\in[a,b]$, where Γ is the gamma function.

The last integral is of Bochner type. Since $f\in C([a,b],X)$, then $f\in L_\infty([a,b],X)$. By [3] we get that $I_{a+;g}^{\nu}f\in C([a,b],X)$. Above we set $I_{a+;g}^{0}f:=f$ and see that $\left(I_{a+;g}^{\nu}f\right)(a)=0$.

When g is the identity function id, we get that $I_{a+;id}^{\nu}=I_{a+}^{\nu}$, the ordinary left Riemann–Liouville fractional integral

$$\left(I_{a+}^{\nu}f\right)(x)=\frac{1}{\Gamma(\nu)}\int_a^x(x-t)^{\nu-1}f(t)\,dt, \tag{11.12}$$

$\forall\, x\in[a,b]$, $\left(I_{a+}^{\nu}f\right)(a)=0$.

We mention

Theorem 11.2 *([3]) Let $\mu,\nu>0$ and $f\in C([a,b],X)$. Then*

$$I_{a+;g}^{\mu}I_{a+;g}^{\nu}f=I_{a+;g}^{\mu+\nu}f=I_{a+;g}^{\nu}I_{a+;g}^{\mu}f. \tag{11.13}$$

We need

Definition 11.3 ([3]) Let $[a,b]\subset\mathbb{R}$, $(X,\|\cdot\|)$ a Banach space, $g\in C^1([a,b])$ and increasing, $f\in C([a,b],X)$, $\nu>0$.

We define the right Riemann–Liouville generalized fractional Bochner integral operator

$$\left(I_{b-;g}^{\nu}f\right)(x) := \frac{1}{\Gamma(\nu)} \int_{x}^{b} (g(z) - g(x))^{\nu-1} g'(z) f(z) \, dz, \tag{11.14}$$

$\forall x \in [a, b]$, where Γ is the gamma function.

The last integral is of Bochner type. Since $f \in C([a, b], X)$, then $f \in L_{\infty}([a, b], X)$. By [3] we get that $I_{b-;g}^{\nu}f \in C([a, b], X)$. Above we set $I_{b-;g}^{0}f := f$ and see that $\left(I_{b-;g}^{\nu}f\right)(b) = 0$.

When g is the identity function id, we get that $I_{b-;id}^{\nu} = I_{b-}^{\nu}$, the ordinary right Riemann–Liouville fractional integral

$$\left(I_{b-}^{\nu}f\right)(x) = \frac{1}{\Gamma(\nu)} \int_{x}^{b} (t - x)^{\nu-1} f(t) \, dt, \tag{11.15}$$

$\forall x \in [a, b]$, with $\left(I_{b-}^{\nu}f\right)(b) = 0$.

We mention

Theorem 11.4 ([3]) *Let $\mu, \nu > 0$ and $f \in C([a, b], X)$. Then*

$$I_{b-;g}^{\mu} I_{b-;g}^{\nu} f = I_{b-;g}^{\mu+\nu} f = I_{b-;g}^{\nu} I_{b-;g}^{\mu} f. \tag{11.16}$$

We will use

Definition 11.5 ([3]) Let $\alpha > 0$, $\lceil \alpha \rceil = n$, $\lceil \cdot \rceil$ the ceiling of the number. Let $f \in C^{n}([a, b], X)$, where $[a, b] \subset \mathbb{R}$, and $(X, \|\cdot\|)$ is a Banach space. Let $g \in C^{1}([a, b])$, strictly increasing, such that $g^{-1} \in C^{n}([g(a), g(b)])$.

We define the left generalized g-fractional derivative X-valued of f of order α as follows:

$$\left(D_{a+;g}^{\alpha}f\right)(x) := \frac{1}{\Gamma(n-\alpha)} \int_{a}^{x} (g(x) - g(t))^{n-\alpha-1} g'(t) \left(f \circ g^{-1}\right)^{(n)} (g(t)) \, dt, \tag{11.17}$$

$\forall x \in [a, b]$. The last integral is of Bochner type.

If $\alpha \notin \mathbb{N}$, by [3], we have that $\left(D_{a+;g}^{\alpha}f\right) \in C([a, b], X)$.

We see that

$$\left(I_{a+;g}^{n-\alpha}\left(\left(f \circ g^{-1}\right)^{(n)} \circ g\right)\right)(x) = \left(D_{a+;g}^{\alpha}f\right)(x), \ \forall x \in [a, b]. \tag{11.18}$$

We set

$$D_{a+;g}^{n}f(x) := \left(\left(f \circ g^{-1}\right)^{(n)} \circ g\right)(x) \in C([a, b], X), \ n \in \mathbb{N}m, \tag{11.19}$$

$$D_{a+;g}^{0}f(x) = f(x), \ \forall x \in [a, b].$$

When $g = id$, then

$$D_{a+;g}^{\alpha} f = D_{a+;id}^{\alpha} f = D_{*a}^{\alpha} f, \tag{11.20}$$

the usual left X-valued Caputo fractional derivative, see [5].

We will use

Definition 11.6 *([3])* Let $\alpha > 0$, $\lceil \alpha \rceil = n$, $\lceil \cdot \rceil$ the ceiling of the number. Let $f \in C^n([a, b], X)$, where $[a, b] \subset \mathbb{R}$, and $(X, \|\cdot\|)$ is a Banach space. Let $g \in C^1([a, b])$, strictly increasing, such that $g^{-1} \in C^n([g(a), g(b)])$.

We define the right generalized g-fractional derivative X-valued of f of order α as follows:

$$\left(D_{b-;g}^{\alpha} f\right)(x) := \frac{(-1)^n}{\Gamma(n-\alpha)} \int_x^b (g(t) - g(x))^{n-\alpha-1} g'(t) \left(f \circ g^{-1}\right)^{(n)} (g(t))\, dt, \tag{11.21}$$

$\forall\, x \in [a, b]$. The last integral is of Bochner type.

If $\alpha \notin \mathbb{N}$, by [3], we have that $\left(D_{b-;g}^{\alpha} f\right) \in C([a, b], X)$.

We see that

$$I_{b-;g}^{n-\alpha} \left((-1)^n \left(f \circ g^{-1}\right)^{(n)} \circ g\right)(x) = \left(D_{b-;g}^{\alpha} f\right)(x), \ a \le x \le b. \tag{11.22}$$

We set

$$D_{b-;g}^{n} f(x) := (-1)^n \left(\left(f \circ g^{-1}\right)^n \circ g\right)(x) \in C([a, b], X), \ n \in \mathbb{N}, \tag{11.23}$$

$$D_{b-;g}^{0} f(x) := f(x), \ \forall x \in [a, b].$$

When $g = id$, then

$$D_{b-;g}^{\alpha} f(x) = D_{b-;id}^{\alpha} f(x) = D_{b-}^{\alpha} f, \tag{11.24}$$

the usual right X-valued Caputo fractional derivative, see [5].

We make

Remark 11.7 All as in Definition 11.5. We have (by [8, Theorem 2.5, p. 7] and [10])

$$\left\|\left(D_{a+;g}^{\alpha} f\right)(x)\right\| \le \frac{1}{\Gamma(n-\alpha)} \int_a^x (g(x) - g(t))^{n-\alpha-1} g'(t) \left\|\left(f \circ g^{-1}\right)^{(n)} (g(t))\right\| dt$$

$$\le \frac{\left\|\left(f \circ g^{-1}\right)^{(n)} \circ g\right\|_{\infty,[a,b]}}{\Gamma(n-\alpha)} \int_{g(a)}^{g(x)} (g(x) - g(t))^{n-\alpha-1} dg(t) =$$

$$\frac{\left\|\left(f \circ g^{-1}\right)^{(n)} \circ g\right\|_{\infty,[a,b]}}{\Gamma(n-\alpha+1)} (g(x) - g(a))^{n-\alpha}. \tag{11.25}$$

That is

$$\left\| \left(D^{\alpha}_{a+;g} f \right)(x) \right\| \leq \frac{\left\| \left(f \circ g^{-1} \right)^{(n)} \circ g \right\|_{\infty,[a,b]}}{\Gamma(n-\alpha+1)} (g(x) - g(a))^{n-\alpha}, \tag{11.26}$$

$\forall\, x \in [a,b]$.

If $\alpha \notin \mathbb{N}$, then $\left(D^{\alpha}_{a+;g} f \right)(a) = 0$.

Similarly, by Definition 11.6 we derive

$$\left\| \left(D^{\alpha}_{b-;g} f \right)(x) \right\| \leq \frac{1}{\Gamma(n-\alpha)} \int_x^b (g(t) - g(x))^{n-\alpha-1} g'(t) \left\| \left(f \circ g^{-1} \right)^{(n)} (g(t)) \right\| dt$$

$$\leq \frac{\left\| \left(f \circ g^{-1} \right)^{(n)} \circ g \right\|_{\infty,[a,b]}}{\Gamma(n-\alpha)} \int_{g(x)}^{g(b)} (g(t) - g(x))^{n-\alpha-1} dg(t) =$$

$$\frac{\left\| \left(f \circ g^{-1} \right)^{(n)} \circ g \right\|_{\infty,[a,b]}}{\Gamma(n-\alpha+1)} (g(b) - g(x))^{n-\alpha}. \tag{11.27}$$

That is

$$\left\| \left(D^{\alpha}_{b-;g} f \right)(x) \right\| \leq \frac{\left\| \left(f \circ g^{-1} \right)^{(n)} \circ g \right\|_{\infty,[a,b]}}{\Gamma(n-\alpha+1)} (g(b) - g(x))^{n-\alpha}, \tag{11.28}$$

$\forall\, x \in [a,b]$.

If $\alpha \notin \mathbb{N}$, then $\left(D^{\alpha}_{b-;g} f \right)(b) = 0$.

Notation 11.8 *We denote by*

$$D^{n\alpha}_{a+;g} := D^{\alpha}_{a+;g} D^{\alpha}_{a+;g} \ldots D^{\alpha}_{a+;g} (n \text{ times}),\ n \in \mathbb{N}, \tag{11.29}$$

$$I^{n\alpha}_{a+;g} := I^{\alpha}_{a+;g} I^{\alpha}_{a+;g} \ldots I^{\alpha}_{a+;g}, \tag{11.30}$$

$$D^{n\alpha}_{b-;g} := D^{\alpha}_{b-;g} D^{\alpha}_{b-;g} \ldots {}^{\alpha}_{b-;g}, \tag{11.31}$$

and

$$I^{n\alpha}_{b-;g} := I^{\alpha}_{b-;g} I^{\alpha}_{b-;g} \ldots I^{\alpha}_{b-;g}, \tag{11.32}$$

(n times), $n \in \mathbb{N}$.

We mention the following g-left generalized modified X-valued Taylor's formula.

Theorem 11.9 ([3]) *Let* $0 < \alpha \leq 1$, $n \in \mathbb{N}$, $f \in C^1([a,b], X)$, $g \in C^1([a,b])$, *strictly increasing, such that* $g^{-1} \in C^1([g(a), g(b)])$. *Let* $F_k := D_{a+;g}^{k\alpha} f$, $k = 1, \dots, n$, *that fulfill* $F_k \in C^1([a,b], X)$.

Then

$$f(x) = \sum_{i=0}^{n} \frac{(g(x) - g(a))^{i\alpha}}{\Gamma(i\alpha + 1)} \left(D_{a+;g}^{i\alpha} f\right)(a) +$$

$$\frac{1}{\Gamma((n+1)\alpha)} \int_a^x (g(x) - g(t))^{(n+1)\alpha - 1} g'(t) \left(D_{a+;g}^{(n+1)\alpha} f\right)(t)\, dt, \tag{11.33}$$

$\forall\, x \in [a,b]$.

Clearly here $D_{a+;g}^{(n+1)\alpha} f \in C([a,b], X)$.

We also mention the following g-right generalized modified X-valued Taylor's formula.

Theorem 11.10 ([3]) *Let* $f \in C^1([a,b], X)$, $g \in C^1([a,b])$, *strictly increasing, such that* $g^{-1} \in C^1([g(a), g(b)])$. *Suppose that* $F_k := D_{b-;g}^{k\alpha} f$, $k = 1, \dots, n$, *fulfill* $F_k \in C^1([a,b], X)$, *where* $0 < \alpha \leq 1$, $n \in \mathbb{N}$.

Then

$$f(x) = \sum_{i=0}^{n} \frac{(g(b) - g(x))^{i\alpha}}{\Gamma(i\alpha + 1)} \left(D_{b-;g}^{i\alpha} f\right)(b) +$$

$$\frac{1}{\Gamma((n+1)\alpha)} \int_x^b (g(t) - g(x))^{(n+1)\alpha - 1} g'(t) \left(D_{b-;g}^{(n+1)\alpha} f\right)(t)\, dt, \tag{11.34}$$

$\forall\, x \in [a,b]$.

Clearly here $D_{b-;g}^{(n+1)\alpha} f \in C([a,b], X)$.

For differentiation of functions from real numbers to normed linear spaces the definition is the same as for the real valued functions, however the limit and convergence is in the norm of linear space $(X, \|\cdot\|)$.

We state

Corollary 11.11 (to Theorem 11.9) *Let* $0 < \alpha < 1$, $n \in \mathbb{N}$, $f \in C^1([a,b], X)$, $g \in C^1([a,b])$, *strictly increasing, such that* $g^{-1} \in C^1([g(a), g(b)])$. *Assume* $D_{a+;g}^{k\alpha} f \in C^1([a,b], X)$, $k = 1, \dots, n$, *and* $\left(D_{a+;g}^{i\alpha} f\right)(a) = 0$, $i = 0, 2, 3, \dots, n$.

Then

$$f(x) = \frac{1}{\Gamma((n+1)\alpha)} \int_a^x (g(x) - g(t))^{(n+1)\alpha - 1} g'(t) \left(D_{a+;g}^{(n+1)\alpha} f\right)(t)\, dt, \tag{11.35}$$

$\forall\, x \in [a,b]$.

We state

Corollary 11.12 (to Theorem 11.10) *Let* $f \in C^1([a,b], X)$, $g \in C^1([a,b])$, *strictly increasing, such that* $g^{-1} \in C^1([g(a), g(b)])$. *Suppose that* $D_{b-;g}^{k\alpha} f \in C^1([a,b], X)$, *for* $k = 1, \ldots, n$; *where* $0 < \alpha < 1$, $n \in \mathbb{N}$. *We further assume that* $\left(D_{b-;g}^{i\alpha} f\right)(b) = 0$, $i = 0, 2, 3, \ldots, n$.

Then

$$f(x) = \frac{1}{\Gamma((n+1)\alpha)} \int_x^b (g(t) - g(x))^{(n+1)\alpha - 1} g'(t) \left(D_{b-;g}^{(n+1)\alpha} f\right)(t)\, dt, \tag{11.36}$$

$\forall\, x \in [a,b]$.

We mention the following g-left generalized X-valued Taylor's formula:

Theorem 11.13 ([3]) *Let* $\alpha > 0$, $n = \lceil \alpha \rceil$, *and* $f \in C^n([a,b], X)$, *where* $[a,b] \subset \mathbb{R}$ *and* $(X, \|\cdot\|)$ *is a Banach space. Let* $g \in C^1([a,b])$, *strictly increasing, such that* $g^{-1} \in C^n([g(a), g(b)])$. *Then*

$$f(x) = f(a) + \sum_{i=1}^{n-1} \frac{(g(x) - g(a))^i}{i!} \left(f \circ g^{-1}\right)^{(i)} (g(a)) +$$

$$\frac{1}{\Gamma(\alpha)} \int_a^x (g(x) - g(t))^{\alpha - 1} g'(t) \left(D_{a+;g}^{\alpha} f\right)(t)\, dt =$$

$$f(a) + \sum_{i=1}^{n-1} \frac{(g(x) - g(a))^i}{i!} \left(f \circ g^{-1}\right)^{(i)} (g(a)) + \tag{11.37}$$

$$\frac{1}{\Gamma(\alpha)} \int_{g(a)}^{g(x)} (g(x) - z)^{\alpha - 1} \left(\left(D_{a+;g}^{\alpha} f\right) \circ g^{-1}\right)(z)\, dz, \ \forall\, x \in [a,b].$$

We also mention the following g-right generalized X-valued Taylor's formula:

Theorem 11.14 ([3]) *Let* $\alpha > 0$, $n = \lceil \alpha \rceil$, *and* $f \in C^n([a,b], X)$, *where* $[a,b] \subset \mathbb{R}$ *and* $(X, \|\cdot\|)$ *is a Banach space. Let* $g \in C^1([a,b])$, *strictly increasing, such that* $g^{-1} \in C^n([g(a), g(b)])$. *Then*

$$f(x) = f(b) + \sum_{i=1}^{n-1} \frac{(g(x) - g(b))^i}{i!} \left(f \circ g^{-1}\right)^{(i)} (g(b)) +$$

$$\frac{1}{\Gamma(\alpha)} \int_x^b (g(t) - g(x))^{\alpha - 1} g'(t) \left(D_{b-;g}^{\alpha} f\right)(t)\, dt =$$

$$f(b) + \sum_{i=1}^{n-1} \frac{(g(x) - g(b))^i}{i!} \left(f \circ g^{-1}\right)^{(i)} (g(b)) + \tag{11.38}$$

$$\frac{1}{\Gamma(\alpha)} \int_{g(x)}^{g(b)} (z - g(x))^{\alpha - 1} \left(\left(D_{b-;g}^{\alpha} f \right) \circ g^{-1} \right) (z) \, dz, \ \forall \, x \in [a, b].$$

Integrals in (11.33)–(11.38) are of Bochner type. For the Bochner integral excellent resources are [7, 9, 10] and [1, pp. 422–428].

We need

Definition 11.15 ([4]) Let $f \in C([a,b], X)$, $[a,b] \subset \mathbb{R}$, $(X, \|\cdot\|)$ a Banach space. We define the first modulus of continuity of f as

$$\omega_1(f, \delta) := \sup_{\substack{x, y \in [a,b]: \\ |x-y| \leq \delta}} \|f(x) - f(y)\|, \ 0 < \delta \leq b - a. \tag{11.39}$$

If $\delta > b - a$, then $\omega_1(f, \delta) = \omega_1(f, b - a)$.

Notice $\omega_1(f, \delta)$ is increasing in $\delta > 0$.

Clearly f is uniformly continuous and $\omega_1(f, \delta) < \infty$. For $f \in B([a,b], X)$ (bounded functions) $\omega_1(f, \delta)$ is defined the same way.

Lemma 11.16 ([4]) *We have $\omega_1(f, \delta) \to 0$ as $\delta \downarrow 0$ iff $f \in C([a,b], X)$.*

We need

Theorem 11.17 ([4]) *Let $f : [a,b]^2 \to X$ be jointly continuous, X is a Banach space. Consider*

$$G(x) = \omega_1(f(\cdot, x), \delta, [x, b]), \tag{11.40}$$

$\delta > 0$, $x \in [a,b]$.

Then G is continuous on $[a,b]$.

Theorem 11.18 ([4]) *Let $f : [a,b]^2 \to X$ be jointly continuous, X is a Banach space. Then*

$$H(x) = \omega_1(f(\cdot, x), \delta, [a, x]), \tag{11.41}$$

$x \in [a,b]$, is continuous in $x \in [a,b]$, $\delta > 0$.

We need

Lemma 11.19 ([2, p. 208, Lemma 7.1.1]) *Let $f \in B([a,b], X)$, $(X, \|\cdot\|)$ is a Banach space. Then*

$$\|f(x) - f(x_0)\| \leq \omega_1(f, h) \left\lceil \frac{|x - x_0|}{h} \right\rceil \leq \omega_1(f, h) \left(1 + \frac{|x - x_0|}{h} \right), \tag{11.42}$$

$\forall \, x, x_0 \in [a,b]$, $h > 0$.

We make

Remark 11.20 All here as in (11.7)–(11.10). Let $f, g \in C([a,b], X)$. We have that

$$\|L_N(f)(x_0) - L_N(g)(x_0)\| = \|(L_N(f-g))(x_0)\| \overset{(11.7)}{\leq} \left(\widetilde{L}_N(\|f-g\|)\right)(x_0), \tag{11.43}$$

$\forall N \in \mathbb{N}, \forall x_0 \in [a,b]$.

In this work, for simplicity, from now on we will assume that

$$\widetilde{L}_N(1) = 1, \ \forall N \in \mathbb{N}. \tag{11.44}$$

We observe that

$$\|(L_N(f))(x_0) - f(x_0)\| \overset{(11.10)}{=} \|(L_N(f))(x_0) - (L_N(f(x_0)))(x_0)\| = \tag{11.45}$$

$$\|(L_N(f - f(x_0)))(x_0)\| \overset{(11.43.)}{\leq} \widetilde{L}_N(\|f(\cdot) - f(x_0)\|)(x_0),$$

$\forall x_0 \in [a,b], \forall N \in \mathbb{N}$.

We have proved that

$$\|(L_N(f))(x_0) - f(x_0)\| \leq \widetilde{L}_N(\|f(\cdot) - f(x_0)\|)(x_0), \tag{11.46}$$

$\forall x_0 \in [a,b], \forall N \in \mathbb{N}$.

11.3 Main Results

We need

Definition 11.21 Let $D_{x_0;g}^{(n+1)\alpha} f$ denote any of $D_{x_0-;g}^{(n+1)\alpha} f$, $D_{x_0+;g}^{(n+1)\alpha} f$, $n \in \mathbb{N}$ and $\delta > 0$. We set

$$\omega_1\left(\left(D_{x_0;g}^{(n+1)\alpha} f\right) \circ g^{-1}, \delta\right) :=$$

$$\max\left\{\omega_1\left(\left(D_{x_0-;g}^{(n+1)\alpha} f\right) \circ g^{-1}, \delta\right)_{[g(a),g(x_0)]}, \omega_1\left(\left(D_{x_0+;g}^{(n+1)\alpha} f\right) \circ g^{-1}, \delta\right)_{[g(x_0),g(b)]}\right\}, \tag{11.47}$$

where $x_0 \in [a,b]$. Here the moduli of continuity are considered over $[g(a), g(x_0)]$ and $[g(x_0), g(b)]$, respectively.

We will use the following:

Theorem 11.22 *Let $0 < \alpha < 1$, $f \in C^1([a,b], X)$, $g \in C^1([a,b])$, g is strictly increasing and $g^{-1} \in C^1([g(a), g(b)])$. Assume that $D_{x_0-;g}^{k\alpha} f \in C^1([a,x_0], X)$ and $D_{x_0+;g}^{k\alpha} f \in C^1([x_0,b], X)$, for $k = 1, \ldots, n$; where $x_0 \in [a,b]$ is fixed. Further assume that $\left(D_{x_0\pm;g}^{i\alpha} f\right)(x_0) = 0$, $i = 2, \ldots, n+1$, $n \in \mathbb{N}$. Then*

$$\|f(x) - f(x_0)\| \leq \frac{\omega_1\left(\left(D_{x_0;g}^{(n+1)\alpha} f\right) \circ g^{-1}, \delta\right)}{\Gamma((n+1)\alpha + 1)} \tag{11.48}$$

$$\left[|g(x) - g(x_0)|^{(n+1)\alpha} + \frac{|g(x) - g(x_0)|^{(n+1)\alpha+1}}{\delta((n+1)\alpha + 1)}\right],$$

$\forall\, x \in [a, b]$, $\delta > 0$.

Proof By $\left(D_{x_0+;g}^{i\alpha} f\right)(x_0) = 0$, for $i = 2, \ldots, n+1$, and (11.33) we have

$$f(x) - f(x_0) = \frac{1}{\Gamma((n+1)\alpha)} \int_{x_0}^{x} (g(x) - g(t))^{(n+1)\alpha - 1} g'(t)$$

$$\left[\left(D_{x_0+;g}^{(n+1)\alpha} f\right)(t) - \left(D_{x_0+;g}^{(n+1)\alpha} f\right)(x_0)\right] dt, \tag{11.49}$$

$\forall\, x \in [x_0, b]$.

Here we apply the change of variables method for Bochner integrals, see [8, Theorem 2.5, p. 7] and [10].

Hence ($z := g(t)$)

$$f(x) - f(x_0) = \frac{1}{\Gamma((n+1)\alpha)} \int_{g(x_0)}^{g(x)} (g(x) - z)^{(n+1)\alpha - 1}$$

$$\left[\left(\left(D_{x_0+;g}^{(n+1)\alpha} f\right) \circ g^{-1}\right)(z) - \left(\left(D_{x_0+;g}^{(n+1)\alpha} f\right) \circ g^{-1}\right)(g(x_0))\right] dz, \tag{11.50}$$

$\forall\, x \in [x_0, b]$.

By $\left(D_{x_0-;g}^{i\alpha} f\right)(x_0) = 0$, for $i = 2, \ldots, n+1 \in \mathbb{N}$, and (11.34) we have

$$f(x) - f(x_0) = \frac{1}{\Gamma((n+1)\alpha)} \int_{x}^{x_0} (g(t) - g(x))^{(n+1)\alpha - 1} g'(t)$$

$$\left[\left(D_{x_0-;g}^{(n+1)\alpha} f\right)(t) - \left(D_{x_0-;g}^{(n+1)\alpha} f\right)(x_0)\right] dt, \tag{11.51}$$

$\forall\, x \in [a, x_0]$.

Hence ($z := g(t)$)

$$f(x) - f(x_0) = \frac{1}{\Gamma((n+1)\alpha)} \int_{g(x)}^{g(x_0)} (z - g(x))^{(n+1)\alpha - 1}$$

$$\left[\left(\left(D_{x_0-;g}^{(n+1)\alpha} f\right) \circ g^{-1}\right)(z) - \left(\left(D_{x_0-;g}^{(n+1)\alpha} f\right) \circ g^{-1}\right)(g(x_0))\right] dz, \tag{11.52}$$

$\forall\, x \in [a, x_0]$.

We have that ($x_0 \leq x \leq b$)

$$\|f(x) - f(x_0)\| \leq \frac{1}{\Gamma((n+1)\alpha)} \int_{g(x_0)}^{g(x)} (g(x) - z)^{(n+1)\alpha - 1}$$

$$\left\| \left(\left(D_{x_0+;g}^{(n+1)\alpha} f \right) \circ g^{-1} \right)(z) - \left(\left(D_{x_0+;g}^{(n+1)\alpha} f \right) \circ g^{-1} \right)(g(x_0)) \right\| dz \underset{(\delta_1 > 0)}{\leq}$$

$$\frac{1}{\Gamma((n+1)\alpha)} \int_{g(x_0)}^{g(x)} (g(x) - z)^{(n+1)\alpha - 1} \tag{11.53}$$

$$\omega_1 \left(\left(D_{x_0+;g}^{(n+1)\alpha} f \right) \circ g^{-1}, \frac{\delta_1 |z - g(x_0)|}{\delta_1} \right)_{[g(x_0), g(b)]} dz \overset{(11.42.)}{\leq}$$

$$\frac{\omega_1 \left(\left(D_{x_0+;g}^{(n+1)\alpha} f \right) \circ g^{-1}, \delta_1 \right)_{[g(x_0), g(b)]}}{\Gamma((n+1)\alpha)}$$

$$\int_{g(x_0)}^{g(x)} (g(x) - z)^{(n+1)\alpha - 1} \left(1 + \frac{(z - g(x_0))}{\delta_1} \right) dz =$$

$$\frac{\omega_1 \left(\left(D_{x_0+;g}^{(n+1)\alpha} f \right) \circ g^{-1}, \delta_1 \right)_{[g(x_0), g(b)]}}{\Gamma((n+1)\alpha)}$$

$$\left[\frac{(g(x) - g(x_0))^{(n+1)\alpha}}{(n+1)\alpha} + \frac{1}{\delta_1} \int_{g(x_0)}^{g(x)} (g(x) - z)^{(n+1)\alpha - 1} (z - g(x_0))^{2-1} dz \right] = \tag{11.54}$$

$$\frac{\omega_1 \left(\left(D_{x_0+;g}^{(n+1)\alpha} f \right) \circ g^{-1}, \delta_1 \right)_{[g(x_0), g(b)]}}{\Gamma((n+1)\alpha)}$$

$$\left[\frac{(g(x) - g(x_0))^{(n+1)\alpha}}{(n+1)\alpha} + \frac{1}{\delta_1} \frac{\Gamma((n+1)\alpha)\,\Gamma(2)}{\Gamma((n+1)\alpha + 2)} (g(x) - g(x_0))^{(n+1)\alpha + 1} \right] =$$

$$\frac{\omega_1 \left(\left(D_{x_0+;g}^{(n+1)\alpha} f \right) \circ g^{-1}, \delta_1 \right)_{[g(x_0), g(b)]}}{\Gamma((n+1)\alpha)} \tag{11.55}$$

$$\left[\frac{(g(x) - g(x_0))^{(n+1)\alpha}}{(n+1)\alpha} + \frac{1}{\delta_1} \frac{(g(x) - g(x_0))^{(n+1)\alpha + 1}}{(n+1)\alpha((n+1)\alpha + 1)} \right].$$

We have proved that

$$\|f(x)-f(x_0)\| \leq \frac{\omega_1\left(\left(D_{x_0+;g}^{(n+1)\alpha}f\right)\circ g^{-1},\delta_1\right)_{[g(x_0),g(b)]}}{\Gamma((n+1)\alpha+1)} \tag{11.56}$$

$$\left[(g(x)-g(x_0))^{(n+1)\alpha}+\frac{(g(x)-g(x_0))^{(n+1)\alpha+1}}{\delta_1((n+1)\alpha+1)}\right],$$

$\forall\, x \in [x_0, b]$, $\delta_1 > 0$.

We have that $(a \leq x \leq x_0)$

$$\|f(x)-f(x_0)\| \overset{(11.52)}{\leq} \frac{1}{\Gamma((n+1)\alpha)}\int_{g(x)}^{g(x_0)}(z-g(x))^{(n+1)\alpha-1}$$

$$\left\|\left(\left(D_{x_0-;g}^{(n+1)\alpha}f\right)\circ g^{-1}\right)(z)-\left(\left(D_{x_0-;g}^{(n+1)\alpha}f\right)\circ g^{-1}\right)(g(x_0))\right\|dz \underset{(\delta_2>0)}{\leq}$$

$$\frac{1}{\Gamma((n+1)\alpha)}\int_{g(x)}^{g(x_0)}(z-g(x))^{(n+1)\alpha-1}$$

$$\omega_1\left(\left(D_{x_0-;g}^{(n+1)\alpha}f\right)\circ g^{-1},\frac{\delta_2\,|z-g(x_0)|}{\delta_2}\right)_{[g(a),g(x_0)]}dz \overset{(11.42.)}{\leq} \tag{11.57}$$

$$\frac{\omega_1\left(\left(D_{x_0-;g}^{(n+1)\alpha}f\right)\circ g^{-1},\delta_2\right)_{[g(a),g(x_0)]}}{\Gamma((n+1)\alpha)}$$

$$\int_{g(x)}^{g(x_0)}(z-g(x))^{(n+1)\alpha-1}\left(1+\frac{g(x_0)-z}{\delta_2}\right)dz=$$

$$\frac{\omega_1\left(\left(D_{x_0-;g}^{(n+1)\alpha}f\right)\circ g^{-1},\delta_2\right)_{[g(a),g(x_0)]}}{\Gamma((n+1)\alpha)}$$

$$\left[\frac{(g(x_0)-g(x))^{(n+1)\alpha}}{(n+1)\alpha}+\frac{1}{\delta_2}\int_{g(x)}^{g(x_0)}(g(x_0)-z)^{2-1}(z-g(x))^{(n+1)\alpha-1}dz\right]= \tag{11.58}$$

$$\frac{\omega_1\left(\left(D_{x_0-;g}^{(n+1)\alpha}f\right)\circ g^{-1},\delta_2\right)_{[g(a),g(x_0)]}}{\Gamma((n+1)\alpha)}$$

$$\left[\frac{(g(x_0)-g(x))^{(n+1)\alpha}}{(n+1)\alpha}+\frac{1}{\delta_2}\frac{\Gamma(2)\Gamma((n+1)\alpha)}{\Gamma((n+1)\alpha+2)}(g(x_0)-g(x))^{(n+1)\alpha+1}\right]=$$

$$\frac{\omega_1\left(\left(D_{x_0-;g}^{(n+1)\alpha}f\right)\circ g^{-1},\delta_2\right)_{[g(a),g(x_0)]}}{\Gamma((n+1)\alpha)}$$

$$\left[\frac{(g(x_0)-g(x))^{(n+1)\alpha}}{(n+1)\alpha}+\frac{1}{\delta_2}\frac{(g(x_0)-g(x))^{(n+1)\alpha+1}}{(n+1)\alpha((n+1)\alpha+1)}\right]= \tag{11.59}$$

$$\frac{\omega_1\left(\left(D_{x_0-;g}^{(n+1)\alpha}f\right)\circ g^{-1},\delta_2\right)_{[g(a),g(x_0)]}}{\Gamma((n+1)\alpha+1)}$$

$$\left[(g(x_0)-g(x))^{(n+1)\alpha}+\frac{(g(x_0)-g(x))^{(n+1)\alpha+1}}{\delta_2((n+1)\alpha+1)}\right].$$

We have proved that

$$\|f(x)-f(x_0)\|\leq\frac{\omega_1\left(\left(D_{x_0-;g}^{(n+1)\alpha}f\right)\circ g^{-1},\delta_2\right)_{[g(a),g(x_0)]}}{\Gamma((n+1)\alpha+1)}$$

$$\left[(g(x_0)-g(x))^{(n+1)\alpha}+\frac{(g(x_0)-g(x))^{(n+1)\alpha+1}}{\delta_2((n+1)\alpha+1)}\right], \tag{11.60}$$

$\forall\, x\in[a,x_0]$, $\delta_2>0$.

By (11.56) and (11.60), setting $\delta=\delta_1=\delta_2>0$, we derive (11.48). ■

We state

Corollary 11.23 (to Theorem 11.22) *All as in Theorem 11.22. Then*

$$\|f(\cdot)-f(x_0)\|\leq\frac{\omega_1\left(\left(D_{x_0;g}^{(n+1)\alpha}f\right)\circ g^{-1},\delta\right)}{\Gamma((n+1)\alpha+1)}$$

$$\left[|g(\cdot)-g(x_0)|^{(n+1)\alpha}+\frac{|g(\cdot)-g(x_0)|^{(n+1)\alpha+1}}{\delta((n+1)\alpha+1)}\right], \tag{11.61}$$

$\delta>0$, *true over* $[a,b]$.

We need

Definition 11.24 Let $D_{x_0;g}^{\alpha}f$ denote any of $D_{x_0\pm;g}^{\alpha}f$ and $\delta>0$. We set

$$\omega_1\left(\left(D^{\alpha}_{x_0;g}f\right)\circ g^{-1},\delta\right) := \tag{11.62}$$

$$\max\left\{\omega_1\left(\left(D^{\alpha}_{x+;g}f\right)\circ g^{-1},\delta\right)_{[g(x_0),g(b)]},\omega_1\left(\left(D^{\alpha}_{x_0-;g}f\right)\circ g^{-1},\delta\right)_{[g(a),g(x_0)]}\right\},$$

where $x_0 \in [a,b]$. Here the moduli of continuity are considered over $[g(x_0), g(b)]$ and $[g(a), g(x_0)]$, respectively.

We will use:

Theorem 11.25 *Let $\alpha > 0$, $n = \lceil\alpha\rceil$, $\alpha \notin \mathbb{N}$, and $f \in C^n([a,b], X)$, where $[a,b] \subset \mathbb{R}$ and $(X, \|\cdot\|)$ is a Banach space. Let $g \in C^1([a,b])$, strictly increasing such that $g^{-1} \in C^n([g(a), g(b)])$. Let $x_0 \in [a,b]$, be fixed such that $\left(f\circ g^{-1}\right)^{(i)}(g(x_0)) = 0$, for, $i = 1, \ldots, n-1$. Then*

$$\|f(x) - f(x_0)\| \leq \frac{\omega_1\left(\left(D^{\alpha}_{x_0;g}f\right)\circ g^{-1},\delta\right)}{\Gamma(\alpha+1)} \tag{11.63}$$

$$\left[|g(x) - g(x_0)|^{\alpha} + \frac{|g(x) - g(x_0)|^{\alpha+1}}{\delta(\alpha+1)}\right],$$

$\forall\, \delta > 0$, $\forall\, x \in [a,b]$.

If $0 < \alpha < 1$, then we do not need any initial conditions.

Proof By Remark 11.7 we get that $\left(D^{\alpha}_{x_0\pm;g}f\right)(x_0) = 0$, for $\alpha \notin \mathbb{N}$. By (11.37) we obtain

$$f(x) - f(x_0) = \frac{1}{\Gamma(\alpha)}\int_{x_0}^{x}(g(x) - g(t))^{\alpha-1}g'(t)\left(D^{\alpha}_{x_0+;g}f\right)(t)\,dt, \tag{11.64}$$

$\forall\, x \in [x_0, b]$.

And, by (11.38) we get

$$f(x) - f(x_0) = \frac{1}{\Gamma(\alpha)}\int_{x}^{x_0}(g(t) - g(x))^{\alpha-1}g'(t)\left(D^{\alpha}_{x_0-;g}f\right)(t)\,dt, \tag{11.65}$$

$\forall\, x \in [a, x_0]$.

When $0 < \alpha < 1$, i.e. $n = 1$, then (11.64), (11.65) are valid without any initial conditions.

We can write

$$f(x) - f(x_0) = \frac{1}{\Gamma(\alpha)}\int_{x_0}^{x}(g(x) - g(t))^{\alpha-1}g'(t)\left[\left(D^{\alpha}_{x_0+;g}f\right)(t) - \left(D^{\alpha}_{x_0+;g}f\right)(x_0)\right]dt, \tag{11.66}$$

$\forall\, x \in [x_0, b]$,

$$f(x) - f(x_0) = \frac{1}{\Gamma(\alpha)} \int_x^{x_0} (g(t) - g(x))^{\alpha-1} g'(t) \left[\left(D^{\alpha}_{x_0-;g} f \right)(t) - \left(D^{\alpha}_{x_0-;g} f \right)(x_0) \right] dt, \tag{11.67}$$

$\forall\, x \in [a, x_0]$.

We can rewrite (11.66) and (11.67) as follows (by $z := g(t)$):

$$f(x) - f(x_0) = \frac{1}{\Gamma(\alpha)} \int_{g(x_0)}^{g(x)} (g(x) - z)^{\alpha-1}$$

$$\left[\left(\left(D^{\alpha}_{x_0+;g} f \right) \circ g^{-1} \right)(z) - \left(\left(D^{\alpha}_{x_0+;g} f \right) \circ g^{-1} \right)(g(x_0)) \right] dz, \tag{11.68}$$

$\forall\, x \in [x_0, b]$, and

$$f(x) - f(x_0) = \frac{1}{\Gamma(\alpha)} \int_{g(x)}^{g(x_0)} (z - g(x))^{\alpha-1}$$

$$\left[\left(\left(D^{\alpha}_{x_0-;g} f \right) \circ g^{-1} \right)(z) - \left(\left(D^{\alpha}_{x_0-;g} f \right) \circ g^{-1} \right)(g(x_0)) \right] dz, \tag{11.69}$$

$\forall\, x \in [a, x_0]$.

We have that ($x_0 \leq x \leq b$)

$$\| f(x) - f(x_0) \| = \frac{1}{\Gamma(\alpha)} \int_{g(x_0)}^{g(x)} (g(x) - z)^{\alpha-1}$$

$$\left\| \left(\left(D^{\alpha}_{x_0+;g} f \right) \circ g^{-1} \right)(z) - \left(\left(D^{\alpha}_{x_0+;g} f \right) \circ g^{-1} \right)(g(x_0)) \right\| dz \leq$$

($\delta_1 > 0$)

$$\frac{1}{\Gamma(\alpha)} \int_{g(x_0)}^{g(x)} (g(x) - z)^{\alpha-1} \omega_1 \left(\left(D^{\alpha}_{x_0+;g} f \right) \circ g^{-1}, \frac{\delta_1 |z - g(x_0)|}{\delta_1} \right)_{[g(x_0), g(b)]} dz \overset{(11.42.)}{\leq}$$

$$\frac{\omega_1 \left(\left(D^{\alpha}_{x_0+;g} f \right) \circ g^{-1}, \delta_1 \right)_{[g(x_0), g(b)]}}{\Gamma(\alpha)} \int_{g(x_0)}^{g(x)} (g(x) - z)^{\alpha-1} \left(1 + \frac{(z - g(x_0))}{\delta_1} \right) dz = \tag{11.70}$$

$$\frac{\omega_1 \left(\left(D^{\alpha}_{x_0+;g} f \right) \circ g^{-1}, \delta_1 \right)_{[g(x_0), g(b)]}}{\Gamma(\alpha)}$$

$$\left[\frac{(g(x) - g(x_0))^{\alpha}}{\alpha} + \frac{1}{\delta_1} \int_{g(x_0)}^{g(x)} (g(x) - z)^{\alpha-1} (z - g(x_0))^{2-1} dz \right] =$$

$$\frac{\omega_1\left(\left(D^{\alpha}_{x_0+;g}f\right)\circ g^{-1},\delta_1\right)_{[g(x_0),g(b)]}}{\Gamma(\alpha)}$$

$$\left[\frac{(g(x)-g(x_0))^{\alpha}}{\alpha}+\frac{1}{\delta_1}\frac{\Gamma(\alpha)\Gamma(2)}{\Gamma(\alpha+2)}(g(x)-g(x_0))^{\alpha+1}\right]=$$

$$\frac{\omega_1\left(\left(D^{\alpha}_{x_0+;g}f\right)\circ g^{-1},\delta_1\right)_{[g(x_0),g(b)]}}{\Gamma(\alpha)}$$

$$\left[\frac{(g(x)-g(x_0))^{\alpha}}{\alpha}+\frac{1}{\delta_1}\frac{1}{\alpha(\alpha+1)}(g(x)-g(x_0))^{\alpha+1}\right], \tag{11.71}$$

$\forall\, x\in[x_0,b]$.

We have proved that

$$\|f(x)-f(x_0)\|\leq\frac{\omega_1\left(\left(D^{\alpha}_{x_0+;g}f\right)\circ g^{-1},\delta_1\right)_{[g(x_0),g(b)]}}{\Gamma(\alpha+1)}$$

$$\left[(g(x)-g(x_0))^{\alpha}+\frac{(g(x)-g(x_0))^{\alpha+1}}{\delta_1(\alpha+1)}\right], \tag{11.72}$$

$\forall\, x\in[x_0,b]$, $\delta_1>0$.

By (11.69) we obtain

$$\|f(x)-f(x_0)\|\leq\frac{1}{\Gamma(\alpha)}\int_{g(x)}^{g(x_0)}(z-g(x))^{\alpha-1}$$

$$\left\|\left(\left(D^{\alpha}_{x_0-;g}f\right)\circ g^{-1}\right)(z)-\left(\left(D^{\alpha}_{x_0-;g}f\right)\circ g^{-1}\right)(g(x_0))\right\|dz\leq$$

$(\delta_2>0)$

$$\frac{1}{\Gamma(\alpha)}\int_{g(x)}^{g(x_0)}(z-g(x))^{\alpha-1}\omega_1\left(\left(D^{\alpha}_{x_0-;g}f\right)\circ g^{-1},\frac{\delta_2(g(x_0)-z)}{\delta_2}\right)_{[g(a),g(x_0)]}dz\overset{(11.42.)}{\leq}$$

$$\frac{\omega_1\left(\left(D^{\alpha}_{x_0-;g}f\right)\circ g^{-1},\delta_2\right)_{[g(a),g(x_0)]}}{\Gamma(\alpha)}\int_{g(x)}^{g(x_0)}(z-g(x))^{\alpha-1}\left(1+\frac{g(x_0)-z}{\delta_2}\right)dz= \tag{11.73}$$

$$\frac{\omega_1\left(\left(D^{\alpha}_{x_0-;g}f\right)\circ g^{-1},\delta_2\right)_{[g(a),g(x_0)]}}{\Gamma(\alpha)}$$

$$\left[\frac{(g(x_0)-g(x))^{\alpha}}{\alpha}+\frac{1}{\delta_2}\int_{g(x)}^{g(x_0)}(g(x_0)-z)^{2-1}(z-g(x))^{\alpha-1}dz\right]=$$

$$\frac{\omega_1\left(\left(D^{\alpha}_{x_0-;g}f\right)\circ g^{-1},\delta_2\right)_{[g(a),g(x_0)]}}{\Gamma(\alpha)}$$

$$\left[\frac{(g(x_0)-g(x))^{\alpha}}{\alpha}+\frac{1}{\delta_2}\frac{\Gamma(2)\Gamma(\alpha)}{\Gamma(\alpha+2)}(g(x_0)-g(x))^{\alpha+1}\right]=$$

$$\frac{\omega_1\left(\left(D^{\alpha}_{x_0-;g}f\right)\circ g^{-1},\delta_2\right)_{[g(a),g(x_0)]}}{\Gamma(\alpha)}$$

$$\left[\frac{(g(x_0)-g(x))^{\alpha}}{\alpha}+\frac{1}{\delta_2}\frac{1}{\alpha(\alpha+1)}(g(x_0)-g(x))^{\alpha+1}\right]= \tag{11.74}$$

$$\frac{\omega_1\left(\left(D^{\alpha}_{x_0-;g}f\right)\circ g^{-1},\delta_2\right)_{[g(a),g(x_0)]}}{\Gamma(\alpha+1)}$$

$$\left[(g(x_0)-g(x))^{\alpha}+\frac{1}{\delta_2(\alpha+1)}(g(x_0)-g(x))^{\alpha+1}\right], \tag{11.75}$$

$\forall\, x\in[a,x_0]$.

In conclusion we derive

$$\|f(x)-f(x_0)\|\le\frac{\omega_1\left(\left(D^{\alpha}_{x_0-;g}f\right)\circ g^{-1},\delta_2\right)_{[g(a),g(x_0)]}}{\Gamma(\alpha+1)}$$

$$\left[(g(x_0)-g(x))^{\alpha}+\frac{(g(x_0)-g(x))^{\alpha+1}}{\delta_2(\alpha+1)}\right], \tag{11.76}$$

$\forall\, x\in[a,x_0]$, $\delta_2>0$.

Choosing $\delta=\delta_1=\delta_2>0$ by (11.72) and (11.76), we get (11.63). ■

We state

Corollary 11.26 *All as in Theorem 11.25. Then*

$$\|f(\cdot)-f(x_0)\|\le\frac{\omega_1\left(\left(D^{\alpha}_{x_0;g}f\right)\circ g^{-1},\delta\right)}{\Gamma(\alpha+1)}$$

$$\left[|g(\cdot)-g(x_0)|^{\alpha}+\frac{|g(\cdot)-g(x_0)|^{\alpha+1}}{\delta(\alpha+1)}\right], \tag{11.77}$$

true over $[a,b]$, $\delta>0$.

If $0<\alpha<1$, *no initial conditions are needed.*

We make

Assumption 11.27 Let $L_N : C([a,b], X) \hookrightarrow C([a,b], X)$, $(X, \|\cdot\|)$ a Banach space, L_N is a linear operator, $\forall\ N \in \mathbb{N}$, $x_0 \in [a,b]$. Let also $\widetilde{L}_N : C([a,b]) \hookrightarrow C([a,b])$, a sequence of positive linear operators, $\forall\ N \in \mathbb{N}$.

We assume that

$$\|(L_N(f))(x_0)\| \leq \left(\widetilde{L}_N(\|f\|)\right)(x_0), \tag{11.78}$$

$\forall\ N \in \mathbb{N}, \forall\ x_0 \in [a,b], \forall\ f \in C([a,b], X)$.

When $g \in C([a,b])$, $c \in X$, we assume that

$$(L_N(cg)) = c\widetilde{L}_N(g),\ \forall\ N \in \mathbb{N}. \tag{11.79}$$

We treat the special case of

$$\widetilde{L}_N(1) = 1,\ \forall\ N \in \mathbb{N}, \tag{11.80}$$

which implies

$$L_N(c) = c,\ \forall\ c \in X,\ \forall\ N \in \mathbb{N}. \tag{11.81}$$

We call $\widetilde{L}_N$ the companion operator of L_N.

Operators $\widetilde{B}_N$ see (1), and B_N see (11.2), fulfill Assumption 11.27.

We make

Remark 11.28 Let $\widetilde{L}_N : C([a,b]) \to C([a,b])$, $N \in \mathbb{N}$, be a sequence of positive linear operators. By Riesz representation theorem (see [11, p. 304]) we have

$$\widetilde{L}_N(f, x_0) = \int_{[a,b]} f(t)\, d\mu_{N_{x_0}}(t), \tag{11.82}$$

$\forall\ x_0 \in [a,b]$, where $\mu_{N_{x_0}}$ is a unique positive finite measure on σ-Borel algebra of $[a,b]$. Call

$$\widetilde{L}_N(1, x_0) = \mu_{N_{x_0}}([a,b]) = M_{N_{x_0}}. \tag{11.83}$$

In our case of $\widetilde{L}_N(1, x_0) = 1$, we have that $M_{N_{x_0}} = 1$, and $\mu_{N_{x_0}}$ is a probability measure.

By Hölder's inequality we obtain

$$\widetilde{L}_N\left(|g(\cdot) - g(x_0)|^{\alpha}\right)(x_0) = \int_{[a,b]} |g(t) - g(x_0)|^{\alpha}\, d\mu_{N_{x_0}}(t) \leq$$

$$\left(\int_{[a,b]} |g(t) - g(x_0)|^{\alpha+1}\, d\mu_{N_{x_0}}(t)\right)^{\frac{\alpha}{\alpha+1}} = \left(\widetilde{L}_N\left(|g(\cdot) - g(x_0)|^{\alpha+1}\right)(x_0)\right)^{\frac{\alpha}{\alpha+1}}. \tag{11.84}$$

Similarly, it holds

$$\widetilde{L}_N\left(|g(\cdot)-g(x_0)|^{(n+1)\alpha}\right)(x_0) \leq \left(\widetilde{L}_N\left(|g(\cdot)-g(x_0)|^{(n+1)\alpha+1}\right)(x_0)\right)^{\frac{(n+1)\alpha}{(n+1)\alpha+1}}. \tag{11.85}$$

We present

Theorem 11.29 *Under Assumption 11.27 and all as in Theorem 11.22, we derive*

$$\|(L_N(f))(x_0)-f(x_0)\| \leq \frac{\omega_1\left(\left(D_{x_0;g}^{(n+1)\alpha}f\right)\circ g^{-1},\delta\right)}{\Gamma((n+1)\alpha+1)} \tag{11.86}$$

$$\left[\widetilde{L}_N\left(|g(\cdot)-g(x_0)|^{(n+1)\alpha}\right)(x_0)+\frac{\widetilde{L}_N\left(|g(\cdot)-g(x_0)|^{(n+1)\alpha+1}\right)(x_0)}{\delta((n+1)\alpha+1)}\right],$$

$\forall\, N \in \mathbb{N}$, $\delta > 0$.

Proof By (11.61) we get

$$\widetilde{L}_N(\|f(\cdot)-f(x_0)\|)(x_0) \leq \frac{\omega_1\left(\left(D_{x_0;g}^{(n+1)\alpha}f\right)\circ g^{-1},\delta\right)}{\Gamma((n+1)\alpha+1)} \tag{11.87}$$

$$\left[\widetilde{L}_N\left(|g(\cdot)-g(x_0)|^{(n+1)\alpha}\right)(x_0)+\frac{\widetilde{L}_N\left(|g(\cdot)-g(x_0)|^{(n+1)\alpha+1}\right)(x_0)}{\delta((n+1)\alpha+1)}\right],$$

along with (11.46) proving the claim. ∎

We also give

Theorem 11.30 *Under Assumption 11.27 and all as in Theorem 11.25, we derive*

$$\|(L_N(f))(x_0)-f(x_0)\| \leq \frac{\omega_1\left(\left(D_{x_0;g}^{\alpha}f\right)\circ g^{-1},\delta\right)}{\Gamma(\alpha+1)}$$

$$\left[\widetilde{L}_N\left(|g(\cdot)-g(x_0)|^{\alpha}\right)(x_0)+\frac{\widetilde{L}_N\left(|g(\cdot)-g(x_0)|^{\alpha+1}\right)(x_0)}{\delta(\alpha+1)}\right], \tag{11.88}$$

$\forall\, N \in \mathbb{N}$, $\delta > 0$.

Proof By (11.77) we get

$$\widetilde{L}_N\left(\|f(\cdot)-f(x_0)\|\right)(x_0) \leq \frac{\omega_1\left(\left(D^{\alpha}_{x_0;g}f\right)\circ g^{-1},\delta\right)}{\Gamma(\alpha+1)}$$

$$\left[\widetilde{L}_N\left(|g(\cdot)-g(x_0)|^{\alpha}\right)(x_0)+\frac{\widetilde{L}_N\left(|g(\cdot)-g(x_0)|^{\alpha+1}\right)(x_0)}{\delta(\alpha+1)}\right], \tag{11.89}$$

along with (11.46) proving the claim. ∎

We give

Theorem 11.31 *Under Assumption 11.27 and all as in Theorem 11.22, we obtain*

$$\|(L_N(f))(x_0)-f(x_0)\| \leq \frac{((n+1)\alpha+2)}{\Gamma((n+1)\alpha+2)}$$

$$\omega_1\left(\left(D^{(n+1)\alpha}_{x_0;g}f\right)\circ g^{-1},\left(\widetilde{L}_N\left(|g(\cdot)-g(x_0)|^{(n+1)\alpha+1}\right)(x_0)\right)^{\frac{1}{(n+1)\alpha+1}}\right) \tag{11.90}$$

$$\left(\widetilde{L}_N\left(|g(\cdot)-g(x_0)|^{(n+1)\alpha+1}\right)(x_0)\right)^{\frac{(n+1)\alpha}{(n+1)\alpha+1}},\ \forall\, N\in\mathbb{N}.$$

If $\widetilde{L}_N\left(|g(\cdot)-g(x_0)|^{(n+1)\alpha+1}\right)(x_0)\to 0$, *then* $L_N(f)(x_0)\to f(x_0)$, *as* $N\to\infty$.

Proof By (11.87) and (11.85) we obtain that

$$\text{Right hand side (11.87)} \leq \frac{\omega_1\left(\left(D^{(n+1)\alpha}_{x_0;g}f\right)\circ g^{-1},\delta\right)}{\Gamma((n+1)\alpha+1)}$$

$$\left[\left(\widetilde{L}_N\left(|g(\cdot)-g(x_0)|^{(n+1)\alpha+1}\right)(x_0)\right)^{\frac{(n+1)\alpha}{(n+1)\alpha+1}}+\frac{\widetilde{L}_N\left(|g(\cdot)-g(x_0)|^{(n+1)\alpha+1}\right)(x_0)}{\delta((n+1)\alpha+1)}\right]=:(*). \tag{11.91}$$

Momentarily we assume that $\widetilde{L}_N\left(|g(\cdot)-g(x_0)|^{(n+1)\alpha+1}\right)(x_0)>0$, and choose

$$\delta:=\left(\widetilde{L}_N\left(|g(\cdot)-g(x_0)|^{(n+1)\alpha+1}\right)(x_0)\right)^{\frac{1}{(n+1)\alpha+1}}>0. \tag{11.92}$$

Thus, we have

$$(*)=\frac{\omega_1\left(\left(D^{(n+1)\alpha}_{x_0;g}f\right)\circ g^{-1},\left(\widetilde{L}_N\left(|g(\cdot)-g(x_0)|^{(n+1)\alpha+1}\right)(x_0)\right)^{\frac{1}{(n+1)\alpha+1}}\right)}{\Gamma((n+1)\alpha+1)}$$

$$\left[\left(\widetilde{L}_N\left(|g(\cdot)-g(x_0)|^{(n+1)\alpha+1}\right)(x_0)\right)^{\frac{(n+1)\alpha}{(n+1)\alpha+1}}+\frac{\delta^{(n+1)\alpha+1}}{\delta((n+1)\alpha+1)}\right]= \tag{11.93}$$

$$\frac{\omega_1\left(\left(D_{x_0;g}^{(n+1)\alpha}f\right)\circ g^{-1},\left(\widetilde{L}_N\left(|g(\cdot)-g(x_0)|^{(n+1)\alpha+1}\right)(x_0)\right)^{\frac{1}{(n+1)\alpha+1}}\right)}{\Gamma((n+1)\alpha+1)}$$

$$\left(\widetilde{L}_N\left(|g(\cdot)-g(x_0)|^{(n+1)\alpha+1}\right)(x_0)\right)^{\frac{(n+1)\alpha}{(n+1)\alpha+1}}\left[1+\frac{1}{(n+1)\alpha+1}\right]=$$

$$\frac{((n+1)\alpha+2)}{\Gamma((n+1)\alpha+2)}\omega_1\left(\left(D_{x_0;g}^{(n+1)\alpha}f\right)\circ g^{-1},\left(\widetilde{L}_N\left(|g(\cdot)-g(x_0)|^{(n+1)\alpha+1}\right)(x_0)\right)^{\frac{1}{(n+1)\alpha+1}}\right) \tag{11.94}$$

$$\left(\widetilde{L}_N\left(|g(\cdot)-g(x_0)|^{(n+1)\alpha+1}\right)(x_0)\right)^{\frac{(n+1)\alpha}{(n+1)\alpha+1}}.$$

So far we have proved that

$$\widetilde{L}_N(\|f(\cdot)-f(x_0)\|)(x_0)\leq\frac{((n+1)\alpha+2)}{\Gamma((n+1)\alpha+2)}$$

$$\omega_1\left(\left(D_{x_0;g}^{(n+1)\alpha}f\right)\circ g^{-1},\left(\widetilde{L}_N\left(|g(\cdot)-g(x_0)|^{(n+1)\alpha+1}\right)(x_0)\right)^{\frac{1}{(n+1)\alpha+1}}\right) \tag{11.95}$$

$$\left(\widetilde{L}_N\left(|g(\cdot)-g(x_0)|^{(n+1)\alpha+1}\right)(x_0)\right)^{\frac{(n+1)\alpha}{(n+1)\alpha+1}},$$

when $\delta>0$.

If $\delta=0$, we get R.H.S.(11.95)$=0$, and

$$\widetilde{L}_N\left(|g(\cdot)-g(x_0)|^{(n+1)\alpha+1}\right)(x_0)=\int_{[a,b]}|g(t)-g(x_0)|^{(n+1)\alpha+1}d\mu_{N_{x_0}}(t)=0. \tag{11.96}$$

Hence $|g(t)-g(x_0)|=0$, a.e. on $[a,b]$. That is $g(t)=g(x_0)$, a.e. on $[a,b]$ and $t=x_0$, a.e. on $[a,b]$, because g is strictly increasing. More precisely $\mu_{N_{x_0}}\{t\in[a,b]:t\neq x_0\}=0$. Therefore $\mu_{N_{x_0}}$ concentrates on x_0, that is $\mu_{N_{x_0}}=\delta_{x_0}$, the unit Dirac measure.

Consequently, it holds

$$\begin{aligned}\widetilde{L}_N(\|f(\cdot)-f(x_0)\|)(x_0)&=\int_{[a,b]}\|f(t)-f(x_0)\|\,d\delta_{x_0}(t)\\&=\|f(x_0)-f(x_0)\|=0.\end{aligned} \tag{11.97}$$

Therefore (11.95) is true always, in both cases, completing the proof of the theorem, by the use of (11.46). ■

We also give

Theorem 11.32 *Under Assumption 11.27 and all as in Theorem 11.25, we obtain*

$$\|(L_N(f))(x_0) - f(x_0)\| \leq \frac{(\alpha+2)}{\Gamma(\alpha+2)}$$

$$\omega_1\left(\left(D^{\alpha}_{x_0;g}f\right)\circ g^{-1}, \left(\widetilde{L}_N\left(|g(\cdot) - g(x_0)|^{\alpha+1}\right)(x_0)\right)^{\frac{1}{\alpha+1}}\right) \tag{11.98}$$

$$\left(\widetilde{L}_N\left(|g(\cdot) - g(x_0)|^{\alpha+1}\right)(x_0)\right)^{\frac{\alpha}{\alpha+1}}, \ \forall\, N \in \mathbb{N}.$$

If $\widetilde{L}_N\left(|g(\cdot) - g(x_0)|^{\alpha+1}\right)(x_0) \to 0$*, then* $L_N(f)(x_0) \to f(x_0)$*, as* $N \to \infty$*.*

Proof Similar to the proof of Theorem 11.31. It is based on (11.46), (11.84) and (11.89). We omit the details. ■

We give

Corollary 11.33 (to Theorem 11.31) *All as in Theorem 11.31. Then*

$$\|(L_N(f))(x_0) - f(x_0)\| \leq \frac{((n+1)\alpha+2)}{\Gamma((n+1)\alpha+2)}$$

$$\omega_1\left(\left(D^{(n+1)\alpha}_{x_0;g}f\right)\circ g^{-1}, \left\|g'\right\|_{\infty,[a,b]}\left(\widetilde{L}_N\left(|\cdot - x_0|^{(n+1)\alpha+1}\right)(x_0)\right)^{\frac{1}{(n+1)\alpha+1}}\right)$$

$$\left(\left\|g'\right\|_{\infty,[a,b]}\right)^{(n+1)\alpha}\left(\widetilde{L}_N\left(|\cdot - x_0|^{(n+1)\alpha+1}\right)(x_0)\right)^{\frac{(n+1)\alpha}{(n+1)\alpha+1}}, \ \forall\, N \in \mathbb{N}. \tag{11.99}$$

If $\widetilde{L}_N\left(|\cdot - x_0|^{(n+1)\alpha+1}\right)(x_0) \to 0$*, then* $L_N(f)(x_0) \to f(x_0)$*, as* $N \to \infty$*.*

Proof By Theorem 11.31 and

$$|g(x) - g(x_0)| \leq \left\|g'\right\|_{\infty,[a,b]}|x - x_0|, \ \forall\, x,\ x_0 \in [a,b]. \tag{11.100}$$

■

We give

Corollary 11.34 (to Theorem 11.32) *All as in Theorem 11.32. Then*

$$\|(L_N(f))(x_0) - f(x_0)\| \leq \frac{(\alpha+2)}{\Gamma(\alpha+2)}$$

$$\omega_1\left(\left(D^{\alpha}_{x_0;g}f\right)\circ g^{-1}, \left\|g'\right\|_{\infty,[a,b]}\left(\widetilde{L}_N\left(|\cdot - x_0|^{\alpha+1}\right)(x_0)\right)^{\frac{1}{\alpha+1}}\right)$$

$$\left(\left\|g'\right\|_{\infty,[a,b]}\right)^{\alpha}\left(\widetilde{L}_N\left(|\cdot - x_0|^{\alpha+1}\right)(x_0)\right)^{\frac{\alpha}{\alpha+1}}, \ \forall\, N \in \mathbb{N}. \tag{11.101}$$

If $\widetilde{L}_N\left(|\cdot - x_0|^{\alpha+1}\right)(x_0) \to 0$*, then* $L_N(f)(x_0) \to f(x_0)$*, as* $N \to \infty$*.*

Proof By Theorem 11.32 and

$$\left|g(x)-g(x_0)\right| \leq \left\|g'\right\|_{\infty,[a,b]}\left|x-x_0\right|, \ \forall\, x,\ x_0 \in [a,b]. \tag{11.102}$$

■

We make

Remark 11.35 Theorems 11.31, 11.32 and Corollaries 11.33, 11.34 have direct applications to B_N, $\widetilde{B}_N$ Bernstein operators over $[0,1]$. We obtain there $B_N(f)(x_0) \to f(x_0)$, as $N \to \infty$.

We notice that

$$|\cdot - x_0|^{\alpha+1} = |\cdot - x_0|^{\alpha}\,|\cdot - x_0| \leq |\cdot - x_0|, \text{ over } [0,1].$$

Hence

$$\widetilde{B}_N\left(|\cdot - x_0|^{\alpha+1}\right)(x_0) \leq \widetilde{B}_N\left(|\cdot - x_0|\right)(x_0),\, x_0 \in [0,1].$$

Similarly, it holds

$$|\cdot - x_0|^{(n+1)\alpha+1} = |\cdot - x_0|^{(n+1)\alpha}\,|\cdot - x_0| \leq |\cdot - x_0|, \text{over } [0,1].$$

Hence

$$\widetilde{B}_N\left(|\cdot - x_0|^{(n+1)\alpha+1}\right)(x_0) \leq \widetilde{B}_N\left(|\cdot - x_0|\right)(x_0),\ x_0 \in [0,1].$$

We have that

$$\widetilde{B}_N\left(|\cdot - x_0|\right)(x_0) \leq \left(\widetilde{B}_N\left((\cdot - x_0)^2\right)(x_0)\right)^{\frac{1}{2}}$$

$$= \sqrt{\frac{x_0(1-x_0)}{N}} \leq \frac{1}{2\sqrt{N}}, \ \forall\, N \in \mathbb{N}. \tag{11.103}$$

We give

Corollary 11.36 *Here $[a,b]=[0,1]$ and all as in Theorem 11.22. Then*

$$\left\|\left(B_N(f)\right)(x_0) - f(x_0)\right\| \leq \frac{((n+1)\alpha+2)}{\Gamma((n+1)\alpha+2)}$$

$$\omega_1\left(\left(D_{x_0;g}^{(n+1)\alpha} f\right)\circ g^{-1}, \left\|g'\right\|_{\infty,[a,b]}\left(\frac{1}{2\sqrt{N}}\right)^{\frac{1}{(n+1)\alpha+1}}\right)$$

$$\left(\left\|g'\right\|_{\infty,[a,b]}\right)^{(n+1)\alpha}\left(\frac{1}{2\sqrt{N}}\right)^{\frac{(n+1)\alpha}{(n+1)\alpha+1}}, \ \forall\, N \in \mathbb{N},\ x_0 \in [0,1]. \tag{11.104}$$

As $N \to \infty$, we get $B_N(f)(x_0) \to f(x_0)$.

Proof By Corollary 11.33 and (11.103). ■

We give

Corollary 11.37 *Here* $[a,b]=[0,1]$ *and all as in Theorem 11.25. Then*

$$\|(B_N(f))(x_0)-f(x_0)\|\leq\frac{(\alpha+2)}{\Gamma(\alpha+2)}$$

$$\omega_1\left(\left(D^{\alpha}_{x_0;g}f\right)\circ g^{-1},\left\|g'\right\|_{\infty,[a,b]}\left(\frac{1}{2\sqrt{N}}\right)^{\frac{1}{\alpha+1}}\right)$$

$$\left(\left\|g'\right\|_{\infty,[a,b]}\right)^{\alpha}\left(\frac{1}{2\sqrt{N}}\right)^{\frac{\alpha}{\alpha+1}},\ \forall\,N\in\mathbb{N}. \tag{11.105}$$

As $N\to\infty$, *we get* $B_N(f)(x_0)\to f(x_0)$, $x_0\in[0,1]$.

Proof By Corollary 11.34 and (11.103). ■

We continue with the following

Corollary 11.38 (to Corollary 11.37) *Here* $[a,b]=[0,1]$, *all as in Theorem 11.25, and* $g(x)=e^x$, $x\in[0,1]$. *Then*

$$\|(B_N(f))(x_0)-f(x_0)\|\leq\frac{(\alpha+2)}{\Gamma(\alpha+2)}$$

$$\omega_1\left(\left(D^{\alpha}_{x_0;e^x}f\right)\circ\ln x,e\left(\frac{1}{2\sqrt{N}}\right)^{\frac{1}{\alpha+1}}\right)e^{\alpha}\left(\frac{1}{2\sqrt{N}}\right)^{\frac{\alpha}{\alpha+1}},\ \forall\,N\in\mathbb{N}. \tag{11.106}$$

As $N\to\infty$, *we get* $B_N(f)(x_0)\to f(x_0)$, $x_0\in[0,1]$.

Proof By Corollary 11.37. ■

We finish with

Corollary 11.39 (to Corollary 11.38) *Here* $[a,b]=[0,1]$, *all as in Theorem 11.25,* $g(x)=e^x$, $x\in[0,1]$ *and* $\alpha=\frac{1}{2}$ *(no initial conditions needed). Then*

$$\|(B_N(f))(x_0)-f(x_0)\|\leq\frac{10}{3}\sqrt{\frac{e}{\pi}}$$

$$\omega_1\left(\left(D^{\frac{1}{2}}_{x_0;e^x}f\right)\circ\ln x,e\left(\frac{1}{2\sqrt{N}}\right)^{\frac{2}{3}}\right)\left(\frac{1}{2\sqrt{N}}\right)^{\frac{1}{3}}\ \forall\,N\in\mathbb{N},\ \forall\,x_0\in[0,1]. \tag{11.107}$$

As $N\to\infty$, *we get* $B_N(f)(x_0)\to f(x_0)$, *any* $x_0\in[0,1]$.

Proof By Corollary 11.38. ■

References

1. Aliprantis, C.D., Border, K.C.: Infinite Dimensional Analysis. Springer, New York (2006)
2. Anastassiou, G.A.: Moments in Probability and Approximation Theory. Pitman Research Notes in Mathematics, vol. 287. Longman Sci. & Tech., Harlow (1993)
3. Anastassiou, G.A.: Principles of General Fractional Analysis for Banach space valued functions. Bull. Allahabad Math. Soc. **32**(1), 71–145 (2017)
4. Anastassiou, G.A.: Vector fractional Korovkin type Approximation. Dyn. Syst. Appl. **26**, 81–104 (2017)
5. Anastassiou, G.: Intelligent Computations: Abstract Fractional Calculus, Inequalities, Approximations. Springer, Heidelberg, New York (2018)
6. Anastassiou, G.: Vectorial generalized g-fractional direct and iterated approximations by linear operators. Fasciculi Mathematici (2020). Accepted
7. Bochner integral. Encyclopedia of Mathematics. http://www.encyclopediaofmath.org/index.php?title=Bochner_integral&oldid=38659
8. Kreuter, M.: Sobolev space of vector-valued functions. Ulm University, Master thesis in Mathematics, Ulm, Germany (2015)
9. Mikkola, K.: Appendix B integration and differentiation in Banach spaces. http://math.aalto.fi/~kmikkola/research/thesis/contents/thesisb.pdf
10. Mikusinski, J.: The Bochner Integral. Academic, New York (1978)
11. Royden, H.L.: Real Analysis, 2nd edn. Macmillan, New York (1968)

Chapter 12
Quantitative Multivariate Complex Korovkin Approximation Theory

Let K be a compact convex subset of $\mathbb{C}^k$, $k \geq 2$, and $C(K, \mathbb{C})$ be the space of continuous functions from K into $\mathbb{C}$. We consider bounded linear operators from $C(K, \mathbb{C})$ into itself. We assume that these are bounded by companion positive linear operators from $C(K, \mathbb{R})$ into itself. We study quantitatively the rate of convergence of the approximation and high order approximation of these multivariate complex operators to the unit operators. Our results are inequalities of Korovkin type involving the multivariate complex modulus of continuity of the engaged function or its partial derivatives and basic test functions.

On the way to prove our main results we establish a complex multivariate Taylor's formula and results on complex multivariate modulus of continuity. We consider related approximations under convexity and we finish with important applications. See also [5].

12.1 Introduction

The study of the convergence of positive linear operators became more intensive and attractive when Korovkin (1953) proved his famous theorem (see [7, p. 14]).

Korovkin's First Theorem. Let $[a, b]$ be a compact interval in $\mathbb{R}$ and $(L_n)_{n\in\mathbb{N}}$ be a sequence of positive linear operators L_n mapping $C([a, b])$ into itself. Assume that $(L_n f)$ converges uniformly to f for the three test functions $f = 1, x, x^2$. Then $(L_n f)$ converges uniformly to f on $[a, b]$ for all functions of $f \in C([a, b])$.

So a lot of authors since then have worked on the theoretical aspects of the above convergence. But Mamedov (1959) (see [10]) was the first to put Korovkin's theorem in a quantitative scheme.

Mamedov's Theorem. Let $\{L_n\}_{n\in\mathbb{N}}$ be a sequence of positive linear operators in the space $C([a, b])$, for which $L_n 1 = 1$, $L_n(t, x) = x + \alpha_n(x)$, $L_n(t^2, x) = x^2 + \beta_n(x)$. Then

G. A. Anastassiou, *Generalized Fractional Calculus*, Studies in Systems, Decision and Control 305, https://doi.org/10.1007/978-3-030-56962-4_12

$$\|L_n(f, x) - f(x)\|_\infty \leq 3\omega_1\left(f, \sqrt{d_n}\right), \tag{12.1}$$

where ω_1 is the first modulus of continuity and $d_n = \|\beta_n(x) - 2x\alpha_n(x)\|_\infty$.

An improvement of the last result was the following.

Shisha and Mond's Theorem. (1968, see [13]). Let $[a, b] \subset \mathbb{R}$ be a compact interval. Let $\{L_n\}_{n\in\mathbb{N}}$ be a sequence of positive linear operators acting on $C([a, b])$. For $n = 1, 2, \ldots,$ suppose $L_n(1)$ is bounded. Let $f \in C([a, b])$. Then for $n = 1, 2, \ldots,$ it holds

$$\|L_n f - f\|_\infty \leq \|f\|_\infty \cdot \|L_n 1 - 1\|_\infty + \|L_n(1) + 1\|_\infty \cdot \omega_1(f, \mu_n), \tag{12.2}$$

where

$$\mu_n := \left\|\left(L_n\left((t-x)^2\right)\right)(x)\right\|_\infty^{\frac{1}{2}}. \tag{12.3}$$

Shisha-Mond inequality generated and inspired a lot of research done by many authors worldwide on the rate of convergence of a sequence of positive linear operators to the unit operator, always producing similar inequalities however in many different directions.

The author (see [1]) in his 1993 research monograph, produces in many directions best upper bounds for $|(L_n f)(x_0) - f(x_0)|$, $x_0 \in Q \subseteq \mathbb{R}^n$, $n \geq 1$, compact and convex, which lead for the first time to sharp/attained inequalities of Shisha-Mond type. The method of proving is probabilistic from the theory of moments. His pointwise approach is closely related to the study of the weak convergence with rates of a sequence of finite positive measures to the unit measure at a specific point.

The author in [3, pp. 383–412] continued this work in an abstract setting: Let X be a normed vector space, Y be a Banach lattice; $M \subset X$ is a compact and convex subset. Consider the space of continuous functions from M into Y, denoted by $C(M, Y)$; also consider the space of bounded functions $B(M, Y)$. He studied the rate of the uniform convergence of lattice homomorphisms $T : C(M, Y) \to C(M, Y)$ or $T : C(M, Y) \to B(M, Y)$ to the unit operator I. See also [2].

Also the author in [4, pp. 175–188] continued the last abstract work for bounded linear operators that are bounded by companion real positive linear operators. Here the involved functions are from $[a, b] \subset \mathbb{R}$ into $(X, \|\cdot\|)$ a Banach space.

All the above have inspired and motivated the work of this chapter. Our results are of Shisha-Mond type, i.e., of Korovkin type.

Namely here let K be a convex and compact subset of $\mathbb{C}^k$, $k \geq 2$, and L be a linear operator from $C(K, \mathbb{C})$ into itself, and let $\widetilde{L}$ be a positive linear operator from $C(K, \mathbb{R})$ into itself, such that $|L(f)| \leq \lambda\widetilde{L}(|f|)$, $\forall\, f \in C(K, \mathbb{C})$, where $\lambda > 0$.

Clearly then L is a bounded linear operator. Here we create a complete quantitative Korovkin type theory over the last described setting.

12.2 Background

We need

Theorem 12.1 *Let $K \subseteq (\mathbb{C}^k, \|\cdot\|)$, $k \in \mathbb{N} - \{1\}$, and f a function from K into $\mathbb{C}$. Consider the first multivariate complex modulus of continuity*

$$\omega_1(f, \delta) := \sup_{\substack{x,y \in K \\ \|x-y\| < \delta}} |f(x) - f(y)|, \ \delta > 0. \tag{12.4}$$

We have:

(1)' If K is open convex or compact convex, then $\omega_1(f, \delta) < \infty$, $\forall \delta > 0$, where $f \in UC(K, \mathbb{C})$ (uniformly continuous functions).

(2)' If K is open convex or compact convex, then $\omega_1(f, \delta)$ is continuous on $\mathbb{R}_+$ in δ, for $f \in UC(K, \mathbb{C})$.

(3)' If K is convex, then

$$\omega_1(f, t_1 + t_2) \leq \omega_1(f, t_1) + \omega_1(f, t_2), \ t_1, t_2 > 0, \tag{12.5}$$

that is the subadditivity property is true. Also it holds

$$\omega_1(f, n\delta) \leq n\omega_1(f, \delta) \tag{12.6}$$

and

$$\omega_1(f, \lambda\delta) \leq \lceil\lambda\rceil \omega_1(f, \delta) \leq (\lambda + 1)\omega_1(f, \delta), \tag{12.7}$$

where $n \in \mathbb{N}$, $\lambda > 0$, $\delta > 0$, $\lceil\cdot\rceil$ is the ceiling of the number.

(4)' Clearly in general $\omega_1(f, \delta) \geq 0$ and is increasing in $\delta > 0$ and $\omega_1(f, 0) = 0$.

(5)' If K is open or compact, then $\omega_1(f, \delta) \to 0$ as $\delta \downarrow 0$, iff $f \in UC(K, \mathbb{C})$.

(6)' It holds

$$\omega_1(f + g, \delta) \leq \omega_1(f, \delta) + \omega_1(g, \delta), \tag{12.8}$$

for $\delta > 0$, any $f, g : K \to \mathbb{C}$, $K \subset \mathbb{C}^k$ is arbitrary.

Proof (1)' Here K is open convex. Let here $f \in UC(K, \mathbb{C})$, iff $\forall \varepsilon > 0, \exists \delta > 0$: $\|x - y\| < \delta$ implies $|f(x) - f(y)| < \varepsilon$. Let $\varepsilon_0 > 0$ then $\exists \delta_0 > 0 : \|x - y\| \leq \delta_0$ with $|f(x) - f(y)| < \varepsilon_0$, hence $\omega_1(f, \delta_0) \leq \varepsilon_0 < \infty$.

Let $\delta > 0$ arbitrary and $x, y \in K : \|x - y\| \leq \delta$. Choose $n \in \mathbb{N} : n\delta_0 > \delta$, and set $x_i = x + \frac{i}{n}(y - x)$, $0 \leq i \leq n$. Notice that all $x_i \in K$. Then

$$|f(x) - f(y)| = \left|\sum_{i=0}^{n-1} (f(x_i) - f(x_{i+1}))\right| \leq$$

$$|f(x) - f(x_1)| + |f(x_1) - f(x_2)| + |f(x_2) - f(x_3)| + \cdots + |f(x_{n-1}) - f(y)| \leq$$

$$n\omega_1(f, \delta_0) \leq n\varepsilon_0 < \infty,$$

since $\|x_i - x_{i+1}\| = \frac{1}{n}\|x - y\| \leq \frac{1}{n}\delta < \delta_0$.

Thus $\omega_1(f, \delta) \leq n\varepsilon_0 < \infty$, proving the claim. If K is compact convex, then claim is obvious.

(2)' Let $x, y \in K$ and let $\|x - y\| \leq t_1 + t_2$, then there exists a point $z \in \overline{xy}$, $z \in K : \|x - z\| \leq t_1$ and $\|y - z\| \leq t_2$, where $t_1, t_2 > 0$.

Notice that

$$|f(x) - f(y)| \leq |f(x) - f(z)| + |f(z) - f(y)| \leq \omega_1(f, t_1) + \omega_1(f, t_2).$$

Hence

$$\omega_1(f, t_1 + t_2) \leq \omega_1(f, t_1) + \omega_1(f, t_2),$$

proving (3)'. Then by the obvious property (4)' we get

$$0 \leq \omega_1(f, t_1 + t_2) - \omega_1(f, t_1) \leq \omega_1(f, t_2),$$

and

$$|\omega_1(f, t_1 + t_2) - \omega_1(f, t_1)| \leq \omega_1(f, t_2).$$

Let $f \in UC(K, \mathbb{C})$, then $\lim_{t_2 \downarrow 0} \omega_1(f, t_2) = 0$, by property (5)'. Hence $\omega_1(f, \cdot)$ is continuous on $\mathbb{R}_+$.

(5)' ($\Rightarrow$) Let $\omega_1(f, \delta) \to 0$ as $\delta \downarrow 0$. Then $\forall\, \varepsilon > 0$, $\exists\, \delta > 0$ with $\omega_1(f, \delta) \leq \varepsilon$. I.e. $\forall\, x, y \in K : \|x - y\| \leq \delta$ we get $|f(x) - f(y)| \leq \varepsilon$. That is $f \in UC(K, \mathbb{C})$.

($\Leftarrow$) Let $f \in UC(K, \mathbb{C})$. Then $\forall\ \varepsilon > 0$, $\exists\ \delta > 0 :$ whenever $\|x - y\| \leq \delta$, $x, y \in K$, it implies $|f(x) - f(y)| \leq \varepsilon$. I.e. $\forall\, \varepsilon > 0$, $\exists\, \delta > 0 : \omega_1(f, \delta) \leq \varepsilon$. That is $\omega_1(f, \delta) \to 0$ as $\delta \downarrow 0$.

(6)' Notice that

$$|(f(x) + g(x)) - (f(y) + g(y))| \leq |f(x) - f(y)| + |g(x) - g(y)|.$$

That is property (6)' now is clear. ∎

We need

Theorem 12.2 ([1, p. 208]) *Let* $(V_1, \|\cdot\|)$, $(V_2, \|\cdot\|)$ *be real normed vector spaces and* $Q \subseteq V_1$ *which is star-shaped relative to the fixed point* x_0. *Consider* $f : Q \to V_2$ *with the properties:*

$$f(x_0) = 0, \text{ and } \|s - t\| \leq h \text{ implies } \|f(s) - f(t)\| \leq w;\ w, h > 0. \tag{12.9}$$

Then, there exists a maximal such function Φ, *namely*

$$\Phi(t) := \left\lceil \frac{\|t - x_0\|}{h} \right\rceil \cdot w \cdot \vec{i}, \tag{12.10}$$

where $\vec{i}$ *is any unit vector in* V_2.

That is

$$\|f(t)\| \leq \|\Phi(t)\|, \text{ all } t \in Q. \tag{12.11}$$

Corollary 12.3 *Let* $K \subseteq \left(\mathbb{C}^k, \|\cdot\|\right)$ *be a compact convex subset, and* $f \in C(K, \mathbb{C})$. *Then*

$$|f(x) - f(x_0)| \leq \omega_1(f, \delta) \left\lceil \frac{\|x - x_0\|}{\delta} \right\rceil, \ \delta > 0, \tag{12.12}$$

$\forall\, x, x_0 \in K$.

We make

Remark 12.4 Let $K \subseteq \left(\mathbb{C}^k, \|\cdot\|\right)$ be a compact subset and $g \in C(K, \mathbb{R})$.

A linear functional I from $C(K, \mathbb{R})$ into $\mathbb{R}$ is positive, iff $I(g_1) \geq I(g_2)$, whenever $g_1 \geq g_2$, where $g_1, g_2 \in C(K, \mathbb{R})$.

Let us assume that I is a positive linear functional. Then by Riesz representation theorem, [11, p. 304], there exists a unique Borel measure μ on K such that

$$I(g) = \int_K g(t)\, d\mu(t), \tag{12.13}$$

$\forall\, g \in C(K, \mathbb{R})$, where $\mu(K) < \infty$.

Next the vector derivatives and vector Riemann integral are as the numerical ones, see [12, pp. 83–95].

We have the following vector Taylor type formula:

Theorem 12.5 *Let* $n \in \mathbb{N}$ *and* $f \in C^n([a, b], X)$, *where* $[a, b] \subset \mathbb{R}$ *and* $(X, \|\cdot\|)$ *is a Banach space. Then*

$$f(b) = f(a) + \sum_{i=1}^{n-1} \frac{(b-a)^i}{i!} f^{(i)}(a) + \int_a^b \left(\int_a^{t_1} \cdots \left(\int_a^{t_{n-1}} f^{(n)}(t_n)\, dt_n \right) \cdots \right) dt_1. \tag{12.14}$$

We also have that

$$f(b) = f(a) + \sum_{i=1}^{n} \frac{(b-a)^i}{i!} f^{(i)}(a) +$$

$$\int_a^b \left(\int_a^{t_1} \cdots \left(\int_a^{t_{n-1}} \left(f^{(n)}(t_n) - f^{(n)}(a) \right) dt_n \right) \ldots \right) dt_1. \tag{12.15}$$

Proof By [12, p. 92] we have that

$$f^{(n-1)}(t_{n-1}) = f^{(n-1)}(a) + \int_a^{t_{n-1}} f^{(n)}(t_n) \, dt_n, \tag{12.16}$$

and

$$f^{(n-2)}(t_{n-2}) = f^{(n-2)}(a) + \int_a^{t_{n-2}} f^{(n-1)}(t_{n-1}) \, dt_{n-1}, \tag{12.17}$$

where $a \leq t_{n-1} \leq t_{n-2} \leq b$.

Hence it holds

$$f^{(n-2)}(t_{n-2}) = f^{(n-2)}(a) + \int_a^{t_{n-2}} \left(f^{(n-1)}(a) + \int_a^{t_{n-1}} f^{(n)}(t_n) \, dt_n \right) dt_{n-1}$$

$$= f^{(n-2)}(a) + f^{(n-1)}(a)(t_{n-2} - a) + \int_a^{t_{n-2}} \left(\int_a^{t_{n-1}} f^{(n)}(t_n) \, dt_n \right) dt_{n-1}. \tag{12.18}$$

Similarly, we have

$$f^{(n-3)}(t_{n-3}) = f^{(n-3)}(a) + \int_a^{t_{n-3}} f^{(n-2)}(t_{n-2}) \, dt_{n-2}, \tag{12.19}$$

where $a \leq t_{n-1} \leq t_{n-2} \leq t_{n-3} \leq b$.

Consequently we get

$$f^{(n-3)}(t_{n-3}) = f^{(n-3)}(a) + \int_a^{t_{n-3}} \left[f^{(n-2)}(a) + f^{(n-1)}(a)(t_{n-2} - a) \right.$$

$$\left. + \int_a^{t_{n-2}} \left(\int_a^{t_{n-1}} f^{(n)}(t_n) \, dt_n \right) dt_{n-1} \right] dt_{n-2} =$$

$$f^{(n-3)}(a) + f^{(n-2)}(a)(t_{n-3} - a) + f^{(n-1)}(a) \frac{(t_{n-3} - a)^2}{2} + \tag{12.20}$$

$$\int_a^{t_{n-3}} \left(\int_a^{t_{n-2}} \left(\int_a^{t_{n-1}} f^{(n)}(t_n) \, dt_n \right) dt_{n-1} \right) dt_{n-2}.$$

Similarly, we have

$$f^{(n-4)}(t_{n-4}) = f^{(n-4)}(a) + \int_a^{t_{n-4}} f^{(n-3)}(t_{n-3})\, dt_{n-3}, \tag{12.21}$$

where $a \le t_{n-1} \le t_{n-2} \le t_{n-3} \le t_{n-4} \le b$.

Thus, we find

$$f^{(n-4)}(t_{n-4}) = f^{(n-4)}(a) + \int_a^{t_{n-4}} \Big[f^{(n-3)}(a) +$$

$$f^{(n-2)}(a)(t_{n-3}-a) + \frac{f^{(n-1)}(a)}{2}(t_{n-3}-a)^2 +$$

$$\int_a^{t_{n-3}} \left(\int_a^{t_{n-2}} \left(\int_a^{t_{n-1}} f^{(n)}(t_n)\, dt_n \right) dt_{n-1} \right) dt_{n-2} \Big] dt_{n-3} \tag{12.22}$$

$$= f^{(n-4)}(a) + f^{(n-3)}(a)(t_{n-4}-a) + f^{(n-2)}(a)\frac{(t_{n-4}-a)^2}{2}$$

$$+ \frac{f^{(n-1)}(a)}{3!}(t_{n-4}-a)^3 +$$

$$\int_a^{t_{n-4}} \left(\int_a^{t_{n-3}} \left(\int_a^{t_{n-2}} \left(\int_a^{t_{n-1}} f^{(n)}(t_n)\, dt_n \right) dt_{n-1} \right) dt_{n-2} \right) dt_{n-3}.$$

We have proved that

$$f^{(n-4)}(t_{n-4}) = f^{(n-4)}(a) + f^{(n-3)}(a)(t_{n-4}-a) + \tag{12.23}$$

$$\frac{f^{(n-2)}(a)}{2}(t_{n-4}-a)^2 + \frac{f^{(n-1)}(a)}{3!}(t_{n-4}-a)^3 +$$

$$\int_a^{t_{n-4}} \left(\int_a^{t_{n-3}} \left(\int_a^{t_{n-2}} \left(\int_a^{t_{n-1}} f^{(n)}(t_n)\, dt_n \right) dt_{n-1} \right) dt_{n-2} \right) dt_{n-3},$$

where $a \le t_{n-1} \le t_{n-2} \le t_{n-3} \le t_{n-4} \le b$.

Etc.

Working inductively we have proved the claim. ■

We make

Remark 12.6 Here Q is an open convex subset of $\mathbb{C}^k$, $k \ge 2$; $z := (z_1, \ldots, z_k)$, $x_0 := (x_{01}, \ldots, x_{0k}) \in Q$. Let $f : Q \to \mathbb{C}$ be a coordinate-wise holomorphic function. Then, by the famous Hartog's fundamental theorem [6, 8] f is jointly holomorphic and jointly continuous on Q. Let $n \in \mathbb{N}$. Each nth order complex partial derivative is denoted by $f_\alpha := \frac{\partial^\alpha f}{\partial x^\alpha}$, where $\alpha := (\alpha_1, \ldots, \alpha_k)$, $\alpha_i \in \mathbb{Z}_+$, $i = 1, \ldots, k$ and $|\alpha| := \sum_{i=1}^{k} \alpha_i = n$.

Consider

$$g_z(t) := f(x_0 + t(z - x_0)), \; 0 \leq t \leq 1. \tag{12.24}$$

Clearly it holds that $x_0 + t(z - x_0) \in Q$ and $g_z(t) \in \mathbb{C}$, $\forall\, t \in [0, 1]$.

Then we derive

$$g_z^{(j)}(t) = \left[\left(\sum_{i=1}^{k} (z_i - x_{0i}) \frac{\partial}{\partial x_i}\right)^j f\right](x_{01} + t(z_1 - x_{01}), \ldots, x_{0k} + t(z_k - x_{0k})), \tag{12.25}$$

for all $j = 0, 1, \ldots, n$.

Notice here that any mixed partials commute. We remind that $(\mathbb{C}, |\cdot|)$ is a Banach space.

We also have for $j = 0, 1, \ldots, n$:

$$g_z^{(j)}(0) = \sum_{\substack{\alpha := (\alpha_1, \ldots, \alpha_k), \alpha_j \in \mathbb{Z}^+ \\ i = 1, \ldots, k; |\alpha| := \sum_{i=1}^{k} \alpha_i = j}} \left(\frac{j!}{\prod_{i=1}^{k} \alpha_i!}\right) \left(\prod_{i=1}^{k} (z_i - x_{0i})^{\alpha_i}\right) f_\alpha(x_0). \tag{12.26}$$

Furthermore it holds

$$g_z^{(n)}(\theta) = \sum_{\substack{\alpha := (\alpha_1, \ldots, \alpha_k), \alpha_j \in \mathbb{Z}^+ \\ i = 1, \ldots, k; |\alpha| := \sum_{i=1}^{k} \alpha_i = n}} \left(\frac{n!}{\prod_{i=1}^{k} \alpha_i!}\right) \left(\prod_{i=1}^{k} (z_i - x_{0i})^{\alpha_i}\right) f_\alpha(x_0 + \theta(z - x_0)), \tag{12.27}$$

$0 \leq \theta \leq 1$.

Notice that $g_z(0) = f(x_0)$.

By (12.26), if all $f_\alpha(x_0) = 0$, then $g_z^{(j)}(0) = 0$, where $j \in \{0, 1, \ldots, n\}$.

We give

Example 12.7 Let $n = k = 2$. Then

$$g_z(t) = f(x_{01} + t(z_1 - x_{01}), x_{02} + t(z_2 - x_{02})), \; t \in [0, 1],$$

and

$$g_z'(t) = (z_1 - x_{01}) \frac{\partial f}{\partial x_1}(x_0 + t(z - x_0)) + (z_2 - x_{02}) \frac{\partial f}{\partial x_2}(x_0 + t(z - x_0)). \tag{12.28}$$

In addition,

$$g_z''(t) = (z_1 - x_{01}) \left(\frac{\partial f}{\partial x_1}(x_0 + t(z - x_0)) \right)' + (z_2 - x_{02}) \left(\frac{\partial f}{\partial x_2}(x_0 + t(z - x_0)) \right)'$$

$$= (z_1 - x_{01}) \left\{ (z_1 - x_{01}) \frac{\partial f^2}{\partial x_1^2}(*) + (z_2 - x_{02}) \frac{\partial f^2}{\partial x_2 \partial x_1}(*) \right\} + \tag{12.29}$$

$$(z_2 - x_{02}) \left\{ (z_1 - x_{01}) \frac{\partial f^2}{\partial x_1 \partial x_2}(*) + (z_2 - x_{02}) \frac{\partial f^2}{\partial x_2^2}(*) \right\}.$$

Hence

$$g_z''(t) = (z_1 - x_{01})^2 \frac{\partial f^2}{\partial x_1^2}(*) + (z_1 - x_{01})(z_2 - x_{02}) \frac{\partial f^2}{\partial x_2 \partial x_1}(*)$$

$$+ (z_1 - x_{01})(z_2 - x_{02}) \frac{\partial f^2}{\partial x_1 \partial x_2}(*) + (z_2 - x_{02})^2 \frac{\partial f^2}{\partial x_2^2}(*), \tag{12.30}$$

where $* := x_0 + t(z - x_0)$.

Notice that $g_z(t), g_z'(t), g_z''(t) \in \mathbb{C}$. Etc.

We give the following Taylor's formula on $\mathbb{C}^k$, $k \geq 2$.

Theorem 12.8 *It holds*

$$f(z_1, \ldots, z_k) = g_z(1) = \sum_{j=0}^{n} \frac{g_z^{(j)}(0)}{j!} + R_n(z, 0), \tag{12.31}$$

where

$$R_n(z, 0) := \int_0^1 \left(\int_0^{t_1} \cdots \left(\int_0^{t_{n-1}} \left(g_z^{(n)}(t_n) - g_z^{(n)}(0) \right) dt_n \right) \cdots \right) dt_1, \ \forall z \in Q. \tag{12.32}$$

Proof By Theorem 12.5 (12.15), it is applied to $g_z(t)$, $t \in [0, 1]$. ■

Next we will estimate $R_n(z, 0)$.

We need

Remark 12.9 See [1, p. 210]. Let $h>0$ be fixed, $x\in\mathbb{R}$. Denote by

$$\phi_n(x)=\int_0^{|x|}\int_0^{x_1}\cdots\left(\int_0^{x_{n-1}}\left\lceil\frac{x_n}{h}\right\rceil dx_n\right)\dots dx_1. \tag{12.33}$$

Clearly $\phi_n(0)=0$.

It holds

$$\phi_n(x)\le\frac{|x|^{n+1}}{(n+1)!h}+\frac{|x|^n}{2n!}+\frac{h|x|^{n-1}}{8(n-1)!}, \tag{12.34}$$

with equality only at $x=0$.

Theorem 12.10 *Let $Q\subseteq\left(\mathbb{C}^k,\|\cdot\|_1\right)$ be a compact convex subset, $k\ge 2$, where $\|\cdot\|_1$ is the l_1-norm. Here $f:Q\to\mathbb{C}$ is a coordinate-wise holomorphic function and $R_n(z,0)$ as in (12.32). Then*

$$|R_n(z,0)|\le\left(\max_{|\alpha|=n}\omega_1(f_\alpha,h)\right)\phi_n\left(\|z-x_0\|_1\right), \tag{12.35}$$

$\forall\, z,x_0\in Q$, $n\in\mathbb{N}$.

Proof By Corollary 12.3 we have that

$$|f_\alpha(x_0+t(z-x_0))-f_\alpha(x_0)|\le\omega_1(f_\alpha,h)\left\lceil\frac{t\|z-x_0\|_1}{h}\right\rceil, \tag{12.36}$$

$\forall\, t\in[0,1]$, $h>0$, all $\alpha:|\alpha|=n$.

For $\|z-x_0\|_1\neq 0$, it follows from (12.27), using (12.33), that

$$|R_n(z,0)|\le\int_0^1\left(\int_0^{t_1}\cdots\left(\int_0^{t_{n-1}}\left(\sum_{|\alpha|=n}\frac{n!}{\alpha_1!\dots\alpha_k!}|z_1-x_{01}|^{\alpha_1}\dots|z_k-x_{0k}|^{\alpha_k}\right.\right.\right.$$

$$\left.\left.\left.\left(\max_{|\alpha|=n}\omega_1(f_\alpha,h)\right)\right)\left\lceil\frac{t_n\|z-x_0\|_1}{h}\right\rceil dt_n\right)\dots\right)dt_1=$$

$$\sum_{|\alpha|=n}\frac{n!}{\alpha_1!\dots\alpha_k!}\frac{\prod_{i=1}^k|z_i-x_{0i}|^{\alpha_i}}{\|z-x_0\|_1}\left(\max_{|\alpha|=n}\omega_1(f_\alpha,h)\right)\phi_n\left(\|z-x_0\|_1\right)= \tag{12.37}$$

$$\left(\max_{|\alpha|=n}\omega_1(f_\alpha,h)\right)\phi_n\left(\|z-x_0\|_1\right),$$

by $\|z - x_0\|_1 = \sum_{i=1}^{k} |z_i - x_{0i}|$. Therefore

$$|R_n(z, 0)| \leq \left(\max_{|\alpha|=n} \omega_1(f_\alpha, h) \right) \phi_n(\|z - x_0\|_1), \tag{12.38}$$

for all $z \in Q$.

If $z = x_0$, then $\|z - x_0\|_1 = 0$, and (12.38) is trivially true. ■

We state the following main result:

Theorem 12.11 *Let $Q \subseteq (\mathbb{C}^k, \|\cdot\|_1)$ be a compact convex subset, $k \geq 2$, where $\|\cdot\|_1$ is the l_1-norm. Here $f : Q \to \mathbb{C}$ is a coordinate-wise holomorphic function and $z, x_0 \in Q$, $n \in \mathbb{N}$. Assume that $f_\alpha(x_0) = 0$, for all $\alpha : |\alpha| = j$, $j = 1, \ldots, n$. Then*

(1)

$$f(z) = f(x_0) + R_n(z, 0), \tag{12.39}$$

where $R_n(z, 0)$ as in (12.32),
and

(2)

$$|f(z) - f(x_0)| \leq \left(\max_{|\alpha|=n} \omega_1(f_\alpha, h) \right) \phi_n(\|z - x_0\|_1) \leq$$

$$\left(\max_{|\alpha|=n} \omega_1(f_\alpha, h) \right) \left[\frac{\|z - x_0\|_1^{n+1}}{(n+1)!h} + \frac{\|z - x_0\|_1^n}{2n!} + \frac{h\|z - x_0\|_1^{n-1}}{8(n-1)!} \right], \tag{12.40}$$

where $h > 0$.

Proof By (12.32), (12.34) and (12.35). ■

We give

Corollary 12.12 *All as in Theorem 12.11. Then*

$$|f(\cdot) - f(x_0)| \leq \left(\max_{|\alpha|=n} \omega_1(f_\alpha, h) \right) \left[\frac{\|\cdot - x_0\|_1^{n+1}}{(n+1)!h} + \frac{\|\cdot - x_0\|_1^n}{2n!} + \frac{h\|\cdot - x_0\|_1^{n-1}}{8(n-1)!} \right], \tag{12.41}$$

$h > 0$, true over Q.

Above ω_1 is defined with respect to $\|\cdot\|_1$.

12.3 Main Results

Let K be a compact and convex subset of $\mathbb{C}^k$, $k \geq 2$. Consider $L : C(K, \mathbb{C}) \to C(K, \mathbb{C})$ a linear operator and $\widetilde{L} : C(K, \mathbb{R}) \to C(K, \mathbb{R})$ a positive linear operator (i.e. for $f_1 . f_2 \in C(K, \mathbb{R})$ with $f_1 \geq f_2$ we get $\widetilde{L}(f_1) \geq \widetilde{L}(f_2)$) both over the field of $\mathbb{R}$.

We assume that

$$|L(f)| \leq \lambda \widetilde{L}(|f|), \ \forall f \in C(K, \mathbb{C}), \text{ where } \lambda > 0, \tag{12.42}$$

(i.e. $|L(f)(z)| \leq \lambda \widetilde{L}(|f|)(z), \forall z \in K$).

We call $\widetilde{L}$ the companion operator of L.

Let $z_0 \in K$. Clearly, then $L(\cdot)(z_0)$ is a linear functional from $C(K, \mathbb{C})$ into $\mathbb{C}$, and $\widetilde{L}(\cdot)(z_0)$ is a positive linear functional from $C(K, \mathbb{R})$ into $\mathbb{R}$. Notice $L(f)(z) \in \mathbb{C}$ and $\widetilde{L}(|f|)(z) \in \mathbb{R}, \forall f \in C(K, \mathbb{C})$ (thus $|f| \in C(K, \mathbb{R})$). Here $L(f) \in C(K, \mathbb{C})$, and $\widetilde{L}(|f|) \in C(K, \mathbb{R}), \forall f \in C(K, \mathbb{C})$.

Notice that $C(K, \mathbb{C}) = UC(K, \mathbb{C})$, also $C(K, \mathbb{R}) = UC(K, \mathbb{R})$ (uniformly continuous functions).

By [3, p. 388], we have that $\widetilde{L}(\|\cdot - z_0\|^r)(z_0)$, $r > 0$, is a continuous function in $z_0 \in K$. Here $\|\cdot\|$ is an arbitrary norm.

We have the following approximation result with rates of Korovkin type.

Theorem 12.13 *Here K is a convex and compact subset of $(\mathbb{C}^k, \|\cdot\|)$, $k \geq 2$, and L_n is a sequence of linear operators from $C(K, \mathbb{C})$ into itself, $n \in \mathbb{N}$. There is a sequence of companion positive linear operators $\widetilde{L}_n$ from $C(K, \mathbb{R})$ into itself, such that*

$$|L_n(f)| \leq \lambda \widetilde{L}_n(|f|), \ \lambda > 0, \forall f \in C(K, \mathbb{C}), \ \forall n \in \mathbb{N} \tag{12.43}$$

(i.e. $|L_n(f)(z_0)| \leq \lambda \left(\widetilde{L}_n(|f|)\right)(z_0)$, $\forall z_0 \in K$).

Additionally, we assume that

$$L_n(cg) = c\widetilde{L}_n(g), \ \forall g \in C(K, \mathbb{R}), \ \forall c \in \mathbb{C} \tag{12.44}$$

(i.e. $(L_n(cg))(z_0) = c\left(\widetilde{L}_n(g)\right)(z_0), \forall z_0 \in K$).

Then, for any $f \in C(K, \mathbb{C})$, we have

$$|(L_n(f))(z_0) - f(z_0)| \leq |f(z_0)| \left|\widetilde{L}_n(1(\cdot))(z_0) - 1\right| +$$

$$\lambda \left(\widetilde{L}_n(1(\cdot))(z_0) + 1\right) \omega_1\left(f, \widetilde{L}_n(\|\cdot - z_0\|)(z_0)\right), \tag{12.45}$$

$\forall z_0 \in K$, $\forall n \in \mathbb{N}$.

If $\widetilde{L}_n(1(\cdot))(z_0) = 1$, $\forall z_0 \in K$, then

$$|(L_n(f))(z_0) - f(z_0)| \leq 2\lambda\omega_1\left(f, \widetilde{L}_n(\|\cdot - z_0\|)(z_0)\right), \tag{12.46}$$

$\forall\, z_0 \in K$, $\forall\, n \in \mathbb{N}$.

If $\widetilde{L}_n(1(\cdot))(z_0) \to 1$, *and* $\widetilde{L}_n(\|\cdot - z_0\|)(z_0) \to 0$, *as* $n \to \infty$, *then* $L_n(f)(z_0) \to f(z_0)$, $\forall\, f \in C(K, \mathbb{C})$. *Here* $\widetilde{L}_n(1(\cdot))(z_0)$ *is bounded.*

Proof We notice that

$$|(L_n(f))(z_0) - f(z_0)| =$$

$$|(L_n(f))(z_0) - L_n(f(z_0)(\cdot))(z_0) + L_n(f(z_0)(\cdot))(z_0) - f(z_0)| \overset{(12.44)}{=}$$

$$\left|(L_n(f))(z_0) - L_n(f(z_0)(\cdot))(z_0) + f(z_0)\widetilde{L}_n(1(\cdot))(z_0) - f(z_0)\right| \leq$$

$$|(L_n(f))(z_0) - L_n(f(z_0)(\cdot))(z_0)| + |f(z_0)|\left|\widetilde{L}_n(1(\cdot))(z_0) - 1\right| = \tag{12.47}$$

$$|L_n(f(\cdot) - f(z_0))(z_0)| + |f(z_0)|\left|\widetilde{L}_n(1(\cdot))(z_0) - 1\right| \overset{(12.43)}{\leq}$$

$$|f(z_0)|\left|\widetilde{L}_n(1(\cdot))(z_0) - 1\right| + \lambda\left(\widetilde{L}_n(|f(\cdot) - f(z_0)|)\right)(z_0) \overset{(12.4.)}{\leq}$$

$$|f(z_0)|\left|\widetilde{L}_n(1(\cdot))(z_0) - 1\right| + \lambda\left(\widetilde{L}_n\left(\omega_1\left(f, \frac{\delta\,\|\cdot - z_0\|}{\delta}\right)\right)\right)(z_0) \overset{(12.7.)}{\leq}$$

$$|f(z_0)|\left|\widetilde{L}_n(1(\cdot))(z_0) - 1\right| + \lambda\left(\widetilde{L}_n\left(\omega_1(f, \delta)\left(1(\cdot) + \frac{1}{\delta}\|\cdot - z_0\|\right)\right)\right)(z_0) =$$

$$|f(z_0)|\left|\widetilde{L}_n(1(\cdot))(z_0) - 1\right| + \lambda\omega_1(f, \delta)\left[\widetilde{L}_n(1(\cdot))(z_0) + \frac{1}{\delta}\widetilde{L}_n(\|\cdot - z_0\|)(z_0)\right] \overset{(12.48)}{=}$$

$$|f(z_0)|\left|\widetilde{L}_n(1(\cdot))(z_0) - 1\right| + \lambda\omega_1\left(f, \widetilde{L}_n(\|\cdot - z_0\|)(z_0)\right)\left[\widetilde{L}_n(1(\cdot))(z_0) + 1\right],$$

by choosing

$$\delta := \widetilde{L}_n(\|\cdot - z_0\|)(z_0), \tag{12.49}$$

if $\widetilde{L}_n(\|\cdot - z_0\|)(z_0) > 0$.

Next we consider the case of

$$\widetilde{L}_n(\|\cdot - z_0\|)(z_0) = 0. \tag{12.50}$$

By Riesz representation theorem [11, p. 304] there exists a positive finite measure μ_{z_0} such that

$$\widetilde{L}_n(g)(z_0) = \int_K g(t)\, d\mu_{z_0}(t), \ \forall\, g \in C(K, \mathbb{R}). \tag{12.51}$$

That is

$$\int_K \|t - z_0\| \, d\mu_{z_0}(t) = 0,$$

which implies $\|t - z_0\| = 0$, a.e., hence $t - z_0 = 0$, a.e., and $t = z_0$, a.e. on K. Consequently $\mu_{z_0}(\{t \in K : t \neq z_0\}) = 0$.

That is $\mu_{z_0} = \delta_{z_0} M$ (where $0 < M := \mu_{z_0}(K) = \widetilde{L}_n(1(\cdot))(z_0)$). Hence, in that case $\widetilde{L}_n(g)(z_0) = g(z_0) M$. Consequently, it holds $\omega_1\left(f, \widetilde{L}_n(\|\cdot - z_0\|)(z_0)\right) = 0$, and the right hand side of (12.45) equals $|f(z_0)| \, |M - 1|$.

Also, it is $\widetilde{L}_n(|f(\cdot) - f(z_0)(\cdot)|)(z_0) = |f(z_0) - f(z_0)| \, M = 0$.

And by (12.43) we obtain

$$|(L_n(f(\cdot) - f(z_0)(\cdot)))(z_0)| = 0,$$

that is

$$|L_n(f)(z_0) - L_n(f(z_0)(\cdot))(z_0)| = 0.$$

The last says that

$$L_n(f)(z_0) = L_n(f(z_0)(\cdot))(z_0) \overset{(12.44)}{=} f(z_0) \widetilde{L}_n(1(\cdot))(z_0) = Mf(z_0).$$

Consequently the left hand side of (12.45) becomes

$$|L_n(f)(z_0) - f(z_0)| = |Mf(z_0) - f(z_0)| = |f(z_0)| \, |M - 1|.$$

So that (12.45) becomes an equality, and both sides equal $|f(z_0)| \, |M - 1|$ in the extreme case of $\widetilde{L}_n(\|\cdot - z_0\|)(z_0) = 0$. Thus inequality (12.45) is proved completely in both cases. ■

A similar result follows:

Theorem 12.14 *Here all as in Theorem 12.13. Then, for any $f \in C(K, \mathbb{C})$, we have*

$$|(L_n(f))(z_0) - f(z_0)| \leq |f(z_0)| \left|\widetilde{L}_n(1(\cdot))(z_0) - 1\right| + \tag{12.52}$$

$$\lambda\left(\widetilde{L}_n(1(\cdot))(z_0) + 1\right) \omega_1\left(f, \left(\widetilde{L}_n\left(\|\cdot - z_0\|^2\right)(z_0)\right)^{\frac{1}{2}}\right),$$

$\forall\, z_0 \in K$, $\forall\, n \in \mathbb{N}$.

If $\widetilde{L}_n(1(\cdot))(z_0) = 1$, $\forall\, z_0 \in K$, *then*

$$|(L_n(f))(z_0) - f(z_0)| \leq 2\lambda\omega_1\left(f, \left(\widetilde{L}_n\left(\|\cdot - z_0\|^2\right)(z_0)\right)^{\frac{1}{2}}\right), \tag{12.53}$$

$\forall\, z_0 \in K$, $\forall\, n \in \mathbb{N}$.

Proof Let $t, z_0 \in K$ and $\delta > 0$. If $\|t - z_0\| > \delta$, then

$$|f(t) - f(z_0)| \leq \omega_1 (f, \|t - z_0\|) = \omega_1 \left(f, \|t - z_0\| \, \delta^{-1}\delta\right) \leq \tag{12.54}$$

$$\left(1 + \frac{\|t - z_0\|}{\delta}\right) \omega_1 (f, \delta) \leq \left(1 + \frac{\|t - z_0\|^2}{\delta^2}\right) \omega_1 (f, \delta).$$

The estimate

$$|f(t) - f(z_0)| \leq \left(1 + \frac{\|t - z_0\|^2}{\delta^2}\right) \omega_1 (f, \delta) \tag{12.55}$$

also holds trivially when $\|t - z_0\| \leq \delta$.

So (12.55) is true always, $\forall\, t \in K$, for any $z_0 \in K$.

As in the proof of Theorem 12.13 we have

$$|(L_n(f))(z_0) - f(z_0)| \leq \cdots \leq |f(z_0)| \left|\widetilde{L}_n (1(\cdot))(z_0) - 1\right|$$

$$+\lambda \left(\widetilde{L}_n (|f(\cdot) - f(z_0)|)\right)(z_0) \overset{(12.55)}{\leq} |f(z_0)| \left|\widetilde{L}_n (1(\cdot))(z_0) - 1\right| +$$

$$\lambda \left(\widetilde{L}_n \left(\left(1(\cdot) + \frac{\|\cdot - z_0\|^2}{\delta^2}\right) \omega_1 (f, \delta)\right)\right)(z_0) =$$

$$|f(z_0)| \left|\widetilde{L}_n (1(\cdot))(z_0) - 1\right| + \lambda \omega_1 (f, \delta) \left[\widetilde{L}_n (1(\cdot))(z_0) + \frac{1}{\delta^2} \widetilde{L}_n \left(\|\cdot - z_0\|^2\right)(z_0)\right] =$$

$$|f(z_0)| \left|\widetilde{L}_n (1(\cdot))(z_0) - 1\right| + \lambda \omega_1 \left(f, \left(\widetilde{L}_n \left(\|\cdot - z_0\|^2\right)(z_0)\right)^{\frac{1}{2}}\right) \left[\widetilde{L}_n (1(\cdot))(z_0) + 1\right], \tag{12.56}$$

by choosing

$$\delta := \left(\widetilde{L}_n \left(\|\cdot - z_0\|^2\right)(z_0)\right)^{\frac{1}{2}}, \tag{12.57}$$

if $\widetilde{L}_n \left(\|\cdot - z_0\|^2\right)(z_0) > 0$.

Next we consider the case of

$$\widetilde{L}_n \left(\|\cdot - z_0\|^2\right)(z_0) = 0.$$

By Riesz representation theorem there exists a positive finite measure μ_{z_0} such that

$$\widetilde{L}_n (g)(z_0) = \int_K g(t) \, d\mu_{z_0}(t), \ \forall\, g \in C(K, \mathbb{R}). \tag{12.58}$$

That is

$$\int_K \|t - z_0\|^2 \, d\mu_{z_0}(t) = 0,$$

which implies $\|t - z_0\|^2 = 0$, a.e., hence $\|t - z_0\| = 0$, a.e., thus $t - z_0 = 0$, a.e., and $t = z_0$, a.e. on K. Consequently $\mu_{z_0}(\{t \in K : t \neq z_0\}) = 0$.

That is $\mu_{z_0} = \delta_{z_0} M$ (where $0 < M := \mu_{z_0}(K) = \widetilde{L}_n(1(\cdot))(z_0)$). Hence, in that case $\widetilde{L}_n(g)(z_0) = g(z_0) M$.

Consequently, it holds $\omega_1\left(f, \left(\widetilde{L}_n\left(\|\cdot - z_0\|^2\right)(z_0)\right)^{\frac{1}{2}}\right) = 0$, and the right hand side of (12.52) equals $|f(z_0)| |M - 1|$. Also, it is $\widetilde{L}_n(|f(\cdot) - f(z_0)(\cdot)|)(z_0) = |f(z_0) - f(z_0)| M = 0$.

And by (12.43) we obtain

$$|(L_n(f(\cdot) - f(z_0)(\cdot)))(z_0)| = 0,$$

that is

$$|L_n(f)(z_0) - L_n(f(z_0)(\cdot))(z_0)| = 0.$$

The last says that

$$L_n(f)(z_0) = L_n(f(z_0)(\cdot))(z_0) \overset{(12.44)}{=} f(z_0)\widetilde{L}_n(1(\cdot))(z_0) = Mf(z_0).$$

Consequently the left hand side of (12.52) becomes

$$|L_n(f)(z_0) - f(z_0)| = |Mf(z_0) - f(z_0)| = |f(z_0)| |M - 1|.$$

So that (12.52) becomes an equality, and both sides equal $|f(z_0)| |M - 1|$ in the extreme case of $\widetilde{L}_n\left(\|\cdot - z_0\|^2\right)(z_0) = 0$. Thus inequality (12.52) is proved completely in both cases. ■

We give

Corollary 12.15 *All as in Theorem 12.13, $z_0 \in K$. Then*

$$\|L_n(f) - f\|_\infty \leq \|f\|_\infty \left\|\widetilde{L}_n(1(\cdot)) - 1\right\|_\infty + \tag{12.59}$$

$$\lambda \left\|\widetilde{L}_n(1(\cdot)) + 1\right\|_\infty \omega_1\left(f, \left\|\widetilde{L}_n(\|\cdot - z_0\|)(z_0)\right\|_{\infty, z_0}\right),$$

$\forall\, n \in \mathbb{N}$.

If $\widetilde{L}_n(1(\cdot)) = 1$, then

$$\|L_n(f) - f\|_\infty \leq 2\lambda\omega_1\left(f, \left\|\widetilde{L}_n(\|\cdot - z_0\|)(z_0)\right\|_{\infty, z_0}\right), \tag{12.60}$$

$\forall\, n \in \mathbb{N}$.

As $\widetilde{L}_n(1) \overset{u}{\to} 1$, and $\widetilde{L}_n(\|\cdot - z_0\|)(z_0) \overset{u}{\to} 0$ (u is uniformly), as $n \to \infty$, then $L_n(f) \overset{u}{\to} f, \forall\, f \in C(K, \mathbb{C})$. Notice $\widetilde{L}_n(1)$ is bounded, and all suprema in (12.59) are finite.

Corollary 12.16 *All as in Theorem 12.14, $z_0 \in K$. Then*

$$\|L_n(f) - f\|_\infty \le \|f\|_\infty \left\|\widetilde{L}_n(1(\cdot)) - 1\right\|_\infty + \tag{12.61}$$

$$\lambda \left\|\widetilde{L}_n(1(\cdot)) + 1\right\|_\infty \omega_1\left(f, \left\|\widetilde{L}_n\left(\|\cdot - z_0\|^2\right)(z_0)\right\|_{\infty, z_0}^{\frac{1}{2}}\right),$$

$\forall\, n \in \mathbb{N}$.

If $\widetilde{L}_n(1(\cdot)) = 1$, then

$$\|L_n(f) - f\|_\infty \le 2\lambda\omega_1\left(f, \left\|\widetilde{L}_n\left(\|\cdot - z_0\|^2\right)(z_0)\right\|_{\infty, z_0}^{\frac{1}{2}}\right), \tag{12.62}$$

$\forall\, n \in \mathbb{N}$.

As $\widetilde{L}_n(1) \xrightarrow{u} 1$, and $\widetilde{L}_n\left(\|\cdot - z_0\|^2\right)(z_0) \xrightarrow{u} 0$, then $L_n(f) \xrightarrow{u} f$, as $n \to \infty$, $\forall$ $f \in C(K, \mathbb{C})$.

We need

Theorem 12.17 *Let a convex subset $K \subseteq \left(\mathbb{C}^k, \|\cdot\|\right)$, $k \ge 2$, where $\|\cdot\|$ is an arbitrary norm, $x_0 \in K^0$ (interior of K) and $f : K \to \mathbb{C}$ such that $|f(t) - f(x_0)|$ is convex in $t \in K$. Furthermore let $\delta > 0$ so that the ball $B(x_0, \delta) \subset K$. Then*

$$|f(t) - f(x_0)| \le \frac{\omega_1(f, \delta)}{\delta}\|t - x_0\|, \ \forall\, t \in K. \tag{12.63}$$

Proof Let $g(t) := |f(t) - f(x_0)|$, $t \in K$, which is convex in $t \in K$ and $g(x_0) = 0$.

Then, by Lemma 8.1.1, p. 243 of [1], we obtain

$$g(t) \le \frac{\omega_1(g, \delta)}{\delta}\|t - x_0\|, \ \forall\, t \in K. \tag{12.64}$$

For any $t_1, t_2 \in K : \|t_1 - t_2\| \le \delta$ we get

$$\left||f(t_1) - f(x_0)| - |f(t_2) - f(x_0)|\right| \le |f(t_1) - f(t_2)| \le \omega_1(f, \delta). \tag{12.65}$$

That is

$$\omega_1(g, \delta) \le \omega_1(f, \delta). \tag{12.66}$$

The last (12.66) and (12.64) imply (12.63). ■

We present a convex Korovkin type result:

Theorem 12.18 *Here all as in Theorem 12.13. Let a fixed $z_0 \in K^0$ and assume that $|f(t) - f(z_0)|$ is convex in $t \in K$. Assume the closed ball $B\left(z_0, \widetilde{L}_n(\|\cdot - z_0\|)(z_0)\right)$ $\subset K$. Then*

$$|(L_n(f))(z_0) - f(z_0)| \leq |f(z_0)| \left|\widetilde{L}_n(1(\cdot))(z_0) - 1\right|$$

$$+ \lambda\omega_1\left(f, \widetilde{L}_n(\|\cdot - z_0\|)(z_0)\right), \ \forall\, n \in \mathbb{N}. \tag{12.67}$$

As $\widetilde{L}_n(1(\cdot))(z_0) \to 1$, *and* $\widetilde{L}_n(\|\cdot - z_0\|)(z_0) \to 0$, *then* $(L_n(f))(z_0) \to f(z_0)$, *as* $n \to \infty$.

Proof As in the proof of Theorem 12.13 we have

$$|(L_n(f))(z_0) - f(z_0)| \leq$$

$$|f(z_0)| \left|\widetilde{L}_n(1(\cdot))(z_0) - 1\right| + \lambda\left(\widetilde{L}_n(|f(\cdot) - f(z_0)|)\right)(z_0) \overset{(12.63)}{\leq}$$

$(\delta > 0 : B(z_0, \delta) \subset K)$

$$|f(z_0)| \left|\widetilde{L}_n(1(\cdot))(z_0) - 1\right| + \frac{\lambda\omega_1(f, \delta)}{\delta}\widetilde{L}_n(\|\cdot - z_0\|)(z_0) =$$

$$|f(z_0)| \left|\widetilde{L}_n(1(\cdot))(z_0) - 1\right| + \lambda\omega_1\left(f, \widetilde{L}_n(\|\cdot - z_0\|)(z_0)\right), \tag{12.68}$$

by choosing

$$\delta := \widetilde{L}_n(\|\cdot - z_0\|)(z_0),$$

if $\widetilde{L}_n(\|\cdot - z_0\|)(z_0) > 0$.

The case $\widetilde{L}_n(\|\cdot - z_0\|)(z_0) = 0$ is treated similarly as in the proof of Theorem 12.13. The theorem is proved. ∎

We make

Assumption 12.19 Let $Q \subseteq \left(\mathbb{C}^k, \|\cdot\|_1\right)$ be a compact convex subset, $k \geq 2$, where $\|\cdot\|_1$ is the l_1-norm. Here we consider $f : Q \to \mathbb{C}$ that are coordinate-wise holomorphic functions and $x_0 \in Q$, $n \in \mathbb{N}$. Assume that $f_\alpha(x_0) = 0$ for all $\alpha : |\alpha| = j$, $j = 1, \ldots, n$. Let $\{L_N\}_{N \in \mathbb{N}}$ be a sequence of linear operators from $C(Q, \mathbb{C})$ into itself. There is a sequence of companion positive linear operators $\left\{\widetilde{L}_N\right\}_{N \in \mathbb{N}}$ from $C(Q, \mathbb{R})$ into itself, such that $\widetilde{L}_N(1) = 1$, $\forall\, N \in \mathbb{N}$, and

$$|L_N(F)| \leq \lambda\widetilde{L}_N(|F|), \ \lambda > 0, \tag{12.69}$$

$\forall\, F \in C(Q, \mathbb{C})$, $\forall\, N \in \mathbb{N}$,

and

$$L_N(cg) = c\widetilde{L}_N(g), \ \forall\, g \in C(Q, \mathbb{R}), \forall\, c \in \mathbb{C}, \ \forall\, N \in \mathbb{N}. \tag{12.70}$$

We give

Theorem 12.20 *All as in Assumption 12.19. Then*

$$|(L_N(f))(x_0) - f(x_0)| \leq \lambda \left(\max_{|\alpha|=n} \omega_1(f_\alpha, h) \right) \tag{12.71}$$

$$\left[\frac{\widetilde{L}_N\left(\|\cdot - x_0\|_1^{n+1}\right)(x_0)}{(n+1)!h} + \frac{\widetilde{L}_N\left(\|\cdot - x_0\|_1^{n}\right)(x_0)}{2n!} + \frac{h}{8(n-1)!}\widetilde{L}_N\left(\|\cdot - x_0\|_1^{n-1}\right)(x_0) \right],$$

$\forall\, N \in \mathbb{N}$, *where* $h > 0$.

Proof We have that

$$|(L_N(f))(x_0) - f(x_0)| = \left|(L_N(f))(x_0) - f(x_0)\widetilde{L}_N(1)\right| \overset{(12.70)}{=}$$

$$|L_N(f)(x_0) - L_N(f(x_0))(x_0)| = |L_N(f(\cdot) - f(x_0))(x_0)| \overset{(12.69)}{\leq} \tag{12.72}$$

$$\lambda \widetilde{L}_N(|f(\cdot) - f(x_0)|)(x_0) \overset{(12.41)}{\leq} \lambda \left(\max_{|\alpha|=n} \omega_1(f_\alpha, h) \right)$$

$$\left[\frac{\widetilde{L}_N\left(\|\cdot - x_0\|_1^{n+1}\right)(x_0)}{(n+1)!h} + \frac{\widetilde{L}_N\left(\|\cdot - x_0\|_1^{n}\right)(x_0)}{2n!} + \frac{h}{8(n-1)!}\widetilde{L}_N\left(\|\cdot - x_0\|_1^{n-1}\right)(x_0) \right],$$

$h > 0$, $\forall\, N \in \mathbb{N}$. ■

We present

Theorem 12.21 *All as in Assumption 12.19. Then*

$$|(L_N(f))(x_0) - f(x_0)| \leq \frac{\lambda}{n!} \left[\frac{1}{(n+1)} + \frac{1}{2} + \frac{n}{8} \right]$$

$$\left(\max_{|\alpha|=n} \omega_1 \left(f_\alpha, \left(\widetilde{L}_N\left(\|\cdot - x_0\|_1^{n+1}\right)(x_0) \right)^{\frac{1}{n+1}} \right) \right) \left(\widetilde{L}_N\left(\|\cdot - x_0\|_1^{n+1}\right)(x_0) \right)^{\frac{n}{n+1}}, \tag{12.73}$$

$\forall\, N \in \mathbb{N}$, $n \in \mathbb{N}$.

If $\widetilde{L}_N\left(\|\cdot - x_0\|_1^{n+1}\right)(x_0) \to 0$, *then* $L_N(f)(x_0) \to f(x_0)$, *as* $N \to +\infty$.

Proof (i) Assume that $\widetilde{L}_N\left(\|\cdot - x_0\|_1^{n+1}\right)(x_0) = 0$. Then the right hand side of (12.73) equals zero. By Riesz representation theorem there exists a probability measure μ_{x_0} such that

$$\widetilde{L}_n(g)(z_0) = \int_Q g(t)\, d\mu_{x_0}(t), \ \forall\, g \in C(Q, \mathbb{R}).$$

That is

$$\int_Q \|t - x_0\|_1^{n+1} d\mu_{x_0}(t) = 0, \tag{12.74}$$

which implies $\|t - x_0\|_1^{n+1} = 0$, a.e., hence $t = x_0$, a.e. on Q. Consequently $\mu_{x_0}(\{t \in Q : t \neq x_0\}) = 0$.
That is $\mu_{x_0} = \delta_{x_0}$ (the Dirac measure).
Thus $\widetilde{L}_n(g)(x_0) = g(x_0)$, $\forall\ g \in C(Q, \mathbb{R})$ and therefore $\widetilde{L}_N(|f(\cdot) - f(x_0)|)(x_0) = 0$.
By (12.69) we have $|L_N(f(\cdot) - f(x_0))(x_0)| = 0$, which implies $|(L_N(f))(x_0) - f(x_0)| = 0$. Therefore inequality (12.73) is valid trivially.

(ii) Assume that $\widetilde{L}_N\left(\|\cdot - x_0\|_1^{n+1}\right)(x_0) > 0$. Choose

$$h := \left(\widetilde{L}_N\left(\|\cdot - x_0\|_1^{n+1}\right)(x_0)\right)^{\frac{1}{n+1}} > 0, \tag{12.75}$$

then

$$h^{n+1} = \widetilde{L}_N\left(\|\cdot - x_0\|_1^{n+1}\right)(x_0).$$

By Riesz representation theorem and Hölder's inequality we obtain

$$\widetilde{L}_N\left(\|\cdot - x_0\|_1^{n}\right)(x_0) \leq \left(\widetilde{L}_N\left(\|\cdot - x_0\|_1^{n+1}\right)(x_0)\right)^{\frac{n}{n+1}} = h^n, \tag{12.76}$$

and

$$\widetilde{L}_N\left(\|\cdot - x_0\|_1^{n-1}\right)(x_0) \leq \left(\widetilde{L}_N\left(\|\cdot - x_0\|_1^{n+1}\right)(x_0)\right)^{\frac{n-1}{n+1}} = h^{n-1}, \tag{12.77}$$

for $n \in \mathbb{N}$.
The right hand side of (12.71) is less equal to

$$\lambda\left(\max_{|\alpha|=n}\omega_1\left(f_\alpha, \left(\widetilde{L}_N\left(\|\cdot - x_0\|_1^{n+1}\right)(x_0)\right)^{\frac{1}{n+1}}\right)\right)$$

$$\left[\frac{h^n}{(n+1)!} + \frac{h^n}{2n!} + \frac{h^n}{8(n-1)!}\right] =$$

$$\lambda\left[\frac{1}{(n+1)!} + \frac{1}{2n!} + \frac{1}{8(n-1)!}\right]$$

$$\left(\max_{|\alpha|=n}\omega_1\left(f_\alpha, \left(\widetilde{L}_N\left(\|\cdot - x_0\|_1^{n+1}\right)(x_0)\right)^{\frac{1}{n+1}}\right)\right)\left(\widetilde{L}_N\left(\|\cdot - x_0\|_1^{n+1}\right)(x_0)\right)^{\frac{n}{n+1}}$$

$$= \frac{\lambda}{n!}\left[\frac{1}{(n+1)} + \frac{1}{2} + \frac{n}{8}\right] \tag{12.78}$$

$$\left(\max_{|\alpha|=n}\omega_1\left(f_\alpha, \left(\widetilde{L}_N\left(\|\cdot - x_0\|_1^{n+1}\right)(x_0)\right)^{\frac{1}{n+1}}\right)\right)\left(\widetilde{L}_N\left(\|\cdot - x_0\|_1^{n+1}\right)(x_0)\right)^{\frac{n}{n+1}},$$

$\forall\, N \in \mathbb{N}$, proving the claim. ■

We give

Corollary 12.22 *As in Assumption 12.19 with $n = 1$. Then*

$$|(L_N(f))(x_0) - f(x_0)| \leq \frac{9\lambda}{8}$$

$$\left(\max_{i \in \{1,\dots,k\}} \omega_1 \left(\frac{\partial f}{\partial x_i}, \left(\widetilde{L}_N \left(\|\cdot - x_0\|_1^2 \right)(x_0) \right)^{\frac{1}{2}} \right) \right) \left(\widetilde{L}_N \left(\|\cdot - x_0\|_1^2 \right)(x_0) \right)^{\frac{1}{2}}, \quad (12.79)$$

$\forall\, N \in \mathbb{N}$.

If $\widetilde{L}_N \left(\|\cdot - x_0\|_1^2 \right)(x_0) \to 0$, then $L_N(f)(x_0) \to f(x_0)$, as $N \to +\infty$.

Proof By Theorem 12.21. ■

To go further we need

Theorem 12.23 *Let $Q \subseteq \left(\mathbb{C}^k, \|\cdot\|_1 \right)$ be a compact convex subset, $k \geq 2$, where $\|\cdot\|_1$ is the l_1-norm. Here $f : Q \to \mathbb{C}$ is a coordinate-wise holomorphic function and $x_0 \in Q^0$, $n \in \mathbb{N}$. Assume that $f_\alpha(x_0) = 0$ for all $\alpha : |\alpha| = j$, $j = 1, \dots, n$. For all $\alpha : |\alpha| = n$, suppose that $|f_\alpha(z) - f_\alpha(x_0)|$ is convex in $z \in Q$. For $h > 0$ assume that the ball $B(x_0, h) \subset Q$. Then*

$$|f(z) - f(x_0)| \leq \frac{\max\limits_{|\alpha|=n} \omega_1(f_\alpha, h)}{h} \frac{\|z - x_0\|_1^{n+1}}{(n+1)!}, \quad \forall\, z \in Q. \quad (12.80)$$

That is

$$|f(\cdot) - f(x_0)| \leq \frac{\max\limits_{|\alpha|=n} \omega_1(f_\alpha, h)}{h} \frac{\|\cdot - x_0\|_1^{n+1}}{(n+1)!}, \quad (12.81)$$

valid over Q.

Proof We apply Theorem 12.8. Thus, we must estimate

$$R_n(z, 0) = \int_0^1 \left(\int_0^{t_1} \cdots \left(\int_0^{t_{n-1}} \left(g_z^{(n)}(t_n) - g_z^{(n)}(0) \right) dt_n \right) \cdots \right) dt_1, \ z \in Q. \quad (12.82)$$

Here when $|\alpha| = n$, by assumption $|f_\alpha(z) - f_\alpha(x_0)|$ is convex in $z \in Q$. Furthermore for $h > 0$, the ball $B(x_0, h) \subset Q$, where $x_0 \in Q^0$. By Theorem 12.17 we get

$$|f_\alpha(z) - f_\alpha(x_0)| \leq \frac{\omega_1(f_\alpha, h)}{h} \|z - x_0\|_1 \leq \frac{\max\limits_{|\alpha|=n} \omega_1(f_\alpha, h)}{h} \|z - x_0\|_1, \quad (12.83)$$

$\forall\, z \in Q$.

In particular, we have that

$$|f_\alpha(x_0 + t(z - x_0)) - f_\alpha(x_0)| \le \frac{\max\limits_{|\alpha|=n} \omega_1(f_\alpha, h)}{h} t \|z - x_0\|_1, \text{ all } t \in [0,1]. \tag{12.84}$$

It follows, from (12.27) and (12.84), that

$$|R_n(z,0)| \le \int_0^1 \left(\int_0^{t_1} \cdots \left(\int_0^{t_{n-1}} \left(\sum_{|\alpha|=n} \frac{n! \prod_{i=1}^k |z_i - x_{0i}|^{\alpha_i}}{\alpha_1! \ldots \alpha_k!} \right.\right.\right.\tag{12.85}$$

$$\left.\left.\left. \frac{\max\limits_{|\alpha|=n} \omega_1(f_\alpha, h)}{h} \|z - x_0\|_1 t_n \right) dt_n \right) \ldots \right) dt_1 =$$

$$\frac{\max\limits_{|\alpha|=n} \omega_1(f_\alpha, h)}{h} \frac{\|z - x_0\|_1^{n+1}}{(n+1)!}.$$

Therefore, it holds

$$|R_n(z,0)| \le \frac{\max\limits_{|\alpha|=n} \omega_1(f_\alpha, h)}{h} \frac{\|z - x_0\|_1^{n+1}}{(n+1)!}, \ \forall z \in Q, \tag{12.86}$$

proving the claim. ∎

We make

Assumption 12.24 Let $Q \subseteq (\mathbb{C}^k, \|\cdot\|_1)$ be a compact convex subset, $k \ge 2$, where $\|\cdot\|_1$ is the l_1-norm. Here we consider $f : Q \to \mathbb{C}$ to be coordinate-wise holomorphic functions and $x_0 \in Q^0$, $n \in \mathbb{N}$. Assume that $f_\alpha(x_0) = 0$ for all $\alpha : |\alpha| = j$, $j = 1, \ldots, n$. For all $\alpha : |\alpha| = n$, suppose that $|f_\alpha(z) - f_\alpha(x_0)|$ is convex in $z \in Q$. For $h > 0$ assume that the ball $B(x_0, h) \subset Q$.

Let $\{L_N\}_{N\in\mathbb{N}}$ be a sequence of linear operators from $C(Q,\mathbb{C})$ into itself. There is a sequence of companion positive linear operators $\{\widetilde{L}_N\}_{N\in\mathbb{N}}$ from $C(Q,\mathbb{R})$ into itself, such that $\widetilde{L}_N(1) = 1$, $\forall N \in \mathbb{N}$, and

$$|L_N(F)| \le \lambda \widetilde{L}_N(|F|), \ \lambda > 0, \tag{12.87}$$

$\forall F \in C(Q,\mathbb{C})$, $\forall N \in \mathbb{N}$,

and

$$L_N(cg) = c\widetilde{L}_N(g), \ \forall g \in C(Q,\mathbb{R}), \forall c \in \mathbb{C}, \ \forall N \in \mathbb{N}. \tag{12.88}$$

We give

Theorem 12.25 *All as in Assumption 12.24. Then*

$$\left|\left(L_N(f)\right)(x_0)-f(x_0)\right| \leq \frac{\lambda\left(\max\limits_{|\alpha|=n}\omega_1\left(f_\alpha,h\right)\right)}{h\,(n+1)!}\widetilde{L}_N\left(\|\cdot-x_0\|_1^{n+1}\right)(x_0), \quad (12.89)$$

$\forall\, N \in \mathbb{N}$.

Proof We have that

$$\left|\left(L_N(f)\right)(x_0)-f(x_0)\right| = \left|\left(L_N(f)\right)(x_0)-f(x_0)\,\widetilde{L}_N(1)\right| \overset{(12.88)}{=}$$

$$\left|L_N(f)(x_0)-L_N\left(f(x_0)\right)(x_0)\right| = \left|L_N\left(f(\cdot)-f(x_0)\right)(x_0)\right| \overset{(12.87)}{\leq}$$

$$\lambda\widetilde{L}_N\left(\left|f(\cdot)-f(x_0)\right|\right)(x_0) \overset{(12.81)}{\leq} \frac{\lambda\left(\max\limits_{|\alpha|=n}\omega_1\left(f_\alpha,h\right)\right)}{h\,(n+1)!}\widetilde{L}_N\left(\|\cdot-x_0\|_1^{n+1}\right)(x_0), \quad (12.90)$$

$\forall\, N \in \mathbb{N}$. ■

We also give

Theorem 12.26 *All as in Assumption 12.24, with* $h=\widetilde{L}_N\left(\|\cdot-x_0\|_1^{n+1}\right)(x_0)$, $n \in \mathbb{N}$. *Then*

$$\left|\left(L_N(f)\right)(x_0)-f(x_0)\right| \leq \frac{\lambda}{(n+1)!}\left(\max\limits_{|\alpha|=n}\omega_1\left(f_\alpha,\widetilde{L}_N\left(\|\cdot-x_0\|_1^{n+1}\right)(x_0)\right)\right), \quad (12.91)$$

$\forall\, N \in \mathbb{N}$.

If $\widetilde{L}_N\left(\|\cdot-x_0\|_1^{n+1}\right)(x_0)\to 0$, *then* $L_N(f)(x_0)\to f(x_0)$, *as* $N\to+\infty$.

Proof If $\widetilde{L}_N\left(\|\cdot-x_0\|_1^{n+1}\right)(x_0)>0$, then (12.91) comes directly from (12.89). If $\widetilde{L}_N\left(\|\cdot-x_0\|_1^{n+1}\right)(x_0)=0$, then (12.91) is trivially true, by acting as in the proof of Theorem 12.21 ■.

It comes

Corollary 12.27 *All as in Assumption 12.24 with* $n=1$, $h=\widetilde{L}_N\left(\|\cdot-x_0\|_1^{2}\right)(x_0)$. *Then*

$$\left|\left(L_N(f)\right)(x_0)-f(x_0)\right| \leq \frac{\lambda}{2}\left(\max\limits_{i\in\{1,\ldots,k\}}\omega_1\left(\frac{\partial f}{\partial x_i},\widetilde{L}_N\left(\|\cdot-x_0\|_1^{2}\right)(x_0)\right)\right), \quad (12.92)$$

$\forall\, N \in \mathbb{N}$.

If $\widetilde{L}_N\left(\|\cdot-x_0\|_1^{2}\right)(x_0)\to 0$, *then* $L_N(f)(x_0)\to f(x_0)$, *as* $N\to+\infty$.

Proof By Theorem 12.26, for $n=1$. ■

12.4 Applications

Example 12.28 Let $z_j = x_j + iy_j; x_j, y_j \in \mathbb{R}, i = \sqrt{-1}, j = 1, \ldots, k \in \mathbb{N} - \{1\}$. Here $z = (z_1, \ldots, z_k) \in K$, which is a compact convex subset of $\mathbb{C}^k$. Let $f \in C(K, \mathbb{C})$, which is $f(z) = u(x, y) + iv(x, y) = u + iv$, where $x = (x_1, \ldots, x_k)$, $y = (y_1, \ldots, y_k)$.

Consider $\widetilde{L} : C(K, \mathbb{R}) \to C(K, \mathbb{R})$ a positive linear operator. Also consider $L : C(K, \mathbb{C}) \to C(K, \mathbb{C})$ the linear operator such that:

$$L(f)(z) := \left(\widetilde{L}(u)\right)(x, y) + i\left(\widetilde{L}(v)\right)(x, y), \tag{12.93}$$

as $f \in C(K, \mathbb{C})$ iff $u, v \in C(K, \mathbb{R})$.

Here $u = u(x, y)$, $v = v(x, y) \in \mathbb{R}$, and from $|u| \leq |u| \Leftrightarrow -|u| \leq u \leq |u| \Leftrightarrow -\widetilde{L}(|u|) \leq \widetilde{L}(u) \leq \widetilde{L}(|u|) \Leftrightarrow \left|\widetilde{L}(u)\right| \leq \widetilde{L}(|u|)$.

We have that

$$\begin{gathered} |L(f)(z)| \leq \left|\left(\widetilde{L}(u)\right)(x, y)\right| + \left|\left(\widetilde{L}(v)\right)(x, y)\right| \leq \\ \left(\widetilde{L}(|u|)\right)(x, y) + \left(\widetilde{L}(|v|)\right)(x, y) = \left(\widetilde{L}(|u| + |v|)\right)(x, y) \leq \\ \sqrt{2}\left(\widetilde{L}\left(\sqrt{u^2 + v^2}\right)\right)(x, y) = \sqrt{2}\widetilde{L}(|f(z)|) = \sqrt{2}\left(\widetilde{L}(|f|)\right)(z). \end{gathered} \tag{12.94}$$

We have proved that

(1)

$$|L(f)|(z) \leq \sqrt{2}\widetilde{L}(|f|)(z), \ \forall z \in K. \tag{12.95}$$

Next, let $g \in C(K, \mathbb{R})$, and $c \in \mathbb{C}$, i.e. $c = a + bi$; $a, b \in \mathbb{R}$. Then $cg = ag + ibg$.

Clearly, $L(cg) = \widetilde{L}(ag) + i\widetilde{L}(bg) = a\widetilde{L}(g) + ib\widetilde{L}(g) = c\widetilde{L}(g)$.

Thus, it is true

(2)

$$L(cg) = c\widetilde{L}(g), \ \forall c \in \mathbb{C} \text{ and } \forall g \in C(K, \mathbb{R}). \tag{12.96}$$

So our main assumptions in this work are natural and possible.

Example 12.29 *(a realization of Example* 12.28*)* Let $F \in C\left([0, 1]^k\right)$, $k \in \mathbb{N} - \{1\}$, and $n_1, \ldots, n_k \in \mathbb{N}$.

The multivariate Bernstein polynomials ($(x_1, \ldots, x_k) \in [0, 1]^k$)

$$\left(B_{n_1,\ldots,n_k}(F)\right)(x_1, \ldots, x_k) := \sum_{\nu_1=0}^{n_1} \cdots \sum_{\nu_k=0}^{n_k} \binom{n_1}{\nu_1} \cdots \binom{n_k}{\nu_k}$$

$$f\left(\frac{\nu_1}{n_1},\ldots,\frac{\nu_k}{n_k}\right)x^{\nu_1}(1-x)^{n_1-\nu_1}\ldots x_k^{\nu_k}(1-x_k)^{n_k-\nu_k} \tag{12.97}$$

converge to F, uniformly, as $n_1,\ldots,n_k\to\infty$, see [9, p. 51]. These polynomials are positive linear operators.

Notice that $B_{n_1,\ldots,n_k}(1)=1$.

Here $z_j=x_j+iy_j$; $x_j,y_j\in[0,1]$, $j=1,\ldots,k\in\mathbb{N}-\{1\}$; where $z=(z_1,\ldots,z_k)$.

We take $K=[0,1]^{2k}$, which is a compact and convex subset of $\left(\mathbb{C}^k,\|\cdot\|\right)$, where $\|\cdot\|$ is an arbitrary norm.

Thus, for $f\in C\left([0,1]^{2k},\mathbb{C}\right): f(z)=u(x,y)+iv(x,y)=u+iv$, we define

$$\left(B_{n_1,\ldots,n_{2k}}^{\mathbb{C}^k}(f)\right)(z):=\left(B_{n_1,\ldots,n_{2k}}(u)\right)(x,y)+i\left(B_{n_1,\ldots,n_{2k}}(v)\right)(x,y), \tag{12.98}$$

the multivariate complex Bernstein polynomial operators; where $x=(x_1,\ldots,x_k)$ and $y=(y_1,\ldots,y_k)$.

By (12.95) we get that

$$\left|B_{n_1,\ldots,n_{2k}}^{\mathbb{C}^k}(f)\right|\leq\sqrt{2}B_{n_1,\ldots,n_{2k}}(|f|), \tag{12.99}$$

true over $[0,1]^{2k}$,

and by (12.96) we have that

$$B_{n_1,\ldots,n_{2k}}^{\mathbb{C}^k}(cg)=cB_{n_1,\ldots,n_{2k}}(g), \tag{12.100}$$

$\forall\, c\in\mathbb{C}$ and $\forall\, g\in C\left([0,1]^{2k},\mathbb{R}\right)$.

So the main assumptions of our work are fulfilled.

Notice that

$$\left|\left(B_{n_1,\ldots,n_{2k}}^{\mathbb{C}^k}(f)\right)(z)-f(z)\right|=$$

$$\left|\left(B_{n_1,\ldots,n_{2k}}(u)\right)(x,y)+i\left(B_{n_1,\ldots,n_{2k}}(v)(x,y)\right)-u(x,y)-iv(x,y)\right|=$$

$$\left|\left(\left(B_{n_1,\ldots,n_{2k}}(u)\right)(x,y)-u(x,y)\right)+i\left(\left(B_{n_1,\ldots,n_{2k}}(v)\right)(x,y)-v(x,y)\right)\right|=$$

$$\sqrt{\left(\left(B_{n_1,\ldots,n_{2k}}(u)\right)(x,y)-u(x,y)\right)^2+\left(\left(B_{n_1,\ldots,n_{2k}}(v)\right)(x,y)-v(x,y)\right)^2}=:(*). \tag{12.101}$$

We have that $\left|\left(B_{n_1,\ldots,n_{2k}}(u)\right)(x,y)-u(x,y)\right|<\varepsilon_1,\forall x,y\in[0,1]^k,\forall n_1,\ldots,n_{2k}\geq N_1$, and $\left|\left(B_{n_1,\ldots,n_{2k}}(v)\right)(x,y)-v(x,y)\right|<\varepsilon_2,\forall x,y\in[0,1]^k,\forall n_1,\ldots,n_{2k}\geq N_2$; $N_1,N_2\in\mathbb{N}$, where $\varepsilon_1,\varepsilon_2>0$.

Thus, it holds

$$(*)\leq\sqrt{\varepsilon_1^2+\varepsilon_2^2}=:\varepsilon, \tag{12.102}$$

$\forall\, x, y \in [0,1]^k$, $\forall\, n_1, \ldots, n_{2k} \geq \max(N_1, N_2) =: N^*$, $\varepsilon > 0$.

Hence

$$\left| \left(B^{\mathbb{C}^k}_{n_1,\ldots,n_{2k}}(f) \right)(z) - f(z) \right| \leq \varepsilon, \forall\, z \in [0,1]^{2k}, \tag{12.103}$$

$\forall\, n_1, \ldots, n_{2k} \geq N^* \in \mathbb{N}$, where $\varepsilon > 0$.

Therefore $B^{\mathbb{C}^k}_{n_1,\ldots,n_{2k}}(f) \to f$, converges uniformly as $n_1, \ldots, n_{2k} \to \infty$.

Example 12.30 *(continuation of Example* 12.29*)* For $f \in C\left([0,1]^{2k}, \mathbb{C}\right)$, by (12.46), we have that

$$\left| \left(B^{\mathbb{C}^k}_{n_1,\ldots,n_{2k}}(f) \right)(z_0) - f(z_0) \right| \leq 2\sqrt{2}\omega_1\left(f, B_{n_1,\ldots,n_{2k}}\left(\|\cdot - z_0\|_{l_1}\right)(z_0)\right), \tag{12.104}$$

$\forall\, z_0 \in [0,1]^{2k}$, $\forall\, n_1, \ldots, n_{2k} \in \mathbb{N}$, see also (12.99).

Here $z_0 = x_0 + iy_0$, where $x_0 = (x_{01}, \ldots, x_{0k})$, $y_0 = (y_{01}, \ldots, y_{0k})$. It is $z_0 = (x_0, y_0)$. We notice that:

$$\left(B_{n_1,\ldots,n_{2k}}\left(\|\cdot - z_0\|_{l_1}\right)\right)(z_0) = \left(B_{n_1,\ldots,n_{2k}}\left(\sum_{j=1}^{k} \left|\cdot - z_{0j}\right| \right) \right)(z_0) =$$

$$\sum_{j=1}^{k} \left(B_{n_1,\ldots,n_{2k}}\left(\left|\cdot - z_{0j}\right|\right)\right)(z_0) \leq \sum_{j=1}^{k} \left(\left(B_{n_1,\ldots,n_{2k}}\left(\left|\cdot - z_{0j}\right|^2 \right) \right)(z_0) \right)^{\frac{1}{2}} =$$

$$\sum_{j=1}^{k} \left(\left(B_{n_1,\ldots,n_{2k}}\left(\left(\cdot - x_{0j}\right)^2 + \left(\cdot - y_{0j}\right)^2 \right) \right)(x_0, y_0) \right)^{\frac{1}{2}} = \tag{12.105}$$

$$\sum_{j=1}^{k} \left[\left(B_{n_1,\ldots,n_{2k}}\left(\left(\cdot - x_{0j}\right)^2 \right) \right)(x_0, y_0) + \left(B_{n_1,\ldots,n_{2k}}\left(\left(\cdot - y_{0j}\right)^2 \right) \right)(x_0, y_0) \right]^{\frac{1}{2}} =$$

$$\sum_{j=1}^{k} \left[\left(B_{n_j}\left(\left(\cdot - x_{0j}\right)^2 \right) \right)(x_{0j}) + \left(B_{n_{j+k}}\left(\left(\cdot - y_{0j}\right)^2 \right) \right)(y_{0j}) \right]^{\frac{1}{2}}$$

(where B_{n_j} is the basic univariate real Bernstein polynomial on $[0,1]$)

$$= \sum_{j=1}^{k} \left[\frac{x_{0j}\left(1 - x_{0j}\right)}{n_j} + \frac{y_{0j}\left(1 - y_{0j}\right)}{n_{j+k}} \right]^{\frac{1}{2}} \leq$$

$$\sum_{j=1}^{k} \left[\frac{1}{4n_j} + \frac{1}{4n_{j+k}} \right]^{\frac{1}{2}} = \frac{1}{2} \sum_{j=1}^{k} \left(\sqrt{\frac{1}{n_j} + \frac{1}{n_{j+k}}} \right). \tag{12.106}$$

We have found that

$$\left(B_{n_1,\ldots,n_{2k}}\left(\left\|\cdot - z_0\right\|_{l_1}\right)\right)(z_0) \leq \frac{1}{2}\sum_{j=1}^{k}\left(\sqrt{\frac{1}{n_j}+\frac{1}{n_{j+k}}}\right). \tag{12.107}$$

By (12.104) and (12.107) we derive that

$$\left|\left(B_{n_1,\ldots,n_{2k}}^{\mathbb{C}^k}(f)\right)(z_0) - f(z_0)\right| \leq 2\sqrt{2}\omega_1\left(f, \frac{1}{2}\left(\sum_{j=1}^{k}\sqrt{\frac{1}{n_j}+\frac{1}{n_{j+k}}}\right)\right), \tag{12.108}$$

$\forall\, z_0 \in [0,1]^{2k}$, $\forall\, n_1,\ldots,n_{2k} \in \mathbb{N}$, and $\forall\, f \in C\left([0,1]^{2k}, \mathbb{C}\right)$.

Clearly, it holds

$$\left\|B_{n_1,\ldots,n_{2k}}^{\mathbb{C}^k}(f) - f\right\|_{\infty} \leq 2\sqrt{2}\omega_1\left(f, \frac{1}{2}\left(\sum_{j=1}^{k}\sqrt{\frac{1}{n_j}+\frac{1}{n_{j+k}}}\right)\right), \tag{12.109}$$

$\forall\, n_1,\ldots,n_{2k} \in \mathbb{N}$.

We get $B_{n_1,\ldots,n_{2k}}^{\mathbb{C}^k}(f) \to f$, uniformly, as $n_1,\ldots,n_{2k} \to \infty$.

References

1. Anastassiou, G.A.: Moments in Probability and Approximation Theory. Pitman Research Notes in Math, vol. 287. Longman Sci. & Tech., Harlow (1993)
2. Anastassiou, G.A.: Lattice homomorphism - Korovkin type inequalities for vector valued functions. Hokkaido Math. J. **26**, 337–364 (1997)
3. Anastassiou, G.A.: Quantitative Approximations. Chapman & Hall/CRC, Boca Raton (2001)
4. Anastassiou, G.A.: Intelligent Computations: Abstract Fractional Calculus, Inequalities, Approximations. Springer, Heidelberg (2018)
5. Anastassiou, G.: Quantitative multivariate complex Korovkin theory. RACSAM 114 (2020). Article 59
6. Caratheodory, C.: Theory of Functions of a complex variable, vol. 2. Chelsea publishing Company, New York (1954)
7. Korovkin, P.P.: Linear Operators and Approximation Theory. Hindustan Publ. Corp., Delhi (1960)
8. Krantz, S.G.: Function Theory of Several Complex Variables, 2nd edn. AMS Chelsea publishing, Providence (2001)
9. Lorentz, G.G.: Bernstein Polynomials. Chelsea Publishing Company, New York (1986)
10. Mamedov, R.G.: On the order of the approximation of functions by linear positive operators. Dokl. Akad. Nauk USSR **128**, 674–676 (1959)
11. Royden, H.L.: Real Analysis, 2nd edn. Macmillan, New York (1968)
12. Shilov, G.E.: Elementary Functional Analysis. Dover Publications Inc., New York (1996)
13. Shisha, O., Mond, B.: The degree of convergence of sequences of linear positive operators. Natl. Acad. Sci. **60**, 1196–1200 (1968)

Chapter 13
M-Fractional Integral Type Inequalities

Here we present M-fractional integral inequalities of Ostrowski and Polya types. See also [5].

13.1 Introduction

We are inspired by the following results:

Theorem 13.1 ([2, p. 498], [1, 6] Ostrowski inequality)
Let $f \in C^1([a,b])$, $x \in [a,b]$. Then

$$\left| \frac{1}{b-a} \int_a^b f(z)\,dz - f(x) \right| \leq \left(\frac{(x-a)^2 + (b-x)^2}{2(b-a)} \right) \|f'\|_\infty . \tag{13.1}$$

Inequality (13.1) is sharp. In particular the optimal function is

$$f^*(z) := |z-x|^\alpha (b-a), \ \alpha > 1. \tag{13.2}$$

Theorem 13.2 ([7], [8, p. 62], [9], [10, p. 83] Polya integral inequality)
Let $f(x)$ be differentiable and not identically a constant on $[a,b]$ with $f(a) = f(b) = 0$. Then there exists at least one point $\xi \in [a,b]$ such that

$$|f'(\xi)| > \frac{4}{(b-a)^2} \int_a^b f(x)\,dx. \tag{13.3}$$

In this short work we present inequalities of types (13.1) and (13.3) involving the left and right fractional local general M-derivatives, see [3, 4].

G. A. Anastassiou, *Generalized Fractional Calculus*, Studies in Systems, Decision and Control 305, https://doi.org/10.1007/978-3-030-56962-4_13

13.2 Background

We need

Definition 13.3 *([3])* Let $f : [a, \infty) \to \mathbb{R}$ and $t > a$, $a \in \mathbb{R}$. For $0 < \alpha \leq 1$ we define the left local general M-derivative of order α of function f, denoted by $D_{M,a}^{\alpha,\beta} f(t)$, by

$$D_{M,a}^{\alpha,\beta} f(t) := \lim_{\varepsilon \to 0} \frac{f\left(t\mathbb{E}_\beta\left(\varepsilon(t-a)^{-\alpha}\right)\right) - f(t)}{\varepsilon}, \tag{13.4}$$

$\forall\, t > a$, where $\mathbb{E}_\beta(t) = \sum_{k=0}^{\infty} \frac{t^k}{\Gamma(\beta k+1)}$, $\beta > 0$, is the Mittag-Leffler function with one parameter.

If $D_{M,a}^{\alpha,\beta} f(t)$ exists over (a, γ), $\gamma \in \mathbb{R}$ and $\lim_{t \to a+} D_{M,a}^{\alpha,\beta} f(t)$ exists, then

$$D_{M,a}^{\alpha,\beta} f(a) = \lim_{t \to a+} D_{M,a}^{\alpha,\beta} f(t). \tag{13.5}$$

Theorem 13.4 ([3]) *If a function $f : [a, \infty) \to \mathbb{R}$ has the left local general M-derivative of order $\alpha \in (0, 1]$, $\beta > 0$, at $t_0 > a$, then f is continuous at t_0.*

We need

Theorem 13.5 ([3] Mean value theorem) *Let $f : [\gamma, \delta] \to \mathbb{R}$ with $\gamma > a$, $0 \notin [\gamma, \delta]$, such that*

(1) f is continuous on $[\gamma, \delta]$,
(2) there exists $D_{M,a}^{\alpha,\beta} f$ on (γ, δ) for some $\alpha \in (0, 1]$.

Then, there exists $c \in (\gamma, \delta)$ such that

$$f(\delta) - f(\gamma) = \left(D_{M,a}^{\alpha,\beta} f(c)\right) \frac{\Gamma(\beta+1)(c-a)^\alpha}{c} (\delta - \gamma). \tag{13.6}$$

We need

Definition 13.6 *([4])* Let $f : (-\infty, b] \to \mathbb{R}$ and $t < b$, $b \in \mathbb{R}$. For $0 < \alpha \leq 1$ we define the right local general M-derivative of order α of function f, denoted as ${}_{M,b}^{\alpha,\beta}Df(t)$, by

$${}_{M,b}^{\alpha,\beta}Df(t) := -\lim_{\varepsilon \to 0} \frac{f\left(t\mathbb{E}_\beta\left(\varepsilon(b-t)^{-\alpha}\right)\right) - f(t)}{\varepsilon}, \tag{13.7}$$

$\forall\, t < b$.

If ${}_{M,b}^{\alpha,\beta}Df(t)$ exists over (γ, b), $\gamma \in \mathbb{R}$ and $\lim_{t\to b-} {}_{M,b}^{\alpha,\beta}Df(t)$ exists, then

$$ {}_{M,b}^{\alpha,\beta}Df(b) = \lim_{t\to b-} {}_{M,b}^{\alpha,\beta}Df(t). \tag{13.8} $$

Theorem 13.7 ([4]) *If a function $f : (-\infty, b] \to \mathbb{R}$ has the right local general M-derivative of order $\alpha \in (0, 1]$, $\beta > 0$, at $t_0 < b$, then f is continuous at t_0.*

We also need

Theorem 13.8 ([4] Mean value theorem) *Let $f : [\gamma, \delta] \to \mathbb{R}$ with $\delta < b$, $0 \notin [\gamma, \delta]$, such that*

(1) f is continuous on $[\gamma, \delta]$,
(2) there exists ${}_{M,b}^{\alpha,\beta}Df$ on (γ, δ) for some $\alpha \in (0, 1]$.

Then, there exists $c \in (\gamma, \delta)$ such that

$$ f(\delta) - f(\gamma) = \left(-{}_{M,b}^{\alpha,\beta}Df(c)\right)\left(\frac{\Gamma(\beta+1)(b-c)^{\alpha}}{c}\right)(\delta-\gamma). \tag{13.9} $$

Fractional derivatives $D_{M,a}^{\alpha,\beta}$ and ${}_{M,b}^{\alpha,\beta}D$ possess all basic properties of the ordinary derivatives and beyond, see [3, 4].

13.3 Main Results

We present the following M-fractional Ostrowski type inequality:

Theorem 13.9 *Let $a < \gamma < \delta < b$, $0 \notin [\gamma, \delta]$, $f : [a, b] \to \mathbb{R}$, which is continuous over $[\gamma, \delta]$. We assume that $D_{M,a}^{\alpha,\beta}$, ${}_{M,b}^{\alpha,\beta}D$ exist and are continuous over $[\gamma, x_0]$ and $[x_0, \delta]$, respectively, where $x_0 \in [\gamma, \delta]$, for some $\alpha \in (0, 1]$. Then*

$$ \left|\frac{1}{\delta-\gamma}\int_{\gamma}^{\delta} f(x)\,dx - f(x_0)\right| \leq \frac{\Gamma(\beta+1)}{2(\delta-\gamma)} $$
$$ \left[\left\|\frac{D_{M,a}^{\alpha,\beta}f(x)}{x}\right\|_{\infty,[\gamma,x_0]}(x_0-a)^{\alpha}(x_0-\gamma)^2 + \left\|\frac{{}_{M,b}^{\alpha,\beta}Df(x)}{x}\right\|_{\infty,[x_0,\delta]}(b-x_0)^{\alpha}(\delta-x_0)^2\right]. \tag{13.10} $$

Proof Let $x \in [\gamma, x_0]$, the by Theorem 13.5, there exists $c_1 \in (x, x_0)$, such that

$$ f(x_0) - f(x) = \left(\frac{D_{M,a}^{\alpha,\beta}f(c_1)}{c_1}\right)\Gamma(\beta+1)(c_1-a)^{\alpha}(x_0-x). \tag{13.11} $$

Thus

$$|f(x) - f(x_0)| = \left| \frac{D_{M,a}^{\alpha,\beta} f(c_1)}{c_1} \right| \Gamma(\beta+1)(c_1 - a)^{\alpha} |x - x_0| \leq$$

$$\left\| \frac{D_{M,a}^{\alpha,\beta} f(x)}{x} \right\|_{\infty,[\gamma,x_0]} \Gamma(\beta+1)(x_0 - a)^{\alpha} |x - x_0|, \tag{13.12}$$

$\forall\, x \in [\gamma, x_0]$.

Let now $x \in [x_0, \delta]$, then by Theorem 13.8, there exists $c_2 \in (x_0, x)$, such that

$$f(x) - f(x_0) = -\left(\frac{{}_{M,b}^{\alpha,\beta}Df(c_2)}{c_2} \right) \Gamma(\beta+1)(b - c_2)^{\alpha}(x - x_0). \tag{13.13}$$

Thus

$$|f(x) - f(x_0)| = \left| \frac{{}_{M,b}^{\alpha,\beta}Df(c_2)}{c_2} \right| \Gamma(\beta+1)(b - x_0)^{\alpha} |x - x_0| \leq$$

$$\left\| \frac{{}_{M,b}^{\alpha,\beta}Df(x)}{x} \right\|_{\infty,[x_0,\delta]} \Gamma(\beta+1)(b - x_0)^{\alpha} |x - x_0|, \tag{13.14}$$

$\forall\, x \in [x_0, \delta]$.

We have that

$$\left| \frac{1}{\delta - \gamma} \int_{\gamma}^{\delta} f(x)\, dx - f(x_0) \right| = \frac{1}{\delta - \gamma} \left| \int_{\gamma}^{\delta} (f(x) - f(x_0))\, dx \right| \leq$$

$$\frac{1}{\delta - \gamma} \int_{\gamma}^{\delta} |f(x) - f(x_0)|\, dx = \tag{13.15}$$

$$\frac{1}{\delta - \gamma} \left[\int_{\gamma}^{x_0} |f(x) - f(x_0)|\, dx + \int_{x_0}^{\delta} |f(x) - f(x_0)|\, dx \right] \overset{\text{(by (13.12), (13.14))}}{\leq}$$

$$\frac{1}{\delta - \gamma} \left[\left\| \frac{D_{M,a}^{\alpha,\beta} f(x)}{x} \right\|_{\infty,[\gamma,x_0]} \Gamma(\beta+1)(x_0 - a)^{\alpha} \int_{\gamma}^{x_0} (x_0 - x)\, dx \right.$$

$$\left. + \left\| \frac{{}_{M,b}^{\alpha,\beta}Df(x)}{x} \right\|_{\infty,[x_0,\delta]} \Gamma(\beta+1)(b - x_0)^{\alpha} \int_{x_0}^{\delta} (x - x_0)\, dx \right] =$$

$$\frac{\Gamma(\beta+1)}{2(\delta-\gamma)}\left[\left\|\frac{D_{M,a}^{\alpha,\beta}f(x)}{x}\right\|_{\infty,[\gamma,x_0]}(x_0-a)^{\alpha}(x_0-\gamma)^2+\right. \tag{13.16}$$

$$\left.\left\|\frac{{}_{M,b}^{\alpha,\beta}Df(x)}{x}\right\|_{\infty,[x_0,\delta]}(b-x_0)^{\alpha}(\delta-x_0)^2\right].$$

The theorem is proved. ■

Next we give two M-fractional Polya type inequalities:

Theorem 13.10 *All as in Theorem 13.9 and* $f(x_0)=0$. *Then*

$$\left|\int_{\gamma}^{\delta}f(x)\,dx\right|\leq\int_{\gamma}^{\delta}|f(x)|\,dx\leq\frac{\Gamma(\beta+1)}{2}$$

$$\left[\left\|\frac{D_{M,a}^{\alpha,\beta}f(x)}{x}\right\|_{\infty,[\gamma,x_0]}(x_0-a)^{\alpha}(x_0-\gamma)^2+\left\|\frac{{}_{M,b}^{\alpha,\beta}Df(x)}{x}\right\|_{\infty,[x_0,\delta]}(b-x_0)^{\alpha}(\delta-x_0)^2\right]. \tag{13.17}$$

Proof Same as in the proof of Theorem 13.9, by setting $f(x_0)=0$. ■

Corollary 13.11 (to Theorem 13.10, case of $x_0=\frac{\gamma+\delta}{2}$) *All as in Theorem 13.9 and* $f\left(\frac{\gamma+\delta}{2}\right)=0$. *Then*

$$\int_{\gamma}^{\delta}|f(x)|\,dx\leq\frac{\Gamma(\beta+1)(\delta-\gamma)^2}{8}$$

$$\left[\left\|\frac{D_{M,a}^{\alpha,\beta}f(x)}{x}\right\|_{\infty,\left[\gamma,\frac{\gamma+\delta}{2}\right]}\left(\left(\frac{\gamma+\delta}{2}\right)-a\right)^{\alpha}+\left\|\frac{{}_{M,b}^{\alpha,\beta}Df(x)}{x}\right\|_{\infty,\left[\frac{\gamma+\delta}{2},\delta\right]}\left(b-\left(\frac{\gamma+\delta}{2}\right)\right)^{\alpha}\right]. \tag{13.18}$$

Proof Apply (13.17) for $x_0=\frac{\gamma+\delta}{2}$. ■

References

1. Anastassiou, G.A.: Ostrowski type inequalities. Proc. AMS **123**, 3775–3781 (1995)
2. Anastassiou, G.A.: Quantitative Approximations. Chapmann & Hall / CRC, Boca Raton (2001)
3. Anastassiou, G.: On the left fractional local general M-derivative (2019). Submitted for publication
4. Anastassiou, G.: About the right fractional local general M-derivative. Analele Univ. Oradea, Fasc. Mate., Tom **XXVII**(1), 87–94 (2020)

5. Anastassiou, G.: M-fractional integral inequalities. J. Comput. Anal. Appl. **29**(6), 1153–1158 (2021)
6. Ostrowski, A.: Über die Absolutabweichung einer differtentiebaren Funktion von ihrem Integralmittelwert. Comment. Math. Helv. **10**, 226–227 (1938)
7. Polya, G.: Ein mittelwertsatz für Funktionen mehrerer Veränderlichen. Tohoku Math. J. **19**, 1–3 (1921)
8. Polya, G., Szegö, G.: Aufgaben und Lehrsätze aus der Analysis, vol. I. Springer, Berlin (1925). (German)
9. Polya, G., Szegö, G.: Problems and Theorems in Analysis. Classics in Mathematics, vol. I. Springer, Berlin (1972)
10. Polya, G., Szegö, G.: Problems and Theorems in Analysis, vol. I, Chinese edn. (1984)

Chapter 14
Principles of Stochastic Caputo Fractional Calculus with Fractional Approximation of Stochastic Processes

Here we consider and study very general stochastic positive linear operators induced by general positive linear operators that are acting on continuous functions. These are acting on the space of real fractionally differentiable stochastic processes. Under some very mild, general and natural assumptions on the stochastic processes we produce related fractional stochastic Shisha-Mond type inequalities of L^q-type $1 \leq q < \infty$ and corresponding fractional stochastic Korovkin type theorems. These are regarding the stochastic q-mean fractional convergence of a sequence of stochastic positive linear operators to the stochastic unit operator for various cases. All convergences are produced with rates and are given via the fractional stochastic inequalities involving the stochastic modulus of continuity of the αth fractional derivatives of the engaged stochastic process, $\alpha > 0$, $\alpha \notin \mathbb{N}$. The impressive fact is that the basic real Korovkin test functions assumptions are enough for the conclusions of our fractional stochastic Korovkin theory. We give applications to stochastic Bernstein operators. See also [10].

14.1 Introduction

Motivation for this work comes from [2, 3, 15, 16]. This work continues our earlier work [5], now at the stochastic fractional level. First we lay down the foundations of Stochastic fractional calculus in the direct analytical sense, see Sect. 14.2, this is in the Caputo fractional direction. In the Sect. 14.3, about background, we talk about the q-mean $(1 \leq q < \infty)$ first modulus of continuity of a stochastic process and its upper bounds. There we describe completely our setting by introducing our stochastic positive linear operator M, see (14.15), which is based on the positive linear operator $\widetilde{L}$ from $C([a, b])$ into itself. The operator M is acting on a wide space of Caputo fractional differentiable real valued stochastic processes X. See there Assumptions 14.10, 14.13, 14.14. We first give the main pointwise fractional

G. A. Anastassiou, *Generalized Fractional Calculus*, Studies in Systems, Decision and Control 305, https://doi.org/10.1007/978-3-030-56962-4_14

stochastic Shisha-Mond type inequalities [14], see Theorems 14.16, 14.17, and their several corollaries covering important special cases.

We continue with fractional q-mean uniform Shisha-Mond type inequalities, see Theorems 14.23, 14.24, and their interesting corollaries. All this theory is regarding the fractional stochastic convergence of operators M to I (stochastic unit operator) given quantitatively with rates. An extensive application about the stochastic Bernstein operators follows in full details. Based on our Shisha-Mond type inequalities of our main Theorems 14.16, 14.17, 14.23, 14.24 we derive pointwise and uniform Stochastic Korovkin theorems [12] on stochastic processes, see Theorems 14.33–14.36. The amazing fact here is, that basic conditions on operator $\widetilde{L}$ regarding two simple real valued functions that are not stochastic, are able to enforce fractional stochastic convergence on all stochastic processes we are dealing with; see Concepts 14.9 and Assumptions 14.10, 14.13, 14.14.

14.2 Foundation of Stochastic Fractional Calculus

Let $t \in [a, b] \subset \mathbb{R}$, $\omega \in \Omega$, where $(\Omega, \mathcal{F}, P)$ is a probability space. Here $X(t, \omega)$ stands for a stochastic process. Case of $X(\cdot, \omega)$ being continuous on $[a, b]$, $\forall\, \omega \in \Omega$. Then by Caratheodory Theorem 20.15, p.156, [1], we get that $X(t, \omega)$ is jointly measurable.

Next we define the left and right respectively, Riemann–Liouville stochastic fractional integrals, where $\alpha > 0$ is not an integer:

$$I_{a+}^{\alpha} X(x, \omega) = \frac{1}{\Gamma(\alpha)} \int_a^x (x - t)^{\alpha - 1} X(t, \omega)\, dt, \tag{14.1}$$

and

$$I_{b-}^{\alpha} X(x, \omega) = \frac{1}{\Gamma(\alpha)} \int_x^b (t - x)^{\alpha - 1} X(t, \omega)\, dt, \tag{14.2}$$

$\forall\, x \in [a, b]$, $\forall\, \omega \in \Omega$, where Γ is the gamma function.

In the following important cases we prove that $I_{a+}^{\alpha} X$, $I_{b-}^{\alpha} X$ are stochastic processes:

(i) Assume that $(\Omega, \mathcal{F}, P)$ is a complete probability space, and that $(x - t)^{\alpha - 1} X(t, \omega)$ is an integrable function on $[a, x] \times \Omega$, $\forall\, x \in [a, b]$, then by Fubini's theorem, [13], p. 269, $I_{a+}^{\alpha} X(x, \cdot)$ is an integrable function on Ω, $\forall\, x \in [a, b]$. Similarly, if $(t - x)^{\alpha - 1} X(t, \omega)$ is an integrable function on $[x, b] \times \Omega$, $\forall\, x \in [a, b]$, then again by Fubini's theorem $I_{b-}^{\alpha} X(x, \cdot)$ is an integrable function on Ω, $\forall\, x \in [a, b]$. That is $I_{a+}^{\alpha} X(x, \omega)$ and $I_{b-}^{\alpha} X(x, \omega)$ are stochastic processes.

(ii) Assume a general probability space $(\Omega, \mathcal{F}, P)$ and the Lebesgue measure spaces on $[a, x]$, $[x, b]$, $\forall\, x \in [a, b]$. These are clearly σ-finite measure spaces. We assume that the jointly measurable stochastic process $X(t, \omega) \geq 0$ on $[a, b] \times \Omega$, hence $(x-t)^{\alpha-1} X(t,\omega) \geq 0$ on $[a,x] \times \Omega$, and $(t-x)^{\alpha-1} X(t,\omega) \geq 0$ on $[x,b] \times \Omega$, $\forall\, x \in [a,b]$, and both are jointly measurable. Then by Tonelli's theorem, [13, p. 270], we get that $I_{a+}^{\alpha} X(x,\cdot)$, $I_{b-}^{\alpha} X(x,\cdot)$ are measurable functions on Ω, $\forall\, x \in [a,b]$. That is $I_{a+}^{\alpha} X$, $I_{b-}^{\alpha} X$ are stochastic processes. The above facts provide the foundation of stochastic fractional calculus in the direct analytical sense. So it is not unusual to consider that $I_{a+}^{\alpha} X$, $I_{b-}^{\alpha} X$ are stochastic processes.

(iii) Given that $X(\cdot,\omega)$ is in $L_1([a,b])$ then $I_{a+}^{\alpha} X(\cdot,\omega) \in L_1([a,b])$, $\forall\, \omega \in \Omega$, see [11, p. 13], and $I_{b-}^{\alpha} X(\cdot,\omega) \in L_1([a,b])$, $\forall\, \omega \in \Omega$, see [8, p. 334].
And given that $X(\cdot,\omega) \in L_{\infty}([a,b])$, then $I_{a+}^{\alpha} X(\cdot,\omega) \in C([a,b])$, when $0 < \alpha < 1$, and $I_{a+}^{\alpha} X(\cdot,\omega) \in AC([a,b])$ (absolutely continuous functions), when $\alpha \geq 1$, $\forall\, \omega \in \Omega$, see [6, p. 388]. Similarly, if $X(\cdot,\omega) \in L_{\infty}([a,b])$, then $I_{b-}^{\alpha} X(\cdot,\omega) \in C([a,b])$, when $0 < \alpha < 1$, and $I_{b-}^{\alpha} X(\cdot,\omega) \in AC([a,b])$, when $\alpha \geq 1$, $\forall\, \omega \in \Omega$, see [9].

We need

Definition 14.1 Let non-integer $\alpha > 0$, $n = \lceil \alpha \rceil$ ($\lceil \cdot \rceil$ is the ceiling of the number), $t \in [a,b] \subset \mathbb{R}$, $\omega \in \Omega$, where $(\Omega, \mathcal{F}, P)$ is a general probability space. Here $X(t,\omega)$ stands for a stochastic process. Assume that $X(\cdot,\omega) \in AC^n([a,b])$ (spaces of functions $X(\cdot,\omega)$ with $X^{(n-1)}(\cdot,\omega) \in AC([a,b])$), $\forall\, \omega \in \Omega$.

We call stochastic left Caputo fractional derivative

$$D_{*a}^{\alpha} X(x,\omega) = \frac{1}{\Gamma(n-\alpha)} \int_a^x (x-t)^{n-\alpha-1} X^{(n)}(t,\omega)\, dt, \tag{14.3}$$

$\forall\, x \in [a,b]$, $\forall\, \omega \in \Omega$.

And, we call stochastic right Caputo fractional derivative

$$D_{b-}^{\alpha} X(x,\omega) = \frac{(-1)^n}{\Gamma(n-\alpha)} \int_x^b (z-x)^{n-\alpha-1} X^{(n)}(z,\omega)\, dz, \tag{14.4}$$

$\forall\, x \in [a,b]$, $\forall\, \omega \in \Omega$.

Remark 14.2 *(to Definition* 14.1*)* We further assume here that

$$\left| X^{(n)}(t,\omega) \right| \leq M^*, \forall\, (t,\omega) \in [a,b] \times \Omega,$$

where $M^* > 0$. Then, by (14.3), we have

$$\left| D_{*a}^{\alpha} X(x,\omega) \right| \leq \frac{1}{\Gamma(n-\alpha)} \int_a^x (x-t)^{n-\alpha-1} \left| X^{(n)}(t,\omega) \right| dt$$

$$\leq \frac{M^*}{\Gamma(n-\alpha)} \int_a^x (x-t)^{n-\alpha-1}\, dt = \frac{M^* (x-a)^{n-\alpha}}{\Gamma(n-\alpha+1)}.$$

That is

$$\left|D_{*a}^{\alpha}X(x,\omega)\right| \leq \frac{M^*(x-a)^{n-\alpha}}{\Gamma(n-\alpha+1)}, \ \forall x \in [a,b]. \tag{14.5}$$

Also, from (14.4) we get

$$\left|D_{b-}^{\alpha}X(x,\omega)\right| \leq \frac{1}{\Gamma(n-\alpha)}\int_x^b (z-x)^{n-\alpha-1}\left|X^{(n)}(z,\omega)\right| dz \leq$$

$$\frac{M^*}{\Gamma(n-\alpha)}\int_x^b (z-x)^{n-\alpha-1} dz = \frac{M^*(b-x)^{n-\alpha}}{\Gamma(n-\alpha+1)}.$$

That is

$$\left|D_{b-}^{\alpha}X(x,\omega)\right| \leq \frac{M^*(b-x)^{n-\alpha}}{\Gamma(n-\alpha+1)}, \ \forall x \in [a,b]. \tag{14.6}$$

By (14.1)–(14.4), it is not strange to assume that $D_{*a}^{\alpha}X$, $D_{b-}^{\alpha}X$ are stochastic processes.

14.3 Background

We need

Definition 14.3 We define the relative q-mean first modulus of continuity of stochastic process $X(t,\omega)$ by

$$\Omega_1(X,\delta)_{L^q,[c,d]} :=$$

$$\sup\left\{\left(\int_{\Omega}|X(x,\omega)-X(y,\omega)|^q P(d\omega)\right)^{\frac{1}{q}} : x,y \in [c,d] \subseteq [a,b], |x-y| \leq \delta\right\}, \tag{14.7}$$

$\delta > 0$, $1 \leq q < \infty$.

Definition 14.4 Let $1 \leq q < \infty$. Let $X(x,\omega)$ be a stochastic process. We call X a q-mean uniformly continuous stochastic process over $[a,b]$, iff $\forall\, \varepsilon > 0 \ \exists\, \delta > 0 :$ whenever $|x-y| \leq \delta$; $x,y \in [a,b]$ implies that

$$\int_{\Omega}|X(x,s)-X(y,s)|^q P(ds) \leq \varepsilon. \tag{14.8}$$

We denote it as $X \in C_{\mathbb{R}}^{U_q}([a,b])$.

It holds

Proposition 14.5 ([5]) *Let* $X \in C_{\mathbb{R}}^{U_q}([a,b])$, *then* $\Omega_1(X,\delta)_{L^q,[a,b]} < \infty$, *any* $\delta > 0$.

Also it holds

Proposition 14.6 ([5]) *Let* $X(t,\omega)$ *be a stochastic process from* $[a,b] \times (\Omega, \mathcal{F}, P)$ *into* $\mathbb{R}$. *Then following are true* ($[c,d] \subseteq [a,b]$)*:*

(i) $\Omega_1(X,\delta)_{L^q,[c,d]}$ *is nonnegative and nondecreasing in* $\delta > 0$,
(ii) $\lim_{\delta \downarrow 0} \Omega_1(X,\delta)_{L^q,[c,d]} = \Omega_1(X,0)_{L^q,[c,d]} = 0$, *iff* $X \in C_{\mathbb{R}}^{U_q}([c,d])$,
(iii) $\Omega_1(X,\delta_1+\delta_2)_{L^q,[c,d]} \le \Omega_1(X,\delta_1)_{L^q,[c,d]} + \Omega_1(X,\delta_2)_{L^q,[c,d]}$, $\delta_1,\delta_2 > 0$,
(iv) $\Omega_1(X,m\delta)_{L^q,[c,d]} \le m\Omega_1(X,\delta)_{L^q,[c,d]}$, $\delta > 0$, $m \in \mathbb{N}$,
(v) $\Omega_1(X,\lambda\delta)_{L^q,[c,d]} \le \lceil \lambda \rceil \Omega_1(X,\delta)_{L^q,[c,d]} \le (\lambda+1)\Omega_1(X,\delta)_{L^q,[c,d]}$, $\lambda > 0$, $\delta > 0$,
(vi) $\Omega_1(X+Y,\delta)_{L^q,[c,d]} \le \Omega_1(X,\delta)_{L^q,[c,d]} + \Omega_1(Y,\delta)_{L^q,[c,d]}$, $\delta > 0$,
(vii) $\Omega_1(X,\cdot)_{L^q,[c,d]}$ *is continuous on* $\mathbb{R}_+$ *for* $X \in C_{\mathbb{R}}^{U_q}([c,d])$.

We give

Remark 14.7 *(to Proposition* 14.6*)* By Proposition 14.6(v) we get

$$\Omega_1(X,|x-y|)_{L^q,[c,d]} \le \left\lceil \frac{|x-y|}{\delta} \right\rceil \Omega_1(X,\delta)_{L^q,[c,d]}, \tag{14.9}$$

$\forall\, x,y \in [c,d]$, any $\delta > 0$.

We give

Remark 14.8 *(continuation of Remark* 14.2*)* We assume again that $\left|X^{(n)}(t,\omega)\right| \le M^*$, $\forall\,(t,\omega) \in [a,b] \times \Omega$, where $M^* > 0$. Let $\delta > 0$, $1 \le q < \infty$. Then

$$\Omega_1\left(D_{*c}^{\alpha}X,\delta\right)_{L^q,[c,d]} = \sup\left\{\left(\int_{\Omega}\left|\left(D_{*c}^{\alpha}X\right)(x,\omega) - \left(D_{*c}^{\alpha}X\right)(y,\omega)\right|^q P(d\omega)\right)^{\frac{1}{q}} : \right.$$

$$\left. x,y \in [c,d] \subseteq [a,b], |x-y| \le \delta \right\} \le$$

$$\sup\left\{\left(\int_{\Omega}\left(\left|\left(D_{*c}^{\alpha}X\right)(x,\omega)\right| + \left|\left(D_{*c}^{\alpha}X\right)(y,\omega)\right|\right)^q P(d\omega)\right)^{\frac{1}{q}} : \right.$$

$$\left. x,y \in [c,d] \subseteq [a,b], |x-y| \le \delta \right\} \overset{([5])}{\le} \tag{14.10}$$

$$\frac{M^*}{\Gamma(n-\alpha+1)} \sup\left\{\left(\int_{\Omega}\left((x-c)^{n-\alpha} + (y-c)^{n-\alpha}\right)^q P(d\omega)\right)^{\frac{1}{q}} : \right.$$

$$x, y \in [c, d] \subseteq [a, b], |x - y| \leq \delta\} =$$

$$\frac{M^*}{\Gamma(n-\alpha+1)} \sup\left\{\left((x-c)^{n-\alpha} + (y-c)^{n-\alpha}\right) :\right.$$

$$x, y \in [c, d] \subseteq [a, b], |x - y| \leq \delta\} \leq \frac{2M^*}{\Gamma(n-\alpha+1)} (d-c)^{n-\alpha}.$$

That is

$$\Omega_1 \left(D_{*c}^{\alpha} X, \delta\right)_{L^q,[c,d]} \leq \frac{2M^*}{\Gamma(n-\alpha+1)} (d-c)^{n-\alpha}, \tag{14.11}$$

where $a \leq c < d \leq b$.

Similarly, it holds

$$\Omega_1 \left(D_{d-}^{\alpha} X, \delta\right)_{L^q,[c,d]} \leq \frac{2M^*}{\Gamma(n-\alpha+1)} (d-c)^{n-\alpha}, \tag{14.12}$$

where $a \leq c < d \leq b$.

Next let $x_0 \in [a, b]$, then

$$\sup_{x_0 \in [a,b]} \Omega_1 \left(D_{*x_0}^{\alpha} X, \delta\right)_{L^q,[x_0,b]} \leq \frac{2M^*}{\Gamma(n-\alpha+1)} \sup_{x_0 \in [a,b]} (b-x_0)^{n-\alpha}$$

$$= \frac{2M^*}{\Gamma(n-\alpha+1)} (b-a)^{n-\alpha}. \tag{14.13}$$

Similarly, it holds

$$\sup_{x_0 \in [a,b]} \Omega_1 \left(D_{x_0-}^{\alpha} X, \delta\right)_{L^q,[a,x_0]} \leq \frac{2M^*}{\Gamma(n-\alpha+1)} \sup_{x_0 \in [a,b]} (x_0-a)^{n-\alpha}$$

$$= \frac{2M^*}{\Gamma(n-\alpha+1)} (b-a)^{n-\alpha}. \tag{14.14}$$

We need

Concepts 14.9 *Let $\widetilde{L}$ be a positive linear operator from $C([a, b])$ into itself. Let $X(t, \omega)$ be a stochastic process from $[a, b] \times (\Omega, \mathcal{F}, P)$ into $\mathbb{R}$, where $(\Omega, \mathcal{F}, P)$ is a probability space. Here we assume that for non-integer $\alpha > 0$, $\lceil \alpha \rceil = n$, $X(\cdot, \omega) \in AC^n([a, b])$ with $X^{(n)}(\cdot, \omega) \in L_\infty([a, b])$, $\forall\, \omega \in \Omega$.*

Also we assume for each $t \in [a, b]$ that $X^{(k)}(t, \cdot)$ is measurable for all $k = 1, \ldots, n-1$. Further we assume that $D_{*t}^{\alpha} X(z, \omega)$ is a stochastic process for $z \in [t, b]$, $\omega \in \Omega$, and $D_{t-}^{\alpha} X(z, \omega)$ is a stochastic process for $z \in [a, t]$, $\omega \in \Omega$; $\forall\, t \in [a, b]$.

Define

$$M(X)(t,\omega) := \widetilde{L}(X(\cdot,\omega))(t), \ \forall \omega \in \Omega, \forall t \in [a,b], \tag{14.15}$$

and assume that it is a random variable in ω. *Clearly* M *is a positive linear operator on stochastic processes.*

We make

Assumption 14.10 Let non-integer $\alpha > 0$.

(i) For any $t \in [a,b]$ we assume that $D_{*t}^{\alpha}X(z,\omega)$ is continuous in $z \in [t,b]$, uniformly with respect to $\omega \in \Omega$. I.e. $\forall\ \varepsilon > 0\ \exists\ \delta > 0$: whenever $|z_1 - z_2| \leq \delta$; $z_1, z_2 \in [t,b]$, then $\left|D_{*t}^{\alpha}X(z_1,\omega) - D_{*t}^{\alpha}X(z_2,\omega)\right| \leq \varepsilon, \forall\ \omega \in \Omega$.
We denote this by $D_{*t}^{\alpha}X \in C_{\mathbb{R}}^{U}([t,b])$, the space of continuous in x, uniformly with respect to ω, stochastic processes over $[t,b]$.

(ii) For any $t \in [a,b]$ we assume that $D_{t-}^{\alpha}X(z,\omega)$ is continuous in $z \in [a,t]$, uniformly with respect to $\omega \in \Omega$. I.e. $\forall\ \varepsilon > 0\ \exists\ \delta > 0$: whenever $|z_1 - z_2| \leq \delta$; $z_1, z_2 \in [a,t]$, then $\left|D_{t-}^{\alpha}X(z_1,\omega) - D_{t-}^{\alpha}X(z_2,\omega)\right| \leq \varepsilon, \forall\ \omega \in \Omega$.
We denote this by $D_{t-}^{\alpha}X \in C_{\mathbb{R}}^{U}([a,t])$, the space of continuous in x, uniformly with respect to ω, stochastic processes over $[a,t]$.

Remark 14.11 Assumption 14.10 implies:

(i) $D_{*t}^{\alpha}X(\cdot,\omega) \in C([t,b]), \forall\ \omega \in \Omega$, and $D_{*t}^{\alpha}X$ is q-mean uniformly continuous in $z \in [t,b]$, that is $D_{*t}^{\alpha}X \in C_{\mathbb{R}}^{U_q}([t,b])$, for any $1 \leq q < \infty$.

(ii) $D_{t-}^{\alpha}X(\cdot,\omega) \in C([a,t]), \forall\ \omega \in \Omega$, and $D_{t-}^{\alpha}X$ is q-mean uniformly continuous in $z \in [a,t]$, that is $D_{t-}^{\alpha}X \in C_{\mathbb{R}}^{U_q}([a,t])$, for any $1 \leq q < \infty$.

We need

Definition 14.12 Denote by

$$(EX)(t) := \int_{\Omega} X(t,\omega)\, P(d\omega), \ \forall t \in [a,b], \tag{14.16}$$

the expectation operator.

We make

Assumption 14.13 We assume that

$$\left(E\left|X^{(k)}\right|^{q}\right)(t) < \infty, \ \forall t \in [a,b], \tag{14.17}$$

$q > 1$, for all $k = 0, 1, \ldots, n-1$, $n = \lceil \alpha \rceil$; $\alpha > 0$ non-integer.

We make

Assumption 14.14 We assume that

$$\left(E\left|X^{(k)}\right|\right)(t) < \infty, \ \forall t \in [a,b], \tag{14.18}$$

for all $k = 0, 1, \ldots, n-1$, $n = \lceil \alpha \rceil$; $\alpha > 0$ non-integer.

We give

Remark 14.15 By the Riesz representation theorem [13] we have that there exists μ_t unique, completed Borel measure on $[a,b]$ with

$$m_t := \mu_t([a,b]) = \widetilde{L}(1)(t) > 0, \tag{14.19}$$

such that

$$\widetilde{L}(f)(t) = \int_{[a,b]} f(x)\, d\mu_t(x), \ \forall t \in [a,b], \forall f \in C([a,b]). \tag{14.20}$$

Consequently we have that

$$M(X)(t,\omega) = \int_{[a,b]} X(x,\omega)\, d\mu_t(x), \ \forall (t,\omega) \in [a,b] \times \Omega, \tag{14.21}$$

and X as in Concepts 14.9.

Here $\chi_{[\gamma,\delta]}(s)$ stands for the characteristic function on $[\gamma,\delta] \subseteq [a,b]$.

Notice that $(r > 0)$

$$\int_{[t,b]} (s-t)^r \mu_t(ds) = \int_{[a,b]} \chi_{[t,b]}(s)\, |s-t|^r \mu_t(ds) = \widetilde{L}\left(|\cdot - t|^r \chi_{[t,b]}(\cdot)\right)(t), \tag{14.22}$$

and

$$\int_{[a,t]} (t-s)^r \mu_t(ds) = \int_{[a,b]} \chi_{[a,t]}(s)\, |s-t|^r \mu_t(ds) = \widetilde{L}\left(|\cdot - t|^r \chi_{[a,t]}(\cdot)\right)(t). \tag{14.23}$$

Let now $n = \lceil \alpha \rceil$, $\alpha \notin \mathbb{N}$, $\alpha > 0$, $k = 1, \ldots, n-1$. Then by Hölder's inequality we obtain

$$\left|\int_{[a,b]} (x-t)^k d\mu_t(x)\right| \leq \int_{[a,b]} |x-t|^k d\mu_t(x) \leq$$

$$\left(\int_{[a,b]} |x-t|^{\alpha+1} d\mu_t(x)\right)^{\left(\frac{k}{\alpha+1}\right)} (\mu_t([a,b]))^{\left(\frac{\alpha+1-k}{\alpha+1}\right)}. \tag{14.24}$$

Therefore it holds

$$\left\|\widetilde{L}\left((\cdot - t)^k\right)(t)\right\|_{\infty,[a,b]} \leq \left\|\widetilde{L}\left(|\cdot - t|^k\right)(t)\right\|_{\infty,[a,b]} \leq$$

$$\left\| \widetilde{L}\left(|\cdot - t|^{\alpha+1}\right)(t) \right\|_{\infty,[a,b]}^{\left(\frac{k}{\alpha+1}\right)} \left\| \widetilde{L}(1) \right\|_{\infty,[a,b]}^{\left(\frac{\alpha+1-k}{\alpha+1}\right)}, \tag{14.25}$$

all $k = 1, \ldots, n-1$.

Also, we observe that

$$C([a,b]) \ni |\cdot - t|^{\alpha+1} \chi_{[a,t]}(\cdot) \leq |\cdot - t|^{\alpha+1}, \ \forall t \in [a,b], \tag{14.26}$$

and

$$C([a,b]) \ni |\cdot - t|^{\alpha+1} \chi_{[t,b]}(\cdot) \leq |\cdot - t|^{\alpha+1}, \ \forall t \in [a,b]. \tag{14.27}$$

By positivity of $\widetilde{L}$ we obtain

$$\left\| \widetilde{L}\left(|\cdot - t|^{\alpha+1} \chi_{[a,t]}(\cdot)\right)(t) \right\|_{\infty,[a,b]} \leq \left\| \widetilde{L}\left(|\cdot - t|^{\alpha+1}\right)(t) \right\|_{\infty,[a,b]} < \infty, \tag{14.28}$$

by $\widetilde{L}\left(|\cdot - t|^{\alpha+1}\right)(t)$ being continuous in $t \in [a,b]$, see p. 388 of [4],
and

$$\left\| \widetilde{L}\left(|\cdot - t|^{\alpha+1} \chi_{[t,b]}(\cdot)\right)(t) \right\|_{\infty,[a,b]} \leq \left\| \widetilde{L}\left(|\cdot - t|^{\alpha+1}\right)(t) \right\|_{\infty,[a,b]}. \tag{14.29}$$

Take $a \leq s \leq t$, then $(t-s)^{\alpha+1} \leq (t-s)^{\alpha+1} \cdot 1 + 0$, and for $t \leq s \leq b$, we get $(s-t)^{\alpha+1} \leq 0 + (s-t)^{\alpha+1} \cdot 1$.

So we have

$$|\cdot - t|^{\alpha+1} \leq |\cdot - t|^{\alpha+1} \chi_{[a,t]}(\cdot) + |\cdot - t|^{\alpha+1} \chi_{[t,b]}(\cdot), \ \forall t \in [a,b]. \tag{14.30}$$

Thus, by positivity of $\widetilde{L}$ we obtain

$$\begin{aligned} \left\| \widetilde{L}\left(|\cdot - t|^{\alpha+1}\right)(t) \right\|_{\infty,[a,b]} &\leq \left\| \widetilde{L}\left(|\cdot - t|^{\alpha+1} \chi_{[a,t]}(\cdot)\right)(t) \right\|_{\infty,[a,b]} \\ &+ \left\| \widetilde{L}\left(|\cdot - t|^{\alpha+1} \chi_{[t,b]}(\cdot)\right)(t) \right\|_{\infty,[a,b]}. \end{aligned} \tag{14.31}$$

In this work we denote by $\widetilde{L}\left(\chi_{[a,t]}(\cdot)\right)(t) := \mu_t([a,t]) \leq \widetilde{L}(1)(t)$, and by $\widetilde{L}\left(\chi_{[t,b]}(\cdot)\right)(t) := \mu_t([t,b]) \leq \widetilde{L}(1)(t)$.

14.4 Main Results

Next we present our first main result on the quantitative stochastic fractional approximation regarding stochastic processes:

Theorem 14.16 *Suppose Concepts 14.9, Assumptions 14.10 and 14.13. Here the non-integer $\alpha > 0$ is such that $\alpha > \frac{1}{q}$, where $p, q > 1$ such that $\frac{1}{p} + \frac{1}{q} = 1$. Then, $\forall\, t \in [a,b]$, we have:*

$$\left(E\left(|M(X)-X|^q\right)(t)\right)^{\frac{1}{q}} \leq \left(\left(E|X|^q\right)(t)\right)^{\frac{1}{q}}\left|\widetilde{L}(1)(t)-1\right|+$$

$$\sum_{k=1}^{n-1}\frac{\left(\left(E\left|X^{(k)}\right|^q\right)(t)\right)^{\frac{1}{q}}}{k!}\left|\widetilde{L}\left((\cdot-t)^k\right)(t)\right|+\frac{2^{\frac{1}{p}}}{\Gamma(\alpha)(p(\alpha-1)+1)^{\frac{1}{p}}(q+1)^{\frac{\alpha}{q(\alpha+1)}}} \tag{14.32}$$

$$\left\{\left\{\left(\widetilde{L}\left(\chi_{[t,b]}(\cdot)\right)(t)\right)^{\frac{1}{p}}\left(\widetilde{L}\left(|\cdot-t|^{q(\alpha+1)}\chi_{[t,b]}(\cdot)\right)(t)\right)^{\frac{\alpha}{q(\alpha+1)}}\right.\right.$$

$$\left[\left(\widetilde{L}\left(\chi_{[t,b]}(\cdot)\right)(t)\right)^{\frac{1}{\alpha+1}}(q+1)^{\frac{\alpha}{(\alpha+1)}}+1\right]^{\frac{1}{q}}$$

$$\left.\Omega_1\left(D_{*t}^{\alpha}X,\left(\frac{1}{(q+1)}\widetilde{L}\left(|\cdot-t|^{q(\alpha+1)}\chi_{[t,b]}(\cdot)\right)(t)\right)^{\frac{1}{q(\alpha+1)}}\right)_{L^q,[t,b]}\right\}+$$

$$\left\{\left(\widetilde{L}\left(\chi_{[a,t]}(\cdot)\right)(t)\right)^{\frac{1}{p}}\left(\widetilde{L}\left(|\cdot-t|^{q(\alpha+1)}\chi_{[a,t]}(\cdot)\right)(t)\right)^{\frac{\alpha}{q(\alpha+1)}}\right.$$

$$\left[\left(\widetilde{L}\left(\chi_{[a,t]}(\cdot)\right)(t)\right)^{\frac{1}{\alpha+1}}(q+1)^{\frac{\alpha}{(\alpha+1)}}+1\right]^{\frac{1}{q}}$$

$$\left.\left.\Omega_1\left(D_{t-}^{\alpha}X,\left(\frac{1}{(q+1)}\widetilde{L}\left(|\cdot-t|^{q(\alpha+1)}\chi_{[a,t]}(\cdot)\right)(t)\right)^{\frac{1}{q(\alpha+1)}}\right)_{L^q,[a,t]}\right\}\right\}.$$

Proof Here non-integer $\alpha>0$, $\lceil\alpha\rceil=n$, $\alpha>\frac{1}{q}$, $p,q>1:\frac{1}{p}+\frac{1}{q}=1$, $X(\cdot,\omega)\in AC^n([a,b])$, with $X^{(n)}(\cdot,\omega)\in L_{\infty}([a,b])$, $\forall\,\omega\in\Omega$.

We get by left Caputo fractional Taylor's formula that [11, p. 54]

$$X(s,\omega)=\sum_{k=0}^{n-1}\frac{X^{(k)}(t,\omega)}{k!}(s-t)^k+$$

$$\frac{1}{\Gamma(\alpha)}\int_t^s(s-z)^{\alpha-1}\left(D_{*t}^{\alpha}X(z,\omega)-D_{*t}^{\alpha}X(t,\omega)\right)dz, \tag{14.33}$$

for all $t\leq s\leq b$, $\forall\,\omega\in\Omega$.

Also from [8, p. 341], using the right Caputo fractional Taylor formula we get

$$X(s,\omega)=\sum_{k=0}^{n-1}\frac{X^{(k)}(t,\omega)}{k!}(s-t)^k+$$

$$\frac{1}{\Gamma(\alpha)}\int_s^t(z-s)^{\alpha-1}\left(D_{t-}^{\alpha}X(z,\omega)-D_{t-}^{\alpha}X(t,\omega)\right)dz, \tag{14.34}$$

for all $a \leq s \leq t, \forall\, \omega \in \Omega$.

Above notice that

$$D_{*t}^{\alpha} X(t,\omega) = D_{t-}^{\alpha} X(t,\omega) = 0, \tag{14.35}$$

$\forall\, \omega \in \Omega$, see [8, pp.358–359].

We assume that

$$D_{*t}^{\alpha} X(s,\omega) = 0, \text{ for } s < t,$$

and

$$D_{t-}^{\alpha} X(s,\omega) = 0, \text{ for } s > t,$$

$\forall\, \omega \in \Omega$.

Then working on the remainder related to (14.33) ($t \leq s \leq b$), we have

$$\int_t^s (s-z)^{\alpha-1} \left| D_{*t}^{\alpha} X(z,\omega) - D_{*t}^{\alpha} X(t,\omega) \right| dz \leq$$

$$\left(\int_t^s \left| D_{*t}^{\alpha} X(z,\omega) - D_{*t}^{\alpha} X(t,\omega) \right|^q dz \right)^{\frac{1}{q}} \left(\int_t^s \left((s-z)^{\alpha-1} \right)^p dz \right)^{\frac{1}{p}} = \tag{14.36}$$

$$\left(\int_t^s \left| D_{*t}^{\alpha} X(z,\omega) - D_{*t}^{\alpha} X(t,\omega) \right|^q dz \right)^{\frac{1}{q}} \frac{(s-t)^{\frac{q\alpha-1}{q}}}{(p(\alpha-1)+1)^{\frac{1}{p}}}.$$

Hence we have

$$\left(\int_t^s (s-z)^{\alpha-1} \left| D_{*t}^{\alpha} X(z,\omega) - D_{*t}^{\alpha} X(t,\omega) \right| dz \right)^q \leq$$

$$\left(\int_t^s \left| D_{*t}^{\alpha} X(z,\omega) - D_{*t}^{\alpha} X(t,\omega) \right|^q dz \right) \frac{(s-t)^{(q\alpha-1)}}{(p(\alpha-1)+1)^{(q-1)}}, \tag{14.37}$$

$t \leq s \leq b$.

Applying again Hölder's inequality we obtain:

$$\Gamma_1 := \frac{1}{\Gamma(\alpha)}$$

$$\left(\int_{\Omega} \left\{ \int_t^b \left(\int_t^s (s-z)^{\alpha-1} \left| D_{*t}^{\alpha} X(z,\omega) - D_{*t}^{\alpha} X(t,\omega) \right| dz \right) \mu_t(ds) \right\}^q P(d\omega) \right)^{\frac{1}{q}}$$

$$\leq \frac{(\mu_t([t,b]))^{\frac{1}{p}}}{\Gamma(\alpha)}$$

$$\left(\int_\Omega \left\{\int_t^b \left(\int_t^s (s-z)^{\alpha-1} \left|D_{*t}^\alpha X(z,\omega) - D_{*t}^\alpha X(t,\omega)\right| dz\right)^q \mu_t(ds)\right\} P(d\omega)\right)^{\frac{1}{q}} \tag{14.38}$$

$$\overset{(14.37)}{\leq} \frac{(\mu_t([t,b]))^{\frac{1}{p}}}{\Gamma(\alpha)(p(\alpha-1)+1)^{\frac{1}{p}}}$$

$$\left(\int_\Omega \left\{\int_t^b \left[\left(\int_t^s \left|D_{*t}^\alpha X(z,\omega) - D_{*t}^\alpha X(t,\omega)\right|^q dz\right)(s-t)^{(q\alpha-1)}\right] \mu_t(ds)\right\} P(d\omega)\right)^{\frac{1}{q}}$$

$$=: (*),$$

where $\varphi(z,\omega) := \left|D_{*t}^\alpha X(z,\omega) - D_{*t}^\alpha X(t,\omega)\right|^q \geq 0$, is a real valued random variable for each $z \in [t,b]$ (we assumed that $D_{*t}^\alpha X(z,\omega)$, $t \leq z \leq b$, is continuous in z and measurable in ω). Hence $\varphi(z,\omega)$ is continuous in z and jointly measurable, see [7, p. 353]. Hence $\lambda(s,\omega) := \left(\int_t^s \varphi(z,\omega)\,dz\right)(s-t)^{q\alpha-1}$ is a real valued random variable in ω and continuous in $s \in [t,b]$. Again by [7, p. 353], $\lambda(s,\omega)$ is jointly measurable in (s,ω).

Therefore by applying twice Tonelli's theorem, [13, p. 270], we get

$$(*) = \frac{(\mu_t([t,b]))^{\frac{1}{p}}}{\Gamma(\alpha)(p(\alpha-1)+1)^{\frac{1}{p}}}$$

$$\left\{\int_t^b \left(\int_\Omega \left[\int_t^s \left|D_{*t}^\alpha X(z,\omega) - D_{*t}^\alpha X(t,\omega)\right|^q dz\right] P(d\omega)\right)(s-t)^{(q\alpha-1)} \mu_t(ds)\right\}^{\frac{1}{q}} =$$

$$\frac{(\mu_t([t,b]))^{\frac{1}{p}}}{\Gamma(\alpha)(p(\alpha-1)+1)^{\frac{1}{p}}}$$

$$\left\{\int_t^b \left(\int_t^s \left(\int_\Omega \left|D_{*t}^\alpha X(z,\omega) - D_{*t}^\alpha X(t,\omega)\right|^q P(d\omega)\right) dz\right)(s-t)^{(q\alpha-1)} \mu_t(ds)\right\}^{\frac{1}{q}}. \tag{14.39}$$

We have proved that

$$\Gamma_1 := \frac{1}{\Gamma(\alpha)}$$

$$\left(\int_\Omega \left\{\int_t^b \left(\int_t^s (s-z)^{\alpha-1} \left|D_{*t}^\alpha X(z,\omega) - D_{*t}^\alpha X(t,\omega)\right| dz\right) \mu_t(ds)\right\}^q P(d\omega)\right)^{\frac{1}{q}} \tag{14.40}$$

$$\leq \frac{(\mu_t([t,b]))^{\frac{1}{p}}}{\Gamma(\alpha)(p(\alpha-1)+1)^{\frac{1}{p}}}$$

$$\left\{\int_t^b \left(\int_t^s \left(\int_\Omega \left|D_{*t}^\alpha X(z,\omega) - D_{*t}^\alpha X(t,\omega)\right|^q P(d\omega)\right) dz\right) (s-t)^{(q\alpha-1)} \mu_t(ds)\right\}^{\frac{1}{q}}$$

(continuing estimation)

$$\leq \frac{(\mu_t([t,b]))^{\frac{1}{p}}}{\Gamma(\alpha)(p(\alpha-1)+1)^{\frac{1}{p}}}$$

$$\left\{\int_t^b \left(\int_t^s \Omega_1^q \left(D_{*t}^\alpha X, z-t\right)_{L^q,[t,b]} dz\right) (s-z)^{(q\alpha-1)} \mu_t(ds)\right\}^{\frac{1}{q}} \overset{\text{(by (14.9))}}{\leq}$$

(let $h_1 > 0$)

$$\frac{\Omega_1\left(D_{*t}^\alpha X, h_1\right)_{L^q,[t,b]} (\mu_t([t,b]))^{\frac{1}{p}}}{\Gamma(\alpha)(p(\alpha-1)+1)^{\frac{1}{p}}}$$

$$\left\{\int_t^b \left(\int_t^s \left\lceil \frac{z-t}{h_1} \right\rceil^q dz\right) (s-t)^{(q\alpha-1)} \mu_t(ds)\right\}^{\frac{1}{q}} \leq \tag{14.41}$$

$$\frac{\Omega_1\left(D_{*t}^\alpha X, h_1\right)_{L^q,[t,b]} (\mu_t([t,b]))^{\frac{1}{p}}}{\Gamma(\alpha)(p(\alpha-1)+1)^{\frac{1}{p}}}$$

$$\left\{\int_t^b \left(\int_t^s \left(1+\frac{z-t}{h_1}\right)^q dz\right) (s-t)^{(q\alpha-1)} \mu_t(ds)\right\}^{\frac{1}{q}} \leq \tag{14.42}$$

$$\tau\left\{\int_t^b \left(\int_t^s \left(1+\frac{(z-t)^q}{h_1^q}\right) dz\right) (s-t)^{(q\alpha-1)} \mu_t(ds)\right\}^{\frac{1}{q}} =$$

(where

$$\tau = \frac{2^{\frac{1}{p}}\Omega_1\left(D_{*t}^\alpha X, h_1\right)_{L^q,[t,b]} (\mu_t([t,b]))^{\frac{1}{p}}}{\Gamma(\alpha)(p(\alpha-1)+1)^{\frac{1}{p}}}$$

)

$$\tau\left\{\int_{[t,b]} \left[(s-t)^{q\alpha} + \frac{1}{h_1^q}\left(\int_t^s (z-t)^q dz\right)(s-t)^{(q\alpha-1)}\right] \mu_t(ds)\right\}^{\frac{1}{q}} =$$

$$\tau\left\{\int_{[t,b]} \left[(s-t)^{q\alpha} + \frac{1}{h_1^q}\frac{(s-t)^{q+1}}{(q+1)}(s-t)^{(q\alpha-1)}\right] \mu_t(ds)\right\}^{\frac{1}{q}} = \tag{14.43}$$

$$\tau\left\{\int_{[t,b]} \left[(s-t)^{q\alpha} + \frac{1}{h_1^q}\frac{(s-t)^{q(\alpha+1)}}{(q+1)}\right] \mu_t(ds)\right\}^{\frac{1}{q}} =$$

$$\tau\left\{\int_{[t,b]}(s-t)^{q\alpha}\mu_t(ds)+\frac{1}{h_1^q(q+1)}\int_{[t,b]}(s-t)^{q(\alpha+1)}\mu_t(ds)\right\}^{\frac{1}{q}}\leq$$

$$\tau\left\{\left(\int_{[t,b]}(s-t)^{q(\alpha+1)}\mu_t(ds)\right)^{\frac{\alpha}{\alpha+1}}(\mu_t([t,b]))^{\frac{1}{\alpha+1}}+\right.$$

$$\left.\frac{1}{h_1^q(q+1)}\int_{[t,b]}(s-t)^{q(\alpha+1)}\mu_t(ds)\right\}^{\frac{1}{q}}=:(**).\tag{14.44}$$

We set and assume

$$h_1:=\left(\frac{1}{(q+1)}\int_{[t,b]}(s-t)^{q(\alpha+1)}\mu_t(ds)\right)^{\frac{1}{q(\alpha+1)}}>0.\tag{14.45}$$

I.e.

$$h_1^{q(\alpha+1)}=\frac{1}{(q+1)}\int_{[t,b]}(s-t)^{q(\alpha+1)}\mu_t(ds)>0.\tag{14.46}$$

Therefore we get

$$(**)=\tau\left\{(\mu_t([t,b]))^{\frac{1}{\alpha+1}}h_1^{q\alpha}(q+1)^{\frac{\alpha}{\alpha+1}}+h_1^{q\alpha}\right\}^{\frac{1}{q}}=$$

$$\tau h_1^{\alpha}\left\{(\mu_t([t,b]))^{\frac{1}{\alpha+1}}(q+1)^{\frac{\alpha}{\alpha+1}}+1\right\}^{\frac{1}{q}}.\tag{14.47}$$

We have proved that

$$\Gamma_1\leq\tau h_1^{\alpha}\left\{(\mu_t([t,b]))^{\frac{1}{\alpha+1}}(q+1)^{\frac{\alpha}{\alpha+1}}+1\right\}^{\frac{1}{q}}.\tag{14.48}$$

The last means that

$$\Gamma_1:=\frac{1}{\Gamma(\alpha)}$$

$$\left(\int_{\Omega}\left\{\int_t^b\left(\int_t^s(s-z)^{\alpha-1}\left|D_{*t}^{\alpha}X(z,\omega)-D_{*t}^{\alpha}X(t,\omega)\right|dz\right)\mu_t(ds)\right\}^qP(d\omega)\right)^{\frac{1}{q}}$$

$$\leq 2^{\frac{1}{p}}\Omega_1\left(D_{*t}^{\alpha}X,\left(\frac{1}{(q+1)}\int_{[t,b]}(s-t)^{q(\alpha+1)}\mu_t(ds)\right)^{\frac{1}{q(\alpha+1)}}\right)_{L^q,[t,b]}$$

$$\left(\frac{(\mu_t([t,b]))^{\frac{1}{p}}}{\Gamma(\alpha)(p(\alpha-1)+1)^{\frac{1}{p}}}\right)\left(\frac{1}{(q+1)}\int_{[t,b]}(s-t)^{q(\alpha+1)}\mu_t(ds)\right)^{\frac{\alpha}{q(\alpha+1)}}\tag{14.49}$$

$$\left[(\mu_t ([t, b]))^{\frac{1}{\alpha+1}} (q+1)^{\frac{\alpha}{\alpha+1}} + 1 \right]^{\frac{1}{q}}.$$

Next we work on the remainder related to (14.34) ($a \le s \le t$), we have

$$\int_s^t (z-s)^{\alpha-1} \left| D_{t-}^{\alpha} X(z,\omega) - D_{t-}^{\alpha} X(t,\omega) \right| dz \le$$

$$\left(\int_s^t \left| D_{t-}^{\alpha} X(z,\omega) - D_{t-}^{\alpha} X(t,\omega) \right|^q dz \right)^{\frac{1}{q}} \left(\int_s^t (z-s)^{p(\alpha-1)} dz \right)^{\frac{1}{p}} =$$

$$\left(\int_s^t \left| D_{t-}^{\alpha} X(z,\omega) - D_{t-}^{\alpha} X(t,\omega) \right|^q dz \right)^{\frac{1}{q}} \frac{(t-s)^{\frac{q\alpha-1}{q}}}{(p(\alpha-1)+1)^{\frac{1}{p}}}. \tag{14.50}$$

Hence we have

$$\int_s^t (z-s)^{\alpha-1} \left| D_{t-}^{\alpha} X(z,\omega) - D_{t-}^{\alpha} X(t,\omega) \right| dz \le$$

$$\left(\int_s^t \left| D_{t-}^{\alpha} X(z,\omega) - D_{t-}^{\alpha} X(t,\omega) \right|^q dz \right)^{\frac{1}{q}} \frac{(t-s)^{\frac{(q\alpha-1)}{q}}}{(p(\alpha-1)+1)^{\frac{1}{p}}}, \tag{14.51}$$

$a \le s \le t$.

Applying again Hölder's inequality we obtain:

$$\Gamma_2 := \frac{1}{\Gamma(\alpha)}$$

$$\left(\int_\Omega \left\{ \int_a^t \left(\int_s^t (z-s)^{\alpha-1} \left| D_{t-}^{\alpha} X(z,\omega) - D_{t-}^{\alpha} X(t,\omega) \right| dz \right) \mu_t(ds) \right\}^q P(d\omega) \right)^{\frac{1}{q}}$$

$$\le \frac{(\mu_t([a,t]))^{\frac{1}{p}}}{\Gamma(\alpha)}$$

$$\left(\int_\Omega \left\{ \int_a^t \left(\int_s^t (z-s)^{\alpha-1} \left| D_{t-}^{\alpha} X(z,\omega) - D_{t-}^{\alpha} X(t,\omega) \right| dz \right)^q \mu_t(ds) \right\} P(d\omega) \right)^{\frac{1}{q}} \tag{14.52}$$

$$\overset{(14.51)}{\le} \frac{(\mu_t([a,t]))^{\frac{1}{p}}}{\Gamma(\alpha)(p(\alpha-1)+1)^{\frac{1}{p}}}$$

$$\left(\int_\Omega \left\{ \int_a^t \left[\left(\int_s^t \left| D_{t-}^{\alpha} X(z,\omega) - D_{t-}^{\alpha} X(t,\omega) \right|^q dz \right) (t-s)^{(q\alpha-1)} \right] \mu_t(ds) \right\} P(d\omega) \right)^{\frac{1}{q}}$$

(as before)

$$= \frac{(\mu_t([a,t]))^{\frac{1}{p}}}{\Gamma(\alpha)(p(\alpha-1)+1)^{\frac{1}{p}}}$$

$$\left\{\int_a^t \left(\int_\Omega \left[\int_s^t \left|D_{t-}^{\alpha} X(z,\omega) - D_{t-}^{\alpha} X(t,\omega)\right|^q dz\right] P(d\omega)\right)(t-s)^{(q\alpha-1)} \mu_t(ds)\right\}^{\frac{1}{q}}$$

$$= \frac{(\mu_t([a,t]))^{\frac{1}{p}}}{\Gamma(\alpha)(p(\alpha-1)+1)^{\frac{1}{p}}}$$

$$\left\{\int_a^t \left(\int_s^t \left(\int_\Omega \left|D_{t-}^{\alpha} X(z,\omega) - D_{t-}^{\alpha} X(t,\omega)\right|^q P(d\omega)\right) dz\right)(t-s)^{(q\alpha-1)} \mu_t(ds)\right\}^{\frac{1}{q}} \tag{14.53}$$

$$\leq \frac{(\mu_t([a,t]))^{\frac{1}{p}}}{\Gamma(\alpha)(p(\alpha-1)+1)^{\frac{1}{p}}}$$

$$\left\{\int_a^t \left(\int_s^t \Omega_1^q\left(D_{t-}^{\alpha} X, t-z\right)_{L^q,[a,t]} dz\right)(t-s)^{(q\alpha-1)} \mu_t(ds)\right\}^{\frac{1}{q}} \leq$$

(let $h_2 > 0$)

$$\Omega_1\left(D_{t-}^{\alpha} X, h_2\right)_{L^q,[a,t]} \frac{(\mu_t([a,t]))^{\frac{1}{p}}}{\Gamma(\alpha)(p(\alpha-1)+1)^{\frac{1}{p}}}$$

$$\left\{\int_a^t \left(\int_s^t \left\lceil \frac{t-z}{h_2} \right\rceil^q dz\right)(t-s)^{(q\alpha-1)} \mu_t(ds)\right\}^{\frac{1}{q}} \leq$$

$$\Omega_1\left(D_{t-}^{\alpha} X, h_2\right)_{L^q,[a,t]} \frac{(\mu_t([a,t]))^{\frac{1}{p}}}{\Gamma(\alpha)(p(\alpha-1)+1)^{\frac{1}{p}}} \tag{14.54}$$

$$\left\{\int_a^t \left(\int_s^t \left(1+\frac{t-z}{h_2}\right)^q dz\right)(t-s)^{(q\alpha-1)} \mu_t(ds)\right\}^{\frac{1}{q}} \leq$$

$$\rho\left\{\int_a^t \left(\int_s^t \left(1+\frac{(t-z)^q}{h_2^q}\right) dz\right)(t-s)^{(q\alpha-1)} \mu_t(ds)\right\}^{\frac{1}{q}} =$$

(where

$$\rho := 2^{\frac{1}{p}} \Omega_1\left(D_{t-}^{\alpha} X, h_2\right)_{L^q,[a,t]} \frac{(\mu_t([a,t]))^{\frac{1}{p}}}{\Gamma(\alpha)(p(\alpha-1)+1)^{\frac{1}{p}}}$$

)

$$\rho\left\{\int_{[a,t]}\left[(t-s)^{q\alpha}+\frac{1}{h_2^q}\left(\int_s^t(t-z)^q\,dz\right)(t-s)^{(q\alpha-1)}\right]\mu_t\,(ds)\right\}^{\frac{1}{q}}=$$

$$\rho\left\{\int_{[a,t]}\left[(t-s)^{q\alpha}+\frac{1}{h_2^q}\frac{(t-s)^{q+1}}{(q+1)}(t-s)^{(q\alpha-1)}\right]\mu_t\,(ds)\right\}^{\frac{1}{q}}=$$

$$\rho\left\{\int_{[a,t]}\left[(t-s)^{q\alpha}+\frac{1}{h_2^q}\frac{(t-s)^{q(\alpha+1)}}{(q+1)}\right]\mu_t\,(ds)\right\}^{\frac{1}{q}}= \tag{14.55}$$

$$\rho\left\{\int_{[a,t]}(t-s)^{q\alpha}\,\mu_t\,(ds)+\frac{1}{h_2^q\,(q+1)}\int_{[a,t]}(t-s)^{q(\alpha+1)}\,\mu_t\,(ds)\right\}^{\frac{1}{q}}\leq$$

$$\rho\left\{\left(\int_{[a,t]}(t-s)^{q(\alpha+1)}\,\mu_t\,(ds)\right)^{\frac{\alpha}{\alpha+1}}(\mu_t\,([a,t]))^{\frac{1}{\alpha+1}}+\right.$$

$$\left.\frac{1}{h_2^q\,(q+1)}\int_{[a,t]}(t-s)^{q(\alpha+1)}\,\mu_t\,(ds)\right\}^{\frac{1}{q}}=:(***)\,. \tag{14.56}$$

We set and assume

$$h_2:=\left(\frac{1}{(q+1)}\int_{[a,t]}(t-s)^{q(\alpha+1)}\,\mu_t\,(ds)\right)^{\frac{1}{q(\alpha+1)}}>0. \tag{14.57}$$

I.e.

$$h_2^{q(\alpha+1)}=\frac{1}{(q+1)}\int_{[a,t]}(t-s)^{q(\alpha+1)}\,\mu_t\,(ds)>0. \tag{14.58}$$

Therefore we get

$$(***)=\rho\left\{(\mu_t\,([a,t]))^{\frac{1}{\alpha+1}}\,h_2^{q\alpha}\,(q+1)^{\frac{\alpha}{\alpha+1}}+h_2^{q\alpha}\right\}^{\frac{1}{q}}=$$

$$\rho h_2^{\alpha}\left\{(\mu_t\,([a,t]))^{\frac{1}{\alpha+1}}\,(q+1)^{\frac{\alpha}{\alpha+1}}+1\right\}^{\frac{1}{q}}. \tag{14.59}$$

That is

$$\Gamma_2\leq\rho h_2^{\alpha}\left\{(\mu_t\,([a,t]))^{\frac{1}{\alpha+1}}\,(q+1)^{\frac{\alpha}{\alpha+1}}+1\right\}^{\frac{1}{q}}. \tag{14.60}$$

The last means that

$$\Gamma_2:=\frac{1}{\Gamma(\alpha)}$$

$$\left(\int_{\Omega}\left\{\int_{a}^{t}\left(\int_{s}^{t}(z-s)^{\alpha-1}\left|D_{t-}^{\alpha}X(z,\omega)-D_{t-}^{\alpha}X(t,\omega)\right|dz\right)\mu_t(ds)\right\}^{q}P(d\omega)\right)^{\frac{1}{q}}$$

$$\leq 2^{\frac{1}{p}}\Omega_1\left(D_{t-}^{\alpha}X,\left(\frac{1}{(q+1)}\int_{[a,t]}(t-s)^{q(\alpha+1)}\mu_t(ds)\right)^{\frac{1}{q(\alpha+1)}}\right)_{L^q,[a,t]}$$

$$\left(\frac{(\mu_t([a,t]))^{\frac{1}{p}}}{\Gamma(\alpha)(p(\alpha-1)+1)^{\frac{1}{p}}}\right)\left(\frac{1}{(q+1)}\int_{[a,t]}(t-s)^{q(\alpha+1)}\mu_t(ds)\right)^{\frac{\alpha}{q(\alpha+1)}} \quad (14.61)$$

$$\left[(\mu_t([a,t]))^{\frac{1}{\alpha+1}}(q+1)^{\frac{\alpha}{\alpha+1}}+1\right]^{\frac{1}{q}}.$$

Next we connect facts together.

We have that

$$M(X)(t,\omega)-X(t,\omega)\widetilde{L}(1)(t)=$$

$$\int_{[a,b]}X(s,\omega)\mu_t(ds)-X(t,\omega)\widetilde{L}(1)(t)=$$

$$\int_{[a,t]}X(s,\omega)\mu_t(ds)+\int_{(t,b]}X(s,\omega)\mu_t(ds)-X(t,\omega)\widetilde{L}(1)(t)\overset{\text{(by (14.33), (14.34))}}{=}$$

$$\int_{[a,t]}\left[\sum_{k=0}^{n-1}\frac{X^{(k)}(t,\omega)}{k!}(s-t)^k+\right.$$

$$\left.\frac{1}{\Gamma(\alpha)}\int_{s}^{t}(z-s)^{\alpha-1}\left(D_{t-}^{\alpha}X(z,\omega)-D_{t-}^{\alpha}X(t,\omega)\right)dz\right]\mu_t(ds)+ \quad (14.62)$$

$$\int_{(t,b]}\left[\sum_{k=0}^{n-1}\frac{X^{(k)}(t,\omega)}{k!}(s-t)^k+\right.$$

$$\left.\frac{1}{\Gamma(\alpha)}\int_{t}^{s}(s-z)^{\alpha-1}\left(D_{*t}^{\alpha}X(z,\omega)-D_{*t}^{\alpha}X(t,\omega)\right)dz\right]\mu_t(ds)$$

$$-X(t,\omega)\widetilde{L}(1)(t)=$$

$$\sum_{k=0}^{n-1}\frac{X^{(k)}(t,\omega)}{k!}\int_{[a,t]}(s-t)^k\mu_t(ds)+$$

$$\frac{1}{\Gamma(\alpha)}\int_{[a,t]}\left(\int_{s}^{t}(z-s)^{\alpha-1}\left(D_{t-}^{\alpha}X(z,\omega)-D_{t-}^{\alpha}X(t,\omega)\right)dz\right)\mu_t(ds)+$$

$$\sum_{k=0}^{n-1} \frac{X^{(k)}(t,\omega)}{k!} \int_{(t,b]} (s-t)^k \mu_t(ds) +$$

$$\frac{1}{\Gamma(\alpha)} \int_{(t,b]} \left(\int_t^s (s-z)^{\alpha-1} \left(D_{*t}^{\alpha} X(z,\omega) - D_{*t}^{\alpha} X(t,\omega) \right) dz \right) \mu_t(ds)$$

$$- X(t,\omega) \widetilde{L}(1)(t) = \tag{14.63}$$

$$\sum_{k=1}^{n-1} \frac{X^{(k)}(t,\omega)}{k!} \left(\widetilde{L}\left((\cdot - t)^k \right)(t) \right) +$$

$$\frac{1}{\Gamma(\alpha)} \left[\int_{[a,t]} \left(\int_s^t (z-s)^{\alpha-1} \left(D_{t-}^{\alpha} X(z,\omega) - D_{t-}^{\alpha} X(t,\omega) \right) dz \right) \mu_t(ds) + \right.$$

$$\left. \int_{(t,b]} \left(\int_t^s (s-z)^{\alpha-1} \left(D_{*t}^{\alpha} X(z,\omega) - D_{*t}^{\alpha} X(t,\omega) \right) dz \right) \mu_t(ds) \right].$$

Furthermore we have

$$\left| M(X)(t,\omega) - X(t,\omega) \widetilde{L}(1)(t) \right| \leq$$

$$\sum_{k=1}^{n-1} \frac{\left| X^{(k)}(t,\omega) \right|}{k!} \left| \widetilde{L}\left((\cdot - t)^k \right)(t) \right| +$$

$$\frac{1}{\Gamma(\alpha)} \left[\int_{[a,t]} \left(\int_s^t (z-s)^{\alpha-1} \left| D_{t-}^{\alpha} X(z,\omega) - D_{t-}^{\alpha} X(t,\omega) \right| dz \right) \mu_t(ds) + \right.$$

$$\left. \int_{[t,b]} \left(\int_t^s (s-z)^{\alpha-1} \left| D_{*t}^{\alpha} X(z,\omega) - D_{*t}^{\alpha} X(t,\omega) \right| dz \right) \mu_t(ds) \right]. \tag{14.64}$$

We have eventually that

$$|M(X)(t,\omega) - X(t,\omega)| \leq$$

$$\left| M(X)(t,\omega) - X(t,\omega) \widetilde{L}(1)(t) \right| + |X(t,\omega)| \left| \widetilde{L}(1)(t) - 1 \right| \leq \tag{14.65}$$

$$|X(t,\omega)| \left| \widetilde{L}(1)(t) - 1 \right| + \sum_{k=1}^{n-1} \frac{\left| X^{(k)}(t,\omega) \right|}{k!} \left| \widetilde{L}\left((\cdot - t)^k \right)(t) \right| +$$

$$\frac{1}{\Gamma(\alpha)} \left[\int_{[a,t]} \left(\int_s^t (z-s)^{\alpha-1} \left| D_{t-}^{\alpha} X(z,\omega) - D_{t-}^{\alpha} X(t,\omega) \right| dz \right) \mu_t(ds) + \right.$$

$$\int_{[t,b]}\left(\int_{t}^{s}(s-z)^{\alpha-1}\left|D_{*t}^{\alpha}X(z,\omega)-D_{*t}^{\alpha}X(t,\omega)\right|dz\right)\mu_{t}(ds)\Bigg],$$

$\forall\, t\in[a,b]$.

Hence it holds

$$\left(E\left(|M(X)-X|^{q}\right)(t)\right)^{\frac{1}{q}}\leq\left(\left(E|X|^{q}\right)(t)\right)^{\frac{1}{q}}\left|\widetilde{L}(1)(t)-1\right|+$$

$$\sum_{k=1}^{n-1}\frac{\left(\left(E\left|X^{(k)}\right|^{q}\right)(t)\right)^{\frac{1}{q}}}{k!}\left|\widetilde{L}\left((\cdot-t)^{k}\right)(t)\right|+$$

$$\frac{1}{\Gamma(\alpha)}\Bigg\{\left[\int_{\Omega}\left(\int_{[a,t]}\left(\int_{s}^{t}(z-s)^{\alpha-1}\left|D_{t-}^{\alpha}X(z,\omega)-D_{t-}^{\alpha}X(t,\omega)\right|dz\right)\mu_{t}(ds)\right)^{q}P(d\omega)\right]^{\frac{1}{q}}+ \tag{14.66}$$

$$\left[\int_{\Omega}\left(\int_{[t,b]}\left(\int_{t}^{s}(s-z)^{\alpha-1}\left|D_{*t}^{\alpha}X(z,\omega)-D_{*t}^{\alpha}X(t,\omega)\right|dz\right)\mu_{t}(ds)\right)^{q}P(d\omega)\right]^{\frac{1}{q}}\Bigg\}.$$

By (14.49) and (14.61) we obtain

$$\left(E\left(|M(X)-X|^{q}\right)(t)\right)^{\frac{1}{q}}\leq\left(\left(E|X|^{q}\right)(t)\right)^{\frac{1}{q}}\left|\widetilde{L}(1)(t)-1\right|+$$

$$\sum_{k=1}^{n-1}\frac{\left(\left(E\left|X^{(k)}\right|^{q}\right)(t)\right)^{\frac{1}{q}}}{k!}\left|\widetilde{L}\left((\cdot-t)^{k}\right)(t)\right|+$$

$$\Bigg\{2^{\frac{1}{p}}\Omega_{1}\left(D_{*t}^{\alpha}X,\left(\frac{1}{(q+1)}\int_{[t,b]}(s-t)^{q(\alpha+1)}\mu_{t}(ds)\right)^{\frac{1}{q(\alpha+1)}}\right)_{L^{q},[t,b]}$$

$$\left(\frac{\left(\mu_{t}([t,b])\right)^{\frac{1}{p}}}{\Gamma(\alpha)\left(p(\alpha-1)+1\right)^{\frac{1}{p}}}\right)\left(\frac{1}{(q+1)}\int_{[t,b]}(s-t)^{q(\alpha+1)}\mu_{t}(ds)\right)^{\frac{\alpha}{q(\alpha+1)}} \tag{14.67}$$

$$\left[\left(\mu_{t}([t,b])\right)^{\frac{1}{\alpha+1}}(q+1)^{\frac{\alpha}{\alpha+1}}+1\right]^{\frac{1}{q}}\Bigg\}+$$

$$\Bigg\{2^{\frac{1}{p}}\Omega_{1}\left(D_{t-}^{\alpha}X,\left(\frac{1}{(q+1)}\int_{[a,t]}(t-s)^{q(\alpha+1)}\mu_{t}(ds)\right)^{\frac{1}{q(\alpha+1)}}\right)_{L^{q},[a,t]}$$

$$\left(\frac{\left(\mu_{t}([a,t])\right)^{\frac{1}{p}}}{\Gamma(\alpha)\left(p(\alpha-1)+1\right)^{\frac{1}{p}}}\right)\left(\frac{1}{(q+1)}\int_{[a,t]}(t-s)^{q(\alpha+1)}\mu_{t}(ds)\right)^{\frac{\alpha}{q(\alpha+1)}}$$

$$\left[(\mu_t([a,t]))^{\frac{1}{\alpha+1}} (q+1)^{\frac{\alpha}{\alpha+1}} + 1 \right]^{\frac{1}{q}} \Bigg\}.$$

Clearly (14.67) is trivially valid when $t = a$ or $t = b$, from what follows, see (14.69) and (14.73), just replace $[t, b]$ by $[a, b]$ or $[a, t]$ by $[a, b]$, respectively.

Next assume that

$$\int_{[t,b]} (s-t)^{q(\alpha+1)} \mu_t(ds) = 0, \tag{14.68}$$

then $(s-t)^{q(\alpha+1)} = 0$, a.e. on $[t, b]$, thus $s = t$, a.e. on $[t, b]$, i.e. $\mu_t \{s \in [t, b] : s \neq t\} = 0$, hence $\mu_t(t, b] = 0$.

Therefore μ_t concentrates on $[a, t]$. In that case inequality (14.67) becomes

$$\left(E\left(|M(X) - X|^q\right)(t)\right)^{\frac{1}{q}} \leq \left(\left(E|X|^q\right)(t)\right)^{\frac{1}{q}} \left|\widetilde{L}(1)(t) - 1\right| +$$

$$\sum_{k=1}^{n-1} \frac{\left(\left(E\left|X^{(k)}\right|^q\right)(t)\right)^{\frac{1}{q}}}{k!} \left|\widetilde{L}\left((\cdot - t)^k\right)(t)\right| +$$

$$2^{\frac{1}{p}} \Omega_1 \left(D_{t-}^{\alpha} X, \left(\frac{1}{(q+1)} \int_{[a,t]} (t-s)^{q(\alpha+1)} \mu_t(ds) \right)^{\frac{1}{q(\alpha+1)}} \right)_{L^q,[a,t]} \tag{14.69}$$

$$\left(\frac{(\mu_t([a,t]))^{\frac{1}{p}}}{\Gamma(\alpha)(p(\alpha-1)+1)^{\frac{1}{p}}} \right) \left(\frac{1}{(q+1)} \int_{[a,t]} (t-s)^{q(\alpha+1)} \mu_t(ds) \right)^{\frac{\alpha}{q(\alpha+1)}}$$

$$\left[(\mu_t([a,t]))^{\frac{1}{\alpha+1}} (q+1)^{\frac{\alpha}{\alpha+1}} + 1 \right]^{\frac{1}{q}}.$$

So far inequality (14.69) is valid when

$$\int_{[a,t]} (t-s)^{q(\alpha+1)} \mu_t(ds) > 0.$$

In the case of $\int_{[a,t]} (t-s)^{q(\alpha+1)} \mu_t(ds) = 0$, we have $(t-s)^{q(\alpha+1)} = 0$, a.e. on $[a, t]$, thus $s = t$, a.e. on $[a, t]$, i.e. $\mu_t \{s \in [a, t] : s \neq t\} = 0$, hence $\mu_t[a, t) = 0$.

Therefore $\mu_t = \delta_t \mu_t([a, b]) = \delta_t m_t$, where δ_t is the unit Dirac measure at t. Clearly then $\widetilde{L}\left((\cdot - t)^k\right)(t) = 0$, all $k = 1, \ldots, n-1$. Then (14.69) collapses to

$$\left(E\left(|M(X) - X|^q\right)(t)\right)^{\frac{1}{q}} \leq \left(\left(E|X|^q\right)(t)\right)^{\frac{1}{q}} \left|\widetilde{L}(1)(t) - 1\right|, \tag{14.70}$$

equivalently,

$$E\left(|M(X) - X|^q\right)(t) \leq \left(E|X|^q\right)(t) \left|\widetilde{L}(1)(t) - 1\right|^q. \tag{14.71}$$

By (14.21) we have that $M(X)(t,\omega) = X(t,\omega)\widetilde{L}(1)(t)$. Therefore $M(X)(t,\omega) - X(t,\omega) = X(t,\omega)\left(\widetilde{L}(1)(t) - 1\right)$.

Hence, obviously, (14.71) holds as equality. Thus inequality (14.69) is valid trivially.

Finally, let us go the other way around. Let us assume that

$$\int_{[a,t]} (t-s)^{q(\alpha+1)} \mu_t(ds) = 0, \tag{14.72}$$

then $(t-s)^{q(\alpha+1)} = 0$, a.e. on $[a,t]$, thus $s = t$, a.e. on $[a,t]$, i.e. $\mu_t \{s \in [a,t] : s \neq t\} = 0$, hence $\mu_t[a,t) = 0$. Therefore μ_t concentrates on $[t,b]$. In that case inequality (14.67) becomes

$$\left(E\left(|M(X) - X|^q\right)(t)\right)^{\frac{1}{q}} \leq \left(\left(E|X|^q\right)(t)\right)^{\frac{1}{q}} \left|\widetilde{L}(1)(t) - 1\right| +$$

$$\sum_{k=1}^{n-1} \frac{\left(\left(E\left|X^{(k)}\right|^q\right)(t)\right)^{\frac{1}{q}}}{k!} \left|\widetilde{L}\left((\cdot - t)^k\right)(t)\right| +$$

$$2^{\frac{1}{p}} \Omega_1 \left(D^{\alpha}_{*t} X, \left(\frac{1}{(q+1)} \int_{[t,b]} (s-t)^{q(\alpha+1)} \mu_t(ds) \right)^{\frac{1}{q(\alpha+1)}} \right)_{L^q,[t,b]}$$

$$\left(\frac{(\mu_t([t,b]))^{\frac{1}{p}}}{\Gamma(\alpha)(p(\alpha-1)+1)^{\frac{1}{p}}} \right) \left(\frac{1}{(q+1)} \int_{[t,b]} (s-t)^{q(\alpha+1)} \mu_t(ds) \right)^{\frac{\alpha}{q(\alpha+1)}} \tag{14.73}$$

$$\left[(\mu_t([t,b]))^{\frac{1}{\alpha+1}} (q+1)^{\frac{\alpha}{\alpha+1}} + 1 \right]^{\frac{1}{q}}.$$

So inequality (14.73) is valid when

$$\int_{[t,b]} (s-t)^{q(\alpha+1)} \mu_t(ds) > 0.$$

In the case of $\int_{[t,b]} (s-t)^{q(\alpha+1)} \mu_t(ds) = 0$, we have $(s-t)^{q(\alpha+1)} = 0$, a.e. on $[t,b]$, hence $s = t$, a.e. on $[t,b]$. Therefore $\mu_t \{s \in [t,b] : s \neq t\} = 0$, thus $\mu_t(t,b] = 0$.

Consequently $\mu_t = \delta_t \mu_t([a,b]) = \delta_t m_t$. Clearly then $\widetilde{L}\left((\cdot - t)^k\right)(t) = 0$, all $k = 1, \ldots, n-1$. Therefore (14.73) collapses to

$$\left(E\left(|M(X) - X|^q\right)(t)\right)^{\frac{1}{q}} \leq \left(\left(E|X|^q\right)(t)\right)^{\frac{1}{q}} \left|\widetilde{L}(1)(t) - 1\right|. \tag{14.74}$$

The last is an equality as earlier, see (14.71), etc.

Thus inequality (14.73) is valid trivially. Inequality (14.67) has been proved in all possible cases. Inequality (14.32) is the equivalent of (14.67) via (14.20).

Theorem 14.16 now is proved completely. ■

It follows our second main results, the L_1-quantitative stochastic fractional approximation of stochastic processes, it is the $q = 1$ analog of Theorem 14.16. Inequality (14.75) is much simpler than (14.32).

Theorem 14.17 *Suppose Concepts 14.9, Assumptions 14.10 and 14.14. Here* $\alpha > 0$, $\alpha \notin \mathbb{N}$, $n = \lceil \alpha \rceil$. *Then,* $\forall\, t \in [a, b]$, *we have*

$$E\left(|M(X) - X|\right)(t) \le \left(E|X|\right)(t)\left|\widetilde{L}(1) - 1\right| +$$

$$\sum_{k=1}^{n-1} \frac{\left(E\left|X^{(k)}\right|\right)(t)}{k!}\left|\widetilde{L}\left((\cdot - t)^k\right)(t)\right| + \frac{1}{\Gamma(\alpha+1)}$$

$$\left\{\left[\left(\widetilde{L}\left(\chi_{[a,t]}(\cdot)\right)(t)\right)^{\frac{1}{\alpha+1}} + \frac{1}{(\alpha+1)}\right]\left(\widetilde{L}\left(|\cdot - t|^{\alpha+1}\chi_{[a,t]}(\cdot)\right)(t)\right)^{\frac{\alpha}{\alpha+1}}\right.$$

$$\Omega_1\left(D_{t-}^{\alpha}X, \left(\widetilde{L}\left(|\cdot - t|^{\alpha+1}\chi_{[a,t]}(\cdot)\right)(t)\right)^{\frac{1}{\alpha+1}}\right)_{L_1,[a,t]} +$$

$$\left[\left(\widetilde{L}\left(\chi_{[t,b]}(\cdot)\right)(t)\right)^{\frac{1}{\alpha+1}} + \frac{1}{(\alpha+1)}\right]\left(\widetilde{L}\left(|\cdot - t|^{\alpha+1}\chi_{[t,b]}(\cdot)\right)(t)\right)^{\frac{\alpha}{\alpha+1}} \tag{14.75}$$

$$\left.\Omega_1\left(D_{*t}^{\alpha}X, \left(\widetilde{L}\left(|\cdot - t|^{\alpha+1}\chi_{[t,b]}(\cdot)\right)(t)\right)^{\frac{1}{\alpha+1}}\right)_{L_1,[t,b]}\right\}.$$

Proof We have that

$$E\left(|M(X) - X|\right)(t) \le \left(E|X|\right)(t)\left|\widetilde{L}(1)(t) - 1\right| +$$

$$\sum_{k=1}^{n-1} \frac{\left(E\left|X^{(k)}\right|\right)(t)}{k!}\left|\widetilde{L}\left((\cdot - t)^k\right)(t)\right| +$$

$$\frac{1}{\Gamma(\alpha)}\left\{\left[\int_{\Omega}\left(\int_a^t\left(\int_s^t (z-s)^{\alpha-1}\left|D_{t-}^{\alpha}X(z,\omega) - D_{t-}^{\alpha}X(t,\omega)\right|dz\right)\mu_t(ds)\right)P(d\omega)\right] + \right. \tag{14.76}$$

$$\left.\left[\int_{\Omega}\left(\int_t^b\left(\int_t^s (s-z)^{\alpha-1}\left|D_{*t}^{\alpha}X(z,\omega) - D_{*t}^{\alpha}X(t,\omega)\right|dz\right)\mu_t(ds)\right)P(d\omega)\right]\right\} =$$

(by applying Tonelli's theorem twice)

$$\left(E|X|\right)(t)\left|\widetilde{L}(1)(t) - 1\right| + \sum_{k=1}^{n-1} \frac{\left(E\left|X^{(k)}\right|\right)(t)}{k!}\left|\widetilde{L}\left((\cdot - t)^k\right)(t)\right| +$$

$$\frac{1}{\Gamma(\alpha)}\left\{\left[\int_a^t\left(\int_s^t\left(\int_\Omega\left|D_{t-}^\alpha X(z,\omega)-D_{t-}^\alpha X(t,\omega)\right|P(d\omega)\right)(z-s)^{\alpha-1}dz\right)\mu_t(ds)\right]+\right.$$

$$\left.\left[\int_t^b\left(\int_t^s\left(\int_\Omega\left|D_{*t}^\alpha X(z,\omega)-D_{*t}^\alpha X(t,\omega)\right|P(d\omega)\right)(s-z)^{\alpha-1}dz\right)\mu_t(ds)\right]\right\}\leq$$

$$(E|X|)(t)\left|\widetilde{L}(1)(t)-1\right|+\sum_{k=1}^{n-1}\frac{\left(E\left|X^{(k)}\right|\right)(t)}{k!}\left|\widetilde{L}\left((\cdot-t)^k\right)(t)\right|+\tag{14.77}$$

$$\frac{1}{\Gamma(\alpha)}\left\{\left[\int_a^t\left(\int_s^t\Omega_1\left(D_{t-}^\alpha X,t-z\right)_{L^1,[a,t]}(z-s)^{\alpha-1}dz\right)\mu_t(ds)\right]+\right.$$

$$\left.\left[\int_t^b\left(\int_t^s\Omega_1\left(D_{*t}^\alpha X,z-t\right)_{L^1,[t,b]}(s-z)^{\alpha-1}dz\right)\mu_t(ds)\right]\right\}\underset{(h_1,h_2>0)}{\leq}$$

$$(E|X|)(t)\left|\widetilde{L}(1)(t)-1\right|+\sum_{k=1}^{n-1}\frac{\left(E\left|X^{(k)}\right|\right)(t)}{k!}\left|\widetilde{L}\left((\cdot-t)^k\right)(t)\right|+$$

$$\frac{1}{\Gamma(\alpha)}\left\{\Omega_1\left(D_{t-}^\alpha X,h_1\right)_{L^1,[a,t]}\left[\int_a^t\left(\int_s^t\left(1+\frac{t-z}{h_1}\right)(z-s)^{\alpha-1}dz\right)\mu_t(ds)\right]+\right.\tag{14.78}$$

$$\left.\Omega_1\left(D_{*t}^\alpha X,h_2\right)_{L^1,[t,b]}\left[\int_t^b\left(\int_t^s\left(1+\frac{z-t}{h_2}\right)(s-z)^{\alpha-1}dz\right)\mu_t(ds)\right]\right\}=$$

(set $\lambda:=(E|X|)(t)\left|\widetilde{L}(1)(t)-1\right|+\sum_{k=1}^{n-1}\frac{(E|X^{(k)}|)(t)}{k!}\left|\widetilde{L}\left((\cdot-t)^k\right)(t)\right|$)

$$\lambda+\frac{1}{\Gamma(\alpha)}\left\{\Omega_1\left(D_{t-}^\alpha X,h_1\right)_{L^1,[a,t]}\right.$$

$$\left[\int_a^t\left(\frac{(t-s)^\alpha}{\alpha}+\frac{1}{h_1}\int_s^t(t-z)^{2-1}(z-s)^{\alpha-1}dz\right)\mu_t(ds)\right]+$$

$$\left.\Omega_1\left(D_{*t}^\alpha X,h_2\right)_{L^1,[t,b]}\left[\int_t^b\left(\frac{(s-t)^\alpha}{\alpha}+\frac{1}{h_2}\int_t^s(s-z)^{\alpha-1}(z-t)^{2-1}dz\right)\mu_t(ds)\right]\right\}\tag{14.79}$$

$$=\lambda+\frac{1}{\Gamma(\alpha)}\left\{\Omega_1\left(D_{t-}^\alpha X,h_1\right)_{L^1,[a,t]}\right.$$

$$\left[\int_a^t\left(\frac{(t-s)^\alpha}{\alpha}+\frac{1}{h_1}\frac{\Gamma(2)\Gamma(\alpha)}{\Gamma(\alpha+2)}(t-s)^{\alpha+1}\right)\mu_t(ds)\right]+$$

$$\Omega_1\left(D_{*t}^{\alpha}X,h_2\right)_{L^1,[t,b]}\left[\int_t^b\left(\frac{(s-t)^{\alpha}}{\alpha}+\frac{1}{h_2}\frac{\Gamma(\alpha)\Gamma(2)}{\Gamma(\alpha+2)}(s-t)^{\alpha+1}\right)\mu_t(ds)\right]\Bigg\}=$$

$$\lambda+\frac{1}{\Gamma(\alpha+1)}\left\{\Omega_1\left(D_{t-}^{\alpha}X,h_1\right)_{L^1,[a,t]}\left[\int_a^t\left((t-s)^{\alpha}+\frac{(t-s)^{\alpha+1}}{h_1(\alpha+1)}\right)\mu_t(ds)\right]+\right. \tag{14.80}$$

$$\Omega_1\left(D_{*t}^{\alpha}X,h_2\right)_{L^1,[t,b]}\left[\int_t^b\left((s-t)^{\alpha}+\frac{(s-t)^{\alpha+1}}{h_2(\alpha+1)}\right)\mu_t(ds)\right]\Bigg\}=$$

$$\lambda+\frac{1}{\Gamma(\alpha+1)}\left\{\Omega_1\left(D_{t-}^{\alpha}X,h_1\right)_{L^1,[a,t]}\right.$$

$$\left[\int_a^t(t-s)^{\alpha}\mu_t(ds)+\frac{1}{h_1(\alpha+1)}\int_a^t(t-s)^{\alpha+1}\mu_t(ds)\right]+$$

$$\Omega_1\left(D_{*t}^{\alpha}X,h_2\right)_{L^1,[t,b]}\left[\int_t^b(s-t)^{\alpha}\mu_t(ds)+\frac{1}{h_2(\alpha+1)}\int_t^b(s-t)^{\alpha+1}\mu_t(ds)\right]\Bigg\}=:(\xi). \tag{14.81}$$

(Choose

$$h_1=\left(\int_a^t(t-s)^{\alpha+1}\mu_t(ds)\right)^{\frac{1}{\alpha+1}}>0,$$

and

$$h_2=\left(\int_t^b(s-t)^{\alpha+1}\mu_t(ds)\right)^{\frac{1}{\alpha+1}}>0.$$

Then

$$h_1^{\alpha+1}=\int_a^t(t-s)^{\alpha+1}\mu_t(ds)>0,$$

and

$$h_2^{\alpha+1}=\int_t^b(s-t)^{\alpha+1}\mu_t(ds)>0.$$

)

Hence it holds

$$(\xi)=\lambda+\frac{1}{\Gamma(\alpha+1)}\left\{\Omega_1\left(D_{t-}^{\alpha}X,\left(\int_a^t(t-s)^{\alpha+1}\mu_t(ds)\right)^{\frac{1}{\alpha+1}}\right)_{L^1,[a,t]}\right.$$

$$\left[(\mu_t([a,t]))^{\frac{1}{\alpha+1}}\left(\int_a^t(t-s)^{\alpha+1}\mu_t(ds)\right)^{\frac{\alpha}{\alpha+1}}+\frac{h_1^{\alpha}}{(\alpha+1)}\right]+$$

$$\Omega_1\left(D^{\alpha}_{*t}X, \left(\int_t^b (s-t)^{\alpha+1}\mu_t(ds)\right)^{\frac{1}{\alpha+1}}\right)_{L^1,[t,b]} \tag{14.82}$$

$$\left[(\mu_t([t,b]))^{\frac{1}{\alpha+1}}\left(\int_t^b (s-t)^{\alpha+1}\mu_t(ds)\right)^{\frac{\alpha}{\alpha+1}} + \frac{h_2^{\alpha}}{(\alpha+1)}\right]\Bigg\} =$$

$$\lambda + \frac{1}{\Gamma(\alpha+1)}\Bigg\{\Omega_1\left(D^{\alpha}_{t-}X, \left(\int_a^t (t-s)^{\alpha+1}\mu_t(ds)\right)^{\frac{1}{\alpha+1}}\right)_{L^1,[a,t]}$$

$$\left(\int_a^t (t-s)^{\alpha+1}\mu_t(ds)\right)^{\frac{\alpha}{\alpha+1}}\left[(\mu_t([a,t]))^{\frac{1}{\alpha+1}} + \frac{1}{(\alpha+1)}\right] +$$

$$\Omega_1\left(D^{\alpha}_{*t}X, \left(\int_t^b (s-t)^{\alpha+1}\mu_t(ds)\right)^{\frac{1}{\alpha+1}}\right)_{L^1,[t,b]} \tag{14.83}$$

$$\left(\int_t^b (s-t)^{\alpha+1}\mu_t(ds)\right)^{\frac{\alpha}{\alpha+1}}\left[(\mu_t([t,b]))^{\frac{1}{\alpha+1}} + \frac{1}{(\alpha+1)}\right]\Bigg\}.$$

We have proved that

$$E(|M(X)-X|)(t) \le (E|X|)(t)\left|\widetilde{L}(1)(t)-1\right| +$$

$$\sum_{k=1}^{n-1}\frac{\left(E\left|X^{(k)}\right|\right)(t)}{k!}\left|\widetilde{L}\left((\cdot-t)^k\right)(t)\right| +$$

$$\frac{1}{\Gamma(\alpha+1)}\Bigg\{\left[(\mu_t([a,t]))^{\frac{1}{\alpha+1}} + \frac{1}{(\alpha+1)}\right]$$

$$\left(\int_a^t (t-s)^{\alpha+1}\mu_t(ds)\right)^{\frac{\alpha}{\alpha+1}}\Omega_1\left(D^{\alpha}_{t-}X, \left(\int_a^t (t-s)^{\alpha+1}\mu_t(ds)\right)^{\frac{1}{\alpha+1}}\right)_{L^1,[a,t]} + \tag{14.84}$$

$$\left[(\mu_t([t,b]))^{\frac{1}{\alpha+1}} + \frac{1}{(\alpha+1)}\right]$$

$$\left(\int_t^b (s-t)^{\alpha+1}\mu_t(ds)\right)^{\frac{\alpha}{\alpha+1}}\Omega_1\left(D^{\alpha}_{*t}X, \left(\int_t^b (s-t)^{\alpha+1}\mu_t(ds)\right)^{\frac{1}{\alpha+1}}\right)_{L^1,[t,b]}\Bigg\}.$$

The trivial cases of $h_1=0$ or $h_2=0$ or $h_1=h_2=0$ are treated similarly as in the proof of Theorem 14.16.

Theorem 14.17 is now proved. ■

Next we give a series of corollaries:

Corollary 14.18 *All as in Theorem 14.16. Further assume* $\widetilde{L}(1) = 1$ *and* $X^{(k)}(t_0, \omega) = 0$, $\forall\, \omega \in \Omega$, *all* $k = 1, \ldots, n-1$, *for a fixed* $t_0 \in [a, b]$. *Then*

$$\left(E\left(|M(X) - X|^q\right)(t_0)\right)^{\frac{1}{q}} \leq \frac{2^{\frac{1}{p}}}{\Gamma(\alpha)\,(p(\alpha-1)+1)^{\frac{1}{p}}\,(q+1)^{\frac{\alpha}{q(\alpha+1)}}}$$

$$\Bigg\{\bigg\{\left(\widetilde{L}\left(\chi_{[t_0,b]}(\cdot)\right)(t_0)\right)^{\frac{1}{p}}\left(\widetilde{L}\left(|\cdot - t_0|^{q(\alpha+1)}\chi_{[t_0,b]}(\cdot)\right)(t_0)\right)^{\frac{\alpha}{q(\alpha+1)}} \tag{14.85}$$

$$\left[\left(\widetilde{L}\left(\chi_{[t_0,b]}(\cdot)\right)(t_0)\right)^{\frac{1}{\alpha+1}}(q+1)^{\frac{\alpha}{(\alpha+1)}} + 1\right]^{\frac{1}{q}}$$

$$\Omega_1\left(D^{\alpha}_{*t_0}X, \left(\frac{1}{(q+1)}\widetilde{L}\left(|\cdot - t_0|^{q(\alpha+1)}\chi_{[t_0,b]}(\cdot)\right)(t_0)\right)^{\frac{1}{q(\alpha+1)}}\right)_{L^q,[t_0,b]}\bigg\} +$$

$$\bigg\{\left(\widetilde{L}\left(\chi_{[a,t_0]}(\cdot)\right)(t_0)\right)^{\frac{1}{p}}\left(\widetilde{L}\left(|\cdot - t_0|^{q(\alpha+1)}\chi_{[a,t_0]}(\cdot)\right)(t_0)\right)^{\frac{\alpha}{q(\alpha+1)}}$$

$$\left[\left(\widetilde{L}\left(\chi_{[a,t_0]}(\cdot)\right)(t_0)\right)^{\frac{1}{\alpha+1}}(q+1)^{\frac{\alpha}{(\alpha+1)}} + 1\right]^{\frac{1}{q}}$$

$$\Omega_1\left(D^{\alpha}_{t_0-}X, \left(\frac{1}{(q+1)}\widetilde{L}\left(|\cdot - t_0|^{q(\alpha+1)}\chi_{[a,t_0]}(\cdot)\right)(t_0)\right)^{\frac{1}{q(\alpha+1)}}\right)_{L^q,[a,t_0]}\bigg\}\Bigg\}.$$

Corollary 14.19 *All as in Theorem 14.16. Further assume that* $\widetilde{L}(1) = 1$ *and* $\frac{1}{q} < \alpha < 1$ *(i.e* $n = 1$*). Then,* $\forall\, t \in [a, b]$, *we have*

$$\left(E\left(|M(X) - X|^q\right)(t)\right)^{\frac{1}{q}} \leq \frac{2^{\frac{1}{p}}}{\Gamma(\alpha)\,(p(\alpha-1)+1)^{\frac{1}{p}}\,(q+1)^{\frac{\alpha}{q(\alpha+1)}}} \tag{14.86}$$

$$\Bigg\{\bigg\{\left(\widetilde{L}\left(\chi_{[t,b]}(\cdot)\right)(t)\right)^{\frac{1}{p}}\left(\widetilde{L}\left(|\cdot - t|^{q(\alpha+1)}\chi_{[t,b]}(\cdot)\right)(t)\right)^{\frac{\alpha}{q(\alpha+1)}}$$

$$\left[\left(\widetilde{L}\left(\chi_{[t,b]}(\cdot)\right)(t)\right)^{\frac{1}{\alpha+1}}(q+1)^{\frac{\alpha}{(\alpha+1)}} + 1\right]^{\frac{1}{q}}$$

$$\Omega_1\left(D^{\alpha}_{*t}X, \left(\frac{1}{(q+1)}\widetilde{L}\left(|\cdot - t|^{q(\alpha+1)}\chi_{[t,b]}(\cdot)\right)(t)\right)^{\frac{1}{q(\alpha+1)}}\right)_{L^q,[t,b]}\bigg\} +$$

$$\left\{\left(\widetilde{L}\left(\chi_{[a,t]}(\cdot)\right)(t)\right)^{\frac{1}{p}}\left(\widetilde{L}\left(|\cdot-t|^{q(\alpha+1)}\chi_{[a,t]}(\cdot)\right)(t)\right)^{\frac{\alpha}{q(\alpha+1)}}\right.$$

$$\left[\left(\widetilde{L}\left(\chi_{[a,t]}(\cdot)\right)(t)\right)^{\frac{1}{\alpha+1}}(q+1)^{\frac{\alpha}{(\alpha+1)}}+1\right]^{\frac{1}{q}}$$

$$\left.\Omega_1\left(D_{t-}^{\alpha}X,\left(\frac{1}{(q+1)}\widetilde{L}\left(|\cdot-t|^{q(\alpha+1)}\chi_{[a,t]}(\cdot)\right)(t)\right)^{\frac{1}{q(\alpha+1)}}\right)_{L^q,[a,t]}\right\}\Bigg\}.$$

Of great interest in Corollary 14.19 is the case of $\alpha=\frac{1}{2}$. We require there $q>2$. Due to lack of space we omit this statement.

Corollary 14.20 *All as in Theorem 14.17. Further assume* $\widetilde{L}(1)=1$ *and* $X^{(k)}(t_0,\omega)=0$, $\forall\,\omega\in\Omega$, *all* $k=1,\ldots,n-1$, *for a fixed* $t_0\in[a,b]$. *Then*

$$E\left(|M(X)-X|\right)(t_0)\le\frac{1}{\Gamma(\alpha+1)}$$

$$\left\{\left[\left(\widetilde{L}\left(\chi_{[a,t_0]}(\cdot)\right)(t_0)\right)^{\frac{1}{\alpha+1}}+\frac{1}{(\alpha+1)}\right]\left(\widetilde{L}\left(|\cdot-t_0|^{\alpha+1}\chi_{[a,t_0]}(\cdot)\right)(t_0)\right)^{\frac{\alpha}{\alpha+1}}\right.$$

$$\Omega_1\left(D_{t_0-}^{\alpha}X,\left(\widetilde{L}\left(|\cdot-t_0|^{\alpha+1}\chi_{[a,t_0]}(\cdot)\right)(t_0)\right)^{\frac{1}{\alpha+1}}\right)_{L^1,[a,t_0]}+\tag{14.87}$$

$$\left[\left(\widetilde{L}\left(\chi_{[t_0,b]}(\cdot)\right)(t_0)\right)^{\frac{1}{\alpha+1}}+\frac{1}{(\alpha+1)}\right]\left(\widetilde{L}\left(|\cdot-t_0|^{\alpha+1}\chi_{[t_0,b]}(\cdot)\right)(t_0)\right)^{\frac{\alpha}{\alpha+1}}$$

$$\left.\Omega_1\left(D_{*t_0}^{\alpha}X,\left(\widetilde{L}\left(|\cdot-t_0|^{\alpha+1}\chi_{[t_0,b]}(\cdot)\right)(t_0)\right)^{\frac{1}{\alpha+1}}\right)_{L^1,[t_0,b]}\right\}.$$

Corollary 14.21 *All as in Theorem 14.17. Further assume that* $\widetilde{L}(1)=1$, $0<\alpha<1$. *Then,* $\forall\,t\in[a,b]$, *we have*

$$E\left(|M(X)-X|\right)(t)\le\frac{1}{\Gamma(\alpha+1)}$$

$$\left\{\left[\left(\widetilde{L}\left(\chi_{[a,t]}(\cdot)\right)(t)\right)^{\frac{1}{\alpha+1}}+\frac{1}{(\alpha+1)}\right]\left(\widetilde{L}\left(|\cdot-t|^{\alpha+1}\chi_{[a,t]}(\cdot)\right)(t)\right)^{\frac{\alpha}{\alpha+1}}\right.$$

$$\Omega_1\left(D_{t-}^{\alpha}X,\left(\widetilde{L}\left(|\cdot-t|^{\alpha+1}\chi_{[a,t]}(\cdot)\right)(t)\right)^{\frac{1}{\alpha+1}}\right)_{L^1,[a,t]}+\tag{14.88}$$

$$\left[\left(\widetilde{L}\left(\chi_{[t,b]}(\cdot)\right)(t)\right)^{\frac{1}{\alpha+1}}+\frac{1}{(\alpha+1)}\right]\left(\widetilde{L}\left(|\cdot-t|^{\alpha+1}\chi_{[t,b]}(\cdot)\right)(t)\right)^{\frac{\alpha}{\alpha+1}}$$

$$\Omega_1\left(D_{*t}^{\alpha}X,\left(\widetilde{L}\left(|\cdot-t|^{\alpha+1}\chi_{[t,b]}(\cdot)\right)(t)\right)^{\frac{1}{\alpha+1}}\right)_{L^1,[t,b]}\Bigg\}.$$

The case $\alpha=\frac{1}{2}$ follows:

Corollary 14.22 *All as in Theorem 14.17. Further assume $\widetilde{L}(1)=1$. Then, $\forall\, t\in[a,b]$, we have*

$$E\left(|M(X)-X|\right)(t)\leq\frac{2}{\sqrt{\pi}}$$

$$\left\{\left[\left(\widetilde{L}\left(\chi_{[a,t]}(\cdot)\right)(t)\right)^{\frac{2}{3}}+\frac{2}{3}\right]\left(\widetilde{L}\left(|\cdot-t|^{\frac{3}{2}}\chi_{[a,t]}(\cdot)\right)(t)\right)^{\frac{1}{3}}\right.$$

$$\Omega_1\left(D_{t-}^{\frac{1}{2}}X,\left(\widetilde{L}\left(|\cdot-t|^{\frac{3}{2}}\chi_{[a,t]}(\cdot)\right)(t)\right)^{\frac{2}{3}}\right)_{L^1,[a,t]}+$$

$$\left[\left(\widetilde{L}\left(\chi_{[t,b]}(\cdot)\right)(t)\right)^{\frac{2}{3}}+\frac{2}{3}\right]\left(\widetilde{L}\left(|\cdot-t|^{\frac{3}{2}}\chi_{[t,b]}(\cdot)\right)(t)\right)^{\frac{1}{3}}\tag{14.89}$$

$$\left.\Omega_1\left(D_{*t}^{\frac{1}{2}}X,\left(\widetilde{L}\left(|\cdot-t|^{\frac{3}{2}}\chi_{[t,b]}(\cdot)\right)(t)\right)^{\frac{2}{3}}\right)_{L^1,[t,b]}\right\}.$$

We give a uniform norm result (denote $\|\cdot\|_{\infty,[a,b]}:=\|f\|_{\infty}$). It is based on Theorem 14.16.

Theorem 14.23 *Suppose Concepts 14.9, Assumptions 14.10 and 14.13. Here the non-integer $\alpha>0$ is such that $\alpha>\frac{1}{q}$, where $p,q>1:\frac{1}{p}+\frac{1}{q}=1$. Additionally assume that $\left|X^{(n)}(t,\omega)\right|\leq M^*$, $\forall\,(t,\omega)\in[a,b]\times\Omega$, where $M^*>0$. Then*

$$\left\|E\left(|M(X)-X|^q\right)\right\|_{\infty}^{\frac{1}{q}}\leq\left\|E\left(|X|^q\right)\right\|_{\infty}^{\frac{1}{q}}\left\|\widetilde{L}(1)-1\right\|_{\infty}+$$

$$\sum_{k=1}^{n-1}\frac{\left\|E\left(\left|X^{(k)}\right|^q\right)\right\|_{\infty}^{\frac{1}{q}}}{k!}\left\|\widetilde{L}\left((\cdot-t)^k\right)(t)\right\|_{\infty}+\frac{2^{\frac{1}{p}}}{\Gamma(\alpha)\left(p(\alpha-1)+1\right)^{\frac{1}{p}}(q+1)^{\frac{\alpha}{q(\alpha+1)}}}$$

$$\left[\left\|\widetilde{L}(1)\right\|_{\infty}^{\frac{1}{\alpha+1}}(q+1)^{\frac{\alpha}{(\alpha+1)}}+1\right]^{\frac{1}{q}}$$

$$\left\{\left(\left\|\widetilde{L}(1)\right\|_{\infty}\right)^{\frac{1}{p}}\left\|\widetilde{L}\left(|\cdot-t|^{q(\alpha+1)}\chi_{[t,b]}(\cdot)\right)(t)\right\|_{\infty}^{\frac{\alpha}{q(\alpha+1)}}\right.$$

$$\left.\sup_{t\in[a,b]}\Omega_1\left(D_{*t}^{\alpha}X,\left(\frac{1}{(q+1)}\left\|\widetilde{L}\left(|\cdot-t|^{q(\alpha+1)}\chi_{[t,b]}(\cdot)\right)(t)\right\|_{\infty}^{\frac{1}{q(\alpha+1)}}\right)\right)_{L^q,[t,b]}\right\}+$$

$$\left\{\left(\left\|\widetilde{L}(1)\right\|_{\infty}\right)^{\frac{1}{p}}\left\|\widetilde{L}\left(|\cdot-t|^{q(\alpha+1)}\chi_{[a,t]}(\cdot)\right)(t)\right\|_{\infty}^{\frac{\alpha}{q(\alpha+1)}}\right.$$

$$\left.\left.\sup_{t\in[a,b]}\Omega_1\left(D_{t-}^{\alpha}X,\left(\frac{1}{(q+1)}\left\|\widetilde{L}\left(|\cdot-t|^{q(\alpha+1)}\chi_{[a,t]}(\cdot)\right)(t)\right\|_{\infty}^{\frac{1}{q(\alpha+1)}}\right)\right)_{L^q,[a,t]}\right\}\right\}. \tag{14.90}$$

Based on Theorem 14.17 we give the following uniform estimate:

Theorem 14.24 *Suppose Concepts 14.9, Assumptions 14.10 and 14.14. Here* $\alpha > 0$, $\alpha \notin \mathbb{N}$, $n = \lceil \alpha \rceil$. *Additionally assume that* $\left|X^{(n)}(t,\omega)\right| \leq M^*$, $\forall (t,\omega) \in [a,b] \times \Omega$, *where* $M^* > 0$. *Then*

$$\|E(|M(X)-X|)\|_{\infty} \leq \|E(|X|)\|_{\infty}\left\|\widetilde{L}(1)-1\right\|_{\infty}+$$

$$\sum_{k=1}^{n-1}\frac{\left\|E\left(\left|X^{(k)}\right|\right)\right\|_{\infty}}{k!}\left\|\widetilde{L}\left((\cdot-t)^k\right)(t)\right\|_{\infty}+\frac{\left[\left\|\widetilde{L}(1)\right\|_{\infty}^{\frac{1}{\alpha+1}}+\frac{1}{\alpha+1}\right]}{\Gamma(\alpha+1)}$$

$$\left\{\left\|\widetilde{L}\left(|\cdot-t|^{\alpha+1}\chi_{[a,t]}(\cdot)\right)(t)\right\|_{\infty}^{\frac{\alpha}{\alpha+1}}\right.$$

$$\sup_{t\in[a,b]}\Omega_1\left(D_{t-}^{\alpha}X,\left\|\widetilde{L}\left(|\cdot-t|^{\alpha+1}\chi_{[a,t]}(\cdot)\right)(t)\right\|_{\infty}^{\frac{1}{\alpha+1}}\right)_{L^1,[a,t]}+$$

$$\left\|\widetilde{L}\left(|\cdot-t|^{\alpha+1}\chi_{[t,b]}(\cdot)\right)(t)\right\|_{\infty}^{\frac{\alpha}{\alpha+1}}$$

$$\left.\sup_{t\in[a,b]}\Omega_1\left(D_{*t}^{\alpha}X,\left\|\widetilde{L}\left(|\cdot-t|^{\alpha+1}\chi_{[t,b]}(\cdot)\right)(t)\right\|_{\infty}^{\frac{1}{\alpha+1}}\right)_{L^1,[t,b]}\right\}. \tag{14.91}$$

Corollary 14.25 (to Theorem 14.24) *All as in Theorem 14.24. Case of* $0 < \alpha < 1$ *and* $\widetilde{L}(1) = 1$. *Then*

$$\|E(|M(X)-X|)\|_{\infty} \leq \frac{\alpha+2}{\Gamma(\alpha+2)}$$

$$\left\{\left\|\widetilde{L}\left(|\cdot-t|^{\alpha+1}\chi_{[a,t]}(\cdot)\right)(t)\right\|_{\infty}^{\frac{\alpha}{\alpha+1}}\right.$$

$$\sup_{t\in[a,b]}\Omega_1\left(D_{t-}^{\alpha}X,\left\|\widetilde{L}\left(|\cdot-t|^{\alpha+1}\chi_{[a,t]}(\cdot)\right)(t)\right\|_{\infty}^{\frac{1}{\alpha+1}}\right)_{L^1,[a,t]}+$$

$$\left\|\widetilde{L}\left(|\cdot-t|^{\alpha+1}\chi_{[t,b]}(\cdot)\right)(t)\right\|_{\infty}^{\frac{\alpha}{\alpha+1}}$$

$$\sup_{t\in[a,b]} \Omega_1 \left(D_{*t}^{\alpha} X, \left\| \widetilde{L} \left(|\cdot - t|^{\alpha+1} \chi_{[t,b]} (\cdot) \right) (t) \right\|_{\infty}^{\frac{1}{\alpha+1}} \right)_{L^1,[t,b]} \Bigg\} . \quad (14.92)$$

Corollary 14.26 (to Theorem 14.24, $\alpha = \frac{1}{2}$) *All as in Theorem 14.24. Assume* $\widetilde{L}(1) = 1$. *Then*

$$\| E(|M(X) - X|) \|_{\infty} \leq \frac{10}{3\sqrt{\pi}}$$

$$\Bigg\{ \left\| \widetilde{L} \left(|\cdot - t|^{\frac{3}{2}} \chi_{[a,t]} (\cdot) \right) (t) \right\|_{\infty}^{\frac{1}{3}}$$

$$\sup_{t\in[a,b]} \Omega_1 \left(D_{t-}^{\alpha} X, \left\| \widetilde{L} \left(|\cdot - t|^{\frac{3}{2}} \chi_{[a,t]} (\cdot) \right) (t) \right\|_{\infty}^{\frac{2}{3}} \right)_{L^1,[a,t]} +$$

$$\left\| \widetilde{L} \left(|\cdot - t|^{\frac{3}{2}} \chi_{[t,b]} (\cdot) \right) (t) \right\|_{\infty}^{\frac{1}{3}} \quad (14.93)$$

$$\sup_{t\in[a,b]} \Omega_1 \left(D_{*t}^{\alpha} X, \left\| \widetilde{L} \left(|\cdot - t|^{\frac{3}{2}} \chi_{[t,b]} (\cdot) \right) (t) \right\|_{\infty}^{\frac{2}{3}} \right)_{L^1,[t,b]} \Bigg\} .$$

14.5 Application

Let $f \in C([0, 1])$ and the Bernstein polynomials

$$B_N(f)(t) := \sum_{k=0}^{N} f\left(\frac{k}{N}\right) \binom{N}{k} t^k (1-t)^{N-k}, \quad (14.94)$$

$\forall\, t \in [0, 1]$, $\forall\, N \in \mathbb{N}$.

We have that $B_N 1 = 1$ and B_N is a positive linear operator.

We have that

$$B_N\left((\cdot - t)^2\right)(t) = \frac{t(1-t)}{N}, \ \forall\, t \in [0, 1], \quad (14.95)$$

and

$$\left\| B_N\left((\cdot - t)^2\right)(t) \right\|_{\infty}^{\frac{1}{2}} \leq \frac{1}{2\sqrt{N}}, \ \forall\, N \in \mathbb{N}. \quad (14.96)$$

So B_N is like $\widetilde{L}$ in Concepts 14.9.

Define the corresponding application of M by

$$\widetilde{B}_N(X)(t,\omega) := B_N(X(\cdot,\omega))(t) = \sum_{k=0}^{N} X\left(\frac{k}{N},\omega\right)\binom{N}{k} t^k (1-t)^{N-k}, \tag{14.97}$$

$\forall\, t \in [0,1]$, $\forall\, \omega \in \Omega$, $N \in \mathbb{N}$, where X is a stochastic process.

Clearly $\widetilde{B}_N$ is a stochastic process.

We give

Proposition 14.27 *Let $X(t,\omega)$ be a stochastic process from $[0,1] \times (\Omega, \mathcal{F}, P)$ into $\mathbb{R}$, where $(\Omega, \mathcal{F}, P)$ is a probability space. Here $0 < \alpha < 1$ (i.e. $n = 1$) and $X(\cdot,\omega) \in AC([0,1])$ with $X^{(1)}(\cdot,\omega) \in L_\infty([0,1])$, $\forall\, \omega \in \Omega$. Further we assume that $D_{*t}^{\alpha} X(z,\omega)$ is a stochastic process for $z \in [t,1]$, $\omega \in \Omega$, and $D_{t-}^{\alpha} X(z,\omega)$ is a stochastic process for $z \in [0,t]$, $\omega \in \Omega$; $\forall\, t \in [0,1]$. For any $t \in [0,1]$ we assume that $D_{*t}^{\alpha} X(z,\omega)$ is continuous in $z \in [t,1]$, uniformly with respect to $\omega \in \Omega$. And for any $t \in [0,1]$ we assume that $D_{t-}^{\alpha} X(z,\omega)$ is continuous in $z \in [0,t]$, uniformly with respect to $\omega \in \Omega$. Finally, we assume that $(E|X|)(t) < \infty$, $\forall\, t \in [0,1]$. Then, for any $t \in [0,1]$, we have:*

$$E\left(\left|\widetilde{B}_N(X) - X\right|\right)(t) \leq \frac{1}{\Gamma(\alpha+1)}$$

$$\left\{\left[\left(B_N\left(\chi_{[0,t]}(\cdot)\right)(t)\right)^{\frac{1}{\alpha+1}} + \frac{1}{(\alpha+1)}\right]\left(B_N\left(|\cdot - t|^{\alpha+1}\chi_{[0,t]}(\cdot)\right)(t)\right)^{\frac{\alpha}{\alpha+1}}\right.$$

$$\Omega_1\left(D_{t-}^{\alpha} X, \left(B_N\left(|\cdot - t|^{\alpha+1}\chi_{[0,t]}(\cdot)\right)(t)\right)^{\frac{1}{\alpha+1}}\right)_{L^1,[0,t]} +$$

$$\left[\left(B_N\left(\chi_{[t,1]}(\cdot)\right)(t)\right)^{\frac{1}{\alpha+1}} + \frac{1}{(\alpha+1)}\right]\left(B_N\left(|\cdot - t|^{\alpha+1}\chi_{[t,1]}(\cdot)\right)(t)\right)^{\frac{\alpha}{\alpha+1}} \tag{14.98}$$

$$\left.\Omega_1\left(D_{*t}^{\alpha} X, \left(B_N\left(|\cdot - t|^{\alpha+1}\chi_{[t,1]}(\cdot)\right)(t)\right)^{\frac{1}{\alpha+1}}\right)_{L^1,[t,1]}\right\}.$$

Proof By Corollary 14.21. ■

We give

Proposition 14.28 *All as in Proposition 14.27 with $\alpha = \frac{1}{2}$. Then, for any $t \in [0,1]$, we have:*

$$E\left(\left|\widetilde{B}_N(X) - X\right|\right)(t) \leq \frac{2}{\sqrt{\pi}}$$

$$\left\{\left[\left(B_N\left(\chi_{[0,t]}(\cdot)\right)(t)\right)^{\frac{2}{3}} + \frac{2}{3}\right]\left(B_N\left(|\cdot - t|^{\frac{3}{2}}\chi_{[0,t]}(\cdot)\right)(t)\right)^{\frac{1}{3}}\right. \tag{14.99}$$

$$\Omega_1\left(D_{t-}^{\frac{1}{2}} X, \left(B_N\left(|\cdot - t|^{\frac{3}{2}}\chi_{[0,t]}(\cdot)\right)(t)\right)^{\frac{2}{3}}\right)_{L^1,[0,t]} +$$

$$\left[\left(B_N\left(\chi_{[t,1]}(\cdot)\right)(t)\right)^{\frac{2}{3}}+\frac{2}{3}\right]\left(B_N\left(|\cdot-t|^{\frac{3}{2}}\chi_{[t,1]}(\cdot)\right)(t)\right)^{\frac{1}{3}}$$

$$\Omega_1\left(D_{*t}^{\frac{1}{2}}X,\left(B_N\left(|\cdot-t|^{\frac{3}{2}}\chi_{[t,1]}(\cdot)\right)(t)\right)^{\frac{2}{3}}\right)_{L^1,[t,1]}\Bigg\}.$$

Proof By Corollary 14.22. ■

We give

Proposition 14.29 *All as in Proposition 14.27, $\alpha=\frac{1}{2}$. Then*

$$E\left(\left|\widetilde{B}_N(X)-X\right|\right)(t)\le\frac{10}{3\sqrt{\pi}}$$

$$\left\{\left(B_N\left(|\cdot-t|^{\frac{3}{2}}\right)(t)\right)^{\frac{1}{3}}\Omega_1\left(D_{t-}^{\frac{1}{2}}X,\left(B_N\left(|\cdot-t|^{\frac{3}{2}}\right)(t)\right)^{\frac{2}{3}}\right)_{L^1,[0,t]}+\right. \tag{14.100}$$

$$\left.\left(B_N\left(|\cdot-t|^{\frac{3}{2}}\right)(t)\right)^{\frac{1}{3}}\Omega_1\left(D_{*t}^{\frac{1}{2}}X,\left(B_N\left(|\cdot-t|^{\frac{3}{2}}\right)(t)\right)^{\frac{2}{3}}\right)_{L^1,[t,1]}\right\},$$

$\forall\, t\in[0,1]$, $\forall\, N\in\mathbb{N}$.

Proof By (14.99) and the positivity of B_N. ■

We further have

Proposition 14.30 *All as in Proposition 14.27, $\alpha=\frac{1}{2}$. Then*

$$E\left(\left|\widetilde{B}_N(X)-X\right|\right)(t)\le\frac{10}{3\sqrt{2\pi}\sqrt[4]{N}}$$

$$\left\{\Omega_1\left(D_{t-}^{\frac{1}{2}}X,\frac{1}{2\sqrt{N}}\right)_{L^1,[0,t]}+\Omega_1\left(D_{*t}^{\frac{1}{2}}X,\frac{1}{2\sqrt{N}}\right)_{L^1,[t,1]}\right\}, \tag{14.101}$$

$\forall\, t\in[0,1]$, $\forall\, N\in\mathbb{N}$.

Proof By discrete Hölder's inequality we get that

$$B_N\left(|\cdot-t|^{\frac{3}{2}}\right)(t)\le\left(B_N\left((\cdot-t)^2\right)(t)\right)^{\frac{3}{4}}\overset{(14.95)}{=} \tag{14.102}$$

$$\left(\frac{1}{N}t(1-t)\right)^{\frac{3}{4}}\le\frac{1}{(4N)^{\frac{3}{4}}},$$

$\forall\, t \in [0,1]$, $\forall\, N \in \mathbb{N}$.

Hence,

$$\text{the right hand side of (14.100)} \leq$$

$$\frac{10}{3\sqrt{\pi}}\left\{\frac{1}{(4N)^{\frac{1}{4}}}\Omega_1\left(D_{t-}^{\frac{1}{2}}X, \frac{1}{(4N)^{\frac{1}{2}}}\right)_{L^1,[0,t]} + \frac{1}{(4N)^{\frac{1}{4}}}\Omega_1\left(D_{*t}^{\frac{1}{2}}X, \frac{1}{(4N)^{\frac{1}{2}}}\right)_{L^1,[t,1]}\right\} =$$

$$\frac{10}{3\sqrt{2\pi}\sqrt[4]{N}}\left\{\Omega_1\left(D_{t-}^{\frac{1}{2}}X, \frac{1}{2\sqrt{N}}\right)_{L^1,[0,t]} + \Omega_1\left(D_{*t}^{\frac{1}{2}}X, \frac{1}{2\sqrt{N}}\right)_{L^1,[t,1]}\right\}, \quad (14.103)$$

$\forall\, t \in [0,1]$, $\forall\, N \in \mathbb{N}$. ■

Consequently we obtain

Proposition 14.31 *All as in Proposition 14.27,* $\alpha = \frac{1}{2}$. *Additionally assume that* $\left|X^{(1)}(t,\omega)\right| \leq M^*$, $\forall\,(t,\omega) \in [0,1] \times \Omega$, *where* $M^* > 0$. *Then*

$$\left\| E\left(\left|\widetilde{B}_N(X) - X\right|\right)\right\|_{\infty} \leq \frac{10}{3\sqrt{2\pi}\sqrt[4]{N}}$$

$$\left\{\sup_{t\in[0,1]}\Omega_1\left(D_{t-}^{\frac{1}{2}}X, \frac{1}{2\sqrt{N}}\right)_{L^1,[0,t]} + \sup_{t\in[0,1]}\Omega_1\left(D_{*t}^{\frac{1}{2}}X, \frac{1}{2\sqrt{N}}\right)_{L^1,[t,1]}\right\} < \infty, \quad (14.104)$$

$\forall\, N \in \mathbb{N}$.

As $N \to +\infty$, *we obtain*

$$\left\| E\left(\left|\widetilde{B}_N(X) - X\right|\right)\right\|_{\infty} \to 0. \quad (14.105)$$

That is $\widetilde{B}_N \overset{\text{"1-mean"}}{\to} I$ *(unit operator) with rates.*

Proof By (14.101) and (14.13) with (14.14). ■

Remark 14.32 *(to Proposition* 14.31*)* Assume that

$$\Omega_1\left(D_{t-}^{\frac{1}{2}}X, \frac{1}{2\sqrt{N}}\right)_{L^1,[0,t]} \leq \frac{K_1}{2\sqrt{N}}, \ \forall\, t \in [0,1], \ \forall\, N \in \mathbb{N}, \quad (14.106)$$

where $K_1 > 0$,

and assume that

$$\Omega_1\left(D_{*t}^{\frac{1}{2}}X, \frac{1}{2\sqrt{N}}\right)_{L^1,[t,1]} \leq \frac{K_2}{2\sqrt{N}}, \ \forall\, t \in [0,1], \ \forall\, N \in \mathbb{N}, \quad (14.107)$$

where $K_2 > 0$.

Conditions (14.106), (14.107) are of Lipschitz type of order 1.
Then, by (14.104), (14.106) and (14.107), we get:

$$\left\| E\left(\left|\widetilde{B}_N(X) - X\right|\right)\right\|_\infty \leq \frac{5}{3\sqrt{2\pi}}(K_1 + K_2)\frac{1}{\sqrt[4]{N}}\frac{1}{\sqrt{N}}$$

$$= \frac{5(K_1+K_2)}{3\sqrt{2\pi}}\frac{1}{N^{\frac{3}{4}}}, \ \forall N \in \mathbb{N}. \tag{14.108}$$

Thus, at smoothness of order $\frac{1}{2}$ we achieve speed of convergence $\frac{1}{N^{\frac{3}{4}}}$, $N \in \mathbb{N}$. Without any smoothness, in [5], we proved that the speed of convergence was $\frac{1}{\sqrt{N}}$, $N \in \mathbb{N}$.

In the presence of the ordinary first derivative, see [5], in deterministic approximation the rate of convergence was $\frac{1}{N}$, $N \in \mathbb{N}$.

Naturally, we have as expected:

$$\frac{1}{N} < \frac{1}{N^{\frac{3}{4}}} < \frac{1}{\sqrt{N}}, \ \forall N \in \mathbb{N} - \{1\}. \tag{14.109}$$

14.6 Stochastic Korovkin Results

We give at first pointwise results.

Theorem 14.33 *Assume all as in Theorem 14.16. Here $\widetilde{L}$, M are meant as sequences of operators. Assume further that $\widetilde{L}(1)(t) \to 1$ and $\widetilde{L}\left(|\cdot - t|^{q(\alpha+1)}\right)(t) \to 0$, pointwise in $t \in [a,b]$. Then $E\left(|M(X) - X|^q\right)(t) \to 0$, pointwise in $t \in [a,b]$, that is $M \to I$ (stochastic unit operator) in q-mean-pointwise with rates, quantitatively.*

Proof From now on R.H.S denotes the "right hand side". By Hölder's inequality we get that

$$\text{R.H.S. (14.32)} \leq \left(\left(E|X|^q\right)(t)\right)^{\frac{1}{q}}\left|\widetilde{L}(1)(t) - 1\right| +$$

$$\sum_{k=1}^{n-1}\frac{\left(\left(E\left|X^{(k)}\right|^q\right)(t)\right)^{\frac{1}{q}}}{k!}\left(\widetilde{L}\left(|\cdot - t|^{q(\alpha+1)}\right)(t)\right)^{\frac{k}{q(\alpha+1)}}\left(\widetilde{L}(1)(t)\right)^{\frac{q(\alpha+1)-k}{q(\alpha+1)}}$$

$$+\frac{2^{\frac{1}{p}}}{\Gamma(\alpha)(p(\alpha-1)+1)^{\frac{1}{p}}(q+1)^{\frac{\alpha}{q(\alpha+1)}}}$$

$$\left[\left(\widetilde{L}(1)(t)\right)^{\frac{1}{\alpha+1}}(q+1)^{\frac{\alpha}{(\alpha+1)}} + 1\right]^{\frac{1}{q}} \tag{14.110}$$

$$\left(\widetilde{L}(1)(t)\right)^{\frac{1}{p}}\left(\widetilde{L}\left(|\cdot - t|^{q(\alpha+1)}\right)(t)\right)^{\frac{\alpha}{q(\alpha+1)}}$$

$$\left\{\Omega_1\left(D^{\alpha}_{*t}X, \left(\frac{1}{(q+1)}\widetilde{L}\left(|\cdot - t|^{q(\alpha+1)}\right)(t)\right)^{\frac{1}{q(\alpha+1)}}\right)_{L^q,[t,b]} + \right.$$

$$\left.\Omega_1\left(D^{\alpha}_{t-}X, \left(\frac{1}{(q+1)}\widetilde{L}\left(|\cdot - t|^{q(\alpha+1)}\right)(t)\right)^{\frac{1}{q(\alpha+1)}}\right)_{L^q,[a,t]}\right\}.$$

Because $\widetilde{L}(1)(t)$ converges pointwise in t to 1 it is bounded.

Assuming also that $\widetilde{L}\left(|\cdot - t|^{q(\alpha+1)}\right)(t)$ converges pointwise in t to zero, we get that R.H.S. (14.110) converges to zero pointwise in t, any $t \in [a,b]$.

Therefore $\left(E\left(|M(X) - X|^q\right)(t)\right)^{\frac{1}{q}}$ (by (14.32)) converges to zero pointwise in t, any $t \in [a,b]$. ■

We continue with

Theorem 14.34 *Assume all as in Theorem 14.17. Here $\widetilde{L}$, M are meant as sequences of operators. Assume further that $\widetilde{L}(1)(t) \to 1$ and $\widetilde{L}\left(|\cdot - t|^{(\alpha+1)}\right)(t) \to 0$, pointwise in $t \in [a,b]$. Then $E(|M(X) - X|)(t) \to 0$, pointwise in $t \in [a,b]$, that is $M \to I$ in 1-mean-pointwise with rates, quantitatively.*

Proof We observe that

$$\text{R.H.S. (14.75)} \overset{(14.24)}{\leq} (E|X|)(t)\left|\widetilde{L}(1)(t) - 1\right| +$$

$$\sum_{k=1}^{n-1}\frac{\left(E\left|X^{(k)}\right|\right)(t)}{k!}\left(\widetilde{L}\left(|\cdot - t|^{\alpha+1}\right)(t)\right)^{\frac{k}{\alpha+1}}\left(\widetilde{L}(1)(t)\right)^{\frac{\alpha+1-k}{\alpha+1}} \tag{14.111}$$

$$+\frac{1}{\Gamma(\alpha+1)}\left[\left(\widetilde{L}(1)(t)\right)^{\frac{1}{\alpha+1}} + \frac{1}{\alpha+1}\right]$$

$$\left(\widetilde{L}\left(|\cdot - t|^{\alpha+1}\right)(t)\right)^{\frac{\alpha}{\alpha+1}}\left[\Omega_1\left(D^{\alpha}_{t-}X, \left(\widetilde{L}\left(|\cdot - t|^{\alpha+1}\right)(t)\right)^{\frac{1}{\alpha+1}}\right)_{L^1,[a,t]} + \right.$$

$$\left.\Omega_1\left(D^{\alpha}_{*t}X, \left(\widetilde{L}\left(|\cdot - t|^{\alpha+1}\right)(t)\right)^{\frac{1}{\alpha+1}}\right)_{L^1,[t,b]}\right].$$

Because $\widetilde{L}(1)(t)$ converges pointwise in t to 1 it is bounded.

Assuming also that $\widetilde{L}\left(|\cdot - t|^{\alpha+1}\right)(t)$ converges pointwise in t to zero, we get that R.H.S. (14.111) converges to zero pointwise in t, any $t \in [a,b]$.

Therefore $E(|M(X) - X|)(t)$ (by (14.75)) converges to zero pointwise in t, any $t \in [a,b]$. ■

We continue with uniform convergence results.

Theorem 14.35 *All as in Theorem 14.23. Here* $\widetilde{L}$, M *are meant as sequences of operators. Assume further that* $\widetilde{L}(1) \to 1$ *uniformly, and* $\left\| \widetilde{L}\left(|\cdot - t|^{q(\alpha+1)}\right)(t) \right\|_{\infty} \to 0$. *Then* $\left\| E\left(|M(X) - X|^{q}\right) \right\|_{\infty} \to 0$ *over* $[a, b]$, *that is* $M \overset{\text{“}q\text{-mean”}}{\to} I$, *the stochastic unit operator with rates, quantitatively.*

Proof We have that

$$\text{R.H.S. (14.90)} \leq \left\| E\left(|X|^{q}\right) \right\|_{\infty}^{\frac{1}{q}} \left\| \widetilde{L}(1) - 1 \right\|_{\infty} +$$

$$\sum_{k=1}^{n-1} \frac{\left\| E\left(\left|X^{(k)}\right|^{q}\right) \right\|_{\infty}^{\frac{1}{q}}}{k!} \left\| \widetilde{L}\left(|\cdot - t|^{q(\alpha+1)}\right)(t) \right\|_{\infty}^{\frac{k}{q(\alpha+1)}} \left\| \widetilde{L}(1) \right\|_{\infty}^{\frac{q(\alpha+1)-k}{q(\alpha+1)}} +$$

$$\frac{2^{\frac{1}{p}}}{\Gamma(\alpha)\left(p(\alpha-1)+1\right)^{\frac{1}{p}}(q+1)^{\frac{\alpha}{q(\alpha+1)}}} \left[\left\| \widetilde{L}(1) \right\|_{\infty}^{\frac{1}{\alpha+1}} (q+1)^{\frac{\alpha}{(\alpha+1)}} + 1 \right]^{\frac{1}{q}} \tag{14.112}$$

$$\left\| \widetilde{L}(1) \right\|_{\infty}^{\frac{1}{p}} \left\| \widetilde{L}\left(|\cdot - t|^{q(\alpha+1)}\right)(t) \right\|_{\infty}^{\frac{\alpha}{q(\alpha+1)}}$$

$$\left\{ \sup_{t\in[a,b]} \Omega_1\left(D_{*t}^{\alpha} X, \left(\frac{1}{(q+1)} \left\| \widetilde{L}\left(|\cdot - t|^{q(\alpha+1)}\right)(t) \right\|_{\infty}^{\frac{1}{q(\alpha+1)}} \right) \right)_{L^q,[t,b]} + \right.$$

$$\left. \sup_{t\in[a,b]} \Omega_1\left(D_{t-}^{\alpha} X, \left(\frac{1}{(q+1)} \left\| \widetilde{L}\left(|\cdot - t|^{q(\alpha+1)}\right)(t) \right\|_{\infty}^{\frac{1}{q(\alpha+1)}} \right) \right)_{L^q,[a,t]} \right\}.$$

Because $\widetilde{L}(1) \to 1$ uniformly, it is bounded. Assuming also that $\left\| \widetilde{L}\left(|\cdot - t|^{q(\alpha+1)}\right)(t) \right\|_{\infty} \to 0$, we get that R.H.S. (14.112) converges uniformly to zero over $[a, b]$.

Therefore $\left\| E\left(|M(X) - X|^{q}\right) \right\|_{\infty}^{\frac{1}{q}}$ (by (14.90)) converges to zero over $[a, b]$. ■

We also present:

Theorem 14.36 *All as in Theorem 14.24. Here* $\widetilde{L}$, M *are meant as sequences of operators. Assume further that* $\widetilde{L}(1) \to 1$ *uniformly, and* $\left\| \widetilde{L}\left(|\cdot - t|^{(\alpha+1)}\right)(t) \right\|_{\infty} \to 0$. *Then* $\| E(|M(X) - X|) \|_{\infty} \to 0$ *over* $[a, b]$, *that is* $M \overset{\text{“}1\text{-mean”}}{\to} I$ *with rates quantitatively.*

Proof We have that

$$\text{R.H.S. (14.91)} \leq \|E(|X|)\|_{\infty} \left\|\widetilde{L}(1) - 1\right\|_{\infty} +$$

$$\sum_{k=1}^{n-1} \frac{\left\|E\left(\left|X^{(k)}\right|\right)\right\|_{\infty}}{k!} \left\|\widetilde{L}\left(|\cdot - t|^{\alpha+1}\right)(t)\right\|_{\infty}^{\frac{k}{\alpha+1}} \left\|\widetilde{L}(1)\right\|_{\infty}^{\frac{\alpha+1-k}{\alpha+1}} +$$

$$\frac{1}{\Gamma(\alpha+1)} \left[\left\|\widetilde{L}(1)\right\|_{\infty}^{\frac{1}{\alpha+1}} + \frac{1}{\alpha+1}\right] \left\|\widetilde{L}\left(|\cdot - t|^{\alpha+1}\right)(t)\right\|_{\infty}^{\frac{\alpha}{\alpha+1}} \tag{14.113}$$

$$\left[\sup_{t\in[a,b]} \Omega_1\left(D_{t-}^{\alpha}X, \left\|\widetilde{L}\left(|\cdot - t|^{\alpha+1}\right)(t)\right\|_{\infty}^{\frac{1}{\alpha+1}}\right)_{L^1,[a,t]} +\right.$$

$$\left.\sup_{t\in[a,b]} \Omega_1\left(D_{*t}^{\alpha}X, \left\|\widetilde{L}\left(|\cdot - t|^{\alpha+1}\right)(t)\right\|_{\infty}^{\frac{1}{\alpha+1}}\right)_{L^1,[t,b]}\right].$$

Because $\widetilde{L}(1) \to 1$ uniformly, it is bounded. Assuming also $\left\|\widetilde{L}\left(|\cdot - t|^{\alpha+1}\right)(t)\right\|_{\infty} \to 0$, we get that R.H.S. (14.113) converges uniformly to zero over $[a, b]$.

Therefore $\|E(|M(X) - X|)\|_{\infty}$ (by (14.91)) converges to zero over $[a, b]$. ■

We finish with

Remark 14.37 An amazing fact observation follows: In all convergence results here, see Theorems 14.33–14.36, the forcing conditions for convergences are based only on $\widetilde{L}$ and basic real valued continuous functions on $[a, b]$ and are not related to stochastic processes, but they are giving convergence results on stochastic processes!

References

1. Aliprantis, C., Burkinshaw, O.: Principles of Real Analysis, 3rd edn. Academic, San Diego (1998)
2. Anastassiou, G.A.: Korovkin inequalities for stochastic processes. J. Math. Anal. Appl. **157**(2), 366–384 (1991)
3. Anastassiou, G.A.: Moments in Probability and Approximation Theory. Pitman/Longman, # 287, UK (1993)
4. Anastassiou, G.: Quantitative Approximations. Chapman & Hall/CRC, Boca Raton (2001)
5. Anastassiou, G.: Stochastic Korovkin theory given quantitatively. Facta Univ. (Nis) Ser. Math. Inform. **22**(1), 43–60 (2007)
6. Anastassiou, G.: Fractional Differentiation Inequalities. Springer, Heildelberg (2009)
7. Anastassiou, G.A.: Fuzzy Mathematics: Approximation Theory. Springer, Heildelberg (2010)
8. Anastassiou, G.: Intelligent Mathematics: Computational Analysis. Springer, Heidelberg (2011)
9. Anastassiou, G.: Fractional representation formulae and right fractional inequalities. Math. Comput. Model. **54**(11–12), 3098–3115 (2011)

10. Anastassiou, G.: Foundation of stochastic fractional calculus with fractional approximation of stochastic processes. RACSAM **114** (2020). Article 89
11. Diethelm, K.: The Analysis of Fractional Differential Equations. Springer, New York (2010)
12. Korovkin, P.P.: Linear Operators and Approximation Theory. Hindustan Publ. Corp., Delhi (1960)
13. Royden, H.L.: Real Analysis, 2nd edn. MacMillan Publishing Co., Inc., New York (1968)
14. Shisha, O., Mond, B.: The degree of convergence of sequences of linear positive operators. Natl. Acad. Sci. U.S. **60**, 1196–1200 (1968)
15. Weba, M.: Korovkin systems of stochastic processes. Math. Z. **192**, 73–80 (1986)
16. Weba, M.: A quantitative Korovkin theorem for random functions with multivariate domains. J. Approx. Theory **61**, 74–87 (1990)

Chapter 15
Trigonometric Caputo Fractional Approximation of Stochastic Processes

Here we encounter and study very general stochastic positive linear operators induced by general positive linear operators that are acting on continuous functions in the trigonometric sense. These are acting on the space of real fractionally differentiable stochastic processes. Under some very mild, general and natural assumptions on the stochastic processes we produce related trigonometric fractional stochastic Shisha-Mond type inequalities of L^q-type $1 \leq q < \infty$ and corresponding trigonometric fractional stochastic Korovkin type theorems. These are regarding the trigonometric stochastic q-mean fractional convergence of a sequence of stochastic positive linear operators to the stochastic unit operator for various cases. All convergences are produced with rates and are given via the trigonometric fractional stochastic inequalities involving the stochastic modulus of continuity of the αth fractional derivatives of the engaged stochastic process, $\alpha > 0, \alpha \notin \mathbb{N}$. The impressive fact is that only two basic real Korovkin test functions assumptions, one of them trigonometric, are enough for the conclusions of our trigonometric fractional stochastic Korovkin theory. We give applications to stochastic Bernstein operators in the trigonometric sense. See also [11].

15.1 Introduction

Inspiration for this work comes from [2, 3, 15–17]. This work continues our earlier work [5], now at the stochastic fractional level. First we mention the foundations of Stochastic fractional calculus in the direct analytical sense, see [10], in Sect. 15.2, this is in the Caputo fractional direction. In the Sect. 15.3, about background, we talk about the q-mean ($1 \leq q < \infty$) first modulus of continuity of a stochastic process and its upper bounds. There we describe completely our setting by introducing our stochastic positive linear operator M, see (15.15), which is based on the positive linear operator $\widetilde{L}$ from $C([-\pi, \pi])$ into itself. The operator M is acting on a wide space of Caputo fractional differentiable real valued stochastic processes X. See there

G. A. Anastassiou, *Generalized Fractional Calculus*, Studies in Systems, Decision and Control 305, https://doi.org/10.1007/978-3-030-56962-4_15

Assumptions 15.10, 15.13, 15.14. We first give the main trigonometric pointwise fractional stochastic Shisha-Mond type inequalities [15], see Theorems 15.21, 15.22, and their several corollaries covering important trigonometric special cases.

We continue with trigonometric fractional q-mean uniform Shisha-Mond type inequalities, see Theorems 15.23, 15.24, and their interesting corollaries. All this theory is regarding the trigonometric fractional stochastic convergence of operators M to I (stochastic unit operator) given quantitatively with rates. An extensive trigonometric application about the stochastic Bernstein operators follows in full details. Based on our Shisha-Mond type inequalities of our main Theorems 15.21–15.24 we derive trigonometric pointwise and uniform Stochastic Korovkin theorems [13] on stochastic processes, see Theorems 15.40–15.43. The amazing fact here is, that basic conditions on operator $\widetilde{L}$ regarding two simple real valued functions, one of them trigonometric, that are not stochastic, are able to enforce fractional stochastic convergence on all stochastic processes we are dealing with; see Concepts 15.9 and Assumptions 15.10, 15.13, 15.14 on $[-\pi, \pi]$.

15.2 Foundation of Stochastic Fractional Calculus ([10])

Let $t \in [a, b] \subset \mathbb{R}$, $\omega \in \Omega$, where $(\Omega, \mathcal{F}, P)$ is a probability space. Here $X(t, \omega)$ stands for a stochastic process. Case of $X(\cdot, \omega)$ being continuous on $[a, b]$, $\forall\, \omega \in \Omega$. Then by Caratheodory [1, Theorem 20.15, p.156], we get that $X(t, \omega)$ is jointly measurable.

Next we define the left and right respectively, Riemann-Liouville stochastic fractional integrals [10], where $\alpha > 0$ is not an integer:

$$I_{a+}^{\alpha} X(x, \omega) = \frac{1}{\Gamma(\alpha)} \int_a^x (x-t)^{\alpha-1} X(t, \omega)\, dt, \tag{15.1}$$

and

$$I_{b-}^{\alpha} X(x, \omega) = \frac{1}{\Gamma(\alpha)} \int_x^b (t-x)^{\alpha-1} X(t, \omega)\, dt, \tag{15.2}$$

$\forall\, x \in [a, b]$, $\forall\, \omega \in \Omega$, where Γ is the gamma function.

In the following important cases we prove that $I_{a+}^{\alpha} X$, $I_{b-}^{\alpha} X$ are stochastic processes:

(i) Assume that $(\Omega, \mathcal{F}, P)$ is a complete probability space, and that $(x-t)^{\alpha-1} X(t, \omega)$ is an integrable function on $[a, x] \times \Omega$, $\forall\, x \in [a, b]$, then by Fubini's theorem, [14], p. 269, $I_{a+}^{\alpha} X(x, \cdot)$ is an integrable function on Ω, $\forall\, x \in [a, b]$. Similarly, if $(t-x)^{\alpha-1} X(t, \omega)$ is an integrable function on $[x, b] \times \Omega$, $\forall\, x \in [a, b]$, then again by Fubini's theorem $I_{b-}^{\alpha} X(x, \cdot)$ is an integrable function on Ω, $\forall\, x \in [a, b]$. That is $I_{a+}^{\alpha} X(x, \omega)$ and $I_{b-}^{\alpha} X(x, \omega)$ are stochastic processes.

(ii) Assume a general probability space $(\Omega, \mathcal{F}, P)$ and the Lebesgue measure spaces on $[a, x]$, $[x, b]$, $\forall\, x \in [a, b]$. These are clearly σ-finite measure spaces. We assume that the jointly measurable stochastic process $X(t, \omega) \geq 0$ on $[a, b] \times \Omega$, hence $(x-t)^{\alpha-1} X(t, \omega) \geq 0$ on $[a, x] \times \Omega$, and $(t-x)^{\alpha-1} X(t, \omega) \geq 0$ on $[x, b] \times \Omega$, $\forall\, x \in [a, b]$, and both are jointly measurable. Then by Tonelli's theorem, [14, p. 270], we get that $I_{a+}^{\alpha} X(x, \cdot)$, $I_{b-}^{\alpha} X(x, \cdot)$ are measurable functions on Ω, $\forall\, x \in [a, b]$. That is $I_{a+}^{\alpha} X$, $I_{b-}^{\alpha} X$ are stochastic processes. The above facts provide the foundation of stochastic fractional calculus in the direct analytical sense. So it is not unusual to consider that $I_{a+}^{\alpha} X$, $I_{b-}^{\alpha} X$ are stochastic processes.

(iii) Given that $X(\cdot, \omega)$ is in $L_1([a, b])$ then $I_{a+}^{\alpha} X(\cdot, \omega) \in L_1([a, b])$, $\forall\, \omega \in \Omega$, see [12, p. 13], and $I_{b-}^{\alpha} X(\cdot, \omega) \in L_1([a, b])$, $\forall\, \omega \in \Omega$, see [8, p. 334].

And given that $X(\cdot, \omega) \in L_{\infty}([a, b])$, then $I_{a+}^{\alpha} X(\cdot, \omega) \in C([a, b])$, when $0 < \alpha < 1$, and $I_{a+}^{\alpha} X(\cdot, \omega) \in AC([a, b])$ (absolutely continuous functions), when $\alpha \geq 1$, $\forall\, \omega \in \Omega$, see [6, p. 388]. Similarly, if $X(\cdot, \omega) \in L_{\infty}([a, b])$, then $I_{b-}^{\alpha} X(\cdot, \omega) \in C([a, b])$, when $0 < \alpha < 1$, and $I_{b-}^{\alpha} X(\cdot, \omega) \in AC([a, b])$, when $\alpha \geq 1$, $\forall\, \omega \in \Omega$, see [9].

We need

Definition 15.1 *([10])* Let non-integer $\alpha > 0$, $n = \lceil \alpha \rceil$ ($\lceil \cdot \rceil$ is the ceiling of the number), $t \in [a, b] \subset \mathbb{R}$, $\omega \in \Omega$, where $(\Omega, \mathcal{F}, P)$ is a general probability space. Here $X(t, \omega)$ stands for a stochastic process. Assume that $X(\cdot, \omega) \in AC^n([a, b])$ (spaces of functions $X(\cdot, \omega)$ with $X^{(n-1)}(\cdot, \omega) \in AC([a, b])$), $\forall\, \omega \in \Omega$.

We call stochastic left Caputo fractional derivative

$$D_{*a}^{\alpha} X(x, \omega) = \frac{1}{\Gamma(n-\alpha)} \int_a^x (x-t)^{n-\alpha-1} X^{(n)}(t, \omega)\, dt, \tag{15.3}$$

$\forall\, x \in [a, b]$, $\forall\, \omega \in \Omega$.

And, we call stochastic right Caputo fractional derivative

$$D_{b-}^{\alpha} X(x, \omega) = \frac{(-1)^n}{\Gamma(n-\alpha)} \int_x^b (z-x)^{n-\alpha-1} X^{(n)}(z, \omega)\, dz, \tag{15.4}$$

$\forall\, x \in [a, b]$, $\forall\, \omega \in \Omega$.

Remark 15.2 *(to Definition 15.1)* We further assume here that

$$\left|X^{(n)}(t, \omega)\right| \leq M^*, \forall\, (t, \omega) \in [a, b] \times \Omega,$$

where $M^* > 0$. Then, by (15.3), we have

$$\left|D_{*a}^{\alpha}X(x,\omega)\right| \leq \frac{1}{\Gamma(n-\alpha)}\int_{a}^{x}(x-t)^{n-\alpha-1}\left|X^{(n)}(t,\omega)\right|dt$$

$$\leq \frac{M^{*}}{\Gamma(n-\alpha)}\int_{a}^{x}(x-t)^{n-\alpha-1}dt = \frac{M^{*}(x-a)^{n-\alpha}}{\Gamma(n-\alpha+1)}.$$

That is

$$\left|D_{*a}^{\alpha}X(x,\omega)\right| \leq \frac{M^{*}(x-a)^{n-\alpha}}{\Gamma(n-\alpha+1)},\ \forall x\in[a,b]. \tag{15.5}$$

Also, from (15.4) we get

$$\left|D_{b-}^{\alpha}X(x,\omega)\right| \leq \frac{1}{\Gamma(n-\alpha)}\int_{x}^{b}(z-x)^{n-\alpha-1}\left|X^{(n)}(z,\omega)\right|dz \leq$$

$$\frac{M^{*}}{\Gamma(n-\alpha)}\int_{x}^{b}(z-x)^{n-\alpha-1}dz = \frac{M^{*}(b-x)^{n-\alpha}}{\Gamma(n-\alpha+1)}.$$

That is

$$\left|D_{b-}^{\alpha}X(x,\omega)\right| \leq \frac{M^{*}(b-x)^{n-\alpha}}{\Gamma(n-\alpha+1)},\ \forall x\in[a,b]. \tag{15.6}$$

By (15.1)–(15.4), it is not strange to assume that $D_{*a}^{\alpha}X$, $D_{b-}^{\alpha}X$ are stochastic processes.

15.3 Background (See Also [10])

We need

Definition 15.3 We define the relative q-mean first modulus of continuity of stochastic process $X(t,\omega)$ by

$$\Omega_1(X,\delta)_{L^q,[c,d]} :=$$

$$\sup\left\{\left(\int_{\Omega}|X(x,\omega)-X(y,\omega)|^{q}P(d\omega)\right)^{\frac{1}{q}} : x,y\in[c,d]\subseteq[a,b],\ |x-y|\leq\delta\right\}, \tag{15.7}$$

$\delta>0$, $1\leq q<\infty$.

Definition 15.4 Let $1\leq q<\infty$. Let $X(x,\omega)$ be a stochastic process. We call X a q-mean uniformly continuous stochastic process over $[a,b]$, iff $\forall\ \varepsilon>0\ \exists\ \delta>0$: whenever $|x-y|\leq\delta$; $x,y\in[a,b]$ implies that

$$\int_{\Omega}|X(x,s)-X(y,s)|^{q}P(ds)\leq\varepsilon. \tag{15.8}$$

We denote it as $X \in C_{\mathbb{R}}^{U_q}([a,b])$.

It holds

Proposition 15.5 ([5]) *Let* $X \in C_{\mathbb{R}}^{U_q}([a,b])$, *then* $\Omega_1(X,\delta)_{L^q,[a,b]} < \infty$, *any* $\delta > 0$.

Also it holds

Proposition 15.6 ([5]) *Let* $X(t,\omega)$ *be a stochastic process from* $[a,b] \times (\Omega, \mathcal{F}, P)$ *into* $\mathbb{R}$. *Then following are true* ($[c,d] \subseteq [a,b]$):

(i) $\Omega_1(X,\delta)_{L^q,[c,d]}$ *is nonnegative and nondecreasing in* $\delta > 0$,
(ii) $\lim_{\delta \downarrow 0} \Omega_1(X,\delta)_{L^q,[c,d]} = \Omega_1(X,0)_{L^q,[c,d]} = 0$, *iff* $X \in C_{\mathbb{R}}^{U_q}([c,d])$,
(iii) $\Omega_1(X,\delta_1+\delta_2)_{L^q,[c,d]} \leq \Omega_1(X,\delta_1)_{L^q,[c,d]} + \Omega_1(X,\delta_2)_{L^q,[c,d]}$, $\delta_1, \delta_2 > 0$,
(iv) $\Omega_1(X,m\delta)_{L^q,[c,d]} \leq m\Omega_1(X,\delta)_{L^q,[c,d]}$, $\delta > 0$, $m \in \mathbb{N}$,
(v) $\Omega_1(X,\lambda\delta)_{L^q,[c,d]} \leq \lceil \lambda \rceil \Omega_1(X,\delta)_{L^q,[c,d]} \leq (\lambda+1)\Omega_1(X,\delta)_{L^q,[c,d]}$, $\lambda > 0$, $\delta > 0$,
(vi) $\Omega_1(X+Y,\delta)_{L^q,[c,d]} \leq \Omega_1(X,\delta)_{L^q,[c,d]} + \Omega_1(Y,\delta)_{L^q,[c,d]}$, $\delta > 0$,
(vii) $\Omega_1(X,\cdot)_{L^q,[c,d]}$ *is continuous on* $\mathbb{R}_+$ *for* $X \in C_{\mathbb{R}}^{U_q}([c,d])$.

We give

Remark 15.7 *(to Proposition* 15.6*)* By Proposition 15.6(v) we get

$$\Omega_1(X,|x-y|)_{L^q,[c,d]} \leq \left\lceil \frac{|x-y|}{\delta} \right\rceil \Omega_1(X,\delta)_{L^q,[c,d]}, \tag{15.9}$$

$\forall\, x, y \in [c,d]$, any $\delta > 0$.

We give

Remark 15.8 *(continuation of Remark* 15.2*)* We assume again that $\left|X^{(n)}(t,\omega)\right| \leq M^*$, $\forall\,(t,\omega) \in [a,b] \times \Omega$, where $M^* > 0$. Let $\delta > 0$, $1 \leq q < \infty$. Then

$$\Omega_1\left(D_{*c}^{\alpha}X,\delta\right)_{L^q,[c,d]} = \sup\left\{\left(\int_{\Omega}\left|\left(D_{*c}^{\alpha}X\right)(x,\omega) - \left(D_{*c}^{\alpha}X\right)(y,\omega)\right|^q P(d\omega)\right)^{\frac{1}{q}} :\right.$$

$$\left. x, y \in [c,d] \subseteq [a,b],\ |x-y| \leq \delta\right\} \leq$$

$$\sup\left\{\left(\int_{\Omega}\left(\left|\left(D_{*c}^{\alpha}X\right)(x,\omega)\right| + \left|\left(D_{*c}^{\alpha}X\right)(y,\omega)\right|\right)^q P(d\omega)\right)^{\frac{1}{q}} :\right.$$

$$\left. x, y \in [c,d] \subseteq [a,b],\ |x-y| \leq \delta\right\} \overset{([5])}{\leq} \tag{15.10}$$

$$\frac{M^*}{\Gamma(n-\alpha+1)} \sup \left\{ \left(\int_{\Omega} \left((x-c)^{n-\alpha} + (y-c)^{n-\alpha} \right)^q P(d\omega) \right)^{\frac{1}{q}} : \right.$$

$$\left. x, y \in [c,d] \subseteq [a,b], |x-y| \leq \delta \right\} =$$

$$\frac{M^*}{\Gamma(n-\alpha+1)} \sup \left\{ \left((x-c)^{n-\alpha} + (y-c)^{n-\alpha} \right) : \right.$$

$$\left. x, y \in [c,d] \subseteq [a,b], |x-y| \leq \delta \right\} \leq \frac{2M^*}{\Gamma(n-\alpha+1)} (d-c)^{n-\alpha}.$$

That is

$$\Omega_1 \left(D_{*c}^{\alpha} X, \delta \right)_{L^q,[c,d]} \leq \frac{2M^*}{\Gamma(n-\alpha+1)} (d-c)^{n-\alpha}, \tag{15.11}$$

where $a \leq c < d \leq b$.

Similarly, it holds

$$\Omega_1 \left(D_{d-}^{\alpha} X, \delta \right)_{L^q,[c,d]} \leq \frac{2M^*}{\Gamma(n-\alpha+1)} (d-c)^{n-\alpha}, \tag{15.12}$$

where $a \leq c < d \leq b$.

Next let $x_0 \in [a,b]$, then

$$\sup_{x_0 \in [a,b]} \Omega_1 \left(D_{*x_0}^{\alpha} X, \delta \right)_{L^q,[x_0,b]} \leq \frac{2M^*}{\Gamma(n-\alpha+1)} \sup_{x_0 \in [a,b]} (b-x_0)^{n-\alpha}$$

$$= \frac{2M^*}{\Gamma(n-\alpha+1)} (b-a)^{n-\alpha}. \tag{15.13}$$

Similarly, it holds

$$\sup_{x_0 \in [a,b]} \Omega_1 \left(D_{x_0-}^{\alpha} X, \delta \right)_{L^q,[a,x_0]} \leq \frac{2M^*}{\Gamma(n-\alpha+1)} \sup_{x_0 \in [a,b]} (x_0-a)^{n-\alpha}$$

$$= \frac{2M^*}{\Gamma(n-\alpha+1)} (b-a)^{n-\alpha}. \tag{15.14}$$

We need

Concepts 15.9 (see also [10]) *Let $\widetilde{L}$ be a positive linear operator from $C([a,b])$ into itself. Let $X(t,\omega)$ be a stochastic process from $[a,b] \times (\Omega, \mathcal{F}, P)$ into $\mathbb{R}$, where $(\Omega, \mathcal{F}, P)$ is a probability space. Here we assume that for non-integer $\alpha > 0$, $\lceil \alpha \rceil = n$, $X(\cdot,\omega) \in AC^n([a,b])$ with $X^{(n)}(\cdot,\omega) \in L_{\infty}([a,b])$, $\forall\, \omega \in \Omega$.*

Also we assume for each $t \in [a,b]$ that $X^{(k)}(t,\cdot)$ is measurable for all $k = 1,\ldots,n-1$. Further we assume that $D_{*t}^{\alpha} X(z,\omega)$ is a stochastic process for

$z \in [t, b]$, $\omega \in \Omega$, and $D_{t-}^{\alpha} X(z, \omega)$ is a stochastic process for $z \in [a, t]$, $\omega \in \Omega$; $\forall\, t \in [a, b]$.

Define

$$M(X)(t, \omega) := \widetilde{L}(X(\cdot, \omega))(t), \forall \omega \in \Omega, \forall t \in [a, b], \tag{15.15}$$

and assume that it is a random variable in ω. Clearly M is a positive linear operator on stochastic processes.

We mention

Assumption 15.10 *(as in* [10]*)* Let non-integer $\alpha > 0$.

(i) For any $t \in [a, b]$ we assume that $D_{*t}^{\alpha} X(z, \omega)$ is continuous in $z \in [t, b]$, uniformly with respect to $\omega \in \Omega$. I.e. $\forall\, \varepsilon > 0\ \exists\, \delta > 0$: whenever $|z_1 - z_2| \leq \delta$; $z_1, z_2 \in [t, b]$, then $\left|D_{*t}^{\alpha} X(z_1, \omega) - D_{*t}^{\alpha} X(z_2, \omega)\right| \leq \varepsilon, \forall\, \omega \in \Omega$.
We denote this by $D_{*t}^{\alpha} X \in C_{\mathbb{R}}^{U}([t, b])$, the space of continuous in x, uniformly with respect to ω, stochastic processes over $[t, b]$.
(ii) For any $t \in [a, b]$ we assume that $D_{t-}^{\alpha} X(z, \omega)$ is continuous in $z \in [a, t]$, uniformly with respect to $\omega \in \Omega$. I.e. $\forall\, \varepsilon > 0\ \exists\, \delta > 0$: whenever $|z_1 - z_2| \leq \delta$; $z_1, z_2 \in [a, t]$, then $\left|D_{t-}^{\alpha} X(z_1, \omega) - D_{t-}^{\alpha} X(z_2, \omega)\right| \leq \varepsilon, \forall\, \omega \in \Omega$.
We denote this by $D_{t-}^{\alpha} X \in C_{\mathbb{R}}^{U}([a, t])$, the space of continuous in x, uniformly with respect to ω, stochastic processes over $[a, t]$.

Remark 15.11 Assumption 15.10 implies:

(i) $D_{*t}^{\alpha} X(\cdot, \omega) \in C([t, b]), \forall\, \omega \in \Omega$, and $D_{*t}^{\alpha} X$ is q-mean uniformly continuous in $z \in [t, b]$, that is $D_{*t}^{\alpha} X \in C_{\mathbb{R}}^{U_q}([t, b])$, for any $1 \leq q < \infty$.
(ii) $D_{t-}^{\alpha} X(\cdot, \omega) \in C([a, t]), \forall\, \omega \in \Omega$, and $D_{t-}^{\alpha} X$ is q-mean uniformly continuous in $z \in [a, t]$, that is $D_{t-}^{\alpha} X \in C_{\mathbb{R}}^{U_q}([a, t])$, for any $1 \leq q < \infty$.

We need

Definition 15.12 Denote by

$$(EX)(t) := \int_{\Omega} X(t, \omega) P(d\omega), \forall t \in [a, b], \tag{15.16}$$

the expectation operator.

We make

Assumption 15.13 *(as in* [10]*)* We assume that

$$\left(E\left|X^{(k)}\right|^{q}\right)(t) < \infty, \forall t \in [a, b], \tag{15.17}$$

$q > 1$, for all $k = 0, 1, \ldots, n - 1$, $n = \lceil \alpha \rceil$; $\alpha > 0$ non-integer.

We make

Assumption 15.14 *(as in* [10]*)* We assume that

$$\left(E\left|X^{(k)}\right|\right)(t) < \infty, \forall t \in [a,b], \tag{15.18}$$

for all $k = 0, 1, \ldots, n-1$, $n = \lceil \alpha \rceil$; $\alpha > 0$ non-integer.

We give

Remark 15.15 By the Riesz representation theorem [14] we have that there exists μ_t unique, completed Borel measure on $[a, b]$ with

$$m_t := \mu_t([a,b]) = \widetilde{L}(1)(t) > 0, \tag{15.19}$$

such that

$$\widetilde{L}(f)(t) = \int_{[a,b]} f(x)\, d\mu_t(x), \forall t \in [a,b], \forall f \in C([a,b]). \tag{15.20}$$

Consequently we have that

$$M(X)(t,\omega) = \int_{[a,b]} X(x,\omega)\, d\mu_t(x), \ \forall\ (t,\omega) \in [a,b] \times \Omega, \tag{15.21}$$

and X as in Concepts 15.9.

Here $\chi_{[\gamma,\delta]}(s)$ stands for the characteristic function on $[\gamma, \delta] \subseteq [a,b]$.

Notice that $(r > 0)$

$$\int_{[t,b]} (s-t)^r \mu_t(ds) = \int_{[a,b]} \chi_{[t,b]}(s)\, |s-t|^r \mu_t(ds) = \widetilde{L}\left(|\cdot - t|^r \chi_{[t,b]}(\cdot)\right)(t), \tag{15.22}$$

and

$$\int_{[a,t]} (t-s)^r \mu_t(ds) = \int_{[a,b]} \chi_{[a,t]}(s)\, |s-t|^r \mu_t(ds) = \widetilde{L}\left(|\cdot - t|^r \chi_{[a,t]}(\cdot)\right)(t). \tag{15.23}$$

Let now $n = \lceil \alpha \rceil$, $\alpha \notin \mathbb{N}$, $\alpha > 0$, $k = 1, \ldots, n-1$. Then by Hölder's inequality we obtain

$$\left|\int_{[a,b]} (x-t)^k\, d\mu_t(x)\right| \leq \int_{[a,b]} |x-t|^k\, d\mu_t(x) \leq$$

$$\left(\int_{[a,b]} |x-t|^{\alpha+1}\, d\mu_t(x)\right)^{\left(\frac{k}{\alpha+1}\right)} \left(\mu_t([a,b])\right)^{\left(\frac{\alpha+1-k}{\alpha+1}\right)}. \tag{15.24}$$

Therefore it holds

$$\left\| \widetilde{L}\left((\cdot - t)^k \right)(t) \right\|_{\infty,[a,b]} \leq \left\| \widetilde{L}\left(|\cdot - t|^k \right)(t) \right\|_{\infty,[a,b]} \leq$$

$$\left\| \widetilde{L}\left(|\cdot - t|^{\alpha+1} \right)(t) \right\|_{\infty,[a,b]}^{\left(\frac{k}{\alpha+1}\right)} \left\| \widetilde{L}(1) \right\|_{\infty,[a,b]}^{\left(\frac{\alpha+1-k}{\alpha+1}\right)}, \quad (15.25)$$

all $k = 1, \ldots, n-1$.

Also, we observe that

$$C([a,b]) \ni |\cdot - t|^{\alpha+1} \chi_{[a,t]}(\cdot) \leq |\cdot - t|^{\alpha+1}, \ \forall t \in [a,b], \quad (15.26)$$

and

$$C([a,b]) \ni |\cdot - t|^{\alpha+1} \chi_{[t,b]}(\cdot) \leq |\cdot - t|^{\alpha+1}, \ \forall t \in [a,b]. \quad (15.27)$$

By positivity of $\widetilde{L}$ we obtain

$$\left\| \widetilde{L}\left(|\cdot - t|^{\alpha+1} \chi_{[a,t]}(\cdot) \right)(t) \right\|_{\infty,[a,b]} \leq \left\| \widetilde{L}\left(|\cdot - t|^{\alpha+1} \right)(t) \right\|_{\infty,[a,b]} < \infty, \quad (15.28)$$

by $\widetilde{L}\left(|\cdot - t|^{\alpha+1} \right)(t)$ being continuous in $t \in [a,b]$, see p. 388 of [4],

and

$$\left\| \widetilde{L}\left(|\cdot - t|^{\alpha+1} \chi_{[t,b]}(\cdot) \right)(t) \right\|_{\infty,[a,b]} \leq \left\| \widetilde{L}\left(|\cdot - t|^{\alpha+1} \right)(t) \right\|_{\infty,[a,b]}. \quad (15.29)$$

Above (15.22)–(15.25) and (15.28), (15.29) can be used to derive convergence from (15.30), (15.32)–(15.34) next.

In this work we denote by $\widetilde{L}\left(\chi_{[a,t]}(\cdot) \right)(t) := \mu_t([a,t]) \leq \widetilde{L}(1)(t)$, and by $\widetilde{L}\left(\chi_{[t,b]}(\cdot) \right)(t) := \mu_t([t,b]) \leq \widetilde{L}(1)(t)$.

Next we mention the first main result from [10], on the quantitative stochastic fractional approximation regarding stochastic processes:

Theorem 15.16 ([10]) *Suppose Concepts 15.9, Assumptions 15.10 and 15.13. Here the non-integer $\alpha > 0$ is such that $\alpha > \frac{1}{q}$, where $p, q > 1$ such that $\frac{1}{p} + \frac{1}{q} = 1$. Then, $\forall\, t \in [a,b]$, we have:*

$$\left(E\left(|M(X) - X|^q \right)(t) \right)^{\frac{1}{q}} \leq \left(\left(E|X|^q \right)(t) \right)^{\frac{1}{q}} \left| \widetilde{L}(1)(t) - 1 \right| +$$

$$\sum_{k=1}^{n-1} \frac{\left(\left(E\left| X^{(k)} \right|^q \right)(t) \right)^{\frac{1}{q}}}{k!} \left| \widetilde{L}\left((\cdot - t)^k \right)(t) \right| + \frac{2^{\frac{1}{p}}}{\Gamma(\alpha)\left(p(\alpha-1)+1 \right)^{\frac{1}{p}} (q+1)^{\frac{\alpha}{q(\alpha+1)}}} \quad (15.30)$$

$$\left\{ \left\{ \left(\widetilde{L}\left(\chi_{[t,b]}(\cdot) \right)(t) \right)^{\frac{1}{p}} \left(\widetilde{L}\left(|\cdot - t|^{q(\alpha+1)} \chi_{[t,b]}(\cdot) \right)(t) \right)^{\frac{\alpha}{q(\alpha+1)}} \right. \right.$$

$$\left[\left(\widetilde{L}\left(\chi_{[t,b]}(\cdot) \right)(t) \right)^{\frac{1}{\alpha+1}} (q+1)^{\frac{\alpha}{(\alpha+1)}} + 1 \right]^{\frac{1}{q}}$$

$$\Omega_1\left(D^{\alpha}_{*t}X,\left(\frac{1}{(q+1)}\widetilde{L}\left(|\cdot-t|^{q(\alpha+1)}\chi_{[t,b]}(\cdot)\right)(t)\right)^{\frac{1}{q(\alpha+1)}}\right)_{L^q,[t,b]}\Bigg\}+$$

$$\left\{\left(\widetilde{L}\left(\chi_{[a,t]}(\cdot)\right)(t)\right)^{\frac{1}{p}}\left(\widetilde{L}\left(|\cdot-t|^{q(\alpha+1)}\chi_{[a,t]}(\cdot)\right)(t)\right)^{\frac{\alpha}{q(\alpha+1)}}\right.$$

$$\left[\left(\widetilde{L}\left(\chi_{[a,t]}(\cdot)\right)(t)\right)^{\frac{1}{\alpha+1}}(q+1)^{\frac{\alpha}{(\alpha+1)}}+1\right]^{\frac{1}{q}}$$

$$\Omega_1\left(D^{\alpha}_{t-}X,\left(\frac{1}{(q+1)}\widetilde{L}\left(|\cdot-t|^{q(\alpha+1)}\chi_{[a,t]}(\cdot)\right)(t)\right)^{\frac{1}{q(\alpha+1)}}\right)_{L^q,[a,t]}\Bigg\}\Bigg\}.$$

Here we have that

$$D^{\alpha}_{*t}X(t,\omega)=D^{\alpha}_{t-}X(t,\omega)=0, \tag{15.31}$$

$\forall\,\omega\in\Omega$, see [8, pp. 358–359].

We assume also

$$D^{\alpha}_{*t}X(s,\omega)=0,\text{ for } s<t,$$

and

$$D^{\alpha}_{t-}X(s,\omega)=0,\text{ for } s>t,$$

$\forall\,\omega\in\Omega.$

We mention the second main result from [10], the L_1 -quantitative stochastic fractional approximation of stochastic processes, it is the $q=1$ analog of Theorem 15.16. Inequality (15.32) is much simpler than (15.30).

Theorem 15.17 ([10]) *Suppose Concepts 15.9, Assumptions 15.10 and 15.14. Here* $\alpha>0$, $\alpha\notin\mathbb{N}$, $n=\lceil\alpha\rceil$. *Then,* $\forall\,t\in[a,b]$, *we have*

$$E\left(|M(X)-X|\right)(t)\leq\left(E|X|\right)(t)\left|\widetilde{L}(1)(t)-1\right|+$$

$$\sum_{k=1}^{n-1}\frac{\left(E\left|X^{(k)}\right|\right)(t)}{k!}\left|\widetilde{L}\left((\cdot-t)^k\right)(t)\right|+\frac{1}{\Gamma(\alpha+1)}$$

$$\left\{\left[\left(\widetilde{L}\left(\chi_{[a,t]}(\cdot)\right)(t)\right)^{\frac{1}{\alpha+1}}+\frac{1}{(\alpha+1)}\right]\left(\widetilde{L}\left(|\cdot-t|^{\alpha+1}\chi_{[a,t]}(\cdot)\right)(t)\right)^{\frac{\alpha}{\alpha+1}}\right.$$

$$\Omega_1\left(D^{\alpha}_{t-}X,\left(\widetilde{L}\left(|\cdot-t|^{\alpha+1}\chi_{[a,t]}(\cdot)\right)(t)\right)^{\frac{1}{\alpha+1}}\right)_{L^1,[a,t]}+$$

$$\left[\left(\widetilde{L}\left(\chi_{[t,b]}(\cdot)\right)(t)\right)^{\frac{1}{\alpha+1}}+\frac{1}{(\alpha+1)}\right]\left(\widetilde{L}\left(|\cdot-t|^{\alpha+1}\chi_{[t,b]}(\cdot)\right)(t)\right)^{\frac{\alpha}{\alpha+1}} \tag{15.32}$$

$$\Omega_1 \left(D^{\alpha}_{*t} X, \left(\widetilde{L} \left(|\cdot - t|^{\alpha+1} \chi_{[t,b]} (\cdot) \right) (t) \right)^{\frac{1}{\alpha+1}} \right)_{L^1,[t,b]} \Bigg\} .$$

We also mention a uniform norm result (denote $\|\cdot\|_{\infty,[a,b]} := \|f\|_{\infty}$). It is based on Theorem 15.16.

Theorem 15.18 ([10]) *Suppose Concepts 15.9, Assumptions 15.10 and 15.13. Here the non-integer $\alpha > 0$ is such that $\alpha > \frac{1}{q}$, where $p, q > 1 : \frac{1}{p} + \frac{1}{q} = 1$. Additionally assume that $\left|X^{(n)}(t,\omega)\right| \le M^*$, $\forall (t,\omega) \in [a,b] \times \Omega$, where $M^* > 0$. Then*

$$\left\| E\left(|M(X) - X|^q \right) \right\|_{\infty}^{\frac{1}{q}} \le \left\| E\left(|X|^q \right) \right\|_{\infty}^{\frac{1}{q}} \left\| \widetilde{L}(1) - 1 \right\|_{\infty} +$$

$$\sum_{k=1}^{n-1} \frac{\left\| E\left(\left|X^{(k)}\right|^q \right) \right\|_{\infty}^{\frac{1}{q}}}{k!} \left\| \widetilde{L}\left((\cdot - t)^k \right)(t) \right\|_{\infty} + \frac{2^{\frac{1}{p}}}{\Gamma(\alpha)\left(p(\alpha-1)+1\right)^{\frac{1}{p}} (q+1)^{\frac{\alpha}{q(\alpha+1)}}}$$

$$\left[\left\| \widetilde{L}(1) \right\|_{\infty}^{\frac{1}{\alpha+1}} (q+1)^{\frac{\alpha}{(\alpha+1)}} + 1 \right]^{\frac{1}{q}} \left\| \widetilde{L}(1) \right\|_{\infty}^{\frac{1}{p}}$$

$$\left\{ \left\{ \left\| \widetilde{L}\left(|\cdot - t|^{q(\alpha+1)} \chi_{[t,b]}(\cdot) \right)(t) \right\|_{\infty}^{\frac{\alpha}{q(\alpha+1)}} \right. \right. \tag{15.33}$$

$$\sup_{t \in [a,b]} \Omega_1 \left(D^{\alpha}_{*t} X, \left(\frac{1}{(q+1)} \left\| \widetilde{L}\left(|\cdot - t|^{q(\alpha+1)} \chi_{[t,b]}(\cdot) \right)(t) \right\|_{\infty} \right)^{\frac{1}{q(\alpha+1)}} \right)_{L^q,[t,b]} \Bigg\}$$

$$+ \left\{ \left\| \widetilde{L}\left(|\cdot - t|^{q(\alpha+1)} \chi_{[a,t]}(\cdot) \right)(t) \right\|_{\infty}^{\frac{\alpha}{q(\alpha+1)}} \right.$$

$$\sup_{t \in [a,b]} \Omega_1 \left(D^{\alpha}_{t-} X, \left(\frac{1}{(q+1)} \left\| \widetilde{L}\left(|\cdot - t|^{q(\alpha+1)} \chi_{[a,t]}(\cdot) \right)(t) \right\|_{\infty} \right)^{\frac{1}{q(\alpha+1)}} \right)_{L^q,[a,t]} \Bigg\} \Bigg\} .$$

Based on Theorem 15.17 we mention the following uniform estimate:

Theorem 15.19 ([10]) *Suppose Concepts 15.9, Assumptions 15.10 and 15.14. Here $\alpha > 0$, $\alpha \notin \mathbb{N}$, $n = \lceil \alpha \rceil$. Additionally assume that $\left|X^{(n)}(t,\omega)\right| \le M^*$, $\forall (t,\omega) \in [a,b] \times \Omega$, where $M^* > 0$. Then*

$$\left\| E\left(|M(X) - X| \right) \right\|_{\infty} \le \left\| E\left(|X| \right) \right\|_{\infty} \left\| \widetilde{L}(1) - 1 \right\|_{\infty} +$$

$$\sum_{k=1}^{n-1} \frac{\left\| E\left(\left|X^{(k)}\right| \right) \right\|_{\infty}}{k!} \left\| \widetilde{L}\left((\cdot - t)^k \right)(t) \right\|_{\infty} + \frac{\left[\left\| \widetilde{L}(1) \right\|_{\infty}^{\frac{1}{\alpha+1}} + \frac{1}{\alpha+1} \right]}{\Gamma(\alpha+1)}$$

$$\left\{ \left\| \widetilde{L} \left(|\cdot - t|^{\alpha+1} \chi_{[a,t]} (\cdot) \right) (t) \right\|_{\infty}^{\frac{\alpha}{\alpha+1}} \right.$$

$$\sup_{t \in [a,b]} \Omega_1 \left(D_{t-}^{\alpha} X, \left\| \widetilde{L} \left(|\cdot - t|^{\alpha+1} \chi_{[a,t]} (\cdot) \right) (t) \right\|_{\infty}^{\frac{1}{\alpha+1}} \right)_{L^1,[a,t]} + \tag{15.34}$$

$$\left\| \widetilde{L} \left(|\cdot - t|^{\alpha+1} \chi_{[t,b]} (\cdot) \right) (t) \right\|_{\infty}^{\frac{\alpha}{\alpha+1}}$$

$$\left. \sup_{t \in [a,b]} \Omega_1 \left(D_{*t}^{\alpha} X, \left\| \widetilde{L} \left(|\cdot - t|^{\alpha+1} \chi_{[t,b]} (\cdot) \right) (t) \right\|_{\infty}^{\frac{1}{\alpha+1}} \right)_{L^1,[t,b]} \right\}.$$

We make

Remark 15.20 Next we specify $[a, b]$ as $[-\pi, \pi]$. Clearly then $\widetilde{L} : C([-\pi, \pi]) \to C([-\pi, \pi])$ is the positive linear operator on hand.

Here $n = \lceil \alpha \rceil, \alpha \notin \mathbb{N}, \alpha > 0, k = 1, \ldots, n-1$. Next we use Hölder's inequality. We notice that

$$\widetilde{L} \left(\left(\sin \left(\frac{|\cdot - t|}{4} \right) \right)^k \right) (t) = \int_{[-\pi,\pi]} \left(\sin \left(\frac{|x - t|}{4} \right)^k \right) d\mu_t (x) \leq$$

$$\left(\int_{[-\pi,\pi]} \left(\sin \left(\frac{|x - t|}{4} \right)^{\alpha+1} \right) d\mu_t (x) \right)^{\frac{k}{\alpha+1}} (\mu_t ([-\pi, \pi]))^{\frac{\alpha+1-k}{\alpha+1}} = \tag{15.35}$$

$$\left(\widetilde{L} \left(\left(\sin \left(\frac{|\cdot - t|}{4} \right) \right)^{\alpha+1} \right) (t) \right)^{\frac{k}{\alpha+1}} \left(\widetilde{L} (1) (t) \right)^{\frac{\alpha+1-k}{\alpha+1}}.$$

That is

$$\widetilde{L} \left(\left(\sin \left(\frac{|\cdot - t|}{4} \right) \right)^k \right) (t) \leq \tag{15.36}$$

$$\left(\widetilde{L} \left(\left(\sin \left(\frac{|\cdot - t|}{4} \right) \right)^{\alpha+1} \right) (t) \right)^{\frac{k}{\alpha+1}} \left(\widetilde{L} (1) (t) \right)^{\frac{\alpha+1-k}{\alpha+1}},$$

for $k = 1, \ldots, n-1$; true also for $q(\alpha + 1)$ instead of $(\alpha + 1)$, for any $1 < q < \infty$.

Next $\|\cdot\|_{\infty}$ denotes $\|\cdot\|_{\infty,[-\pi,\pi]}$.

Consequently, it holds

$$\left\| \widetilde{L} \left(\left(\sin \left(\frac{|\cdot - t|}{4} \right) \right)^k \right) (t) \right\|_{\infty} \leq \tag{15.37}$$

$$\left\| \widetilde{L}\left(\left(\sin\left(\frac{|\cdot-t|}{4}\right)\right)^{\alpha+1}\right)(t)\right\|_{\infty}^{\frac{k}{\alpha+1}} \left\|\widetilde{L}(1)(t)\right\|_{\infty}^{\frac{\alpha+1-k}{\alpha+1}},$$

for $k = 1, \ldots, n-1$; true also for $q(\alpha+1)$ instead of $(\alpha+1)$, for any $1 < q < \infty$.

In this work we use a lot the following well known inequality:

$$|z| \leq \pi \sin\left(\frac{|z|}{2}\right), \forall z \in [-\pi, \pi]. \tag{15.38}$$

Notice that, for any $t \in [-\pi, \pi]$, we have $C([-\pi,\pi]) \ni |\cdot - t| \chi_{[-\pi,t]}(\cdot) \leq |\cdot - t| \in C([-\pi,\pi])$, therefore

$$C([-\pi,\pi]) \ni \left(\sin\left(\frac{|\cdot-t|\chi_{[-\pi,t]}(\cdot)}{4}\right)\right)^{\alpha+1} \leq \left(\sin\left(\frac{|\cdot-t|}{4}\right)\right)^{\alpha+1} \in C([-\pi,\pi]). \tag{15.39}$$

Consequently, by positivity of $\widetilde{L}$ we obtain

$$\left\| \widetilde{L}\left(\left(\sin\left(\frac{|\cdot-t|\chi_{[-\pi,t]}(\cdot)}{4}\right)\right)^{\alpha+1}\right)(t)\right\|_{\infty} \leq \left\| \widetilde{L}\left(\left(\sin\left(\frac{|\cdot-t|}{4}\right)\right)^{\alpha+1}\right)(t)\right\|_{\infty}. \tag{15.40}$$

Similarly, for any $t \in [-\pi, \pi]$, we have $C([-\pi,\pi]) \ni |\cdot - t| \chi_{[t,\pi]}(\cdot) \leq |\cdot - t| \in C([-\pi,\pi])$, thus

$$C([-\pi,\pi]) \ni \left(\sin\left(\frac{|\cdot-t|\chi_{[t,\pi]}(\cdot)}{4}\right)\right)^{\alpha+1} \leq \left(\sin\left(\frac{|\cdot-t|}{4}\right)\right)^{\alpha+1} \in C([-\pi,\pi]). \tag{15.41}$$

Hence

$$\left\| \widetilde{L}\left(\left(\sin\left(\frac{|\cdot-t|\chi_{[t,\pi]}(\cdot)}{4}\right)\right)^{\alpha+1}\right)(t)\right\|_{\infty} \leq \left\| \widetilde{L}\left(\left(\sin\left(\frac{|\cdot-t|}{4}\right)\right)^{\alpha+1}\right)(t)\right\|_{\infty}. \tag{15.42}$$

So, if the right hand side of (15.40), (15.42) goes to zero, so do their left hand sides. Above in (15.39)–(15.42), one can use $q(\alpha+1)$ instead of $(\alpha+1)$, $1 < q < \infty$.

A further detailed analysis reveals:

We have that $(1 \leq q < \infty)$

$$\widetilde{L}\left(|\cdot - t|^{q(\alpha+1)} \chi_{[t,\pi]}(\cdot)\right)(t) \overset{(15.20)}{=} \int_{[t,\pi]} (x-t)^{q(\alpha+1)} d\mu_t(x) =$$

$$2^{q(\alpha+1)} \int_{[t,\pi]} \left(\frac{x-t}{2}\right)^{q(\alpha+1)} d\mu_t(x) \overset{(15.38)}{\leq}$$

$$(2\pi)^{q(\alpha+1)} \int_{[t,\pi]} \left(\sin\left(\frac{x-t}{4}\right)\right)^{q(\alpha+1)} d\mu_t(x) = \tag{15.43}$$

$$(2\pi)^{q(\alpha+1)} \int_{[t,\pi]} \left(\sin\left(\frac{|x-t|}{4}\right)\right)^{q(\alpha+1)} d\mu_t(x) =$$

$$(2\pi)^{q(\alpha+1)} \int_{[-\pi,\pi]} \left(\sin\left(\frac{|x-t|}{4}\right)\right)^{q(\alpha+1)} \chi_{[t,\pi]}(x)\, d\mu_t(x) =$$

$$(2\pi)^{q(\alpha+1)} \int_{[-\pi,\pi]} \left(\sin\left(\frac{|x-t|}{4}\chi_{[t,\pi]}(x)\right)\right)^{q(\alpha+1)} d\mu_t(x) \overset{(15.20)}{=}$$

$$(2\pi)^{q(\alpha+1)} \widetilde{L}\left(\left(\sin\left(\frac{|\cdot-t|}{4}\chi_{[t,\pi]}(\cdot)\right)\right)^{q(\alpha+1)}\right)(t).$$

That is, we have obtained

$$\widetilde{L}\left(|\cdot-t|^{q(\alpha+1)}\chi_{[t,\pi]}(\cdot)\right)(t) \leq (2\pi)^{q(\alpha+1)} \widetilde{L}\left(\left(\sin\left(\frac{|\cdot-t|}{4}\chi_{[t,\pi]}(\cdot)\right)\right)^{q(\alpha+1)}\right)(t). \tag{15.44}$$

Similarly, it holds

$$\widetilde{L}\left(|\cdot-t|^{q(\alpha+1)}\chi_{[-\pi,t]}(\cdot)\right)(t) \leq (2\pi)^{q(\alpha+1)} \widetilde{L}\left(\left(\sin\left(\frac{|\cdot-t|}{4}\chi_{[-\pi,t]}(\cdot)\right)\right)^{q(\alpha+1)}\right)(t). \tag{15.45}$$

Above inequalities (15.44), (15.45) are valid for any $1 \leq q < \infty$.

Furthermore, we observe that

$$\left|\widetilde{L}\left((\cdot-t)^k\right)(t)\right| = \left|\int_{[-\pi,\pi]} (x-t)^k\, d\mu_t(x)\right| \leq \int_{[-\pi,\pi]} |x-t|^k\, d\mu_t(x) =$$

$$2^k \int_{[-\pi,\pi]} \left(\frac{|x-t|}{2}\right)^k d\mu_t(x) \overset{(15.38)}{\leq} (2\pi)^k \int_{[-\pi,\pi]} \left(\sin\left(\frac{|x-t|}{4}\right)\right)^k d\mu_t(x) \tag{15.46}$$

$$= (2\pi)^k \widetilde{L}\left(\left(\sin\left(\frac{|\cdot-t|}{4}\right)\right)^k\right)(t).$$

That is

$$\left|\widetilde{L}\left((\cdot-t)^k\right)(t)\right| \leq (2\pi)^k \widetilde{L}\left(\left(\sin\left(\frac{|\cdot-t|}{4}\right)\right)^k\right)(t), \tag{15.47}$$

$\forall\, t \in [-\pi, \pi]$, all $k = 1, \ldots, n-1$.

15.4 Main Results

Next we present our first main result on the trigonometric quantitative stochastic fractional approximation of stochastic processes.

It is a pointwise result.

Theorem 15.21 *Here* $[a, b] = [-\pi, \pi]$. *Suppose Concepts 15.9, Assumptions 15.10 and 15.13. Let* $\alpha > 0$, $\alpha \notin \mathbb{N}$, *such that* $\alpha > \frac{1}{q}$, *where* $p, q > 1$ *and* $\frac{1}{p} + \frac{1}{q} = 1$. *Then,* $\forall\, t \in [-\pi, \pi]$, *we have*

$$\left(E\left(|M(X) - X|^q\right)(t)\right)^{\frac{1}{q}} \leq \left(\left(E|X|^q\right)(t)\right)^{\frac{1}{q}} \left|\widetilde{L}(1)(t) - 1\right| +$$

$$\sum_{k=1}^{n-1} \frac{\left(\left(E\left|X^{(k)}\right|^q\right)(t)\right)^{\frac{1}{q}}}{k!} (2\pi)^k \widetilde{L}\left(\left(\sin\left(\frac{|\cdot - t|}{4}\right)\right)^k\right)(t) +$$

$$\frac{2^{\alpha+\frac{1}{p}}\pi^{\alpha}}{\Gamma(\alpha)(p(\alpha-1)+1)^{\frac{1}{p}}(q+1)^{\frac{\alpha}{q(\alpha+1)}}} \tag{15.48}$$

$$\left\{\left\{\left(\widetilde{L}\left(\chi_{[t,\pi]}(\cdot)\right)(t)\right)^{\frac{1}{p}} \left(\widetilde{L}\left(\left(\sin\left(\frac{|\cdot - t|}{4}\chi_{[t,\pi]}(\cdot)\right)\right)^{q(\alpha+1)}\right)(t)\right)^{\frac{\alpha}{q(\alpha+1)}}\right.\right.$$

$$\left[\left(\widetilde{L}\left(\chi_{[t,\pi]}(\cdot)\right)(t)\right)^{\frac{1}{\alpha+1}} (q+1)^{\frac{\alpha}{(\alpha+1)}} + 1\right]^{\frac{1}{q}}$$

$$\left.\Omega_1\left(D^{\alpha}_{*t}X, 2\pi\left(\frac{1}{(q+1)}\widetilde{L}\left(\left(\sin\left(\frac{|\cdot - t|}{4}\chi_{[t,\pi]}(\cdot)\right)\right)^{q(\alpha+1)}\right)(t)\right)^{\frac{1}{q(\alpha+1)}}\right)_{L^q,[t,\pi]}\right\} +$$

$$\left\{\left(\widetilde{L}\left(\chi_{[-\pi,t]}(\cdot)\right)(t)\right)^{\frac{1}{p}} \left(\widetilde{L}\left(\left(\sin\left(\frac{|\cdot - t|}{4}\chi_{[-\pi,t]}(\cdot)\right)\right)^{q(\alpha+1)}\right)(t)\right)^{\frac{\alpha}{q(\alpha+1)}}\right.$$

$$\left[\left(\widetilde{L}\left(\chi_{[-\pi,t]}(\cdot)\right)(t)\right)^{\frac{1}{\alpha+1}} (q+1)^{\frac{\alpha}{(\alpha+1)}} + 1\right]^{\frac{1}{q}}$$

$$\left.\left.\Omega_1\left(D^{\alpha}_{t-}X, 2\pi\left(\frac{1}{(q+1)}\widetilde{L}\left(\left(\sin\left(\frac{|\cdot - t|}{4}\chi_{[-\pi,t]}(\cdot)\right)\right)^{q(\alpha+1)}\right)(t)\right)^{\frac{1}{q(\alpha+1)}}\right)_{L^q,[-\pi,t]}\right\}\right\}.$$

Proof By Theorem 15.16 on $[-\pi, \pi]$ we get ($\forall\, t \in [-\pi, \pi]$) that

$$\left(E\left(|M(X) - X|^q\right)(t)\right)^{\frac{1}{q}} \leq \left(\left(E|X|^q\right)(t)\right)^{\frac{1}{q}} \left|\widetilde{L}(1)(t) - 1\right| +$$

$$\sum_{k=1}^{n-1} \frac{\left(\left(E\left|X^{(k)}\right|^q\right)(t)\right)^{\frac{1}{q}}}{k!} \left|\widetilde{L}\left((\cdot - t)^k\right)(t)\right| + \frac{2^{\frac{1}{p}}}{\Gamma(\alpha)\left(p(\alpha-1)+1\right)^{\frac{1}{p}} (q+1)^{\frac{\alpha}{q(\alpha+1)}}}$$

$$\left\{\left\{\left(\widetilde{L}\left(\chi_{[t,\pi]}(\cdot)\right)(t)\right)^{\frac{1}{p}} \left(\widetilde{L}\left(|\cdot - t|^{q(\alpha+1)} \chi_{[t,\pi]}(\cdot)\right)(t)\right)^{\frac{\alpha}{q(\alpha+1)}}\right.\right.$$

$$\left[\left(\widetilde{L}\left(\chi_{[t,\pi]}(\cdot)\right)(t)\right)^{\frac{1}{\alpha+1}} (q+1)^{\frac{\alpha}{(\alpha+1)}} + 1\right]^{\frac{1}{q}}$$

$$\left.\Omega_1\left(D_{*t}^{\alpha} X, \left(\frac{1}{(q+1)} \widetilde{L}\left(|\cdot - t|^{q(\alpha+1)} \chi_{[t,\pi]}(\cdot)\right)(t)\right)^{\frac{1}{q(\alpha+1)}}\right)_{L^q,[t,\pi]}\right\} +$$

$$\left\{\left(\widetilde{L}\left(\chi_{[-\pi,t]}(\cdot)\right)(t)\right)^{\frac{1}{p}} \left(\widetilde{L}\left(|\cdot - t|^{q(\alpha+1)} \chi_{[-\pi,t]}(\cdot)\right)(t)\right)^{\frac{\alpha}{q(\alpha+1)}}\right. \tag{15.49}$$

$$\left[\left(\widetilde{L}\left(\chi_{[-\pi,t]}(\cdot)\right)(t)\right)^{\frac{1}{\alpha+1}} (q+1)^{\frac{\alpha}{(\alpha+1)}} + 1\right]^{\frac{1}{q}}$$

$$\left.\left.\Omega_1\left(D_{t-}^{\alpha} X, \left(\frac{1}{(q+1)} \widetilde{L}\left(|\cdot - t|^{q(\alpha+1)} \chi_{[-\pi,t]}(\cdot)\right)(t)\right)^{\frac{1}{q(\alpha+1)}}\right)_{L^q,[-\pi,t]}\right\}\right\} =: (*).$$

Using (15.44), (15.45) and (15.47) we obtain:

$$(*) \leq \left(\left(E|X|^q\right)(t)\right)^{\frac{1}{q}} \left|\widetilde{L}(1)(t) - 1\right| +$$

$$\sum_{k=1}^{n-1} \frac{\left(\left(E\left|X^{(k)}\right|^q\right)(t)\right)^{\frac{1}{q}}}{k!} (2\pi)^k \widetilde{L}\left(\left(\sin\left(\frac{|\cdot - t|}{4}\right)\right)^k\right)(t) +$$

$$\frac{2^{\frac{1}{p}}}{\Gamma(\alpha)\left(p(\alpha-1)+1\right)^{\frac{1}{p}} (q+1)^{\frac{\alpha}{q(\alpha+1)}}}$$

$$\left\{\left\{\left(\widetilde{L}\left(\chi_{[t,\pi]}(\cdot)\right)(t)\right)^{\frac{1}{p}} (2\pi)^{\alpha} \left(\widetilde{L}\left(\left(\sin\left(\frac{|\cdot - t|}{4} \chi_{[t,\pi]}(\cdot)\right)\right)^{q(\alpha+1)}\right)(t)\right)^{\frac{\alpha}{q(\alpha+1)}}\right.\right.$$

$$\left[\left(\widetilde{L}\left(\chi_{[t,\pi]}(\cdot)\right)(t)\right)^{\frac{1}{\alpha+1}} (q+1)^{\frac{\alpha}{(\alpha+1)}} + 1\right]^{\frac{1}{q}} \tag{15.50}$$

$$\Omega_1\left(D^{\alpha}_{*t}X, 2\pi\left(\frac{1}{(q+1)}\widetilde{L}\left(\left(\sin\left(\frac{|\cdot-t|}{4}\chi_{[t,\pi]}(\cdot)\right)\right)^{q(\alpha+1)}\right)(t)\right)^{\frac{1}{q(\alpha+1)}}\right)_{L^q,[t,\pi]}\Bigg\}+$$

$$\Bigg\{\left(\widetilde{L}\left(\chi_{[-\pi,t]}(\cdot)\right)(t)\right)^{\frac{1}{p}}(2\pi)^{\alpha}\left(\widetilde{L}\left(\left(\sin\left(\frac{|\cdot-t|}{4}\chi_{[-\pi,t]}(\cdot)\right)\right)^{q(\alpha+1)}\right)(t)\right)^{\frac{\alpha}{q(\alpha+1)}}$$

$$\left[\left(\widetilde{L}\left(\chi_{[-\pi,t]}(\cdot)\right)(t)\right)^{\frac{1}{\alpha+1}}(q+1)^{\frac{\alpha}{(\alpha+1)}}+1\right]^{\frac{1}{q}}$$

$$\Omega_1\left(D^{\alpha}_{t-}X, 2\pi\left(\frac{1}{(q+1)}\widetilde{L}\left(\left(\sin\left(\frac{|\cdot-t|}{4}\chi_{[-\pi,t]}(\cdot)\right)\right)^{q(\alpha+1)}\right)(t)\right)^{\frac{1}{q(\alpha+1)}}\right)_{L^q,[-\pi,t]}\Bigg\}\Bigg\},$$

proving the claim. ■

We continue with the trigonometric L_1-quantitative stochastic fractional pointwise approximation of stochastic processes.

Theorem 15.22 *Here $[a,b]=[-\pi,\pi]$. Suppose Concepts 15.9, Assumptions 15.10 and 15.14. Here $\alpha>0$, $\alpha\notin\mathbb{N}$, $n=\lceil\alpha\rceil$. Then, $\forall\, t\in[-\pi,\pi]$, we have*

$$E\left(|M(X)-X|\right)(t)\leq\left(E|X|\right)(t)\left|\widetilde{L}(1)(t)-1\right|+$$

$$\sum_{k=1}^{n-1}\frac{\left(E\left|X^{(k)}\right|\right)(t)}{k!}(2\pi)^k\,\widetilde{L}\left(\left(\sin\left(\frac{|\cdot-t|}{4}\right)\right)^k\right)(t)+\frac{(2\pi)^{\alpha}}{\Gamma(\alpha+1)}$$

$$\Bigg\{\left[\left(\widetilde{L}\left(\chi_{[-\pi,t]}(\cdot)\right)(t)\right)^{\frac{1}{\alpha+1}}+\frac{1}{(\alpha+1)}\right]\left(\widetilde{L}\left(\left(\sin\left(\frac{|\cdot-t|}{4}\chi_{[-\pi,t]}(\cdot)\right)\right)^{\alpha+1}\right)(t)\right)^{\frac{\alpha}{\alpha+1}}$$

$$\Omega_1\left(D^{\alpha}_{t-}X, 2\pi\left(\widetilde{L}\left(\left(\sin\left(\frac{|\cdot-t|}{4}\chi_{[-\pi,t]}(\cdot)\right)\right)^{\alpha+1}\right)(t)\right)^{\frac{1}{\alpha+1}}\right)_{L^1,[-\pi,t]}+$$

$$\left[\left(\widetilde{L}\left(\chi_{[t,\pi]}(\cdot)\right)(t)\right)^{\frac{1}{\alpha+1}}+\frac{1}{(\alpha+1)}\right]\left(\widetilde{L}\left(\left(\sin\left(\frac{|\cdot-t|}{4}\chi_{[t,\pi]}(\cdot)\right)\right)^{\alpha+1}\right)(t)\right)^{\frac{\alpha}{\alpha+1}}$$

$$\Omega_1\left(D^{\alpha}_{*t}X, 2\pi\left(\widetilde{L}\left(\left(\sin\left(\frac{|\cdot-t|}{4}\chi_{[t,\pi]}(\cdot)\right)\right)^{\alpha+1}\right)(t)\right)^{\frac{1}{\alpha+1}}\right)_{L^1,[t,\pi]}\Bigg\}. \quad (15.51)$$

Proof By Theorem 15.17 for $[a,b]=[-\pi,\pi]$, and $\forall\, t\in[-\pi,\pi]$, we get

$$E\left(|M(X)-X|\right)(t)\leq (E|X|)(t)\left|\widetilde{L}(1)(t)-1\right|+$$

$$\sum_{k=1}^{n-1}\frac{\left(E\left|X^{(k)}\right|\right)(t)}{k!}\left|\widetilde{L}\left((\cdot-t)^{k}\right)(t)\right|+\frac{1}{\Gamma(\alpha+1)}$$

$$\left\{\left[\left(\widetilde{L}\left(\chi_{[-\pi,t]}(\cdot)\right)(t)\right)^{\frac{1}{\alpha+1}}+\frac{1}{(\alpha+1)}\right]\left(\widetilde{L}\left(|\cdot-t|^{\alpha+1}\chi_{[-\pi,t]}(\cdot)\right)(t)\right)^{\frac{\alpha}{\alpha+1}}\right.$$

$$\Omega_1\left(D_{t-}^{\alpha}X,\left(\widetilde{L}\left(|\cdot-t|^{\alpha+1}\chi_{[-\pi,t]}(\cdot)\right)(t)\right)^{\frac{1}{\alpha+1}}\right)_{L^1,[-\pi,t]}+ \tag{15.52}$$

$$\left[\left(\widetilde{L}\left(\chi_{[t,\pi]}(\cdot)\right)(t)\right)^{\frac{1}{\alpha+1}}+\frac{1}{(\alpha+1)}\right]\left(\widetilde{L}\left(|\cdot-t|^{\alpha+1}\chi_{[t,\pi]}(\cdot)\right)(t)\right)^{\frac{\alpha}{\alpha+1}}$$

$$\left.\Omega_1\left(D_{*t}^{\alpha}X,\left(\widetilde{L}\left(|\cdot-t|^{\alpha+1}\chi_{[t,\pi]}(\cdot)\right)(t)\right)^{\frac{1}{\alpha+1}}\right)_{L^1,[t,\pi]}\right\}=:(**).$$

Using (15.44), (15.45) and (15.47) we obtain:

$$(**)\leq (E|X|)(t)\left|\widetilde{L}(1)(t)-1\right|+$$

$$\sum_{k=1}^{n-1}\frac{\left(E\left|X^{(k)}\right|\right)(t)}{k!}(2\pi)^{k}\,\widetilde{L}\left(\left(\sin\left(\frac{|\cdot-t|}{4}\right)\right)^{k}\right)(t)+\frac{(2\pi)^{\alpha}}{\Gamma(\alpha+1)}$$

$$\left\{\left[\left(\widetilde{L}\left(\chi_{[-\pi,t]}(\cdot)\right)(t)\right)^{\frac{1}{\alpha+1}}+\frac{1}{(\alpha+1)}\right]\left(\widetilde{L}\left(\left(\sin\left(\frac{|\cdot-t|}{4}\chi_{[-\pi,t]}(\cdot)\right)\right)^{\alpha+1}\right)(t)\right)^{\frac{\alpha}{\alpha+1}}\right.$$

$$\Omega_1\left(D_{t-}^{\alpha}X,2\pi\left(\widetilde{L}\left(\left(\sin\left(\frac{|\cdot-t|}{4}\chi_{[-\pi,t]}(\cdot)\right)\right)^{\alpha+1}\right)(t)\right)^{\frac{1}{\alpha+1}}\right)_{L^1,[-\pi,t]}+$$

$$\left[\left(\widetilde{L}\left(\chi_{[t,\pi]}(\cdot)\right)(t)\right)^{\frac{1}{\alpha+1}}+\frac{1}{(\alpha+1)}\right]\left(\widetilde{L}\left(\left(\sin\left(\frac{|\cdot-t|}{4}\chi_{[t,\pi]}(\cdot)\right)\right)^{\alpha+1}\right)(t)\right)^{\frac{\alpha}{\alpha+1}}$$

$$\left.\Omega_1\left(D_{*t}^{\alpha}X,2\pi\left(\widetilde{L}\left(\left(\sin\left(\frac{|\cdot-t|}{4}\chi_{[t,\pi]}(\cdot)\right)\right)^{\alpha+1}\right)(t)\right)^{\frac{1}{\alpha+1}}\right)_{L^1,[t,\pi]}\right\}. \tag{15.53}$$

The claim is proved. ■

We continue with trigonometric fractional uniform estimates ($\|\cdot\|_{\infty,[-\pi,\pi]} := \|\cdot\|_{\infty}$) in L_q-mean ($1 \leq q < \infty$).

Theorem 15.23 *Here $[a,b] = [-\pi,\pi]$. Suppose Concepts 15.9, Assumptions 15.10 and 15.13. Let $\alpha > 0$, $\alpha \notin \mathbb{N}$, $n = \lceil \alpha \rceil$, such that $\alpha > \frac{1}{q}$, where $p, q > 1$ and $\frac{1}{p} + \frac{1}{q} = 1$. Additionally assume that $\left|X^{(n)}(t,\omega)\right| \leq M^*$, $\forall (t,\omega) \in [-\pi,\pi] \times \Omega$, where $M^* > 0$. Then*

$$\left\|E\left(|M(X) - X|^q\right)\right\|_{\infty}^{\frac{1}{q}} \leq \left\|E\left(|X|^q\right)\right\|_{\infty}^{\frac{1}{q}} \left\|\widetilde{L}(1) - 1\right\|_{\infty} +$$

$$\sum_{k=1}^{n-1} \frac{\left\|E\left(\left|X^{(k)}\right|^q\right)\right\|_{\infty}^{\frac{1}{q}}}{k!} (2\pi)^k \left\|\widetilde{L}\left(\left(\sin\left(\frac{|\cdot - t|}{4}\right)\right)^k\right)(t)\right\|_{\infty} +$$

$$\frac{\pi^{\alpha} 2^{\alpha + \frac{1}{p}} \left[\left\|\widetilde{L}(1)\right\|_{\infty}^{\frac{1}{\alpha+1}} (q+1)^{\frac{\alpha}{(\alpha+1)}} + 1\right]^{\frac{1}{q}} \left\|\widetilde{L}(1)\right\|_{\infty}^{\frac{1}{p}}}{\Gamma(\alpha)(p(\alpha - 1) + 1)^{\frac{1}{p}} (q+1)^{\frac{\alpha}{q(\alpha+1)}}} \tag{15.54}$$

$$\left\{\left\{\left\|\widetilde{L}\left(\left(\sin\left(\frac{|\cdot - t|}{4}\chi_{[t,\pi]}(\cdot)\right)\right)^{q(\alpha+1)}\right)(t)\right\|_{\infty}^{\frac{\alpha}{q(\alpha+1)}}\right.\right.$$

$$\left.\sup_{t\in[-\pi,\pi]} \Omega_1\left(D_{*t}^{\alpha} X, 2\pi\left(\frac{1}{(q+1)}\left\|\widetilde{L}\left(\left(\sin\left(\frac{|\cdot - t|}{4}\chi_{[t,\pi]}(\cdot)\right)\right)^{q(\alpha+1)}\right)(t)\right\|_{\infty}\right)^{\frac{1}{q(\alpha+1)}}\right)_{L^q,[t,\pi]}\right\}$$

$$+\left\{\left\|\widetilde{L}\left(\left(\sin\left(\frac{|\cdot - t|}{4}\chi_{[-\pi,t]}(\cdot)\right)\right)^{q(\alpha+1)}\right)(t)\right\|_{\infty}^{\frac{\alpha}{q(\alpha+1)}}\right.$$

$$\left.\left.\sup_{t\in[-\pi,\pi]} \Omega_1\left(D_{t-}^{\alpha} X, 2\pi\left(\frac{1}{(q+1)}\left\|\widetilde{L}\left(\left(\sin\left(\frac{|\cdot - t|}{4}\chi_{[-\pi,t]}(\cdot)\right)\right)^{q(\alpha+1)}\right)(t)\right\|_{\infty}\right)^{\frac{1}{q(\alpha+1)}}\right)_{L^q,[-\pi,t]}\right\}\right\}.$$

Proof Based on Theorem 15.21. We take into account (15.13), (15.14) and the positivity of $\widetilde{L}$. ∎

We continue with the L_1-mean uniform stochastic fractional result.

Theorem 15.24 *Here $[a,b] = [-\pi,\pi]$. Suppose Concepts 15.9, Assumptions 15.10 and 15.14. Let $\alpha > 0$, $\alpha \notin \mathbb{N}$, $n = \lceil \alpha \rceil$. Additionally assume that $\left|X^{(n)}(t,\omega)\right| \leq M^*$, $\forall (t,\omega) \in [-\pi,\pi] \times \Omega$, where $M^* > 0$. Then*

$$\|E(|M(X) - X|)\|_{\infty} \leq \|E(|X|)\|_{\infty} \left\|\widetilde{L}(1) - 1\right\|_{\infty} +$$

$$\sum_{k=1}^{n-1} \frac{\left\|E\left(\left|X^{(k)}\right|\right)\right\|_{\infty}}{k!} (2\pi)^k \left\|\widetilde{L}\left(\left(\sin\left(\frac{|\cdot - t|}{4}\right)\right)^k\right)(t)\right\|_{\infty} +$$

$$\frac{(2\pi)^{\alpha}}{\Gamma(\alpha+1)}\left[\left\|\widetilde{L}(1)\right\|_{\infty}^{\frac{1}{\alpha+1}}+\frac{1}{\alpha+1}\right]$$

$$\left\{\left\|\widetilde{L}\left(\left(\sin\left(\frac{|\cdot-t|}{4}\chi_{[-\pi,t]}(\cdot)\right)\right)^{\alpha+1}\right)(t)\right\|_{\infty}^{\frac{\alpha}{\alpha+1}}\right.$$

$$\sup_{t\in[-\pi,\pi]}\left\|\Omega_1\left(D_{t-}^{\alpha}X,2\pi\left\|\widetilde{L}\left(\left(\sin\left(\frac{|\cdot-t|}{4}\chi_{[-\pi,t]}(\cdot)\right)\right)^{\alpha+1}\right)(t)\right\|_{\infty}^{\frac{1}{\alpha+1}}\right)\right\|_{L^1,[-\pi,t]}$$

$$+\left\|\widetilde{L}\left(\left(\sin\left(\frac{|\cdot-t|}{4}\chi_{[t,\pi]}(\cdot)\right)\right)^{\alpha+1}\right)(t)\right\|_{\infty}^{\frac{\alpha}{\alpha+1}}$$

$$\left.\sup_{t\in[-\pi,\pi]}\left\|\Omega_1\left(D_{*t}^{\alpha}X,2\pi\left\|\widetilde{L}\left(\left(\sin\left(\frac{|\cdot-t|}{4}\chi_{[t,\pi]}(\cdot)\right)\right)^{\alpha+1}\right)(t)\right\|_{\infty}^{\frac{1}{\alpha+1}}\right)\right\|_{L^1,[t,\pi]}\right\}. \tag{15.55}$$

Proof Based on Theorem 15.22. We take into account (15.13), (15.14) and the positivity of $\widetilde{L}$. ■

We continue with interesting corollaries.

Corollary 15.25 *All as in Theorem 15.21. Further assume* $\widetilde{L}(1)=1$ *and* $X^{(k)}(t_0,\omega)=0$, $\forall\,\omega\in\Omega$, *all* $k=1,\ldots,n-1$, *for a fixed* $t_0\in[-\pi,\pi]$. *Then*

$$\left(E\left(|M(X)-X|^{q}\right)(t_0)\right)^{\frac{1}{q}}\le\frac{2^{\alpha+\frac{1}{p}}\pi^{\alpha}}{\Gamma(\alpha)(p(\alpha-1)+1)^{\frac{1}{p}}(q+1)^{\frac{\alpha}{q(\alpha+1)}}} \tag{15.56}$$

$$\left\{\left\{\left(\widetilde{L}\left(\chi_{[t_0,\pi]}(\cdot)\right)(t_0)\right)^{\frac{1}{p}}\left(\widetilde{L}\left(\left(\sin\left(\frac{|\cdot-t_0|}{4}\chi_{[t_0,\pi]}(\cdot)\right)\right)^{q(\alpha+1)}\right)(t_0)\right)^{\frac{\alpha}{q(\alpha+1)}}\right.\right.$$

$$\left[\left(\widetilde{L}\left(\chi_{[t_0,\pi]}(\cdot)\right)(t_0)\right)^{\frac{1}{\alpha+1}}(q+1)^{\frac{\alpha}{(\alpha+1)}}+1\right]^{\frac{1}{q}}$$

$$\left.\left\|\Omega_1\left(D_{*t_0}^{\alpha}X,2\pi\left(\frac{1}{(q+1)}\widetilde{L}\left(\left(\sin\left(\frac{|\cdot-t_0|}{4}\chi_{[t_0,\pi]}(\cdot)\right)\right)^{q(\alpha+1)}\right)(t_0)\right)^{\frac{1}{q(\alpha+1)}}\right)\right\|_{L^q,[t_0,\pi]}\right\}$$

$$+\left\{\left(\widetilde{L}\left(\chi_{[-\pi,t_0]}(\cdot)\right)(t_0)\right)^{\frac{1}{p}}\left(\widetilde{L}\left(\left(\sin\left(\frac{|\cdot-t_0|}{4}\chi_{[-\pi,t_0]}(\cdot)\right)\right)^{q(\alpha+1)}\right)(t_0)\right)^{\frac{\alpha}{q(\alpha+1)}}\right.$$

$$\left[\left(\widetilde{L}\left(\chi_{[-\pi,t_0]}(\cdot)\right)(t_0)\right)^{\frac{1}{\alpha+1}}(q+1)^{\frac{\alpha}{(\alpha+1)}}+1\right]^{\frac{1}{q}}$$

$$\Omega_1\left(D^{\alpha}_{t_0-}X, 2\pi\left(\frac{1}{(q+1)}\widetilde{L}\left(\left(\sin\left(\frac{|\cdot-t_0|}{4}\chi_{[-\pi,t_0]}(\cdot)\right)\right)^{q(\alpha+1)}\right)(t_0)\right)^{\frac{1}{q(\alpha+1)}}\right)_{L^q,[-\pi,t_0]}\Bigg\}\Bigg\}.$$

Corollary 15.26 *All as in Theorem 15.21. Further assume that $\widetilde{L}(1)=1$ and $\frac{1}{q}<\alpha<1$ (i.e. $n=1$). Then, $\forall\, t\in[-\pi,\pi]$, we have*

$$\left(E\left(|M(X)-X|^q\right)(t)\right)^{\frac{1}{q}}\leq\frac{2^{\alpha+\frac{1}{p}}\pi^{\alpha}}{\Gamma(\alpha)(p(\alpha-1)+1)^{\frac{1}{p}}(q+1)^{\frac{\alpha}{q(\alpha+1)}}}\tag{15.57}$$

$$\Bigg\{\Bigg\{\left(\widetilde{L}\left(\chi_{[t,\pi]}(\cdot)\right)(t)\right)^{\frac{1}{p}}\left(\widetilde{L}\left(\left(\sin\left(\frac{|\cdot-t|}{4}\chi_{[t,\pi]}(\cdot)\right)\right)^{q(\alpha+1)}\right)(t)\right)^{\frac{\alpha}{q(\alpha+1)}}$$

$$\left[\left(\widetilde{L}\left(\chi_{[t,\pi]}(\cdot)\right)(t)\right)^{\frac{1}{\alpha+1}}(q+1)^{\frac{\alpha}{(\alpha+1)}}+1\right]^{\frac{1}{q}}$$

$$\Omega_1\left(D^{\alpha}_{*t}X, 2\pi\left(\frac{1}{(q+1)}\widetilde{L}\left(\left(\sin\left(\frac{|\cdot-t|}{4}\chi_{[t,\pi]}(\cdot)\right)\right)^{q(\alpha+1)}\right)(t)\right)^{\frac{1}{q(\alpha+1)}}\right)_{L^q,[t,\pi]}\Bigg\}$$

$$+\Bigg\{\left(\widetilde{L}\left(\chi_{[-\pi,t]}(\cdot)\right)(t)\right)^{\frac{1}{p}}\left(\widetilde{L}\left(\left(\sin\left(\frac{|\cdot-t|}{4}\chi_{[-\pi,t]}(\cdot)\right)\right)^{q(\alpha+1)}\right)(t)\right)^{\frac{\alpha}{q(\alpha+1)}}$$

$$\left[\left(\widetilde{L}\left(\chi_{[-\pi,t]}(\cdot)\right)(t)\right)^{\frac{1}{\alpha+1}}(q+1)^{\frac{\alpha}{(\alpha+1)}}+1\right]^{\frac{1}{q}}$$

$$\Omega_1\left(D^{\alpha}_{t-}X, 2\pi\left(\frac{1}{(q+1)}\widetilde{L}\left(\left(\sin\left(\frac{|\cdot-t|}{4}\chi_{[-\pi,t]}(\cdot)\right)\right)^{q(\alpha+1)}\right)(t)\right)^{\frac{1}{q(\alpha+1)}}\right)_{L^q,[-\pi,t]}\Bigg\}\Bigg\}.$$

Of great interest in Corollary 15.26 is the case of $\alpha=\frac{1}{2}$. We require there $q>2$. Due to lack of space we omit this statement.

Corollary 15.27 *All as in Theorem 15.22. Further assume that $\widetilde{L}(1)=1$ and $X^{(k)}(t_0,\omega)=0$, $\forall\,\omega\in\Omega$, all $k=1,\dots,n-1$, for a fixed $t_0\in[-\pi,\pi]$. Then*

$$E\left(|M(X)-X|\right)(t_0)\leq\frac{(2\pi)^{\alpha}}{\Gamma(\alpha+1)}$$

$$\Bigg\{\left[\left(\widetilde{L}\left(\chi_{[-\pi,t_0]}(\cdot)\right)(t_0)\right)^{\frac{1}{\alpha+1}}+\frac{1}{(\alpha+1)}\right]\left(\widetilde{L}\left(\left(\sin\left(\frac{|\cdot-t_0|}{4}\chi_{[-\pi,t_0]}(\cdot)\right)\right)^{\alpha+1}\right)(t_0)\right)^{\frac{\alpha}{\alpha+1}}$$

$$\Omega_1\left(D^{\alpha}_{t_0-}X, 2\pi\left(\widetilde{L}\left(\left(\sin\left(\frac{|\cdot - t_0|}{4}\chi_{[-\pi,t_0]}(\cdot)\right)\right)^{\alpha+1}\right)(t_0)\right)^{\frac{1}{\alpha+1}}\right)_{L^1,[-\pi,t_0]} +$$

$$\left[\left(\widetilde{L}\left(\chi_{[t_0,\pi]}(\cdot)\right)(t_0)\right)^{\frac{1}{\alpha+1}} + \frac{1}{(\alpha+1)}\right]\left(\widetilde{L}\left(\left(\sin\left(\frac{|\cdot - t_0|}{4}\chi_{[t_0,\pi]}(\cdot)\right)\right)^{\alpha+1}\right)(t_0)\right)^{\frac{\alpha}{\alpha+1}}$$

$$\left.\Omega_1\left(D^{\alpha}_{*t_0}X, 2\pi\left(\widetilde{L}\left(\left(\sin\left(\frac{|\cdot - t_0|}{4}\chi_{[t_0,\pi]}(\cdot)\right)\right)^{\alpha+1}\right)(t_0)\right)^{\frac{1}{\alpha+1}}\right)_{L^1,[t_0,\pi]}\right\}. \tag{15.58}$$

Corollary 15.28 *All as in Theorem 15.22. Further assume that* $\widetilde{L}(1) = 1$ *and* $0 < \alpha < 1$. *Then,* $\forall\, t \in [-\pi, \pi]$, *we have*

$$E(|M(X) - X|)(t) \leq \frac{(2\pi)^{\alpha}}{\Gamma(\alpha+1)}$$

$$\left\{\left[\left(\widetilde{L}\left(\chi_{[-\pi,t]}(\cdot)\right)(t)\right)^{\frac{1}{\alpha+1}} + \frac{1}{(\alpha+1)}\right]\left(\widetilde{L}\left(\left(\sin\left(\frac{|\cdot - t|}{4}\chi_{[-\pi,t]}(\cdot)\right)\right)^{\alpha+1}\right)(t)\right)^{\frac{\alpha}{\alpha+1}}\right.$$

$$\Omega_1\left(D^{\alpha}_{t-}X, 2\pi\left(\widetilde{L}\left(\left(\sin\left(\frac{|\cdot - t|}{4}\chi_{[-\pi,t]}(\cdot)\right)\right)^{\alpha+1}\right)(t)\right)^{\frac{1}{\alpha+1}}\right)_{L^1,[-\pi,t]} +$$

$$\left[\left(\widetilde{L}\left(\chi_{[t,\pi]}(\cdot)\right)(t)\right)^{\frac{1}{\alpha+1}} + \frac{1}{(\alpha+1)}\right]\left(\widetilde{L}\left(\left(\sin\left(\frac{|\cdot - t|}{4}\chi_{[t,\pi]}(\cdot)\right)\right)^{\alpha+1}\right)(t)\right)^{\frac{\alpha}{\alpha+1}}$$

$$\left.\Omega_1\left(D^{\alpha}_{*t}X, 2\pi\left(\widetilde{L}\left(\left(\sin\left(\frac{|\cdot - t|}{4}\chi_{[t,\pi]}(\cdot)\right)\right)^{\alpha+1}\right)(t)\right)^{\frac{1}{\alpha+1}}\right)_{L^1,[t,\pi]}\right\}. \tag{15.59}$$

Corollary 15.29 *All as in Theorem 15.22. Further assume* $\widetilde{L}(1) = 1$. *Then* $\forall\, t \in [-\pi, \pi]$, *we have*

$$E(|M(X) - X|)(t) \leq 2\sqrt{2}$$

$$\left\{\left[\left(\widetilde{L}\left(\chi_{[-\pi,t]}(\cdot)\right)(t)\right)^{\frac{2}{3}} + \frac{2}{3}\right]\left(\widetilde{L}\left(\left(\sin\left(\frac{|\cdot - t|}{4}\chi_{[-\pi,t]}(\cdot)\right)\right)^{\frac{3}{2}}\right)(t)\right)^{\frac{1}{3}}\right.$$

$$\Omega_1\left(D_{t-}^{\frac{1}{2}}X, 2\pi\left(\widetilde{L}\left(\left(\sin\left(\frac{|\cdot-t|}{4}\chi_{[-\pi,t]}(\cdot)\right)\right)^{\frac{3}{2}}\right)(t)\right)^{\frac{2}{3}}\right)_{L^1,[-\pi,t]} +$$

$$\left[\left(\widetilde{L}\left(\chi_{[t,\pi]}(\cdot)\right)(t)\right)^{\frac{2}{3}}+\frac{2}{3}\right]\left(\widetilde{L}\left(\left(\sin\left(\frac{|\cdot-t|}{4}\chi_{[t,\pi]}(\cdot)\right)\right)^{\frac{3}{2}}\right)(t)\right)^{\frac{1}{3}}$$

$$\left.\Omega_1\left(D_{*t}^{\frac{1}{2}}X, 2\pi\left(\widetilde{L}\left(\left(\sin\left(\frac{|\cdot-t|}{4}\chi_{[t,\pi]}(\cdot)\right)\right)^{\frac{3}{2}}\right)(t)\right)^{\frac{2}{3}}\right)_{L^1,[t,\pi]}\right\}. \tag{15.60}$$

Next we give uniform norm results. They are based on Theorems 15.23, 15.24.

Corollary 15.30 *All as in Theorem 15.23. Assume further* $\frac{1}{q} < \alpha < 1$ *and* $\widetilde{L}(1) = 1$. *Then*

$$\left\|E\left(|M(X)-X|^q\right)\right\|_{\infty}^{\frac{1}{q}} \leq \frac{\pi^{\alpha}2^{\alpha+\frac{1}{p}}\left[(q+1)^{\frac{\alpha}{(\alpha+1)}}+1\right]^{\frac{1}{q}}}{\Gamma(\alpha)(p(\alpha-1)+1)^{\frac{1}{p}}(q+1)^{\frac{\alpha}{q(\alpha+1)}}} \tag{15.61}$$

$$\left\{\left\{\left\|\widetilde{L}\left(\left(\sin\left(\frac{|\cdot-t|}{4}\chi_{[t,\pi]}(\cdot)\right)\right)^{q(\alpha+1)}\right)(t)\right\|_{\infty}^{\frac{\alpha}{q(\alpha+1)}}\right.\right.$$

$$\left.\sup_{t\in[-\pi,\pi]}\Omega_1\left(D_{*t}^{\alpha}X, 2\pi\left(\frac{1}{(q+1)}\left\|\widetilde{L}\left(\left(\sin\left(\frac{|\cdot-t|}{4}\chi_{[t,\pi]}(\cdot)\right)\right)^{q(\alpha+1)}\right)(t)\right\|_{\infty}\right)^{\frac{1}{q(\alpha+1)}}\right)_{L^q,[t,\pi]}\right\}$$

$$+\left\{\left\|\widetilde{L}\left(\left(\sin\left(\frac{|\cdot-t|}{4}\chi_{[-\pi,t]}(\cdot)\right)\right)^{q(\alpha+1)}\right)(t)\right\|_{\infty}^{\frac{\alpha}{q(\alpha+1)}}\right.$$

$$\left.\left.\sup_{t\in[-\pi,\pi]}\Omega_1\left(D_{t-}^{\alpha}X, 2\pi\left(\frac{1}{(q+1)}\left\|\widetilde{L}\left(\left(\sin\left(\frac{|\cdot-t|}{4}\chi_{[-\pi,t]}(\cdot)\right)\right)^{q(\alpha+1)}\right)(t)\right\|_{\infty}\right)^{\frac{1}{q(\alpha+1)}}\right)_{L^q,[-\pi,t]}\right\}\right\}.$$

Corollary 15.31 *All as in Theorem 15.24. Assume further* $0 < \alpha < 1$ *and* $\widetilde{L}(1) = 1$. *Then*

$$\|E(|M(X)-X|)\|_{\infty} \leq \frac{(2\pi)^{\alpha}(\alpha+2)}{\Gamma(\alpha+2)}$$

$$\left\{\left\|\widetilde{L}\left(\left(\sin\left(\frac{|\cdot-t|}{4}\chi_{[-\pi,t]}(\cdot)\right)\right)^{\alpha+1}\right)(t)\right\|_{\infty}^{\frac{\alpha}{\alpha+1}}\right.$$

$$\sup_{t\in[-\pi,\pi]} \Omega_1\left(D_{t-}^{\alpha}X, 2\pi\left\|\widetilde{L}\left(\left(\sin\left(\frac{|\cdot-t|}{4}\chi_{[-\pi,t]}(\cdot)\right)\right)^{\alpha+1}\right)(t)\right\|_{\infty}^{\frac{1}{\alpha+1}}\right)_{L^1,[-\pi,t]}$$

$$+\left\|\widetilde{L}\left(\left(\sin\left(\frac{|\cdot-t|}{4}\chi_{[t,\pi]}(\cdot)\right)\right)^{\alpha+1}\right)(t)\right\|_{\infty}^{\frac{\alpha}{\alpha+1}}$$

$$\left.\sup_{t\in[-\pi,\pi]} \Omega_1\left(D_{*t}^{\alpha}X, 2\pi\left\|\widetilde{L}\left(\left(\sin\left(\frac{|\cdot-t|}{4}\chi_{[t,\pi]}(\cdot)\right)\right)^{\alpha+1}\right)(t)\right\|_{\infty}^{\frac{1}{\alpha+1}}\right)_{L^1,[t,\pi]}\right\}. \tag{15.62}$$

Corollary 15.32 *All as in Theorem 15.24. Here* $\alpha = \frac{1}{2}$ *and* $\widetilde{L}(1) = 1$. *Then*

$$\|E(|M(X) - X|)\|_{\infty} \le \frac{10\sqrt{2}}{3}$$

$$\left\{\left\|\widetilde{L}\left(\left(\sin\left(\frac{|\cdot-t|}{4}\chi_{[-\pi,t]}(\cdot)\right)\right)^{\frac{3}{2}}\right)(t)\right\|_{\infty}^{\frac{1}{3}}\right. \tag{15.63}$$

$$\sup_{t\in[-\pi,\pi]} \Omega_1\left(D_{t-}^{\frac{1}{2}}X, 2\pi\left\|\widetilde{L}\left(\left(\sin\left(\frac{|\cdot-t|}{4}\chi_{[-\pi,t]}(\cdot)\right)\right)^{\frac{3}{2}}\right)(t)\right\|_{\infty}^{\frac{2}{3}}\right)_{L^1,[-\pi,t]}$$

$$+\left\|\widetilde{L}\left(\left(\sin\left(\frac{|\cdot-t|}{4}\chi_{[t,\pi]}(\cdot)\right)\right)^{\frac{3}{2}}\right)(t)\right\|_{\infty}^{\frac{1}{3}}$$

$$\left.\sup_{t\in[-\pi,\pi]} \Omega_1\left(D_{*t}^{\frac{1}{2}}X, 2\pi\left\|\widetilde{L}\left(\left(\sin\left(\frac{|\cdot-t|}{4}\chi_{[t,\pi]}(\cdot)\right)\right)^{\frac{3}{2}}\right)(t)\right\|_{\infty}^{\frac{2}{3}}\right)_{L^1,[t,\pi]}\right\}.$$

15.5 Application

Consider the Bernstein polynomials on $[-\pi, \pi]$ for $f \in C([-\pi, \pi])$:

$$B_N(f)(x) = \sum_{k=0}^{N}\binom{N}{k} f\left(-\pi + \frac{2\pi k}{N}\right)\left(\frac{x+\pi}{2\pi}\right)^{k}\left(\frac{\pi - x}{2\pi}\right)^{N-k}, \tag{15.64}$$

$N \in \mathbb{N}$, any $x \in [-\pi, \pi]$. There are positive linear operators from $C([-\pi, \pi])$ into itself.

Setting $g(t) = f(2\pi t - \pi)$, $t \in [0, 1]$, we have $g(0) = f(-\pi)$, $g(1) = f(\pi)$, and

$$(B_N g)(t) = \sum_{k=0}^{N} \binom{N}{k} g\left(\frac{k}{N}\right) t^k (1-t)^{N-k} = (B_N f)(x), \ x \in [-\pi, \pi]. \tag{15.65}$$

Here $x = \varphi(t) = 2\pi t - \pi$ is an $1-1$ and onto map from $[0, 1]$ onto $[-\pi, \pi]$. Clearly here $g \in C([0, 1])$.

Notice also that

$$\left(B_N\left((\cdot - x)^2\right)\right)(x) = \left[\left(B_N\left((\cdot - t)^2\right)\right)(t)\right](2\pi)^2 = \frac{(2\pi)^2}{N} t(1-t)$$

$$= \frac{(2\pi)^2}{N}\left(\frac{x+\pi}{2\pi}\right)\left(\frac{\pi - x}{2\pi}\right) = \frac{1}{N}(x+\pi)(\pi - x) \leq \frac{\pi^2}{N}, \ \forall x \in [-\pi, \pi].$$

I.e.

$$\left(B_N\left((\cdot - x)^2\right)\right)(x) \leq \frac{\pi^2}{N}, \ \forall x \in [-\pi, \pi]. \tag{15.66}$$

In particular

$$(B_N 1)(x) = 1, \ \forall x \in [-\pi, \pi]. \tag{15.67}$$

Define the corresponding application of M by

$$\widetilde{B}_N(X)(t, \omega) := B_N(X(\cdot, \omega))(t) =$$

$$\sum_{k=0}^{N} \binom{N}{k} X\left(-\pi + \frac{2\pi k}{N}, \omega\right)\left(\frac{t+\pi}{2\pi}\right)^k \left(\frac{\pi - t}{2\pi}\right)^{N-k}, \tag{15.68}$$

$\forall N \in \mathbb{N}$, $\forall t \in [-\pi, \pi]$, $\forall \omega \in \Omega$, where X is a stochastic process. Clearly $\widetilde{B}_N$ is a stochastic process.

We give

Proposition 15.33 *Let $X(t, \omega)$ be a stochastic process from $[-\pi, \pi] \times (\Omega, \mathcal{F}, P)$ into $\mathbb{R}$, where $(\Omega, \mathcal{F}, P)$ is a probability space. Here $0 < \alpha < 1$ (i.e. $n = 1$) and $X(\cdot, \omega) \in AC([-\pi, \pi])$ with $X^{(1)}(\cdot, \omega) \in L_\infty([-\pi, \pi])$, $\forall \omega \in \Omega$. Further we assume that $D_{*t}^{\alpha} X(z, \omega)$ is a stochastic process for $z \in [t, \pi]$, $\omega \in \Omega$, and $D_{t-}^{\alpha} X(z, \omega)$ is a stochastic process for $z \in [-\pi, t]$, $\omega \in \Omega$; $\forall t \in [-\pi, \pi]$. For any $t \in [-\pi, \pi]$ we assume that $D_{*t}^{\alpha} X(z, \omega)$ is continuous in $z \in [t, \pi]$, uniformly with respect to $\omega \in \Omega$. And for any $t \in [-\pi, \pi]$ we assume that $D_{t-}^{\alpha} X(z, \omega)$ is continuous in $z \in [-\pi, t]$, uniformly with respect to $\omega \in \Omega$. Finally, we assume that $(E|X|)(t) < \infty$, $\forall t \in [-\pi, \pi]$. Then, for any $t \in [-\pi, \pi]$, we have:*

$$E\left(\left|\widetilde{B}_N(X)-X\right|\right)(t)\leq\frac{(2\pi)^{\alpha}}{\Gamma(\alpha+1)}$$

$$\left\{\left[\left(B_N\left(\chi_{[-\pi,t]}(\cdot)\right)(t)\right)^{\frac{1}{\alpha+1}}+\frac{1}{(\alpha+1)}\right]\left(B_N\left(\left(\sin\left(\frac{|\cdot-t|}{4}\chi_{[-\pi,t]}(\cdot)\right)\right)^{\alpha+1}\right)(t)\right)^{\frac{\alpha}{\alpha+1}}\right.$$

$$\Omega_1\left(D_{t-}^{\alpha}X,2\pi\left(B_N\left(\left(\sin\left(\frac{|\cdot-t|}{4}\chi_{[-\pi,t]}(\cdot)\right)\right)^{\alpha+1}\right)(t)\right)^{\frac{1}{\alpha+1}}\right)_{L^1,[-\pi,t]}+$$

$$\left[\left(B_N\left(\chi_{[t,\pi]}(\cdot)\right)(t)\right)^{\frac{1}{\alpha+1}}+\frac{1}{(\alpha+1)}\right]\left(B_N\left(\left(\sin\left(\frac{|\cdot-t|}{4}\chi_{[t,\pi]}(\cdot)\right)\right)^{\alpha+1}\right)(t)\right)^{\frac{\alpha}{\alpha+1}}$$

$$\left.\Omega_1\left(D_{*t}^{\alpha}X,2\pi\left(B_N\left(\left(\sin\left(\frac{|\cdot-t|}{4}\chi_{[t,\pi]}(\cdot)\right)\right)^{\alpha+1}\right)(t)\right)^{\frac{1}{\alpha+1}}\right)_{L^1,[t,\pi]}\right\},\quad(15.69)$$

$\forall\, N\in\mathbb{N}$.

Proof By Corollary 15.28. ∎

We give

Proposition 15.34 *All as in Proposition 15.33 with* $\alpha=\frac{1}{2}$. *Then, for any* $t\in[-\pi,\pi]$, *we have:*

$$E\left(\left|\widetilde{B}_N(X)-X\right|\right)(t)\leq 2\sqrt{2}$$

$$\left\{\left[\left(B_N\left(\chi_{[-\pi,t]}(\cdot)\right)(t)\right)^{\frac{2}{3}}+\frac{2}{3}\right]\left(B_N\left(\left(\sin\left(\frac{|\cdot-t|}{4}\chi_{[-\pi,t]}(\cdot)\right)\right)^{\frac{3}{2}}\right)(t)\right)^{\frac{1}{3}}\right.$$

$$\Omega_1\left(D_{t-}^{\frac{1}{2}}X,2\pi\left(B_N\left(\left(\sin\left(\frac{|\cdot-t|}{4}\chi_{[-\pi,t]}(\cdot)\right)\right)^{\frac{3}{2}}\right)(t)\right)^{\frac{2}{3}}\right)_{L^1,[-\pi,t]}+$$

$$\left[\left(B_N\left(\chi_{[t,\pi]}(\cdot)\right)(t)\right)^{\frac{2}{3}}+\frac{2}{3}\right]\left(B_N\left(\left(\sin\left(\frac{|\cdot-t|}{4}\chi_{[t,\pi]}(\cdot)\right)\right)^{\frac{3}{2}}\right)(t)\right)^{\frac{1}{3}}$$

$$\left.\Omega_1\left(D_{*t}^{\frac{1}{2}}X,2\pi\left(B_N\left(\left(\sin\left(\frac{|\cdot-t|}{4}\chi_{[t,\pi]}(\cdot)\right)\right)^{\frac{3}{2}}\right)(t)\right)^{\frac{2}{3}}\right)_{L^1,[t,\pi]}\right\},\quad(15.70)$$

$\forall\, N\in\mathbb{N}$.

Proof By Proposition 15.33. ∎

We continue with

Proposition 15.35 *All as in Proposition 15.33 with* $\alpha = \frac{1}{2}$. *Then, for any* $t \in [-\pi, \pi]$, *we have:*

$$E\left(\left|\widetilde{B}_N(X) - X\right|\right)(t) \leq \frac{10\sqrt{2}}{3}\left(B_N\left(\left(\sin\left(\frac{|\cdot - t|}{4}\right)\right)^{\frac{3}{2}}\right)(t)\right)^{\frac{1}{3}}$$

$$\left[\Omega_1\left(D_{t-}^{\frac{1}{2}}X, 2\pi\left(B_N\left(\left(\sin\left(\frac{|\cdot - t|}{4}\right)\right)^{\frac{3}{2}}\right)(t)\right)^{\frac{2}{3}}\right)_{L^1,[-\pi,t]} + \right.$$

$$\left.\Omega_1\left(D_{*t}^{\frac{1}{2}}X, 2\pi\left(B_N\left(\left(\sin\left(\frac{|\cdot - t|}{4}\right)\right)^{\frac{3}{2}}\right)(t)\right)^{\frac{2}{3}}\right)_{L^1,[t,\pi]}\right], \tag{15.71}$$

$\forall\, N \in \mathbb{N}$.

Proof By (15.70) and the positivity of B_N, see also (15.39) and (15.41). ∎

We make

Remark 15.36 By $|\sin x| < |x|$, $\forall\, x \in \mathbb{R} - \{0\}$, in particular $\sin x \leq x$, for $x \geq 0$, we get

$$\left(\sin\left(\frac{|\cdot - t|}{4}\right)\right)^{\frac{3}{2}} \leq \left(\frac{|\cdot - t|}{4}\right)^{\frac{3}{2}} = \frac{1}{8}|\cdot - t|^{\frac{3}{2}}.$$

Hence

$$\left\|B_N\left(\left(\sin\left(\frac{|\cdot - t|}{4}\right)\right)^{\frac{3}{2}}\right)(t)\right\|_\infty \leq \frac{1}{8}\left\|B_N\left(|\cdot - t|^{\frac{3}{2}}\right)(t)\right\|_\infty. \tag{15.72}$$

We observe that

$$B_N\left(|\cdot - t|^{\frac{3}{2}}\right)(t) = \sum_{k=0}^{N}\left|t + \pi - \frac{2\pi k}{N}\right|^{\frac{3}{2}}\binom{N}{k}\left(\frac{t+\pi}{2\pi}\right)^k\left(\frac{\pi - t}{2\pi}\right)^{N-k}$$

(by discrete Hölder's inequality)

$$\leq \left[\sum_{k=0}^{N} \left(t + \pi - \frac{2\pi k}{N} \right)^2 \binom{N}{k} \left(\frac{t+\pi}{2\pi} \right)^k \left(\frac{\pi - t}{2\pi} \right)^{N-k} \right]^{\frac{3}{4}}$$

$$= \left(B_N \left((\cdot - t)^2 \right) (t) \right)^{\frac{3}{4}} \overset{(15.66)}{\leq} \frac{\pi^{\frac{3}{2}}}{N^{\frac{3}{4}}}, \forall t \in [-\pi, \pi]. \tag{15.73}$$

Consequently it holds

$$\left\| B_N \left(|\cdot - t|^{\frac{3}{2}} \right) (t) \right\|_{\infty} \leq \frac{\pi^{\frac{3}{2}}}{N^{\frac{3}{4}}}, \tag{15.74}$$

and

$$\left\| B_N \left(\left(\sin \left(\frac{|\cdot - t|}{4} \right) \right)^{\frac{3}{2}} \right) (t) \right\|_{\infty} \leq \frac{\pi^{\frac{3}{2}}}{8N^{\frac{3}{4}}}, \forall N \in \mathbb{N}. \tag{15.75}$$

We further have

Proposition 15.37 *All as in Proposition 15.33 with $\alpha = \frac{1}{2}$. Then, for any $t \in [-\pi, \pi]$ we have:*

$$E \left(\left| \widetilde{B}_N (X) - X \right| \right) (t) \leq \frac{5\sqrt{2\pi}}{\sqrt[4]{N}}$$

$$\left[\Omega_1 \left(D_{t-}^{\frac{1}{2}} X, \frac{\pi^2}{2\sqrt{N}} \right)_{L^1, [-\pi, t]} + \Omega_1 \left(D_{*t}^{\frac{1}{2}} X, \frac{\pi^2}{2\sqrt{N}} \right)_{L^1, [t, \pi]} \right], \tag{15.76}$$

$\forall N \in \mathbb{N}$.

As $N \to +\infty$, we get $E \left(\left| \widetilde{B}_N (X) - X \right| \right) (t) \to 0$.

Proof By (15.71), (15.75) and positivity of B_N. See also Proposition 15.5. ■

Consequently we obtain

Proposition 15.38 *All as in Proposition 15.33 with $\alpha = \frac{1}{2}$. Assume further that $\left| X^{(1)} (t, \omega) \right| \leq M^*$, $\forall (t, \omega) \in [-\pi, \pi] \times \Omega$, where $M^* > 0$. Then*

$$\left\| E \left(\left| \widetilde{B}_N (X) - X \right| \right) \right\|_{\infty} \leq \frac{5\sqrt{2\pi}}{\sqrt[4]{N}}$$

$$\left[\sup_{t \in [-\pi, \pi]} \Omega_1 \left(D_{t-}^{\frac{1}{2}} X, \frac{\pi^2}{2\sqrt{N}} \right)_{L^1, [-\pi, t]} + \sup_{t \in [-\pi, \pi]} \Omega_1 \left(D_{*t}^{\frac{1}{2}} X, \frac{\pi^2}{2\sqrt{N}} \right)_{L^1, [t, \pi]} \right], \tag{15.77}$$

$\forall N \in \mathbb{N}$.

As $N \to +\infty$, then $\left\| E \left(\left| \widetilde{B}_N (X) - X \right| \right) \right\|_{\infty} \to 0$, i.e. $\widetilde{B}_N \to I$ (stochastic unit operator) in 1-mean.

Proof By (15.76) and Remark 15.8, see (15.13), (15.14). ∎

Remark 15.39 *(to Proposition* 15.38*)* Assume that

$$\Omega_1\left(D_{t-}^{\frac{1}{2}}X, \frac{\pi^2}{2\sqrt{N}}\right)_{L^1,[-\pi,t]} \leq \frac{K_1\pi^2}{2\sqrt{N}}, \forall t\in[-\pi,\pi], \forall N\in\mathbb{N}, \tag{15.78}$$

where $K_1 > 0$.

And assume that

$$\Omega_1\left(D_{*t}^{\frac{1}{2}}X, \frac{\pi^2}{2\sqrt{N}}\right)_{L^1,[t,\pi]} \leq \frac{K_2\pi^2}{2\sqrt{N}}, \forall t\in[-\pi,\pi], \forall N\in\mathbb{N}, \tag{15.79}$$

where $K_2 > 0$.

Conditions (15.78), (15.79) are of Lipschitz type of order 1.

By (15.77) and (15.78), (15.79), we easily derive that

$$\left\| E\left(\left|\widetilde{B}_N(X) - X\right|\right)\right\|_\infty \leq \frac{\theta}{N^{\frac{3}{4}}}, \tag{15.80}$$

$\forall N\in\mathbb{N}$, where $\theta > 0$.

Thus, at smoothness of order $\frac{1}{2}$ we achieve speed of convergence $\frac{1}{N^{\frac{3}{4}}}$, $N\in\mathbb{N}$. Without any smoothness, in [5], we proved that the speed of convergence was $\frac{1}{\sqrt{N}}$, $N\in\mathbb{N}$.

In the presence of the ordinary first derivative, see [5], in deterministic approximation the rate of convergence was $\frac{1}{N}$, $N\in\mathbb{N}$.

Naturally, we have as expected:

$$\frac{1}{N} < \frac{1}{N^{\frac{3}{4}}} < \frac{1}{\sqrt{N}}, \forall N\in\mathbb{N}-\{1\}. \tag{15.81}$$

15.6 Trigonometric Stochastic Korovkin Results

In this section $\widetilde{L}$, M are meant as sequences of operators.

We give first pointwise results:

Theorem 15.40 *Here all as in Theorem 15.21. Assume further that* $\widetilde{L}(1)(t)\to 1$ *and* $\widetilde{L}\left(\left(\sin\left(\frac{|\cdot-t|}{4}\right)\right)^{q(\alpha+1)}\right)(t)\to 0$, *pointwise in* $t\in[-\pi,\pi]$.

Then $E\left(|M(X)-X|^q\right)(t)\to 0$, *pointwise in* $t\in[-\pi,\pi]$, *that is* $M\to I$ *(stochastic unit operator) in* q*-mean-pointwise with rates, quantitatively.*

Proof We use (15.48), we take into account $\widetilde{L}(1)(t) \to 1$, (15.36); and $\widetilde{L}(\chi_{[t,\pi]}(\cdot))(t)$, $\widetilde{L}(\chi_{[-\pi,t]}(\cdot))(t) \leq \widetilde{L}(1)(t)$, which $\widetilde{L}(1)(t)$ is bounded, and by (15.39), (15.41) and positivity of $\widetilde{L}$ we get that $\widetilde{L}\left(\left(\sin\left(\frac{|\cdot-t|}{4}\chi_{[t,\pi]}(\cdot)\right)\right)^{q(\alpha+1)}\right)(t)$, $\widetilde{L}\left(\left(\sin\left(\frac{|\cdot-t|}{4}\chi_{[-\pi,t]}(\cdot)\right)\right)^{q(\alpha+1)}\right)(t) \leq \widetilde{L}\left(\left(\sin\left(\frac{|\cdot-t|}{4}\right)\right)^{q(\alpha+1)}\right)(t) \to 0.$

Finally, we use Proposition 15.6(ii) for the $\Omega_1(\cdot,\cdot)$'s to go to zero. ■

We continue with

Theorem 15.41 *Here all as in Theorem 15.22. Assume further that* $\widetilde{L}(1)(t) \to 1$ *and* $\widetilde{L}\left(\left(\sin\left(\frac{|\cdot-t|}{4}\right)\right)^{(\alpha+1)}\right)(t) \to 0$, *pointwise in* $t \in [-\pi, \pi]$.

Then $E(|M(X) - X|)(t) \to 0$, *pointwise in* $t \in [-\pi, \pi]$, *that is* $M \to I$ *in* 1-*mean-pointwise with rates, quantitatively.*

Proof Based on (15.51), similar to the proof of Theorem 15.40, just take $q = 1$ there. ■

Next we give uniform results:

Theorem 15.42 *Here all as in Theorem 15.23. Assume further that* $\widetilde{L}(1) \to 1$, *uniformly, and* $\left\|\widetilde{L}\left(\left(\sin\left(\frac{|\cdot-t|}{4}\right)\right)^{q(\alpha+1)}\right)(t)\right\|_{\infty} \to 0.$

Then $\left\|E\left(|M(X) - X|^q\right)\right\|_{\infty} \to 0$ *over* $[-\pi, \pi]$, *that is* $M \to I$ *in the* q-*mean, quantitatively with rates.*

Proof We use (15.54), we take into account $\widetilde{L}(1) \to 1$, uniformly, (15.37); $\left\|\widetilde{L}(1)\right\|_{\infty}$ is bounded, use of (15.40), (15.42) and Remark 15.8, see there (15.13), (15.14). ■

Next we give the L_1-mean uniform result

Theorem 15.43 *Here all as in Theorem 15.24. Assume further that* $\widetilde{L}(1) \to 1$, *uniformly, and* $\left\|\widetilde{L}\left(\left(\sin\left(\frac{|\cdot-t|}{4}\right)\right)^{(\alpha+1)}\right)(t)\right\|_{\infty} \to 0.$

Then $\|E(|M(X) - X|)\|_{\infty} \to 0$ *over* $[-\pi, \pi]$, *that is* $M \to I$ *in the* 1-*mean, quantitatively with rates.*

Proof Use of (15.55), similar to the proof of Theorem 15.42, just take $q = 1$ there. ■

We finish with

Remark 15.44 An amazing fact/observation follows: In all trigonometric convergence results here, see Theorems 15.40–15.43, the forcing conditions for convergences are based only on $\widetilde{L}$ and basic real valued continuous functions on $[-\pi, \pi]$ and are not related to stochastic processes, but they are giving trigonometric convergence results on stochastic processes!

References

1. Aliprantis, C., Burkinshaw, O.: Principles of Real Analysis, 3rd edn. Academic, San Diego (1998)
2. Anastassiou, G.A.: Korovkin inequalities for stochastic processes. J. Math. Anal. Appl. **157**(2), 366–384 (1991)
3. Anastassiou, G.A.: Moments in Probability and Approximation Theory. Pitman/Longman, # 287, UK (1993)
4. Anastassiou, G.: Quantitative Approximations. Chapman & Hall/CRC, Boca Raton (2001)
5. Anastassiou, G.: Stochastic Korovkin theory given quantitatively, Facta Universitatis (Nis). Ser. Math. Inform. **22**(1), 43–60 (2007)
6. Anastassiou, G.: Fractional Differentiation Inequalities. Springer, Heildelberg (2009)
7. Anastassiou, G.A.: Fuzzy Mathematics: Approximation Theory. Springer, Heildelberg (2010)
8. Anastassiou, G.: Intelligent Mathematics: Computational Analysis. Springer, Heidelberg (2011)
9. Anastassiou, G.: Fractional representation formulae and right fractional inequalities. Math. Comput. Model. **54**(11–12), 3098–3115 (2011)
10. Anastassiou, G.A.: Foundation of stochastic fractional calculus with fractional approximation of Stochastic processes. RACSAM 114 (2020). Article 89
11. Anastassiou, G.A.: Trigonometric fractional approximation of stochastic processes (2020). Submitted
12. Diethelm, K.: The Analysis of Fractional Differential Equations. Springer, New York (2010)
13. Korovkin, P.P.: Linear Operators and Approximation Theory. Hindustan Publ. Corp., Delhi (1960)
14. Royden, H.L.: Real Analysis, 2nd edn. MacMillan Publishing Co., Inc., New York (1968)
15. Shisha, O., Mond, B.: The degree of convergence of sequences of linear positive operators. Natl. Acad. Sci. U.S. **60**, 1196–1200 (1968)
16. Weba, M.: Korovkin systems of stochastic processes. Math. Z. **192**, 73–80 (1986)
17. Weba, M.: A quantitative Korovkin theorem for random functions with multivariate domains. J. Approx. Theory **61**, 74–87 (1990)

Chapter 16
Conformable Fractional Quantitative Approximation of Stochastic Processes

Here we consider and study very general stochastic positive linear operators induced by general positive linear operators that are acting on continuous functions. These are acting on the space of real conformable fractionally differentiable stochastic processes. Under some very mild, general and natural assumptions on the stochastic processes we produce related conformable fractional stochastic Shisha-Mond type inequalities of L^q-type $1 \leq q < \infty$ and corresponding conformable fractional stochastic Korovkin type theorems. These are regarding the stochastic q -mean conformable fractional convergence of a sequence of stochastic positive linear operators to the stochastic unit operator for various cases. All convergences are produced with rates and are given via the conformable fractional stochastic inequalities involving the stochastic modulus of continuity of the αth conformable fractional derivatives of the engaged stochastic process, $\alpha > 0$, $\alpha \notin \mathbb{N}$. The impressive fact is that the basic real Korovkin test functions assumptions are enough for the conclusions of our conformable fractional stochastic Korovkin theory. We give conformable fractional applications to stochastic Bernstein operators. See also [9].

16.1 Introduction

Inspiration for this work comes from [2, 3, 14, 15]. This work continues our earlier work [5], now at the stochastic conformable fractional level. First we mention the basics of conformable fractional calculus. In the Sect. 16.3, about background, we talk about the q-mean $(1 \leq q < \infty)$ first modulus of continuity of a stochastic process and its upper bounds. There we describe completely our setting by introducing our stochastic positive linear operator M, see (16.31), which is based on the positive linear operator $\widetilde{L}$ from $C([a, b])$ into itself. The operator M is acting on a wide space of conformable fractional differentiable real valued stochastic processes X. See there Assumptions 16.16, 16.19, 16.20, 16.22. We first give the main

G. A. Anastassiou, *Generalized Fractional Calculus*, Studies in Systems, Decision and Control 305, https://doi.org/10.1007/978-3-030-56962-4_16

pointwise conformable fractional stochastic Shisha-Mond type inequalities [13], see Theorems 16.23, 16.24, and their several corollaries covering important special cases.

We continue with conformable fractional q-mean uniform Shisha-Mond type inequalities, see Theorems 16.30, 16.31, and their interesting corollaries. All this theory is regarding the conformable fractional stochastic convergence of operators M to I (stochastic unit operator) given quantitatively with rates. An extensive fractional conformable application about the stochastic Bernstein operators follows in full details. Based on our Shisha-Mond type inequalities of our main Theorems 16.23, 16.24, 16.30, 16.31 we derive pointwise and uniform Stochastic Korovkin theorems [11] on stochastic processes, see Theorems 16.41–16.44. The amazing fact here is, that basic conditions on operator $\widetilde{L}$ regarding two simple real valued functions that are not stochastic, are able to enforce conformable fractional stochastic convergence on all stochastic processes we are dealing with; see Concepts 16.15 and Assumptions 16.16, 16.19, 16.20, 16.22.

16.2 Background—I

Here we follow [1], see also [10].

We need

Definition 16.1 *([1])* Let $a, b \in \mathbb{R}$. The left conformable fractional derivative starting from a of a function $f : [a, \infty) \to \mathbb{R}$ of order $0 < \alpha \leq 1$ is defined by

$$\left(T_{\alpha}^{a} f\right)(t) = \lim_{\varepsilon \to 0} \frac{f\left(t + \varepsilon (t - a)^{1-\alpha}\right) - f(t)}{\varepsilon}. \tag{16.1}$$

If $\left(T_{\alpha}^{a} f\right)(t)$ exists on (a, b), then

$$\left(T_{\alpha}^{a} f\right)(a) = \lim_{t \to a+} \left(T_{\alpha}^{a} f\right)(t). \tag{16.2}$$

The right conformable fractional derivative of order $0 < \alpha \leq 1$ terminating at b of $f : (-\infty, b] \to \mathbb{R}$ is defined by

$$\left({}_{\alpha}^{b}Tf\right)(t) = -\lim_{\varepsilon \to 0} \frac{f\left(t + \varepsilon (b - t)^{1-\alpha}\right) - f(t)}{\varepsilon}. \tag{16.3}$$

If $\left({}_{\alpha}^{b}Tf\right)(t)$ exists on (a, b), then

$$\left({}_{\alpha}^{b}Tf\right)(b) = \lim_{t \to b-} \left({}_{\alpha}^{b}Tf\right)(t). \tag{16.4}$$

Note that if f is differentiable then

$$\left(T_{\alpha}^{a}f\right)(t) = (t-a)^{1-\alpha} f'(t), \tag{16.5}$$

and

$$\left({}_{\alpha}^{b}Tf\right)(t) = -(b-t)^{1-\alpha} f'(t). \tag{16.6}$$

Denote by

$$\left(I_{\alpha}^{a}f\right)(t) = \int_{a}^{t} (x-a)^{\alpha-1} f(x)\,dx, \tag{16.7}$$

and

$$\left({}^{b}I_{\alpha}f\right)(t) = \int_{t}^{b} (b-x)^{\alpha-1} f(x)\,dx, \tag{16.8}$$

these are the left and right conformable fractional integrals of order $0 < \alpha \leq 1$.

In the higher order case we can generalize things as follows:

Definition 16.2 *([1])* Let $\alpha \in (n, n+1]$, and set $\beta = \alpha - n$. Then, the left conformable fractional derivative starting from a of a function $f : [a, \infty) \to \mathbb{R}$ of order α, where $f^{(n)}(t)$ exists, is defined by

$$\left(\mathbf{T}_{\alpha}^{a}f\right)(t) = \left(T_{\beta}^{a}f^{(n)}\right)(t). \tag{16.9}$$

The right conformable fractional derivative of order α terminating at b of $f : (-\infty, b] \to \mathbb{R}$, where $f^{(n)}(t)$ exists, is defined by

$$\left({}_{\alpha}^{b}\mathbf{T}f\right)(t) = (-1)^{n+1}\left({}_{\beta}^{b}Tf^{(n)}\right)(t). \tag{16.10}$$

If $\alpha = n+1$ then $\beta = 1$ and $\mathbf{T}_{n+1}^{a}f = f^{(n+1)}$.

If n is odd, then ${}_{n+1}^{b}\mathbf{T}f = -f^{(n+1)}$, and if n is even, then ${}_{n+1}^{b}\mathbf{T}f = f^{(n+1)}$.

When $n = 0$ (or $\alpha \in (0, 1]$), then $\beta = \alpha$, and (16.9), (16.10) collapse to {(16.1), (16.2)}, {(16.3), (16.4)} respectively.

Lemma 16.3 ([1]) *Let $f : (a, b) \to \mathbb{R}$ be continuously differentiable and $0 < \alpha \leq 1$. Then, for all $t > a$ we have*

$$I_{\alpha}^{a}T_{\alpha}^{a}(f)(t) = f(t) - f(a). \tag{16.11}$$

We need

Definition 16.4 *(see also [1])* If $\alpha \in (n, n+1]$, then the left fractional integral of order α starting at a is defined by

$$\left(\mathbf{I}_{\alpha}^{a}f\right)(t) = \frac{1}{n!}\int_{a}^{t} (t-x)^{n} (x-a)^{\beta-1} f(x)\,dx. \tag{16.12}$$

Similarly, (author's definition) the right fractional integral of order α terminating at b is defined by

$$\left({}^{b}\mathbf{I}_{\alpha} f\right)(t) = \frac{1}{n!} \int_{t}^{b} (x-t)^{n} (b-x)^{\beta-1} f(x)\, dx. \tag{16.13}$$

We need

Proposition 16.5 ([1]) *Let $\alpha \in (n, n+1]$ and $f : [a, \infty) \to \mathbb{R}$ be $(n+1)$ times continuously differentiable for $t > a$. Then, for all $t > a$ we have*

$$\mathbf{I}_{\alpha}^{a} \mathbf{T}_{\alpha}^{a} (f)(t) = f(t) - \sum_{k=0}^{n} \frac{f^{(k)}(a)(t-a)^{k}}{k!}. \tag{16.14}$$

We also have

Proposition 16.6 ([8, p. 154]) *Let $\alpha \in (n, n+1]$ and $f : (-\infty, b] \to \mathbb{R}$ be $(n+1)$ times continuously differentiable for $t < b$. Then, for all $t < b$ we have*

$$-{}^{b}\mathbf{I}_{\alpha}\, {}_{\alpha}^{b}\mathbf{T}(f)(t) = f(t) - \sum_{k=0}^{n} \frac{f^{(k)}(b)(t-b)^{k}}{k!}. \tag{16.15}$$

If $n = 0$ or $0 < \alpha \leq 1$, then (see also [1])

$${}^{b}I_{\alpha}\, {}_{\alpha}^{b}T(f)(t) = f(t) - f(b). \tag{16.16}$$

In conclusion we derive

Theorem 16.7 ([8, p. 155]) *Let $\alpha \in (n, n+1]$ and $f \in C^{n+1}([a, b])$, $n \in \mathbb{N}$. Then*

(1)

$$f(t) - \sum_{k=0}^{n} \frac{f^{(k)}(a)(t-a)^{k}}{k!} = \frac{1}{n!} \int_{a}^{t} (t-x)^{n} (x-a)^{\beta-1} \left(\mathbf{T}_{\alpha}^{a}(f)\right)(x)\, dx, \tag{16.17}$$

and

(2)

$$f(t) - \sum_{k=0}^{n} \frac{f^{(k)}(b)(t-b)^{k}}{k!} = -\frac{1}{n!} \int_{t}^{b} (b-x)^{\beta-1} (x-t)^{n} \left({}_{\alpha}^{b}\mathbf{T}(f)\right)(x)\, dx, \tag{16.18}$$

$\forall\, t \in [a, b]$.

We make

Remark 16.8 ([8, p. 155]) We notice the following: let $\alpha \in (n, n+1]$ and $f \in C^{n+1}([a, b])$, $n \in \mathbb{N}$. Then ($\beta := \alpha - n$, $0 < \beta \leq 1$)

$$\left(\mathbf{T}_{\alpha}^{a}(f)\right)(x) = \left(T_{\beta}^{a} f^{(n)}\right)(x) = (x-a)^{1-\beta} f^{(n+1)}(x), \tag{16.19}$$

and

$$\left({}_{\alpha}^{b}\mathbf{T}(f)\right)(x) = (-1)^{n+1}\left({}_{\beta}^{b}T f^{(n)}\right)(x) =$$

$$(-1)^{n+1}(-1)(b-x)^{1-\beta} f^{(n+1)}(x) = (-1)^{n}(b-x)^{1-\beta} f^{(n+1)}(x). \tag{16.20}$$

Consequently we get that

$$\left(\mathbf{T}_{\alpha}^{a}(f)\right)(x),\ \left({}_{\alpha}^{b}\mathbf{T}(f)\right)(x) \in C([a, b]).$$

Furthermore it is obvious that

$$\left(\mathbf{T}_{\alpha}^{a}(f)\right)(a) = \left({}_{\alpha}^{b}\mathbf{T}(f)\right)(b) = 0, \tag{16.21}$$

when $0 < \beta < 1$, i.e. when $\alpha \in (n, n+1)$.

If $f^{(k)}(a) = 0$, $k = 1, \ldots, n$, then

$$f(t) - f(a) = \frac{1}{n!}\int_{a}^{t}(t-x)^{n}(x-a)^{\beta-1}\left(\mathbf{T}_{\alpha}^{a}(f)\right)(x)\,dx, \tag{16.22}$$

$\forall\, t \in [a, b]$.

If $f^{(k)}(b) = 0$, $k = 1, \ldots, n$, then

$$f(t) - f(b) = -\frac{1}{n!}\int_{t}^{b}(b-x)^{\beta-1}(x-t)^{n}\left({}_{\alpha}^{b}\mathbf{T}(f)\right)(x)\,dx, \tag{16.23}$$

$\forall\, t \in [a, b]$.

16.3 Background—II

We need

Definition 16.9 Here $(\Omega, \mathcal{F}, P)$ is a probability space, $\omega \in \Omega$. We define the relative q-mean first modulus of continuity of stochastic process $X(t, \omega)$ by

$$\Omega_1(X, \delta)_{L^q,[c,d]} :=$$

$$\sup \left\{ \left(\int_{\Omega} |X(x,\omega) - X(y,\omega)|^q P(d\omega) \right)^{\frac{1}{q}} : x, y \in [c,d] \subseteq [a,b], \ |x-y| \leq \delta \right\}, \tag{16.24}$$

where $\delta > 0$, $1 \leq q < \infty$, $t \in [a,b] \subset \mathbb{R}$.

Definition 16.10 Let $1 \leq q < \infty$. Let $X(x,\omega)$ be a stochastic process. We call X a q-mean uniformly continuous stochastic process over $[a,b]$, iff $\forall\, \varepsilon > 0\ \exists\, \delta > 0$: whenever $|x-y| \leq \delta$; $x, y \in [a,b]$ implies that

$$\int_{\Omega} |X(x,s) - X(y,s)|^q P(ds) \leq \varepsilon. \tag{16.25}$$

We denote it as $X \in C_{\mathbb{R}}^{U_q}([a,b])$.

It holds

Proposition 16.11 ([5]) *Let* $X \in C_{\mathbb{R}}^{U_q}([a,b])$, *then* $\Omega_1(X,\delta)_{L^q,[a,b]} < \infty$, *any* $\delta > 0$.

Also it holds

Proposition 16.12 ([5]) *Let* $X(t,\omega)$ *be a stochastic process from* $[a,b] \times (\Omega, \mathcal{F}, P)$ *into* $\mathbb{R}$. *Then following are true* ($[c,d] \subseteq [a,b]$):

(i) $\Omega_1(X,\delta)_{L^q,[c,d]}$ *is nonnegative and nondecreasing in* $\delta > 0$,
(ii) $\lim_{\delta \downarrow 0} \Omega_1(X,\delta)_{L^q,[c,d]} = \Omega_1(X,0)_{L^q,[c,d]} = 0$, *iff* $X \in C_{\mathbb{R}}^{U_q}([c,d])$,
(iii) $\Omega_1(X,\delta_1+\delta_2)_{L^q,[c,d]} \leq \Omega_1(X,\delta_1)_{L^q,[c,d]} + \Omega_1(X,\delta_2)_{L^q,[c,d]}$, $\delta_1, \delta_2 > 0$,
(iv) $\Omega_1(X,m\delta)_{L^q,[c,d]} \leq m\Omega_1(X,\delta)_{L^q,[c,d]}$, $\delta > 0$, $m \in \mathbb{N}$,
(v) $\Omega_1(X,\lambda\delta)_{L^q,[c,d]} \leq \lceil \lambda \rceil \Omega_1(X,\delta)_{L^q,[c,d]} \leq (\lambda+1)\Omega_1(X,\delta)_{L^q,[c,d]}$, $\lambda > 0$, $\delta > 0$, $\lceil \cdot \rceil$ *is the ceiling of the number,*
(vi) $\Omega_1(X+Y,\delta)_{L^q,[c,d]} \leq \Omega_1(X,\delta)_{L^q,[c,d]} + \Omega_1(Y,\delta)_{L^q,[c,d]}$, $\delta > 0$,
(vii) $\Omega_1(X,\cdot)_{L^q,[c,d]}$ *is continuous on* $\mathbb{R}_+$ *for* $X \in C_{\mathbb{R}}^{U_q}([c,d])$.

We give

Remark 16.13 *(*to Proposition 16.12*)* By Proposition 16.12(v) we get

$$\Omega_1(X,|x-y|)_{L^q,[c,d]} \leq \left\lceil \frac{|x-y|}{\delta} \right\rceil \Omega_1(X,\delta)_{L^q,[c,d]}, \tag{16.26}$$

$\forall\, x, y \in [c,d]$, any $\delta > 0$.

We give

Remark 16.14 Let $\alpha \in (n, n+1]$, $n \in \mathbb{Z}_+$, $X(\cdot,\omega) \in C^{n+1}([a,b])$, $\forall\, \omega \in \Omega$. We assume that $\left|X^{(n+1)}(t,\omega)\right| \leq M^*$, $\forall\, (t,\omega) \in [a,b] \times \Omega$, where $M^* > 0$. Let $\delta > 0$, $1 \leq q < \infty$. Then

$$\Omega_1\left(\mathbf{T}_\alpha^c X,\delta\right)_{L^q,[c,d]} = \sup\left\{\left(\int_\Omega \left|\left(\mathbf{T}_\alpha^c X\right)(x,\omega) - \left(\mathbf{T}_\alpha^c X\right)(y,\omega)\right|^q P(d\omega)\right)^{\frac{1}{q}} :\right.$$

$$\left. x, y \in [c,d] \subseteq [a,b],\ |x-y| \leq \delta\right\} \leq$$

$$\sup\left\{\left(\int_\Omega \left(\left|\left(\mathbf{T}_\alpha^c X\right)(x,\omega)\right| + \left|\left(\mathbf{T}_\alpha^c X\right)(y,\omega)\right|\right)^q P(d\omega)\right)^{\frac{1}{q}} :\right.$$

$$\left. x, y \in [c,d] \subseteq [a,b],\ |x-y| \leq \delta\right\} \underset{(\beta:=\alpha-n)}{\overset{(16.19.)}{\leq}}$$

$$\sup\left\{\left(\int_\Omega \left[(x-c)^{1-\beta}\left|X^{(n+1)}(x,\omega)\right| + (y-c)^{1-\beta}\left|X^{(n+1)}(y,\omega)\right|\right]^q P(d\omega)\right)^{\frac{1}{q}} :\right.$$

$$\left. x, y \in [c,d] \subseteq [a,b],\ |x-y| \leq \delta\right\} \leq$$

$$M^* \sup\left\{\left(\int_\Omega \left((x-c)^{1-\beta} + (y-c)^{1-\beta}\right)^q P(d\omega)\right)^{\frac{1}{q}} :\right.$$

$$\left. x, y \in [c,d] \subseteq [a,b],\ |x-y| \leq \delta\right\} \leq 2M^*(b-a)^{1-\beta}.$$

That is

$$\Omega_1\left(\mathbf{T}_\alpha^c X,\delta\right)_{L^q,[c,d]} \leq 2M^*(b-a)^{1-\beta}, \tag{16.27}$$

for any $[c,d] \subseteq [a,b]$.

Similarly, it holds

$$\Omega_1\left({}_\alpha^d\mathbf{T}X,\delta\right)_{L^q,[c,d]} \leq 2M^*(b-a)^{1-\beta}, \tag{16.28}$$

for any $[c,d] \subseteq [a,b]$.

In particular it holds

$$\sup_{t\in[a,b]} \Omega_1\left(\mathbf{T}_\alpha^t X,\delta\right)_{L^q,[t,b]} \leq 2M^*(b-a)^{1-\beta}, \tag{16.29}$$

and

$$\sup_{t\in[a,b]} \Omega_1\left({}_\alpha^t\mathbf{T}X,\delta\right)_{L^q,[a,t]} \leq 2M^*(b-a)^{1-\beta}. \tag{16.30}$$

Above, it is not strange to assume that $\mathbf{T}_\alpha^c X, {}_\alpha^d\mathbf{T}X, \mathbf{T}_\alpha^t X, {}_\alpha^t\mathbf{T}X$ are stochastic processes, see Remark 16.8.

We need

Concepts 16.15 *Let $\widetilde{L}$ be a positive linear operator from $C([a,b])$ into itself. Let $X(t,\omega)$ be a stochastic process from $[a,b] \times (\Omega, \mathcal{F}, P)$ into $\mathbb{R}$, where $(\Omega, \mathcal{F}, P)$ is a probability space. Let $\alpha \in (n, n+1]$, $n \in \mathbb{Z}_+$, and $X(\cdot,\omega) \in C^{n+1}([a,b])$, $\forall$ $\omega \in \Omega$. Also we assume for each $t \in [a,b]$ that $X^{(k)}(t,\cdot)$ is measurable over $(\Omega, \mathcal{F})$ for all $k = 1, \ldots, n+1$. Define*

$$M(X)(t,\omega) := \widetilde{L}(X(\cdot,\omega))(t), \ \forall \omega \in \Omega, \forall t \in [a,b], \tag{16.31}$$

and assume that it is a random variable in ω. Clearly M is a positive linear operator on stochastic processes.

We make

Assumption 16.16 (i) For any $t \in [a,b]$ we assume that $\mathbf{T}_\alpha^t X(z,\omega)$ is continuous in $z \in [t,b]$, uniformly with respect to $\omega \in \Omega$. I.e. $\forall \varepsilon > 0 \ \exists \delta > 0$: whenever $|z_1 - z_2| \leq \delta$; $z_1, z_2 \in [t,b]$, then $\left|\mathbf{T}_\alpha^t X(z_1,\omega) - \mathbf{T}_\alpha^t X(z_2,\omega)\right| \leq \varepsilon$, $\forall$ $\omega \in \Omega$.
We denote this by $\mathbf{T}_\alpha^t X \in C_{\mathbb{R}}^U([t,b])$, the space of continuous in x, uniformly with respect to ω, stochastic processes over $[t,b]$.

(ii) For any $t \in [a,b]$ we assume that ${}_\alpha^t\mathbf{T}X(z,\omega)$ is continuous in $z \in [a,t]$, uniformly with respect to $\omega \in \Omega$. I.e. $\forall \varepsilon > 0 \ \exists \delta > 0$: whenever $|z_1 - z_2| \leq \delta$; $z_1, z_2 \in [a,t]$, then $\left|{}_\alpha^t\mathbf{T}X(z_1,\omega) - {}_\alpha^t\mathbf{T}X(z_2,\omega)\right| \leq \varepsilon, \forall \omega \in \Omega$.
We denote this by ${}_\alpha^t\mathbf{T}X \in C_{\mathbb{R}}^U([a,t])$, the space of continuous in x, uniformly with respect to ω, stochastic processes over $[a,t]$.

Remark 16.17 Assumption 16.16 implies:

(i) $\mathbf{T}_\alpha^t X(\cdot,\omega) \in C([t,b])$, $\forall \omega \in \Omega$, and $\mathbf{T}_\alpha^t X$ is q-mean uniformly continuous in $z \in [t,b]$, that is $\mathbf{T}_\alpha^t X \in C_{\mathbb{R}}^{U_q}([t,b])$, for any $1 \leq q < \infty$.

(ii) ${}_\alpha^t\mathbf{T}X(\cdot,\omega) \in C([a,t])$, $\forall \omega \in \Omega$, and ${}_\alpha^t\mathbf{T}X$ is q-mean uniformly continuous in $z \in [a,t]$, that is ${}_\alpha^t\mathbf{T}X \in C_{\mathbb{R}}^{U_q}([a,t])$, for any $1 \leq q < \infty$.

We need

Definition 16.18 Denote by

$$(EX)(t) := \int_\Omega X(t,\omega) P(d\omega), \ \forall t \in [a,b], \tag{16.32}$$

the expectation operator.

We make

Assumption 16.19 We assume that

$$\left(E\left|X^{(k)}\right|^q\right)(t) < \infty, \ \forall t \in [a,b], \tag{16.33}$$

$q > 1$, for all $k = 0, 1, \ldots, n$.

We make

Assumption 16.20 We assume that

$$\left(E\left|X^{(k)}\right|\right)(t)<\infty,\ \forall\, t\in[a,b], \tag{16.34}$$

for all $k=0,1,\ldots,n$.

We give

Remark 16.21 By the Riesz representation theorem [12] we have that there exists μ_t unique, completed Borel measure on $[a,b]$ with

$$m_t:=\mu_t([a,b])=\widetilde{L}(1)(t)>0, \tag{16.35}$$

such that

$$\widetilde{L}(f)(t)=\int_{[a,b]} f(x)\,d\mu_t(x),\ \forall\, t\in[a,b],\ \forall\, f\in C([a,b]). \tag{16.36}$$

Consequently we have that

$$M(X)(t,\omega)=\int_{[a,b]} X(x,\omega)\,d\mu_t(x),\ \forall\ (t,\omega)\in[a,b]\times\Omega, \tag{16.37}$$

and X as in Concepts 16.15.

Here $\chi_{[\gamma,\delta]}(s)$ stands for the characteristic function on $[\gamma,\delta]\subseteq[a,b]$.

Notice that $(r>0)$

$$\int_{[t,b]}(s-t)^r\,\mu_t(ds)=\int_{[a,b]}\chi_{[t,b]}(s)\,|s-t|^r\,\mu_t(ds)=\widetilde{L}\left(|\cdot-t|^r\,\chi_{[t,b]}(\cdot)\right)(t), \tag{16.38}$$

and

$$\int_{[a,t]}(t-s)^r\,\mu_t(ds)=\int_{[a,b]}\chi_{[a,t]}(s)\,|s-t|^r\,\mu_t(ds)=\widetilde{L}\left(|\cdot-t|^r\,\chi_{[a,t]}(\cdot)\right)(t). \tag{16.39}$$

Let $\alpha\in(n,n+1]$, $n\in\mathbb{N}$ and $k=1,\ldots,n$. Then by Hö lder's inequality we obtain

$$\left|\int_{[a,b]}(x-t)^k\,d\mu_t(x)\right|\leq\int_{[a,b]}|x-t|^k\,d\mu_t(x)$$

$$\leq\left(\int_{[a,b]}|x-t|^{\alpha+1}\,d\mu_t(x)\right)^{\left(\frac{k}{\alpha+1}\right)}\left(\mu_t([a,b])\right)^{\left(\frac{\alpha+1-k}{\alpha+1}\right)}. \tag{16.40}$$

Therefore it holds

$$\left\| \widetilde{L}\left((\cdot - t)^k \right)(t) \right\|_{\infty,[a,b]} \leq \left\| \widetilde{L}\left(|\cdot - t|^k \right)(t) \right\|_{\infty,[a,b]} \leq$$

$$\left\| \widetilde{L}\left(|\cdot - t|^{\alpha+1} \right)(t) \right\|_{\infty,[a,b]}^{\left(\frac{k}{\alpha+1}\right)} \left\| \widetilde{L}(1) \right\|_{\infty,[a,b]}^{\left(\frac{\alpha+1-k}{\alpha+1}\right)}, \tag{16.41}$$

all $k = 1, \ldots, n$.

Also, we observe that

$$C([a,b]) \ni |\cdot - t|^{\alpha+1} \chi_{[a,t]}(\cdot) \leq |\cdot - t|^{\alpha+1}, \quad \forall\, t \in [a,b], \tag{16.42}$$

and

$$C([a,b]) \ni |\cdot - t|^{\alpha+1} \chi_{[t,b]}(\cdot) \leq |\cdot - t|^{\alpha+1}, \quad \forall\, t \in [a,b]. \tag{16.43}$$

By positivity of $\widetilde{L}$ we obtain

$$\left\| \widetilde{L}\left(|\cdot - t|^{\alpha+1} \chi_{[a,t]}(\cdot) \right)(t) \right\|_{\infty,[a,b]} \leq \left\| \widetilde{L}\left(|\cdot - t|^{\alpha+1} \right)(t) \right\|_{\infty,[a,b]} < \infty, \tag{16.44}$$

by $\widetilde{L}\left(|\cdot - t|^{\alpha+1} \right)(t)$ being continuous in $t \in [a,b]$, see p. 388 of [4],

and

$$\left\| \widetilde{L}\left(|\cdot - t|^{\alpha+1} \chi_{[t,b]}(\cdot) \right)(t) \right\|_{\infty,[a,b]} \leq \left\| \widetilde{L}\left(|\cdot - t|^{\alpha+1} \right)(t) \right\|_{\infty,[a,b]}. \tag{16.45}$$

Above (16.38)–(16.41) and (16.44), (16.45) can be used to derive convergences from our main results next.

In this work we denote by $\widetilde{L}\left(\chi_{[a,t]}(\cdot)\right)(t) := \mu_t([a,t]) \leq \widetilde{L}(1)(t)$, and by $\widetilde{L}\left(\chi_{[t,b]}(\cdot)\right)(t) := \mu_t([t,b]) \leq \widetilde{L}(1)(t)$.

We make

Assumption 16.22 Assume that $\left| X^{(n+1)}(t,\omega) \right| \leq M^*, \forall\, (t,\omega) \in [a,b] \times \Omega$, where $M^* > 0$. And suppose that $X^{(n+1)}(\cdot,\omega)$ is continuous over $[a,b]$, uniformly with respect to $\omega \in \Omega$.

Clearly, Assumption 16.22 implies Assumption 16.16, see also Remark 16.8.

16.4 Main Results

Next we present our first main result on the quantitative stochastic fractional approximation regarding stochastic processes:

Theorem 16.23 *Let* $n \in \mathbb{Z}_+$, $\alpha \in (n, n+1)$; $p, q > 1 : \frac{1}{p} + \frac{1}{q} = 1$, *with* $\beta := \alpha - n > \frac{1}{q}$; *set* $\lambda_1 := \left(\frac{\Gamma(np+1)\Gamma(p(\beta-1)+1)}{\Gamma(p(\alpha-1)+2)} \right)^{(q-1)}$. *Suppose Concepts 16.15, Assumptions 16.16 and 16.19.*

Then, $\forall\, t \in [a, b]$, *we have:*

$$\left(E\left(|M(X) - X|^q\right)(t)\right)^{\frac{1}{q}} \leq \left(\left(E|X|^q\right)(t)\right)^{\frac{1}{q}} \left|\widetilde{L}(1)(t) - 1\right| +$$

$$\sum_{k=1}^{n} \frac{\left(\left(E\left|X^{(k)}\right|^q\right)(t)\right)^{\frac{1}{q}}}{k!} \left|\widetilde{L}\left((\cdot - t)^k\right)(t)\right| + \frac{2^{\frac{1}{p}} \lambda_1^{\frac{1}{q}}}{(q+1)^{\frac{\alpha}{q(\alpha+1)}}\, n!}$$

$$\left\{\left[\Omega_1\left(\mathbf{T}_\alpha^t X, \left(\frac{1}{(q+1)} \widetilde{L}\left(|\cdot - t|^{q(\alpha+1)} \chi_{[t,b]}(\cdot)\right)(t)\right)^{\frac{1}{q(\alpha+1)}}\right)_{L^q,[t,b]}\right.\right.$$

$$\left(\widetilde{L}\left(\chi_{[t,b]}(\cdot)\right)(t)\right)^{\frac{1}{p}} \left(\widetilde{L}\left(|\cdot - t|^{q(\alpha+1)} \chi_{[t,b]}(\cdot)\right)(t)\right)^{\frac{\alpha}{q(\alpha+1)}}$$

$$\left.\left[\left(\widetilde{L}\left(\chi_{[t,b]}(\cdot)\right)(t)\right)^{\frac{1}{\alpha+1}} (q+1)^{\frac{\alpha}{(\alpha+1)}} + 1\right]^{\frac{1}{q}}\right] + \tag{16.46}$$

$$\left[\Omega_1\left({}_\alpha^t\mathbf{T} X, \left(\frac{1}{(q+1)} \widetilde{L}\left(|\cdot - t|^{q(\alpha+1)} \chi_{[a,t]}(\cdot)\right)(t)\right)^{\frac{1}{q(\alpha+1)}}\right)_{L^q,[a,t]}\right.$$

$$\left(\widetilde{L}\left(\chi_{[a,t]}(\cdot)\right)(t)\right)^{\frac{1}{p}} \left(\widetilde{L}\left(|\cdot - t|^{q(\alpha+1)} \chi_{[a,t]}(\cdot)\right)(t)\right)^{\frac{\alpha}{q(\alpha+1)}}$$

$$\left.\left.\left[\left(\widetilde{L}\left(\chi_{[a,t]}(\cdot)\right)(t)\right)^{\frac{1}{\alpha+1}} (q+1)^{\frac{\alpha}{(\alpha+1)}} + 1\right]^{\frac{1}{q}}\right]\right\}$$

Proof By Theorem 16.7 we have

$$X(s, \omega) = \sum_{k=0}^{n} \frac{X^{(k)}(t, \omega)}{k!} (s - t)^k +$$

$$\frac{1}{n!} \int_t^s (s - z)^n (z - t)^{\beta - 1} \left(\mathbf{T}_\alpha^t X(z, \omega) - \mathbf{T}_\alpha^t X(t, \omega)\right) dz, \tag{16.47}$$

for all $t \leq s \leq b$, $\forall\, \omega \in \Omega$; $\alpha \in (n, n+1)$, see also (16.21).

Furthermore it holds from Theorem 16.7 that

$$X(s, \omega) = \sum_{k=0}^{n} \frac{X^{(k)}(t, \omega)}{k!} (s - t)^k -$$

$$\frac{1}{n!} \int_s^t (t - z)^{\beta - 1} (z - s)^n \left({}_\alpha^t\mathbf{T} X(z, \omega) - {}_\alpha^t \mathbf{T} X(t, \omega)\right) dz, \tag{16.48}$$

for all $a \leq s \leq t$, $\forall\, \omega \in \Omega$; $\alpha \in (n, n+1)$, see also (16.21).

Let $p, q > 1 : \frac{1}{p} + \frac{1}{q} = 1$, and $\alpha > n + \frac{1}{q}$, i.e. $\beta := \alpha - n > \frac{1}{q}$, implying $p(\beta - 1) + 1 > 0$.

Next, we work on the remainder of (16.47). Using Hölder's inequality we have

$$\int_t^s (s-z)^n (z-t)^{\beta-1} \left| \mathbf{T}_\alpha^t X(z,\omega) - \mathbf{T}_\alpha^t X(t,\omega) \right| dz \leq$$

$$\left(\int_t^s \left| \mathbf{T}_\alpha^t X(z,\omega) - \mathbf{T}_\alpha^t X(t,\omega) \right|^q dz \right)^{\frac{1}{q}}$$

$$\left(\int_t^s (s-z)^{(np+1)-1} (z-t)^{(p(\beta-1)+1)-1} dz \right)^{\frac{1}{p}} = \tag{16.49}$$

$$\left(\int_t^s \left| \mathbf{T}_\alpha^t X(z,\omega) - \mathbf{T}_\alpha^t X(t,\omega) \right|^q dz \right)^{\frac{1}{q}}$$

$$\left(\frac{\Gamma(np+1)\,\Gamma(p(\beta-1)+1)}{\Gamma(p(\alpha-1)+2)} \right)^{\frac{1}{p}} (s-t)^{\frac{q\alpha-1}{q}},$$

$t \leq s \leq b$.

Hence we have

$$\left(\int_t^s (s-z)^n (z-t)^{\beta-1} \left| \mathbf{T}_\alpha^t X(z,\omega) - \mathbf{T}_\alpha^t X(t,\omega) \right| dz \right)^q \leq$$

$$\left(\int_t^s \left| \mathbf{T}_\alpha^t X(z,\omega) - \mathbf{T}_\alpha^t X(t,\omega) \right|^q dz \right)$$

$$\left(\frac{\Gamma(np+1)\,\Gamma(p(\beta-1)+1)}{\Gamma(p(\alpha-1)+2)} \right)^{(q-1)} (s-t)^{(q\alpha-1)}, \tag{16.50}$$

$t \leq s \leq b$.

We call

$$\lambda_1 := \left(\frac{\Gamma(np+1)\,\Gamma(p(\beta-1)+1)}{\Gamma(p(\alpha-1)+2)} \right)^{(q-1)}. \tag{16.51}$$

Applying again Hölder's inequality we obtain:

$$\Gamma_1 := \frac{1}{n!} \left(\int_\Omega \left\{ \int_t^b \left(\int_t^s (s-z)^n (z-t)^{\beta-1} \right.\right.\right.$$

$$\left.\left.\left. \left| \mathbf{T}_\alpha^t X(z,\omega) - \mathbf{T}_\alpha^t X(t,\omega) \right| dz \right) \mu_t(ds) \right\}^q P(d\omega) \right)^{\frac{1}{q}}$$

$$\leq \frac{(\mu_t([t,b]))^{\frac{1}{p}}}{n!} \left(\int_{\Omega} \left\{ \int_t^b \left(\int_t^s (s-z)^n (z-t)^{\beta-1} \right.\right.\right.$$

$$\left.\left.\left. \left| \mathbf{T}_{\alpha}^t X(z,\omega) - \mathbf{T}_{\alpha}^t X(t,\omega) \right| dz \right)^q \mu_t(ds) \right\} P(d\omega) \right)^{\frac{1}{q}} \tag{16.52}$$

$$\overset{((16.50.),(16.51.))}{\leq} \frac{\lambda_1^{\frac{1}{q}} (\mu_t([t,b]))^{\frac{1}{p}}}{n!} \left(\int_{\Omega} \left\{ \int_t^b \left[\left(\int_t^s \left| \mathbf{T}_{\alpha}^t X(z,\omega) - \mathbf{T}_{\alpha}^t X(t,\omega) \right|^q dz \right) \right.\right.\right.$$

$$\left.\left.\left. (s-t)^{(q\alpha-1)} \right] \mu_t(ds) \right\} P(d\omega) \right)^{\frac{1}{q}} =: (*),$$

where $\varphi(z,\omega) := \left| \mathbf{T}_{\alpha}^t X(z,\omega) - \mathbf{T}_{\alpha}^t X(t,\omega) \right|^q \geq 0$, is a real valued random variable for each $z \in [t,b]$ (by assumptions $\mathbf{T}_{\alpha}^t X(z,\omega)$, $t \leq z \leq b$, is continuous in z and measurable in ω). Hence $\varphi(z,\omega)$ is continuous in z and jointly measurable, see [6, p. 353]. Thus $\delta(s,\omega) := \left(\int_t^s \varphi(z,\omega) dz \right) (s-t)^{q\alpha-1}$ is a real valued random variable in ω and continuous in $s \in [t,b]$. Again by [6, p. 353], $\delta(s,\omega)$ is jointly measurable in (s,ω).

Therefore by applying twice Tonelli's theorem, [12, p. 270], we get

$$(*) = \frac{\lambda_1^{\frac{1}{q}} (\mu_t([t,b]))^{\frac{1}{p}}}{n!}$$

$$\left\{ \int_t^b \left(\int_t^s \left(\int_{\Omega} \left| \mathbf{T}_{\alpha}^t X(z,\omega) - \mathbf{T}_{\alpha}^t X(t,\omega) \right|^q P(d\omega) \right) dz \right) (s-t)^{(q\alpha-1)} \mu_t(ds) \right\}^{\frac{1}{q}}. \tag{16.53}$$

We have proved that

$$\Gamma_1 := \frac{1}{n!} \left(\int_{\Omega} \left\{ \int_t^b \left(\int_t^s (s-z)^n (z-t)^{\beta-1} \right.\right.\right.$$

$$\left.\left.\left. \left| \mathbf{T}_{\alpha}^t X(z,\omega) - \mathbf{T}_{\alpha}^t X(t,\omega) \right| dz \right) \mu_t(ds) \right\}^q P(d\omega) \right)^{\frac{1}{q}}$$

$$\leq \frac{\lambda_1^{\frac{1}{q}} (\mu_t([t,b]))^{\frac{1}{p}}}{n!}$$

$$\left\{ \int_t^b \left(\int_t^s \left(\int_{\Omega} \left| \mathbf{T}_{\alpha}^t X(z,\omega) - \mathbf{T}_{\alpha}^t X(t,\omega) \right|^q P(d\omega) \right) dz \right) (s-t)^{(q\alpha-1)} \mu_t(ds) \right\}^{\frac{1}{q}} \tag{16.54}$$

(continuing estimation)

$$\leq \frac{\lambda_1^{\frac{1}{q}} (\mu_t([t,b]))^{\frac{1}{p}}}{n!}$$

$$\left\{\int_t^b\left(\int_t^s\Omega_1^q\left(\mathbf{T}_\alpha^tX,z-t\right)_{L^q,[t,b]}dz\right)(s-t)^{(q\alpha-1)}\mu_t(ds)\right\}^{\frac{1}{q}}\overset{(16.26.)}{\leq}$$

(let $h_1>0$)

$$\Omega_1\left(\mathbf{T}_\alpha^tX,h_1\right)_{L^q,[t,b]}\frac{\lambda_1^{\frac{1}{q}}\left(\mu_t([t,b])\right)^{\frac{1}{p}}}{n!}$$

$$\left\{\int_t^b\left(\int_t^s\left\lceil\frac{z-t}{h_1}\right\rceil^q dz\right)(s-t)^{(q\alpha-1)}\mu_t(ds)\right\}^{\frac{1}{q}}\leq \tag{16.55}$$

$$\Omega_1\left(\mathbf{T}_\alpha^tX,h_1\right)_{L^q,[t,b]}\frac{\lambda_1^{\frac{1}{q}}\left(\mu_t([t,b])\right)^{\frac{1}{p}}}{n!}$$

$$\left\{\int_t^b\left(\int_t^s\left(1+\frac{z-t}{h_1}\right)^q dz\right)(s-t)^{(q\alpha-1)}\mu_t(ds)\right\}^{\frac{1}{q}}\leq$$

$$\tau\left\{\int_t^b\left(\int_t^s\left(1+\frac{(z-t)^q}{h_1^q}\right)dz\right)(s-t)^{(q\alpha-1)}\mu_t(ds)\right\}^{\frac{1}{q}}=$$

(where

$$\tau:=2^{\frac{1}{p}}\Omega_1\left(\mathbf{T}_\alpha^tX,h_1\right)_{L^q,[t,b]}\frac{\lambda_1^{\frac{1}{q}}\left(\mu_t([t,b])\right)^{\frac{1}{p}}}{n!} \tag{16.56}$$

)

$$\tau\left\{\int_{[t,b]}\left[(s-t)^{q\alpha}+\frac{1}{h_1^q}\left(\int_t^s(z-t)^q dz\right)(s-t)^{(q\alpha-1)}\right]\mu_t(ds)\right\}^{\frac{1}{q}}=$$

$$\tau\left\{\int_{[t,b]}\left[(s-t)^{q\alpha}+\frac{1}{h_1^q}\frac{(s-t)^{q+1}}{(q+1)}(s-t)^{(q\alpha-1)}\right]\mu_t(ds)\right\}^{\frac{1}{q}}=$$

$$\tau\left\{\int_{[t,b]}\left[(s-t)^{q\alpha}+\frac{1}{h_1^q}\frac{(s-t)^{q(\alpha+1)}}{(q+1)}\right]\mu_t(ds)\right\}^{\frac{1}{q}}= \tag{16.57}$$

$$\tau\left\{\int_{[t,b]}(s-t)^{q\alpha}\mu_t(ds)+\frac{1}{h_1^q(q+1)}\int_{[t,b]}(s-t)^{q(\alpha+1)}\mu_t(ds)\right\}^{\frac{1}{q}}\leq$$

$$\tau\left\{\left(\int_{[t,b]}(s-t)^{q(\alpha+1)}\mu_t(ds)\right)^{\frac{\alpha}{\alpha+1}}\left(\mu_t([t,b])\right)^{\frac{1}{\alpha+1}}+\right.$$

$$\left.\frac{1}{h_1^q (q+1)} \int_{[t,b]} (s-t)^{q(\alpha+1)} \mu_t (ds)\right\}^{\frac{1}{q}} =: (**) .$$

We set and assume

$$h_1 := \left(\frac{1}{(q+1)} \int_{[t,b]} (s-t)^{q(\alpha+1)} \mu_t (ds) \right)^{\frac{1}{q(\alpha+1)}} > 0. \tag{16.58}$$

I.e.

$$h_1^{q(\alpha+1)} = \frac{1}{(q+1)} \int_{[t,b]} (s-t)^{q(\alpha+1)} \mu_t (ds) > 0. \tag{16.59}$$

Therefore we get

$$(**) = \tau \left\{ (\mu_t ([t,b]))^{\frac{1}{\alpha+1}} h_1^{q\alpha} (q+1)^{\frac{\alpha}{\alpha+1}} + h_1^{q\alpha} \right\}^{\frac{1}{q}} =$$

$$\tau h_1^{\alpha} \left\{ (\mu_t ([t,b]))^{\frac{1}{\alpha+1}} (q+1)^{\frac{\alpha}{\alpha+1}} + 1 \right\}^{\frac{1}{q}} . \tag{16.60}$$

We have proved that

$$\Gamma_1 \leq \tau h_1^{\alpha} \left\{ (\mu_t ([t,b]))^{\frac{1}{\alpha+1}} (q+1)^{\frac{\alpha}{\alpha+1}} + 1 \right\}^{\frac{1}{q}} . \tag{16.61}$$

The last means that

$$\Gamma_1 := \frac{1}{n!} \left(\int_{\Omega} \left\{ \int_t^b \left(\int_t^s (s-z)^n (z-t)^{\beta-1} \right.\right.\right.$$

$$\left.\left.\left. \left| \mathbf{T}_{\alpha}^t X (z,\omega) - \mathbf{T}_{\alpha}^t X (t,\omega) \right| dz \right) \mu_t (ds) \right\}^q P (d\omega) \right)^{\frac{1}{q}}$$

$$\leq 2^{\frac{1}{p}} \Omega_1 \left(\mathbf{T}_{\alpha}^t X, \left(\frac{1}{(q+1)} \int_{[t,b]} (s-t)^{q(\alpha+1)} \mu_t (ds) \right)^{\frac{1}{q(\alpha+1)}} \right)_{L^q,[t,b]}$$

$$\frac{\lambda_1^{\frac{1}{q}} (\mu_t ([t,b]))^{\frac{1}{p}}}{n!} \left(\frac{1}{(q+1)} \int_{[t,b]} (s-t)^{q(\alpha+1)} \mu_t (ds) \right)^{\frac{\alpha}{q(\alpha+1)}}$$

$$\left\{ (\mu_t ([t,b]))^{\frac{1}{\alpha+1}} (q+1)^{\frac{\alpha}{\alpha+1}} + 1 \right\}^{\frac{1}{q}} . \tag{16.62}$$

Next we work on the remainder related to (16.48) ($a \leq s \leq t$), we have

$$\int_s^t (t-z)^{\beta-1} (z-s)^n \left| {}_\alpha^t \mathbf{T}X(z,\omega) - {}_\alpha^t \mathbf{T}X(t,\omega) \right| dz \leq$$

$$\left(\int_s^t \left| {}_\alpha^t \mathbf{T}X(z,\omega) - {}_\alpha^t \mathbf{T}X(t,\omega) \right|^q dz \right)^{\frac{1}{q}}$$

$$\left(\int_s^t (t-z)^{(p(\beta-1)+1)-1} (z-s)^{(pn+1)-1} dz \right)^{\frac{1}{p}} = \tag{16.63}$$

$$\left(\int_s^t \left| {}_\alpha^t \mathbf{T}X(z,\omega) - {}_\alpha^t \mathbf{T}X(t,\omega) \right|^q dz \right)^{\frac{1}{q}}$$

$$\left(\frac{\Gamma(p(\beta-1)+1)\Gamma(pn+1)}{\Gamma(p(\alpha-1)+2)} \right)^{\frac{1}{p}} (t-s)^{\frac{q\alpha-1}{q}},$$

$a \leq s \leq t$.

Hence we have

$$\left(\int_s^t (t-z)^{\beta-1} (z-s)^n \left| {}_\alpha^t \mathbf{T}X(z,\omega) - {}_\alpha^t \mathbf{T}X(t,\omega) \right| dz \right)^q \leq$$

$$\left(\int_s^t \left| {}_\alpha^t \mathbf{T}X(z,\omega) - {}_\alpha^t \mathbf{T}X(t,\omega) \right|^q dz \right) \tag{16.64}$$

$$\left(\frac{\Gamma(p(\beta-1)+1)\Gamma(pn+1)}{\Gamma(p(\alpha-1)+2)} \right)^{(q-1)} (t-s)^{(q\alpha-1)},$$

$a \leq s \leq t$.

Applying again Hölder's inequality we obtain

$$\Gamma_2 := \frac{1}{n!} \left(\int_\Omega \left\{ \int_a^t \left(\int_s^t (t-z)^{\beta-1} (z-s)^n \right.\right.\right.$$

$$\left.\left.\left. \left| {}_\alpha^t \mathbf{T}X(z,\omega) - {}_\alpha^t \mathbf{T}X(t,\omega) \right| dz \right) \mu_t(ds) \right\}^q P(d\omega) \right)^{\frac{1}{q}}$$

$$\leq \frac{(\mu_t([a,t]))^{\frac{1}{p}}}{n!} \left(\int_\Omega \left\{ \int_a^t \left(\int_s^t (t-z)^{\beta-1} (z-s)^n \right.\right.\right.$$

$$\left.\left.\left. \left| {}_\alpha^t \mathbf{T}X(z,\omega) - {}_\alpha^t \mathbf{T}X(t,\omega) \right| dz \right)^q \mu_t(ds) \right\} P(d\omega) \right)^{\frac{1}{q}}$$

$$\overset{((16.51.),(16.64.))}{\leq} \frac{\lambda_1^{\frac{1}{q}} (\mu_t([a,t]))^{\frac{1}{p}}}{n!}$$

$$\left(\int_{\Omega}\left\{\int_a^t\left[\left(\int_s^t \left|{}_{\alpha}^{t}\mathbf{T}X(z,\omega)-{}_{\alpha}^{t}\mathbf{T}X(t,\omega)\right|^q dz\right)(t-s)^{(q\alpha-1)}\right]\mu_t(ds)\right\}P(d\omega)\right)^{\frac{1}{q}} \tag{16.65}$$

(as before)

$$=\frac{\lambda_1^{\frac{1}{q}}\left(\mu_t([a,t])\right)^{\frac{1}{p}}}{n!}$$

$$\left\{\int_a^t\left(\int_s^t\left(\int_{\Omega}\left|{}_{\alpha}^{t}\mathbf{T}X(z,\omega)-{}_{\alpha}^{t}\mathbf{T}X(t,\omega)\right|^q P(d\omega)\right)dz\right)(t-s)^{(q\alpha-1)}\mu_t(ds)\right\}^{\frac{1}{q}}$$

$$\leq\frac{\lambda_1^{\frac{1}{q}}\left(\mu_t([a,t])\right)^{\frac{1}{p}}}{n!}$$

$$\left\{\int_a^t\left(\int_s^t\Omega_1^q\left({}_{\alpha}^{t}\mathbf{T}X,t-z\right)_{L^q,[a,t]}dz\right)(t-s)^{(q\alpha-1)}\mu_t(ds)\right\}^{\frac{1}{q}}\leq$$

(let $h_2>0$)

$$\Omega_1\left({}_{\alpha}^{t}\mathbf{T}X,h_2\right)_{L^q,[a,t]}\frac{\lambda_1^{\frac{1}{q}}\left(\mu_t([a,t])\right)^{\frac{1}{p}}}{n!}$$

$$\left\{\int_a^t\left(\int_s^t\left\lceil\frac{t-z}{h_2}\right\rceil^q dz\right)(t-s)^{(q\alpha-1)}\mu_t(ds)\right\}^{\frac{1}{q}}\leq$$

$$\Omega_1\left({}_{\alpha}^{t}\mathbf{T}X,h_2\right)_{L^q,[a,t]}\frac{\lambda_1^{\frac{1}{q}}\left(\mu_t([a,t])\right)^{\frac{1}{p}}}{n!}$$

$$\left\{\int_a^t\left(\int_s^t\left(1+\frac{t-z}{h_2}\right)^q dz\right)(t-s)^{(q\alpha-1)}\mu_t(ds)\right\}^{\frac{1}{q}}\leq$$

$$\rho\left\{\int_a^t\left(\int_s^t\left(1+\frac{(t-z)^q}{h_2^q}\right)dz\right)(t-s)^{(q\alpha-1)}\mu_t(ds)\right\}^{\frac{1}{q}}=$$

(where

$$\rho:=2^{\frac{1}{p}}\Omega_1\left({}_{\alpha}^{t}\mathbf{T}X,h_2\right)_{L^q,[a,t]}\frac{\lambda_1^{\frac{1}{q}}\left(\mu_t([a,t])\right)^{\frac{1}{p}}}{n!} \tag{16.66}$$

)

$$\rho\left\{\int_{[a,t]}\left[(t-s)^{q\alpha}+\frac{1}{h_2^q}\left(\int_s^t (t-z)^q\,dz\right)(t-s)^{(q\alpha-1)}\right]\mu_t\,(ds)\right\}^{\frac{1}{q}}=$$

$$\rho\left\{\int_{[a,t]}\left[(t-s)^{q\alpha}+\frac{1}{h_2^q}\frac{(t-s)^{q+1}}{(q+1)}(t-s)^{(q\alpha-1)}\right]\mu_t\,(ds)\right\}^{\frac{1}{q}}=$$

$$\rho\left\{\int_{[a,t]}\left[(t-s)^{q\alpha}+\frac{1}{h_2^q}\frac{(t-s)^{q(\alpha+1)}}{(q+1)}\right]\mu_t\,(ds)\right\}^{\frac{1}{q}}= \tag{16.67}$$

$$\rho\left\{\int_{[a,t]}(t-s)^{q\alpha}\,\mu_t\,(ds)+\frac{1}{h_2^q\,(q+1)}\int_{[a,t]}(t-s)^{q(\alpha+1)}\,\mu_t\,(ds)\right\}^{\frac{1}{q}}\leq$$

$$\rho\left\{\left(\int_{[a,t]}(t-s)^{q(\alpha+1)}\,\mu_t\,(ds)\right)^{\frac{\alpha}{\alpha+1}}(\mu_t\,([a,t]))^{\frac{1}{\alpha+1}}+\right.$$

$$\left.\frac{1}{h_2^q\,(q+1)}\int_{[a,t]}(t-s)^{q(\alpha+1)}\,\mu_t\,(ds)\right\}^{\frac{1}{q}}=:(***)\,.$$

We set and assume

$$h_2:=\left(\frac{1}{(q+1)}\int_{[a,t]}(t-s)^{q(\alpha+1)}\,\mu_t\,(ds)\right)^{\frac{1}{q(\alpha+1)}}>0. \tag{16.68}$$

I.e.

$$h_2^{q(\alpha+1)}=\frac{1}{(q+1)}\int_{[a,t]}(t-s)^{q(\alpha+1)}\,\mu_t\,(ds)>0. \tag{16.69}$$

Therefore we get

$$(***)=\rho\left\{(\mu_t\,([a,t]))^{\frac{1}{\alpha+1}}\,h_2^{q\alpha}\,(q+1)^{\frac{\alpha}{\alpha+1}}+h_2^{q\alpha}\right\}^{\frac{1}{q}}= \tag{16.70}$$

$$\rho h_2^{\alpha}\left\{(\mu_t\,([a,t]))^{\frac{1}{\alpha+1}}\,(q+1)^{\frac{\alpha}{\alpha+1}}+1\right\}^{\frac{1}{q}}.$$

That is

$$\Gamma_2\leq\rho h_2^{\alpha}\left\{(\mu_t\,([a,t]))^{\frac{1}{\alpha+1}}\,(q+1)^{\frac{\alpha}{\alpha+1}}+1\right\}^{\frac{1}{q}}. \tag{16.71}$$

The last means

$$\Gamma_2:=\frac{1}{n!}\left(\int_{\Omega}\left\{\int_a^t\left(\int_s^t (t-z)^{\beta-1}\,(z-s)^n\right.\right.\right.$$

$$\left|{}_{\alpha}^{t}\mathbf{T}X(z,\omega)-{}_{\alpha}^{t}\mathbf{T}X(t,\omega)\right|dz\Big)\mu_t(ds)\Big\}^q P(d\omega)\Big)^{\frac{1}{q}}$$

$$\leq 2^{\frac{1}{p}}\Omega_1\left({}_{\alpha}^{t}\mathbf{T}X,\left(\frac{1}{(q+1)}\int_{[a,t]}(t-s)^{q(\alpha+1)}\mu_t(ds)\right)^{\frac{1}{q(\alpha+1)}}\right)_{L^q,[a,t]}$$

$$\frac{\lambda_1^{\frac{1}{q}}(\mu_t([a,t]))^{\frac{1}{p}}}{n!}\left(\frac{1}{(q+1)}\int_{[a,t]}(t-s)^{q(\alpha+1)}\mu_t(ds)\right)^{\frac{\alpha}{q(\alpha+1)}}$$

$$\left\{(\mu_t([a,t]))^{\frac{1}{\alpha+1}}(q+1)^{\frac{\alpha}{\alpha+1}}+1\right\}^{\frac{1}{q}}. \tag{16.72}$$

Next we connect facts together.

We have that

$$M(X)(t,\omega)-X(t,\omega)\widetilde{L}(1)(t)=$$

$$\int_{[a,b]}X(s,\omega)\mu_t(ds)-X(t,\omega)\widetilde{L}(1)(t)=$$

$$\int_{[a,t]}X(s,\omega)\mu_t(ds)+\int_{(t,b]}X(s,\omega)\mu_t(ds)-X(t,\omega)\widetilde{L}(1)(t)\overset{((16.47.),(16.48.))}{=} \tag{16.73}$$

$$\int_{[a,t]}\left[\sum_{k=0}^{n}\frac{X^{(k)}(t,\omega)}{k!}(s-t)^k-\right.$$

$$\left.\frac{1}{n!}\int_s^t(t-z)^{\beta-1}(z-s)^n\left({}_{\alpha}^{t}\mathbf{T}X(z,\omega)-{}_{\alpha}^{t}\mathbf{T}X(t,\omega)\right)dz\right]\mu_t(ds)+$$

$$\int_{(t,b]}\left[\sum_{k=0}^{n}\frac{X^{(k)}(t,\omega)}{k!}(s-t)^k+\right.$$

$$\left.\frac{1}{n!}\int_t^s(s-z)^n(z-t)^{\beta-1}\left(\mathbf{T}_{\alpha}^{t}X(z,\omega)-\mathbf{T}_{\alpha}^{t}X(t,\omega)\right)dz\right]\mu_t(ds)$$

$$-X(t,\omega)\widetilde{L}(1)(t)=$$

$$\sum_{k=0}^{n}\frac{X^{(k)}(t,\omega)}{k!}\int_{[a,t]}(s-t)^k\mu_t(ds)-$$

$$\frac{1}{n!}\int_{[a,t]}\left(\int_s^t(t-z)^{\beta-1}(z-s)^n\left({}_{\alpha}^{t}\mathbf{T}X(z,\omega)-{}_{\alpha}^{t}\mathbf{T}X(t,\omega)\right)dz\right)\mu_t(ds)+$$

$$\sum_{k=0}^{n} \frac{X^{(k)}(t,\omega)}{k!} \int_{(t,b]} (s-t)^k \mu_t(ds) +$$

$$\frac{1}{n!} \int_{(t,b]} \left(\int_t^s (s-z)^n (z-t)^{\beta-1} \left(\mathbf{T}_\alpha^t X(z,\omega) - \mathbf{T}_\alpha^t X(t,\omega) \right) dz \right) \mu_t(ds)$$

$$- X(t,\omega) \widetilde{L}(1)(t) = \tag{16.74}$$

$$\sum_{k=1}^{n} \frac{X^{(k)}(t,\omega)}{k!} \left(\widetilde{L}\left((\cdot - t)^k \right)(t) \right) +$$

$$\frac{1}{n!} \left[\int_{(t,b]} \left(\int_t^s (s-z)^n (z-t)^{\beta-1} \left(\mathbf{T}_\alpha^t X(z,\omega) - \mathbf{T}_\alpha^t X(t,\omega) \right) dz \right) \mu_t(ds) - \right.$$

$$\left. \int_{[a,t]} \left(\int_s^t (t-z)^{\beta-1} (z-s)^n \left({}_\alpha^t \mathbf{T} X(z,\omega) - {}_\alpha^t \mathbf{T} X(t,\omega) \right) dz \right) \mu_t(ds) \right]. \tag{16.75}$$

Furthermore we have

$$\left| M(X)(t,\omega) - X(t,\omega) \widetilde{L}(1)(t) \right| \leq$$

$$\sum_{k=1}^{n} \frac{\left| X^{(k)}(t,\omega) \right|}{k!} \left| \widetilde{L}\left((\cdot - t)^k \right)(t) \right| +$$

$$\frac{1}{n!} \left[\int_{[t,b]} \left(\int_t^s (s-z)^n (z-t)^{\beta-1} \left| \mathbf{T}_\alpha^t X(z,\omega) - \mathbf{T}_\alpha^t X(t,\omega) \right| dz \right) \mu_t(ds) + \right.$$

$$\left. \int_{[a,t]} \left(\int_s^t (t-z)^{\beta-1} (z-s)^n \left| {}_\alpha^t \mathbf{T} X(z,\omega) - {}_\alpha^t \mathbf{T} X(t,\omega) \right| dz \right) \mu_t(ds) \right]. \tag{16.76}$$

We have eventually that

$$|M(X)(t,\omega) - X(t,\omega)| \leq$$

$$\left| M(X)(t,\omega) - X(t,\omega) \widetilde{L}(1)(t) \right| + |X(t,\omega)| \left| \widetilde{L}(1)(t) - 1 \right| \leq$$

$$|X(t,\omega)| \left| \widetilde{L}(1)(t) - 1 \right| + \sum_{k=1}^{n} \frac{\left| X^{(k)}(t,\omega) \right|}{k!} \left| \widetilde{L}\left((\cdot - t)^k \right)(t) \right| +$$

$$\frac{1}{n!} \left[\int_{[t,b]} \left(\int_t^s (s-z)^n (z-t)^{\beta-1} \left| \mathbf{T}_\alpha^t X(z,\omega) - \mathbf{T}_\alpha^t X(t,\omega) \right| dz \right) \mu_t(ds) + \right. \tag{16.77}$$

$$\int_{[a,t]}\left(\int_s^t (t-z)^{\beta-1}(z-s)^n \left|{}_{\alpha}^{t}\mathbf{T}X(z,\omega) -{}_{\alpha}^{t}\mathbf{T}X(t,\omega)\right| dz\right)\mu_t(ds)\Bigg].$$

Hence it holds

$$\left(E\left(|M(X)-X|^q\right)(t)\right)^{\frac{1}{q}} \leq \left(\left(E|X|^q\right)(t)\right)^{\frac{1}{q}}\left|\widetilde{L}(1)(t)-1\right| +$$

$$\sum_{k=1}^{n}\frac{\left(\left(E\left|X^{(k)}\right|^q\right)(t)\right)^{\frac{1}{q}}}{k!}\left|\widetilde{L}\left((\cdot-t)^k\right)(t)\right| +$$

$$\frac{1}{n!}\Bigg\{\Bigg[\int_\Omega\Bigg(\int_{[t,b]}\Bigg(\int_t^s (s-z)^n (z-t)^{\beta-1}$$

$$\left|\mathbf{T}_{\alpha}^{t}X(z,\omega)-\mathbf{T}_{\alpha}^{t}X(t,\omega)\right| dz\Big)\mu_t(ds)\Big)^q P(d\omega)\Bigg]^{\frac{1}{q}} \tag{16.78}$$

$$+\Bigg[\int_\Omega\Bigg(\int_{[a,t]}\Bigg(\int_s^t (t-z)^{\beta-1}(z-s)^n$$

$$\left|{}_{\alpha}^{t}\mathbf{T}X(z,\omega) -{}_{\alpha}^{t}\mathbf{T}X(t,\omega)\right| dz\Big)\mu_t(ds)\Big)^q P(d\omega)\Bigg]^{\frac{1}{q}}\Bigg\}.$$

By (16.62) and (16.72) we obtain

$$\left(E\left(|M(X)-X|^q\right)(t)\right)^{\frac{1}{q}} \leq \left(\left(E|X|^q\right)(t)\right)^{\frac{1}{q}}\left|\widetilde{L}(1)(t)-1\right| +$$

$$\sum_{k=1}^{n}\frac{\left(\left(E\left|X^{(k)}\right|^q\right)(t)\right)^{\frac{1}{q}}}{k!}\left|\widetilde{L}\left((\cdot-t)^k\right)(t)\right| +$$

$$\frac{2^{\frac{1}{p}}\lambda_1^{\frac{1}{q}}}{(q+1)^{\frac{\alpha}{q(\alpha+1)}}\,n!}\left[\Omega_1\left(\mathbf{T}_{\alpha}^{t}X,\left(\frac{1}{(q+1)}\int_{[t,b]}(s-t)^{q(\alpha+1)}\mu_t(ds)\right)^{\frac{1}{q(\alpha+1)}}\right)_{L^q,[t,b]}\right.$$

$$\left(\mu_t([t,b])\right)^{\frac{1}{p}}\left(\int_{[t,b]}(s-t)^{q(\alpha+1)}\mu_t(ds)\right)^{\frac{\alpha}{q(\alpha+1)}}$$

$$\left\{\left(\mu_t([t,b])\right)^{\frac{1}{\alpha+1}}(q+1)^{\frac{\alpha}{\alpha+1}}+1\right\}^{\frac{1}{q}} +$$

$$\Omega_1\left({}_{\alpha}^{t}\mathbf{T}X,\left(\frac{1}{(q+1)}\int_{[a,t]}(t-s)^{q(\alpha+1)}\mu_t(ds)\right)^{\frac{1}{q(\alpha+1)}}\right)_{L^q,[a,t]}$$

$$(\mu_t([a,t]))^{\frac{1}{p}} \left(\int_{[a,t]} (t-s)^{q(\alpha+1)} \mu_t(ds) \right)^{\frac{\alpha}{q(\alpha+1)}}$$

$$\left. \left\{ (\mu_t([a,t]))^{\frac{1}{\alpha+1}} (q+1)^{\frac{\alpha}{\alpha+1}} + 1 \right\}^{\frac{1}{q}} \right]. \tag{16.79}$$

Clearly (16.79) is trivially valid when $t = a$ or $t = b$, from what follows, see (16.81) and (16.85), just replace $[t, b]$ by $[a, b]$ or $[a, t]$ by $[a, b]$, respectively.

Next assume that

$$\int_{[t,b]} (s-t)^{q(\alpha+1)} \mu_t(ds) = 0, \tag{16.80}$$

then $(s-t)^{q(\alpha+1)} = 0$, a.e. on $[t, b]$, thus $s = t$, a.e. on $[t, b]$, i.e. $\mu_t \{s \in [t, b] : s \neq t\} = 0$, hence $\mu_t(t, b] = 0$.

Therefore μ_t concentrates on $[a, t]$. In that case inequality (16.79) becomes

$$\left(E\left(|M(X) - X|^q\right)(t)\right)^{\frac{1}{q}} \leq \left(\left(E|X|^q\right)(t)\right)^{\frac{1}{q}} \left|\widetilde{L}(1)(t) - 1\right| +$$

$$\sum_{k=1}^{n} \frac{\left(\left(E\left|X^{(k)}\right|^q\right)(t)\right)^{\frac{1}{q}}}{k!} \left|\widetilde{L}\left((\cdot - t)^k\right)(t)\right| +$$

$$\frac{2^{\frac{1}{p}} \lambda_1^{\frac{1}{q}}}{(q+1)^{\frac{\alpha}{q(\alpha+1)}} n!} \Omega_1 \left({}_{\alpha}^{t}\mathbf{T}X, \left(\frac{1}{(q+1)} \int_{[a,t]} (t-s)^{q(\alpha+1)} \mu_t(ds) \right)^{\frac{1}{q(\alpha+1)}} \right)_{L^q,[a,t]} \tag{16.81}$$

$$(\mu_t([a,t]))^{\frac{1}{p}} \left(\int_{[a,t]} (t-s)^{q(\alpha+1)} \mu_t(ds) \right)^{\frac{\alpha}{q(\alpha+1)}}$$

$$\left[(\mu_t([a,t]))^{\frac{1}{\alpha+1}} (q+1)^{\frac{\alpha}{\alpha+1}} + 1 \right]^{\frac{1}{q}}.$$

So far inequality (16.81) is valid when

$$\int_{[a,t]} (t-s)^{q(\alpha+1)} \mu_t(ds) > 0.$$

In the case of $\int_{[a,t]} (t-s)^{q(\alpha+1)} \mu_t(ds) = 0$, we have $(t-s)^{q(\alpha+1)} = 0$, a.e. on $[a, t]$, thus $s = t$, a.e. on $[a, t]$, i.e. $\mu_t \{s \in [a, t] : s \neq t\} = 0$, hence $\mu_t[a, t) = 0$.

Therefore $\mu_t = \delta_t \mu_t([a, b]) = \delta_t m_t$, where δ_t is the unit Dirac measure at t. Clearly then $\widetilde{L}\left((\cdot - t)^k\right)(t) = 0$, all $k = 1, \ldots, n$. Then (16.81) collapses to

$$\left(E\left(|M(X) - X|^q\right)(t)\right)^{\frac{1}{q}} \leq \left(\left(E|X|^q\right)(t)\right)^{\frac{1}{q}} \left|\widetilde{L}(1)(t) - 1\right|, \tag{16.82}$$

equivalently,

$$E\left(|M(X)-X|^{q}\right)(t)\leq\left(E|X|^{q}\right)(t)\left|\widetilde{L}(1)(t)-1\right|^{q}. \tag{16.83}$$

By (16.37) we have that $M(X)(t,\omega)=X(t,\omega)\widetilde{L}(1)(t)$. Therefore $M(X)(t,\omega)-X(t,\omega)=X(t,\omega)\left(\widetilde{L}(1)(t)-1\right)$.

Hence, obviously, (16.83) holds as equality. Thus inequality (16.81) is valid trivially.

Finally, let us go the other way around. Let us assume that

$$\int_{[a,t]}(t-s)^{q(\alpha+1)}\mu_t(ds)=0, \tag{16.84}$$

then $(t-s)^{q(\alpha+1)}=0$, a.e. on $[a,t]$, thus $s=t$, a.e. on $[a,t]$, i.e. $\mu_t\{s\in[a,t]:s\neq t\}=0$, hence $\mu_t[a,t)=0$. Therefore μ_t concentrates on $[t,b]$. In that case inequality (16.79) becomes

$$\left(E\left(|M(X)-X|^{q}\right)(t)\right)^{\frac{1}{q}}\leq\left(\left(E|X|^{q}\right)(t)\right)^{\frac{1}{q}}\left|\widetilde{L}(1)(t)-1\right|+$$

$$\sum_{k=1}^{n}\frac{\left(\left(E\left|X^{(k)}\right|^{q}\right)(t)\right)^{\frac{1}{q}}}{k!}\left|\widetilde{L}\left((\cdot-t)^{k}\right)(t)\right|+$$

$$\frac{2^{\frac{1}{p}}\lambda_1^{\frac{1}{q}}}{(q+1)^{\frac{\alpha}{q(\alpha+1)}}n!}\Omega_1\left(\mathbf{T}_{\alpha}^{t}X,\left(\frac{1}{(q+1)}\int_{[t,b]}(s-t)^{q(\alpha+1)}\mu_t(ds)\right)^{\frac{1}{q(\alpha+1)}}\right)_{L^q,[t,b]}$$

$$\left(\mu_t([t,b])\right)^{\frac{1}{p}}\left(\int_{[t,b]}(s-t)^{q(\alpha+1)}\mu_t(ds)\right)^{\frac{\alpha}{q(\alpha+1)}} \tag{16.85}$$

$$\left[\left(\mu_t([t,b])\right)^{\frac{1}{\alpha+1}}(q+1)^{\frac{\alpha}{\alpha+1}}+1\right]^{\frac{1}{q}}.$$

So inequality (16.85) is valid when

$$\int_{[t,b]}(s-t)^{q(\alpha+1)}\mu_t(ds)>0.$$

In the case of $\int_{[t,b]}(s-t)^{q(\alpha+1)}\mu_t(ds)=0$, we have $(s-t)^{q(\alpha+1)}=0$, a.e. on $[t,b]$, hence $s=t$, a.e. on $[t,b]$. Therefore $\mu_t\{s\in[t,b]:s\neq t\}=0$, thus $\mu_t(t,b]=0$.

Consequently $\mu_t=\delta_t\mu_t([a,b])=\delta_t m_t$. Clearly then $\widetilde{L}\left((\cdot-t)^{k}\right)(t)=0$, all $k=1,\ldots,n$. Therefore (16.85) collapses to

$$\left(E\left(|M(X)-X|^{q}\right)(t)\right)^{\frac{1}{q}}\leq\left(\left(E|X|^{q}\right)(t)\right)^{\frac{1}{q}}\left|\widetilde{L}(1)(t)-1\right|. \tag{16.86}$$

The last is an equality as earlier, see (16.83), etc.

Thus inequality (16.85) is valid trivially. Inequality (16.79) has been proved in all possible cases. Inequality (16.46) is equivalent to (16.79) via (16.36).

Theorem 16.23 now is proved completely. ■

It follows our second main result, the L_1-quantitative stochastic fractional approximation of stochastic processes.

Theorem 16.24 *Let* $n \in \mathbb{Z}_+$ *and* $\alpha \in (n, n+1)$, $\beta := \alpha - n$. *Suppose Concepts 16.15, Assumptions 16.16, 16.20. Then,* $\forall\, t \in [a, b]$, *we have:*

$$E\left(|M(X) - X|\right)(t) \leq \left(E|X|\right)(t)\left|\widetilde{L}(1) - 1\right| +$$

$$\sum_{k=1}^{n} \frac{\left(E\left|X^{(k)}\right|\right)(t)}{k!}\left|\widetilde{L}\left((\cdot - t)^k\right)(t)\right| + \frac{1}{\prod_{j=1}^{n}(\beta + j)}$$

$$\left\{\left[\frac{\left(\widetilde{L}\left(\chi_{[a,t]}(\cdot)\right)(t)\right)^{\frac{1}{\alpha+1}}}{\alpha - n} + \frac{1}{\alpha + 1}\right]\left(\widetilde{L}\left(|\cdot - t|^{\alpha+1}\chi_{[a,t]}(\cdot)\right)(t)\right)^{\frac{\alpha}{\alpha+1}}\right.$$

$$\Omega_1\left({}_{\alpha}^{t}\mathbf{T}X, \left(\widetilde{L}\left(|\cdot - t|^{\alpha+1}\chi_{[a,t]}(\cdot)\right)(t)\right)^{\frac{1}{\alpha+1}}\right)_{L^1,[a,t]} +$$

$$\left[\frac{\left(\widetilde{L}\left(\chi_{[t,b]}(\cdot)\right)(t)\right)^{\frac{1}{\alpha+1}}}{\alpha - n} + \frac{1}{(\alpha + 1)}\right]\left(\widetilde{L}\left(|\cdot - t|^{\alpha+1}\chi_{[t,b]}(\cdot)\right)(t)\right)^{\frac{\alpha}{\alpha+1}}$$

$$\left.\Omega_1\left(\mathbf{T}_{\alpha}^{t}X, \left(\widetilde{L}\left(|\cdot - t|^{\alpha+1}\chi_{[t,b]}(\cdot)\right)(t)\right)^{\frac{1}{\alpha+1}}\right)_{L^1,[t,b]}\right\}. \tag{16.87}$$

Proof We have that

$$E\left(|M(X) - X|\right)(t) \leq \left(E|X|\right)(t)\left|\widetilde{L}(1)(t) - 1\right| +$$

$$\sum_{k=1}^{n} \frac{\left(E\left|X^{(k)}\right|\right)(t)}{k!}\left|\widetilde{L}\left((\cdot - t)^k\right)(t)\right| +$$

$$\frac{1}{n!}\left\{\left[\int_{\Omega}\left(\int_{a}^{t}\left(\int_{s}^{t}(t - z)^{\beta-1}(z - s)^n\right.\right.\right.\right. \tag{16.88}$$

$$\left.\left.\left.\left|{}_{\alpha}^{t}\mathbf{T}X(z, \omega) - {}_{\alpha}^{t}\mathbf{T}X(t, \omega)\right| dz\right)\mu_t(ds)\right)P(d\omega)\right] +$$

$$\left[\int_{\Omega}\left(\int_t^b\left(\int_t^s (s-z)^n (z-t)^{\beta-1}\left|\mathbf{T}_{\alpha}^t X(z,\omega)-\mathbf{T}_{\alpha}^t X(t,\omega)\right| dz\right)\right.\right.$$

$$\left.\left.\mu_t(ds)\right)P(d\omega)\right]\Big\} =$$

(by applying Tonelli's theorem twice)

$$(E|X|)(t)\left|\widetilde{L}(1)(t)-1\right|+\sum_{k=1}^{n}\frac{\left(E\left|X^{(k)}\right|\right)(t)}{k!}\left|\widetilde{L}\left((\cdot-t)^k\right)(t)\right|+$$

$$\frac{1}{n!}\left\{\left[\int_a^t\left(\int_s^t\left(\int_{\Omega}\left|{}_{\alpha}^{t}\mathbf{T}X(z,\omega)-{}_{\alpha}^{t}\mathbf{T}X(t,\omega)\right|P(d\omega)\right)\right.\right.\right.$$

$$\left.\left.(t-z)^{\beta-1}(z-s)^n dz\right)\mu_t(ds)\right]+$$

$$\left[\int_t^b\left(\int_t^s\left(\int_{\Omega}\left|\mathbf{T}_{\alpha}^t X(z,\omega)-\mathbf{T}_{\alpha}^t X(t,\omega)\right|P(d\omega)\right)\right.\right.$$

$$\left.\left.(s-z)^n(z-t)^{\beta-1}dz\right)\mu_t(ds)\right]\Big\}\leq$$

$$(E|X|)(t)\left|\widetilde{L}(1)(t)-1\right|+\sum_{k=1}^{n}\frac{\left(E\left|X^{(k)}\right|\right)(t)}{k!}\left|\widetilde{L}\left((\cdot-t)^k\right)(t)\right|+ \tag{16.89}$$

$$\frac{1}{n!}\left\{\left[\int_a^t\left(\int_s^t \Omega_1\left({}_{\alpha}^{t}\mathbf{T}X,t-z\right)_{L^1,[a,t]}(t-z)^{\beta-1}(z-s)^n dz\right)\mu_t(ds)\right]+\right.$$

$$\left.\left[\int_t^b\left(\int_t^s \Omega_1\left(\mathbf{T}_{\alpha}^t X,z-t\right)_{L^1,[t,b]}(s-z)^n(z-t)^{\beta-1}dz\right)\mu_t(ds)\right]\right\}\underset{(h_1,h_2>0)}{\leq}$$

$$(E|X|)(t)\left|\widetilde{L}(1)(t)-1\right|+\sum_{k=1}^{n}\frac{\left(E\left|X^{(k)}\right|\right)(t)}{k!}\left|\widetilde{L}\left((\cdot-t)^k\right)(t)\right|+$$

$$\frac{1}{n!}\left\{\Omega_1\left({}_{\alpha}^{t}\mathbf{T}X,h_1\right)_{L^1,[a,t]}\left[\int_a^t\left(\int_s^t\left(1+\frac{t-z}{h_1}\right)(t-z)^{\beta-1}(z-s)^n dz\right)\mu_t(ds)\right]+\right. \tag{16.90}$$

$$\left.\Omega_1\left(\mathbf{T}_{\alpha}^t X,h_2\right)_{L^1,[t,b]}\left[\int_t^b\left(\int_t^s\left(1+\frac{z-t}{h_2}\right)(s-z)^n(z-t)^{\beta-1}dz\right)\mu_t(ds)\right]\right\}=$$

(set $\lambda_2 := (E|X|)(t)\left|\widetilde{L}(1)(t)-1\right|+\sum_{k=1}^{n}\frac{(E|X^{(k)}|)(t)}{k!}\left|\widetilde{L}\left((\cdot-t)^k\right)(t)\right|$)

$$\lambda_2 + \frac{1}{n!}\left\{\Omega_1\left({}_{\alpha}^{t}\mathbf{T}X, h_1\right)_{L^1,[a,t]}\right.$$

$$\left[\int_a^t \left[\int_s^t (t-z)^{\beta-1}(z-s)^{(n+1)-1}dz+\right.\right.$$

$$\left.\left.\frac{1}{h_1}\int_s^t (t-z)^{(\beta+1)-1}(z-s)^{(n+1)-1}dz\right]\mu_t(ds)\right]$$

$$+\Omega_1\left(\mathbf{T}_{\alpha}^{t}X, h_2\right)_{L^1,[t,b]}\left[\int_t^b\left[\int_t^s (s-z)^{(n+1)-1}(z-t)^{\beta-1}dz+\right.\right.$$

$$\left.\left.\left.\frac{1}{h_2}\int_t^s (s-z)^{(n+1)-1}(z-t)^{(\beta+1)-1}dz\right]\mu_t(ds)\right]\right\}$$

$$= \lambda_2 + \frac{1}{n!}\left\{\Omega_1\left({}_{\alpha}^{t}\mathbf{T}X, h_1\right)_{L^1,[a,t]}\right.$$

$$\left[\int_a^t\left[\frac{\Gamma(\beta)\,n!}{\Gamma(\beta+1+1)}(t-s)^{\alpha}+\frac{1}{h_1}\frac{\Gamma(\beta+1)\,n!}{\Gamma(\beta+n+2)}(t-s)^{\alpha+1}\right]\mu_t(ds)\right]+$$

$$\Omega_1\left(\mathbf{T}_{\alpha}^{t}X, h_2\right)_{L^1,[t,b]}$$

$$\left.\left[\int_t^b\left(\frac{n!\Gamma(\beta)}{\Gamma(\beta+n+1)}(s-t)^{\alpha}+\frac{1}{h_2}\frac{n!\Gamma(\beta+1)}{\Gamma(\beta+n+2)}(s-t)^{\alpha+1}\right)\mu_t(ds)\right]\right\}$$

$$= \lambda_2 + \left\{\Omega_1\left({}_{\alpha}^{t}\mathbf{T}X, h_1\right)_{L^1,[a,t]}\right.$$

$$\left.\left[\int_a^t\left[\frac{1}{\prod_{j=0}^{n}(\beta+j)}(t-s)^{\alpha}+\frac{1}{h_1}\frac{1}{\prod_{j=0}^{n}(\beta+1+j)}(t-s)^{\alpha+1}\right]\mu_t(ds)\right]\right\}+ \tag{16.91}$$

$$\left\{\Omega_1\left(\mathbf{T}_{\alpha}^{t}X, h_2\right)_{L^1,[t,b]}\right.$$

$$\left.\left[\int_t^b\left[\frac{1}{\prod_{j=0}^{n}(\beta+j)}(s-t)^{\alpha}+\frac{1}{h_2}\frac{1}{\prod_{j=0}^{n}(\beta+1+j)}(s-t)^{\alpha+1}\right]\mu_t(ds)\right]\right\}$$

$$= \lambda_2 + \left\{\Omega_1\left({}_{\alpha}^{t}\mathbf{T}X, h_1\right)_{L^1,[a,t]}\right.$$

$$\left[\frac{1}{\prod_{j=0}^{n}(\beta+j)} \int_a^t (t-s)^{\alpha} \mu_t(ds) + \frac{1}{\prod_{j=0}^{n}(\beta+1+j)} \frac{1}{h_1} \int_a^t (t-s)^{\alpha+1} \mu_t(ds) \right] \Bigg\}$$

$$+ \Big\{ \Omega_1 \left(\mathbf{T}_{\alpha}^{t} X, h_2 \right)_{L^1,[t,b]}$$

$$\left[\frac{1}{\prod_{j=0}^{n}(\beta+j)} \int_t^b (s-t)^{\alpha} \mu_t(ds) + \frac{1}{\prod_{j=0}^{n}(\beta+1+j)} \frac{1}{h_2} \int_t^b (s-t)^{\alpha+1} \mu_t(ds) \right] \Bigg\} =: (\xi) . \tag{16.92}$$

(Choose

$$h_1 := \left(\int_a^t (t-s)^{\alpha+1} \mu_t(ds) \right)^{\frac{1}{\alpha+1}} > 0,$$

and

$$h_2 := \left(\int_t^b (s-t)^{\alpha+1} \mu_t(ds) \right)^{\frac{1}{\alpha+1}} > 0.$$

Then

$$h_1^{\alpha+1} = \int_a^t (t-s)^{\alpha+1} \mu_t(ds) > 0,$$

and

$$h_2^{\alpha+1} = \int_t^b (s-t)^{\alpha+1} \mu_t(ds) > 0.$$

)

Hence it holds

$$(\xi) \leq \lambda_2 + \Bigg\{ \Omega_1 \left({}_{\alpha}^{t}\mathbf{T} X, \left(\int_a^t (t-s)^{\alpha+1} \mu_t(ds) \right)^{\frac{1}{\alpha+1}} \right)_{L^1,[a,t]}$$

$$\left[\frac{(\mu_t([a,t]))^{\frac{1}{\alpha+1}}}{\prod_{j=0}^{n}(\beta+j)} \left(\int_a^t (t-s)^{\alpha+1} \mu_t(ds) \right)^{\frac{\alpha}{\alpha+1}} + \frac{h_1^{\alpha}}{\prod_{j=0}^{n}(\beta+1+j)} \right] +$$

$$\Omega_1 \left(\mathbf{T}_{\alpha}^{t} X, \left(\int_t^b (s-t)^{\alpha+1} \mu_t(ds) \right)^{\frac{1}{\alpha+1}} \right)_{L^1,[t,b]} \tag{16.93}$$

$$\left.\left.\left[\frac{(\mu_t([t,b]))^{\frac{1}{\alpha+1}}}{\prod_{j=0}^{n}(\beta+j)}\left(\int_t^b (s-t)^{\alpha+1}\mu_t(ds)\right)^{\frac{\alpha}{\alpha+1}}+\frac{h_2^{\alpha}}{\prod_{j=0}^{n}(\beta+1+j)}\right]\right]\right\}=$$

$$\lambda_2+\left\{\Omega_1\left({}_{\alpha}^{t}\mathbf{T}X,\left(\int_a^t (t-s)^{\alpha+1}\mu_t(ds)\right)^{\frac{1}{\alpha+1}}\right)_{L^1,[a,t]}\right.$$

$$\left(\int_a^t (t-s)^{\alpha+1}\mu_t(ds)\right)^{\frac{\alpha}{\alpha+1}}\left[\frac{(\mu_t([a,t]))^{\frac{1}{\alpha+1}}}{\prod_{j=0}^{n}(\beta+j)}+\frac{1}{\prod_{j=0}^{n}(\beta+1+j)}\right]+$$

$$\Omega_1\left(\mathbf{T}_{\alpha}^{t}X,\left(\int_t^b (s-t)^{\alpha+1}\mu_t(ds)\right)^{\frac{1}{\alpha+1}}\right)_{L^1,[t,b]} \tag{16.94}$$

$$\left.\left(\int_t^b (s-t)^{\alpha+1}\mu_t(ds)\right)^{\frac{\alpha}{\alpha+1}}\left[\frac{(\mu_t([t,b]))^{\frac{1}{\alpha+1}}}{\prod_{j=0}^{n}(\beta+j)}+\frac{1}{\prod_{j=0}^{n}(\beta+1+j)}\right]\right\}.$$

We have proved that

$$E(|M(X)-X|)(t)\leq(E|X|)(t)\left|\widetilde{L}(1)(t)-1\right|+$$

$$\sum_{k=1}^{n}\frac{\left(E\left|X^{(k)}\right|\right)(t)}{k!}\left|\widetilde{L}\left((\cdot-t)^k\right)(t)\right|+\frac{1}{\prod_{j=1}^{n}(\beta+j)}$$

$$\left\{\left[\frac{(\mu_t([a,t]))^{\frac{1}{\alpha+1}}}{\alpha-n}+\frac{1}{(\alpha+1)}\right]\right.$$

$$\left(\int_a^t (t-s)^{\alpha+1}\mu_t(ds)\right)^{\frac{\alpha}{\alpha+1}}\Omega_1\left({}_{\alpha}^{t}\mathbf{T}X,\left(\int_a^t (t-s)^{\alpha+1}\mu_t(ds)\right)^{\frac{1}{\alpha+1}}\right)_{L^1,[a,t]} \tag{16.95}$$

$$+\left[\frac{(\mu_t([t,b]))^{\frac{1}{\alpha+1}}}{\alpha-n}+\frac{1}{(\alpha+1)}\right]$$

$$\left(\int_t^b (s-t)^{\alpha+1} \mu_t(ds)\right)^{\frac{\alpha}{\alpha+1}} \Omega_1\left(\mathbf{T}_\alpha^t X, \left(\int_t^b (s-t)^{\alpha+1} \mu_t(ds)\right)^{\frac{1}{\alpha+1}}\right)_{L^1,[t,b]}\Bigg\}.$$

The trivial cases of $h_1 = 0$ or $h_2 = 0$ or $h_1 = h_2 = 0$ are treated similarly as in the proof of Theorem 16.23.

Theorem 16.24 is now proved. ■

Next we give several corollaries:

Corollary 16.25 (to Theorem 16.23, case of $n = 0$) *Let* $\alpha \in (0, 1)$; $p, q > 1 : \frac{1}{p} + \frac{1}{q} = 1$, *with* $\alpha > \frac{1}{q}$. *Suppose Concepts 16.15, Assumptions 16.16 and 16.19. Then,* $\forall\, t \in [a, b]$, *we have:*

$$\left(E\left(|M(X) - X|^q\right)(t)\right)^{\frac{1}{q}} \leq \left(\left(E|X|^q\right)(t)\right)^{\frac{1}{q}} \left|\widetilde{L}(1)(t) - 1\right| +$$

$$\frac{2^{\frac{1}{p}}}{(q+1)^{\frac{\alpha}{q(\alpha+1)}} (p(\alpha-1)+1)^{\frac{1}{p}}}$$

$$\Bigg\{\Bigg[\Omega_1\left(T_\alpha^t X, \left(\frac{1}{(q+1)}\widetilde{L}\left(|\cdot - t|^{q(\alpha+1)} \chi_{[t,b]}(\cdot)\right)(t)\right)^{\frac{1}{q(\alpha+1)}}\right)_{L^q,[t,b]}$$

$$\left(\widetilde{L}\left(\chi_{[t,b]}(\cdot)\right)(t)\right)^{\frac{1}{p}} \left(\widetilde{L}\left(|\cdot - t|^{q(\alpha+1)} \chi_{[t,b]}(\cdot)\right)(t)\right)^{\frac{\alpha}{q(\alpha+1)}}$$

$$\left[\left(\widetilde{L}\left(\chi_{[t,b]}(\cdot)\right)(t)\right)^{\frac{1}{\alpha+1}} (q+1)^{\frac{\alpha}{(\alpha+1)}} + 1\right]^{\frac{1}{q}}\Bigg] + \tag{16.96}$$

$$\Bigg[\Omega_1\left({}_\alpha^t T X, \left(\frac{1}{(q+1)}\widetilde{L}\left(|\cdot - t|^{q(\alpha+1)} \chi_{[a,t]}(\cdot)\right)(t)\right)^{\frac{1}{q(\alpha+1)}}\right)_{L^q,[a,t]}$$

$$\left(\widetilde{L}\left(\chi_{[a,t]}(\cdot)\right)(t)\right)^{\frac{1}{p}} \left(\widetilde{L}\left(|\cdot - t|^{q(\alpha+1)} \chi_{[a,t]}(\cdot)\right)(t)\right)^{\frac{\alpha}{q(\alpha+1)}}$$

$$\left[\left(\widetilde{L}\left(\chi_{[a,t]}(\cdot)\right)(t)\right)^{\frac{1}{\alpha+1}} (q+1)^{\frac{\alpha}{(\alpha+1)}} + 1\right]^{\frac{1}{q}}\Bigg]\Bigg\}.$$

Corollary 16.26 (to Theorem 16.24, case of $n = 0$) *Let* $\alpha \in (0, 1)$. *Suppose Concepts 16.15, Assumptions 16.16, 16.20. Then,* $\forall\, t \in [a, b]$, *we have:*

$$E(|M(X) - X|)(t) \leq (E|X|)(t) \left|\widetilde{L}(1) - 1\right| +$$

$$\Bigg\{\left[\frac{\left(\widetilde{L}\left(\chi_{[a,t]}(\cdot)\right)(t)\right)^{\frac{1}{\alpha+1}}}{\alpha} + \frac{1}{\alpha+1}\right] \left(\widetilde{L}\left(|\cdot - t|^{\alpha+1} \chi_{[a,t]}(\cdot)\right)(t)\right)^{\frac{\alpha}{\alpha+1}} \tag{16.97}$$

$$\Omega_1\left({}_{\alpha}^{t}TX,\left(\widetilde{L}\left(|\cdot-t|^{\alpha+1}\chi_{[a,t]}(\cdot)\right)(t)\right)^{\frac{1}{\alpha+1}}\right)_{L^1,[a,t]}+$$

$$\left[\frac{\left(\widetilde{L}\left(\chi_{[t,b]}(\cdot)\right)(t)\right)^{\frac{1}{\alpha+1}}}{\alpha}+\frac{1}{(\alpha+1)}\right]\left(\widetilde{L}\left(|\cdot-t|^{\alpha+1}\chi_{[t,b]}(\cdot)\right)(t)\right)^{\frac{\alpha}{\alpha+1}}$$

$$\left.\Omega_1\left(T_{\alpha}^{t}X,\left(\widetilde{L}\left(|\cdot-t|^{\alpha+1}\chi_{[t,b]}(\cdot)\right)(t)\right)^{\frac{1}{\alpha+1}}\right)_{L^1,[t,b]}\right\}.$$

Corollary 16.27 *All as in Theorem 16.23. Case of $\widetilde{L}(1)=1$ and $X^{(k)}(t_0,\omega)=0$, $\forall\,\omega\in\Omega$, all $k=1,\dots,n$, for a fixed $t_0\in[a,b]$. Then*

$$\left(E\left(|M(X)-X|^q\right)(t_0)\right)^{\frac{1}{q}}\leq\frac{2^{\frac{1}{p}}\lambda_1^{\frac{1}{q}}}{(q+1)^{\frac{\alpha}{q(\alpha+1)}}\,n!}$$

$$\left\{\left[\Omega_1\left(\mathbf{T}_{\alpha}^{t_0}X,\left(\frac{1}{(q+1)}\widetilde{L}\left(|\cdot-t_0|^{q(\alpha+1)}\chi_{[t_0,b]}(\cdot)\right)(t_0)\right)^{\frac{1}{q(\alpha+1)}}\right)_{L^q,[t_0,b]}\right.\right.$$

$$\left(\widetilde{L}\left(\chi_{[t_0,b]}(\cdot)\right)(t_0)\right)^{\frac{1}{p}}\left(\widetilde{L}\left(|\cdot-t_0|^{q(\alpha+1)}\chi_{[t_0,b]}(\cdot)\right)(t_0)\right)^{\frac{\alpha}{q(\alpha+1)}}$$

$$\left.\left[\left(\widetilde{L}\left(\chi_{[t_0,b]}(\cdot)\right)(t_0)\right)^{\frac{1}{\alpha+1}}(q+1)^{\frac{\alpha}{(\alpha+1)}}+1\right]^{\frac{1}{q}}\right]+\tag{16.98}$$

$$\left[\Omega_1\left({}_{\alpha}^{t_0}\mathbf{T}X,\left(\frac{1}{(q+1)}\widetilde{L}\left(|\cdot-t_0|^{q(\alpha+1)}\chi_{[a,t_0]}(\cdot)\right)(t_0)\right)^{\frac{1}{q(\alpha+1)}}\right)_{L^q,[a,t_0]}\right.$$

$$\left(\widetilde{L}\left(\chi_{[a,t_0]}(\cdot)\right)(t_0)\right)^{\frac{1}{p}}\left(\widetilde{L}\left(|\cdot-t_0|^{q(\alpha+1)}\chi_{[a,t_0]}(\cdot)\right)(t_0)\right)^{\frac{\alpha}{q(\alpha+1)}}$$

$$\left.\left.\left[\left(\widetilde{L}\left(\chi_{[a,t_0]}(\cdot)\right)(t_0)\right)^{\frac{1}{\alpha+1}}(q+1)^{\frac{\alpha}{(\alpha+1)}}+1\right]^{\frac{1}{q}}\right]\right\}.$$

Corollary 16.28 *All as in Theorem 16.24. Case of $\widetilde{L}(1)=1$ and $X^{(k)}(t_0,\omega)=0$, $\forall\,\omega\in\Omega$, all $k=1,\dots,n$, for a fixed $t_0\in[a,b]$. Then*

$$E\left(|M(X)-X|\right)(t_0)\leq\frac{1}{\prod_{j=1}^{n}(\beta+j)}$$

$$\left\{\left[\frac{\left(\widetilde{L}\left(\chi_{[a,t_0]}(\cdot)\right)(t_0)\right)^{\frac{1}{\alpha+1}}}{\alpha-n}+\frac{1}{\alpha+1}\right]\left(\widetilde{L}\left(|\cdot-t_0|^{\alpha+1}\chi_{[a,t_0]}(\cdot)\right)(t_0)\right)^{\frac{\alpha}{\alpha+1}}\right.$$

$$\Omega_1\left({}_{\alpha}^{t_0}\mathbf{T}X,\left(\widetilde{L}\left(|\cdot-t_0|^{\alpha+1}\chi_{[a,t_0]}(\cdot)\right)(t_0)\right)^{\frac{1}{\alpha+1}}\right)_{L^1,[a,t_0]}+$$

$$\left[\frac{\left(\widetilde{L}\left(\chi_{[t_0,b]}(\cdot)\right)(t_0)\right)^{\frac{1}{\alpha+1}}}{\alpha-n}+\frac{1}{(\alpha+1)}\right]\left(\widetilde{L}\left(|\cdot-t_0|^{\alpha+1}\chi_{[t_0,b]}(\cdot)\right)(t_0)\right)^{\frac{\alpha}{\alpha+1}}$$

$$\left.\Omega_1\left(\mathbf{T}_{\alpha}^{t_0}X,\left(\widetilde{L}\left(|\cdot-t_0|^{\alpha+1}\chi_{[t_0,b]}(\cdot)\right)(t_0)\right)^{\frac{1}{\alpha+1}}\right)_{L^1,[t_0,b]}\right\}. \tag{16.99}$$

Corollary 16.29 *All as in Theorem 16.24, $\alpha=\frac{1}{2}$ and $\widetilde{L}(1)=1$. Then, $\forall\, t\in[a,b]$, we have:*

$$E\left(|M(X)-X|\right)(t)\leq$$

$$\left\{\left[2\left(\widetilde{L}\left(\chi_{[a,t]}(\cdot)\right)(t)\right)^{\frac{2}{3}}+\frac{2}{3}\right]\left(\widetilde{L}\left(|\cdot-t|^{\frac{3}{2}}\chi_{[a,t]}(\cdot)\right)(t)\right)^{\frac{1}{3}}\right.$$

$$\Omega_1\left({}_{\frac{1}{2}}^{t}TX,\left(\widetilde{L}\left(|\cdot-t|^{\frac{3}{2}}\chi_{[a,t]}(\cdot)\right)(t)\right)^{\frac{2}{3}}\right)_{L^1,[a,t]}+$$

$$\left[2\left(\widetilde{L}\left(\chi_{[t,b]}(\cdot)\right)(t)\right)^{\frac{2}{3}}+\frac{2}{3}\right]\left(\widetilde{L}\left(|\cdot-t|^{\frac{3}{2}}\chi_{[t,b]}(\cdot)\right)(t)\right)^{\frac{1}{3}}$$

$$\left.\Omega_1\left(T_{\frac{1}{2}}^{t}X,\left(\widetilde{L}\left(|\cdot-t|^{\frac{3}{2}}\chi_{[t,b]}(\cdot)\right)(t)\right)^{\frac{2}{3}}\right)_{L^1,[t,b]}\right\}. \tag{16.100}$$

We continue with uniform estimates. First is an application of Theorem 16.23.

Theorem 16.30 *Let $n\in\mathbb{Z}_+$, $\alpha\in(n,n+1)$; $p,q>1:\frac{1}{p}+\frac{1}{q}=1$, with $\beta:=\alpha-n>\frac{1}{q}$; set $\lambda_1:=\left(\frac{\Gamma(np+1)\Gamma(p(\beta-1)+1)}{\Gamma(p(\alpha-1)+2)}\right)^{(q-1)}$. Suppose Concepts 16.15, Assumptions 16.19 and 16.22. Then*

$$\left\|E\left(|M(X)-X|^q\right)\right\|_{\infty}^{\frac{1}{q}}\leq\left\|E\left(|X|^q\right)\right\|_{\infty}^{\frac{1}{q}}\left\|\widetilde{L}(1)-1\right\|_{\infty}+$$

$$\sum_{k=1}^{n}\frac{\left\|E\left(\left|X^{(k)}\right|^q\right)\right\|_{\infty}^{\frac{1}{q}}}{k!}\left\|\widetilde{L}\left((\cdot-t)^k\right)(t)\right\|_{\infty}+\frac{\lambda_1^{\frac{1}{q}}\left(2\left\|\widetilde{L}(1)\right\|_{\infty}\right)^{\frac{1}{p}}}{(q+1)^{\frac{\alpha}{q(\alpha+1)}}\,n!}$$

$$\left[\left\|\widetilde{L}(1)\right\|_{\infty}^{\frac{1}{\alpha+1}}(q+1)^{\frac{\alpha}{(\alpha+1)}}+1\right]^{\frac{1}{q}}$$

$$\left\{\left[\sup_{t\in[a,b]}\left\|\Omega_1\left(\mathbf{T}^t_\alpha X, \frac{1}{(q+1)}\left\|\widetilde{L}\left(|\cdot - t|^{q(\alpha+1)}\chi_{[t,b]}(\cdot)\right)(t)\right\|_\infty^{\frac{1}{q(\alpha+1)}}\right)\right\|_{L^q,[t,b]}\right.\right.$$

$$\left.\left\|\widetilde{L}\left(|\cdot - t|^{q(\alpha+1)}\chi_{[t,b]}(\cdot)\right)(t)\right\|_\infty^{\frac{\alpha}{q(\alpha+1)}}\right]+$$

$$\left[\sup_{t\in[a,b]}\left\|\Omega_1\left({}^t_\alpha\mathbf{T}X, \frac{1}{(q+1)}\left\|\widetilde{L}\left(|\cdot - t|^{q(\alpha+1)}\chi_{[a,t]}(\cdot)\right)(t)\right\|_\infty^{\frac{1}{q(\alpha+1)}}\right)\right\|_{L^q,[a,t]}\right.$$

$$\left.\left.\left\|\widetilde{L}\left(|\cdot - t|^{q(\alpha+1)}\chi_{[a,t]}(\cdot)\right)(t)\right\|_\infty^{\frac{\alpha}{q(\alpha+1)}}\right]\right\}. \quad (16.101)$$

The next is an application of Theorem 16.24.

Theorem 16.31 *Let $n \in \mathbb{Z}_+$, $\alpha \in (n, n+1)$, $\beta := \alpha - n$. Suppose Concepts 16.15, Assumptions 16.20, 16.22. Then*

$$\|E(|M(X) - X|)\|_\infty \leq \|E(|X|)\|_\infty \left\|\widetilde{L}(1) - 1\right\|_\infty +$$

$$\sum_{k=1}^{n}\frac{\left\|E\left(\left|X^{(k)}\right|\right)\right\|_\infty}{k!}\left\|\widetilde{L}\left((\cdot - t)^k\right)(t)\right\|_\infty +$$

$$\frac{1}{\prod_{j=1}^{n}(\beta + j)}\left[\frac{\left\|\widetilde{L}(1)\right\|_\infty^{\frac{1}{\alpha+1}}}{\alpha - n} + \frac{1}{\alpha+1}\right]$$

$$\left\{\left\|\widetilde{L}\left(|\cdot - t|^{\alpha+1}\chi_{[a,t]}(\cdot)\right)(t)\right\|_\infty^{\frac{\alpha}{\alpha+1}}\right.$$

$$\sup_{t\in[a,b]}\left\|\Omega_1\left({}^t_\alpha\mathbf{T}X, \left\|\widetilde{L}\left(|\cdot - t|^{\alpha+1}\chi_{[a,t]}(\cdot)\right)(t)\right\|_\infty^{\frac{1}{\alpha+1}}\right)\right\|_{L^1,[a,t]} +$$

$$\left\|\widetilde{L}\left(|\cdot - t|^{\alpha+1}\chi_{[t,b]}(\cdot)\right)(t)\right\|_\infty^{\frac{\alpha}{\alpha+1}}$$

$$\left.\sup_{t\in[a,b]}\left\|\Omega_1\left(\mathbf{T}^t_\alpha X, \left\|\widetilde{L}\left(|\cdot - t|^{\alpha+1}\chi_{[t,b]}(\cdot)\right)(t)\right\|_\infty^{\frac{1}{\alpha+1}}\right)\right\|_{L^1,[t,b]}\right\}. \quad (16.102)$$

Some corollaries follow:

Corollary 16.32 (to Theorem 16.30, case of $n = 0$) *Let $\alpha \in (0, 1)$; $p, q > 1 : \frac{1}{p} + \frac{1}{q} = 1$, with $\alpha > \frac{1}{q}$. Suppose Concepts 16.15, Assumptions 16.19, 16.22 and $\widetilde{L}(1) = 1$. Then*

$$\left\| E\left(|M(X) - X|^q\right)\right\|_\infty^{\frac{1}{q}} \leq \frac{2^{\frac{1}{p}}\left[(q+1)^{\frac{\alpha}{\alpha+1}} + 1\right]^{\frac{1}{q}}}{(q+1)^{\frac{\alpha}{q(\alpha+1)}}(p(\alpha-1)+1)^{\frac{1}{p}}}$$

$$\left\{\left[\sup_{t\in[a,b]} \Omega_1\left(T_\alpha^t X, \left(\frac{1}{(q+1)}\left\|\widetilde{L}\left(|\cdot - t|^{q(\alpha+1)}\chi_{[t,b]}(\cdot)\right)(t)\right\|_\infty^{\frac{1}{q(\alpha+1)}}\right)\right)_{L^q,[t,b]}\right.\right.$$

$$\left.\left\|\widetilde{L}\left(|\cdot - t|^{q(\alpha+1)}\chi_{[t,b]}(\cdot)\right)(t)\right\|_\infty^{\frac{\alpha}{q(\alpha+1)}}\right] +$$

$$\left[\sup_{t\in[a,b]} \Omega_1\left({}_\alpha^t TX, \left(\frac{1}{(q+1)}\left\|\widetilde{L}\left(|\cdot - t|^{q(\alpha+1)}\chi_{[a,t]}(\cdot)\right)(t)\right\|_\infty^{\frac{1}{q(\alpha+1)}}\right)\right)_{L^q,[a,t]}\right.$$

$$\left.\left.\left\|\widetilde{L}\left(|\cdot - t|^{q(\alpha+1)}\chi_{[a,t]}(\cdot)\right)(t)\right\|_\infty^{\frac{\alpha}{q(\alpha+1)}}\right]\right\}. \tag{16.103}$$

The next is an application of Theorem 16.31.

Corollary 16.33 (to Theorem 16.31, case of $n = 0$) *Here $\alpha \in (0,1)$. Suppose Concepts 16.15, Assumptions 16.20, 16.22 and $\widetilde{L}(1) = 1$. Then*

$$\|E(|M(X) - X|)\|_\infty \leq \left(\frac{2\alpha + 1}{\alpha(\alpha+1)}\right)$$

$$\left\{\left\|\widetilde{L}\left(|\cdot - t|^{\alpha+1}\chi_{[a,t]}(\cdot)\right)(t)\right\|_\infty^{\frac{\alpha}{\alpha+1}}\right.$$

$$\sup_{t\in[a,b]} \Omega_1\left({}_\alpha^t TX, \left\|\widetilde{L}\left(|\cdot - t|^{\alpha+1}\chi_{[a,t]}(\cdot)\right)(t)\right\|_\infty^{\frac{1}{\alpha+1}}\right)_{L^1,[a,t]} +$$

$$\left\|\widetilde{L}\left(|\cdot - t|^{\alpha+1}\chi_{[t,b]}(\cdot)\right)(t)\right\|_\infty^{\frac{\alpha}{\alpha+1}}$$

$$\left.\sup_{t\in[a,b]} \Omega_1\left(T_\alpha^t X, \left\|\widetilde{L}\left(|\cdot - t|^{\alpha+1}\chi_{[t,b]}(\cdot)\right)(t)\right\|_\infty^{\frac{1}{\alpha+1}}\right)_{L^1,[t,b]}\right\}. \tag{16.104}$$

Corollary 16.34 (to Theorem 16.31, case of $n = 0$) *Here $\alpha = \frac{1}{2}$. Suppose Concepts 16.15, Assumptions 16.20, 16.22 and $\widetilde{L}(1) = 1$. Then*

$$\|E(|M(X) - X|)\|_\infty \leq \frac{8}{3}\left\{\left\|\widetilde{L}\left(|\cdot - t|^{\frac{3}{2}}\chi_{[a,t]}(\cdot)\right)(t)\right\|_\infty^{\frac{1}{3}}\right.$$

$$\sup_{t\in[a,b]} \Omega_1\left({}_\alpha^t TX, \left\|\widetilde{L}\left(|\cdot - t|^{\frac{3}{2}}\chi_{[a,t]}(\cdot)\right)(t)\right\|_\infty^{\frac{2}{3}}\right)_{L^1,[a,t]} +$$

$$\left\| \widetilde{L} \left(|\cdot - t|^{\frac{3}{2}} \chi_{[t,b]} (\cdot) \right) (t) \right\|_{\infty}^{\frac{1}{3}}$$

$$\left. \sup_{t \in [a,b]} \Omega_1 \left(T_{\alpha}^{t} X, \left\| \widetilde{L} \left(|\cdot - t|^{\frac{3}{2}} \chi_{[t,b]} (\cdot) \right) (t) \right\|_{\infty}^{\frac{2}{3}} \right)_{L^1,[t,b]} \right\} . \tag{16.105}$$

16.5 Application

Let $f \in C([0, 1])$ and the Bernstein polynomials

$$B_N(f)(t) := \sum_{k=0}^{N} f\left(\frac{k}{N}\right) \binom{N}{k} t^k (1-t)^{N-k}, \tag{16.106}$$

$\forall\, t \in [0, 1]$, $\forall\, N \in \mathbb{N}$.

We have that $B_N 1 = 1$ and B_N is a positive linear operator.

We have that

$$B_N\left((\cdot - t)^2\right)(t) = \frac{t(1-t)}{N}, \quad \forall\, t \in [0, 1], \tag{16.107}$$

and

$$\left\| B_N\left((\cdot - t)^2\right)(t) \right\|_{\infty}^{\frac{1}{2}} \leq \frac{1}{2\sqrt{N}}, \quad \forall\, N \in \mathbb{N}. \tag{16.108}$$

So B_N is like $\widetilde{L}$ in Concepts 16.15.

Define the corresponding application of M by

$$\widetilde{B}_N(X)(t,\omega) := B_N(X(\cdot,\omega))(t) = \sum_{k=0}^{N} X\left(\frac{k}{N},\omega\right) \binom{N}{k} t^k (1-t)^{N-k}, \tag{16.109}$$

$\forall\, t \in [0, 1]$, $\forall\, \omega \in \Omega$, $N \in \mathbb{N}$, where X is a stochastic process.

Clearly $\widetilde{B}_N X$ is a stochastic process and $\widetilde{B}_N$ is a positive linear operator.

We give

Proposition 16.35 *Let $\alpha \in (0, 1)$. Let $X(t, \omega)$ be a stochastic process from $[0, 1] \times (\Omega, \mathcal{F}, P)$ into $\mathbb{R}$, where $(\Omega, \mathcal{F}, P)$ is a probability space. Here $X(\cdot, \omega) \in C^1([0, 1])$, $\forall\, \omega \in \Omega$. Also we assume for each $t \in [0, 1]$ that $X^{(1)}(t, \cdot)$ is measurable over $(\Omega, \mathcal{F})$. For any $t \in [0, 1]$ we assume that $T_{\alpha}^{t} X(z, \omega)$ is continuous in $z \in [t, 1]$, uniformly with respect to $\omega \in \Omega$, and for any $t \in [0, 1]$ we assume that ${}_{\alpha}^{t}T X(z, \omega)$ is continuous in $z \in [0, t]$, uniformly with respect to $\omega \in \Omega$. Suppose also that $(E|X|)(t) < \infty$, $\forall\, t \in [0, 1]$. Then, $\forall\, t \in [0, 1]$, we have:*

$$E\left(\left|\widetilde{B}_N(X) - X\right|\right)(t) \leq$$

$$\left\{ \left[\frac{\left(B_N\left(\chi_{[0,t]}(\cdot)\right)(t)\right)^{\frac{1}{\alpha+1}}}{\alpha} + \frac{1}{\alpha+1} \right] \left(B_N\left(|\cdot - t|^{\alpha+1}\chi_{[0,t]}(\cdot)\right)(t)\right)^{\frac{\alpha}{\alpha+1}} \right.$$

$$\Omega_1\left({}_{\alpha}^{t}TX, \left(B_N\left(|\cdot - t|^{\alpha+1}\chi_{[0,t]}(\cdot)\right)(t)\right)^{\frac{1}{\alpha+1}}\right)_{L^1,[0,t]} +$$

$$\left[\frac{\left(B_N\left(\chi_{[t,1]}(\cdot)\right)(t)\right)^{\frac{1}{\alpha+1}}}{\alpha} + \frac{1}{\alpha+1} \right] \left(B_N\left(|\cdot - t|^{\alpha+1}\chi_{[t,1]}(\cdot)\right)(t)\right)^{\frac{\alpha}{\alpha+1}} \tag{16.110}$$

$$\left. \Omega_1\left(T_{\alpha}^{t}X, \left(B_N\left(|\cdot - t|^{\alpha+1}\chi_{[t,1]}(\cdot)\right)(t)\right)^{\frac{1}{\alpha+1}}\right)_{L^1,[t,1]} \right\}.$$

Proof By Corollary 16.26. ■

Proposition 16.36 *All as in Proposition 16.35 with $\alpha = \frac{1}{2}$. Then, $\forall\, t \in [0, 1]$, we have:*

$$E\left(\left|\widetilde{B}_N(X) - X\right|\right)(t) \leq$$

$$\left\{ \left[2\left(B_N\left(\chi_{[0,t]}(\cdot)\right)(t)\right)^{\frac{2}{3}} + \frac{2}{3} \right] \left(B_N\left(|\cdot - t|^{\frac{3}{2}}\chi_{[0,t]}(\cdot)\right)(t)\right)^{\frac{1}{3}} \right. \tag{16.111}$$

$$\Omega_1\left({}_{\frac{1}{2}}^{t}TX, \left(B_N\left(|\cdot - t|^{\frac{3}{2}}\chi_{[0,t]}(\cdot)\right)(t)\right)^{\frac{2}{3}}\right)_{L^1,[0,t]} +$$

$$\left[2\left(B_N\left(\chi_{[t,1]}(\cdot)\right)(t)\right)^{\frac{2}{3}} + \frac{2}{3} \right] \left(B_N\left(|\cdot - t|^{\frac{3}{2}}\chi_{[t,1]}(\cdot)\right)(t)\right)^{\frac{1}{3}}$$

$$\left. \Omega_1\left(T_{\frac{1}{2}}^{t}X, \left(B_N\left(|\cdot - t|^{\frac{3}{2}}\chi_{[t,1]}(\cdot)\right)(t)\right)^{\frac{2}{3}}\right)_{L^1,[t,1]} \right\}.$$

Proof By Proposition 16.35. ■

Proposition 16.37 *All as in Proposition 16.35 with $\alpha = \frac{1}{2}$. Then*

$$E\left(\left|\widetilde{B}_N(X) - X\right|\right)(t) \leq \frac{8}{3}\left(B_N\left(|\cdot - t|^{\frac{3}{2}}\right)(t)\right)^{\frac{1}{3}}$$

$$\left[\Omega_1\left({}_{\frac{1}{2}}^{t}TX, \left(B_N\left(|\cdot - t|^{\frac{3}{2}}\right)(t)\right)^{\frac{2}{3}}\right)_{L^1,[0,t]} + \right.$$

$$\left. \Omega_1\left(T_{\frac{1}{2}}^{t}X, \left(B_N\left(|\cdot - t|^{\frac{3}{2}}\right)(t)\right)^{\frac{2}{3}}\right)_{L^1,[t,1]} \right], \tag{16.112}$$

$\forall\, t \in [0,1]$, $\forall\, N \in \mathbb{N}$.

Proof From (16.111) and the positivity of B_N. ∎

We further have

Proposition 16.38 *All as in Proposition 16.35 with $\alpha = \frac{1}{2}$. Then*

$$E\left(\left|\widetilde{B}_N(X) - X\right|\right)(t) \leq \frac{8}{3\sqrt[4]{4N}}$$

$$\left\{\Omega_1\left({}_{\frac{1}{2}}^{t}TX, \frac{1}{2\sqrt{N}}\right)_{L^1,[0,t]} + \Omega_1\left(T_{\frac{1}{2}}^{t}X, \frac{1}{2\sqrt{N}}\right)_{L^1,[t,1]}\right\}, \tag{16.113}$$

$\forall\, t \in [0,1]$, $\forall\, N \in \mathbb{N}$.

Proof By discrete Hölder's inequality we get that

$$B_N\left(|\cdot - t|^{\frac{3}{2}}\right)(t) \leq \left(B_N\left((\cdot - t)^2\right)(t)\right)^{\frac{3}{4}} \overset{(16.107.)}{=}$$

$$\left(\frac{1}{N}t(1-t)\right)^{\frac{3}{4}} \leq \frac{1}{(4N)^{\frac{3}{4}}}, \tag{16.114}$$

$\forall\, t \in [0,1]$, $\forall\, N \in \mathbb{N}$.

Hence,

$$\text{the right hand side of (16.112.)} \leq$$

$$\frac{8}{3}\frac{1}{\sqrt[4]{4N}}\left\{\Omega_1\left({}_{\frac{1}{2}}^{t}TX, \frac{1}{\sqrt{4N}}\right)_{L^1,[0,t]} + \Omega_1\left(T_{\frac{1}{2}}^{t}X, \frac{1}{\sqrt{4N}}\right)_{L^1,[t,1]}\right\}. \tag{16.115}$$

∎

Consequently we obtain

Proposition 16.39 *All as in Proposition 16.35 with $\alpha = \frac{1}{2}$. Additionally assume that $\left|X^{(1)}(t,\omega)\right| \leq M^*$, $\forall\, (t,\omega) \in [0,1] \times \Omega$, where $M^* > 0$. Then*

$$\left\|E\left(\left|\widetilde{B}_N(X) - X\right|\right)\right\|_{\infty} \leq \frac{8}{3\sqrt[4]{4N}}$$

$$\left\{\sup_{t\in[0,1]}\Omega_1\left({}_{\frac{1}{2}}^{t}TX, \frac{1}{2\sqrt{N}}\right)_{L^1,[0,t]} + \sup_{t\in[0,1]}\Omega_1\left(T_{\frac{1}{2}}^{t}X, \frac{1}{2\sqrt{N}}\right)_{L^1,[t,1]}\right\} < \infty, \tag{16.116}$$

$\forall\, N \in \mathbb{N}$.

As $N \to +\infty$, we obtain

$$\left\| E\left(\left|\widetilde{B}_N(X) - X\right|\right)\right\|_{\infty} \to 0. \tag{16.117}$$

That is $\widetilde{B}_N \overset{\text{“1-mean”}}{\to} I$ (unit operator) with rates.

Proof By (16.113); see also Remark 16.14, in particular see (16.29), (16.30). ■

Remark 16.40 *(to Proposition* 16.39*)* Assume that

$$\Omega_1\left(T^t_{\frac{1}{2}}X, \frac{1}{2\sqrt{N}}\right)_{L^1,[0,t]} \leq \frac{K_1}{2\sqrt{N}}, \ \forall\, t \in [0,1], \ \forall\, N \in \mathbb{N}, \tag{16.118}$$

where $K_1 > 0$,
and

$$\Omega_1\left({}^{t}_{\frac{1}{2}}TX, \frac{1}{2\sqrt{N}}\right)_{L^1,[t,1]} \leq \frac{K_2}{2\sqrt{N}}, \ \forall\, t \in [0,1], \ \forall\, N \in \mathbb{N}, \tag{16.119}$$

where $K_2 > 0$.

Conditions (16.118), (16.119) are of Lipschitz type of order 1.

Then, by (16.116), (16.118), (16.119), we get:

$$\left\| E\left(\left|\widetilde{B}_N(X) - X\right|\right)\right\|_{\infty} \leq \frac{8}{3\sqrt[4]{4N}}\left[\frac{K_1}{2\sqrt{N}} + \frac{K_2}{2\sqrt{N}}\right]$$

$$= \frac{4\left(K_1 + K_2\right)}{3\sqrt[4]{4}N^{\frac{3}{4}}}, \ \forall\, N \in \mathbb{N}. \tag{16.120}$$

That is

$$\left\| E\left(\left|\widetilde{B}_N(X) - X\right|\right)\right\|_{\infty} \leq \frac{4\left(K_1 + K_2\right)}{3\sqrt[4]{4}N^{\frac{3}{4}}}, \ \forall\, N \in \mathbb{N}. \tag{16.121}$$

Thus, at fractional smoothness of order $\frac{1}{2}$ we achieve speed of convergence $\frac{1}{N^{\frac{3}{4}}}$, $N \in \mathbb{N}$. Without any smoothness, in [5], we proved that the speed of convergence was $\frac{1}{\sqrt{N}}$, $N \in \mathbb{N}$.

In the presence of the ordinary first derivative, see [5], in deterministic approximation the rate of convergence was $\frac{1}{N}$, $N \in \mathbb{N}$.

Naturally, we have as expected:

$$\frac{1}{N} < \frac{1}{N^{\frac{3}{4}}} < \frac{1}{\sqrt{N}}, \ \forall\, N \in \mathbb{N} - \{1\}.$$

16.6 Stochastic Korovkin Results

We give at first pointwise results.

Theorem 16.41 *Assume all as in Theorem 16.23. Here* $\widetilde{L}$, M *are meant as sequences of operators. Assume further that* $\widetilde{L}(1)(t) \to 1$ *and* $\widetilde{L}\left(|\cdot - t|^{q(\alpha+1)}\right)(t) \to 0$, *pointwise in* $t \in [a,b]$. *Then* $E\left(|M(X) - X|^q\right)(t) \to 0$, *pointwise in* $t \in [a,b]$, *that is* $M \to I$ *(stochastic unit operator) in* q*-mean-pointwise with rates, quantitatively.*

Proof We use (16.46). First let $n \in \mathbb{N}$. We take into account $\widetilde{L}(1) \to 1$ and that

$$\left|\widetilde{L}\left((\cdot - t)^k\right)(t)\right| \leq \left(\widetilde{L}\left(|\cdot - t|^{q(\alpha+1)}\right)(t)\right)^{\frac{k}{q(\alpha+1)}} \left(\widetilde{L}(1)(t)\right)^{\frac{q(\alpha+1)-k}{q(\alpha+1)}}, \tag{16.122}$$

$k = 1, \ldots, n$.

Here it is $\widetilde{L}\left(\chi_{[t,b]}(\cdot)\right)(t), \widetilde{L}\left(\chi_{[a,t]}(\cdot)\right)(t) \leq \widetilde{L}(1)(t)$, which $\widetilde{L}(1)(t)$ is bounded, and similarly to (16.42), (16.43) and by the positivity of $\widetilde{L}$ we get that $\widetilde{L}\left(|\cdot - t|^{q(\alpha+1)} \chi_{[t,b]}(\cdot)\right)(t)$, $\widetilde{L}\left(|\cdot - t|^{q(\alpha+1)} \chi_{[a,t]}(\cdot)\right)(t) \leq \widetilde{L}\left(|\cdot - t|^{q(\alpha+1)}\right)(t) \to 0$.

Finally, we use Proposition 16.12(ii) for the $\Omega_1(\cdot,\cdot)$'s to go to zero.

The case of $n = 0$ is now obvious. ■

We continue with

Theorem 16.42 *Assume all as in Theorem 16.24. Here* $\widetilde{L}$, M *are meant as sequences of operators. Assume further that* $\widetilde{L}(1)(t) \to 1$ *and* $\widetilde{L}\left(|\cdot - t|^{(\alpha+1)}\right)(t) \to 0$, *pointwise in* $t \in [a,b]$. *Then* $E(|M(X) - X|)(t) \to 0$, *pointwise in* $t \in [a,b]$, *that is* $M \to I$ *in* 1*-mean-pointwise with rates, quantitatively.*

Proof We use (16.87). First let $n \in \mathbb{N}$. We take into account $\widetilde{L}(1) \to 1$, also (16.36) and (16.40). Notice $\widetilde{L}\left(\chi_{[t,b]}(\cdot)\right)(t)$, $\widetilde{L}\left(\chi_{[a,t]}(\cdot)\right)(t) \leq \widetilde{L}(1)(t)$, which $\widetilde{L}(1)(t)$ is bounded, and by (16.42), (16.43) and positivity of $\widetilde{L}$ we get that $\widetilde{L}\left(|\cdot - t|^{\alpha+1} \chi_{[t,b]}(\cdot)\right)(t)$, $\widetilde{L}\left(|\cdot - t|^{\alpha+1} \chi_{[a,t]}(\cdot)\right)(t) \leq \widetilde{L}\left(|\cdot - t|^{\alpha+1}\right)(t) \to 0$.

Finally, we use Proposition 16.12(ii) for the $\Omega_1(\cdot,\cdot)$'s to go to zero. The case of $n = 0$ is now obvious. ■

We continue with uniform convergence results.

Theorem 16.43 *All as in Theorem 16.30. Here* $\widetilde{L}$, M *are meant as sequences of operators. Assume further that* $\widetilde{L}(1) \to 1$ *uniformly, and* $\left\|\widetilde{L}\left(|\cdot - t|^{q(\alpha+1)}\right)(t)\right\|_\infty \to 0$. *Then* $\left\|E\left(|M(X) - X|^q\right)\right\|_\infty \to 0$ *over* $[a,b]$, *that is* $M \overset{\text{"q-mean"}}{\to} I$, *the stochastic unit operator with rates, quantitatively.*

Proof Use of (16.101). Here $\left\|\widetilde{L}(1) - 1\right\|_\infty \to 0$ and

$$\left\|\widetilde{L}\left((\cdot - t)^k\right)(t)\right\|_\infty \leq \left\|\widetilde{L}\left(|\cdot - t|^{q(\alpha+1)}\right)(t)\right\|_\infty^{\frac{k}{q(\alpha+1)}} \left\|\widetilde{L}(1)\right\|_\infty^{\frac{q(\alpha+1)-k}{q(\alpha+1)}}, \tag{16.123}$$

$k = 1, \ldots, n$.

Also $\left\| \widetilde{L}(1) \right\|_{\infty}$ is bounded, see further that $\left\| \widetilde{L}\left(|\cdot - t|^{q(\alpha+1)} \chi_{[t,b]}(\cdot) \right)(t) \right\|_{\infty}$, $\left\| \widetilde{L}\left(|\cdot - t|^{q(\alpha+1)} \chi_{[a,t]}(\cdot) \right)(t) \right\|_{\infty} \leq \left\| \widetilde{L}\left(|\cdot - t|^{q(\alpha+1)} \right)(t) \right\|_{\infty} < \infty$, similarly to (16.44), (16.45). The rest are clear. ■

We continue with

Theorem 16.44 *All as in Theorem 16.31. Here $\widetilde{L}$, M are meant as sequences of operators. Assume further that $\widetilde{L}(1) \to 1$ uniformly, and $\left\| \widetilde{L}\left(|\cdot - t|^{(\alpha+1)} \right)(t) \right\|_{\infty} \to 0$. Then $\| E(|M(X) - X|) \|_{\infty} \to 0$ over $[a, b]$, that is $M \overset{\text{“1-mean”}}{\to} I$ with rates quantitatively.*

Proof Use of (16.102) and $\left\| \widetilde{L}(1) - 1 \right\|_{\infty} \to 0$, (16.41), (16.44), (16.45), also $\left\| \widetilde{L}(1) \right\|_{\infty}$ is bounded. The rest are clear. ■

We finish with

Remark 16.45 An amazing fact/observation follows: In all convergence results here, see Theorems 16.41–16.44, the forcing conditions for convergences are based only on $\widetilde{L}$ and basic real valued continuous functions on $[a, b]$ and are not related to stochastic processes, but they are giving convergence results on stochastic processes!

As related work also see [7, pp. 355–376].

References

1. Abdeljawad, T.: On conformable fractional calculus. J. Comput. Appl. Math. **279**, 57–66 (2015)
2. Anastassiou, G.A.: Korovkin inequalities for stochastic processes. J. Math. Anal. Appl. **157**(2), 366–384 (1991)
3. Anastassiou, G.A.: Moments in Probability and Approximation Theory. Pitman/Longman, # 287, UK (1993)
4. Anastassiou, G.: Quantitative Approximations. Chapman & Hall/CRC, Boca Raton (2001)
5. Anastassiou, G.: Stochastic Korovkin theory given quantitatively, Facta Universitatis (Nis). Ser. Math. Inform. **22**(1), 43–60 (2007)
6. Anastassiou, G.A.: Fuzzy Mathematics: Approximation Theory. Springer, Heildelberg (2010)
7. Anastassiou, G.: Intelligent Mathematics: Computational Analysis. Springer, Heidelberg (2011)
8. Anastassiou, G.: Nonlinearity: Ordinary and Fractional Approximations by Sublinear and Max-Product Operators. Springer, Heidelberg (2018)
9. Anastassiou, G.: Conformable fractional approximation of stochastic processes. Submitted (2020)
10. Khalil, R., Al Horani, M., Yousef, A., Sababheh, M.: A new definition of fractional derivative. J. Comput. Appl. Math. **264**, 65–70 (2014)
11. Korovkin, P.P.: Linear Operators and Approximation Theory. Hindustan Publ. Corp., Delhi (1960)
12. Royden, H.L.: Real Analysis, 2nd edn. MacMillan Publishing Co., Inc., New York (1968)
13. Shisha, O., Mond, B.: The degree of convergence of sequences of linear positive operators. Natl. Acad. Sci. U.S. **60**, 1196–1200 (1968)
14. Weba, M.: Korovkin systems of stochastic processes. Math. Z. **192**, 73–80 (1986)
15. Weba, M.: A quantitative Korovkin theorem for random functions with multivariate domains. J. Approx. Theory **61**, 74–87 (1990)

Chapter 17
Trigonometric Conformable Fractional Approximation of Stochastic Processes

Here we consider very general stochastic positive linear operators induced by general positive linear operators that are acting on continuous functions in the trigonometric sense. These are acting on the space of real conformable fractionally differentiable stochastic processes. Under some very mild, general and natural assumptions on the stochastic processes we produce related trigonometric conformable fractional stochastic Shisha-Mond type inequalities of L^q-type $1 \leq q < \infty$ and corresponding trigonometric conformable fractional stochastic Korovkin type theorems. These are regarding the trigonometric stochastic q-mean conformable fractional convergence of a sequence of stochastic positive linear operators to the stochastic unit operator for various cases. All convergences are derived with rates and are given via the trigonometric conformable fractional stochastic inequalities involving the stochastic modulus of continuity of the αth conformable fractional derivatives of the engaged stochastic process, $\alpha \in (n, n+1), n \in \mathbb{Z}_+$. The impressive fact is that only two basic real Korovkin test functions assumptions, one of them trigonometric, are enough for the conclusions of our trigonometric conformable fractional stochastic Korovkin theorems. We give applications to stochastic Bernstein operators in the trigonometric sense. See also [8].

17.1 Introduction

Motivation for this work comes from [2, 3, 12–14]. This work continues our earlier work [5], now at the stochastic conformable fractional level.

In Sect. 17.2 we talk briefly about conformable fractional calculus. In the Sect. 17.3, we talk about the q-mean $(1 \leq q < \infty)$ first modulus of continuity of a stochastic process and its upper bounds. There we describe completely our setting by introducing our stochastic positive linear operator M, see (17.19), which is based on the positive linear operator $\widetilde{L}$ from $C([-\pi, \pi])$ into itself. The operator M is acting on a wide space of conformable fractional differentiable real valued stochastic

G. A. Anastassiou, *Generalized Fractional Calculus*, Studies in Systems, Decision and Control 305, https://doi.org/10.1007/978-3-030-56962-4_17

processes X. See there Assumptions 17.11, 17.14, 17.15, 17.17. We first give the main trigonometric pointwise conformable fractional stochastic Shisha-Mond type inequalities [12], see Theorems 17.23, 17.24, and their several corollaries covering important trigonometric special cases.

We continue with trigonometric conformable fractional q-mean uniform Shisha-Mond type inequalities, see Theorems 17.25, 17.26, and their interesting corollaries. All this theory is regarding the trigonometric conformable fractional stochastic convergence of operators M to I (stochastic unit operator) given quantitatively with rates. An extensive trigonometric application about the stochastic Bernstein operators follows in full details. Based on our Shisha-Mond type inequalities of our main Theorems 17.23–17.26 we derive trigonometric pointwise and uniform Stochastic Korovkin theorems [10] on stochastic processes, see Theorems 17.41–17.44. The amazing fact here is, that basic conditions on operator $\widetilde{L}$ regarding two simple real valued functions, one of them trigonometric, that are not stochastic, are able to imply conformable fractional stochastic convergence on all stochastic processes we are dealing with; see Concepts 17.10 and Assumptions 17.11, 17.14, 17.15, 17.17 on $[-\pi, \pi]$.

This work is based on [7].

17.2 Background—I

Here we follow [1], see also [9].

We need

Definition 17.1 ([1]) Let $a, b \in \mathbb{R}$. The left conformable fractional derivative starting from a of a function $f : [a, \infty) \to \mathbb{R}$ of order $0 < \alpha \leq 1$ is defined by

$$\left(T_{\alpha}^{a} f\right)(t) = \lim_{\varepsilon \to 0} \frac{f\left(t + \varepsilon (t - a)^{1-\alpha}\right) - f(t)}{\varepsilon}. \tag{17.1}$$

If $\left(T_{\alpha}^{a} f\right)(t)$ exists on (a, b), then

$$\left(T_{\alpha}^{a} f\right)(a) = \lim_{t \to a+} \left(T_{\alpha}^{a} f\right)(t). \tag{17.2}$$

The right conformable fractional derivative of order $0 < \alpha \leq 1$ terminating at b of $f : (-\infty, b] \to \mathbb{R}$ is defined by

$$\left({}_{\alpha}^{b} T f\right)(t) = -\lim_{\varepsilon \to 0} \frac{f\left(t + \varepsilon (b - t)^{1-\alpha}\right) - f(t)}{\varepsilon}. \tag{17.3}$$

If $\left({}_{\alpha}^{b} T f\right)(t)$ exists on (a, b), then

$$\left({}_{\alpha}^{b} T f\right)(b) = \lim_{t \to b-} \left({}_{\alpha}^{b} T f\right)(t). \tag{17.4}$$

Note that if f is differentiable then

$$\left(T_{\alpha}^{a} f\right)(t) = (t-a)^{1-\alpha} f'(t), \tag{17.5}$$

and

$$\left({}_{\alpha}^{b}Tf\right)(t) = -(b-t)^{1-\alpha} f'(t). \tag{17.6}$$

In the higher order case we can generalize things as follows:

Definition 17.2 *([1])* Let $\alpha \in (n, n+1]$, and set $\beta = \alpha - n$. Then, the left conformable fractional derivative starting from a of a function $f : [a, \infty) \to \mathbb{R}$ of order α, where $f^{(n)}(t)$ exists, is defined by

$$\left(\mathbf{T}_{\alpha}^{a} f\right)(t) = \left(T_{\beta}^{a} f^{(n)}\right)(t). \tag{17.7}$$

The right conformable fractional derivative of order α terminating at b of f : $(-\infty, b] \to \mathbb{R}$, where $f^{(n)}(t)$ exists, is defined by

$$\left({}_{\alpha}^{b}\mathbf{T}f\right)(t) = (-1)^{n+1}\left({}_{\beta}^{b}Tf^{(n)}\right)(t). \tag{17.8}$$

If $\alpha = n+1$ then $\beta = 1$ and $\mathbf{T}_{n+1}^{a} f = f^{(n+1)}$.

If n is odd, then ${}_{n+1}^{b}\mathbf{T}f = -f^{(n+1)}$, and if n is even, then ${}_{n+1}^{b}\mathbf{T}f = f^{(n+1)}$.

When $n = 0$ (or $\alpha \in (0, 1]$), then $\beta = \alpha$, and (17.7), (17.8) collapse to {(17.1), (17.2)}, {(17.3), (17.4)} respectively.

We make

Remark 17.3 *([6, p. 155])* We notice the following: let $\alpha \in (n, n+1]$ and $f \in C^{n+1}([a, b])$, $n \in \mathbb{N}$. Then ($\beta := \alpha - n$, $0 < \beta \leq 1$)

$$\left(\mathbf{T}_{\alpha}^{a}(f)\right)(x) = \left(T_{\beta}^{a} f^{(n)}\right)(x) = (x-a)^{1-\beta} f^{(n+1)}(x), \tag{17.9}$$

and

$$\left({}_{\alpha}^{b}\mathbf{T}(f)\right)(x) = (-1)^{n+1}\left({}_{\beta}^{b}Tf^{(n)}\right)(x) =$$

$$(-1)^{n+1}(-1)(b-x)^{1-\beta} f^{(n+1)}(x) = (-1)^{n}(b-x)^{1-\beta} f^{(n+1)}(x). \tag{17.10}$$

Consequently we get that

$$\left(\mathbf{T}_{\alpha}^{a}(f)\right)(x),\ \left({}_{\alpha}^{b}\mathbf{T}(f)\right)(x) \in C([a, b]).$$

Furthermore it is obvious that

$$\left(\mathbf{T}_{\alpha}^{a}(f)\right)(a) = \left({}_{\alpha}^{b}\mathbf{T}(f)\right)(b) = 0, \tag{17.11}$$

when $0 < \beta < 1$, i.e. when $\alpha \in (n, n+1)$.

17.3 Background—II

We need

Definition 17.4 Here $(\Omega, \mathcal{F}, P)$ is a probability space, $\omega \in \Omega$. We define the relative q-mean first modulus of continuity of stochastic process $X(t, \omega)$ by

$$\Omega_1(X, \delta)_{L^q,[c,d]} :=$$

$$\sup\left\{\left(\int_\Omega |X(x,\omega) - X(y,\omega)|^q P(d\omega)\right)^{\frac{1}{q}} : x, y \in [c,d] \subseteq [a,b], \ |x-y| \le \delta\right\}, \tag{17.12}$$

where $\delta > 0$, $1 \le q < \infty$, $t \in [a,b] \subset \mathbb{R}$.

Definition 17.5 Let $1 \le q < \infty$. Let $X(x, \omega)$ be a stochastic process. We call X a q-mean uniformly continuous stochastic process over $[a, b]$, iff $\forall\, \varepsilon > 0\ \exists\, \delta > 0$: whenever $|x - y| \le \delta$; $x, y \in [a, b]$ implies that

$$\int_\Omega |X(x,s) - X(y,s)|^q P(ds) \le \varepsilon. \tag{17.13}$$

We denote it as $X \in C_{\mathbb{R}}^{U_q}([a,b])$.

It holds

Proposition 17.6 ([5]) *Let* $X \in C_{\mathbb{R}}^{U_q}([a,b])$, *then* $\Omega_1(X,\delta)_{L^q,[a,b]} < \infty$, *any* $\delta > 0$.

Also it holds

Proposition 17.7 ([5]) *Let* $X(t, \omega)$ *be a stochastic process from* $[a, b] \times (\Omega, \mathcal{F}, P)$ *into* $\mathbb{R}$. *Then following are true* ($[c, d] \subseteq [a, b]$):

(i) $\Omega_1(X,\delta)_{L^q,[c,d]}$ *is nonnegative and nondecreasing in* $\delta > 0$,
(ii) $\lim_{\delta \downarrow 0} \Omega_1(X,\delta)_{L^q,[c,d]} = \Omega_1(X,0)_{L^q,[c,d]} = 0$, *iff* $X \in C_{\mathbb{R}}^{U_q}([c,d])$,
(iii) $\Omega_1(X,\delta_1+\delta_2)_{L^q,[c,d]} \le \Omega_1(X,\delta_1)_{L^q,[c,d]} + \Omega_1(X,\delta_2)_{L^q,[c,d]}$, $\delta_1, \delta_2 > 0$,
(iv) $\Omega_1(X,m\delta)_{L^q,[c,d]} \le m\Omega_1(X,\delta)_{L^q,[c,d]}$, $\delta > 0$, $m \in \mathbb{N}$,
(v) $\Omega_1(X,\lambda\delta)_{L^q,[c,d]} \le \lceil\lambda\rceil \Omega_1(X,\delta)_{L^q,[c,d]} \le (\lambda+1)\Omega_1(X,\delta)_{L^q,[c,d]}$, $\lambda > 0$, $\delta > 0$, $\lceil\cdot\rceil$ *is the ceiling of the number,*
(vi) $\Omega_1(X+Y,\delta)_{L^q,[c,d]} \le \Omega_1(X,\delta)_{L^q,[c,d]} + \Omega_1(Y,\delta)_{L^q,[c,d]}$, $\delta > 0$,
(vii) $\Omega_1(X,\cdot)_{L^q,[c,d]}$ *is continuous on* $\mathbb{R}_+$ *for* $X \in C_{\mathbb{R}}^{U_q}([c,d])$.

We give

Remark 17.8 *(to Proposition* 17.7*)* By Proposition 17.7 (v) we get

$$\Omega_1(X,|x-y|)_{L^q,[c,d]} \le \left\lceil \frac{|x-y|}{\delta} \right\rceil \Omega_1(X,\delta)_{L^q,[c,d]}, \tag{17.14}$$

$\forall\, x, y \in [c, d]$, any $\delta > 0$.

We give

Remark 17.9 Let $\alpha \in (n, n+1]$, $n \in \mathbb{Z}_+$, $X(\cdot, \omega) \in C^{n+1}([a,b])$, $\forall\, \omega \in \Omega$. We assume that $\left|X^{(n+1)}(t,\omega)\right| \leq M^*$, $\forall\,(t,\omega) \in [a,b] \times \Omega$, where $M^* > 0$. Let $\delta > 0$, $1 \leq q < \infty$. Then

$$\Omega_1\left(\mathbf{T}^c_\alpha X, \delta\right)_{L^q,[c,d]} = \sup\left\{\left(\int_\Omega \left|\left(\mathbf{T}^c_\alpha X\right)(x,\omega) - \left(\mathbf{T}^c_\alpha X\right)(y,\omega)\right|^q P(d\omega)\right)^{\frac{1}{q}} :\right.$$

$$\left. x, y \in [c,d] \subseteq [a,b],\ |x-y| \leq \delta\right\} \leq$$

$$\sup\left\{\left(\int_\Omega \left(\left|\left(\mathbf{T}^c_\alpha X\right)(x,\omega)\right| + \left|\left(\mathbf{T}^c_\alpha X\right)(y,\omega)\right|\right)^q P(d\omega)\right)^{\frac{1}{q}} :\right.$$

$$\left. x, y \in [c,d] \subseteq [a,b],\ |x-y| \leq \delta\right\} \underset{(\beta:=\alpha-n)}{\overset{(17.9.)}{\leq}}$$

$$\sup\left\{\left(\int_\Omega \left[(x-c)^{1-\beta}\left|X^{(n+1)}(x,\omega)\right| + (y-c)^{1-\beta}\left|X^{(n+1)}(y,\omega)\right|\right]^q P(d\omega)\right)^{\frac{1}{q}} :\right.$$

$$\left. x, y \in [c,d] \subseteq [a,b],\ |x-y| \leq \delta\right\} \leq$$

$$M^* \sup\left\{\left(\int_\Omega \left((x-c)^{1-\beta} + (y-c)^{1-\beta}\right)^q P(d\omega)\right)^{\frac{1}{q}} :\right.$$

$$\left. x, y \in [c,d] \subseteq [a,b],\ |x-y| \leq \delta\right\} \leq 2M^*(b-a)^{1-\beta}.$$

That is

$$\Omega_1\left(\mathbf{T}^c_\alpha X, \delta\right)_{L^q,[c,d]} \leq 2M^*(b-a)^{1-\beta}, \tag{17.15}$$

for any $[c,d] \subseteq [a,b]$.

Similarly, it holds

$$\Omega_1\left({}^d_\alpha\mathbf{T} X, \delta\right)_{L^q,[c,d]} \leq 2M^*(b-a)^{1-\beta}, \tag{17.16}$$

for any $[c,d] \subseteq [a,b]$.

In particular it holds

$$\sup_{t\in[a,b]} \Omega_1\left(\mathbf{T}^t_\alpha X, \delta\right)_{L^q,[t,b]} \leq 2M^*(b-a)^{1-\beta}, \tag{17.17}$$

and

$$\sup_{t\in[a,b]} \Omega_1\left({}^t_\alpha\mathbf{T} X, \delta\right)_{L^q,[a,t]} \leq 2M^*(b-a)^{1-\beta}. \tag{17.18}$$

Above, it is not strange to assume that $\mathbf{T}_{\alpha}^{c}X, {}_{\alpha}^{d}\mathbf{T}X, \mathbf{T}_{\alpha}^{t}X, {}_{\alpha}^{t}\mathbf{T}X$ are stochastic processes, see Remark 17.3.

We need

Concepts 17.10 *Let $\widetilde{L}$ be a positive linear operator from $C([a,b])$ into itself. Let $X(t,\omega)$ be a stochastic process from $[a,b] \times (\Omega, \mathcal{F}, P)$ into $\mathbb{R}$, where $(\Omega, \mathcal{F}, P)$ is a probability space. Let $\alpha \in (n, n+1]$, $n \in \mathbb{Z}_{+}$, and $X(\cdot,\omega) \in C^{n+1}([a,b])$, $\forall$ $\omega \in \Omega$. Also we assume for each $t \in [a,b]$ that $X^{(k)}(t,\cdot)$ is measurable over $(\Omega, \mathcal{F})$ for all $k = 1, \ldots, n+1$. Define*

$$M(X)(t,\omega) := \widetilde{L}(X(\cdot,\omega))(t), \quad \forall\, \omega \in \Omega, \forall\, t \in [a,b], \tag{17.19}$$

and assume that it is a random variable in ω. Clearly M is a positive linear operator on stochastic processes.

We make

Assumption 17.11 (i) For any $t \in [a,b]$ we assume that $\mathbf{T}_{\alpha}^{t}X(z,\omega)$ is continuous in $z \in [t,b]$, uniformly with respect to $\omega \in \Omega$. I.e. $\forall\, \varepsilon > 0\, \exists\, \delta > 0$: whenever $|z_1 - z_2| \leq \delta$; $z_1, z_2 \in [t,b]$, then $\left|\mathbf{T}_{\alpha}^{t}X(z_1,\omega) - \mathbf{T}_{\alpha}^{t}X(z_2,\omega)\right| \leq \varepsilon$, $\forall$ $\omega \in \Omega$.
We denote this by $\mathbf{T}_{\alpha}^{t}X \in C_{\mathbb{R}}^{U}([t,b])$, the space of continuous in x, uniformly with respect to ω, stochastic processes over $[t,b]$.

(ii) For any $t \in [a,b]$ we assume that ${}_{\alpha}^{t}\mathbf{T}X(z,\omega)$ is continuous in $z \in [a,t]$, uniformly with respect to $\omega \in \Omega$. I.e. $\forall\, \varepsilon > 0\, \exists\, \delta > 0$: whenever $|z_1 - z_2| \leq \delta$; $z_1, z_2 \in [a,t]$, then $\left|{}_{\alpha}^{t}\mathbf{T}X(z_1,\omega) - {}_{\alpha}^{t}\mathbf{T}X(z_2,\omega)\right| \leq \varepsilon$, $\forall\, \omega \in \Omega$.
We denote this by ${}_{\alpha}^{t}\mathbf{T}X \in C_{\mathbb{R}}^{U}([a,t])$, the space of continuous in x, uniformly with respect to ω, stochastic processes over $[a,t]$.

Remark 17.12 Assumption 17.11 implies:

(i) $\mathbf{T}_{\alpha}^{t}X(\cdot,\omega) \in C([t,b])$, $\forall\, \omega \in \Omega$, and $\mathbf{T}_{\alpha}^{t}X$ is q-mean uniformly continuous in $z \in [t,b]$, that is $\mathbf{T}_{\alpha}^{t}X \in C_{\mathbb{R}}^{U_q}([t,b])$, for any $1 \leq q < \infty$.

(ii) ${}_{\alpha}^{t}\mathbf{T}X(\cdot,\omega) \in C([a,t])$, $\forall\, \omega \in \Omega$, and ${}_{\alpha}^{t}\mathbf{T}X$ is q-mean uniformly continuous in $z \in [a,t]$, that is ${}_{\alpha}^{t}\mathbf{T}X \in C_{\mathbb{R}}^{U_q}([a,t])$, for any $1 \leq q < \infty$.

We need

Definition 17.13 Denote by

$$(EX)(t) := \int_{\Omega} X(t,\omega)\, P(d\omega), \quad \forall\, t \in [a,b], \tag{17.20}$$

the expectation operator.

We make

Assumption 17.14 We assume that

$$\left(E\left|X^{(k)}\right|^{q}\right)(t)<\infty,\ \forall\, t\in[a,b],\tag{17.21}$$

$q>1$, for all $k=0,1,\ldots,n$.

We make

Assumption 17.15 We assume that

$$\left(E\left|X^{(k)}\right|\right)(t)<\infty,\ \forall\, t\in[a,b],\tag{17.22}$$

for all $k=0,1,\ldots,n$.

We give

Remark 17.16 By the Riesz representation theorem [11] we have that there exists μ_t unique, completed Borel measure on $[a,b]$ with

$$m_t:=\mu_t([a,b])=\widetilde{L}(1)(t)>0,\tag{17.23}$$

such that

$$\widetilde{L}(f)(t)=\int_{[a,b]}f(x)\,d\mu_t(x),\quad \forall\, t\in[a,b],\forall\, f\in C([a,b]).\tag{17.24}$$

Consequently we have that

$$M(X)(t,\omega)=\int_{[a,b]}X(x,\omega)\,d\mu_t(x),\ \ \forall\ (t,\omega)\in[a,b]\times\Omega,\tag{17.25}$$

and X as in Concepts 17.10.

Here $\chi_{[\gamma,\delta]}(s)$ stands for the characteristic function on $[\gamma,\delta]\subseteq[a,b]$.

Notice that $(r>0)$

$$\int_{[t,b]}(s-t)^r\mu_t(ds)=\int_{[a,b]}\chi_{[t,b]}(s)\,|s-t|^r\mu_t(ds)=\widetilde{L}\left(|\cdot-t|^r\chi_{[t,b]}(\cdot)\right)(t),\tag{17.26}$$

and

$$\int_{[a,t]}(t-s)^r\mu_t(ds)=\int_{[a,b]}\chi_{[a,t]}(s)\,|s-t|^r\mu_t(ds)=\widetilde{L}\left(|\cdot-t|^r\chi_{[a,t]}(\cdot)\right)(t).\tag{17.27}$$

Let $\alpha\in(n,n+1]$, $n\in\mathbb{N}$ and $k=1,\ldots,n$. Then by Hölder's inequality we obtain

$$\left|\int_{[a,b]} (x-t)^k \, d\mu_t(x)\right| \le \int_{[a,b]} |x-t|^k \, d\mu_t(x) \le$$

$$\left(\int_{[a,b]} |x-t|^{\alpha+1} \, d\mu_t(x)\right)^{\left(\frac{k}{\alpha+1}\right)} (\mu_t([a,b]))^{\left(\frac{\alpha+1-k}{\alpha+1}\right)}. \tag{17.28}$$

Therefore it holds

$$\left\|\widetilde{L}\left((\cdot-t)^k\right)(t)\right\|_{\infty,[a,b]} \le \left\|\widetilde{L}\left(|\cdot-t|^k\right)(t)\right\|_{\infty,[a,b]} \le$$

$$\left\|\widetilde{L}\left(|\cdot-t|^{\alpha+1}\right)(t)\right\|_{\infty,[a,b]}^{\left(\frac{k}{\alpha+1}\right)} \left\|\widetilde{L}(1)\right\|_{\infty,[a,b]}^{\left(\frac{\alpha+1-k}{\alpha+1}\right)}, \tag{17.29}$$

all $k = 1, \ldots, n$.

Also, we observe that

$$C([a,b]) \ni |\cdot - t|^{\alpha+1} \chi_{[a,t]}(\cdot) \le |\cdot - t|^{\alpha+1}, \quad \forall\, t \in [a,b], \tag{17.30}$$

and

$$C([a,b]) \ni |\cdot - t|^{\alpha+1} \chi_{[t,b]}(\cdot) \le |\cdot - t|^{\alpha+1}, \quad \forall\, t \in [a,b]. \tag{17.31}$$

By positivity of $\widetilde{L}$ we obtain

$$\left\|\widetilde{L}\left(|\cdot-t|^{\alpha+1}\chi_{[a,t]}(\cdot)\right)(t)\right\|_{\infty,[a,b]} \le \left\|\widetilde{L}\left(|\cdot-t|^{\alpha+1}\right)(t)\right\|_{\infty,[a,b]} < \infty, \tag{17.32}$$

by $\widetilde{L}\left(|\cdot-t|^{\alpha+1}\right)(t)$ being continuous in $t \in [a,b]$, see p. 388 of [4],

and

$$\left\|\widetilde{L}\left(|\cdot-t|^{\alpha+1}\chi_{[t,b]}(\cdot)\right)(t)\right\|_{\infty,[a,b]} \le \left\|\widetilde{L}\left(|\cdot-t|^{\alpha+1}\right)(t)\right\|_{\infty,[a,b]}. \tag{17.33}$$

Above (17.26)–(17.29) and (17.32), (17.33) were used to derive convergences in our earlier results mentioned next.

In this work we denote by $\widetilde{L}\left(\chi_{[a,t]}(\cdot)\right)(t) := \mu_t([a,t]) \le \widetilde{L}(1)(t)$, and by $\widetilde{L}\left(\chi_{[t,b]}(\cdot)\right)(t) := \mu_t([t,b]) \le \widetilde{L}(1)(t)$.

We make

Assumption 17.17 Assume that $\left|X^{(n+1)}(t,\omega)\right| \le M^*, \forall\,(t,\omega) \in [a,b] \times \Omega$, where $M^* > 0$. And suppose that $X^{(n+1)}(\cdot,\omega)$ is continuous over $[a,b]$, uniformly with respect to $\omega \in \Omega$.

Clearly, Assumption 17.17 implies Assumption 17.11, see also Remark 17.3.

We mention the following pointwise result on the quantitative stochastic conformable fractional approximation regarding stochastic processes:

Theorem 17.18 *([7]) Let $n \in \mathbb{Z}_+$, $\alpha \in (n, n+1)$; $p, q > 1 : \frac{1}{p} + \frac{1}{q} = 1$, with $\beta := \alpha - n > \frac{1}{q}$; set $\lambda_1 := \left(\frac{\Gamma(np+1)\Gamma(p(\beta-1)+1)}{\Gamma(p(\alpha-1)+2)}\right)^{(q-1)}$. Suppose Concepts 17.10, Assumptions 17.11 and 17.14.*

Then, $\forall\, t \in [a, b]$, we have:

$$\left(E\left(|M(X) - X|^q\right)(t)\right)^{\frac{1}{q}} \leq \left(\left(E|X|^q\right)(t)\right)^{\frac{1}{q}} \left|\widetilde{L}(1)(t) - 1\right| +$$

$$\sum_{k=1}^{n} \frac{\left(\left(E\left|X^{(k)}\right|^q\right)(t)\right)^{\frac{1}{q}}}{k!} \left|\widetilde{L}\left((\cdot - t)^k\right)(t)\right| + \frac{2^{\frac{1}{p}} \lambda_1^{\frac{1}{q}}}{(q+1)^{\frac{\alpha}{q(\alpha+1)}}\, n!}$$

$$\left\{\left[\Omega_1\left(\mathbf{T}_\alpha^t X, \left(\frac{1}{(q+1)}\widetilde{L}\left(|\cdot - t|^{q(\alpha+1)} \chi_{[t,b]}(\cdot)\right)(t)\right)^{\frac{1}{q(\alpha+1)}}\right)_{L^q,[t,b]}\right.\right.$$

$$\left(\widetilde{L}\left(\chi_{[t,b]}(\cdot)\right)(t)\right)^{\frac{1}{p}} \left(\widetilde{L}\left(|\cdot - t|^{q(\alpha+1)} \chi_{[t,b]}(\cdot)\right)(t)\right)^{\frac{\alpha}{q(\alpha+1)}}$$

$$\left.\left[\left(\widetilde{L}\left(\chi_{[t,b]}(\cdot)\right)(t)\right)^{\frac{1}{\alpha+1}} (q+1)^{\frac{\alpha}{(\alpha+1)}} + 1\right]^{\frac{1}{q}}\right] + \tag{17.34}$$

$$\left[\Omega_1\left({}_{\alpha}^{t}\mathbf{T}X, \left(\frac{1}{(q+1)}\widetilde{L}\left(|\cdot - t|^{q(\alpha+1)} \chi_{[a,t]}(\cdot)\right)(t)\right)^{\frac{1}{q(\alpha+1)}}\right)_{L^q,[a,t]}\right.$$

$$\left(\widetilde{L}\left(\chi_{[a,t]}(\cdot)\right)(t)\right)^{\frac{1}{p}} \left(\widetilde{L}\left(|\cdot - t|^{q(\alpha+1)} \chi_{[a,t]}(\cdot)\right)(t)\right)^{\frac{\alpha}{q(\alpha+1)}}$$

$$\left.\left.\left[\left(\widetilde{L}\left(\chi_{[a,t]}(\cdot)\right)(t)\right)^{\frac{1}{\alpha+1}} (q+1)^{\frac{\alpha}{(\alpha+1)}} + 1\right]^{\frac{1}{q}}\right]\right\}$$

We mention also the following pointwise results, the L_1-quantitative stochastic conformable fractional approximation of stochastic processes.

Theorem 17.19 ([7]) *Let $n \in \mathbb{Z}_+$ and $\alpha \in (n, n+1)$, $\beta := \alpha - n$. Suppose Concepts 17.10, Assumptions 17.11, 17.15. Then, $\forall\, t \in [a, b]$, we have:*

$$E(|M(X) - X|)(t) \leq (E|X|)(t) \left|\widetilde{L}(1)(t) - 1\right| +$$

$$\sum_{k=1}^{n} \frac{\left(E\left|X^{(k)}\right|\right)(t)}{k!} \left|\widetilde{L}\left((\cdot - t)^k\right)(t)\right| + \frac{1}{\prod_{j=1}^{n}(\beta + j)}$$

$$\left\{\left[\frac{\left(\widetilde{L}\left(\chi_{[a,t]}(\cdot)\right)(t)\right)^{\frac{1}{\alpha+1}}}{\alpha-n}+\frac{1}{\alpha+1}\right]\left(\widetilde{L}\left(|\cdot-t|^{\alpha+1}\chi_{[a,t]}(\cdot)\right)(t)\right)^{\frac{\alpha}{\alpha+1}}\right.$$

$$\Omega_1\left({}_{\alpha}^{t}\mathbf{T}X,\left(\widetilde{L}\left(|\cdot-t|^{\alpha+1}\chi_{[a,t]}(\cdot)\right)(t)\right)^{\frac{1}{\alpha+1}}\right)_{L^1,[a,t]}+$$

$$\left[\frac{\left(\widetilde{L}\left(\chi_{[t,b]}(\cdot)\right)(t)\right)^{\frac{1}{\alpha+1}}}{\alpha-n}+\frac{1}{(\alpha+1)}\right]\left(\widetilde{L}\left(|\cdot-t|^{\alpha+1}\chi_{[t,b]}(\cdot)\right)(t)\right)^{\frac{\alpha}{\alpha+1}}$$

$$\left.\Omega_1\left(\mathbf{T}_{\alpha}^{t}X,\left(\widetilde{L}\left(|\cdot-t|^{\alpha+1}\chi_{[t,b]}(\cdot)\right)(t)\right)^{\frac{1}{\alpha+1}}\right)_{L^1,[t,b]}\right\}. \tag{17.35}$$

Uniform estimates follow:
Theorem 17.18 implies

Theorem 17.20 ([7]) *Let $n\in\mathbb{Z}_+$, $\alpha\in(n,n+1)$; $p,q>1:\frac{1}{p}+\frac{1}{q}=1$, with $\beta:=\alpha-n>\frac{1}{q}$; set $\lambda_1:=\left(\frac{\Gamma(np+1)\Gamma(p(\beta-1)+1)}{\Gamma(p(\alpha-1)+2)}\right)^{(q-1)}$. Suppose Concepts 17.10, Assumptions 17.14 and 17.17. Then*

$$\left\|E\left(|M(X)-X|^q\right)(t)\right\|_\infty^{\frac{1}{q}}\leq\left\|E\left(|X|^q\right)\right\|_\infty^{\frac{1}{q}}\left\|\widetilde{L}(1)-1\right\|_\infty+$$

$$\sum_{k=1}^{n}\frac{\left\|E\left(\left|X^{(k)}\right|^q\right)\right\|_\infty^{\frac{1}{q}}}{k!}\left\|\widetilde{L}\left((\cdot-t)^k\right)(t)\right\|_\infty+\frac{\lambda_1^{\frac{1}{q}}\left(2\left\|\widetilde{L}(1)\right\|_\infty\right)^{\frac{1}{p}}}{(q+1)^{\frac{\alpha}{q(\alpha+1)}}\,n!}$$

$$\left[\left\|\widetilde{L}(1)\right\|_\infty^{\frac{1}{\alpha+1}}(q+1)^{\frac{\alpha}{(\alpha+1)}}+1\right]^{\frac{1}{q}}$$

$$\left\{\left[\sup_{t\in[a,b]}\Omega_1\left(\mathbf{T}_{\alpha}^{t}X,\frac{1}{(q+1)}\left\|\widetilde{L}\left(|\cdot-t|^{q(\alpha+1)}\chi_{[t,b]}(\cdot)\right)(t)\right\|_\infty^{\frac{1}{q(\alpha+1)}}\right)_{L^q,[t,b]}\right.\right.$$

$$\left.\left\|\widetilde{L}\left(|\cdot-t|^{q(\alpha+1)}\chi_{[t,b]}(\cdot)\right)(t)\right\|_\infty^{\frac{\alpha}{q(\alpha+1)}}\right]+$$

$$\left[\sup_{t\in[a,b]}\Omega_1\left({}_{\alpha}^{t}\mathbf{T}X,\frac{1}{(q+1)}\left\|\widetilde{L}\left(|\cdot-t|^{q(\alpha+1)}\chi_{[a,t]}(\cdot)\right)(t)\right\|_\infty^{\frac{1}{q(\alpha+1)}}\right)_{L^q,[a,t]}\right. \tag{17.36}$$

$$\left.\left.\left\|\widetilde{L}\left(|\cdot-t|^{q(\alpha+1)}\chi_{[a,t]}(\cdot)\right)(t)\right\|_\infty^{\frac{\alpha}{q(\alpha+1)}}\right]\right\}.$$

Theorem 17.19 implies:

Theorem 17.21 ([7]) *Let* $n \in \mathbb{Z}_+$, $\alpha \in (n, n+1)$, $\beta := \alpha - n$. *Suppose Concepts 17.10, Assumptions 17.15, 17.17. Then*

$$\|E(|M(X) - X|)\|_\infty \leq \|E(|X|)\|_\infty \left\|\widetilde{L}(1) - 1\right\|_\infty +$$

$$\sum_{k=1}^{n} \frac{\left\|E\left(\left|X^{(k)}\right|\right)\right\|_\infty}{k!} \left\|\widetilde{L}\left((\cdot - t)^k\right)(t)\right\|_\infty +$$

$$\frac{1}{\prod_{j=1}^{n}(\beta + j)} \left[\frac{\left\|\widetilde{L}(1)\right\|_\infty^{\frac{1}{\alpha+1}}}{\alpha - n} + \frac{1}{\alpha + 1}\right]$$

$$\left\{\left\|\widetilde{L}\left(|\cdot - t|^{\alpha+1} \chi_{[a,t]}(\cdot)\right)(t)\right\|_\infty^{\frac{\alpha}{\alpha+1}}\right.$$

$$\sup_{t \in [a,b]} \Omega_1\left({}_{\alpha}^{t}\mathbf{T}X, \left\|\widetilde{L}\left(|\cdot - t|^{\alpha+1} \chi_{[a,t]}(\cdot)\right)(t)\right\|_\infty^{\frac{1}{\alpha+1}}\right)_{L^1,[a,t]} +$$

$$\left\|\widetilde{L}\left(|\cdot - t|^{\alpha+1} \chi_{[t,b]}(\cdot)\right)(t)\right\|_\infty^{\frac{\alpha}{\alpha+1}}$$

$$\left.\sup_{t \in [a,b]} \Omega_1\left(\mathbf{T}_{\alpha}^{t}X, \left\|\widetilde{L}\left(|\cdot - t|^{\alpha+1} \chi_{[t,b]}(\cdot)\right)(t)\right\|_\infty^{\frac{1}{\alpha+1}}\right)_{L^1,[t,b]}\right\}. \tag{17.37}$$

Remark 17.22 Next we specify $[a, b]$ as $[-\pi, \pi]$. Clearly then $\widetilde{L} : C([-\pi, \pi]) \to C([-\pi, \pi])$ is the positive linear operator on hand.

Here $\alpha \in (n, n+1)$, $n \in \mathbb{Z}_+$, $k = 1, \ldots, n$. Next we use of Hölder's inequality we notice that

$$\widetilde{L}\left(\left(\sin\left(\frac{|\cdot - t|}{4}\right)\right)^k\right)(t) = \int_{[-\pi,\pi]} \left(\sin\left(\frac{|x - t|}{4}\right)\right)^k d\mu_t(x) \leq$$

$$\left(\int_{[-\pi,\pi]} \left(\sin\left(\frac{|x - t|}{4}\right)\right)^{\alpha+1} d\mu_t(x)\right)^{\frac{k}{\alpha+1}} \left(\mu_t([-\pi, \pi])\right)^{\frac{\alpha+1-k}{\alpha+1}} = \tag{17.38}$$

$$\left(\widetilde{L}\left(\left(\sin\left(\frac{|\cdot - t|}{4}\right)\right)^{\alpha+1}\right)(t)\right)^{\frac{k}{\alpha+1}} \left(\widetilde{L}(1)(t)\right)^{\frac{\alpha+1-k}{\alpha+1}}.$$

That is

$$\widetilde{L}\left(\left(\sin\left(\frac{|\cdot - t|}{4}\right)\right)^k\right)(t) \leq \tag{17.39}$$

$$\left(\widetilde{L}\left(\left(\sin\left(\frac{|\cdot - t|}{4}\right)\right)^{\alpha+1}\right)(t)\right)^{\frac{k}{\alpha+1}} (L(1)(t))^{\frac{\alpha+1-k}{\alpha+1}},$$

for $k = 1, \ldots, n$; true also for $q(\alpha+1)$ instead of $(\alpha+1)$, for any $1 < q < \infty$.

Next $\|\cdot\|_{\infty}$ denotes $\|\cdot\|_{\infty,[-\pi,\pi]}$.

Consequently, it holds

$$\left\|\widetilde{L}\left(\left(\sin\left(\frac{|\cdot - t|}{4}\right)\right)^{k}\right)(t)\right\|_{\infty} \leq \tag{17.40}$$

$$\left\|\widetilde{L}\left(\left(\sin\left(\frac{|\cdot - t|}{4}\right)\right)^{\alpha+1}\right)(t)\right\|_{\infty}^{\frac{k}{\alpha+1}} \left\|\widetilde{L}(1)(t)\right\|_{\infty}^{\frac{\alpha+1-k}{\alpha+1}},$$

for $k = 1, \ldots, n$; true also for $q(\alpha+1)$ instead of $(\alpha+1)$, for any $1 < q < \infty$.

In this work we use a lot the following well known inequality:

$$|z| \leq \pi \sin\left(\frac{|z|}{2}\right), \ \forall\, z \in [-\pi, \pi]. \tag{17.41}$$

Notice that, for any $t \in [-\pi, \pi]$, we have $C([-\pi,\pi]) \ni |\cdot - t|\,\chi_{[-\pi,t]}(\cdot) \leq |\cdot - t| \in C([-\pi,\pi])$, therefore

$$C([-\pi,\pi]) \ni \left(\sin\left(\frac{|\cdot - t|\,\chi_{[-\pi,t]}(\cdot)}{4}\right)\right)^{\alpha+1} \leq \left(\sin\left(\frac{|\cdot - t|}{4}\right)\right)^{\alpha+1} \in C([-\pi,\pi]). \tag{17.42}$$

Consequently, by positivity of $\widetilde{L}$ we obtain

$$\left\|\widetilde{L}\left(\left(\sin\left(\frac{|\cdot - t|\,\chi_{[-\pi,t]}(\cdot)}{4}\right)\right)^{\alpha+1}\right)(t)\right\|_{\infty} \leq \left\|\widetilde{L}\left(\left(\sin\left(\frac{|\cdot - t|}{4}\right)\right)^{\alpha+1}\right)(t)\right\|_{\infty}. \tag{17.43}$$

Similarly, for any $t \in [-\pi, \pi]$, we have $C([-\pi,\pi]) \ni |\cdot - t|\,\chi_{[t,\pi]}(\cdot) \leq |\cdot - t| \in C([-\pi,\pi])$, thus

$$C([-\pi,\pi]) \ni \left(\sin\left(\frac{|\cdot - t|\,\chi_{[t,\pi]}(\cdot)}{4}\right)\right)^{\alpha+1} \leq \left(\sin\left(\frac{|\cdot - t|}{4}\right)\right)^{\alpha+1} \in C([-\pi,\pi]). \tag{17.44}$$

Hence

$$\left\|\widetilde{L}\left(\left(\sin\left(\frac{|\cdot - t|\,\chi_{[t,\pi]}(\cdot)}{4}\right)\right)^{\alpha+1}\right)(t)\right\|_{\infty} \leq \left\|\widetilde{L}\left(\left(\sin\left(\frac{|\cdot - t|}{4}\right)\right)^{\alpha+1}\right)(t)\right\|_{\infty}. \tag{17.45}$$

So, if the right hand side of (17.43), (17.45) goes to zero, so do their left hand sides. Above in (17.42)–(17.45), one can use $q\ (\alpha+1)$ instead of $(\alpha+1)$, $1 < q < \infty$.

A further detailed analysis reveals:

We have that $(1 \leq q < \infty)$

$$\widetilde{L}\left(|\cdot - t|^{q(\alpha+1)} \chi_{[t,\pi]}(\cdot)\right)(t) \overset{(17.24.)}{=} \int_{[t,\pi]} (x-t)^{q(\alpha+1)}\, d\mu_t(x) =$$

$$2^{q(\alpha+1)} \int_{[t,\pi]} \left(\frac{x-t}{2}\right)^{q(\alpha+1)} d\mu_t(x) \overset{(17.41.)}{\leq}$$

$$(2\pi)^{q(\alpha+1)} \int_{[t,\pi]} \left(\sin\left(\frac{x-t}{4}\right)\right)^{q(\alpha+1)} d\mu_t(x) = \tag{17.46}$$

$$(2\pi)^{q(\alpha+1)} \int_{[t,\pi]} \left(\sin\left(\frac{|x-t|}{4}\right)\right)^{q(\alpha+1)} d\mu_t(x) =$$

$$(2\pi)^{q(\alpha+1)} \int_{[-\pi,\pi]} \left(\sin\left(\frac{|x-t|}{4}\right)\right)^{q(\alpha+1)} \chi_{[t,\pi]}(x)\, d\mu_t(x) =$$

$$(2\pi)^{q(\alpha+1)} \int_{[-\pi,\pi]} \left(\sin\left(\frac{|x-t|}{4}\chi_{[t,\pi]}(x)\right)\right)^{q(\alpha+1)} d\mu_t(x) \overset{(17.24.)}{=}$$

$$(2\pi)^{q(\alpha+1)}\, \widetilde{L}\left(\left(\sin\left(\frac{|\cdot - t|}{4}\chi_{[t,\pi]}(\cdot)\right)\right)^{q(\alpha+1)}\right)(t).$$

That is, we have obtained

$$\widetilde{L}\left(|\cdot - t|^{q(\alpha+1)} \chi_{[t,\pi]}(\cdot)\right)(t) \leq (2\pi)^{q(\alpha+1)}\, \widetilde{L}\left(\left(\sin\left(\frac{|\cdot - t|}{4}\chi_{[t,\pi]}(\cdot)\right)\right)^{q(\alpha+1)}\right)(t). \tag{17.47}$$

Similarly, it holds

$$\widetilde{L}\left(|\cdot - t|^{q(\alpha+1)} \chi_{[-\pi,t]}(\cdot)\right)(t) \leq (2\pi)^{q(\alpha+1)}\, \widetilde{L}\left(\left(\sin\left(\frac{|\cdot - t|}{4}\chi_{[-\pi,t]}(\cdot)\right)\right)^{q(\alpha+1)}\right)(t). \tag{17.48}$$

Above inequalities (17.47), (17.48) are valid for any $1 \leq q < \infty$.

Furthermore, we observe that

$$\left|\widetilde{L}\left((\cdot - t)^k\right)(t)\right| = \left|\int_{[-\pi,\pi]} (x-t)^k\, d\mu_t(x)\right| \leq \int_{[-\pi,\pi]} |x-t|^k\, d\mu_t(x) =$$

$$2^{k}\int_{[-\pi,\pi]}\left(\frac{|x-t|}{2}\right)^{k}d\mu_{t}(x)\overset{(17.41.)}{\leq}(2\pi)^{k}\int_{[-\pi,\pi]}\left(\sin\left(\frac{|x-t|}{4}\right)\right)^{k}d\mu_{t}(x)$$
$$=(2\pi)^{k}\widetilde{L}\left(\left(\sin\left(\frac{|\cdot-t|}{4}\right)\right)^{k}\right)(t). \tag{17.49}$$

That is

$$\left|\widetilde{L}\left((\cdot-t)^{k}\right)(t)\right|\leq(2\pi)^{k}\widetilde{L}\left(\left(\sin\left(\frac{|\cdot-t|}{4}\right)\right)^{k}\right)(t), \tag{17.50}$$

$\forall\, t\in[-\pi,\pi]$, all $k=1,\ldots,n$.

17.4 Main Results

Next we give our first main result on the trigonometric quantitative stochastic conformable fractional approximation of stochastic processes, a pointwise result.

Theorem 17.23 *All as in Theorem 17.18 for $[a,b]=[-\pi,\pi]$. Then, $\forall\, t\in[-\pi,\pi]$, we have*

$$\left(E\left(|M(X)-X|^{q}\right)(t)\right)^{\frac{1}{q}}\leq\left(\left(E|X|^{q}\right)(t)\right)^{\frac{1}{q}}\left|\widetilde{L}(1)(t)-1\right|+$$
$$\sum_{k=1}^{n}\frac{\left(\left(E\left|X^{(k)}\right|^{q}\right)(t)\right)^{\frac{1}{q}}}{k!}(2\pi)^{k}\widetilde{L}\left(\left(\sin\left(\frac{|\cdot-t|}{4}\right)\right)^{k}\right)(t)+\frac{2^{\alpha+\frac{1}{p}}\lambda_{1}^{\frac{1}{q}}\pi^{\alpha}}{(q+1)^{\frac{\alpha}{q(\alpha+1)}}n!}\cdot \tag{17.51}$$
$$\left\{\left[\left\|\Omega_{1}\left(\mathbf{T}_{\alpha}^{t}X,2\pi\left(\frac{1}{(q+1)}\widetilde{L}\left(\left(\sin\left(\frac{|\cdot-t|}{4}\chi_{[t,\pi]}(\cdot)\right)\right)^{q(\alpha+1)}\right)(t)\right)^{\frac{1}{q(\alpha+1)}}\right)\right\|_{L^{q},[t,\pi]}\right.\right.$$
$$\left(\widetilde{L}\left(\chi_{[t,\pi]}(\cdot)\right)(t)\right)^{\frac{1}{p}}\left(\widetilde{L}\left(\left(\sin\left(\frac{|\cdot-t|}{4}\chi_{[t,\pi]}(\cdot)\right)\right)^{q(\alpha+1)}\right)(t)\right)^{\frac{\alpha}{q(\alpha+1)}}$$
$$\left.\left[\left(\widetilde{L}\left(\chi_{[t,\pi]}(\cdot)\right)(t)\right)^{\frac{1}{\alpha+1}}(q+1)^{\frac{\alpha}{(\alpha+1)}}+1\right]^{\frac{1}{q}}\right]+$$
$$\left[\left\|\Omega_{1}\left({}_{\alpha}^{t}\mathbf{T}X,2\pi\left(\frac{1}{(q+1)}\widetilde{L}\left(\left(\sin\left(\frac{|\cdot-t|}{4}\chi_{[-\pi,t]}(\cdot)\right)\right)^{q(\alpha+1)}\right)(t)\right)^{\frac{1}{q(\alpha+1)}}\right)\right\|_{L^{q},[-\pi,t]}\right.$$

$$\left(\widetilde{L}\left(\chi_{[-\pi,t]}(\cdot)\right)(t)\right)^{\frac{1}{p}}\left(\widetilde{L}\left(\left(\sin\left(\frac{|\cdot-t|}{4}\chi_{[-\pi,t]}(\cdot)\right)\right)^{q(\alpha+1)}\right)(t)\right)^{\frac{\alpha}{q(\alpha+1)}}$$

$$\left[\left(\widetilde{L}\left(\chi_{[-\pi,t]}(\cdot)\right)(t)\right)^{\frac{1}{\alpha+1}}(q+1)^{\frac{\alpha}{(\alpha+1)}}+1\right]^{\frac{1}{q}}\Bigg]\Bigg\}.$$

Proof Apply Theorem 17.18, upper bound right hand side of (17.34) by using (17.50), (17.47) and (17.48). ■

We continue with the trigonometric L_1-quantitative stochastic conformable fractional pointwise approximation of stochastic processes.

Theorem 17.24 *All as in Theorem 17.19 for $[a,b]=[-\pi,\pi]$. Then, $\forall\, t\in[-\pi,\pi]$, we have:*

$$E\left(|M(X)-X|\right)(t)\leq\left(E|X|\right)(t)\left|\widetilde{L}(1)(t)-1\right|+$$

$$\sum_{k=1}^{n}\frac{\left(E\left|X^{(k)}\right|\right)(t)}{k!}(2\pi)^{k}\,\widetilde{L}\left(\left(\sin\left(\frac{|\cdot-t|}{4}\right)\right)^{k}\right)(t)+\frac{(2\pi)^{\alpha}}{\prod_{j=1}^{n}(\beta+j)}$$

$$\left\{\left[\frac{\left(\widetilde{L}\left(\chi_{[-\pi,t]}(\cdot)\right)(t)\right)^{\frac{1}{\alpha+1}}}{\alpha-n}+\frac{1}{(\alpha+1)}\right]\left(\widetilde{L}\left(\left(\sin\left(\frac{|\cdot-t|}{4}\chi_{[-\pi,t]}(\cdot)\right)\right)^{\alpha+1}\right)(t)\right)^{\frac{\alpha}{\alpha+1}}\right.$$

$$\Omega_1\left({}_{\alpha}^{t}\mathbf{T}X,2\pi\left(\widetilde{L}\left(\left(\sin\left(\frac{|\cdot-t|}{4}\chi_{[-\pi,t]}(\cdot)\right)\right)^{\alpha+1}\right)(t)\right)^{\frac{1}{\alpha+1}}\right)_{L^1,[-\pi,t]}+$$

$$\left[\frac{\left(\widetilde{L}\left(\chi_{[t,\pi]}(\cdot)\right)(t)\right)^{\frac{1}{\alpha+1}}}{\alpha-n}+\frac{1}{(\alpha+1)}\right]\left(\widetilde{L}\left(\left(\sin\left(\frac{|\cdot-t|}{4}\chi_{[t,\pi]}(\cdot)\right)\right)^{\alpha+1}\right)(t)\right)^{\frac{\alpha}{\alpha+1}}$$

$$\left.\Omega_1\left(\mathbf{T}_{\alpha}^{t}X,2\pi\left(\widetilde{L}\left(\left(\sin\left(\frac{|\cdot-t|}{4}\chi_{[t,\pi]}(\cdot)\right)\right)^{\alpha+1}\right)(t)\right)^{\frac{1}{\alpha+1}}\right)_{L^1,[t,\pi]}\right\}. \tag{17.52}$$

Proof Apply Theorem 17.19, upper bound right hand side of (17.35) by using (17.50), (17.47) and (17.48) for $q=1$. ■

We continue with trigonometric conformable fractional uniform estimates ($\|\cdot\|_{\infty,[-\pi,\pi]}:=\|\cdot\|_{\infty}$) in L_q-mean ($1\leq q<\infty$).

Theorem 17.25 *All as in Theorem 17.20 for $[a,b]=[-\pi,\pi]$. Then*

$$\left\|E\left(|M(X)-X|^{q}\right)\right\|_{\infty}^{\frac{1}{q}}\leq\left\|E\left(|X|^{q}\right)\right\|_{\infty}^{\frac{1}{q}}\left\|\widetilde{L}(1)-1\right\|_{\infty}+$$

$$\sum_{k=1}^{n} \frac{\left\| E\left(\left| X^{(k)} \right|^{q} \right) \right\|_{\infty}^{\frac{1}{q}}}{k!} (2\pi)^{k} \left\| \widetilde{L}\left(\left(\sin\left(\frac{|\cdot - t|}{4} \right) \right)^{k} \right)(t) \right\|_{\infty} +$$

$$\frac{2^{\alpha+\frac{1}{p}} \lambda_{1}^{\frac{1}{q}} \pi^{\alpha} \left\| \widetilde{L}(1) \right\|_{\infty}^{\frac{1}{p}}}{(q+1)^{\frac{\alpha}{q(\alpha+1)}} n!} \left[\left\| \widetilde{L}(1) \right\|_{\infty}^{\frac{1}{\alpha+1}} (q+1)^{\frac{\alpha}{(\alpha+1)}} + 1 \right]^{\frac{1}{q}} \quad (17.53)$$

$$\left[\sup_{t\in[-\pi,\pi]} \Omega_{1} \left(\mathbf{T}_{\alpha}^{t} X, 2\pi \left\| \frac{1}{(q+1)} \widetilde{L}\left(\left(\sin\left(\frac{|\cdot - t|}{4} \chi_{[t,\pi]}(\cdot) \right) \right)^{q(\alpha+1)} \right)(t) \right\|_{\infty}^{\frac{1}{q(\alpha+1)}} \right)_{L^{q},[t,\pi]} \right.$$

$$\left. \left\| \widetilde{L}\left(\left(\sin\left(\frac{|\cdot - t|}{4} \chi_{[t,\pi]}(\cdot) \right) \right)^{q(\alpha+1)} \right)(t) \right\|_{\infty}^{\frac{\alpha}{q(\alpha+1)}} \right] +$$

$$\left[\sup_{t\in[-\pi,\pi]} \Omega_{1} \left({}_{\alpha}^{t}\mathbf{T} X, 2\pi \left\| \frac{1}{(q+1)} \widetilde{L}\left(\left(\sin\left(\frac{|\cdot - t|}{4} \chi_{[-\pi,t]}(\cdot) \right) \right)^{q(\alpha+1)} \right)(t) \right\|_{\infty}^{\frac{1}{q(\alpha+1)}} \right)_{L^{q},[-\pi,t]} \right.$$

$$\left. \left. \left\| \widetilde{L}\left(\left(\sin\left(\frac{|\cdot - t|}{4} \chi_{[-\pi,t]}(\cdot) \right) \right)^{q(\alpha+1)} \right)(t) \right\|_{\infty}^{\frac{\alpha}{q(\alpha+1)}} \right] \right\}.$$

Proof By (17.51). ■

Theorem 17.26 *All as in Theorem 17.21 for* $[a, b] = [-\pi, \pi]$. *Then*

$$\| E(|M(X) - X|) \|_{\infty} \leq \| E(|X|) \|_{\infty} \left\| \widetilde{L}(1) - 1 \right\|_{\infty} +$$

$$\sum_{k=1}^{n} \frac{\left\| E\left(\left| X^{(k)} \right| \right) \right\|_{\infty}}{k!} (2\pi)^{k} \left\| \widetilde{L}\left(\left(\sin\left(\frac{|\cdot - t|}{4} \right) \right)^{k} \right)(t) \right\|_{\infty} +$$

$$\frac{(2\pi)^{\alpha}}{\prod_{j=1}^{n} (\beta + j)} \left[\frac{\left\| \widetilde{L}(1) \right\|_{\infty}^{\frac{1}{\alpha+1}}}{\alpha - n} + \frac{1}{\alpha+1} \right] \quad (17.54)$$

$$\left\{ \left\| \widetilde{L}\left(\left(\sin\left(\frac{|\cdot - t|}{4} \chi_{[-\pi,t]}(\cdot) \right) \right)^{\alpha+1} \right)(t) \right\|_{\infty}^{\frac{\alpha}{\alpha+1}} \right.$$

$$\sup_{t\in[-\pi,\pi]} \Omega_{1} \left({}_{\alpha}^{t}\mathbf{T} X, 2\pi \left\| \widetilde{L}\left(\left(\sin\left(\frac{|\cdot - t|}{4} \chi_{[-\pi,t]}(\cdot) \right) \right)^{\alpha+1} \right)(t) \right\|_{\infty}^{\frac{1}{\alpha+1}} \right)_{L^{1},[-\pi,t]} +$$

$$\left\| \widetilde{L} \left(\left(\sin \left(\frac{|\cdot - t|}{4} \chi_{[t,\pi]} (\cdot) \right) \right)^{\alpha+1} \right) (t) \right\|_{\infty}^{\frac{\alpha}{\alpha+1}}$$

$$\sup_{t \in [-\pi,\pi]} \Omega_1 \left(\mathbf{T}_{\alpha}^{t} X, 2\pi \left\| \widetilde{L} \left(\left(\sin \left(\frac{|\cdot - t|}{4} \chi_{[t,\pi]} (\cdot) \right) \right)^{\alpha+1} \right) (t) \right\|_{\infty}^{\frac{1}{\alpha+1}} \right)_{L^1,[t,\pi]} \Bigg\}.$$

Proof By (17.52). ■

We continue with interesting pointwise corollaries.

Corollary 17.27 *All as in Theorem 17.18 for $[a, b] = [-\pi, \pi]$. Further assume that $\widetilde{L}(1)(t_0) = 1$, and $X^{(k)}(t_0, \omega) = 0$, $\forall\ \omega \in \Omega$, all $k = 1, \dots, n$, for a fixed $t_0 \in [-\pi, \pi]$. Then*

$$\left(E\left(|M(X) - X|^q\right)(t_0)\right)^{\frac{1}{q}} \leq \frac{2^{\alpha+\frac{1}{p}} \lambda_1^{\frac{1}{q}} \pi^{\alpha}}{(q+1)^{\frac{\alpha}{q(\alpha+1)}}\, n!} \tag{17.55}$$

$$\Bigg\{ \Bigg[\Omega_1 \left(\mathbf{T}_{\alpha}^{t_0} X, 2\pi \left(\frac{1}{(q+1)} \widetilde{L} \left(\left(\sin \left(\frac{|\cdot - t_0|}{4} \chi_{[t_0,\pi]} (\cdot) \right) \right)^{q(\alpha+1)} \right) (t_0) \right)^{\frac{1}{q(\alpha+1)}} \right)_{L^q,[t_0,\pi]}$$

$$\left(\widetilde{L}\left(\chi_{[t_0,\pi]}(\cdot)\right)(t_0)\right)^{\frac{1}{p}} \left(\widetilde{L} \left(\left(\sin \left(\frac{|\cdot - t_0|}{4} \chi_{[t_0,\pi]} (\cdot) \right) \right)^{q(\alpha+1)} \right) (t_0) \right)^{\frac{\alpha}{q(\alpha+1)}}$$

$$\left[\left(\widetilde{L}\left(\chi_{[t_0,\pi]}(\cdot)\right)(t_0)\right)^{\frac{1}{\alpha+1}} (q+1)^{\frac{\alpha}{(\alpha+1)}} + 1 \right]^{\frac{1}{q}} +$$

$$\Omega_1 \left({}_{\alpha}^{t_0}\mathbf{T} X, 2\pi \left(\frac{1}{(q+1)} \widetilde{L} \left(\left(\sin \left(\frac{|\cdot - t_0|}{4} \chi_{[-\pi,t_0]} (\cdot) \right) \right)^{q(\alpha+1)} \right) (t_0) \right)^{\frac{1}{q(\alpha+1)}} \right)_{L^q,[-\pi,t_0]}$$

$$\left(\widetilde{L}\left(\chi_{[-\pi,t_0]}(\cdot)\right)(t_0)\right)^{\frac{1}{p}} \left(\widetilde{L} \left(\left(\sin \left(\frac{|\cdot - t_0|}{4} \chi_{[-\pi,t_0]} (\cdot) \right) \right)^{q(\alpha+1)} \right) (t_0) \right)^{\frac{\alpha}{q(\alpha+1)}}$$

$$\left[\left(\widetilde{L}\left(\chi_{[-\pi,t_0]}(\cdot)\right)(t_0)\right)^{\frac{1}{\alpha+1}} (q+1)^{\frac{\alpha}{(\alpha+1)}} + 1 \right]^{\frac{1}{q}} \Bigg] \Bigg\}.$$

Proof By (17.51). ■

Corollary 17.28 *All as in Theorem 17.18 for $[a, b] = [-\pi, \pi]$. Further assume that $\widetilde{L}(1) = 1$ and $\frac{1}{q} < \alpha < 1$ (i.e. $n = 0$). Then, $\forall\ t \in [-\pi, \pi]$, we have*

$$\left(E\left(|M(X)-X|^{q}\right)(t)\right)^{\frac{1}{q}} \leq \frac{2^{\alpha+\frac{1}{p}}\pi^{\alpha}}{(p(\alpha-1)+1)^{\frac{1}{p}}(q+1)^{\frac{\alpha}{q(\alpha+1)}}} \tag{17.56}$$

$$\left\{\left[\Omega_{1}\left(T_{\alpha}^{t}X, 2\pi\left(\frac{1}{(q+1)}\widetilde{L}\left(\left(\sin\left(\frac{|\cdot-t|}{4}\chi_{[t,\pi]}(\cdot)\right)\right)^{q(\alpha+1)}\right)(t)\right)^{\frac{1}{q(\alpha+1)}}\right)_{L^{q},[t,\pi]}\right.\right.$$

$$\left(\widetilde{L}\left(\chi_{[t,\pi]}(\cdot)\right)(t)\right)^{\frac{1}{p}}\left(\widetilde{L}\left(\left(\sin\left(\frac{|\cdot-t|}{4}\chi_{[t,\pi]}(\cdot)\right)\right)^{q(\alpha+1)}\right)(t)\right)^{\frac{\alpha}{q(\alpha+1)}}$$

$$\left.\left[\left(\widetilde{L}\left(\chi_{[t,\pi]}(\cdot)\right)(t)\right)^{\frac{1}{\alpha+1}}(q+1)^{\frac{\alpha}{(\alpha+1)}}+1\right]^{\frac{1}{q}}+\right.$$

$$+\left[\Omega_{1}\left({}_{\alpha}^{t}TX, 2\pi\left(\frac{1}{(q+1)}\widetilde{L}\left(\left(\sin\left(\frac{|\cdot-t|}{4}\chi_{[-\pi,t]}(\cdot)\right)\right)^{q(\alpha+1)}\right)(t)\right)^{\frac{1}{q(\alpha+1)}}\right)_{L^{q},[-\pi,t]}\right.$$

$$\left(\widetilde{L}\left(\chi_{[-\pi,t]}(\cdot)\right)(t)\right)^{\frac{1}{p}}\left(\widetilde{L}\left(\left(\sin\left(\frac{|\cdot-t|}{4}\chi_{[-\pi,t]}(\cdot)\right)\right)^{q(\alpha+1)}\right)(t)\right)^{\frac{\alpha}{q(\alpha+1)}}$$

$$\left.\left.\left[\left(\widetilde{L}\left(\chi_{[-\pi,t]}(\cdot)\right)(t)\right)^{\frac{1}{\alpha+1}}(q+1)^{\frac{\alpha}{(\alpha+1)}}+1\right]^{\frac{1}{q}}\right]\right\}.$$

Proof By (17.51). ∎

Corollary 17.29 *All as in Theorem 17.19 for $[a,b]=[-\pi,\pi]$. Further assume that $\widetilde{L}(1)(t_0)=1$, and $X^{(k)}(t_0,\omega)=0$, $\forall\ \omega\in\Omega$, all $k=1,\dots,n$, for a fixed $t_0\in[-\pi,\pi]$. Then*

$$E(|M(X)-X|)(t_0) \leq \frac{(2\pi)^{\alpha}}{\prod_{j=1}^{n}(\beta+j)} \tag{17.57}$$

$$\left\{\left[\frac{\left(\widetilde{L}\left(\chi_{[-\pi,t_0]}(\cdot)\right)(t_0)\right)^{\frac{1}{\alpha+1}}}{\alpha-n}+\frac{1}{(\alpha+1)}\right]\right.$$

$$\left(\widetilde{L}\left(\left(\sin\left(\frac{|\cdot-t_0|}{4}\chi_{[-\pi,t_0]}(\cdot)\right)\right)^{\alpha+1}\right)(t_0)\right)^{\frac{\alpha}{\alpha+1}}$$

$$\Omega_{1}\left({}_{\alpha}^{t_0}\mathbf{T}X, 2\pi\left(\widetilde{L}\left(\left(\sin\left(\frac{|\cdot-t_0|}{4}\chi_{[-\pi,t_0]}(\cdot)\right)\right)^{\alpha+1}\right)(t_0)\right)^{\frac{1}{\alpha+1}}\right)_{L^{1},[-\pi,t_0]}+$$

$$\left[\frac{\left(\widetilde{L}\left(\chi_{[t_0,\pi]}(\cdot)\right)(t_0)\right)^{\frac{1}{\alpha+1}}}{\alpha-n}+\frac{1}{(\alpha+1)}\right]$$

$$\left(\widetilde{L}\left(\left(\sin\left(\frac{|\cdot-t_0|}{4}\chi_{[t_0,\pi]}(\cdot)\right)\right)^{\alpha+1}\right)(t_0)\right)^{\frac{\alpha}{\alpha+1}}$$

$$\left.\Omega_1\left(\mathbf{T}_{\alpha}^{t_0}X, 2\pi\left(\widetilde{L}\left(\left(\sin\left(\frac{|\cdot-t_0|}{4}\chi_{[t_0,\pi]}(\cdot)\right)\right)^{\alpha+1}\right)(t_0)\right)^{\frac{1}{\alpha+1}}\right)_{L^1,[t_0,\pi]}\right\}.$$

Proof By (17.52). ■

Corollary 17.30 *All as in Theorem 17.19 for* $[a, b] = [-\pi, \pi]$. *Further assume that* $\widetilde{L}(1) = 1$ *and* $0 < \alpha < 1$. *Then,* $\forall\, t \in [-\pi, \pi]$, *we have*

$$E\left(|M(X)-X|\right)(t) \leq (2\pi)^{\alpha} \tag{17.58}$$

$$\left\{\left[\frac{\left(\widetilde{L}\left(\chi_{[-\pi,t]}(\cdot)\right)(t)\right)^{\frac{1}{\alpha+1}}}{\alpha}+\frac{1}{(\alpha+1)}\right]\left(\widetilde{L}\left(\left(\sin\left(\frac{|\cdot-t|}{4}\chi_{[-\pi,t]}(\cdot)\right)\right)^{\alpha+1}\right)(t)\right)^{\frac{\alpha}{\alpha+1}}\right.$$

$$\Omega_1\left({}_{\alpha}^{t}TX, 2\pi\left(\widetilde{L}\left(\left(\sin\left(\frac{|\cdot-t|}{4}\chi_{[-\pi,t]}(\cdot)\right)\right)^{\alpha+1}\right)(t)\right)^{\frac{1}{\alpha+1}}\right)_{L^1,[-\pi,t]}+$$

$$\left[\frac{\left(\widetilde{L}\left(\chi_{[t,\pi]}(\cdot)\right)(t)\right)^{\frac{1}{\alpha+1}}}{\alpha}+\frac{1}{(\alpha+1)}\right]\left(\widetilde{L}\left(\left(\sin\left(\frac{|\cdot-t|}{4}\chi_{[t,\pi]}(\cdot)\right)\right)^{\alpha+1}\right)(t)\right)^{\frac{\alpha}{\alpha+1}}$$

$$\left.\Omega_1\left(T_{\alpha}^{t}X, 2\pi\left(\widetilde{L}\left(\left(\sin\left(\frac{|\cdot-t|}{4}\chi_{[t,\pi]}(\cdot)\right)\right)^{\alpha+1}\right)(t)\right)^{\frac{1}{\alpha+1}}\right)_{L^1,[t,\pi]}\right\}.$$

Proof By (17.52). ■

Corollary 17.31 *All as in Theorem 17.19 for* $[a, b] = [-\pi, \pi]$. *Further assume that* $\widetilde{L}(1) = 1$ *and* $\alpha = \frac{1}{2}$. *Then,* $\forall\, t \in [-\pi, \pi]$, *we have*

$$E\left(|M(X)-X|\right)(t) \leq \sqrt{2\pi} \tag{17.59}$$

$$\left\{\left[2\left(\widetilde{L}\left(\chi_{[-\pi,t]}(\cdot)\right)(t)\right)^{\frac{2}{3}}+\frac{2}{3}\right]\left(\widetilde{L}\left(\left(\sin\left(\frac{|\cdot-t|}{4}\chi_{[-\pi,t]}(\cdot)\right)\right)^{\frac{3}{2}}\right)(t)\right)^{\frac{1}{3}}\right.$$

$$\Omega_1\left({}_{\frac{1}{2}}^{t}TX, 2\pi\left(\widetilde{L}\left(\left(\sin\left(\frac{|\cdot-t|}{4}\chi_{[-\pi,t]}(\cdot)\right)\right)^{\frac{3}{2}}\right)(t)\right)^{\frac{2}{3}}\right)_{L^1,[-\pi,t]} +$$

$$\left[2\left(\widetilde{L}\left(\chi_{[t,\pi]}(\cdot)\right)(t)\right)^{\frac{2}{3}} + \frac{2}{3}\right]\left(\widetilde{L}\left(\left(\sin\left(\frac{|\cdot-t|}{4}\chi_{[t,\pi]}(\cdot)\right)\right)^{\frac{3}{2}}\right)(t)\right)^{\frac{1}{3}}$$

$$\left.\Omega_1\left(T_{\frac{1}{2}}^{t}X, 2\pi\left(\widetilde{L}\left(\left(\sin\left(\frac{|\cdot-t|}{4}\chi_{[t,\pi]}(\cdot)\right)\right)^{\frac{3}{2}}\right)(t)\right)^{\frac{2}{3}}\right)_{L^1,[t,\pi]}\right\}.$$

Proof By (17.58). ■

We continue with interesting uniform corollaries:

Corollary 17.32 *All as in Theorem 17.20 for* $[a, b] = [-\pi, \pi]$. *Further assume that* $\widetilde{L}(1) = 1$ *and* $\frac{1}{q} < \alpha < 1$ *(i.e.* $n = 0$*). Then*

$$\left\|E\left(|M(X) - X|^q\right)\right\|_{\infty}^{\frac{1}{q}} \leq \frac{2^{\alpha+\frac{1}{p}}\pi^{\alpha}\left\|\widetilde{L}(1)\right\|_{\infty}^{\frac{1}{p}}}{(p(\alpha-1)+1)^{\frac{1}{p}}(q+1)^{\frac{\alpha}{q(\alpha+1)}}}$$

$$\left[\left\|\widetilde{L}(1)\right\|_{\infty}^{\frac{1}{\alpha+1}}(q+1)^{\frac{\alpha}{\alpha+1}} + 1\right]^{\frac{1}{q}} \tag{17.60}$$

$$\left[\sup_{t\in[-\pi,\pi]}\Omega_1\left(T_{\alpha}^{t}X, 2\pi\left\|\frac{1}{(q+1)}\widetilde{L}\left(\left(\sin\left(\frac{|\cdot-t|}{4}\chi_{[t,\pi]}(\cdot)\right)\right)^{q(\alpha+1)}\right)(t)\right\|_{\infty}^{\frac{1}{q(\alpha+1)}}\right)_{L^q,[t,\pi]}\right.$$

$$\left.\left\|\widetilde{L}\left(\left(\sin\left(\frac{|\cdot-t|}{4}\chi_{[t,\pi]}(\cdot)\right)\right)^{q(\alpha+1)}\right)(t)\right\|_{\infty}^{\frac{\alpha}{q(\alpha+1)}}\right] +$$

$$\left[\sup_{t\in[-\pi,\pi]}\Omega_1\left({}_{\alpha}^{t}TX, 2\pi\left\|\frac{1}{(q+1)}\widetilde{L}\left(\left(\sin\left(\frac{|\cdot-t|}{4}\chi_{[-\pi,t]}(\cdot)\right)\right)^{q(\alpha+1)}\right)(t)\right\|_{\infty}^{\frac{1}{q(\alpha+1)}}\right)_{L^q,[-\pi,t]}\right.$$

$$\left.\left.\left\|\widetilde{L}\left(\left(\sin\left(\frac{|\cdot-t|}{4}\chi_{[-\pi,t]}(\cdot)\right)\right)^{q(\alpha+1)}\right)(t)\right\|_{\infty}^{\frac{\alpha}{q(\alpha+1)}}\right]\right\}.$$

Proof By (17.53). ■

Corollary 17.33 *All as in Theorem 17.21 for* $[a, b] = [-\pi, \pi]$. *Further assume that* $\widetilde{L}(1) = 1$, *and* $0 < \alpha < 1$ *(i.e.* $n = 0$*). Then*

$$\|E(|M(X)-X|)\|_\infty \le (2\pi)^\alpha \left[\frac{\|\widetilde{L}(1)\|_\infty^{\frac{1}{\alpha+1}}}{\alpha} + \frac{1}{\alpha+1}\right]$$

$$\left\{ \left\| \widetilde{L}\left(\left(\sin\left(\frac{|\cdot - t|}{4} \chi_{[-\pi,t]}(\cdot) \right) \right)^{\alpha+1} \right)(t) \right\|_\infty^{\frac{\alpha}{\alpha+1}} \right.$$

$$\sup_{t\in[-\pi,\pi]} \Omega_1 \left({}_\alpha^t TX, 2\pi \left\| \widetilde{L}\left(\left(\sin\left(\frac{|\cdot - t|}{4} \chi_{[-\pi,t]}(\cdot) \right) \right)^{\alpha+1} \right)(t) \right\|_\infty^{\frac{1}{\alpha+1}} \right)_{L^1,[-\pi,t]} \tag{17.61}$$

$$+ \left\| \widetilde{L}\left(\left(\sin\left(\frac{|\cdot - t|}{4} \chi_{[t,\pi]}(\cdot) \right) \right)^{\alpha+1} \right)(t) \right\|_\infty^{\frac{\alpha}{\alpha+1}}$$

$$\left. \sup_{t\in[-\pi,\pi]} \Omega_1 \left(T_\alpha^t X, 2\pi \left\| \widetilde{L}\left(\left(\sin\left(\frac{|\cdot - t|}{4} \chi_{[t,\pi]}(\cdot) \right) \right)^{\alpha+1} \right)(t) \right\|_\infty^{\frac{1}{\alpha+1}} \right)_{L^1,[t,\pi]} \right\}.$$

Proof By (17.54). ■

Corollary 17.34 *All as in Theorem 17.21 for* $[a,b]=[-\pi,\pi]$. *Further assume that* $\widetilde{L}(1)=1$, *and* $\alpha=\frac{1}{2}$. *Then*

$$\|E(|M(X)-X|)\|_\infty \le \sqrt{2\pi}\left[2\|\widetilde{L}(1)\|_\infty^{\frac{2}{3}} + \frac{2}{3}\right]$$

$$\left\{ \left\| \widetilde{L}\left(\left(\sin\left(\frac{|\cdot - t|}{4} \chi_{[-\pi,t]}(\cdot) \right) \right)^{\frac{3}{2}} \right)(t) \right\|_\infty^{\frac{1}{3}} \right.$$

$$\sup_{t\in[-\pi,\pi]} \Omega_1 \left({}_{\frac{1}{2}}^t TX, 2\pi \left\| \widetilde{L}\left(\left(\sin\left(\frac{|\cdot - t|}{4} \chi_{[-\pi,t]}(\cdot) \right) \right)^{\frac{3}{2}} \right)(t) \right\|_\infty^{\frac{2}{3}} \right)_{L^1,[-\pi,t]} \tag{17.62}$$

$$+ \left\| \widetilde{L}\left(\left(\sin\left(\frac{|\cdot - t|}{4} \chi_{[t,\pi]}(\cdot) \right) \right)^{\frac{3}{2}} \right)(t) \right\|_\infty^{\frac{1}{3}}$$

$$\left. \sup_{t\in[-\pi,\pi]} \Omega_1 \left(T_{\frac{1}{2}}^t X, 2\pi \left\| \widetilde{L}\left(\left(\sin\left(\frac{|\cdot - t|}{4} \chi_{[t,\pi]}(\cdot) \right) \right)^{\frac{3}{2}} \right)(t) \right\|_\infty^{\frac{2}{3}} \right)_{L^1,[t,\pi]} \right\}.$$

Proof By (17.61). ∎

17.5 Application

Consider the Bernstein polynomials on $[-\pi, \pi]$ for $f \in C([-\pi, \pi])$:

$$B_N(f)(x) = \sum_{k=0}^{N} \binom{N}{k} f\left(-\pi + \frac{2\pi k}{N}\right) \left(\frac{x+\pi}{2\pi}\right)^k \left(\frac{\pi - x}{2\pi}\right)^{N-k}, \tag{17.63}$$

$N \in \mathbb{N}$, any $x \in [-\pi, \pi]$. There are positive linear operators from $C([-\pi, \pi])$ into itself.

Setting $g(t) = f(2\pi t - \pi)$, $t \in [0, 1]$, we have $g(0) = f(-\pi)$, $g(1) = f(\pi)$, and

$$(B_N g)(t) = \sum_{k=0}^{N} \binom{N}{k} g\left(\frac{k}{N}\right) t^k (1-t)^{N-k} = (B_N f)(x), \quad x \in [-\pi, \pi]. \tag{17.64}$$

Here $x = \varphi(t) = 2\pi t - \pi$ is an $1-1$ and onto map from $[0, 1]$ onto $[-\pi, \pi]$. Clearly here $g \in C([0, 1])$.

Notice also that

$$\left(B_N\left((\cdot - x)^2\right)\right)(x) = \left[\left(B_N\left((\cdot - t)^2\right)\right)(t)\right](2\pi)^2 = \frac{(2\pi)^2}{N} t(1-t)$$

$$= \frac{(2\pi)^2}{N}\left(\frac{x+\pi}{2\pi}\right)\left(\frac{\pi - x}{2\pi}\right) = \frac{1}{N}(x+\pi)(\pi - x) \le \frac{\pi^2}{N}, \quad \forall\, x \in [-\pi, \pi].$$

I.e.

$$\left(B_N\left((\cdot - x)^2\right)\right)(x) \le \frac{\pi^2}{N}, \quad \forall\, x \in [-\pi, \pi]. \tag{17.65}$$

In particular

$$(B_N 1)(x) = 1, \quad \forall\, x \in [-\pi, \pi]. \tag{17.66}$$

Define the corresponding application of M by

$$\widetilde{B}_N(X)(t, \omega) := B_N(X(\cdot, \omega))(t) =$$

$$\sum_{k=0}^{N} \binom{N}{k} X\left(-\pi + \frac{2\pi k}{N}, \omega\right) \left(\frac{t+\pi}{2\pi}\right)^k \left(\frac{\pi - t}{2\pi}\right)^{N-k}, \tag{17.67}$$

$\forall\, N \in \mathbb{N}, \forall\, t \in [-\pi, \pi], \forall\, \omega \in \Omega$, where X is a stochastic process. Clearly $\widetilde{B}_N$ is a stochastic process.

We give

Proposition 17.35 *Let $X(t, \omega)$ be a stochastic process from $[-\pi, \pi] \times (\Omega, \mathcal{F}, P)$ into $\mathbb{R}$, where $(\Omega, \mathcal{F}, P)$ is a probability space. Here $0 < \alpha < 1$ (i.e. $n = 0$) and $X(\cdot, \omega) \in C^1([-\pi, \pi])$, $\forall\, \omega \in \Omega$ and $X^{(1)}(t, \cdot)$ is measurable over $(\Omega, \mathcal{F})$, $\forall\, t \in [-\pi, \pi]$. For any $t \in [-\pi, \pi]$ we assume that $T_\alpha^t X(z, \omega)$ is continuous in $z \in [t, \pi]$, uniformly with respect to $\omega \in \Omega$. And for any $t \in [-\pi, \pi]$ we assume that ${}_\alpha^t T X(z, \omega)$ is continuous in $z \in [-\pi, t]$, uniformly with respect to $\omega \in \Omega$. Finally, we assume that $(E|X|)(t) < \infty$, $\forall\, t \in [-\pi, \pi]$. Then, for any $t \in [-\pi, \pi]$, we have:*

$$E\left(\left|\widetilde{B}_N(X) - X\right|\right)(t) \leq (2\pi)^\alpha$$

$$\left\{\left[\frac{\left(B_N\left(\chi_{[-\pi,t]}(\cdot)\right)(t)\right)^{\frac{1}{\alpha+1}}}{\alpha} + \frac{1}{(\alpha+1)}\right]\left(B_N\left(\left(\sin\left(\frac{|\cdot - t|}{4}\chi_{[-\pi,t]}(\cdot)\right)\right)^{\alpha+1}\right)(t)\right)^{\frac{\alpha}{\alpha+1}}\right.$$

$$\Omega_1\left({}_\alpha^t T X, 2\pi\left(B_N\left(\left(\sin\left(\frac{|\cdot - t|}{4}\chi_{[-\pi,t]}(\cdot)\right)\right)^{\alpha+1}\right)(t)\right)^{\frac{1}{\alpha+1}}\right)_{L^1,[-\pi,t]} +$$

$$\left[\frac{\left(B_N\left(\chi_{[t,\pi]}(\cdot)\right)(t)\right)^{\frac{1}{\alpha+1}}}{\alpha} + \frac{1}{(\alpha+1)}\right]\left(B_N\left(\left(\sin\left(\frac{|\cdot - t|}{4}\chi_{[t,\pi]}(\cdot)\right)\right)^{\alpha+1}\right)(t)\right)^{\frac{\alpha}{\alpha+1}}$$

$$\left.\Omega_1\left(T_\alpha^t X, 2\pi\left(B_N\left(\left(\sin\left(\frac{|\cdot - t|}{4}\chi_{[t,\pi]}(\cdot)\right)\right)^{\alpha+1}\right)(t)\right)^{\frac{1}{\alpha+1}}\right)_{L^1,[t,\pi]}\right\}, \tag{17.68}$$

$\forall\, N \in \mathbb{N}$.

Proof By Corollary 17.30. ∎

We give

Proposition 17.36 *All as in Proposition 17.35 with $\alpha = \frac{1}{2}$. Then, for any $t \in [-\pi, \pi]$, we have:*

$$E\left(\left|\widetilde{B}_N(X) - X\right|\right)(t) \leq \sqrt{2\pi}$$

$$\left\{\left[2\left(B_N\left(\chi_{[-\pi,t]}(\cdot)\right)(t)\right)^{\frac{2}{3}} + \frac{2}{3}\right]\left(B_N\left(\left(\sin\left(\frac{|\cdot - t|}{4}\chi_{[-\pi,t]}(\cdot)\right)\right)^{\frac{3}{2}}\right)(t)\right)^{\frac{1}{3}}\right.$$

$$\Omega_1\left({}_{\frac{1}{2}}^t T X, 2\pi\left(B_N\left(\left(\sin\left(\frac{|\cdot - t|}{4}\chi_{[-\pi,t]}(\cdot)\right)\right)^{\frac{3}{2}}\right)(t)\right)^{\frac{2}{3}}\right)_{L^1,[-\pi,t]} +$$

$$\left[2\left(B_N\left(\chi_{[t,\pi]}(\cdot)\right)(t)\right)^{\frac{2}{3}}+\frac{2}{3}\right]\left(B_N\left(\left(\sin\left(\frac{|\cdot-t|}{4}\chi_{[t,\pi]}(\cdot)\right)\right)^{\frac{3}{2}}\right)(t)\right)^{\frac{1}{3}}$$

$$\left.\Omega_1\left(T_{\frac{1}{2}}^{t}X,2\pi\left(B_N\left(\left(\sin\left(\frac{|\cdot-t|}{4}\chi_{[t,\pi]}(\cdot)\right)\right)^{\frac{3}{2}}\right)(t)\right)^{\frac{2}{3}}\right)_{L^1,[t,\pi]}\right\}, \tag{17.69}$$

$\forall\, N\in\mathbb{N}$.

Proof By Proposition 17.35. ∎

We continue with

Proposition 17.37 *All as in Proposition 17.35 with* $\alpha=\frac{1}{2}$ *. Then, for any* $t\in[-\pi,\pi]$ *, we have:*

$$E\left(\left|\widetilde{B}_N(X)-X\right|\right)(t)\leq\frac{5\sqrt{2\pi}}{3}\left(B_N\left(\left(\sin\left(\frac{|\cdot-t|}{4}\right)\right)^{\frac{3}{2}}\right)(t)\right)^{\frac{1}{3}}$$

$$\left[\Omega_1\left({}_{\frac{1}{2}}^{t}TX,2\pi\left(B_N\left(\left(\sin\left(\frac{|\cdot-t|}{4}\right)\right)^{\frac{3}{2}}\right)(t)\right)^{\frac{2}{3}}\right)_{L^1,[-\pi,t]}+\right.$$

$$\left.\Omega_1\left(T_{\frac{1}{2}}^{t}X,2\pi\left(B_N\left(\left(\sin\left(\frac{|\cdot-t|}{4}\right)\right)^{\frac{3}{2}}\right)(t)\right)^{\frac{2}{3}}\right)_{L^1,[t,\pi]}\right], \tag{17.70}$$

$\forall\, N\in\mathbb{N}$.

Proof By (17.69) and the positivity of B_N, see also (17.42) and (17.44). ∎

We make

Remark 17.38 By $|\sin x|<|x|$, $\forall\, x\in\mathbb{R}-\{0\}$, in particular $\sin x\leq x$, for $x\geq 0$, we get

$$\left(\sin\left(\frac{|\cdot-t|}{4}\right)\right)^{\frac{3}{2}}\leq\left(\frac{|\cdot-t|}{4}\right)^{\frac{3}{2}}=\frac{1}{8}|\cdot-t|^{\frac{3}{2}}.$$

Hence

$$\left\|B_N\left(\left(\sin\left(\frac{|\cdot-t|}{4}\right)\right)^{\frac{3}{2}}\right)(t)\right\|_\infty\leq\frac{1}{8}\left\|B_N\left(|\cdot-t|^{\frac{3}{2}}\right)(t)\right\|_\infty. \tag{17.71}$$

We observe that

$$B_N\left(|\cdot - t|^{\frac{3}{2}}\right)(t) = \sum_{k=0}^{N}\left|t+\pi-\frac{2\pi k}{N}\right|^{\frac{3}{2}}\binom{N}{k}\left(\frac{t+\pi}{2\pi}\right)^k\left(\frac{\pi-t}{2\pi}\right)^{N-k}$$

(by discrete Hölder's inequality)

$$\leq \left[\sum_{k=0}^{N}\left(t+\pi-\frac{2\pi k}{N}\right)^2\binom{N}{k}\left(\frac{t+\pi}{2\pi}\right)^k\left(\frac{\pi-t}{2\pi}\right)^{N-k}\right]^{\frac{3}{4}}$$

$$= \left(B_N\left((\cdot - t)^2\right)(t)\right)^{\frac{3}{4}} \overset{(17.65.)}{\leq} \frac{\pi^{\frac{3}{2}}}{N^{\frac{3}{4}}},\ \forall\, t \in [-\pi, \pi]. \tag{17.72}$$

Consequently it holds

$$\left\| B_N\left(|\cdot - t|^{\frac{3}{2}}\right)(t)\right\|_\infty \leq \frac{\pi^{\frac{3}{2}}}{N^{\frac{3}{4}}}, \tag{17.73}$$

and

$$\left\| B_N\left(\left(\sin\left(\frac{|\cdot - t|}{4}\right)\right)^{\frac{3}{2}}\right)(t)\right\|_\infty \leq \frac{\pi^{\frac{3}{2}}}{8N^{\frac{3}{4}}},\quad \forall\, N \in \mathbb{N}. \tag{17.74}$$

We further have

Proposition 17.39 *All as in Proposition 17.35 with $\alpha = \frac{1}{2}$. Then, for any $t \in [-\pi, \pi]$ we have:*

$$E\left(\left|\widetilde{B}_N(X) - X\right|\right)(t) \leq \frac{5\sqrt{2}\pi}{6\sqrt[4]{N}}$$

$$\left[\Omega_1\left({}_{\frac{1}{2}}^{t}TX, \frac{\pi^2}{2\sqrt{N}}\right)_{L^1,[-\pi,t]} + \Omega_1\left(T_{\frac{1}{2}}^{t}X, \frac{\pi^2}{2\sqrt{N}}\right)_{L^1,[t,\pi]}\right], \tag{17.75}$$

$\forall\, N \in \mathbb{N}$.

As $N \to +\infty$, we get $E\left(\left|\widetilde{B}_N(X) - X\right|\right)(t) \to 0$.

Proof By positivity of B_N, (17.70) and (17.74). See also Proposition 17.6. ■

Consequently we obtain

Proposition 17.40 *All as in Proposition 17.35 with $\alpha = \frac{1}{2}$. Assume further that $\left|X^{(1)}(t, \omega)\right| \leq M^*$, $\forall\, (t, \omega) \in [-\pi, \pi] \times \Omega$, where $M^* > 0$. Then*

$$\left\| E\left(\left|\widetilde{B}_N(X) - X\right|\right)\right\|_\infty \leq \frac{5\sqrt{2}\pi}{6\sqrt[4]{N}} \tag{17.76}$$

$$\left[\sup_{t\in[-\pi,\pi]} \Omega_1 \left({}_{\frac{1}{2}}^{t} TX, \frac{\pi^2}{2\sqrt{N}} \right)_{L^1,[-\pi,t]} + \sup_{t\in[-\pi,\pi]} \Omega_1 \left(T_{\frac{1}{2}}^{t} X, \frac{\pi^2}{2\sqrt{N}} \right)_{L^1,[t,\pi]} \right],$$

$\forall\, N \in \mathbb{N}$.

As $N \to +\infty$, *then* $\left\| E\left(\left| \widetilde{B}_N (X) - X \right| \right) \right\|_\infty \to 0$, *i.e.* $\widetilde{B}_N \to I$ *(stochastic unit operator) in* 1*-mean.*

Proof By (17.75) and Remark 17.9, see (17.17), (17.18). ■

17.6 Trigonometric Conformable Fractional Stochastic Korovkin Results

In this section $\widetilde{L}$, M are meant as sequences of operators.

We give first pointwise results:

Theorem 17.41 *Here all as in Theorem 17.23. Assume further that* $\widetilde{L}(1)(t) \to 1$ *and* $\widetilde{L}\left(\left(\sin\left(\frac{|\cdot - t|}{4} \right) \right)^{q(\alpha+1)} \right)(t) \to 0$, *pointwise in* $t \in [-\pi, \pi]$.

Then $E\left(|M(X) - X|^q \right)(t) \to 0$, *pointwise in* $t \in [-\pi, \pi]$, *that is* $M \to I$ *(stochastic unit operator) in* q*-mean-pointwise with rates, quantitatively.*

Proof We use (17.51), we take into account $\widetilde{L}(1)(t) \to 1$, (17.39); and $\widetilde{L}\left(\chi_{[t,\pi]}(\cdot) \right)(t)$, $\widetilde{L}\left(\chi_{[-\pi,t]}(\cdot) \right)(t) \leq \widetilde{L}(1)(t)$, which $\widetilde{L}(1)(t)$ is bounded, and by (17.42), (17.44) and the positivity of $\widetilde{L}$ we get that

$$\widetilde{L}\left(\left(\sin\left(\tfrac{|\cdot - t|}{4} \chi_{[t,\pi]}(\cdot) \right) \right)^{q(\alpha+1)} \right)(t),\ \widetilde{L}\left(\left(\sin\left(\tfrac{|\cdot - t|}{4} \chi_{[-\pi,t]}(\cdot) \right) \right)^{q(\alpha+1)} \right)(t)$$
$$\leq \widetilde{L}\left(\left(\sin\left(\tfrac{|\cdot - t|}{4} \right) \right)^{q(\alpha+1)} \right)(t) \to 0.$$

Finally, we use Proposition 17.7(ii) for the $\Omega_1(\cdot,\cdot)$'s to go to zero. ■

We continue with

Theorem 17.42 *Here all as in Theorem 17.24. Assume further that* $\widetilde{L}(1)(t) \to 1$ *and* $\widetilde{L}\left(\left(\sin\left(\frac{|\cdot - t|}{4} \right) \right)^{(\alpha+1)} \right)(t) \to 0$, *pointwise in* $t \in [-\pi, \pi]$.

Then $E(|M(X) - X|)(t) \to 0$, *pointwise in* $t \in [-\pi, \pi]$, *that is* $M \to I$ *in* 1*-mean-pointwise with rates, quantitatively.*

Proof Based on (17.52), similar to the proof of Theorem 17.41, just take $q = 1$ there. ■

Next we give uniform results:

Theorem 17.43 *Here all as in Theorem 17.25. Assume further that* $\widetilde{L}(1) \to 1$, *uniformly, and* $\left\| \widetilde{L}\left(\left(\sin\left(\frac{|\cdot - t|}{4}\right)\right)^{q(\alpha+1)}\right)(t) \right\|_{\infty} \to 0.$

Then $\left\| E\left(|M(X) - X|^{q}\right) \right\|_{\infty} \to 0$ *over* $[-\pi, \pi]$, *that is* $M \to I$ *in the* q*-mean, quantitatively with rates.*

Proof We use (17.53), we take into account $\widetilde{L}(1) \to 1$ uniformly, (17.40); $\left\| \widetilde{L}(1) \right\|_{\infty}$ is bounded, use of (17.43), (17.45) and Remark 17.9, see there (17.17), (17.18). ∎

Next we give the L_1-mean uniform result

Theorem 17.44 *Here all as in Theorem 17.26. Assume further that* $\widetilde{L}(1) \to 1$, *uniformly, and* $\left\| \widetilde{L}\left(\left(\sin\left(\frac{|\cdot - t|}{4}\right)\right)^{(\alpha+1)}\right)(t) \right\|_{\infty} \to 0.$

Then $\| E(|M(X) - X|) \|_{\infty} \to 0$ *over* $[-\pi, \pi]$, *that is* $M \to I$ *in the* 1*-mean, quantitatively with rates.*

Proof Use of (17.54), similar to the proof of Theorem 17.43, just take $q = 1$ there. ∎

We finish with

Remark 17.45 An amazing fact/observation follows: In all trigonometric convergence results here, see Theorems 17.41–17.44, the forcing conditions for convergences are based only on $\widetilde{L}$ and basic real valued continuous functions on $[-\pi, \pi]$ and are not related to stochastic processes, but they are giving trigonometric convergence results on stochastic processes!

References

1. Abdeljawad, T.: On conformable fractional calculus. J. Comput. Appl. Math. **279**, 57–66 (2015)
2. Anastassiou, G.A.: Korovkin inequalities for stochastic processes. J. Math. Anal. Appl. **157**(2), 366–384 (1991)
3. Anastassiou, G.A.: Moments in Probability and Approximation Theory. Pitman/Longman, # 287, UK (1993)
4. Anastassiou, G.: Quantitative Approximations. Chapman & Hall/CRC, Boca Raton (2001)
5. Anastassiou, G.: Stochastic Korovkin theory given quantitatively, Facta Universitatis (Nis). Ser. Math. Inform. **22**(1), 43–60 (2007)
6. Anastassiou, G.: Nonlinearity: Ordinary and Fractional Approximations by Sublinear and Max-product operators. Springer, Heidelberg (2018)
7. Anastassiou, G.: Conformable fractional approximation of stochastic processes (2020). Submitted for publication
8. Anastassiou, G.: Trigonometric conformable fractional quantitative approximation of stochastic processes. In: Progress in Fractional Differentiation and Applications (2020). Accepted

9. Khalil, R., Al Horani, M., Yousef, A., Sababheh, M.: A new definition of fractional derivative. J. Comput. Appl. Math. **264**, 65–70 (2014)
10. Korovkin, P.P.: Linear Operators and Approximation Theory. Hindustan Publ. Corp., Delhi, India (1960)
11. Royden, H.L.: Real Analysis, 2nd edn. MacMillan Publishing Co., Inc., New York (1968)
12. Shisha, O., Mond, B.: The degree of approximation to periodic functions by linear positive Operators. J. Approx. Theory **1**, 335–339 (1968)
13. Weba, M.: Korovkin systems of stochastic processes. Math. Z. **192**(1), 73–80 (1986)
14. Weba, M.: A quantitative Korovkin theorem for random functions with multivariate domains. J. Approx. Theory **61**(1), 74–87 (1990)

Chapter 18
Commutative Caputo Fractional Korovkin Approximation for Stochastic Processes

Here we consider and study expectation commutative stochastic positive linear operators acting on L^1-continuous stochastic processes which are Caputo fractional differentiable. Under some mild, general and natural assumptions on the stochastic processes we produce related Caputo fractional stochastic Shisha-Mond type inequalities pointwise and uniform. All convergences are produced with rates and are given by the fractional stochastic inequalities involving the first modulus of continuity of the expectation of the αth right and left fractional derivatives of the engaged stochastic process, $\alpha > 0$, $\alpha \notin \mathbb{N}$. The amazing fact here is that the basic real Korovkin test functions assumptions impose the conclusions of our Caputo fractional stochastic Korovkin theory. We include also a detailed application to stochastic Bernstein operators. See also [11].

18.1 Introduction

Our work is motivated by the following:

Korovkin's Theorem [13, 1960] Let $\left(T_j\right)_{j\in\mathbb{N}}$ be a sequence of positive linear operators from $C([a,b])$ into itself, $[a,b] \subset \mathbb{R}$. In order to have $\lim_{j\to\infty} \left(T_j f\right)(t) = f(t)$ (in the sup-norm) for all $f \in C([a,b])$, it is enough to prove it for $f_0(t) = 1$, $f_1(t) = t$ and $f_2(t) = t^2$. The rate of the above convergence for arbitrary $f \in C([a,b])$ can be determined exactly from the rates of convergence for f_0, f_1, f_2.

The above theorem was put in an inequality form:

Shisha-Mond inequality [15] We have

$$\left\|T_j(f) - f\right\| \leq \|f\| \cdot \left\|T_j(1) - 1\right\| + \omega_1\left(f, \rho_j\right) \cdot \left(1 + \left\|T_j(1)\right\|\right),$$

where

$$\rho_j = \left(\left\|T_j\left((x-y)^2\right)(y)\right\|\right)^{\frac{1}{2}}.$$

G. A. Anastassiou, *Generalized Fractional Calculus*, Studies in Systems, Decision and Control 305, https://doi.org/10.1007/978-3-030-56962-4_18

In the last inequality $\|\cdot\|$ stands for the supremum norm and ω_1 for the first modulus of continuity. This inequality gives the rate of convergence of T_j to the unit operator I.

Anastassiou in [2–4] established a series of sharp inequalities for various cases of the parameters of the problem. However, Weba in [16–19] was the first, among many workers in quantitative results of Shisha-Mond type, to produce inequalities for stochastic processes. He assumed that T_j are E-commutative (E means expectation) and stochastically simple. According to his work, if a stochastic process $X(t, \omega)$, $t \in Q$—a compact convex subset of a real normed vector space, $\omega \in Q$—probability space, is to be approximated by positive linear operators T_j, then the maximal error in the qth mean is ($q \geq 1$)

$$\left\|T_j X - X\right\| = \sup_{t \in Q} \left(E\left|\left(T_j X\right)(t, \omega) - X(t, \omega)\right|^q\right)^{\frac{1}{q}}.$$

So, Weba established upper bounds for $\left\|T_j X - X\right\|$ involving his own natural general first modulus of continuity of X with several interesting applications.

Anastassiou met [5] the pointwise case of $q = 1$. Without stochastic simplicity of T_j he found nearly best and best upper bounds for $\left|E\left(T_j X\right)(x_0) - (EX)(x_0)\right|$, $x_0 \in Q$.

The author here continues his above work on the approximation of stochastic processes, now at the Caputo stochastic fractional level. He derives pointwise and uniform Caputo fractional stochastic Shisha-Mond type inequalities, see the main Theorems 18.4, 18.7 and the several related corollaries. He gives an extensive application to stochastic Bernstein operators. He finishes with a pointwise and a uniform fractional stochastic Korovkin theorem, derived by Theorems 18.4, 18.7. The stochastic convergences, about stochastic processes, of our fractional Korovkin Theorems 18.15, 18.16 are enforced only by the convergences of real basic non-stochastic functions.

18.2 Background

We need

Definition 18.1 *([10])* Let non-integer $\alpha > 0$, $n = \lceil \alpha \rceil$ ($\lceil \cdot \rceil$ is the ceiling of the number), $t \in [a, b] \subset \mathbb{R}$, $\omega \in \Omega$, where $(\Omega, \mathcal{F}, P)$ is a general probability space. Here $X(t, \omega)$ stands for a stochastic process. Assume that $X(\cdot, \omega) \in AC^n([a, b])$ (spaces of functions $X(\cdot, \omega)$ with $X^{(n-1)}(\cdot, \omega) \in AC([a, b])$ absolutely continuous functions), $\forall\, \omega \in \Omega$.

We call stochastic left Caputo fractional derivative

$$D_{*a}^{\alpha} X(x, \omega) = \frac{1}{\Gamma(n - \alpha)} \int_a^x (x - t)^{n-\alpha-1} X^{(n)}(t, \omega)\, dt, \tag{18.1}$$

$\forall\, x \in [a, b]$, $\forall\, \omega \in \Omega$.

And, we call stochastic right Caputo fractional derivative

$$D_{b-}^{\alpha}X(x,\omega)=\frac{(-1)^{n}}{\Gamma(n-\alpha)}\int_{x}^{b}(z-x)^{n-\alpha-1}X^{(n)}(z,\omega)\,dz, \tag{18.2}$$

$\forall\, x\in[a,b], \forall\, \omega\in\Omega$. Above Γ stands for the gamma function.

We make

Remark 18.2 *(to Definition* 18.1*)* We further assume here that

$$\left|X^{(n)}(t,\omega)\right|\leq M, \forall\ (t,\omega)\in[a,b]\times\Omega,$$

where $M>0$.

Then, by (18.1), we have

$$\left|D_{*a}^{\alpha}X(x,\omega)\right|\leq\frac{1}{\Gamma(n-\alpha)}\int_{a}^{x}(x-t)^{n-\alpha-1}\left|X^{(n)}(t,\omega)\right|dt\leq$$

$$\frac{M}{\Gamma(n-\alpha)}\int_{a}^{x}(x-t)^{n-\alpha-1}\,dt=\frac{M(x-a)^{n-\alpha}}{\Gamma(n-\alpha+1)}.$$

That is

$$\left|D_{*a}^{\alpha}X(x,\omega)\right|\leq\frac{M(x-a)^{n-\alpha}}{\Gamma(n-\alpha+1)},\quad\forall\, x\in[a,b],\ \text{any}\ \omega\in\Omega. \tag{18.3}$$

Also, from (18.2) we get

$$\left|D_{b-}^{\alpha}X(x,\omega)\right|\leq\frac{1}{\Gamma(n-\alpha)}\int_{x}^{b}(z-x)^{n-\alpha-1}\left|X^{(n)}(z,\omega)\right|dz\leq$$

$$\frac{M}{\Gamma(n-\alpha)}\int_{x}^{b}(z-x)^{n-\alpha-1}\,dz=\frac{M(b-x)^{n-\alpha}}{\Gamma(n-\alpha+1)}.$$

That is

$$\left|D_{b-}^{\alpha}X(x,\omega)\right|\leq\frac{M(b-x)^{n-\alpha}}{\Gamma(n-\alpha+1)},\quad\forall\, x\in[a,b],\ \text{any}\ \omega\in\Omega. \tag{18.4}$$

It is not strange to assume that $D_{*a}^{\alpha}X$, $D_{b-}^{\alpha}X$ are stochastic processes.

By [7, p. 388], we get that $D_{*a}^{\alpha}X(\cdot,\omega)\in C([a,b]), \forall\,\omega\in\Omega$. And by [8], we get that $D_{b-}^{\alpha}X(\cdot,\omega)\in C([a,b]), \forall\,\omega\in\Omega$.

Similarly, we obtain

$$\left|D_{*t}^{\alpha}X(x,\omega)\right| \leq \frac{M(x-t)^{n-\alpha}}{\Gamma(n-\alpha+1)} \leq \frac{M(b-t)^{n-\alpha}}{\Gamma(n-\alpha+1)}, \tag{18.5}$$

$\forall\, x \in [t,b]$, any $t \in [a,b]$, $\forall\, \omega \in \Omega$,

and

$$\left|D_{t-}^{\alpha}X(x,\omega)\right| \leq \frac{M(t-x)^{n-\alpha}}{\Gamma(n-\alpha+1)} \leq \frac{M(t-a)^{n-\alpha}}{\Gamma(n-\alpha+1)}, \tag{18.6}$$

$\forall\, x \in [a,t]$, any $t \in [a,b]$, $\forall\, \omega \in \Omega$.

Above $D_{*t}^{\alpha}X$, $D_{t-}^{\alpha}X$ are assumed to be stochastic processes for any $t \in [a,b]$, and it holds $D_{*t}^{\alpha}X(\cdot,\omega) \in C([t,b])$, $D_{t-}^{\alpha}X(\cdot,\omega) \in C([a,t])$, $\forall\, \omega \in \Omega$.

Clearly, then

$$\left|E\left(D_{*t}^{\alpha}X\right)(x)\right| \leq \frac{M(b-t)^{n-\alpha}}{\Gamma(n-\alpha+1)}, \tag{18.7}$$

$\forall\, x \in [t,b]$, any $t \in [a,b]$, where E is the expectation operator $(EX)(t) = \int_{\Omega} X(t,\omega)\,P(d\omega)$ and similarly,

$$\left|E\left(D_{t-}^{\alpha}X\right)(x)\right| \leq \frac{M(t-a)^{n-\alpha}}{\Gamma(n-\alpha+1)}, \tag{18.8}$$

$\forall\, x \in [a,t]$, any $t \in [a,b]$.

We observe that the first modulus of continuity ($\delta > 0$)

$$\omega_1\left(E\left(D_{*t}^{\alpha}X\right),\delta\right)_{[t,b]} := \sup_{\substack{x,y\in[t,b]:\\ |x-y|\leq\delta}} \left|E\left(D_{*t}^{\alpha}X\right)(x) - E\left(D_{*t}^{\alpha}X\right)(y)\right|$$

$$\overset{(18.7)}{\leq} \frac{2M(b-t)^{n-\alpha}}{\Gamma(n-\alpha+1)}, \tag{18.9}$$

any $t \in [a,b]$.

Hence, it holds ($\delta > 0$)

$$\sup_{t\in[a,b]} \omega_1\left(E\left(D_{*t}^{\alpha}X\right),\delta\right)_{[t,b]} \leq \frac{2M(b-a)^{n-\alpha}}{\Gamma(n-\alpha+1)}. \tag{18.10}$$

Similarly, it holds ($\delta > 0$)

$$\sup_{t\in[a,b]} \omega_1\left(E\left(D_{t-}^{\alpha}X\right),\delta\right)_{[a,t]} \leq \frac{2M(b-a)^{n-\alpha}}{\Gamma(n-\alpha+1)}. \tag{18.11}$$

By [6, p. 209], we have that ($\delta_1 > 0$)

$$\left|E\left(D_{*t}^{\alpha}X\right)(z) - E\left(D_{*t}^{\alpha}X\right)(t)\right| \leq \omega_1\left(E\left(D_{*t}^{\alpha}X\right), \delta_1\right)_{[t,b]} \left\lceil \frac{|z-t|}{\delta_1} \right\rceil \leq$$

$$\omega_1\left(E\left(D_{*t}^{\alpha}X\right), \delta_1\right)_{[t,b]} \left(1 + \frac{(z-t)}{\delta_1}\right), \tag{18.12}$$

$\forall\, z \in [t, b]$,
and similarly ($\delta_2 > 0$),

$$\left|E\left(D_{t-}^{\alpha}X\right)(z) - E\left(D_{t-}^{\alpha}X\right)(t)\right| \leq \omega_1\left(E\left(D_{t-}^{\alpha}X\right), \delta_2\right)_{[a,t]} \left(1 + \frac{t-z}{\delta_2}\right), \tag{18.13}$$

$\forall\, z \in [a, t]$.
We also set

$$\omega_1\left(E\left(D_t^{\alpha}X\right), \delta\right) := \max\left\{\omega_1\left(E\left(D_{*t}^{\alpha}X\right), \delta\right)_{[t,b]}, \omega_1\left(E\left(D_{t-}^{\alpha}X\right), \delta\right)_{[a,t]}\right\}, \tag{18.14}$$

where $\delta > 0$.
We make

Remark 18.3 Let the positive linear operator L mapping $C([a, b])$ into $B([a, b])$ (the bounded functions). By the Riesz representation theorem [14] we have that there exists μ_t unique, completed Borel measure on $[a, b]$ with

$$\mu_t([a, b]) = L(1)(t) > 0, \tag{18.15}$$

such that

$$L(f)(t) = \int_{[a,b]} f(s)\, d\mu_t(s), \quad \forall\, t \in [a, b], \forall\, f \in C([a, b]). \tag{18.16}$$

Let now $n = \lceil \alpha \rceil$, $\alpha \notin \mathbb{N}$, $\alpha > 0$, $k = 1, \ldots, n-1$. Then by Hölder's inequality we obtain

$$\left|\int_{[a,b]} (s-t)^k\, d\mu_t(s)\right| \leq \int_{[a,b]} |s-t|^k\, d\mu_t(s) \leq$$

$$\left(\int_{[a,b]} |s-t|^{\alpha+1}\, d\mu_t(s)\right)^{\left(\frac{k}{\alpha+1}\right)} \left(\mu_t([a, b])\right)^{\left(\frac{\alpha+1-k}{\alpha+1}\right)}. \tag{18.17}$$

The last means

$$\left|L\left((s-t)^k\right)(t)\right| \leq L\left(|s-t|^k\right)(t) \leq \left(L\left(|s-t|^{\alpha+1}\right)(t)\right)^{\left(\frac{k}{\alpha+1}\right)} \left(L(1)(t)\right)^{\left(\frac{\alpha+1-k}{\alpha+1}\right)}, \tag{18.18}$$

all $k = 1, \ldots, n-1$. It is clear that

$$\left\| L\left(|s-t|^{\alpha+1}\right)(t)\right\|_{\infty,[a,b]} < \infty.$$

Furthermore we derive

$$\left\| L\left((s-t)^{k}\right)(t)\right\|_{\infty,[a,b]} \leq \left\| L\left(|s-t|^{k}\right)(t)\right\|_{\infty,[a,b]} \leq$$

$$\left\| L\left(|s-t|^{\alpha+1}\right)(t)\right\|_{\infty,[a,b]}^{\left(\frac{k}{\alpha+1}\right)} \left\| L(1)\right\|_{\infty,[a,b]}^{\left(\frac{\alpha+1-k}{\alpha+1}\right)}, \tag{18.19}$$

all $k = 1, \ldots, n-1$.

From now on we will denote $\|\cdot\|_{\infty,[a,b]} = \|\cdot\|_{\infty}$ the supremum norm.

18.3 Preliminaries

Let $(\Omega, \mathcal{F}, P)$ be a probabilistic space and $L^1(\Omega, \mathcal{F}, P)$ be the space of all real-valued random variables $Y = Y(\omega)$ with

$$\int_{\Omega} |Y(\omega)|\, P(d\omega) < \infty.$$

Let $X = X(t,\omega)$ denote a stochastic process with index set $[a,b] \subset \mathbb{R}$ and real state space $(\mathbb{R}, \mathcal{B})$, where $\mathcal{B}$ is the σ-field of Borel subsets of $\mathbb{R}$. Here $C([a,b])$ is the space of continuous real-valued functions on $[a,b]$ and $\mathcal{B}([a,b])$ is the space of bounded real-valued functions on $[a,b]$. Also $C_{\Omega}([a,b]) = C\left([a,b], L^1(\Omega, \mathcal{F}, P)\right)$ is the space of L^1-continuous stochastic processes in t and $B_{\Omega}([a,b]) = \left\{X : \sup_{t\in[a,b]} \int_{\Omega} |X(t,\omega)|\, P(d\omega) < \infty\right\}$, obviously $C_{\Omega}([a,b]) \subset B_{\Omega}([a,b])$.

Let $\alpha > 0$, $\alpha \notin \mathbb{N}$, $\lceil \alpha \rceil = n$, and consider the subspace of stochastic processes $C_{\Omega}^{\alpha,n}([a,b]) := \{X : X(\cdot,\omega) \in AC^n([a,b]),\ \forall\ \omega \in \Omega$ and $\left|X^{(n)}(t,\omega)\right| \leq M$, $\forall$ $(t,\omega) \in [a,b] \times \Omega$, where $M > 0$; $X^{(k)}(t,\omega) \in C_{\Omega}([a,b])$, $k = 0, 1, \ldots, n-1$; also $D_{*t}^{\alpha} X$, $D_{t-}^{\alpha} X$ are stochastic processes for any $t \in [a,b]\}$. That is, for every $\omega \in \Omega$ we have $X(t,\omega) \in C^{n-1}([a,b])$.

Consider the linear operator

$$L : C_{\Omega}([a,b]) \hookrightarrow B_{\Omega}([a,b]).$$

If $X \in C_{\Omega}([a,b])$ is nonnegative and LX, too, then L is called positive. If $EL = LE$, then L is called E-commutative.

18.4 Main Results

Following 3. Preliminaries we state

Theorem 18.4 *Consider the positive E-commutative linear operator* $L : C_{\Omega}([a,b]) \hookrightarrow B_{\Omega}([a,b])$, *and* $\alpha > 0$, $\alpha \notin \mathbb{N}$, $\lceil \alpha \rceil = n$, *and let* $X \in C_{\Omega}^{\alpha,n}([a,b])$, *with* $\delta > 0$.

Then

$$|(E(LX))(t) - (EX)(t)| \leq |(EX)(t)| \, |(L(1))(t) - 1| + \tag{18.20}$$

$$\sum_{k=1}^{n-1} \frac{\left|\left(EX^{(k)}\right)(t)\right|}{k!} \left|L\left((s-t)^k\right)(t)\right| + \frac{\omega_1\left(E\left(D_t^{\alpha}X\right), \delta\right)}{\Gamma(\alpha+1)}$$

$$\left(L\left(|s-t|^{\alpha+1}\right)(t)\right)^{\frac{\alpha}{\alpha+1}} \left[(L(1)(t))^{\frac{1}{\alpha+1}} + \frac{\left(L\left(|s-t|^{\alpha+1}\right)(t)\right)^{\frac{1}{\alpha+1}}}{\delta(\alpha+1)}\right],$$

$\forall \, t \in [a,b]$.

Above $\omega_1\left(E\left(D_t^{\alpha}X\right), \delta\right)$ *is as in (18.14).*

Proof We have that

$$D_{*t}^{\alpha}X(t,\omega) = D_{t-}^{\alpha}X(t,\omega) = 0, \tag{18.21}$$

$\forall \, \omega \in \Omega$, see [9, pp. 358–359].

We also assume

$$D_{*t}^{\alpha}X(s,\omega) = 0, \text{ for } s < t,$$

and

$$D_{t-}^{\alpha}X(s,\omega) = 0, \text{ for } s > t,$$

$\forall \, \omega \in \Omega$.

We get by left Caputo fractional Taylor's formula that [12, p. 54]

$$X(s,\omega) = \sum_{k=0}^{n-1} \frac{X^{(k)}(t,\omega)}{k!}(s-t)^k + \tag{18.22}$$

$$\frac{1}{\Gamma(\alpha)} \int_t^s (s-z)^{\alpha-1} \left(D_{*t}^{\alpha}X(z,\omega) - D_{*t}^{\alpha}X(t,\omega)\right) dz,$$

for all $t \leq s \leq b$, $\forall \, \omega \in \Omega$.

Also, from [9, p. 341], using the right Caputo fractional Taylor formula we get

$$X(s,\omega)=\sum_{k=0}^{n-1}\frac{X^{(k)}(t,\omega)}{k!}(s-t)^{k}+ \tag{18.23}$$

$$\frac{1}{\Gamma(\alpha)}\int_{s}^{t}(z-s)^{\alpha-1}\left(D_{t-}^{\alpha}X(z,\omega)-D_{t-}^{\alpha}X(t,\omega)\right)dz,$$

for all $a\leq s\leq t$, $\forall\,\omega\in\Omega$.

Therefore we get

$$(EX)(s)=\sum_{k=0}^{n-1}\frac{\left(EX^{(k)}\right)(t)}{k!}(s-t)^{k}+$$

$$\frac{1}{\Gamma(\alpha)}E\left(\int_{t}^{s}(s-z)^{\alpha-1}\left(D_{*t}^{\alpha}X(z,\omega)-D_{*t}^{\alpha}X(t,\omega)\right)dz\right), \tag{18.24}$$

all $t\leq s\leq b$,

and

$$(EX)(s)=\sum_{k=0}^{n-1}\frac{\left(EX^{(k)}\right)(t)}{k!}(s-t)^{k}+$$

$$\frac{1}{\Gamma(\alpha)}E\left(\int_{s}^{t}(z-s)^{\alpha-1}\left(D_{t-}^{\alpha}X(z,\omega)-D_{t-}^{\alpha}X(t,\omega)\right)dz\right), \tag{18.25}$$

all $a\leq s\leq t$.

By [1, p. 156, Theorem 20.15], the functions $D_{*t}^{\alpha}X$ and $D_{t-}^{\alpha}X$ are jointly measurable.

By (18.5) and (18.6) we obtain that $E\left(\int_{t}^{s}(s-z)^{\alpha-1}\left|D_{*t}^{\alpha}X(z,\omega)-D_{*t}^{\alpha}X(t,\omega)\right|dz\right)$, $E\left(\int_{s}^{t}(z-s)^{\alpha-1}\left|D_{t-}^{\alpha}X(z,\omega)-D_{t-}^{\alpha}X(t,\omega)\right|dz\right)$ are finite.

Therefore, by Fubini-Tonelli's theorem [14], we get

$$(EX)(s)=\sum_{k=0}^{n-1}\frac{\left(EX^{(k)}\right)(t)}{k!}(s-t)^{k}+$$

$$\frac{1}{\Gamma(\alpha)}\int_{t}^{s}(s-z)^{\alpha-1}\left(E\left(D_{*t}^{\alpha}X\right)(z)-E\left(D_{*t}^{\alpha}X\right)(t)\right)dz, \tag{18.26}$$

all $t\leq s\leq b$,

and

$$(EX)(s) = \sum_{k=0}^{n-1} \frac{\left(EX^{(k)}\right)(t)}{k!}(s-t)^k +$$

$$\frac{1}{\Gamma(\alpha)} \int_s^t (z-s)^{\alpha-1} \left(E\left(D_{t-}^{\alpha} X\right)(z) - E\left(D_{t-}^{\alpha} X\right)(t)\right) dz, \tag{18.27}$$

all $a \leq s \leq t$.

Call the remainders

$$R_1(t,s) := \frac{1}{\Gamma(\alpha)} \int_t^s (s-z)^{\alpha-1} \left(E\left(D_{*t}^{\alpha} X\right)(z) - E\left(D_{*t}^{\alpha} X\right)(t)\right) dz, \tag{18.28}$$

all $t \leq s \leq b$,

and

$$R_2(s,t) = \frac{1}{\Gamma(\alpha)} \int_s^t (z-s)^{\alpha-1} \left(E\left(D_{t-}^{\alpha} X\right)(z) - E\left(D_{t-}^{\alpha} X\right)(t)\right) dz, \tag{18.29}$$

all $a \leq s \leq t$.

We observe that ($t \leq s \leq b$, $\delta_1 > 0$)

$$|R_1(t,s)| \leq \frac{1}{\Gamma(\alpha)} \int_t^s (s-z)^{\alpha-1} \left|E\left(D_{*t}^{\alpha} X\right)(z) - E\left(D_{*t}^{\alpha} X\right)(t)\right| dz \overset{(18.12)}{\leq}$$

$$\frac{\omega_1\left(E\left(D_{*t}^{\alpha} X\right), \delta_1\right)_{[t,b]}}{\Gamma(\alpha)} \int_t^s (s-z)^{\alpha-1} \left(1 + \frac{(z-t)}{\delta_1}\right) dz =$$

$$\frac{\omega_1\left(E\left(D_{*t}^{\alpha} X\right), \delta_1\right)_{[t,b]}}{\Gamma(\alpha)} \left[\int_t^s (s-z)^{\alpha-1} dz + \frac{1}{\delta_1} \int_t^s (s-z)^{\alpha-1} (z-t)^{2-1} dz\right] =$$

$$\frac{\omega_1\left(E\left(D_{*t}^{\alpha} X\right), \delta_1\right)_{[t,b]}}{\Gamma(\alpha)} \left[\frac{(s-t)^{\alpha}}{\alpha} + \frac{1}{\delta_1} \frac{\Gamma(\alpha)\Gamma(2)}{\Gamma(\alpha+2)} (s-t)^{\alpha+1}\right] = \tag{18.30}$$

$$\frac{\omega_1\left(E\left(D_{*t}^{\alpha} X\right), \delta_1\right)_{[t,b]}}{\Gamma(\alpha+1)} \left[(s-t)^{\alpha} + \frac{(s-t)^{\alpha+1}}{\delta_1(\alpha+1)}\right].$$

That is

$$|R_1(t,s)| \leq \frac{\omega_1\left(E\left(D_{*t}^{\alpha} X\right), \delta_1\right)_{[t,b]}}{\Gamma(\alpha+1)} \left[(s-t)^{\alpha} + \frac{(s-t)^{\alpha+1}}{\delta_1(\alpha+1)}\right], \tag{18.31}$$

where $t \leq s \leq b$, any $t \in [a,b]$, $\delta_1 > 0$.

Similarly, we have that ($a \leq s \leq t$, $\delta_2 > 0$)

$$\left|R_2(s,t)\right| \leq \frac{1}{\Gamma(\alpha)} \int_s^t (z-s)^{\alpha-1} \left|E\left(D_{t-}^{\alpha}X\right)(z) - E\left(D_{t-}^{\alpha}X\right)(t)\right| dz \overset{(18.13)}{\leq}$$

$$\frac{\omega_1\left(E\left(D_{t-}^{\alpha}X\right),\delta_2\right)_{[a,t]}}{\Gamma(\alpha)} \int_s^t (z-s)^{\alpha-1}\left(1+\frac{t-z}{\delta_2}\right) dz =$$

$$\frac{\omega_1\left(E\left(D_{t-}^{\alpha}X\right),\delta_2\right)_{[a,t]}}{\Gamma(\alpha)} \left[\int_s^t (z-s)^{\alpha-1} dz + \frac{1}{\delta_2}\int_s^t (t-z)^{2-1}(z-s)^{\alpha-1} dz\right] \overset{(18.32)}{=}$$

$$\frac{\omega_1\left(E\left(D_{t-}^{\alpha}X\right),\delta_2\right)_{[a,t]}}{\Gamma(\alpha)} \left[\frac{(t-s)^{\alpha}}{\alpha} + \frac{1}{\delta_2}\frac{\Gamma(2)\Gamma(\alpha)}{\Gamma(\alpha+2)}(t-s)^{\alpha+1}\right] =$$

$$\frac{\omega_1\left(E\left(D_{t-}^{\alpha}X\right),\delta_2\right)_{[a,t]}}{\Gamma(\alpha+1)} \left[(t-s)^{\alpha} + \frac{(t-s)^{\alpha+1}}{\delta_2(\alpha+1)}\right].$$

That is

$$\left|R_2(s,t)\right| \leq \frac{\omega_1\left(E\left(D_{t-}^{\alpha}X\right),\delta_2\right)_{[a,t]}}{\Gamma(\alpha+1)} \left[(t-s)^{\alpha} + \frac{(t-s)^{\alpha+1}}{\delta_2(\alpha+1)}\right], \tag{18.33}$$

where $a \leq s \leq t$, any $t \in [a,b]$, $\delta_2 > 0$.

So, we have

$$(EX)(s) - (EX)(t) = \sum_{k=1}^{n-1} \frac{\left(EX^{(k)}\right)(t)}{k!}(s-t)^k + R_1(t,s), \tag{18.34}$$

all $t \leq s \leq b$,

and

$$(EX)(s) - (EX)(t) = \sum_{k=1}^{n-1} \frac{\left(EX^{(k)}\right)(t)}{k!}(s-t)^k + R_2(s,t), \tag{18.35}$$

all $a \leq s \leq t$.

From now on we take $\delta_1 = \delta_2 =: \delta > 0$.

Therefore, it holds

$$\left|R_1(t,s)\right|, \left|R_2(s,t)\right| \leq \frac{\omega_1\left(E\left(D_t^{\alpha}X\right),\delta\right)}{\Gamma(\alpha+1)} \left[|t-s|^{\alpha} + \frac{|t-s|^{\alpha+1}}{\delta(\alpha+1)}\right], \tag{18.36}$$

for any $s,t \in [a,b]$.

We have that

$$L\left(EX\right)(t)-\left(EX\right)(t)\,L\left(1\right)(t)\overset{(18.16)}{=}$$

$$\int_{[a,b]}\left(EX\right)(s)\,d\mu_t\left(s\right)-\left(EX\right)(t)\,L\left(1\right)(t)\overset{(18.15)}{=}$$

$$\int_{[a,t)}\left(EX\right)(s)\,d\mu_t\left(s\right)+\int_{[t,b]}\left(EX\right)(s)\,d\mu_t\left(s\right)-$$

$$\int_{[a,t)}\left(EX\right)(t)\,d\mu_t\left(s\right)-\int_{[t,b]}\left(EX\right)(t)\,d\mu_t\left(s\right)=$$

$$\int_{[a,t)}\left(\left(EX\right)(s)-\left(EX\right)(t)\right)d\mu_t\left(s\right)+\int_{[t,b]}\left(\left(EX\right)(s)-\left(EX\right)(t)\right)d\mu_t\left(s\right)\overset{\text{(by (18.34), (18.35))}}{=}$$

$$\sum_{k=1}^{n-1}\frac{\left(EX^{(k)}\right)(t)}{k!}L\left((s-t)^k\right)(t)+\int_{[a,t)}R_2\left(s,t\right)d\mu_t\left(s\right)+\int_{[t,b]}R_1\left(t,s\right)d\mu_t\left(s\right). \tag{18.37}$$

That is

$$L\left(EX\right)(t)-\left(EX\right)(t)\,L\left(1\right)(t)=\sum_{k=1}^{n-1}\frac{\left(EX^{(k)}\right)(t)}{k!}L\left((s-t)^k\right)(t)$$

$$+\int_{[a,t)}R_2\left(s,t\right)d\mu_t\left(s\right)+\int_{[t,b]}R_1\left(t,s\right)d\mu_t\left(s\right). \tag{18.38}$$

From [18, pp. 3–5] we have the following results

(i) $C\left([a,b]\right)\subset C_\Omega\left([a,b]\right)$,

(ii) if $X\in C_\Omega\left([a,b]\right)$, then $EX\in C\left([a,b]\right)$,

and

(iii) if L is E-commutative, then L maps the subspace $C\left([a,b]\right)$ into $B\left([a,b]\right)$.

One can rewrite

$$L\left(EX\right)(t)-\left(EX\right)(t)=\left(EX\right)(t)\left[\left(L\left(1\right)\right)(t)-1\right]+$$

$$\sum_{k=1}^{n-1}\frac{\left(EX^{(k)}\right)(t)}{k!}L\left((s-t)^k\right)(t)+\int_{[a,t)}R_2\left(s,t\right)d\mu_t\left(s\right) \tag{18.39}$$

$$+\int_{[t,b]}R_1\left(t,s\right)d\mu_t\left(s\right).$$

Consequently, by E-commutativity of L we find

$$|E[(LX)(t,\omega)-X(t,\omega)]| \leq |(EX)(t)|\,|(L(1))(t)-1| +$$

$$\sum_{k=1}^{n-1}\frac{\left|\left(EX^{(k)}\right)(t)\right|}{k!}\left|L\left((s-t)^k\right)(t)\right| + \int_{[a,t)}|R_2(s,t)|\,d\mu_t(s)$$

$$+\int_{[t,b]}|R_1(t,s)|\,d\mu_t(s) \overset{(18.36)}{\leq}$$

$$|(EX)(t)|\,|(L(1))(t)-1| + \sum_{k=1}^{n-1}\frac{\left|\left(EX^{(k)}\right)(t)\right|}{k!}\left|L\left((s-t)^k\right)(t)\right| +$$

$$\frac{\omega_1\left(E\left(D_t^{\alpha}X\right),\delta\right)}{\Gamma(\alpha+1)}\left[\int_{[a,b]}|t-s|^{\alpha}\,d\mu_t(s) + \frac{1}{\delta(\alpha+1)}\int_{[a,b]}|t-s|^{\alpha+1}\,d\mu_t(s)\right] \leq$$

(by Hölder's inequality)

$$|(EX)(t)|\,|(L(1))(t)-1| + \sum_{k=1}^{n-1}\frac{\left|\left(EX^{(k)}\right)(t)\right|}{k!}\left|L\left((s-t)^k\right)(t)\right| +$$

$$\frac{\omega_1\left(E\left(D_t^{\alpha}X\right),\delta\right)}{\Gamma(\alpha+1)}\left[\left(\int_{[a,b]}|t-s|^{\alpha+1}\,d\mu_t(s)\right)^{\frac{\alpha}{\alpha+1}}(L(1)(t))^{\frac{1}{\alpha+1}} \right. \tag{18.40}$$

$$\left. +\frac{1}{\delta(\alpha+1)}\int_{[a,b]}|t-s|^{\alpha+1}\,d\mu_t(s)\right].$$

We have proved that

$$|(E(LX))(t)-(EX)(t)| \leq |(EX)(t)|\,|(L(1))(t)-1| +$$

$$\sum_{k=1}^{n-1}\frac{\left|\left(EX^{(k)}\right)(t)\right|}{k!}\left|L\left((s-t)^k\right)(t)\right| + \frac{\omega_1\left(E\left(D_t^{\alpha}X\right),\delta\right)}{\Gamma(\alpha+1)}$$

$$\left[\left(L\left(|s-t|^{\alpha+1}\right)(t)\right)^{\frac{\alpha}{\alpha+1}}(L(1)(t))^{\frac{1}{\alpha+1}} + \frac{1}{\delta(\alpha+1)}\left(L\left(|s-t|^{\alpha+1}\right)(t)\right)\right] = \tag{18.41}$$

$$|(EX)(t)|\,|(L(1))(t)-1| + \sum_{k=1}^{n-1}\frac{\left|\left(EX^{(k)}\right)(t)\right|}{k!}\left|L\left((s-t)^k\right)(t)\right| +$$

$$\frac{\omega_1\left(E\left(D_t^{\alpha}X\right),\delta\right)}{\Gamma(\alpha+1)}\left(L\left(|s-t|^{\alpha+1}\right)(t)\right)^{\frac{\alpha}{\alpha+1}}$$

$$\left[(L(1)(t))^{\frac{1}{\alpha+1}}+\frac{\left(L\left(|s-t|^{\alpha+1}\right)(t)\right)^{\frac{1}{\alpha+1}}}{\delta(\alpha+1)}\right].$$

The theorem now is valid. ■

We need

Definition 18.5 If $0<\alpha<1$, then $n=1$, and $C_{\Omega}^{\alpha,1}([a,b]):=\{X:X(\cdot,\omega)\in AC([a,b]),\ \forall\,\omega\in\Omega$ and $\left|X^{(1)}(t,\omega)\right|\leq M,\ \forall\,(t,\omega)\in[a,b]\times\Omega$, where $M>0$; $X(t,\omega)\in C_{\Omega}([a,b])$; also $D_{*t}^{\alpha}X, D_{t-}^{\alpha}X$ are stochastic processes for any $t\in[a,b]\}$.

We give

Corollary 18.6 *Consider the positive E-commutative linear operator $L:C_{\Omega}([a,b])\hookrightarrow B_{\Omega}([a,b])$, and $0<\alpha<1$ and let $X\in C_{\Omega}^{\alpha,1}([a,b])$, with $\delta>0$. Then*

$$|(E(LX))(t)-(EX)(t)|\leq|(EX)(t)|\,|(L(1))(t)-1|+$$

$$\frac{\omega_1\left(E\left(D_t^{\alpha}X\right),\delta\right)}{\Gamma(\alpha+1)}\left(L\left(|s-t|^{\alpha+1}\right)(t)\right)^{\frac{\alpha}{\alpha+1}}$$

$$\left[(L(1)(t))^{\frac{1}{\alpha+1}}+\frac{\left(L\left(|s-t|^{\alpha+1}\right)(t)\right)^{\frac{1}{\alpha+1}}}{\delta(\alpha+1)}\right],\tag{18.42}$$

$\forall\,t\in[a,b]$.

Proof By Theorem 18.4. ■

We further present

Theorem 18.7 *All as in Theorem 18.4. Then*

$$\|E(LX)-EX\|_{\infty}\leq\|EX\|_{\infty}\|L(1)-1\|_{\infty}+$$

$$\sum_{k=1}^{n-1}\frac{\left\|EX^{(k)}\right\|_{\infty}}{k!}\left\|L\left((s-t)^k\right)(t)\right\|_{\infty}+\sup_{t\in[a,b]}\frac{\omega_1\left(E\left(D_t^{\alpha}X\right),\delta\right)}{\Gamma(\alpha+1)}$$

$$\left\|L\left(|s-t|^{\alpha+1}\right)(t)\right\|_{\infty}^{\frac{\alpha}{\alpha+1}}\left[\|L(1)\|_{\infty}^{\frac{1}{\alpha+1}}+\frac{\left\|L\left(|s-t|^{\alpha+1}\right)(t)\right\|_{\infty}^{\frac{1}{\alpha+1}}}{\delta(\alpha+1)}\right]<\infty.\tag{18.43}$$

Proof By Theorem 18.4, see (18.10), (18.11) and (18.14). ■

Corollary 18.8 *All as in Corollary 18.6. Then*

$$\|E(LX) - EX\|_\infty \leq \|EX\|_\infty \|L(1) - 1\|_\infty +$$

$$\sup_{t\in[a,b]} \frac{\omega_1\left(E\left(D_t^\alpha X\right), \delta\right)}{\Gamma(\alpha+1)} \left\|L\left(|s-t|^{\alpha+1}\right)(t)\right\|_\infty^{\frac{\alpha}{\alpha+1}}$$

$$\left[\|L(1)\|_\infty^{\frac{1}{\alpha+1}} + \frac{\left\|L\left(|s-t|^{\alpha+1}\right)(t)\right\|_\infty^{\frac{1}{\alpha+1}}}{\delta(\alpha+1)}\right]. \tag{18.44}$$

Proof By Corollary 18.6. ■

Corollary 18.9 *All as in Corollary 18.6, and* $L(1) = 1$. *Then*

$$\|E(LX) - EX\|_\infty \leq \sup_{t\in[a,b]} \frac{\omega_1\left(E\left(D_t^\alpha X\right), \delta\right)}{\Gamma(\alpha+1)}$$

$$\left\|L\left(|s-t|^{\alpha+1}\right)(t)\right\|_\infty^{\frac{\alpha}{\alpha+1}} \left[1 + \frac{\left\|L\left(|s-t|^{\alpha+1}\right)(t)\right\|_\infty^{\frac{1}{\alpha+1}}}{\delta(\alpha+1)}\right]. \tag{18.45}$$

Proof By (18.44). ■

We give

Corollary 18.10 *All as in Corollary 18.6, and* $L(1) = 1$. *Then*

$$\|E(LX) - EX\|_\infty \leq \frac{2\sup\limits_{t\in[a,b]} \omega_1\left(E\left(D_t^\alpha X\right), \frac{1}{(\alpha+1)}\left\|L\left(|s-t|^{\alpha+1}\right)(t)\right\|_\infty^{\frac{1}{\alpha+1}}\right)}{\Gamma(\alpha+1)}$$

$$\left\|L\left(|s-t|^{\alpha+1}\right)(t)\right\|_\infty^{\frac{\alpha}{\alpha+1}}. \tag{18.46}$$

Proof By (18.45): we take there

$$\delta = \frac{1}{(\alpha+1)} \left\|L\left(|s-t|^{\alpha+1}\right)(t)\right\|_\infty^{\frac{1}{\alpha+1}} > 0.$$

In case of $\left\|L\left(|s-t|^{\alpha+1}\right)(t)\right\|_\infty = 0$, we have that $L\left(|s-t|^{\alpha+1}\right)(t) = 0$, $\forall\, t \in [a,b]$. That is, by (18.16) $\int_{[a,b]} |s-t|^{\alpha+1} d\mu_t(s) = 0$, $\forall\, t \in [a,b]$, where μ_t is a probability measure, any $t \in [a,b]$.

The last implies $|s-t|^{\alpha+1}=0$, a.e., hence $|s-t|=0$, a.e., thus $s=t$, a.e., which means $\mu_t\{s\in[a,b]:s\neq t\}=0$, i.e. $\mu_t=\delta_t$, $\forall\, t\in[a,b]$, where δ_t is the unit Dirac measure.

Consequently we have $E(LX)(t)=L(EX)(t)=\int_{[a,b]}(EX)(s)\,d\delta_t(s)=(EX)(t)$, $\forall\, t\in[a,b]$.

That is $E(LX)=EX$ over $[a,b]$. Therefore both sides of inequality (18.46) equal to zero.

Hence (18.46) is always true. ■

18.5 Application

Let $f\in C([0,1])$ and the Bernstein polynomials

$$B_N(f)(t):=\sum_{k=0}^{N}f\left(\frac{k}{N}\right)\binom{N}{k}t^k(1-t)^{N-k}, \tag{18.47}$$

$\forall\, t\in[0,1]$, $\forall\, N\in\mathbb{N}$.

We have that $B_N1=1$ and B_N is a positive linear operator.

We have that

$$B_N\left((\cdot-t)^2\right)(t)=\frac{t(1-t)}{N},\ \forall\, t\in[0,1], \tag{18.48}$$

and

$$\left\|B_N\left((\cdot-t)^2\right)(t)\right\|_\infty^{\frac{1}{2}}\leq\frac{1}{2\sqrt{N}},\ \forall\, N\in\mathbb{N}. \tag{18.49}$$

Define the corresponding stochastic application of B_N by

$$B_N(X)(t,\omega):=B_N(X(\cdot,\omega))(t)=\sum_{k=0}^{N}X\left(\frac{k}{N},\omega\right)\binom{N}{k}t^k(1-t)^{N-k}, \tag{18.50}$$

$\forall\, t\in[0,1]$, $\forall\,\omega\in\Omega$, $N\in\mathbb{N}$, where X is a stochastic process. Clearly $B_N(X)$ is a stochastic process and $B_N:C_\Omega([0,1])\hookrightarrow C_\Omega([0,1])$. Notice that

$$(EB_N(X))(t)=\sum_{k=0}^{N}(EX)\left(\frac{k}{N}\right)\binom{N}{k}t^k(1-t)^{N-k}=(B_N(EX))(t), \tag{18.51}$$

$\forall\, t\in[0,1]$.

That is $EB_N=B_NE$, i.e. B_N is an E-commutative positive linear operator.

We give

Proposition 18.11 *Let* $0 < \alpha < 1$ *and* $X \in C_{\Omega}^{\alpha,1}([0,1])$. *Then*

$$\|E(B_N X) - EX\|_{\infty} \leq \frac{2 \sup_{t\in[0,1]} \omega_1 \left(E\left(D_t^{\alpha} X\right), \frac{1}{(\alpha+1)} \left\| B_N\left(|s-t|^{\alpha+1}\right)(t) \right\|_{\infty}^{\frac{1}{\alpha+1}} \right)}{\Gamma(\alpha+1)}$$

$$\left\| B_N\left(|s-t|^{\alpha+1}\right)(t) \right\|_{\infty}^{\frac{\alpha}{\alpha+1}}. \tag{18.52}$$

Proof By Corollary 18.10. ∎

In particular we get:

Corollary 18.12 *Let* $X \in C_{\Omega}^{\frac{1}{2},1}([0,1])$. *Then*

$$\|E(B_N X) - EX\|_{\infty} \leq \frac{4}{\sqrt{\pi}} \sup_{t\in[0,1]} \omega_1 \left(E\left(D_t^{\frac{1}{2}} X\right), \frac{2}{3} \left\| B_N\left(|s-t|^{\frac{3}{2}}\right)(t) \right\|_{\infty}^{\frac{2}{3}} \right)$$

$$\left\| B_N\left(|s-t|^{\frac{3}{2}}\right)(t) \right\|_{\infty}^{\frac{1}{3}}, \quad \forall\, N \in \mathbb{N}. \tag{18.53}$$

Proof Apply (18.52) for $\alpha = \frac{1}{2}$. ∎

We make

Remark 18.13 We notice that

$$B_N\left(|s-t|^{\frac{3}{2}}\right)(t) = \sum_{k=0}^{N} \left| t - \frac{k}{N} \right|^{\frac{3}{2}} \binom{N}{k} t^k (1-t)^{N-k}$$

(by discrete Hölder's inequality)

$$\leq \left(\sum_{k=0}^{N} \left| t - \frac{k}{N} \right|^{2} \binom{N}{k} t^k (1-t)^{N-k} \right)^{\frac{3}{4}} \tag{18.54}$$

$$\overset{(18.48)}{=} \left(\frac{1}{N} t (1-t) \right)^{\frac{3}{4}} \leq \frac{1}{(4N)^{\frac{3}{4}}}, \quad \forall\, t \in [0,1].$$

That is

$$\left\| B_N\left(|s-t|^{\frac{3}{2}}\right)(t) \right\|_{\infty} \leq \frac{1}{(4N)^{\frac{3}{4}}}, \tag{18.55}$$

and

$$\left\| B_N\left(|s-t|^{\frac{3}{2}}\right)(t) \right\|_{\infty}^{\frac{1}{3}} \leq \frac{1}{(4N)^{\frac{1}{4}}}, \tag{18.56}$$

and

$$\left\| B_N \left(|s-t|^{\frac{3}{2}} \right)(t) \right\|_{\infty}^{\frac{2}{3}} \leq \frac{1}{2\sqrt{N}}, \tag{18.57}$$

$\forall\, N \in \mathbb{N}$.

We derive

Proposition 18.14 *Let $X \in C_{\Omega}^{\frac{1}{2},1}([0,1])$. Then*

$$\|E(B_N X) - EX\|_{\infty} \leq \frac{\left(\sqrt{2}\right)^3}{\sqrt{\pi}\sqrt[4]{N}} \sup_{t\in[0,1]} \omega_1 \left(E\left(D_t^{\frac{1}{2}} X \right), \frac{1}{3\sqrt{N}} \right), \tag{18.58}$$

$\forall\, N \in \mathbb{N}$.

Hence $\lim_{N\to\infty} E(B_N X) = EX$, *uniformly.*

Proof By (18.53) and Remark 18.13. ■

18.6 Caputo Fractional Stochastic Korovkin Theory

Here L is meant as a sequence of positive E-commutative linear operators and all assumptions are as in Theorem 18.4.

We give

Theorem 18.15 *We further assume that $L(1)(t) \to 1$ and $L\left(|s-t|^{\alpha+1}\right)(t) \to 0$, then $(E(LX))(t) \to (EX)(t)$, for any $X \in C_{\Omega}^{\alpha,n}([a,b])$, $\forall\, t \in [a,b]$, a pointwise convergence; where $\alpha > 0$, $\alpha \notin \mathbb{N}$, $\lceil \alpha \rceil = n$.*

Proof Based on (18.20), (18.18), and that $L(1)(t)$ is bounded as a sequence of functions. Also $\omega_1\left(E\left(D_t^{\alpha} X\right), \delta\right)$ is bounded, see (18.10), (18.11) and (18.14). ■

We continue with

Theorem 18.16 *We further assume that $L(1)(t) \to 1$, uniformly and $\left\|L\left(|s-t|^{\alpha+1}\right)(t)\right\|_{\infty} \to 0$, then $E(LX) \to EX$, uniformly over $[a,b]$, for any $X \in C_{\Omega}^{\alpha,n}([a,b])$; where $\alpha > 0$, $\alpha \notin \mathbb{N}$, $\lceil \alpha \rceil = n$.*

Proof Based on (18.43), (18.19), and that $\|L(1)\|_{\infty}$ is bounded. Also it is $\sup_{t\in[a,b]} \omega_1\left(E\left(D_t^{\alpha} X\right), \delta\right) < \infty$, by (18.10), (18.11) and (18.14). ■

We finish with

Remark 18.17 The stochastic convergences of Theorems 18.15, 18.16 are derived by the convergences of the basic and simple real non-stochastic functions $\left\{1, |s-t|^{\alpha+1}\right\}$, an amazing fact!

References

1. Aliprantis, C., Burkinshaw, O.: Principles of Real Analysis, 3rd edn. Academic, San Diego (1998)
2. Anastassiou, G.: A study of positive linear operators by the method of moments, one-dimensional case. J. Approx. Theory **45**, 247–270 (1985)
3. Anastassiou, G.: Korovkin type inequalities in real normed vector spaces. Approx. Theory Appl. **2**, 39–53 (1986)
4. Anastassiou, G.: Multi-dimensional quantitative results for probability measures approximating the unit measure. Approx. Theory Appl. **2**, 93–103 (1986)
5. Anastassiou, G.A.: Korovkin inequalities for stochastic processes. J. Math. Anal. Appl. **157**(2), 366–384 (1991)
6. Anastassiou, G.A.: Moments in Probability and Approximation Theory. Pitman/Longman, # 287, UK (1993)
7. Anastassiou, G.: Fractional Differentiation Inequalities. Springer, Heildelberg (2009)
8. Anastassiou, G.: Fractional representation formulae and right fractional inequalities. Math. Comput. Model. **54**(11–12), 3098–3115 (2011)
9. Anastassiou, G.: Intelligent Mathematics: Computational Analysis. Springer, Heidelberg (2011)
10. Anastassiou, G.A.: Foundation of stochastic fractional calculus with fractional approximation of stochastic processes. Rev. R. Acad. Cienc. Exactas Fis. Nat. Ser. A Mat. RACSAM **114**(2), Paper No. 89 (2020)
11. Anastassiou, G.: Commutative caputo fractional Korovkin inequalities for stochastic processes (2020). Submitted
12. Diethelm, K.: The Analysis of Fractional Differential Equations. Springer, New York (2010)
13. Korovkin, P.P.: Linear Operators and Approximation Theory. Hindustan Publ. Corp., Delhi (1960)
14. Royden, H.L.: Real Analysis, 2nd edn. MacMillan Publishing Co., Inc., New York (1968)
15. Shisha, O., Mond, B.: The degree of convergence of sequences of linear positive operators. Natl. Acad. of Sci. U.S., **60**, 1196–1200 (1968)
16. Weba, M.: Korovkin systems of stochastic processes. Math. Z. **192**(1), 73–80 (1986)
17. Weba, M.: Quantitative results on monotone approximation of stochastic processes. Probab. Math. Stat. **11**(1), 109–120 (1990)
18. Weba, M.: A quantitative Korovkin theorem for random functions with multivariate domains. J. Approx. Theory **61**(1), 74–87 (1990)
19. Weba, M.: Monotone approximation of random functions with multivariate domains in respect of lattice semi-norms. Results Math. **20**(1–2), 554–576 (1991)

Chapter 19
Trigonometric Commutative Caputo Fractional Korovkin Approximation for Stochastic Processes

Here we consider and study from the trigonometric point of view expectation commutative stochastic positive linear operators acting on L^1-continuous stochastic processes which are Caputo fractional differentiable. Under some mild, general and natural assumptions on the stochastic processes we produce related trigonometric Caputo fractional stochastic Shisha-Mond type inequalities pointwise and uniform. All convergences are produced with rates and are given by the trigonometric fractional stochastic inequalities involving the first modulus of continuity of the expectation of the αth right and left fractional derivatives of the engaged stochastic process, $\alpha > 0$, $\alpha \notin \mathbb{N}$. The amazing fact here is that the basic non-stochastic real Korovkin test functions assumptions impose the conclusions of our trigonometric Caputo fractional stochastic Korovkin theory. We include also a detailed trigonometric application to stochastic Bernstein operators. See also [11].

19.1 Introduction

In this work among others we are motivated by the following results.

Theorem A (Korovkin [12], (1960)) *Let* $L_n : C([-\pi, \pi]) \to C([-\pi, \pi])$, $n \in \mathbb{N}$, *be a sequence of positive linear operators. Assume* $L_n(1) \overset{u}{\to} 1$ *(uniformly),* $L_n(\cos t) \overset{u}{\to} \cos t$, $L_n(\sin t) \overset{u}{\to} \sin t$, as $n \to \infty$. *Then* $L_n f \overset{u}{\to} f$, *for every* $f \in C([-\pi, \pi])$ *that is* 2π*-periodic.*

Let $f \in C([a, b])$ *and* $0 \le \delta \le b - a$. *The first modulus of continuity of* f *at* δ *is given by*

$$\omega_1(f, \delta) = \sup\{|f(x) - f(y)|;\, x, y \in [a, b], |x - y| \le \delta\}.$$

If $\delta > b - a$, *then we define*

$$\omega_1(f, \delta) = \omega_1(f, b - a).$$

G. A. Anastassiou, *Generalized Fractional Calculus*, Studies in Systems, Decision and Control 305, https://doi.org/10.1007/978-3-030-56962-4_19

Another motivation is the following.

Theorem B (Shisha and Mond [14], (1968)) *Let $L_1, L_2, \ldots,$ be linear positive operators, whose common domain D consists of real functions with domain $(-\infty, \infty)$. Suppose $1, \cos x, \sin x, f$ belong to D, where f is an everywhere continuous, 2π-periodic function, with modulus of continuity ω_1. Let $-\infty < a < b < \infty$, and suppose that for $n = 1, 2, \ldots, L_n(1)$ is bounded in $[a, b]$. Then for $n = 1, 2, \ldots,$*

$$\|L_n(f) - f\|_\infty \leq \|f\|_\infty \|L_n(1) - 1\|_\infty + \|L_n(1) + 1\|_\infty \omega_1(f, \mu_n), \quad (*)$$

where

$$\mu_n = \pi \left\| \left(L_n \left(\sin^2 \left(\frac{t-x}{2} \right) \right) \right) (x) \right\|_\infty^{\frac{1}{2}},$$

and $\|\cdot\|_\infty$ stands for the sup norm over $[a, b]$.

In particular, if $L_n(1) = 1$, then (*) reduces to

$$\|L_n(f) - f\|_\infty \leq 2\omega_1(f, \mu_n).$$

One can easily see that, for $m = 1, 2, \ldots,$

$$\mu_n^2 \leq \left(\frac{\pi^2}{2} \right) \left[\|L_n(1) - 1\|_\infty + \right.$$

$$\left. \|(L_n(\cos t))(x) - \cos x\|_\infty + \|(L_n(\sin t))(x) - \sin x\|_\infty \right],$$

so the last along with (*) prove Korovkin's Theorem A in a quantitative way and with rates of convergence.

One more motivation follows.

Theorem C (see [5, p. 217]) *Let $f \in C^n([-\pi, \pi])$, $n \geq 1$, and μ a measure on $[-\pi, \pi]$ of mass $m > 0$. Put*

$$\beta := \left(\int \left(\sin \frac{|t|}{2} \right)^{n+1} \cdot \mu(dt) \right)^{\frac{1}{n+1}}$$

and denote by $\omega := \omega_1(f^{(n)}, \beta)$ the modulus of continuity of $f^{(n)}$ at β. Then

$$\left| \int f d\mu - f(0) \right| \leq |f(0)| \cdot |m-1| + \sum_{k=1}^{n} \frac{|f^{(k)}(0)|}{k!} \cdot \left| \int t^k \mu(dt) \right| +$$

$$w \left[m^{\frac{1}{n+1}} + \frac{\pi}{n+1} \right] \cdot \frac{\pi^n \beta^n}{n!}.$$

Anastassiou in [1–3] established a series of sharp inequalities for various cases of the parameters of the problem. However, Weba in [15–18] was the first, among many workers in quantitative results of Shisha-Mond type, to produce inequalities for stochastic processes. He assumed that the positive linear operators L_j are E-commutative (E means expectation) and stochastically simple. According to his work, if a stochastic process $X(t,\omega)$, $t \in Q$—a compact convex subset of a real normed vector space, $\omega \in Q$—probability space, is to be approximated by positive linear operators L_j, then the maximal error in the qth mean is ($q \geq 1$)

$$\left\| L_j X - X \right\| = \sup_{t \in Q} \left(E \left| \left(L_j X \right)(t,\omega) - X(t,\omega) \right|^q \right)^{\frac{1}{q}}.$$

So, Weba established upper bounds for $\left\| L_j X - X \right\|$ involving his own natural general first modulus of continuity of X with several interesting applications.

Anastassiou met [4] the pointwise case of $q = 1$. Without stochastic simplicity of L_j he found nearly best and best upper bounds for $\left| E\left(L_j X\right)(x_0) - (EX)(x_0) \right|$, $x_0 \in Q$.

The author here continues his above work on the trigonometric approximation of stochastic processes, now at the Caputo stochastic fractional level. He derives pointwise and uniform trigonometric Caputo fractional stochastic Shisha-Mond type inequalities, see the main Theorems 19.8, 19.9 and the several related corollaries. He gives an extensive trigonometric application to stochastic Bernstein operators. He finishes with a pointwise and a uniform fractional trigonometric stochastic Korovkin theorem, derived by Theorems 19.8, 19.9. The stochastic convergences, about stochastic processes, of our trigonometric fractional Korovkin Theorems 19.18, 19.19 are implied only by the convergences of real basic non-stochastic functions.

Our results here are built on [10].

19.2 Background—I

We need

Definition 19.1 *([9])* Let non-integer $\alpha > 0$, $n = \lceil \alpha \rceil$ ($\lceil \cdot \rceil$ is the ceiling of the number), $t \in [a,b] \subset \mathbb{R}$, $\omega \in \Omega$, where $(\Omega, \mathcal{F}, P)$ is a general probability space. Here $X(t,\omega)$ stands for a stochastic process. Assume that $X(\cdot,\omega) \in AC^n([a,b])$ (spaces of functions $X(\cdot,\omega)$ with $X^{(n-1)}(\cdot,\omega) \in AC([a,b])$ absolutely continuous functions), $\forall\, \omega \in \Omega$.

We call stochastic left Caputo fractional derivative

$$D_{*a}^{\alpha} X(x,\omega) = \frac{1}{\Gamma(n-\alpha)} \int_a^x (x-t)^{n-\alpha-1} X^{(n)}(t,\omega)\, dt, \tag{19.1}$$

$\forall\, x \in [a,b]$, $\forall\, \omega \in \Omega$.

And, we call stochastic right Caputo fractional derivative

$$D_{b-}^{\alpha} X(x, \omega) = \frac{(-1)^n}{\Gamma(n-\alpha)} \int_x^b (z-x)^{n-\alpha-1} X^{(n)}(z, \omega)\, dz, \tag{19.2}$$

$\forall\, x \in [a, b]$, $\forall\, \omega \in \Omega$. Above Γ stands for the gamma function.

We make

Remark 19.2 *(to Definition* 19.1*)* We further assume here that

$$\left| X^{(n)}(t, \omega) \right| \leq M, \forall\ (t, \omega) \in [a, b] \times \Omega,$$

where $M > 0$.

Then, by (19.1), we have

$$\left| D_{*a}^{\alpha} X(x, \omega) \right| \leq \frac{1}{\Gamma(n-\alpha)} \int_a^x (x-t)^{n-\alpha-1} \left| X^{(n)}(t, \omega) \right| dt \leq$$

$$\frac{M}{\Gamma(n-\alpha)} \int_a^x (x-t)^{n-\alpha-1}\, dt = \frac{M(x-a)^{n-\alpha}}{\Gamma(n-\alpha+1)}.$$

That is

$$\left| D_{*a}^{\alpha} X(x, \omega) \right| \leq \frac{M(x-a)^{n-\alpha}}{\Gamma(n-\alpha+1)}, \forall\, x \in [a, b],\ \text{any}\ \omega \in \Omega. \tag{19.3}$$

Also, from (19.2) we get

$$\left| D_{b-}^{\alpha} X(x, \omega) \right| \leq \frac{1}{\Gamma(n-\alpha)} \int_x^b (z-x)^{n-\alpha-1} \left| X^{(n)}(z, \omega) \right| dz \leq$$

$$\frac{M}{\Gamma(n-\alpha)} \int_x^b (z-x)^{n-\alpha-1}\, dz = \frac{M(b-x)^{n-\alpha}}{\Gamma(n-\alpha+1)}.$$

That is

$$\left| D_{b-}^{\alpha} X(x, \omega) \right| \leq \frac{M(b-x)^{n-\alpha}}{\Gamma(n-\alpha+1)}, \quad \forall\, x \in [a, b], \text{any}\ \omega \in \Omega. \tag{19.4}$$

It is not strange to assume that $D_{*a}^{\alpha} X$, $D_{b-}^{\alpha} X$ are stochastic processes.

By [6], p. 388, we get that $D_{*a}^{\alpha} X(\cdot, \omega) \in C([a, b])$, $\forall\, \omega \in \Omega$. And by [7], we get that $D_{b-}^{\alpha} X(\cdot, \omega) \in C([a, b])$, $\forall\, \omega \in \Omega$.

Similarly, we obtain

$$\left|D_{*t}^{\alpha}X\left(x,\omega\right)\right|\leq\frac{M\left(x-t\right)^{n-\alpha}}{\Gamma\left(n-\alpha+1\right)}\leq\frac{M\left(b-t\right)^{n-\alpha}}{\Gamma\left(n-\alpha+1\right)},\tag{19.5}$$

$\forall\, x\in\left[t,b\right]$, any $t\in\left[a,b\right]$, $\forall\,\omega\in\Omega$,

and

$$\left|D_{t-}^{\alpha}X\left(x,\omega\right)\right|\leq\frac{M\left(t-x\right)^{n-\alpha}}{\Gamma\left(n-\alpha+1\right)}\leq\frac{M\left(t-a\right)^{n-\alpha}}{\Gamma\left(n-\alpha+1\right)},\tag{19.6}$$

$\forall\, x\in\left[a,t\right]$, any $t\in\left[a,b\right]$, $\forall\,\omega\in\Omega$.

Above $D_{*t}^{\alpha}X$, $D_{t-}^{\alpha}X$ are assumed to be stochastic processes for any $t\in[a,b]$, and it holds $D_{*t}^{\alpha}X\left(\cdot,\omega\right)\in C\left([t,b]\right)$, $D_{t-}^{\alpha}X\left(\cdot,\omega\right)\in C\left([a,t]\right)$, $\forall\,\omega\in\Omega$.

Clearly, then

$$\left|E\left(D_{*t}^{\alpha}X\right)(x)\right|\leq\frac{M\left(b-t\right)^{n-\alpha}}{\Gamma\left(n-\alpha+1\right)},\tag{19.7}$$

$\forall\; x\in[t,b]$, any $t\in[a,b]$, where E is the expectation operator $\left(EX\right)(t)=\int_{\Omega}X\left(t,\omega\right)P\left(d\omega\right)$ and similarly,

$$\left|E\left(D_{t-}^{\alpha}X\right)(x)\right|\leq\frac{M\left(t-a\right)^{n-\alpha}}{\Gamma\left(n-\alpha+1\right)},\tag{19.8}$$

$\forall\, x\in[a,t]$, any $t\in\left[a,b\right]$.

We observe that the first modulus of continuity ($\delta>0$)

$$\omega_{1}\left(E\left(D_{*t}^{\alpha}X\right),\delta\right)_{[t,b]}:=\sup_{\substack{x,y\in[t,b]:\\|x-y|\leq\delta}}\left|E\left(D_{*t}^{\alpha}X\right)(x)-E\left(D_{*t}^{\alpha}X\right)(y)\right|$$

$$\overset{(19.7)}{\leq}\frac{2M\left(b-t\right)^{n-\alpha}}{\Gamma\left(n-\alpha+1\right)},\tag{19.9}$$

any $t\in\left[a,b\right]$.

Hence, it holds ($\delta>0$)

$$\sup_{t\in[a,b]}\omega_{1}\left(E\left(D_{*t}^{\alpha}X\right),\delta\right)_{[t,b]}\leq\frac{2M\left(b-a\right)^{n-\alpha}}{\Gamma\left(n-\alpha+1\right)}.\tag{19.10}$$

Similarly, it holds ($\delta>0$)

$$\sup_{t\in[a,b]}\omega_{1}\left(E\left(D_{t-}^{\alpha}X\right),\delta\right)_{[a,t]}\leq\frac{2M\left(b-a\right)^{n-\alpha}}{\Gamma\left(n-\alpha+1\right)}.\tag{19.11}$$

We also set

$$\omega_1\left(E\left(D_t^\alpha X\right),\delta\right):=\max\left\{\omega_1\left(E\left(D_{*t}^\alpha X\right),\delta\right)_{[t,b]},\omega_1\left(E\left(D_{t-}^\alpha X\right),\delta\right)_{[a,t]}\right\}, \tag{19.12}$$

where $\delta>0$.

We make

Remark 19.3 Let the positive linear operator L mapping $C([a,b])$ into $B([a,b])$ (the bounded functions). By the Riesz representation theorem [13] we have that there exists μ_t unique, completed Borel measure on $[a,b]$ with

$$\mu_t([a,b])=L(1)(t)>0, \tag{19.13}$$

such that

$$L(f)(t)=\int_{[a,b]}f(s)\,d\mu_t(s),\quad\forall\,t\in[a,b],\forall\,f\in C([a,b]). \tag{19.14}$$

We denote $\|\cdot\|_\infty=\|\cdot\|_{\infty,[-\pi,\pi]}$ the supremum norm.

Next we specify $[a,b]$ as $[-\pi,\pi]$. Clearly then $L:C([-\pi,\pi])\to B([-\pi,\pi])$ is the positive linear operator on hand.

Here $n=\lceil\alpha\rceil,\alpha\notin\mathbb{N},\alpha>0,k=1,\dots,n-1$. By the use of Hölder's inequality we notice that

$$L\left(\left(\sin\left(\frac{|s-t|}{4}\right)\right)^k\right)(t)=\int_{[-\pi,\pi]}\left(\sin\left(\frac{|s-t|}{4}\right)\right)^k d\mu_t(s)\le$$

$$\left(\int_{[-\pi,\pi]}\left(\sin\left(\frac{|s-t|}{4}\right)\right)^{\alpha+1}d\mu_t(s)\right)^{\frac{k}{\alpha+1}}\left(\mu_t([-\pi,\pi])\right)^{\frac{\alpha+1-k}{\alpha+1}}= \tag{19.15}$$

$$\left(L\left(\left(\sin\left(\frac{|s-t|}{4}\right)\right)^{\alpha+1}\right)(t)\right)^{\frac{k}{\alpha+1}}\left(L(1)(t)\right)^{\frac{\alpha+1-k}{\alpha+1}}.$$

That is

$$L\left(\left(\sin\left(\frac{|s-t|}{4}\right)\right)^k\right)(t)\le \tag{19.16}$$

$$\left(L\left(\left(\sin\left(\frac{|s-t|}{4}\right)\right)^{\alpha+1}\right)(t)\right)^{\frac{k}{\alpha+1}}\left(L(1)(t)\right)^{\frac{\alpha+1-k}{\alpha+1}},$$

for $k=1,\dots,n-1$.

Consequently, it holds

$$\left\| L\left(\left(\sin\left(\frac{|s-t|}{4}\right)\right)^{k}\right)(t)\right\|_{\infty} \leq \tag{19.17}$$

$$\left\| L\left(\left(\sin\left(\frac{|s-t|}{4}\right)\right)^{\alpha+1}\right)(t)\right\|_{\infty}^{\frac{k}{\alpha+1}} \|L(1)\|_{\infty}^{\frac{\alpha+1-k}{\alpha+1}},$$

for $k = 1, \ldots, n-1$.

In this work we will use a lot the following well known inequality:

$$|z| \leq \pi \sin\left(\frac{|z|}{2}\right), \quad \forall\, z \in [-\pi, \pi]. \tag{19.18}$$

Furthermore, we observe that

$$\left|L\left((s-t)^{k}\right)(t)\right| = \left|\int_{[-\pi,\pi]} (s-t)^{k}\, d\mu_t(s)\right| \leq \int_{[-\pi,\pi]} |s-t|^{k}\, d\mu_t(s) =$$

$$2^{k}\int_{[-\pi,\pi]} \left(\frac{|s-t|}{2}\right)^{k} d\mu_t(s) \overset{(19.18.)}{\leq} (2\pi)^{k}\int_{[-\pi,\pi]} \left(\sin\left(\frac{|s-t|}{4}\right)\right)^{k} d\mu_t(s) \tag{19.19}$$

$$= (2\pi)^{k} L\left(\left(\sin\left(\frac{|s-t|}{4}\right)\right)^{k}\right)(t).$$

That is

$$\left|L\left((s-t)^{k}\right)(t)\right| \leq (2\pi)^{k} L\left(\left(\sin\left(\frac{|s-t|}{4}\right)\right)^{k}\right)(t), \tag{19.20}$$

$\forall\, t \in [-\pi, \pi]$, and

$$\left\|L\left((s-t)^{k}\right)(t)\right\|_{\infty} \leq (2\pi)^{k}\left\| L\left(\left(\sin\left(\frac{|s-t|}{4}\right)\right)^{k}\right)(t)\right\|_{\infty}, \tag{19.21}$$

all $k = 1, \ldots, n-1$.

Then, by (19.16) and (19.20) we get

$$\left|L\left((s-t)^{k}\right)(t)\right| \leq (2\pi)^{k}\left(L\left(\left(\sin\left(\frac{|s-t|}{4}\right)\right)^{\alpha+1}\right)(t)\right)^{\frac{k}{\alpha+1}} (L(1)(t))^{\frac{\alpha+1-k}{\alpha+1}}, \tag{19.22}$$

$k = 1, \ldots, n-1$, $\forall\, t \in [-\pi, \pi]$, and by (19.17) and (19.21), we find

$$\left\|L\left((s-t)^{k}\right)(t)\right\|_{\infty} \leq (2\pi)^{k}\left\|L\left(\left(\sin\left(\frac{|s-t|}{4}\right)\right)^{\alpha+1}\right)(t)\right\|^{\frac{k}{\alpha+1}}\|L(1)\|_{\infty}^{\frac{\alpha+1-k}{\alpha+1}}, \tag{19.23}$$

$k=1,\ldots,n-1$.

We also have

$$L\left(|s-t|^{\alpha+1}\right)(t)=\int_{[-\pi,\pi]}|s-t|^{\alpha+1}d\mu_{t}(s)=$$

$$2^{\alpha+1}\int_{[-\pi,\pi]}\left(\frac{|s-t|}{2}\right)^{\alpha+1}d\mu_{t}(s) \overset{(19.18)}{\leq} (2\pi)^{\alpha+1}\int_{[-\pi,\pi]}\left(\sin\left(\frac{|s-t|}{4}\right)\right)^{\alpha+1}d\mu_{t}(s) \tag{19.24}$$

$$=(2\pi)^{\alpha+1}L\left(\left(\sin\left(\frac{|s-t|}{4}\right)\right)^{\alpha+1}\right)(t).$$

That is

$$L\left(|s-t|^{\alpha+1}\right)(t) \leq (2\pi)^{\alpha+1}L\left(\left(\sin\left(\frac{|s-t|}{4}\right)\right)^{\alpha+1}\right)(t), \tag{19.25}$$

$\forall\, t\in[-\pi,\pi]$,

and

$$\left\|L\left(|s-t|^{\alpha+1}\right)(t)\right\|_{\infty} \leq (2\pi)^{\alpha+1}\left\|L\left(\left(\sin\left(\frac{|s-t|}{4}\right)\right)^{\alpha+1}\right)(t)\right\|_{\infty}. \tag{19.26}$$

Also we have

$$D_{*t}^{\alpha}X(t,\omega)=D_{t-}^{\alpha}X(t,\omega)=0, \tag{19.27}$$

$\forall\,\omega\in\Omega$, see [8], pp. 358–359.

We assume that

$$D_{*t}^{\alpha}X(s,\omega)=0,\ \text{for } s<t,$$

and

$$D_{t-}^{\alpha}X(s,\omega)=0,\ \text{for } s>t,$$

$\forall\,\omega\in\Omega$.

19.3 Preliminaries

Let $(\Omega, \mathcal{F}, P)$ be a probabilistic space and $L^1(\Omega, \mathcal{F}, P)$ be the space of all real-valued random variables $Y = Y(\omega)$ with

$$\int_\Omega |Y(\omega)| P(d\omega) < \infty.$$

Let $X = X(t, \omega)$ denote a stochastic process with index set $[a, b] \subset \mathbb{R}$ and real state space $(\mathbb{R}, \mathcal{B})$, where $\mathcal{B}$ is the σ-field of Borel subsets of $\mathbb{R}$. Here $C([a, b])$ is the space of continuous real-valued functions on $[a, b]$ and $\mathcal{B}([a, b])$ is the space of bounded real-valued functions on $[a, b]$. Also $C_\Omega([a, b]) = C\left([a, b], L^1(\Omega, \mathcal{F}, P)\right)$ is the space of L^1-continuous stochastic processes in t and $B_\Omega([a, b]) = \left\{X : \sup_{t\in[a,b]} \int_\Omega |X(t, \omega)| P(d\omega) < \infty\right\}$, obviously $C_\Omega([a, b]) \subset B_\Omega([a, b])$.

Let $\alpha > 0$, $\alpha \notin \mathbb{N}$, $\lceil\alpha\rceil = n$, and consider the subspace of stochastic processes $C_\Omega^{\alpha,n}([a, b]) := \{X : X(\cdot, \omega) \in AC^n([a, b]),\ \forall\ \omega \in \Omega$ and $\left|X^{(n)}(t, \omega)\right| \le M$, $\forall$ $(t, \omega) \in [a, b] \times \Omega$, where $M > 0$; $X^{(k)}(t, \omega) \in C_\Omega([a, b])$, $k = 0, 1, \ldots, n-1$; also $D_{*t}^\alpha X$, $D_{t-}^\alpha X$ are stochastic processes for any $t \in [a, b]\}$. That is, for every $\omega \in \Omega$ we have $X(t, \omega) \in C^{n-1}([a, b])$.

Consider the linear operator

$$L : C_\Omega([a, b]) \hookrightarrow B_\Omega([a, b]).$$

If $X \in C_\Omega([a, b])$ is nonnegative and LX, too, then L is called positive. If $EL = LE$, then L is called E-commutative.

19.4 Background—II

Following 3. Preliminaries we proved

Theorem 19.4 ([10]) *Consider the positive E-commutative linear operator $L : C_\Omega([a, b]) \hookrightarrow B_\Omega([a, b])$, and $\alpha > 0$, $\alpha \notin \mathbb{N}$, $\lceil\alpha\rceil = n$, and let $X \in C_\Omega^{\alpha,n}([a, b])$, with $\delta > 0$.*

Then

$$|(E(LX))(t) - (EX)(t)| \le |(EX)(t)|\, |(L(1))(t) - 1| + \tag{19.28}$$

$$\sum_{k=1}^{n-1} \frac{\left|\left(EX^{(k)}\right)(t)\right|}{k!} \left|L\left((s-t)^k\right)(t)\right| + \frac{\omega_1\left(E\left(D_t^\alpha X\right), \delta\right)}{\Gamma(\alpha+1)}$$

$$\left(L\left(|s-t|^{\alpha+1}\right)(t)\right)^{\frac{\alpha}{\alpha+1}}\left[(L(1)(t))^{\frac{1}{\alpha+1}}+\frac{\left(L\left(|s-t|^{\alpha+1}\right)(t)\right)^{\frac{1}{\alpha+1}}}{\delta(\alpha+1)}\right],$$

$\forall\, t \in [a,b]$.

Above $\omega_1\left(E\left(D_t^{\alpha}X\right),\delta\right)$ *is as in (19.12).*

We also mention

Theorem 19.5 ([10]) *All as in Theorem 19.4. Then*

$$\|E(LX)-EX\|_\infty \le \|EX\|_\infty \|L(1)-1\|_\infty +$$

$$\sum_{k=1}^{n-1}\frac{\left\|EX^{(k)}\right\|_\infty}{k!}\left\|L\left((s-t)^k\right)(t)\right\|_\infty + \sup_{t\in[a,b]}\frac{\omega_1\left(E\left(D_t^{\alpha}X\right),\delta\right)}{\Gamma(\alpha+1)} \tag{19.29}$$

$$\left\|L\left(|s-t|^{\alpha+1}\right)(t)\right\|_\infty^{\frac{\alpha}{\alpha+1}}\left[\|L(1)\|_\infty^{\frac{1}{\alpha+1}}+\frac{\left\|L\left(|s-t|^{\alpha+1}\right)(t)\right\|_\infty^{\frac{1}{\alpha+1}}}{\delta(\alpha+1)}\right]<\infty.$$

We specify:

Definition 19.6 If $0<\alpha<1$, then $n=1$, and $C_\Omega^{\alpha,1}([a,b]) := \{X : X(\cdot,\omega) \in AC([a,b])\,,\ \forall\,\omega\in\Omega$ and $\left|X^{(1)}(t,\omega)\right| \le M\,,\ \forall\,(t,\omega)\in[a,b]\times\Omega$, where $M>0$; $X(t,\omega)\in C_\Omega([a,b])$; also $D_{*t}^{\alpha}X$, $D_{t-}^{\alpha}X$ are stochastic processes for any $t\in[a,b]\}$. We will specialize on $C_\Omega^{\alpha,1}([-\pi,\pi])$.

Note 19.7 From [17, pp. 3–5] we have the following results

(i) $C([a,b]) \subset C_\Omega([a,b])$,
(ii) if $X\in C_\Omega([a,b])$, then $EX\in C([a,b])$,
and
(iii) if L is E-commutative, then L maps the subspace $C([a,b])$ into $B([a,b])$.

19.5 Main Results

We present the following pointwise result over $[-\pi,\pi]$.

Theorem 19.8 *Consider the 3. Preliminaries for* $[a,b]=[-\pi,\pi]$ *and the positive* E*-commutative linear operator* $L: C_\Omega([-\pi,\pi]) \hookrightarrow B_\Omega([-\pi,\pi])$, *and* $\alpha>0$, $\alpha\notin\mathbb{N}$, $\lceil\alpha\rceil=n$, *and let* $X\in C_\Omega^{\alpha,n}([-\pi,\pi])$, *with* $\delta>0$.

Then

$$|(E(LX))(t)-(EX)(t)| \le |(EX)(t)|\,|(L(1))(t)-1| +$$

$$\sum_{k=1}^{n-1}\frac{\left|\left(EX^{(k)}\right)(t)\right|}{k!}(2\pi)^k L\left(\left(\sin\left(\frac{|s-t|}{4}\right)\right)^k\right)(t)+$$

$$\frac{\omega_1\left(E\left(D_t^\alpha X\right),\delta\right)}{\Gamma(\alpha+1)}(2\pi)^\alpha\left(L\left(\left(\sin\left(\frac{|s-t|}{4}\right)\right)^{\alpha+1}\right)(t)\right)^{\frac{\alpha}{\alpha+1}}$$

$$\left[(L(1)(t))^{\frac{1}{\alpha+1}}+\frac{2\pi}{\delta(\alpha+1)}\left(L\left(\left(\sin\left(\frac{|s-t|}{4}\right)\right)^{\alpha+1}\right)(t)\right)^{\frac{1}{\alpha+1}}\right], \tag{19.30}$$

$\forall\, t\in[-\pi,\pi]$.

Above $\omega_1\left(E\left(D_t^\alpha X\right),\delta\right)$ *is over* $[-\pi,\pi]$.

Proof From (19.28) we have ($\forall\, t\in[-\pi,\pi]$)

$$|(E(LX))(t)-(EX)(t)|\leq|(EX)(t)|\,|(L(1))(t)-1|+$$

$$\sum_{k=1}^{n-1}\frac{\left|\left(EX^{(k)}\right)(t)\right|}{k!}\left|L\left((s-t)^k\right)(t)\right|+\frac{\omega_1\left(E\left(D_t^\alpha X\right),\delta\right)}{\Gamma(\alpha+1)}$$

$$\left(L\left(|s-t|^{\alpha+1}\right)(t)\right)^{\frac{\alpha}{\alpha+1}}\left[(L(1)(t))^{\frac{1}{\alpha+1}}+\frac{\left(L\left(|s-t|^{\alpha+1}\right)(t)\right)^{\frac{1}{\alpha+1}}}{\delta(\alpha+1)}\right]$$

$$\overset{\text{(by (19.20), (19.25))}}{\leq}|(EX)(t)|\,|(L(1))(t)-1|+$$

$$\sum_{k=1}^{n-1}\frac{\left|\left(EX^{(k)}\right)(t)\right|}{k!}(2\pi)^k L\left(\left(\sin\left(\frac{|s-t|}{4}\right)\right)^k\right)(t)+$$

$$\frac{\omega_1\left(E\left(D_t^\alpha X\right),\delta\right)}{\Gamma(\alpha+1)}(2\pi)^\alpha\left(L\left(\left(\sin\left(\frac{|s-t|}{4}\right)\right)^{\alpha+1}\right)(t)\right)^{\frac{\alpha}{\alpha+1}}$$

$$\left[(L(1)(t))^{\frac{1}{\alpha+1}}+\frac{2\pi}{\delta(\alpha+1)}\left(L\left(\left(\sin\left(\frac{|s-t|}{4}\right)\right)^{\alpha+1}\right)(t)\right)^{\frac{1}{\alpha+1}}\right], \tag{19.31}$$

proving the claim. ■

By (19.30) we obtain the following uniform estimate

Theorem 19.9 *All as in Theorem 19.8. Then*

$$\|E(LX)-EX\|_{\infty}\leq\|EX\|_{\infty}\|L(1)-1\|_{\infty}+$$

$$\sum_{k=1}^{n-1}\frac{\|EX^{(k)}\|_{\infty}}{k!}(2\pi)^{k}\left\|L\left(\left(\sin\left(\frac{|s-t|}{4}\right)\right)^{k}\right)(t)\right\|_{\infty}+$$

$$\frac{\sup_{t\in[-\pi,\pi]}\omega_{1}\left(E\left(D_{t}^{\alpha}X\right),\delta\right)}{\Gamma(\alpha+1)}(2\pi)^{\alpha}\left\|L\left(\left(\sin\left(\frac{|s-t|}{4}\right)\right)^{\alpha+1}\right)(t)\right\|_{\infty}^{\frac{\alpha}{\alpha+1}}$$

$$\left[\|L(1)\|_{\infty}^{\frac{1}{\alpha+1}}+\frac{2\pi}{\delta(\alpha+1)}\left\|L\left(\left(\sin\left(\frac{|s-t|}{4}\right)\right)^{\alpha+1}\right)(t)\right\|_{\infty}^{\frac{1}{\alpha+1}}\right]<\infty. \quad (19.32)$$

We give

Corollary 19.10 *Consider the positive E-commutative linear operator* $L:C_{\Omega}([-\pi,\pi])\hookrightarrow B_{\Omega}([-\pi,\pi])$, *and* $0<\alpha<1$ *and let* $X\in C_{\Omega}^{\alpha,1}([-\pi,\pi])$, *with* $\delta>0$. *Then*

$$|(E(LX))(t)-(EX)(t)|\leq|(EX)(t)||(L(1))(t)-1|+$$

$$\frac{\omega_{1}\left(E\left(D_{t}^{\alpha}X\right),\delta\right)}{\Gamma(\alpha+1)}(2\pi)^{\alpha}\left(L\left(\left(\sin\left(\frac{|s-t|}{4}\right)\right)^{\alpha+1}\right)(t)\right)^{\frac{\alpha}{\alpha+1}}$$

$$\left[(L(1)(t))^{\frac{1}{\alpha+1}}+\frac{2\pi}{\delta(\alpha+1)}\left(L\left(\left(\sin\left(\frac{|s-t|}{4}\right)\right)^{\alpha+1}\right)(t)\right)^{\frac{1}{\alpha+1}}\right], \quad (19.33)$$

$\forall\, t\in[-\pi,\pi]$.

Proof By (19.30). ■

We continue with

Corollary 19.11 *All as in Corollary 19.10. Then*

$$\|E(LX)-EX\|_{\infty}\leq\|EX\|_{\infty}\|L(1)-1\|_{\infty}+$$

$$\frac{\sup_{t\in[-\pi,\pi]}\omega_{1}\left(E\left(D_{t}^{\alpha}X\right),\delta\right)}{\Gamma(\alpha+1)}(2\pi)^{\alpha}\left\|L\left(\left(\sin\left(\frac{|s-t|}{4}\right)\right)^{\alpha+1}\right)(t)\right\|_{\infty}^{\frac{\alpha}{\alpha+1}}$$

$$\left[\|L(1)\|_{\infty}^{\frac{1}{\alpha+1}}+\frac{2\pi}{\delta(\alpha+1)}\left\|L\left(\left(\sin\left(\frac{|s-t|}{4}\right)\right)^{\alpha+1}\right)(t)\right\|_{\infty}^{\frac{1}{\alpha+1}}\right]. \quad (19.34)$$

Proof From (19.33). ∎

Corollary 19.12 *All as in Corollary 19.10 and L (1) = 1. Then*

$$\|E(LX) - EX\|_\infty \leq$$

$$\frac{\sup_{t\in[-\pi,\pi]} \omega_1\left(E\left(D_t^\alpha X\right), \delta\right)}{\Gamma(\alpha+1)} (2\pi)^\alpha \left\| L\left(\left(\sin\left(\frac{|s-t|}{4}\right)\right)^{\alpha+1}\right)(t)\right\|_\infty^{\frac{\alpha}{\alpha+1}}$$

$$\left[1 + \frac{2\pi}{\delta(\alpha+1)} \left\| L\left(\left(\sin\left(\frac{|s-t|}{4}\right)\right)^{\alpha+1}\right)(t)\right\|_\infty^{\frac{1}{\alpha+1}}\right]. \tag{19.35}$$

Proof From (19.34). ∎

In particular we give

Corollary 19.13 *All as in Corollary 19.10 and L (1) = 1. Then*

$$\|E(LX) - EX\|_\infty \leq$$

$$\frac{2^{\alpha+1}\pi^\alpha}{\Gamma(\alpha+1)} \sup_{t\in[-\pi,\pi]} \omega_1\left(E\left(D_t^\alpha X\right), \frac{2\pi}{\alpha+1}\left\| L\left(\left(\sin\left(\frac{|s-t|}{4}\right)\right)^{\alpha+1}\right)(t)\right\|_\infty^{\frac{1}{\alpha+1}}\right)$$

$$\left\| L\left(\left(\sin\left(\frac{|s-t|}{4}\right)\right)^{\alpha+1}\right)(t)\right\|_\infty^{\frac{\alpha}{\alpha+1}}. \tag{19.36}$$

Proof By (19.35): we take there

$$\delta = \frac{2\pi}{\alpha+1}\left\| L\left(\left(\sin\left(\frac{|s-t|}{4}\right)\right)^{\alpha+1}\right)(t)\right\|_\infty^{\frac{1}{\alpha+1}} > 0.$$

In case of $\left\| L\left(\left(\sin\left(\frac{|s-t|}{4}\right)\right)^{\alpha+1}\right)(t)\right\|_\infty = 0$, we have that $L\left(\left(\sin\left(\frac{|s-t|}{4}\right)\right)^{\alpha+1}\right)(t)$ $= 0, \forall\, t \in [-\pi, \pi]$. That is, by (19.14), $\int_{[-\pi,\pi]}\left(\sin\left(\frac{|s-t|}{4}\right)\right)^{\alpha+1} d\mu_t(s) = 0, \forall\, t \in [-\pi, \pi]$, where μ_t is a probability measure on $[-\pi, \pi]$.

Since $\sin\left(\frac{|s-t|}{4}\right) \geq 0, \forall\, s \in [-\pi, \pi]$, we get $\sin\left(\frac{|s-t|}{4}\right) = 0$, a.e., that is $|s-t| = 0$, a.e., and $s = t$, a.e., which means $\mu_t\{s \in [-\pi, \pi] : s \neq t\} = 0$, i.e. $\mu_t = \delta_t$, $\forall\, t \in [-\pi, \pi]$, where δ_t is the unit Dirac measure.

Consequently we have

$$E(LX)(t) = L(EX)(t) = \int_{[-\pi,\pi]} (EX)(s)\, d\delta_t(s) = (EX)(t),$$

$\forall\, t \in [-\pi, \pi]$.

That is $E(LX) = EX$ over $[-\pi, \pi]$. Therefore both sides of inequality (19.36) equal to zero.

Hence (19.36) is always true. ■

19.6 Application

Consider the Bernstein polynomials on $[-\pi, \pi]$ for $f \in C([-\pi, \pi])$:

$$B_N(f)(x) = \sum_{k=0}^{N} \binom{N}{k} f\left(-\pi + \frac{2\pi k}{N}\right) \left(\frac{x+\pi}{2\pi}\right)^k \left(\frac{\pi - x}{2\pi}\right)^{N-k}, \tag{19.37}$$

$N \in \mathbb{N}$, any $x \in [-\pi, \pi]$. There are positive linear operators from $C([-\pi, \pi])$ into itself.

Setting $g(t) = f(2\pi t - \pi)$, $t \in [0, 1]$, we have $g(0) = f(-\pi)$, $g(1) = f(\pi)$, and

$$(B_N g)(t) = \sum_{k=0}^{N} \binom{N}{k} g\left(\frac{k}{N}\right) t^k (1-t)^{N-k} = (B_N f)(x), \quad x \in [-\pi, \pi]. \tag{19.38}$$

Here $x = \varphi(t) = 2\pi t - \pi$ is an $1-1$ and onto map from $[0, 1]$ onto $[-\pi, \pi]$. Clearly here $g \in C([0, 1])$.

Notice also that

$$\left(B_N\left((\cdot - x)^2\right)\right)(x) = \left[\left(B_N\left((\cdot - t)^2\right)\right)(t)\right](2\pi)^2 = \frac{(2\pi)^2}{N} t(1-t)$$

$$= \frac{(2\pi)^2}{N} \left(\frac{x+\pi}{2\pi}\right)\left(\frac{\pi - x}{2\pi}\right) = \frac{1}{N}(x+\pi)(\pi - x) \leq \frac{\pi^2}{N}, \quad \forall\, x \in [-\pi, \pi].$$

I.e.

$$\left(B_N\left((\cdot - x)^2\right)\right)(x) \leq \frac{\pi^2}{N}, \quad \forall\, x \in [-\pi, \pi]. \tag{19.39}$$

In particular

$$(B_N 1)(x) = 1, \quad \forall\, x \in [-\pi, \pi]. \tag{19.40}$$

Define the corresponding stochastic application of B_N as follows:

$$B_N\left(X\left(\cdot,\omega\right)\right)(t)=\sum_{k=0}^{N}\binom{N}{k}X\left(-\pi+\frac{2\pi k}{N},\omega\right)\left(\frac{t+\pi}{2\pi}\right)^{k}\left(\frac{\pi-t}{2\pi}\right)^{N-k}, \tag{19.41}$$

$\forall\, N\in\mathbb{N}, \forall\, t\in[-\pi,\pi], \forall\, \omega\in\Omega$, where X is a stochastic process. Clearly $B_N X$ is a stochastic process.

Notice that

$$E\left(B_N X\right)(t)=\sum_{k=0}^{N}\binom{N}{k}(EX)\left(-\pi+\frac{2\pi k}{N}\right)\left(\frac{t+\pi}{2\pi}\right)^{k}\left(\frac{\pi-t}{2\pi}\right)^{N-k}$$

$$=B_N\left(EX\right)(t), \tag{19.42}$$

i.e. $EB_N=B_N E$, that is B_N is an E-commutative positive linear operator from $C_\Omega\left([-\pi,\pi]\right)$ into itself.

We give

Corollary 19.14 *Let* $0<\alpha<1$ *and* $X\in C_\Omega^{\alpha,1}\left([-\pi,\pi]\right)$. *Then*

$$\left\|E\left(B_N X\right)-EX\right\|_\infty\leq\frac{2^{\alpha+1}\pi^\alpha}{\Gamma\left(\alpha+1\right)}$$

$$\sup_{t\in[-\pi,\pi]}\omega_1\left(E\left(D_t^\alpha X\right),\frac{2\pi}{\alpha+1}\left\|B_N\left(\left(\sin\left(\frac{|s-t|}{4}\right)\right)^{\alpha+1}\right)(t)\right\|_\infty^{\frac{1}{\alpha+1}}\right)$$

$$\left\|B_N\left(\left(\sin\left(\frac{|s-t|}{4}\right)\right)^{\alpha+1}\right)(t)\right\|_\infty^{\frac{\alpha}{\alpha+1}}, \tag{19.43}$$

$\forall\, N\in\mathbb{N}$.

Proof By (19.36). ■

In particular we get:

Corollary 19.15 *Let* $X\in C_\Omega^{\frac{1}{2},1}\left([-\pi,\pi]\right)$. *Then*

$$\left\|E\left(B_N X\right)-EX\right\|_\infty\leq 4\sqrt{2}$$

$$\sup_{t\in[-\pi,\pi]}\omega_1\left(E\left(D_t^{\frac{1}{2}} X\right),\frac{4\pi}{3}\left\|B_N\left(\left(\sin\left(\frac{|s-t|}{4}\right)\right)^{\frac{3}{2}}\right)(t)\right\|_\infty^{\frac{2}{3}}\right)$$

$$\left\|B_N\left(\left(\sin\left(\frac{|s-t|}{4}\right)\right)^{\frac{3}{2}}\right)(t)\right\|_\infty^{\frac{1}{3}}, \tag{19.44}$$

$\forall\, N \in \mathbb{N}$.

Proof By (19.43) for $\alpha = \frac{1}{2}$. ∎

We make

Remark 19.16 By $|\sin x| < |x|$, $\forall\, x \in \mathbb{R} - \{0\}$, in particular $\sin x \leq x$, for $x \geq 0$, we get

$$\left(\sin\left(\frac{|\cdot - t|}{4}\right)\right)^{\frac{3}{2}} \leq \left(\frac{|\cdot - t|}{4}\right)^{\frac{3}{2}} = \frac{1}{8}\,|\cdot - t|^{\frac{3}{2}}.$$

Hence

$$\left\| B_N\left(\left(\sin\left(\frac{|\cdot - t|}{4}\right)\right)^{\frac{3}{2}}\right)(t)\right\|_{\infty} \leq \frac{1}{8}\left\| B_N\left(|\cdot - t|^{\frac{3}{2}}\right)(t)\right\|_{\infty}. \tag{19.45}$$

We observe that

$$B_N\left(|\cdot - t|^{\frac{3}{2}}\right)(t) = \sum_{k=0}^{N}\left|t + \pi - \frac{2\pi k}{N}\right|^{\frac{3}{2}}\binom{N}{k}\left(\frac{t+\pi}{2\pi}\right)^{k}\left(\frac{\pi - t}{2\pi}\right)^{N-k}$$

(by discrete Hölder's inequality)

$$\leq \left[\sum_{k=0}^{N}\left(t + \pi - \frac{2\pi k}{N}\right)^{2}\binom{N}{k}\left(\frac{t+\pi}{2\pi}\right)^{k}\left(\frac{\pi - t}{2\pi}\right)^{N-k}\right]^{\frac{3}{4}} \tag{19.46}$$

$$= \left(B_N\left((\cdot - t)^2\right)(t)\right)^{\frac{3}{4}} \overset{(19.39)}{\leq} \frac{\pi^{\frac{3}{2}}}{N^{\frac{3}{4}}},\ \ \forall\, t \in [-\pi, \pi].$$

Consequently it holds

$$\left\| B_N\left(|\cdot - t|^{\frac{3}{2}}\right)(t)\right\|_{\infty} \leq \frac{\pi^{\frac{3}{2}}}{N^{\frac{3}{4}}}, \tag{19.47}$$

and

$$\left\| B_N\left(\left(\sin\left(\frac{|\cdot - t|}{4}\right)\right)^{\frac{3}{2}}\right)(t)\right\|_{\infty} \leq \frac{\pi^{\frac{3}{2}}}{8N^{\frac{3}{4}}},\ \ \forall\, N \in \mathbb{N}. \tag{19.48}$$

We derive

Proposition 19.17 *Let* $X \in C_{\Omega}^{\frac{1}{2},1}([-\pi, \pi])$. *Then*

$$\|E(B_N X) - EX\|_{\infty} \leq \frac{2\sqrt{2\pi}}{\sqrt[4]{N}}\sup_{t\in[-\pi,\pi]}\omega_1\left(E\left(D_t^{\frac{1}{2}}X\right), \frac{\pi^2}{3\sqrt{N}}\right), \tag{19.49}$$

$\forall \, N \in \mathbb{N}$.

Hence $\lim_{N\to\infty} E\left(B_N X\right) = EX$, *uniformly.*

Proof By (19.44) and (19.48). ■

19.7 Commutative Trigonometric Caputo Fractional Stochastic Korovkin Results

Here L is meant as a sequence of positive E-commutative linear operators and all assumptions are as in Theorem 19.8.

We give

Theorem 19.18 *We further assume that* $L\left(1\right)\left(t\right) \to 1$ *and* $L\left(\left(\sin\left(\frac{|s-t|}{4}\right)\right)^{\alpha+1}\right)(t)$ $\to 0$, *then* $\left(E\left(LX\right)\right)(t) \to \left(EX\right)(t)$, *for any* $X \in C_{\Omega}^{\alpha,n}\left([-\pi,\pi]\right)$, $\forall \, t \in [-\pi,\pi]$, *a pointwise convergence; where* $\alpha > 0$, $\alpha \notin \mathbb{N}$, $\lceil \alpha \rceil = n$.

Proof Based on (19.30), (19.16), and that $L\left(1\right)(t)$ is bounded as a sequence of functions. Also $\omega_1\left(E\left(D_t^{\alpha}X\right), \delta\right)$ over $[-\pi,\pi]$ is bounded, see (19.10)–(19.12). ■

We continue with

Theorem 19.19 *We further assume that* $L\left(1\right)(t) \to 1$, *uniformly and* $\left\|L\left(\left(\sin\left(\frac{|s-t|}{4}\right)\right)^{\alpha+1}\right)(t)\right\|_{\infty} \to 0$, *then* $E\left(LX\right) \to EX$, *uniformly over* $[-\pi,\pi]$, *for any* $X \in C_{\Omega}^{\alpha,n}\left([-\pi,\pi]\right)$; *where* $\alpha > 0$, $\alpha \notin \mathbb{N}$, $\lceil \alpha \rceil = n$.

Proof Based on (19.32), (19.17), and that $\|L(1)\|_{\infty}$ is bounded. Also it is $\sup_{t\in[-\pi,\pi]} \omega_1\left(E\left(D_t^{\alpha}X\right), \delta\right) < \infty$, by (19.10)–(19.12). ■

We finish with

Remark 19.20 The stochastic convergences of Theorems 19.18, 19.19 are derived by the convergences of the basic and simple real non-stochastic functions $\left\{1, \left(\sin\left(\frac{|s-t|}{4}\right)\right)^{\alpha+1}\right\}$, an amazing fact!

References

1. Anastassiou, G.: A study of positive linear operators by the method of moments, one-dimensional case. J. Approx. Theory **45**, 247–270 (1985)
2. Anastassiou, G.: Korovkin type inequalities in real normed vector spaces. Approx. Theory Appl. **2**, 39–53 (1986)

3. Anastassiou, G.: Multi-dimensional quantitative results for probability measures approximating the unit measure. Approx. Theory Appl. **2**, 93–103 (1986)
4. Anastassiou, G.A.: Korovkin inequalities for stochastic processes. J. Math. Anal. Appl. **157**(2), 366–384 (1991)
5. Anastassiou, G.A.: Moments in Probability and Approximation Theory. Pitman/Longman, # 287, UK (1993)
6. Anastassiou, G.: Fractional Differentiation Inequalities. Springer, Heildelberg (2009)
7. Anastassiou, G.: Fractional representation formulae and right fractional inequalities. Math. Comput. Model. **54**(11–12), 3098–3115 (2011)
8. Anastassiou, G.: Intelligent Mathematics: Computational Analysis. Springer, Heidelberg (2011)
9. Anastassiou, G.A.: Foundation of stochastic fractional calculus with fractional approximation of stochastic processes. Rev. R. Acad. Cienc. Exactas Fis. Nat. Ser. A Mat. RACSAM 114(2), Paper No. 89 (2020)
10. Anastassiou, G.A.: Commutative caputo fractional Korovkin inequalities for Stochastic processes (2020). Submitted for publication
11. Anastassiou, G.A.: Trigonometric commutative caputo fractional Korovkin theory for stochastic processes. Progr. Fract. Differ. Appl. (2020). Accepted
12. Korovkin, P.P.: Linear Operators and Approximation Theory. Hindustan Publ. Corp., Delhi (1960)
13. Royden, H.L.: Real Analysis, 2nd edn. MacMillan Publishing Co., Inc., New York (1968)
14. Shisha, O., Mond, B.: The degree of approximation to periodic functions by linear positive Operators. J. Approx. Theory **1**, 335–339 (1968)
15. Weba, M.: Korovkin systems of stochastic processes. Math. Z. **192**(1), 73–80 (1986)
16. Weba, M.: Quantitative results on monotone approximation of stochastic processes. Probab. Math. Stat. **11**(1), 109–120 (1990)
17. Weba, M.: A quantitative Korovkin theorem for random functions with multivariate domains. J. Approx. Theory **61**(1), 74–87 (1990)
18. Weba, M.: Monotone approximation of random functions with multivariate domains in respect of lattice semi-norms. Results Math. **20**(1–2), 554–576 (1991)

Chapter 20
Commutative Conformable Fractional Korovkin Approximation for Stochastic Processes

Here we research the expectation commutative stochastic positive linear operators acting on L^1-continuous stochastic processes which are conformable fractional differentiable. Under some mild, general and natural assumptions on the stochastic processes we produce related conformable fractional stochastic Shisha-Mond type inequalities pointwise and uniform. All convergences are produced quantitatively and are given by the conformable fractional stochastic inequalities involving the first modulus of continuity of the expectation of the αth right and left conformable fractional derivatives of the engaged stochastic process, $\alpha \in (n, n+1), n \in \mathbb{Z}_+$. The amazing fact here is that the simple real Korovkin test functions assumptions imply the conclusions of our conformable fractional stochastic Korovkin theory. We include also a full details application to stochastic Bernstein operators. See also [9].

20.1 Introduction

Our work is inspired by the following:

Korovkin's Theorem ([11], 1960) Let $(T_j)_{j \in \mathbb{N}}$ be a sequence of positive linear operators from $C([a,b])$ into itself, $[a,b] \subset \mathbb{R}$. In order to have $\lim_{j \to \infty} (T_j f)(t) = f(t)$ (in the sup-norm) for all $f \in C([a,b])$, it is enough to prove it for $f_0(t) = 1$, $f_1(t) = t$ and $f_2(t) = t^2$. The rate of the above convergence for arbitrary $f \in C([a,b])$ can be determined exactly from the rates of convergence for f_0, f_1, f_2.

The above theorem was put in an inequality form:

Shisha-Mond inequality ([13]) We have

$$\left\|T_j(f) - f\right\| \leq \|f\| \cdot \left\|T_j(1) - 1\right\| + \omega_1\left(f, \rho_j\right) \cdot \left(1 + \left\|T_j(1)\right\|\right),$$

where

$$\rho_j = \left(\left\|T_j\left((x-y)^2\right)(y)\right\|\right)^{\frac{1}{2}}.$$

G. A. Anastassiou, *Generalized Fractional Calculus*, Studies in Systems, Decision and Control 305, https://doi.org/10.1007/978-3-030-56962-4_20

In the last inequality $\|\cdot\|$ stands for the supremum norm and ω_1 for the first modulus of continuity. This inequality gives the rate of convergence of T_j to the unit operator I.

Anastassiou in [3–5] established a series of sharp inequalities for various cases of the parameters of the problem. However, Weba in [14–17] was the first, among many workers in quantitative results of Shisha-Mond type, to produce inequalities for stochastic processes. He assumed that T_j are E-commutative (E means expectation) and stochastically simple. According to his work, if a stochastic process $X(t,\omega)$, $t \in Q$ - a compact convex subset of a real normed vector space, $\omega \in Q$ - probability space, is to be approximated by positive linear operators T_j, then the maximal error in the qth mean is ($q \geq 1$)

$$\left\| T_j X - X \right\| = \sup_{t \in Q} \left(E \left| \left(T_j X \right)(t,\omega) - X(t,\omega) \right|^q \right)^{\frac{1}{q}}.$$

So, Weba established upper bounds for $\left\| T_j X - X \right\|$ involving his own natural general first modulus of continuity of X with several interesting applications.

Anastassiou continued [6] with the pointwise case of $q = 1$. Without stochastic simplicity of T_j he presented nearly best and best upper bounds for $\left| E \left(T_j X \right)(x_0) - (EX)(x_0) \right|$, $x_0 \in Q$.

The author here continues his above work on the approximation of stochastic processes, now at the conformable stochastic fractional level. He derives pointwise and uniform conformable fractional stochastic Shisha-Mond type inequalities, see the main Theorems 20.10, 20.11 and the several related corollaries. He gives an extensive application to stochastic Bernstein operators. He finishes with a pointwise and a uniform conformable fractional stochastic Korovkin theorem, derived by Theorems 20.10, 20.11. The stochastic convergences, about stochastic processes, of our conformable fractional Korovkin Theorems 20.21, 20.22 are implied only by the convergences of real basic non-stochastic functions.

20.2 Background—I

Here we follow [1], see also [10].

We need

Definition 20.1 *([1])* Let $a, b \in \mathbb{R}$. The left conformable fractional derivative starting from a of a function $f : [a, \infty) \to \mathbb{R}$ of order $0 < \alpha \leq 1$ is defined by

$$\left(T_\alpha^a f \right)(t) = \lim_{\varepsilon \to 0} \frac{f\left(t + \varepsilon (t-a)^{1-\alpha} \right) - f(t)}{\varepsilon}. \tag{20.1}$$

If $\left(T_\alpha^a f \right)(t)$ exists on (a, b), then

$$\left(T_{\alpha}^{a} f\right)(a) = \lim_{t \to a+} \left(T_{\alpha}^{a} f\right)(t). \tag{20.2}$$

The right conformable fractional derivative of order $0 < \alpha \leq 1$ terminating at b of $f : (-\infty, b] \to \mathbb{R}$ is defined by

$$\left({}_{\alpha}^{b}T f\right)(t) = -\lim_{\varepsilon \to 0} \frac{f\left(t + \varepsilon (b-t)^{1-\alpha}\right) - f(t)}{\varepsilon}. \tag{20.3}$$

If $\left({}_{\alpha}^{b}T f\right)(t)$ exists on (a, b), then

$$\left({}_{\alpha}^{b}T f\right)(b) = \lim_{t \to b-} \left({}_{\alpha}^{b}T f\right)(t). \tag{20.4}$$

Note that if f is differentiable then

$$\left(T_{\alpha}^{a} f\right)(t) = (t-a)^{1-\alpha} f'(t), \tag{20.5}$$

and

$$\left({}_{\alpha}^{b}T f\right)(t) = -(b-t)^{1-\alpha} f'(t). \tag{20.6}$$

Denote by

$$\left(I_{\alpha}^{a} f\right)(t) = \int_{a}^{t} (x-a)^{\alpha-1} f(x)\, dx, \tag{20.7}$$

and

$$\left({}^{b}I_{\alpha} f\right)(t) = \int_{t}^{b} (b-x)^{\alpha-1} f(x)\, dx, \tag{20.8}$$

these are the left and right conformable fractional integrals of order $0 < \alpha \leq 1$.

In the higher order case we can generalize things as follows:

Definition 20.2 *([1])* Let $\alpha \in (n, n+1]$, and set $\beta = \alpha - n$. Then, the left conformable fractional derivative starting from a of a function $f : [a, \infty) \to \mathbb{R}$ of order α, where $f^{(n)}(t)$ exists, is defined by

$$\left(\mathbf{T}_{\alpha}^{a} f\right)(t) = \left(T_{\beta}^{a} f^{(n)}\right)(t). \tag{20.9}$$

The right conformable fractional derivative of order α terminating at b of $f : (-\infty, b] \to \mathbb{R}$, where $f^{(n)}(t)$ exists, is defined by

$$\left({}_{\alpha}^{b}\mathbf{T} f\right)(t) = (-1)^{n+1} \left({}_{\beta}^{b}T f^{(n)}\right)(t). \tag{20.10}$$

If $\alpha = n+1$ then $\beta = 1$ and $\mathbf{T}_{n+1}^{a} f = f^{(n+1)}$.

If n is odd, then ${}_{n+1}^{b}\mathbf{T} f = -f^{(n+1)}$, and if n is even, then ${}_{n+1}^{b}\mathbf{T} f = f^{(n+1)}$.

When $n = 0$ (or $\alpha \in (0, 1]$), then $\beta = \alpha$, and (20.9), (20.10) collapse to (20.1), (20.2), (20.3), (20.4) respectively.

Lemma 20.3 ([1]) *Let $f : (a, b) \to \mathbb{R}$ be continuously differentiable and $0 < \alpha \leq 1$. Then, for all $t > a$ we have*

$$I_{\alpha}^{a} T_{\alpha}^{a} (f) (t) = f (t) - f (a) . \tag{20.11}$$

We need

Definition 20.4 *(see also* [1]*)* If $\alpha \in (n, n + 1]$, then the left fractional integral of order α starting at a is defined by

$$\left(\mathbf{I}_{\alpha}^{a} f\right) (t) = \frac{1}{n!} \int_{a}^{t} (t - x)^{n} (x - a)^{\alpha - 1} f (x) \, dx. \tag{20.12}$$

Similarly, (author's definition) the right fractional integral of order α terminating at b is defined by

$$\left({}^{b}\mathbf{I}_{\alpha} f\right) (t) = \frac{1}{n!} \int_{t}^{b} (x - t)^{n} (b - x)^{\alpha - 1} f (x) \, dx. \tag{20.13}$$

We need

Proposition 20.5 ([1]) *Let $\alpha \in (n, n + 1]$ and $f : [a, \infty) \to \mathbb{R}$ be $(n + 1)$ times continuously differentiable for $t > a$. Then, for all $t > a$ we have*

$$\mathbf{I}_{\alpha}^{a} \mathbf{T}_{\alpha}^{a} (f) (t) = f (t) - \sum_{k=0}^{n} \frac{f^{(k)} (a) (t - a)^{k}}{k!}. \tag{20.14}$$

We also have

Proposition 20.6 ([8], p. 154) *Let $\alpha \in (n, n + 1]$ and $f : (-\infty, b] \to \mathbb{R}$ be $(n + 1)$ times continuously differentiable for $t < b$. Then, for all $t < b$ we have*

$$-{}^{b}\,\mathbf{I}_{\alpha}\, {}_{\alpha}^{b}\mathbf{T} (f) (t) = f (t) - \sum_{k=0}^{n} \frac{f^{(k)} (b) (t - b)^{k}}{k!}. \tag{20.15}$$

If $n = 0$ or $0 < \alpha \leq 1$, then (see also [1])

$${}^{b}I_{\alpha}\, {}_{\alpha}^{b}T (f) (t) = f (t) - f (b) . \tag{20.16}$$

In conclusion we derive

Theorem 20.7 ([8], p. 155) *Let* $\alpha \in (n, n+1]$ *and* $f \in C^{n+1}([a,b])$, $n \in \mathbb{N}$. *Then*

(1)

$$f(t) - \sum_{k=0}^{n} \frac{f^{(k)}(a)(t-a)^k}{k!} = \frac{1}{n!} \int_a^t (t-x)^n (x-a)^{\beta-1} \left(\mathbf{T}_\alpha^a(f)\right)(x)\,dx, \tag{20.17}$$

and

(2)

$$f(t) - \sum_{k=0}^{n} \frac{f^{(k)}(b)(t-b)^k}{k!} = -\frac{1}{n!} \int_t^b (b-x)^{\beta-1} (x-t)^n \left({}_\alpha^b\mathbf{T}(f)\right)(x)\,dx, \tag{20.18}$$

$\forall\, t \in [a,b]$.

We make

Remark 20.8 *([8], p. 155)* We notice the following: let $\alpha \in (n, n+1]$ and $f \in C^{n+1}([a,b])$, $n \in \mathbb{N}$. Then ($\beta := \alpha - n$, $0 < \beta \le 1$)

$$\left(\mathbf{T}_\alpha^a(f)\right)(x) = \left(T_\beta^a f^{(n)}\right)(x) = (x-a)^{1-\beta} f^{(n+1)}(x), \tag{20.19}$$

and

$$\left({}_\alpha^b\mathbf{T}(f)\right)(x) = (-1)^{n+1} \left({}_\beta^b T f^{(n)}\right)(x) =$$

$$(-1)^{n+1}(-1)(b-x)^{1-\beta} f^{(n+1)}(x) = (-1)^n (b-x)^{1-\beta} f^{(n+1)}(x). \tag{20.20}$$

Consequently we get that

$$\left(\mathbf{T}_\alpha^a(f)\right)(x),\ \left({}_\alpha^b\mathbf{T}(f)\right)(x) \in C([a,b]).$$

Furthermore it is obvious that

$$\left(\mathbf{T}_\alpha^a(f)\right)(a) = \left({}_\alpha^b\mathbf{T}(f)\right)(b) = 0, \tag{20.21}$$

when $0 < \beta < 1$, i.e. when $\alpha \in (n, n+1)$.

If $f^{(k)}(a) = 0$, $k = 1, \ldots, n$, then

$$f(t) - f(a) = \frac{1}{n!} \int_a^t (t-x)^n (x-a)^{\beta-1} \left(\mathbf{T}_\alpha^a(f)\right)(x)\,dx, \tag{20.22}$$

$\forall\, t \in [a,b]$.

If $f^{(k)}(b) = 0$, $k = 1, \ldots, n$, then

$$f(t) - f(b) = -\frac{1}{n!} \int_t^b (b-x)^{\beta-1} (x-t)^n \left({}_\alpha^b\mathbf{T}(f)\right)(x)\,dx, \tag{20.23}$$

$\forall\, t \in [a, b]$.

We make

Remark 20.9 Let the positive linear operator L mapping $C([a, b])$ into $B([a, b])$ (the bounded functions). By the Riesz representation theorem [12] we have that there exists μ_t unique, completed Borel measure on $[a, b]$ with

$$\mu_t([a, b]) = L(1)(t) > 0, \tag{20.24}$$

such that

$$L(f)(t) = \int_{[a,b]} f(s)\, d\mu_t(s), \ \forall\, t \in [a, b], \forall\, f \in C([a, b]). \tag{20.25}$$

Let $\alpha \in (n, n+1)$, $n \in \mathbb{Z}_+$, $k = 1, \ldots, n$. Then by Hölder's inequality we obtain

$$\left| \int_{[a,b]} (s-t)^k\, d\mu_t(s) \right| \leq \int_{[a,b]} |s-t|^k\, d\mu_t(s) \leq$$

$$\left(\int_{[a,b]} |s-t|^{\alpha+1}\, d\mu_t(s) \right)^{\left(\frac{k}{\alpha+1}\right)} (\mu_t([a, b]))^{\left(\frac{\alpha+1-k}{\alpha+1}\right)}. \tag{20.26}$$

The last means

$$\left| L\left((s-t)^k\right)(t) \right| \leq L\left(|s-t|^k\right)(t) \leq \left(L\left(|s-t|^{\alpha+1}\right)(t) \right)^{\left(\frac{k}{\alpha+1}\right)} (L(1)(t))^{\left(\frac{\alpha+1-k}{\alpha+1}\right)}, \tag{20.27}$$

all $k = 1, \ldots, n$. It is clear that

$$\left\| L\left(|s-t|^{\alpha+1}\right)(t) \right\|_{\infty,[a,b]} < \infty.$$

Furthermore we derive

$$\left\| L\left((s-t)^k\right)(t) \right\|_{\infty,[a,b]} \leq \left\| L\left(|s-t|^k\right)(t) \right\|_{\infty,[a,b]} \leq$$

$$\left\| L\left(|s-t|^{\alpha+1}\right)(t) \right\|_{\infty,[a,b]}^{\left(\frac{k}{\alpha+1}\right)} \|L(1)\|_{\infty,[a,b]}^{\left(\frac{\alpha+1-k}{\alpha+1}\right)}, \tag{20.28}$$

all $k = 1, \ldots, n$.

From now on we will denote $\|\cdot\|_{\infty,[a,b]} = \|\cdot\|_{\infty}$ the supremum norm.

20.3 Preliminaries

Let $(\Omega, \mathcal{F}, P)$ be a probabilistic space and $L^1(\Omega, \mathcal{F}, P)$ be the space of all real-valued random variables $Y = Y(\omega)$ with

$$\int_{\Omega} |Y(\omega)| \, P(d\omega) < \infty.$$

Let $X = X(t, \omega)$ denote a stochastic process with index set $[a, b] \subset \mathbb{R}$ and real state space $(\mathbb{R}, \mathcal{B})$, where $\mathcal{B}$ is the σ-field of Borel subsets of $\mathbb{R}$. Here $C([a, b])$ is the space of continuous real-valued functions on $[a, b]$ and $\mathcal{B}([a, b])$ is the space of bounded real-valued functions on $[a, b]$. Also $C_{\Omega}([a, b]) = C\left([a, b], L^1(\Omega, \mathcal{F}, P)\right)$ is the space of L^1-continuous stochastic processes in t and $B_{\Omega}([a, b]) = \left\{X : \sup_{t \in [a,b]} \int_{\Omega} |X(t, \omega)| \, P(d\omega) < \infty\right\}$, obviously $C_{\Omega}([a, b]) \subset B_{\Omega}([a, b])$.

Let $\alpha \in (n, n+1)$, $n \in \mathbb{Z}_+$, and consider the subspace of stochastic processes $C_{\Omega}^{\alpha,n+1}([a, b]) := \{X : X(\cdot, \omega) \in C^{n+1}([a, b]), \forall \omega \in \Omega$ and $\left|X^{(n+1)}(t, \omega)\right| \leq M$, $\forall (t, \omega) \in [a, b] \times \Omega$, where $M > 0$; $X^{(k)}(t, \omega) \in C_{\Omega}([a, b]), k = 0, 1, \ldots, n$; also $X^{(n+1)}(t, \omega)$, $(t, \omega) \in [a, b] \times \Omega$, is a stochastic process$\}$. Clearly by (20.19) and (20.20), respectively, we get that $\mathbf{T}_{\alpha}^{t} X(z, \omega)$, $z \in [t, b]$, and ${}_{\alpha}^{t}\mathbf{T}X(z, \omega)$, $z \in [a, t]$, are stochastic processes for any $t \in [a, b]$.

Consider the linear operator

$$L : C_{\Omega}([a, b]) \hookrightarrow B_{\Omega}([a, b]).$$

If $X \in C_{\Omega}([a, b])$ is nonnegative and LX, too, then L is called positive. If $EL = LE$, then L is called E-commutative, where E is the expectation operator.

From [16, pp. 3–5] we have the following results

(i) $C([a, b]) \subset C_{\Omega}([a, b])$,
(ii) if $X \in C_{\Omega}([a, b])$, then $EX \in C([a, b])$,
and
(iii) if L is E-commutative, then L maps the subspace $C([a, b])$ into $B([a, b])$.

20.4 Background—II

Let $X \in C_{\Omega}^{\alpha,n+1}([a, b])$. We have that $\left|X^{(n+1)}(t, \omega)\right| \leq M$, $\forall (t, \omega) \in [a, b] \times \Omega$.

We notice that

$$\mathbf{T}_{\alpha}^{t} X(z, \omega) \overset{(20.19)}{=} (z - t)^{1-\beta} X^{(n+1)}(z, \omega), \; z \in [t, b],$$

and (20.29)

$${}_{\alpha}^{t}\mathbf{T}X(z, \omega) \overset{(20.20)}{=} (-1)^n (t - z)^{1-\beta} X^{(n+1)}(z, \omega), \; z \in [a, t],$$

$\forall\, \omega \in \Omega$, where $\beta = \alpha - n$, $\alpha \in (n, n+1)$, $n \in \mathbb{Z}_+$; $\forall\, t \in [a, b]$.

We have that

$$\mathbf{T}_{\alpha}^{t} X(t, \omega) = {}_{\alpha}^{t}\mathbf{T} X(t, \omega) = 0. \tag{20.30}$$

It is

$$\left|\mathbf{T}_{\alpha}^{t} X(z, \omega)\right| = \left|(z-t)^{1-\beta} X^{(n+1)}(z, \omega)\right| \leq (b-a)^{1-\beta} M, \tag{20.31}$$

$\forall\, (z, \omega) \in [t, b] \times \Omega$,

and

$$\left|{}_{\alpha}^{t}\mathbf{T} X(z, \omega)\right| = \left|(-1)^{n} (t-z)^{1-\beta} X^{(n+1)}(z, \omega)\right| \leq (b-a)^{1-\beta} M, \tag{20.32}$$

$\forall\, (z, \omega) \in [a, t] \times \Omega$.

Here E is the expectation operator

$$(EX)(t) = \int_{\Omega} X(t, \omega)\, P(d\omega).$$

Clearly, we get that

$$\left|\left(E\left(\mathbf{T}_{\alpha}^{t} X\right)\right)(z)\right| \leq (b-a)^{1-\beta} M, \tag{20.33}$$

$\forall\, z \in [t, b]$,

and

$$\left|\left(E\left({}_{\alpha}^{t}\mathbf{T} X\right)\right)(z)\right| \leq (b-a)^{1-\beta} M, \tag{20.34}$$

$\forall\, z \in [a, t]$.

We observe that the first modulus of continuity ($\delta > 0$)

$$\omega_1\left(\left(E\left(\mathbf{T}_{\alpha}^{t} X\right)\right), \delta\right)_{[t,b]} := \sup_{\substack{x,y \in [t,b] \\ |x-y| \leq \delta}} \left|\left(E\left(\mathbf{T}_{\alpha}^{t} X\right)\right)(x) - \left(E\left(\mathbf{T}_{\alpha}^{t} X\right)\right)(y)\right|$$

$$\overset{(20.33)}{\leq} 2(b-a)^{1-\beta} M, \quad \forall\, t \in [a, b]. \tag{20.35}$$

Hence it holds ($\delta > 0$)

$$\sup_{t \in [a,b]} \omega_1\left(\left(E\left(\mathbf{T}_{\alpha}^{t} X\right)\right), \delta\right)_{[t,b]} \leq 2(b-a)^{1-\beta} M, \tag{20.36}$$

and similarly, it holds

$$\sup_{t \in [a,b]} \omega_1\left(\left(E\left({}_{\alpha}^{t}\mathbf{T} X\right)\right), \delta\right)_{[a,t]} \overset{(20.34)}{\leq} 2(b-a)^{1-\beta} M. \tag{20.37}$$

By [7], p. 209, we have that ($\delta_1 > 0$)

$$\left|E\left(\mathbf{T}_{\alpha}^{t}X\right)(z)-E\left(\mathbf{T}_{\alpha}^{t}X\right)(t)\right|\leq\omega_1\left(E\left(\mathbf{T}_{\alpha}^{t}X\right),\delta_1\right)_{[t,b]}\left\lceil\frac{|z-t|}{\delta_1}\right\rceil\leq \tag{20.38}$$

$$\omega_1\left(E\left(\mathbf{T}_{\alpha}^{t}X\right),\delta_1\right)_{[t,b]}\left(1+\frac{(z-t)}{\delta_1}\right),$$

$\forall\, z\in[t,b]$, where $\lceil\cdot\rceil$ is the ceiling of the number, and similarly ($\delta_2>0$)

$$\left|E\left({}_{\alpha}^{t}\mathbf{T}X\right)(z)-E\left({}_{\alpha}^{t}\mathbf{T}X\right)(t)\right|\leq\omega_1\left(E\left({}_{\alpha}^{t}\mathbf{T}X\right),\delta_2\right)_{[a,t]}\left(1+\frac{(t-z)}{\delta_2}\right), \tag{20.39}$$

$\forall\, z\in[a,t]$.

We also set

$$\omega_1\left(E\left({}^{\alpha}\mathbf{T}^{t}X\right),\delta\right):=\max\left\{\omega_1\left(E\left(\mathbf{T}_{\alpha}^{t}X\right),\delta_1\right)_{[t,b]},\omega_1\left(E\left({}_{\alpha}^{t}\mathbf{T}X\right),\delta_2\right)_{[a,t]}\right\}, \tag{20.40}$$

where $\delta>0$.

20.5 Main Results

Following 3. Preliminaries we state

Theorem 20.10 *Let $\alpha\in(n,n+1)$, $n\in\mathbb{Z}_+$. Consider the positive E-commutative linear operator $L:C_{\Omega}([a,b])\hookrightarrow B_{\Omega}([a,b])$, and let $X\in C_{\Omega}^{\alpha,n+1}([a,b])$, $\delta>0$. Here $\omega_1\left(E\left({}^{\alpha}\mathbf{T}^{t}X\right),\delta\right)$ as in (20.40).*

Then

$$|(E(LX))(t)-(EX)(t)|\leq|(EX)(t)|\,|(L(1))(t)-1|+$$

$$\sum_{k=1}^{n}\frac{\left|\left(EX^{(k)}\right)(t)\right|}{k!}\left|L\left((s-t)^{k}\right)(t)\right|+\frac{\omega_1\left(E\left({}^{\alpha}\mathbf{T}^{t}X\right),\delta\right)}{\prod_{j=1}^{n}(\alpha+j-n)}$$

$$\left(L\left(|s-t|^{\alpha+1}\right)(t)\right)^{\frac{\alpha}{\alpha+1}}\left[\frac{(L(1)(t))^{\frac{1}{\alpha+1}}}{\alpha-n}+\frac{\left(L\left(|s-t|^{\alpha+1}\right)(t)\right)^{\frac{1}{\alpha+1}}}{\delta(\alpha+1)}\right], \tag{20.41}$$

$\forall\, t\in[a,b]$.

Proof By Theorem 20.7 we have

$$X(s,\omega)=\sum_{k=0}^{n}\frac{X^{(k)}(t,\omega)}{k!}(s-t)^{k}+$$

$$\frac{1}{n!}\int_{t}^{s}(s-z)^{n}(z-t)^{\beta-1}\left(\mathbf{T}_{\alpha}^{t}X(z,\omega)-\mathbf{T}_{\alpha}^{t}X(t,\omega)\right)dz, \tag{20.42}$$

for all $t\leq s\leq b$, $\forall\,\omega\in\Omega$; $\alpha\in(n,n+1)$, see also (20.30).

Furthermore it holds from Theorem 20.7 that

$$X(s,\omega)=\sum_{k=0}^{n}\frac{X^{(k)}(t,\omega)}{k!}(s-t)^{k}-$$

$$\frac{1}{n!}\int_{s}^{t}(t-z)^{\beta-1}(z-s)^{n}\left({}_{\alpha}^{t}\mathbf{T}X(z,\omega)-{}_{\alpha}^{t}\mathbf{T}X(t,\omega)\right)dz, \tag{20.43}$$

for all $a\leq s\leq t$, $\forall\,\omega\in\Omega$; $\alpha\in(n,n+1)$, see also (20.30).

Therefore we get

$$(EX)(s)=\sum_{k=0}^{n}\frac{\left(EX^{(k)}\right)(t)}{k!}(s-t)^{k}+$$

$$\frac{1}{n!}E\left(\int_{t}^{s}(s-z)^{n}(z-t)^{\beta-1}\left(\mathbf{T}_{\alpha}^{t}X(z,\omega)-\mathbf{T}_{\alpha}^{t}X(t,\omega)\right)dz\right), \tag{20.44}$$

all $t\leq s\leq b$,

and

$$(EX)(s)=\sum_{k=0}^{n}\frac{\left(EX^{(k)}\right)(t)}{k!}(s-t)^{k}-$$

$$\frac{1}{n!}E\left(\int_{s}^{t}(t-z)^{\beta-1}(z-s)^{n}\left({}_{\alpha}^{t}\mathbf{T}X(z,\omega)-{}_{\alpha}^{t}\mathbf{T}X(t,\omega)\right)dz\right), \tag{20.45}$$

all $a\leq s\leq t$.

By [2], p. 156, Theorem 20.15, the functions $\mathbf{T}_{\alpha}^{t}X$ and ${}_{\alpha}^{t}\mathbf{T}X$ are jointly measurable.

By (20.31) and (20.32) we obtain that

$E\left(\int_{t}^{s}(s-z)^{n}(z-t)^{\beta-1}\left(\mathbf{T}_{\alpha}^{t}X(z,\omega)-\mathbf{T}_{\alpha}^{t}X(t,\omega)\right)dz\right)$,

$E\left(\int_{s}^{t}(t-z)^{\beta-1}(z-s)^{n}\left({}_{\alpha}^{t}\mathbf{T}X(z,\omega)-{}_{\alpha}^{t}\mathbf{T}X(t,\omega)\right)dz\right)$ are finite.

Therefore, by Fubini-Tonelli's theorem [12], we get

$$(EX)(s) - (EX)(t) = \sum_{k=0}^{n} \frac{\left(EX^{(k)}\right)(t)}{k!} (s-t)^k + R_1(t,s), \tag{20.46}$$

where

$$R_1(t,s) := \frac{1}{n!} \int_t^s (s-z)^n (z-t)^{\beta-1} \left(E\left(\mathbf{T}_\alpha^t X\right)(z) - E\left(\mathbf{T}_\alpha^t X\right)(t)\right) dz, \tag{20.47}$$

all $t \le s \le b$,

and

$$(EX)(s) - (EX)(t) = \sum_{k=0}^{n} \frac{\left(EX^{(k)}\right)(t)}{k!} (s-t)^k - R_2(s,t), \tag{20.48}$$

where

$$R_2(s,t) := \frac{1}{n!} \int_s^t (t-z)^{\beta-1} (z-s)^n \left(E\left({}_\alpha^t\mathbf{T}X\right)(z) - E\left({}_\alpha^t\mathbf{T}X\right)(t)\right) dz, \tag{20.49}$$

all $a \le s \le t$.

We observe that ($t \le s \le b$, $\delta_1 > 0$)

$$|R_1(t,s)| \le \frac{1}{n!} \int_t^s (s-z)^n (z-t)^{\beta-1} \left|E\left(\mathbf{T}_\alpha^t X\right)(z) - E\left(\mathbf{T}_\alpha^t X\right)(t)\right| dz \overset{(20.38)}{\le}$$

$$\frac{\omega_1\left(E\left(\mathbf{T}_\alpha^t X\right), \delta_1\right)_{[t,b]}}{n!} \left(\int_t^s (s-z)^n (z-t)^{\beta-1} \left(1 + \frac{(z-t)}{\delta_1}\right) dz\right) = \tag{20.50}$$

$$\frac{\omega_1\left(E\left(\mathbf{T}_\alpha^t X\right), \delta_1\right)_{[t,b]}}{n!}$$

$$\left[\int_t^s (s-z)^{(n+1)-1} (z-t)^{\beta-1} dz + \frac{1}{\delta_1} \int_t^s (s-z)^{(n+1)-1} (z-t)^{(\beta+1)-1} dz\right] =$$

$$\frac{\omega_1\left(E\left(\mathbf{T}_\alpha^t X\right), \delta_1\right)_{[t,b]}}{n!}$$

$$\left[\frac{\Gamma(n+1)\Gamma(\beta)}{\Gamma(n+1+\beta)} (s-t)^{n+\beta} + \frac{1}{\delta_1} \frac{\Gamma(n+1)\Gamma(\beta+1)}{\Gamma(n+\beta+2)} (s-t)^{n+\beta+1}\right] =$$

$$\omega_1\left(E\left(\mathbf{T}_\alpha^t X\right), \delta_1\right)_{[t,b]} \left[\frac{1}{\prod_{j=0}^{n}(\beta+j)} (s-t)^{\alpha} + \frac{1}{\delta_1} \frac{1}{\prod_{j=0}^{n}(\beta+1+j)} (s-t)^{\alpha+1}\right]. \tag{20.51}$$

That is

$$|R_1(t,s)| \leq \omega_1\left(E\left(\mathbf{T}_\alpha^t X\right), \delta_1\right)_{[t,b]} \left[\frac{(s-t)^\alpha}{\prod_{j=0}^{n}(\beta+j)} + \frac{(s-t)^{\alpha+1}}{\delta_1 \prod_{j=0}^{n}(\beta+1+j)} \right], \tag{20.52}$$

all $t \leq s \leq b$, $\delta_1 > 0$.

Next we also observe ($a \leq s \leq t$, $\delta_2 > 0$)

$$|R_2(s,t)| \leq \frac{1}{n!}\int_s^t (t-z)^{\beta-1}(z-s)^n \left|E\left({}_\alpha^t\mathbf{T}X\right)(z) - E\left({}_\alpha^t\mathbf{T}X\right)(t)\right| dz \overset{(20.39)}{\leq}$$

$$\frac{\omega_1\left(E\left({}_\alpha^t\mathbf{T}X\right), \delta_2\right)_{[a,t]}}{n!}\int_s^t (t-z)^{\beta-1}(z-s)^n\left(1+\frac{(t-z)}{\delta_2}\right)dz = \tag{20.53}$$

$$\frac{\omega_1\left(E\left({}_\alpha^t\mathbf{T}X\right), \delta_2\right)_{[a,t]}}{n!}$$

$$\left[\int_s^t (t-z)^{\beta-1}(z-s)^{(n+1)-1}dz + \frac{1}{\delta_2}\int_s^t (t-z)^{(\beta+1)-1}(z-s)^{(n+1)-1}dz\right] =$$

$$\frac{\omega_1\left(E\left({}_\alpha^t\mathbf{T}X\right), \delta_2\right)_{[a,t]}}{n!}$$

$$\left[\frac{\Gamma(\beta)\Gamma(n+1)}{\Gamma(\beta+n+1)}(t-s)^{\beta+n} + \frac{1}{\delta_2}\frac{\Gamma(\beta+1)\Gamma(n+1)}{\Gamma(\beta+n+2)}(t-s)^{\beta+n+1}\right] =$$

$$\omega_1\left(E\left({}_\alpha^t\mathbf{T}X\right), \delta_2\right)_{[a,t]}\left[\frac{1}{\prod_{j=0}^{n}(\beta+j)}(t-s)^\alpha + \frac{1}{\delta_2}\frac{1}{\prod_{j=0}^{n}(\beta+1+j)}(t-s)^{\alpha+1}\right]. \tag{20.54}$$

That is

$$|R_2(s,t)| \leq \omega_1\left(E\left({}_\alpha^t\mathbf{T}X\right), \delta_2\right)_{[a,t]}\left[\frac{(t-s)^\alpha}{\prod_{j=0}^{n}(\beta+j)} + \frac{(t-s)^{\alpha+1}}{\delta_2\prod_{j=0}^{n}(\beta+1+j)}\right], \tag{20.55}$$

all $a \leq s \leq t$, $\delta_2 > 0$.

From now on we choose $\delta_1 = \delta_2 = \delta > 0$.

In conclusion we have (see (20.52), (20.55), (20.40))

$$|R_1(t,s)|, |R_2(s,t)| \leq$$

$$\omega_1\left(E\left({}^{\alpha}\mathbf{T}^{t}X\right),\delta\right)\left[\frac{|s-t|^{\alpha}}{\prod_{j=0}^{n}(\alpha+j-n)}+\frac{|s-t|^{\alpha+1}}{\delta\prod_{j=0}^{n}(\alpha+j+1-n)}\right], \tag{20.56}$$

for any $s,t\in[a,b]$.

We further have that

$$L(EX)(t)-(EX)(t)L(1)(t) \overset{(20.25)}{=}$$

$$\int_{[a,b]}(EX)(s)\,d\mu_t(s)-(EX)(t)L(1)(t) \overset{(20.24)}{=}$$

$$\int_{[a,t)}(EX)(s)\,d\mu_t(s)+\int_{[t,b]}(EX)(s)\,d\mu_t(s)-$$

$$\int_{[a,t)}(EX)(t)\,d\mu_t(s)-\int_{[t,b]}(EX)(t)\,d\mu_t(s)=$$

$$\int_{[a,t)}((EX)(s)-(EX)(t))\,d\mu_t(s)+\int_{[t,b]}((EX)(s)-(EX)(t))\,d\mu_t(s) \overset{\text{(by (20.46), (20.48))}}{=} \tag{20.57}$$

$$\sum_{k=1}^{n}\frac{\left(EX^{(k)}\right)(t)}{k!}L\left((s-t)^{k}\right)(t)-\int_{[a,t)}R_2(s,t)\,d\mu_t(s)+\int_{[t,b]}R_1(t,s)\,d\mu_t(s).$$

That is

$$L(EX)(t)-(EX)(t)L(1)(t)=\sum_{k=1}^{n}\frac{\left(EX^{(k)}\right)(t)}{k!}L\left((s-t)^{k}\right)(t)$$

$$-\int_{[a,t)}R_2(s,t)\,d\mu_t(s)+\int_{[t,b]}R_1(t,s)\,d\mu_t(s). \tag{20.58}$$

One can rewrite

$$L(EX)(t)-(EX)(t)=(EX)(t)\left[(L(1))(t)-1\right]+$$

$$\sum_{k=1}^{n}\frac{\left(EX^{(k)}\right)(t)}{k!}L\left((s-t)^{k}\right)(t)-\int_{[a,t)}R_2(s,t)\,d\mu_t(s) \tag{20.59}$$

$$+\int_{[t,b]}R_1(t,s)\,d\mu_t(s).$$

Consequently, by E-commutativity of L we find

$$|E[(LX)(t,\omega) - X(t,\omega)]| \leq |(EX)(t)| |(L(1))(t) - 1| +$$

$$\sum_{k=1}^{n} \frac{\left|\left(EX^{(k)}\right)(t)\right|}{k!} \left|L\left((s-t)^k\right)(t)\right| + \int_{[a,t)} |R_2(s,t)| \, d\mu_t(s) \tag{20.60}$$

$$+ \int_{[t,b]} |R_1(t,s)| \, d\mu_t(s) \overset{(20.56)}{\leq}$$

$$|(EX)(t)| |(L(1))(t) - 1| + \sum_{k=1}^{n} \frac{\left|\left(EX^{(k)}\right)(t)\right|}{k!} \left|L\left((s-t)^k\right)(t)\right| + \omega_1\left(E\left({}^{\alpha}\mathbf{T}^t X\right), \delta\right)$$

$$\left[\frac{1}{\prod_{j=0}^{n}(\alpha+j-n)} \int_{[a,b]} |s-t|^{\alpha} \, d\mu_t(s) + \frac{1}{\delta \prod_{j=0}^{n}(\alpha+j+1-n)} \int_{[a,b]} |s-t|^{\alpha+1} \, d\mu_t(s)\right]$$

(by Hölder's inequality)

$$\leq |(EX)(t)| |(L(1))(t) - 1| + \sum_{k=1}^{n} \frac{\left|\left(EX^{(k)}\right)(t)\right|}{k!} \left|L\left((s-t)^k\right)(t)\right| +$$

$$\omega_1\left(E\left({}^{\alpha}\mathbf{T}^t X\right), \delta\right) \left[\frac{1}{\prod_{j=0}^{n}(\alpha+j-n)} \left(\int_{[a,b]} |t-s|^{\alpha+1} \, d\mu_t(s)\right)^{\frac{\alpha}{\alpha+1}} (L(1)(t))^{\frac{1}{\alpha+1}}\right.$$

$$\left. + \frac{1}{\delta \prod_{j=0}^{n}(\alpha+j+1-n)} \int_{[a,b]} |t-s|^{\alpha+1} \, d\mu_t(s)\right]. \tag{20.61}$$

We have proved that

$$|(E(LX))(t) - (EX)(t)| \leq |(EX)(t)| |(L(1))(t) - 1| +$$

$$\sum_{k=1}^{n} \frac{\left|\left(EX^{(k)}\right)(t)\right|}{k!} \left|L\left((s-t)^k\right)(t)\right| + \omega_1\left(E\left({}^{\alpha}\mathbf{T}^t X\right), \delta\right)$$

$$\left[\frac{1}{\prod_{j=0}^{n}(\alpha+j-n)}\left(L\left(|s-t|^{\alpha+1}\right)(t)\right)^{\frac{\alpha}{\alpha+1}}(L(1)(t))^{\frac{1}{\alpha+1}}+\right.$$

$$\left.\frac{1}{\delta\prod_{j=0}^{n}(\alpha+j+1-n)}\left(L\left(|s-t|^{\alpha+1}\right)(t)\right)\right] \tag{20.62}$$

$$=|(EX)(t)|\,|(L(1))(t)-1|+\sum_{k=1}^{n}\frac{\left|\left(EX^{(k)}\right)(t)\right|}{k!}\left|L\left((s-t)^{k}\right)(t)\right|+$$

$$\frac{\omega_1\left(E\left({}^{\alpha}\mathbf{T}^{t}X\right),\delta\right)}{\prod_{j=1}^{n}(\alpha+j-n)}\left(L\left(|s-t|^{\alpha+1}\right)(t)\right)^{\frac{\alpha}{\alpha+1}}$$

$$\left[\frac{(L(1)(t))^{\frac{1}{\alpha+1}}}{\alpha-n}+\frac{\left(L\left(|s-t|^{\alpha+1}\right)(t)\right)^{\frac{1}{\alpha+1}}}{\delta(\alpha+1)}\right],$$

establishing the claim. ∎

We continue with a uniform estimate:

Theorem 20.11 *All as in Theorem 20.10. Then*

$$\|E(LX)-EX\|_{\infty}\leq\|EX\|_{\infty}\|L(1)-1\|_{\infty}+$$

$$\sum_{k=1}^{n}\frac{\left\|EX^{(k)}\right\|_{\infty}}{k!}\left\|L\left((s-t)^{k}\right)(t)\right\|_{\infty}+\frac{\sup_{t\in[a,b]}\omega_1\left(E\left({}^{\alpha}\mathbf{T}^{t}X\right),\delta\right)}{\prod_{j=1}^{n}(\alpha+j-n)}$$

$$\left\|L\left(|s-t|^{\alpha+1}\right)(t)\right\|_{\infty}^{\frac{\alpha}{\alpha+1}}\left[\frac{\|L(1)\|_{\infty}^{\frac{1}{\alpha+1}}}{\alpha-n}+\frac{\left\|L\left(|s-t|^{\alpha+1}\right)(t)\right\|_{\infty}^{\frac{1}{\alpha+1}}}{\delta(\alpha+1)}\right]<\infty. \tag{20.63}$$

Proof By (20.36), (20.37), (20.40) and (20.41). ∎

Next, we cover the case $n=0$.

Definition 20.12 We call the subspace of stochastic process: $C_{\Omega}^{\alpha,1}([a,b]) := \{X : X(\cdot,\omega) \in C^{1}([a,b]), \forall \omega \in \Omega$ and $\left|X^{(1)}(t,\omega)\right| \leq M, \forall (t,\omega) \in [a,b] \times \Omega$, where $M > 0$; and $X(t,\omega) \in C_{\Omega}([a,b])$; also $X^{(1)}(t,\omega), (t,\omega) \in [a,b] \times \Omega$ is a stochastic process$\}$.

Corollary 20.13 *Let $0 < \alpha < 1$. Consider the positive E-commutative linear operator $L : C_{\Omega}([a,b]) \hookrightarrow B_{\Omega}([a,b])$, and let $X \in C_{\Omega}^{\alpha,1}([a,b])$, $\delta > 0$. Here $\omega_1\left(E\left({}^{\alpha}T^{t}X\right),\delta\right)$ as in (20.40).*

Then

$$|(E(LX))(t) - (EX)(t)| \leq |(EX)(t)|\,|(L(1))(t) - 1| + \tag{20.64}$$

$$\omega_1\left(E\left({}^{\alpha}T^{t}X\right),\delta\right)\left(L\left(|s-t|^{\alpha+1}\right)(t)\right)^{\frac{\alpha}{\alpha+1}}$$

$$\left[\frac{(L(1)(t))^{\frac{1}{\alpha+1}}}{\alpha} + \frac{\left(L\left(|s-t|^{\alpha+1}\right)(t)\right)^{\frac{1}{\alpha+1}}}{\delta(\alpha+1)}\right],$$

$\forall\, t \in [a,b]$.

Proof By (20.41). ∎

Corollary 20.14 *All as in Corollary 20.13. Then*

$$\|E(LX) - EX\|_{\infty} \leq \|EX\|_{\infty}\,\|L(1) - 1\|_{\infty} +$$

$$\sup_{t\in[a,b]} \omega_1\left(E\left({}^{\alpha}T^{t}X\right),\delta\right)\left\|L\left(|s-t|^{\alpha+1}\right)(t)\right\|_{\infty}^{\frac{\alpha}{\alpha+1}} \tag{20.65}$$

$$\left[\frac{\|L(1)\|_{\infty}^{\frac{1}{\alpha+1}}}{\alpha} + \frac{\left\|L\left(|s-t|^{\alpha+1}\right)(t)\right\|_{\infty}^{\frac{1}{\alpha+1}}}{\delta(\alpha+1)}\right] < \infty.$$

Proof By (20.63). ∎

Corollary 20.15 *All as in Corollary 20.13, and $L(1) = 1$. Then*

$$\|E(LX) - EX\|_{\infty} \leq \sup_{t\in[a,b]} \omega_1\left(E\left({}^{\alpha}T^{t}X\right),\delta\right) \tag{20.66}$$

$$\left\|L\left(|s-t|^{\alpha+1}\right)(t)\right\|_{\infty}^{\frac{\alpha}{\alpha+1}}\left[\frac{1}{\alpha} + \frac{\left\|L\left(|s-t|^{\alpha+1}\right)(t)\right\|_{\infty}^{\frac{1}{\alpha+1}}}{\delta(\alpha+1)}\right].$$

Proof From (20.65). ∎

We continue with

Corollary 20.16 *All as in Corollary 20.13, and* $L(1) = 1$. *Then*

$$\|E(LX) - EX\|_\infty \leq \frac{\alpha+1}{\alpha} \sup_{t\in[a,b]} \omega_1 \left(E\left({}^{\alpha}T^{t}X\right), \frac{1}{(\alpha+1)} \left\| L\left(|s-t|^{\alpha+1}\right)(t) \right\|_\infty^{\frac{1}{\alpha+1}} \right)$$

$$\left\| L\left(|s-t|^{\alpha+1}\right)(t) \right\|_\infty^{\frac{\alpha}{\alpha+1}}. \tag{20.67}$$

Proof By (20.66): we take there

$$\delta = \frac{1}{(\alpha+1)} \left\| L\left(|s-t|^{\alpha+1}\right)(t) \right\|_\infty^{\frac{1}{\alpha+1}} > 0.$$

In case of $\left\| L\left(|s-t|^{\alpha+1}\right)(t) \right\|_\infty = 0$, we have that $L\left(|s-t|^{\alpha+1}\right)(t) = 0$, $\forall\, t \in [a,b]$. That is, by (20.25) $\int_{[a,b]} |s-t|^{\alpha+1} d\mu_t(s) = 0$, $\forall\, t \in [a,b]$, where μ_t is a probability measure, any $t \in [a,b]$.

The last implies $|s-t|^{\alpha+1} = 0$, a.e., hence $|s-t| = 0$, a.e., thus $s = t$, a.e., which means $\mu_t\{s \in [a,b] : s \neq t\} = 0$, i.e. $\mu_t = \delta_t$, $\forall\, t \in [a,b]$, where δ_t is the unit Dirac measure.

Consequently we have $E(LX)(t) = L(EX)(t) = \int_{[a,b]} (EX)(s)\, d\delta_t(s) = (EX)(t)$, $\forall\, t \in [a,b]$.

That is $E(LX) = EX$ over $[a,b]$. Therefore both sides of inequality (20.67) equal to zero.

Hence (20.67) is always true. ∎

20.6 Application

Let $f \in C([0,1])$ and the Bernstein polynomials

$$B_N(f)(t) := \sum_{k=0}^{N} f\left(\frac{k}{N}\right) \binom{N}{k} t^k (1-t)^{N-k}, \tag{20.68}$$

$\forall\, t \in [0,1]$, $\forall\, N \in \mathbb{N}$.

We have that $B_N 1 = 1$ and B_N is a positive linear operator.

We have that

$$B_N\left((\cdot - t)^2\right)(t) = \frac{t(1-t)}{N}, \quad \forall\, t \in [0,1], \tag{20.69}$$

and

$$\left\| B_N\left((\cdot - t)^2\right)(t) \right\|_\infty^{\frac{1}{2}} \leq \frac{1}{2\sqrt{N}}, \quad \forall\, N \in \mathbb{N}. \tag{20.70}$$

Define the corresponding stochastic application of B_N by

$$B_N(X)(t,\omega) := B_N(X(\cdot,\omega))(t) = \sum_{k=0}^{N} X\left(\frac{k}{N},\omega\right)\binom{N}{k} t^k (1-t)^{N-k}, \tag{20.71}$$

$\forall\, t \in [0,1], \forall\, \omega \in \Omega$, $N \in \mathbb{N}$, where X is a stochastic process. Clearly $B_N(X)$ is a stochastic process and $B_N : C_\Omega([0,1]) \hookrightarrow C_\Omega([0,1])$. Notice that

$$(EB_N(X))(t) = \sum_{k=0}^{N} (EX)\left(\frac{k}{N}\right)\binom{N}{k} t^k (1-t)^{N-k} = (B_N(EX))(t), \tag{20.72}$$

$\forall\, t \in [0,1]$.

That is $EB_N = B_N E$, i.e. B_N is an E-commutative positive linear operator.

We give

Proposition 20.17 *Let* $0 < \alpha < 1$ *and* $X \in C_\Omega^{\alpha,1}([0,1])$. *Then*

$$\|E(B_N X) - EX\|_\infty \leq$$

$$\frac{\alpha+1}{\alpha} \sup_{t\in[0,1]} \omega_1\left(E\left({}^{\alpha}T^{t}X\right), \frac{1}{(\alpha+1)}\left\|B_N\left(|s-t|^{\alpha+1}\right)(t)\right\|_\infty^{\frac{1}{\alpha+1}}\right)$$

$$\left\|B_N\left(|s-t|^{\alpha+1}\right)(t)\right\|_\infty^{\frac{\alpha}{\alpha+1}}, \quad \forall\, N \in \mathbb{N}. \tag{20.73}$$

Proof By (20.67). ∎

In particular we get:

Proposition 20.18 *Let* $X \in C_\Omega^{\frac{1}{2},1}([0,1])$. *Then*

$$\|E(B_N X) - EX\|_\infty \leq 3 \sup_{t\in[0,1]} \omega_1\left(E\left({}^{\frac{1}{2}}T^{t}X\right), \frac{2}{3}\left\|B_N\left(|s-t|^{\frac{3}{2}}\right)(t)\right\|_\infty^{\frac{2}{3}}\right)$$

$$\left\|B_N\left(|s-t|^{\frac{3}{2}}\right)(t)\right\|_\infty^{\frac{1}{3}}, \quad \forall\, N \in \mathbb{N}. \tag{20.74}$$

Proof By (20.73) and $\alpha = \frac{1}{2}$. ∎

We make

Remark 20.19 We notice that

$$B_N\left(|s-t|^{\frac{3}{2}}\right)(t) = \sum_{k=0}^{N}\left|t-\frac{k}{N}\right|^{\frac{3}{2}}\binom{N}{k} t^k (1-t)^{N-k}$$

(by discrete Hölder's inequality)

$$\leq \left(\sum_{k=0}^{N} \left| t - \frac{k}{N} \right|^2 \binom{N}{k} t^k (1-t)^{N-k} \right)^{\frac{3}{4}} \tag{20.75}$$

$$\overset{(20.69)}{=} \left(\frac{1}{N} t (1-t) \right)^{\frac{3}{4}} \leq \frac{1}{(4N)^{\frac{3}{4}}}, \quad \forall\, t \in [0,1].$$

That is

$$\left\| B_N \left(|s-t|^{\frac{3}{2}} \right)(t) \right\|_{\infty} \leq \frac{1}{(4N)^{\frac{3}{4}}}, \tag{20.76}$$

and

$$\left\| B_N \left(|s-t|^{\frac{3}{2}} \right)(t) \right\|_{\infty}^{\frac{1}{3}} \leq \frac{1}{(4N)^{\frac{1}{4}}}, \tag{20.77}$$

and

$$\left\| B_N \left(|s-t|^{\frac{3}{2}} \right)(t) \right\|_{\infty}^{\frac{2}{3}} \leq \frac{1}{2\sqrt{N}}, \tag{20.78}$$

$\forall\, N \in \mathbb{N}$.

We derive

Proposition 20.20 *Let* $X \in C_{\Omega}^{\frac{1}{2},1}([0,1])$. *Then*

$$\| E(B_N X) - EX \|_{\infty} \leq \frac{3}{\sqrt[4]{4N}} \sup_{t \in [0,1]} \omega_1 \left(E\left({}_{\frac{1}{2}}T^t X \right), \frac{1}{3\sqrt{N}} \right), \tag{20.79}$$

$\forall\, N \in \mathbb{N}$.

Hence $\lim_{N \to \infty} E(B_N X) = EX$, *uniformly.*

Proof By (20.74) and Remark 20.19. ∎

20.7 Conformable Fractional Stochastic Korovkin Theory

Here L is meant as a sequence of positive E-commutative linear operators and all assumptions are as in Theorem 20.10.

We give

Theorem 20.21 *We further assume that* $L(1)(t) \to 1$ *and* $L\left(|s-t|^{\alpha+1}\right)(t) \to 0$, *then* $(E(LX))(t) \to (EX)(t)$, *for any* $X \in C_{\Omega}^{\alpha,n+1}([a,b])$, $\forall\, t \in [a,b]$, *a pointwise convergence; where* $\alpha \in (n, n+1)$, $n \in \mathbb{Z}_+$.

Proof Based on (20.41), (20.27), and that $L(1)(t)$ is bounded as a sequence of functions. Also $\omega_1\left(E\left({}^{\alpha}\mathbf{T}^t X\right), \delta\right)$ is bounded, see (20.36), (20.37) and (20.40). ■

We continue with

Theorem 20.22 *We further assume that* $L(1)(t) \to 1$, *uniformly and* $\left\| L\left(|s-t|^{\alpha+1}\right)(t)\right\|_{\infty} \to 0$, *then* $E(LX) \to EX$, *uniformly over* $[a,b]$, *for any* $X \in C_{\Omega}^{\alpha,n+1}([a,b])$; *where* $\alpha \in (n, n+1)$, $n \in \mathbb{Z}_+$.

Proof Based on (20.63), (20.28), and that $\|L(1)\|_{\infty}$ is bounded. Also it is $\sup_{t\in[a,b]} \omega_1\left(E\left({}^{\alpha}\mathbf{T}^t X\right), \delta\right) < \infty$, by (20.36), (20.37) and (20.40). ■

We finish with

Remark 20.23 The stochastic convergences of Theorems 20.21, 20.22 are derived by the convergences of the basic and simple real non-stochastic functions $\left\{1, |s-t|^{\alpha+1}\right\}$, an amazing fact!

References

1. Abdeljawad, T.: On conformable fractional calculus. J. Comput. Appl. Math. **279**, 57–66 (2015)
2. Aliprantis, C., Burkinshaw, O.: Principles of Real Analysis, 3rd edn. Academic, San Diego (1998)
3. Anastassiou, G.: A study of positive linear operators by the method of moments, one-dimensional case. J. Approx. Theory **45**, 247–270 (1985)
4. Anastassiou, G.: Korovkin type inequalities in real normed vector spaces. Approx. Theory Appl. **2**, 39–53 (1986)
5. Anastassiou, G.: Multi-dimensional quantitative results for probability measures approximating the unit measure. Approx. Theory Appl. **2**, 93–103 (1986)
6. Anastassiou, G.A.: Korovkin inequalities for stochastic processes. J. Math. Anal. Appl. **157**(2), 366–384 (1991)
7. Anastassiou, G.A.: Moments in Probability and Approximation Theory. Pitman/Longman, # 287, UK (1993)
8. Anastassiou, G.: Nonlinearity: Ordinary and Fractional Approximations by Sublinear and Max-product operators. Springer, Heidelberg (2018)
9. Anastassiou, G.: Commutative conformable fractional Korovkin properties for Stochastic processes (2020). Submitted for publication
10. Khalil, R., Al Horani, M., Yousef, A., Sababheh, M.: A new definition of fractional derivative. J. Comput. Appl. Math. **264**, 65–70 (2014)
11. Korovkin, P.P.: Linear Operators and Approximation Theory. Hindustan Publ. Corp., Delhi (1960)
12. Royden, H.L.: Real Analysis, 2nd edn. MacMillan Publishing Co., Inc., New York (1968)
13. Shisha, O., Mond, B.: The degree of convergence of sequences of linear positive operators. Natl. Acad. Sci. U.S. **60**, 1196–1200 (1968)
14. Weba, M.: Korovkin systems of stochastic processes. Math. Z. **192**(1), 73–80 (1986)
15. Weba, M.: Quantitative results on monotone approximation of stochastic processes. Probab. Math. Stat. **11**(1), 109–120 (1990)
16. Weba, M.: A quantitative Korovkin theorem for random functions with multivariate domains. J. Approx. Theory **61**(1), 74–87 (1990)
17. Weba, M.: Monotone approximation of random functions with multivariate domains in respect of lattice semi-norms. Results Math. **20**(1–2), 554–576 (1991)

Chapter 21 Trigonometric Commutative Conformable Fractional Korovkin Approximation for Stochastic Processes

Here we research from the trigonometric point of view expectation commutative stochastic positive linear operators acting on L^1-continuous stochastic processes which are Conformable fractional differentiable. Under some mild, general and natural assumptions on the stochastic processes we produce related trigonometric Conformable fractional stochastic Shisha-Mond type inequalities pointwise and uniform. All convergences are produced with rates and are given by the trigonometric Conformable fractional stochastic inequalities involving the first modulus of continuity of the expectation of the αth right and left Conformable fractional derivatives of the engaged stochastic process, $\alpha \in (n, n+1)$, $n \in \mathbb{Z}_+$. The amazing fact here is that the basic non-stochastic real Korovkin test functions assumptions imply the conclusions of our trigonometric Conformable fractional stochastic Korovkin theory. We include also a detailed trigonometric application to stochastic Bernstein operators. See also [8].

21.1 Introduction

In this work among others we are motivated by the following results.

Theorem A (Korovkin [10], (1960)) *Let $L_n : C([-\pi, \pi]) \to C([-\pi, \pi])$, $n \in \mathbb{N}$, be a sequence of positive linear operators. Assume $L_n(1) \overset{u}{\to} 1$ (uniformly), $L_n(\cos t) \overset{u}{\to} \cos t$, $L_n(\sin t) \overset{u}{\to} \sin t$, as $n \to \infty$. Then $L_n f \overset{u}{\to} f$, for every $f \in C([-\pi, \pi])$ that is 2π-periodic.*

Let $f \in C([a, b])$ and $0 \le \delta \le b - a$. The first modulus of continuity of f at δ is given by

$$\omega_1(f, \delta) = \sup\{|f(x) - f(y)|; x, y \in [a, b], |x - y| \le \delta\}.$$

G. A. Anastassiou, *Generalized Fractional Calculus*, Studies in Systems, Decision and Control 305, https://doi.org/10.1007/978-3-030-56962-4_21

If $\delta > b - a$, then we define

$$\omega_1(f, \delta) = \omega_1(f, b - a).$$

Another motivation is the following.

Theorem B (Shisha and Mond [12], (1968)) *Let $L_1, L_2, \ldots,$ be linear positive operators, whose common domain D consists of real functions with domain $(-\infty, \infty)$. Suppose $1, \cos x, \sin x, f$ belong to D, where f is an everywhere continuous, 2π-periodic function, with modulus of continuity ω_1. Let $-\infty < a < b < \infty$, and suppose that for $n = 1, 2, \ldots, L_n(1)$ is bounded in $[a, b]$. Then for $n = 1, 2, \ldots,$*

$$\|L_n(f) - f\|_\infty \leq \|f\|_\infty \|L_n(1) - 1\|_\infty + \|L_n(1) + 1\|_\infty \omega_1(f, \mu_n), \quad (*)$$

where

$$\mu_n = \pi \left\| \left(L_n \left(\sin^2 \left(\frac{t - x}{2} \right) \right) \right) (x) \right\|_\infty^{\frac{1}{2}},$$

and $\|\cdot\|_\infty$ stands for the sup norm over $[a, b]$.

In particular, if $L_n(1) = 1$, then (*) reduces to

$$\|L_n(f) - f\|_\infty \leq 2\omega_1(f, \mu_n).$$

One can easily see that, for $m = 1, 2, \ldots,$

$$\mu_n^2 \leq \left(\frac{\pi^2}{2}\right) \left[\|L_n(1) - 1\|_\infty + \right.$$

$$\left. \|(L_n(\cos t))(x) - \cos x\|_\infty + \|(L_n(\sin t))(x) - \sin x\|_\infty \right],$$

so the last along with (*) prove Korovkin's Theorem A in a quantitative way and with rates of convergence.

Anastassiou in [2–4] established a series of sharp inequalities for various cases of the parameters of the problem. However, Weba in [13–16] was the first, among many workers in quantitative results of Shisha-Mond type, to produce inequalities for stochastic processes. He assumed that the positive linear operators L_j acting on stochastic processes X, are E-commutative (E means expectation) and stochastically simple.

Anastassiou continued in [5] the pointwise case and without stochastic simplicity of L_j he found nearly best and best upper bounds for $\left|E\left(L_j X\right)(x_0) - (EX)(x_0)\right|$, $x_0 \in Q$—a compact convex subset of a real normed vector space.

The author here continues his above work on the trigonometric approximation of stochastic processes, now at the Conformable stochastic fractional level. He derives pointwise and uniform trigonometric Conformable fractional stochastic Shisha-

Mond type inequalities, see the main Theorems 21.8, 21.9 and the several related corollaries. He gives an extensive trigonometric application to stochastic Bernstein operators. He finishes with a pointwise and a uniform Conformable fractional trigonometric stochastic Korovkin theorem, derived by Theorems 21.8, 21.9. The stochastic convergences, about stochastic processes, of our trigonometric Conformable fractional Korovkin Theorems 21.18, 21.19 are derived only by the convergences of real basic non-stochastic functions.

Our results here are relied on [7].

21.2 Background—I

Here we follow [1], see also [9].

We need

Definition 21.1 *([1])* Let $a, b \in \mathbb{R}$. The left conformable fractional derivative starting from a of a function $f : [a, \infty) \to \mathbb{R}$ of order $0 < \alpha \leq 1$ is defined by

$$\left(T_{\alpha}^{a} f\right)(t) = \lim_{\varepsilon \to 0} \frac{f\left(t + \varepsilon (t-a)^{1-\alpha}\right) - f(t)}{\varepsilon}. \tag{21.1}$$

If $\left(T_{\alpha}^{a} f\right)(t)$ exists on (a, b), then

$$\left(T_{\alpha}^{a} f\right)(a) = \lim_{t \to a+} \left(T_{\alpha}^{a} f\right)(t). \tag{21.2}$$

The right conformable fractional derivative of order $0 < \alpha \leq 1$ terminating at b of $f : (-\infty, b] \to \mathbb{R}$ is defined by

$$\left({}_{\alpha}^{b}T f\right)(t) = -\lim_{\varepsilon \to 0} \frac{f\left(t + \varepsilon (b-t)^{1-\alpha}\right) - f(t)}{\varepsilon}. \tag{21.3}$$

If $\left({}_{\alpha}^{b}T f\right)(t)$ exists on (a, b), then

$$\left({}_{\alpha}^{b}T f\right)(b) = \lim_{t \to b-} \left({}_{\alpha}^{b}T f\right)(t). \tag{21.4}$$

Note that if f is differentiable then

$$\left(T_{\alpha}^{a} f\right)(t) = (t-a)^{1-\alpha} f'(t), \tag{21.5}$$

and

$$\left({}_{\alpha}^{b}T f\right)(t) = -(b-t)^{1-\alpha} f'(t). \tag{21.6}$$

We need:

Definition 21.2 *([1])* Let $\alpha \in (n, n+1]$, and set $\beta = \alpha - n$. Then, the left conformable fractional derivative starting from a of a function $f : [a, \infty) \to \mathbb{R}$ of order α, where $f^{(n)}(t)$ exists, is defined by

$$\left(\mathbf{T}_{\alpha}^{a} f\right)(t) = \left(T_{\beta}^{a} f^{(n)}\right)(t). \tag{21.7}$$

The right conformable fractional derivative of order α terminating at b of $f : (-\infty, b] \to \mathbb{R}$, where $f^{(n)}(t)$ exists, is defined by

$$\left({}_{\alpha}^{b}\mathbf{T} f\right)(t) = (-1)^{n+1}\left({}_{\beta}^{b}T f^{(n)}\right)(t). \tag{21.8}$$

If $\alpha = n+1$ then $\beta = 1$ and $\mathbf{T}_{n+1}^{a} f = f^{(n+1)}$.

If n is odd, then ${}_{n+1}^{b}\mathbf{T} f = -f^{(n+1)}$, and if n is even, then ${}_{n+1}^{b}\mathbf{T} f = f^{(n+1)}$.

When $n = 0$ (or $\alpha \in (0, 1]$), then $\beta = \alpha$, and (21.7), (21.8) collapse to (21.1), (21.2), (21.3), (21.4) respectively.

We make

Remark 21.3 *([6], p. 155)* We notice the following: let $\alpha \in (n, n+1]$ and $f \in C^{n+1}([a, b])$, $n \in \mathbb{N}$. Then ($\beta := \alpha - n$, $0 < \beta \leq 1$)

$$\left(\mathbf{T}_{\alpha}^{a}(f)\right)(x) = \left(T_{\beta}^{a} f^{(n)}\right)(x) = (x-a)^{1-\beta} f^{(n+1)}(x), \tag{21.9}$$

and

$$\left({}_{\alpha}^{b}\mathbf{T}(f)\right)(x) = (-1)^{n+1}\left({}_{\beta}^{b}T f^{(n)}\right)(x) =$$

$$(-1)^{n+1}(-1)(b-x)^{1-\beta} f^{(n+1)}(x) = (-1)^{n}(b-x)^{1-\beta} f^{(n+1)}(x). \tag{21.10}$$

Consequently we get that

$$\left(\mathbf{T}_{\alpha}^{a}(f)\right)(x),\ \left({}_{\alpha}^{b}\mathbf{T}(f)\right)(x) \in C([a, b]).$$

Furthermore it is obvious that

$$\left(\mathbf{T}_{\alpha}^{a}(f)\right)(a) = \left({}_{\alpha}^{b}\mathbf{T}(f)\right)(b) = 0, \tag{21.11}$$

when $0 < \beta < 1$, i.e. when $\alpha \in (n, n+1)$.

21.3 Preliminaries

Let $(\Omega, \mathcal{F}, P)$ be a probabilistic space and $L^1(\Omega, \mathcal{F}, P)$ be the space of all real-valued random variables $Y = Y(\omega)$ with

$$\int_{\Omega} |Y(\omega)| P(d\omega) < \infty.$$

Let $X = X(t, \omega)$ denote a stochastic process with index set $[a, b] \subset \mathbb{R}$ and real state space $(\mathbb{R}, \mathcal{B})$, where $\mathcal{B}$ is the σ-field of Borel subsets of $\mathbb{R}$. Here $C([a, b])$ is the space of continuous real-valued functions on $[a, b]$ and $\mathcal{B}([a, b])$ is the space of bounded real-valued functions on $[a, b]$. Also $C_{\Omega}([a, b]) = C\left([a, b], L^1(\Omega, \mathcal{F}, P)\right)$ is the space of L^1-continuous stochastic processes in t and $B_{\Omega}([a, b]) = \left\{X : \sup_{t \in [a,b]} \int_{\Omega} |X(t, \omega)| P(d\omega) < \infty\right\}$, obviously $C_{\Omega}([a, b]) \subset B_{\Omega}([a, b])$.

Let $\alpha \in (n, n+1)$, $n \in \mathbb{Z}_+$, and consider the subspace of stochastic processes $C_{\Omega}^{\alpha, n+1}([a, b]) := \{X : X(\cdot, \omega) \in C^{n+1}([a, b]), \forall\, \omega \in \Omega$ and $\left|X^{(n+1)}(t, \omega)\right| \leq M$, $\forall\, (t, \omega) \in [a, b] \times \Omega$, where $M > 0$; $X^{(k)}(t, \omega) \in C_{\Omega}([a, b]), k = 0, 1, \ldots, n$; also $X^{(n+1)}(t, \omega)$, $(t, \omega) \in [a, b] \times \Omega$, is a stochastic process$\}$. Clearly by (21.9) and (21.10), respectively, we get that $\mathbf{T}_{\alpha}^{t} X(z, \omega)$, $z \in [t, b]$, and ${}_{\alpha}^{t}\mathbf{T}X(z, \omega)$, $z \in [a, t]$, are stochastic processes for any $t \in [a, b]$.

Consider the linear operator

$$L : C_{\Omega}([a, b]) \hookrightarrow B_{\Omega}([a, b]).$$

If $X \in C_{\Omega}([a, b])$ is nonnegative and LX, too, then L is called positive. If $EL = LE$, then L is called E -commutative, where E is the expectation operator.

From [15, pp. 3–5] we have the following results

(i) $C([a, b]) \subset C_{\Omega}([a, b])$,
(ii) if $X \in C_{\Omega}([a, b])$, then $EX \in C([a, b])$,
and
(iii) if L is E-commutative, then L maps the subspace $C([a, b])$ into $B([a, b])$.

21.4 Background—II

Let $X \in C_{\Omega}^{\alpha, n+1}([a, b])$. We have that $\left|X^{(n+1)}(t, \omega)\right| \leq M, \forall\, (t, \omega) \in [a, b] \times \Omega$.

We notice that

$$\mathbf{T}_{\alpha}^{t} X(z, \omega) \overset{(21.9)}{=} (z - t)^{1-\beta} X^{(n+1)}(z, \omega), \; z \in [t, b],$$

and (21.12)

$${}_{\alpha}^{t}\mathbf{T}X(z, \omega) \overset{(21.10)}{=} (-1)^n (t - z)^{1-\beta} X^{(n+1)}(z, \omega), \; z \in [a, t],$$

$\forall\, \omega \in \Omega$, where $\beta = \alpha - n$, $\alpha \in (n, n+1)$, $n \in \mathbb{Z}_+$; $\forall\, t \in [a, b]$.

We have that

$$\mathbf{T}_{\alpha}^{t} X(t, \omega) = {}_{\alpha}^{t}\mathbf{T}X(t, \omega) = 0. \tag{21.13}$$

It is

$$\left|\mathbf{T}_{\alpha}^{t}X(z,\omega)\right| = \left|(z-t)^{1-\beta}X^{(n+1)}(z,\omega)\right| \leq (b-a)^{1-\beta}M, \tag{21.14}$$

$\forall (z,\omega) \in [t,b] \times \Omega$,

and

$$\left|{}_{\alpha}^{t}\mathbf{T}X(z,\omega)\right| = \left|(-1)^{n}(t-z)^{1-\beta}X^{(n+1)}(z,\omega)\right| \leq (b-a)^{1-\beta}M, \tag{21.15}$$

$\forall (z,\omega) \in [a,t] \times \Omega$.

Here E is the expectation operator

$$(EX)(t) = \int_{\Omega} X(t,\omega)\,P(d\omega).$$

Clearly, we get that

$$\left|\left(E\left(\mathbf{T}_{\alpha}^{t}X\right)\right)(z)\right| \leq (b-a)^{1-\beta}M, \tag{21.16}$$

$\forall z \in [t,b]$,

and

$$\left|\left(E\left({}_{\alpha}^{t}\mathbf{T}X\right)\right)(z)\right| \leq (b-a)^{1-\beta}M, \tag{21.17}$$

$\forall z \in [a,t]$.

We observe that the first modulus of continuity ($\delta > 0$)

$$\omega_1\left(\left(E\left(\mathbf{T}_{\alpha}^{t}X\right)\right),\delta\right)_{[t,b]} := \sup_{\substack{x,y\in[t,b]\\|x-y|\leq\delta}} \left|\left(E\left(\mathbf{T}_{\alpha}^{t}X\right)\right)(x) - \left(E\left(\mathbf{T}_{\alpha}^{t}X\right)\right)(y)\right|$$

$$\overset{(21.16)}{\leq} 2(b-a)^{1-\beta}M, \ \forall t \in [a,b]. \tag{21.18}$$

Hence it holds ($\delta > 0$)

$$\sup_{t\in[a,b]} \omega_1\left(\left(E\left(\mathbf{T}_{\alpha}^{t}X\right)\right),\delta\right)_{[t,b]} \leq 2(b-a)^{1-\beta}M, \tag{21.19}$$

and similarly, it holds

$$\sup_{t\in[a,b]} \omega_1\left(\left(E\left({}_{\alpha}^{t}\mathbf{T}X\right)\right),\delta\right)_{[a,t]} \overset{(21.17)}{\leq} 2(b-a)^{1-\beta}M. \tag{21.20}$$

We also set

$$\omega_1\left(E\left({}^{\alpha}\mathbf{T}^{t}X\right),\delta\right) := \max\left\{\omega_1\left(E\left(\mathbf{T}_{\alpha}^{t}X\right),\delta\right)_{[t,b]}, \omega_1\left(E\left({}_{\alpha}^{t}\mathbf{T}X\right),\delta\right)_{[a,t]}\right\}, \tag{21.21}$$

where $\delta > 0$.

We make

Remark 21.4 Let the positive linear operator L mapping $C([a,b])$ into $B([a,b])$ (the bounded functions). By the Riesz representation theorem [11] we have that there exists μ_t unique, completed Borel measure on $[a,b]$ with

$$\mu_t([a,b]) = L(1)(t) > 0, \tag{21.22}$$

such that

$$L(f)(t) = \int_{[a,b]} f(s)\, d\mu_t(s), \ \forall\, t \in [a,b], \forall\, f \in C([a,b]). \tag{21.23}$$

We denote $\|\cdot\|_\infty = \|\cdot\|_{\infty,[-\pi,\pi]}$ the supremum norm.

Next we specify $[a,b]$ as $[-\pi,\pi]$. Clearly then $L : C([-\pi,\pi]) \to B([-\pi,\pi])$ is the positive linear operator on hand.

Here $\alpha \in (n, n+1)$, $n \in \mathbb{Z}_+$, $k = 1, \ldots, n$. By the use of Hölder's inequality we notice that

$$L\left(\left(\sin\left(\frac{|s-t|}{4}\right)\right)^k\right)(t) = \int_{[-\pi,\pi]} \left(\sin\left(\frac{|s-t|}{4}\right)\right)^k d\mu_t(s) \leq$$

$$\left(\int_{[-\pi,\pi]} \left(\sin\left(\frac{|s-t|}{4}\right)\right)^{\alpha+1} d\mu_t(s)\right)^{\frac{k}{\alpha+1}} (\mu_t([-\pi,\pi]))^{\frac{\alpha+1-k}{\alpha+1}} = \tag{21.24}$$

$$\left(L\left(\left(\sin\left(\frac{|s-t|}{4}\right)\right)^{\alpha+1}\right)(t)\right)^{\frac{k}{\alpha+1}} (L(1)(t))^{\frac{\alpha+1-k}{\alpha+1}}.$$

That is

$$L\left(\left(\sin\left(\frac{|s-t|}{4}\right)\right)^k\right)(t) \leq \tag{21.25}$$

$$\left(L\left(\left(\sin\left(\frac{|s-t|}{4}\right)\right)^{\alpha+1}\right)(t)\right)^{\frac{k}{\alpha+1}} (L(1)(t))^{\frac{\alpha+1-k}{\alpha+1}},$$

for $k = 1, \ldots, n$.

Consequently, it holds

$$\left\| L\left(\left(\sin\left(\frac{|s-t|}{4}\right)\right)^k\right)(t)\right\|_\infty \leq \tag{21.26}$$

$$\left\| L\left(\left(\sin\left(\frac{|s-t|}{4} \right) \right)^{\alpha+1} \right)(t) \right\|_{\infty}^{\frac{k}{\alpha+1}} \|L(1)\|_{\infty}^{\frac{\alpha+1-k}{\alpha+1}},$$

for $k = 1, \ldots, n$.

In this work we will use a lot the following well known inequality:

$$|z| \leq \pi \sin\left(\frac{|z|}{2} \right), \forall z \in [-\pi, \pi]. \tag{21.27}$$

Furthermore, we observe that

$$\left| L\left((s-t)^k \right)(t) \right| = \left| \int_{[-\pi,\pi]} (s-t)^k \, d\mu_t(s) \right| \leq \int_{[-\pi,\pi]} |s-t|^k \, d\mu_t(s) =$$

$$2^k \int_{[-\pi,\pi]} \left(\frac{|s-t|}{2} \right)^k d\mu_t(s) \overset{(21.27)}{\leq} (2\pi)^k \int_{[-\pi,\pi]} \left(\sin\left(\frac{|s-t|}{4} \right) \right)^k d\mu_t(s)$$

$$= (2\pi)^k L\left(\left(\sin\left(\frac{|s-t|}{4} \right) \right)^k \right)(t). \tag{21.28}$$

That is

$$\left| L\left((s-t)^k \right)(t) \right| \leq (2\pi)^k L\left(\left(\sin\left(\frac{|s-t|}{4} \right) \right)^k \right)(t), \tag{21.29}$$

$\forall t \in [-\pi, \pi]$, and

$$\left\| L\left((s-t)^k \right)(t) \right\|_{\infty} \leq (2\pi)^k \left\| L\left(\left(\sin\left(\frac{|s-t|}{4} \right) \right)^k \right)(t) \right\|_{\infty}, \tag{21.30}$$

all $k = 1, \ldots, n$.

Then, by (21.25) and (21.29) we get

$$\left| L\left((s-t)^k \right)(t) \right| \leq (2\pi)^k \left(L\left(\left(\sin\left(\frac{|s-t|}{4} \right) \right)^{\alpha+1} \right)(t) \right)^{\frac{k}{\alpha+1}} (L(1)(t))^{\frac{\alpha+1-k}{\alpha+1}}, \tag{21.31}$$

$k = 1, \ldots, n$, $\forall t \in [-\pi, \pi]$, and by (21.26) and (21.30), we find

$$\left\| L\left((s-t)^k \right)(t) \right\|_{\infty} \leq (2\pi)^k \left\| L\left(\left(\sin\left(\frac{|s-t|}{4} \right) \right)^{\alpha+1} \right)(t) \right\|^{\frac{k}{\alpha+1}} \|L(1)\|_{\infty}^{\frac{\alpha+1-k}{\alpha+1}}, \tag{21.32}$$

$k = 1, \ldots, n$.

We also have

$$L\left(|s-t|^{\alpha+1}\right)(t)=\int_{[-\pi,\pi]}|s-t|^{\alpha+1}\,d\mu_t(s)=$$

$$2^{\alpha+1}\int_{[-\pi,\pi]}\left(\frac{|s-t|}{2}\right)^{\alpha+1}d\mu_t(s)\overset{(21.27)}{\leq}(2\pi)^{\alpha+1}\int_{[-\pi,\pi]}\left(\sin\left(\frac{|s-t|}{4}\right)\right)^{\alpha+1}d\mu_t(s)$$

$$=(2\pi)^{\alpha+1}L\left(\left(\sin\left(\frac{|s-t|}{4}\right)\right)^{\alpha+1}\right)(t). \tag{21.33}$$

That is

$$L\left(|s-t|^{\alpha+1}\right)(t)\leq(2\pi)^{\alpha+1}L\left(\left(\sin\left(\frac{|s-t|}{4}\right)\right)^{\alpha+1}\right)(t), \tag{21.34}$$

$\forall\, t\in[-\pi,\pi]$,
and

$$\left\|L\left(|s-t|^{\alpha+1}\right)(t)\right\|_\infty\leq(2\pi)^{\alpha+1}\left\|L\left(\left(\sin\left(\frac{|s-t|}{4}\right)\right)^{\alpha+1}\right)(t)\right\|_\infty. \tag{21.35}$$

Following 3. Preliminaries we mention

Theorem 21.5 ([7]) *Let $\alpha\in(n,n+1)$, $n\in\mathbb{Z}_+$. Consider the positive E-commutative linear operator $L:C_\Omega([a,b])\hookrightarrow B_\Omega([a,b])$, and let $X\in C_\Omega^{\alpha,n+1}([a,b])$, $\delta>0$. Here $\omega_1\left(E\left({}^{\alpha}\mathbf{T}^{t}X\right),\delta\right)$ as in (21.21).*

Then

$$|(E(LX))(t)-(EX)(t)|\leq|(EX)(t)|\,|(L(1))(t)-1|+$$

$$\sum_{k=1}^{n}\frac{\left|\left(EX^{(k)}\right)(t)\right|}{k!}\left|L\left((s-t)^k\right)(t)\right|+\frac{\omega_1\left(E\left({}^{\alpha}\mathbf{T}^{t}X\right),\delta\right)}{\prod_{j=1}^{n}(\alpha+j-n)}$$

$$\left(L\left(|s-t|^{\alpha+1}\right)(t)\right)^{\frac{\alpha}{\alpha+1}}\left[\frac{(L(1)(t))^{\frac{1}{\alpha+1}}}{\alpha-n}+\frac{\left(L\left(|s-t|^{\alpha+1}\right)(t)\right)^{\frac{1}{\alpha+1}}}{\delta(\alpha+1)}\right], \tag{21.36}$$

$\forall\, t\in[a,b]$.

We also need

Theorem 21.6 ([7]) *All as in Theorem 21.5. Then*

$$\|E(LX) - EX\|_\infty \le \|EX\|_\infty \|L(1) - 1\|_\infty +$$

$$\sum_{k=1}^{n} \frac{\|EX^{(k)}\|_\infty}{k!} \left\|L\left((s-t)^k\right)(t)\right\|_\infty + \frac{\sup_{t\in[a,b]} \omega_1\left(E\left({}^{\alpha}\mathbf{T}^t X\right), \delta\right)}{\prod_{j=1}^{n}(\alpha+j-n)}$$

$$\left\|L\left(|s-t|^{\alpha+1}\right)(t)\right\|_\infty^{\frac{\alpha}{\alpha+1}} \left[\frac{\|L(1)\|_\infty^{\frac{1}{\alpha+1}}}{\alpha-n} + \frac{\left\|L\left(|s-t|^{\alpha+1}\right)(t)\right\|_\infty^{\frac{1}{\alpha+1}}}{\delta(\alpha+1)}\right] < \infty. \quad (21.37)$$

We will use (case of $n=0$):

Definition 21.7 *([7])* We call the subspace of stochastic process: $C_\Omega^{\alpha,1}([a,b]) := \{X : X(\cdot,\omega) \in C^1([a,b]), \forall\, \omega \in \Omega$ and $\left|X^{(1)}(t,\omega)\right| \le M, \forall\,(t,\omega) \in [a,b]\times\Omega$, where $M>0$; and $X(t,\omega) \in C_\Omega([a,b])$; also $X^{(1)}(t,\omega)$, $(t,\omega)\in[a,b]\times\Omega$ is a stochastic process$\}$. Here we will consider $C_\Omega^{\alpha,1}([-\pi,\pi])$.

21.5 Main Results

We present our main pointwise approximation result

Theorem 21.8 *Let $\alpha \in (n, n+1)$, $n \in \mathbb{Z}_+$. Consider the positive E-commutative linear operator $L : C_\Omega([-\pi,\pi]) \hookrightarrow B_\Omega([-\pi,\pi])$, and let $X \in C_\Omega^{\alpha,n+1}([-\pi,\pi])$, $\delta > 0$. Here $\omega_1\left(E\left({}^{\alpha}\mathbf{T}^t X\right), \delta\right)$, as in (21.21) it is over $[-\pi,\pi]$.*

Then

$$|(E(LX))(t) - (EX)(t)| \le |(EX)(t)|\,|(L(1))(t) - 1| +$$

$$\sum_{k=1}^{n} \frac{\left|\left(EX^{(k)}\right)(t)\right|}{k!} (2\pi)^k L\left(\left(\sin\left(\frac{|s-t|}{4}\right)\right)^k\right)(t) +$$

$$\frac{\omega_1\left(E\left({}^{\alpha}\mathbf{T}^t X\right), \delta\right)}{\prod_{j=1}^{n}(\alpha+j-n)} (2\pi)^\alpha \left(L\left(\left(\sin\left(\frac{|s-t|}{4}\right)\right)^{\alpha+1}\right)(t)\right)^{\frac{\alpha}{\alpha+1}}$$

$$\left[\frac{(L(1)(t))^{\frac{1}{\alpha+1}}}{\alpha-n} + \frac{(2\pi)\left(L\left(\left(\sin\left(\frac{|s-t|}{4}\right)\right)^{\alpha+1}\right)(t)\right)^{\frac{1}{\alpha+1}}}{\delta(\alpha+1)}\right], \quad (21.38)$$

$\forall\, t \in [-\pi, \pi]$.

Proof By (21.36), applied over $[-\pi, \pi]$, we get

$$|(E(LX))(t) - (EX)(t)| \leq |(EX)(t)|\,|(L(1))(t) - 1| +$$

$$\sum_{k=1}^{n} \frac{\left|\left(EX^{(k)}\right)(t)\right|}{k!} \left|L\left((s-t)^k\right)(t)\right| + \frac{\omega_1\left(E\left({}^{\alpha}\mathbf{T}^{t}X\right), \delta\right)}{\prod_{j=1}^{n}(\alpha + j - n)}$$

$$\left(L\left(|s-t|^{\alpha+1}\right)(t)\right)^{\frac{\alpha}{\alpha+1}} \left[\frac{(L(1)(t))^{\frac{1}{\alpha+1}}}{\alpha - n} + \frac{\left(L\left(|s-t|^{\alpha+1}\right)(t)\right)^{\frac{1}{\alpha+1}}}{\delta(\alpha+1)}\right]$$

$$\overset{\text{(by (21.29), (21.34))}}{\leq} |(EX)(t)|\,|(L(1))(t) - 1| +$$

$$\sum_{k=1}^{n} \frac{\left|\left(EX^{(k)}\right)(t)\right|}{k!} (2\pi)^k L\left(\left(\sin\left(\frac{|s-t|}{4}\right)\right)^k\right)(t) +$$

$$\frac{\omega_1\left(E\left({}^{\alpha}\mathbf{T}^{t}X\right), \delta\right)}{\prod_{j=1}^{n}(\alpha + j - n)} (2\pi)^{\alpha} \left(L\left(\left(\sin\left(\frac{|s-t|}{4}\right)\right)^{\alpha+1}\right)(t)\right)^{\frac{\alpha}{\alpha+1}}$$

$$\left[\frac{(L(1)(t))^{\frac{1}{\alpha+1}}}{\alpha - n} + \frac{(2\pi)\left(L\left(\left(\sin\left(\frac{|s-t|}{4}\right)\right)^{\alpha+1}\right)(t)\right)^{\frac{1}{\alpha+1}}}{\delta(\alpha+1)}\right], \tag{21.39}$$

$\forall\, t \in [-\pi, \pi]$. ∎

It follows the main uniform related estimate

Theorem 21.9 *All as in Theorem 21.8. Then*

$$\|E(LX) - EX\|_{\infty} \leq \|EX\|_{\infty}\, \|L(1) - 1\|_{\infty} +$$

$$\sum_{k=1}^{n} \frac{\left\|EX^{(k)}\right\|_{\infty}}{k!} (2\pi)^k \left\|L\left(\left(\sin\left(\frac{|s-t|}{4}\right)\right)^k\right)(t)\right\|_{\infty} +$$

$$\frac{\sup_{t\in[-\pi,\pi]} \omega_1\left(E\left({}^{\alpha}\mathbf{T}^{t}X\right), \delta\right)}{\prod_{j=1}^{n}(\alpha + j - n)} (2\pi)^{\alpha} \left\|L\left(\left(\sin\left(\frac{|s-t|}{4}\right)\right)^{\alpha+1}\right)(t)\right\|_{\infty}^{\frac{\alpha}{\alpha+1}} \tag{21.40}$$

$$\left[\frac{\|L(1)\|_{\infty}^{\frac{1}{\alpha+1}}}{\alpha-n}+\frac{(2\pi)\left\|L\left(\left(\sin\left(\frac{|s-t|}{4}\right)\right)^{\alpha+1}\right)(t)\right\|_{\infty}^{\frac{1}{\alpha+1}}}{\delta(\alpha+1)}\right]<+\infty.$$

Proof Directly from (21.38). ∎

We give the following special results:

Corollary 21.10 *Let* $0<\alpha<1$. *Consider the positive E-commutative linear operator* $L: C_{\Omega}([-\pi,\pi]) \hookrightarrow B_{\Omega}([-\pi,\pi])$, *and let* $X \in C_{\Omega}^{\alpha,1}([-\pi,\pi])$, $\delta>0$. *Here* $\omega_1\left(E\left({}^{\alpha}\mathbf{T}^{t}X\right),\delta\right)$, *as in (21.21) it is over* $[-\pi,\pi]$.

Then

$$|(E(LX))(t)-(EX)(t)| \leq |(EX)(t)|\,|(L(1))(t)-1|+$$

$$\omega_1\left(E\left({}^{\alpha}T^{t}X\right),\delta\right)(2\pi)^{\alpha}\left(L\left(\left(\sin\left(\frac{|s-t|}{4}\right)\right)^{\alpha+1}\right)(t)\right)^{\frac{\alpha}{\alpha+1}}$$

$$\left[\frac{(L(1)(t))^{\frac{1}{\alpha+1}}}{\alpha}+\frac{(2\pi)\left(L\left(\left(\sin\left(\frac{|s-t|}{4}\right)\right)^{\alpha+1}\right)(t)\right)^{\frac{1}{\alpha+1}}}{\delta(\alpha+1)}\right], \tag{21.41}$$

$\forall\, t\in[-\pi,\pi]$.

Proof From (21.38). ∎

Corollary 21.11 *All as in Corollary 21.10. Then*

$$\|E(LX)-EX\|_{\infty} \leq \|EX\|_{\infty}\,\|L(1)-1\|_{\infty}+$$

$$\sup_{t\in[-\pi,\pi]}\omega_1\left(E\left({}^{\alpha}T^{t}X\right),\delta\right)(2\pi)^{\alpha}\left\|L\left(\left(\sin\left(\frac{|s-t|}{4}\right)\right)^{\alpha+1}\right)(t)\right\|_{\infty}^{\frac{\alpha}{\alpha+1}}$$

$$\left[\frac{\|L(1)\|_{\infty}^{\frac{1}{\alpha+1}}}{\alpha}+\frac{(2\pi)\left\|L\left(\left(\sin\left(\frac{|s-t|}{4}\right)\right)^{\alpha+1}\right)(t)\right\|_{\infty}^{\frac{1}{\alpha+1}}}{\delta(\alpha+1)}\right]. \tag{21.42}$$

Proof From (21.40). ∎

Corollary 21.12 *All as in Corollary 21.10, plus* $L(1) = 1$. *Then*

$$\|E(LX) - EX\|_\infty \leq$$

$$\sup_{t\in[-\pi,\pi]} \omega_1\left(E\left({}^{\alpha}T^{t}X\right), \delta\right)(2\pi)^{\alpha} \left\| L\left(\left(\sin\left(\frac{|s-t|}{4}\right)\right)^{\alpha+1}\right)(t)\right\|_\infty^{\frac{\alpha}{\alpha+1}}$$

$$\left[\frac{1}{\alpha} + \frac{(2\pi)\left\| L\left(\left(\sin\left(\frac{|s-t|}{4}\right)\right)^{\alpha+1}\right)(t)\right\|_\infty^{\frac{1}{\alpha+1}}}{\delta(\alpha+1)}\right]. \tag{21.43}$$

Proof From (21.42). ■

In particular we have

Corollary 21.13 *All as in Corollary 21.10 and* $L(1) = 1$. *Then*

$$\|E(LX) - EX\|_\infty \leq \left(\frac{\alpha+1}{\alpha}\right)(2\pi)^{\alpha}$$

$$\sup_{t\in[-\pi,\pi]} \omega_1\left(E\left({}^{\alpha}T^{t}X\right), \frac{2\pi}{\alpha+1}\left\| L\left(\left(\sin\left(\frac{|s-t|}{4}\right)\right)^{\alpha+1}\right)(t)\right\|_\infty^{\frac{1}{\alpha+1}}\right)$$

$$\left\| L\left(\left(\sin\left(\frac{|s-t|}{4}\right)\right)^{\alpha+1}\right)(t)\right\|_\infty^{\frac{\alpha}{\alpha+1}}. \tag{21.44}$$

Proof By (21.43): we take there

$$\delta = \frac{2\pi}{\alpha+1}\left\| L\left(\left(\sin\left(\frac{|s-t|}{4}\right)\right)^{\alpha+1}\right)(t)\right\|_\infty^{\frac{1}{\alpha+1}} > 0.$$

In case of $\left\| L\left(\left(\sin\left(\frac{|s-t|}{4}\right)\right)^{\alpha+1}\right)(t)\right\|_\infty = 0$, we have that $L\left(\left(\sin\left(\frac{|s-t|}{4}\right)\right)^{\alpha+1}\right)(t)$ $= 0$, $\forall\, t \in [-\pi,\pi]$. That is, by (21.23), $\int_{[-\pi,\pi]}\left(\sin\left(\frac{|s-t|}{4}\right)\right)^{\alpha+1} d\mu_t(s) = 0$, $\forall\, t \in [-\pi,\pi]$, where μ_t is a probability measure on $[-\pi,\pi]$.

Since $\sin\left(\frac{|s-t|}{4}\right) \geq 0$, $\forall\, s \in [-\pi,\pi]$, we get $\sin\left(\frac{|s-t|}{4}\right) = 0$, a.e., that is $|s-t| = 0$, a.e., and $s = t$, a.e., which means $\mu_t\{s \in [-\pi,\pi] : s \neq t\} = 0$, i.e. $\mu_t = \delta_t$, $\forall\, t \in [-\pi,\pi]$, where δ_t is the unit Dirac measure.

Consequently we have

$$E(LX)(t) = L(EX)(t) = \int_{[-\pi,\pi]} (EX)(s)\, d\delta_t(s) = (EX)(t),$$

$\forall\, t \in [-\pi, \pi]$.

That is $E(LX) = EX$ over $[-\pi, \pi]$. Therefore both sides of inequality (21.44) equal to zero.

Hence (21.44) is always true. ■

21.6 Application

Consider the Bernstein polynomials on $[-\pi, \pi]$ for $f \in C([-\pi, \pi])$:

$$B_N(f)(x) = \sum_{k=0}^{N} \binom{N}{k} f\left(-\pi + \frac{2\pi k}{N}\right)\left(\frac{x+\pi}{2\pi}\right)^k \left(\frac{\pi - x}{2\pi}\right)^{N-k}, \tag{21.45}$$

$N \in \mathbb{N}$, any $x \in [-\pi, \pi]$. There are positive linear operators from $C([-\pi, \pi])$ into itself.

Setting $g(t) = f(2\pi t - \pi)$, $t \in [0, 1]$, we have $g(0) = f(-\pi)$, $g(1) = f(\pi)$, and

$$(B_N g)(t) = \sum_{k=0}^{N} \binom{N}{k} g\left(\frac{k}{N}\right) t^k (1-t)^{N-k} = (B_N f)(x), \ x \in [-\pi, \pi]. \tag{21.46}$$

Here $x = \varphi(t) = 2\pi t - \pi$ is an $1-1$ and onto map from $[0, 1]$ onto $[-\pi, \pi]$. Clearly here $g \in C([0, 1])$.

Notice also that

$$\left(B_N\left((\cdot - x)^2\right)\right)(x) = \left[\left(B_N\left((\cdot - t)^2\right)\right)(t)\right](2\pi)^2 = \frac{(2\pi)^2}{N} t(1-t)$$

$$= \frac{(2\pi)^2}{N}\left(\frac{x+\pi}{2\pi}\right)\left(\frac{\pi - x}{2\pi}\right) = \frac{1}{N}(x+\pi)(\pi - x) \leq \frac{\pi^2}{N}, \ \forall\, x \in [-\pi, \pi].$$

I.e.

$$\left(B_N\left((\cdot - x)^2\right)\right)(x) \leq \frac{\pi^2}{N}, \ \forall\, x \in [-\pi, \pi]. \tag{21.47}$$

In particular

$$(B_N 1)(x) = 1,\ \forall\, x \in [-\pi, \pi]. \tag{21.48}$$

Define the corresponding stochastic application of B_N as follows:

$$B_N(X(\cdot,\omega))(t) = \sum_{k=0}^{N} \binom{N}{k} X\left(-\pi + \frac{2\pi k}{N}, \omega\right)\left(\frac{t+\pi}{2\pi}\right)^k \left(\frac{\pi - t}{2\pi}\right)^{N-k}, \tag{21.49}$$

$\forall\, N \in \mathbb{N}$, $\forall\, t \in [-\pi, \pi]$, $\forall\, \omega \in \Omega$, where X is a stochastic process. Clearly $B_N X$ is a stochastic process.

Notice that

$$E(B_N X)(t) = \sum_{k=0}^{N} \binom{N}{k} (EX)\left(-\pi + \frac{2\pi k}{N}\right)\left(\frac{t+\pi}{2\pi}\right)^k \left(\frac{\pi - t}{2\pi}\right)^{N-k}$$

$$= B_N(EX)(t), \tag{21.50}$$

i.e. $EB_N = B_N E$, that is B_N is an E-commutative positive linear operator from $C_\Omega([-\pi, \pi])$ into itself.

We give

Corollary 21.14 *Let* $0 < \alpha < 1$ *and* $X \in C_\Omega^{\alpha,1}([-\pi, \pi])$. *Then*

$$\|E(B_N X) - EX\|_\infty \leq \left(\frac{\alpha+1}{\alpha}\right)(2\pi)^\alpha$$

$$\sup_{t\in[-\pi,\pi]} \omega_1\left(E\left({}^{\alpha}T^{t}X\right), \frac{2\pi}{\alpha+1}\left\| B_N\left(\left(\sin\left(\frac{|s-t|}{4}\right)\right)^{\alpha+1}\right)(t)\right\|_\infty^{\frac{1}{\alpha+1}}\right)$$

$$\left\| B_N\left(\left(\sin\left(\frac{|s-t|}{4}\right)\right)^{\alpha+1}\right)(t)\right\|_\infty^{\frac{\alpha}{\alpha+1}}, \tag{21.51}$$

$\forall\, N \in \mathbb{N}$.

Proof By (21.44). ∎

In particular we get:

Corollary 21.15 *Let* $X \in C_\Omega^{\frac{1}{2},1}([-\pi, \pi])$. *Then*

$$\|E(B_N X) - EX\|_\infty \le 3\sqrt{2\pi}$$

$$\sup_{t\in[-\pi,\pi]} \omega_1 \left(E\left(\tfrac{1}{2} T^t X\right), \frac{4\pi}{3} \left\| B_N \left(\left(\sin\left(\frac{|s-t|}{4} \right) \right)^{\frac{3}{2}} \right)(t) \right\|_\infty^{\frac{2}{3}} \right)$$

$$\left\| B_N \left(\left(\sin\left(\frac{|s-t|}{4} \right) \right)^{\frac{3}{2}} \right)(t) \right\|_\infty^{\frac{1}{3}}, \tag{21.52}$$

$\forall\, N \in \mathbb{N}$.

Proof By (21.51) for $\alpha = \frac{1}{2}$. ∎

We make

Remark 21.16 By $|\sin x| < |x|$, $\forall\, x \in \mathbb{R} - \{0\}$, in particular $\sin x \le x$, for $x \ge 0$, we get

$$\left(\sin\left(\frac{|\cdot - t|}{4} \right) \right)^{\frac{3}{2}} \le \left(\frac{|\cdot - t|}{4} \right)^{\frac{3}{2}} = \frac{1}{8} |\cdot - t|^{\frac{3}{2}}.$$

Hence

$$\left\| B_N \left(\left(\sin\left(\frac{|\cdot - t|}{4} \right) \right)^{\frac{3}{2}} \right)(t) \right\|_\infty \le \frac{1}{8} \left\| B_N \left(|\cdot - t|^{\frac{3}{2}} \right)(t) \right\|_\infty. \tag{21.53}$$

We observe that

$$B_N \left(|\cdot - t|^{\frac{3}{2}} \right)(t) = \sum_{k=0}^{N} \left| t + \pi - \frac{2\pi k}{N} \right|^{\frac{3}{2}} \binom{N}{k} \left(\frac{t+\pi}{2\pi} \right)^k \left(\frac{\pi - t}{2\pi} \right)^{N-k}$$

(by discrete Hölder's inequality)

$$\le \left[\sum_{k=0}^{N} \left(t + \pi - \frac{2\pi k}{N} \right)^2 \binom{N}{k} \left(\frac{t+\pi}{2\pi} \right)^k \left(\frac{\pi - t}{2\pi} \right)^{N-k} \right]^{\frac{3}{4}} \tag{21.54}$$

$$= \left(B_N \left((\cdot - t)^2 \right)(t) \right)^{\frac{3}{4}} \overset{(21.47)}{\le} \frac{\pi^{\frac{3}{2}}}{N^{\frac{3}{4}}}, \ \forall\, t \in [-\pi, \pi].$$

Consequently it holds

$$\left\| B_N \left(|\cdot - t|^{\frac{3}{2}} \right)(t) \right\|_\infty \le \frac{\pi^{\frac{3}{2}}}{N^{\frac{3}{4}}}, \tag{21.55}$$

and

$$\left\| B_N\left(\left(\sin\left(\frac{|\cdot - t|}{4}\right)\right)^{\frac{3}{2}}\right)(t)\right\|_\infty \leq \frac{\pi^{\frac{3}{2}}}{8N^{\frac{3}{4}}},\ \forall\, N \in \mathbb{N}. \tag{21.56}$$

We derive

Proposition 21.17 *Let* $X \in C_\Omega^{\frac{1}{2},1}([-\pi,\pi])$. *Then*

$$\|E(B_N X) - EX\|_\infty \leq \frac{3\sqrt{2}\pi}{2\sqrt[4]{N}} \sup_{t\in[-\pi,\pi]} \omega_1\left(E\left(\tfrac{1}{2}T^t X\right), \frac{\pi^2}{3\sqrt{N}}\right), \tag{21.57}$$

$\forall\, N \in \mathbb{N}$.

Hence $\lim_{N\to\infty} E(B_N X) = EX$, *uniformly.*

Proof By (21.52) and (21.56). ■

21.7 Commutative Trigonometric Conformable Fractional Stochastic Korovkin Properties

Here L is meant as a sequence of positive E-commutative linear operators and all assumptions are as in Theorem 21.8.

We give

Theorem 21.18 *We further assume that* $L(1)(t) \to 1$ *and* $L\left(\left(\sin\left(\frac{|s-t|}{4}\right)\right)^{\alpha+1}\right)(t) \to 0$, *then* $(E(LX))(t) \to (EX)(t)$, *for any* $X \in C_\Omega^{\alpha,n+1}([-\pi,\pi])$, $\forall\, t \in [-\pi,\pi]$, *a pointwise convergence; where* $\alpha \in (n, n+1)$, $n \in \mathbb{Z}_+$.

Proof Based on (21.38), (21.31), and that $L(1)(t)$ is bounded as a sequence of functions. Also $\omega_1\left(E\left({}^{\alpha}\mathbf{T}^t X\right), \delta\right)$ over $[-\pi,\pi]$ is bounded, see (21.19)–(21.21). ■

We continue with

Theorem 21.19 *We further assume that* $L(1)(t) \to 1$, *uniformly and* $\left\|L\left(\left(\sin\left(\frac{|s-t|}{4}\right)\right)^{\alpha+1}\right)(t)\right\|_\infty \to 0$, *then* $E(LX) \to EX$, *uniformly over* $[-\pi,\pi]$, *for any* $X \in C_\Omega^{\alpha,n+1}([-\pi,\pi])$; *where* $\alpha \in (n, n+1)$, $n \in \mathbb{Z}_+$.

Proof Based on (21.40), (21.32), and that $\|L(1)\|_\infty$ is bounded. Also it is $\sup_{t\in[-\pi,\pi]} \omega_1\left(E\left({}^{\alpha}\mathbf{T}^t X\right), \delta\right) < \infty$, by (21.19)–(21.21). ■

We finish with

Remark 21.20 The stochastic convergences of Theorems 21.18, 21.19 are implied by the convergences of the basic and simple real non-stochastic functions $\left\{1, \left(\sin\left(\frac{|s-t|}{4}\right)\right)^{\alpha+1}\right\}$, an amazing fact!

References

1. Abdeljawad, T.: On conformable fractional calculus. J. Comput. Appl. Math. **279**, 57–66 (2015)
2. Anastassiou, G.: A study of positive linear operators by the method of moments, one-dimensional case. J. Approx. Theory **45**, 247–270 (1985)
3. Anastassiou, G.: Korovkin type inequalities in real normed vector spaces. Approx. Theory Appl. **2**, 39–53 (1986)
4. Anastassiou, G.: Multi-dimensional quantitative results for probability measures approximating the unit measure. Approx. Theory Appl. **2**, 93–103 (1986)
5. Anastassiou, G.A.: Korovkin inequalities for stochastic processes. J. Math. Anal. Appl. **157**(2), 366–384 (1991)
6. Anastassiou, G.: Nonlinearity: Ordinary and Fractional Approximations by Sublinear and Max-product operators. Springer, Heidelberg (2018)
7. Anastassiou, G.A.: Commutative conformable fractional Korovkin properties for Stochastic processes (2020). Submitted for publication
8. Anastassiou, G.: Trigonometric Commutative Conformable fractional Korovkin properties for Stochastic processes, Analele Universitatii Oradea Fasc. Matematica (2020). Accepted for publication
9. Khalil, R., Al Horani, M., Yousef, A., Sababheh, M.: A new definition of fractional derivative. J. Comput. Appl. Math. **264**, 65–70 (2014)
10. Korovkin, P.P.: Linear Operators and Approximation Theory. Hindustan Publ. Corp., Delhi (1960)
11. Royden, H.L.: Real Analysis, 2nd edn. MacMillan Publishing Co., Inc., New York (1968)
12. Shisha, O., Mond, B.: The degree of approximation to periodic functions by linear positive operators. J. Approx. Theory **1**, 335–339 (1968)
13. Weba, M.: Korovkin systems of stochastic processes. Math. Z. **192**(1), 73–80 (1986)
14. Weba, M.: Quantitative results on monotone approximation of stochastic processes. Probab. Math. Stat. **11**(1), 109–120 (1990)
15. Weba, M.: A quantitative Korovkin theorem for random functions with multivariate domains. J. Approx. Theory **61**(1), 74–87 (1990)
16. Weba, M.: Monotone approximation of random functions with multivariate domains in respect of lattice semi-norms. Results Math. **20**(1–2), 554–576 (1991)

Chapter 22
Concluding Remarks

In a 1695 letter to L'Hospital, Leibniz asked, "Can we generalize ordinary\derivatives to ones of arbitrary order?" L'Hospital replied to Leibniz with another question, "What if the order is $1/2$?" Then Leibniz, in a letter dated September 30, 1695 replied, "It will lead to a paradox, from which one day many useful consequences will be drawn." That was the beginning of fractional calculus. The subject has been ongoing for more than 300 years now. Many famous mathematicians contributed to the topic over the years such as Liouville, Euler, Laplace, Lagrange, Riemann, Weyl, Fourier, Abel, Lacroix, Grunwald, and Letnikov.

The first attempt to give a rigorous definition of fractional derivatives and study the subject in depth was by Liouville during 1832–1855. The concept grew out of his earlier work in electromagnetism. Fractional derivatives describe solutions of fractional integral equations, many times arising from physics, as done by Abel in 1823 to solve the brachistochrone problem.

Fractional calculus has a long history but, from the applicative point of view, it fell into oblivion for hundreds of years because of lack of applications to other sciences such as physics and engineering. This oblivion was also due to its complexity and lack of physical and geometric connection. Fortunately, that was then as things have changed dramatically. Fractional calculus has developed a lot in the last fifty years and especially since 1974 when the first international conference in the field took place, organized by B. Ross, in New Haven, Connecticut, USA. Now we frequently see such conferences around the world. There exist a lot of related research activities resulting in many interesting articles and books. AMS/Mathscinet lists about 50,000 fractional publications on 4/19/20. We now see many important applications: in acoustic wave propagation in inhomogeneous porous material, diffusive transport, fluid flow, dynamical processes in self-similar structures, dynamics of earthquakes, optics, geology, viscoelastic materials, biosciences, bioengineering, medicine, economics, probability and statistics, astrophysics, chemical engineering, physics, splines, tomography, fluid mechanics, electromagnetic waves, non-linear control, signal processing, control of electronic power, converters, chaotic dynamics,

G. A. Anastassiou, *Generalized Fractional Calculus*, Studies in Systems, Decision and Control 305, https://doi.org/10.1007/978-3-030-56962-4_22

polymer science, proteins, polymer physics, electrochemistry, statistical physics, rheology, thermodynamics, neural networks, and many others. By now almost all fields of research in science and engineering use fractional calculus to better describe them. The current mathematical theory of fractional calculus seems to lag behind the needs for mathematical modeling of all the above applications. Therefore there exists a lot of room for mathematical theoretical expansion on the subject. The recent development of fractal theory has shown great connections to fractional calculus. In general, fractional calculus currently provides the best description for fractals and chaotic situations. All in all fractional calculus provides the best expression for the complexity of our modern science. In recent years researchers came up with more efficient and practical fractional derivatives, e.g. Caputo (1967) inspired by studies in Geophysics, Canavati (1987) and Khalil et al. (2014) with the conformable fractional derivative. The author was the first to study fractional inequalities and fractional approximation theory by publishing numerous of articles and monographs.

In this monograph the author continuous his studies by employing his generalized fractional calculus (2016) methods. He uses also the, in some cases equivalent, generalized fractional calculus by Almeida (2017). The idea is to cover as many as possible cases in the applications since now the fractional calculus is weighted, that is the produced theory is as rich as possible. In the second part of the book the author lays down for the first time the foundations of stochastic fractional calculus and studies the fractional approximation of stochastic processes in detail. Definitely what is developed in this monograph can have great applications and can lead to more interesting avenues of research.

GPSR Compliance
The European Union's (EU) General Product Safety Regulation (GPSR) is a set of rules that requires consumer products to be safe and our obligations to ensure this.

If you have any concerns about our products, you can contact us on

ProductSafety@springernature.com

In case Publisher is established outside the EU, the EU authorized representative is:

Springer Nature Customer Service Center GmbH
Europaplatz 3
69115 Heidelberg, Germany

www.ingramcontent.com/pod-product-compliance
Lightning Source LLC
Chambersburg PA
CBHW070823200526
45355CB00003B/310
* 9 7 8 3 0 3 0 5 6 9 6 1 7 *